May 3rd, 1993

To the HM Karol Foundation and the Karol family:

With my best regards and thanks for the support provided to me by the HMK Foundation during my training in molecular genetics at Harvard Medical School.

[signature], MD

MOLECULAR GENETICS OF NERVOUS SYSTEM TUMORS

MOLECULAR GENETICS OF NERVOUS SYSTEM TUMORS

Editors

ARNOLD J. LEVINE, PH.D.
Department of Molecular Biology
Princeton University
Princeton, New Jersey

HENRY H. SCHMIDEK, M.D., F.A.C.S.
Marion, Massachusetts

WILEY-LISS
A JOHN WILEY & SONS, INC., PUBLICATION
New York • Chichester • Brisbane • Toronto • Singapore

Address All Inquiries to the Publisher
Wiley-Liss, Inc., 605 Third Avenue, New York, NY 10158-0012

Printed in the United States of America.

Library of Congress Cataloging-in-Publication Data

Molecular genetics of nervous system tumors / edited by Arnold J. Levine, Henry H. Schmidek.
p. cm.
Includes bibliographical references and index.
ISBN 0-471-56179-7
1. Nervous system—Tumor—Genetic aspects. 2. Nervous system--Tumors—Molecular aspects. I. Levine, Arnold J. (Arnold Jay), 1939- . II. Schmidek, Henry H., 1937-
[DNLM: 1. Genetics, Biochemical. 2. Nervous System Neoplasms--genetics. WL 100 M71862]
RC280.N4M654 1993
616.99'28042—dc20
DNLM/DLC
for Library of Congress 92-48694
CIP

On the cover: MR image adapted from* Atlas of Human Cross-Sectional Anatomy, Second Edition, *by Donald R. Cahill, Matthew J. Orland, and Carl C. Reading, published by Wiley-Liss, New York (1990). Used with permission. Detail shows small arm of chromosome 17.

The text of this book was printed on acid-free paper.

Contents

Oncogenes, Growth Factors, and Experimental Neuro-Oncogenesis

Effects of Ionizing Radiation on the Central Nervous System

Identifying and Mapping Genetic Changes in Human Nervous System Tumors

Cytogenetic Analysis of Human Brain Tumors

Molecular Genetic Analysis of Human Nervous System Tumors

Tumors of the Peripheral Nervous System

Tumors of the Central Nervous System

Metastatic Disease to the Nervous System

Experimental Brain Tumor Therapy Using Viral Vectors or Plasmids Expressing Antisense Transcript

About the Editors

Arnold J. Levine, Ph.D., is a world-class molecular biologist who is recognized for his important contributions to the field of cancer research. He is a graduate of Harpur College and of the University of Pennsylvania and has held teaching positions at SUNY Stony Brook School of Medicine and Princeton University.

Dr. Levine is the author of numerous scientific papers and serves as the Editor-in-Chief of the Journal of Virology. He is a member of the National Academy of Sciences and recently won the first Brupbacher Award for his work on p53 tumor suppressor genes.

Dr. Levine is the Harry C. Weiss Professor in Life Sciences and chairman of the Department of Molecular Biology at Princeton University.

Henry H. Schmidek, M.D., is a clinical neurosurgeon who received his medical degree with highest honors from the University of Western Ontario. He trained in general surgery at the Royal Victoria Hospital, McGill University, and completed his residency in neurological surgery at the Massachusetts General Hospital, Harvard Medical School.

He has directed a course involving neurobiology and molecular genetics at the Marine Biological Laboratory, Woods Hole. From 1985 to 1988 he was an NIH Senior Investigator in Molecular Genetics and Visiting Professor in the Department of Neurobiology, Harvard Medical School, and Visiting Scientist at Whitaker College, MIT.

Dr. Schmidek is the author of 146 scientific publications. He has served as the editor, co-editor, or contributor to nine books on a variety of neurosurgical topics including volumes on meningiomas, pineal tumors, and tumors of the spine.

Contributors

V.R. BABU
Department of Medical Genetics, Henry Ford Hospital, Detroit, MI 48202 [265]

MARIANO BARBACID
Department of Molecular Biology, Bristol-Myers Squibb Pharmaceutical Research Institute, Princeton, NJ 08543-4000 [123]

CORNELIA I. BARGMANN
Department of anatomy, Programs in Developmental Biology and Neuroscience, Unirersity of California at San Francisco Medical School, San Francisco, CA 94143-0452 [49]

J. CARL BARRETT
Laboratory of Molecular Carcinogenesis, National Institute of Environmental Health Sciences, National Institutes of Health, Research Triangle Park, NC 27709 [61]

RENE BERNARDS
Division of Molecular Genetics, The Cancer Center of Massachusetts General Hospital, Harvard Medical School, Charlestown, MA 02129 [229]

DAVID A. BOOTHMAN
Department of Radiation Oncology, Division of Cancer Biology, University of Michigan Medical School, Ann Arbor, MI 48109 [163]

AUBREY E. BOYD III
Division of Endocrinology, Metabolism, Diabetes, and Molecular Medicine, Tufts University Medical School, New England Medical Center, Boston, MA 02111 [271]

GARRETT M. BRODEUR
Departments of Pediatrics and Genetics, Division of Hematology-Oncology, Washington University School of Medicine, St. Louis, MO 63110 [229]

MAX M. BURGER
Friedrich Miescher Institute, Basel, Switzerland [357]

ANNE L. CALOF
Department of Biology, University of Iowa, Iowa City, IA 52242 [341]

RICHARD CAWTHON
Department of Human Genetics, Howard Hughes Medical Institute, Eccles Institute of Human Genetics, University of Utah School of Medicine, Salt Lake City, UT 84112 [319]

JACOB CHASON
Department of Pathology, Henry Ford Hospital, Detroit, MI 48202 [265]

BRENT H. COCHRAN
Center for Cancer Research, Massachusetts Institute of Technology, Cambridge, MA 02139 [101]

JOHN COLLINS
Department of Neurosurgery, National Naval Medical Center, Rockville, MD 20852 [73]

V. PETER COLLINS
Ludwig Institute for Cancer Research, Karolinska Hospital, Stockholm S-10401, and Division of Pathology I, Sahlgrenska Hospital, Gotheburg, Sweden [241]

LIONEL FEIGENBAUM
Laboratory of Virology, Jerome H. Holland Laboratory, American Red Cross, Rockville, MD 20855 [153]

ROBERT F. GAGEL
Section of Endocrinology, University of Texas, M.D. Anderson Cancer Center, Houston, TX 77030 [271]

JONATHAN L. HAINES
Molecular Neurogenetics Laboratory, Department of Neurology, Massachusetts General Hospital, Harvard Medical School, Charlestown, MA 02129 [303]

TIMOTHY E. HAYES
Department of Brain and Cognitive Sciences, Massachusetts Institute of Technology, Cambridge, MA 02139 [37]

JOSEPH ILAN
Department of Anatomy, Laboratory of Developmental Genetics, and Molecular Biology, The Cancer Center, Case Western Reserve University, Cleveland, OH 44106 [387]

JUDITH ILAN
Departments of Anatomy and Reproductive Biology, Laboratory of Developmental Genetics and Molecular Biology, The Cancer Center, Case Western Reserve University, Cleveland, OH 44106 [387]

LEE B. JACOBY
Neurosurgical Service, Molecular Neurogenetics Laboratory, Massachusetts General Hospital, Harvard Medical School, Charlestown, MA 02129 [209]

C. DAVID JAMES
Department of Neurosurgery, Emory University School of Medicine, Atlanta, GA 30322 [241]

GILBERT JAY
Laboratory of Virology, Jerome H. Holland Laboratory, American Red Cross, Rockville, MD 20855 [153]

THOMAS R. JOHNSON
Department of Anatomy, Laboratory of Developmental Genetics and Molecular Biology, Case Western Reserve University, Cleveland, OH 44106 [387]

PAMELA S. KARNES
Department of Molecular Genetics, Mayo Clinic, Rochester, MN 55905 [217]

ARTHUR D. LANDER
Department of Brain and Cognitive Sciences, Massachusetts Institute of Technology, Cambridge, MA 02139 [341]

ARNOLD J. LEVINE
Department of Molecular Biology, Lewis Thomas Laboratory, Princeton University, Princeton, NJ 08544-1014 [xv, 83,137,145]

M. BEATRIZ S. LOPES
Department of Pathology, Neuropathology Laboratories, University of Virginia School of Medicine, Health Sciences Center, Charlottesville, VA 22908 [1]

ROBERT L. MARTUZA
Department of Neurosurgery, Georgetown University Medical Center, Washington, DC 20007 [381]

RONALD D.G. MCKAY
Department of Brain and Cognitive Sciences, Massachusetts Institute of Technology, Cambridge, MA 02139 [37]

PAUL E. MCKEEVER
Department of Radiation Oncology, Division of Cancer Biology, Section of Neuropathology, University of Michigan Medical School, Ann Arbor, MI 48109 [163]

PATRICK W. MCLAUGHLIN
Department of Radiation Oncology, Division of Cancer Biology, University of Michigan Medical School, Ann Arbor, MI 48109 [163]

SHLOMO MELMED
Division of Endocrinology and Metabolism, Cedars-Sinai Medical Center, UCLA School of Medicine, Los Angeles, CA 90048 [255]

SARA E. MOLE
Cancer Research Campaign, Human Cancer Genetics Research Group, Department of Pathology, University of Cambridge, Cambridge, CB2 1QP, United Kingdom [195]

LOIS M. MULLIGAN
Cancer Research Campaign, Human Cancer Genetics Research Group, Department of Pathology, University of Cambridge, Cambridge, CB2 1QP, United Kingdom [195]

GARTH L. NICHOLSON
Department of Tumor Biology, University of Texas, M.D. Anderson Cancer Center, Houston, TX 77030 [371]

OLUFUNMILAYO I. OLOPADE
Department of Medicine, Section of Hematology/Oncology, Pritzker School of Medicine, University of Chicago Medical Center, Chicago, IL 60637 [179]

SARA PELEG
Section of Endocrinology, University of Texas, M.D. Anderson Cancer Center, Houston, TX 77030 [271]

MARY ELLEN PERRY
Department of Molecular Biology, Lewis Thomas Laboratory, Princeton University, Princeton, NJ 08544-1014 [83]

LARS POULSGÅRD
Department of Neurosurgery, Odense University Hospital, Odense, Denmark [223,249]

COREY RAFFEL
Department of Neurosurgery, Children's Hospital of Los Angeles, Los Angeles, CA 90054-0700 [217]

LEE RATNER
Divison of Hematology and Oncology, Departments of Medicine and Molecular Microbiology, Washington University School of Medicine, St. Louis, MO 63110 [329]

MARION F. ROBINSON
Baylor College of Medicine, Houston, TX 77030 [271]

JACK P. ROCK
Department of Neurosurgery, Henry Ford Neurological Institute, Henry Ford Hospital, Detroit, MI 48202 [265]

MOGENS RØNNE
Institute for Zoology and Zoophysiology, Zoological Laboratory, Aarhus University, Aarhus, Denmark [223,249]

JANET D. ROWLEY
Departments of Medicine and Molecular Genetics and Cellular Biology, Pritzker School of Medicine, University of Chicago Medical Center, Chicago, IL 60637 [179]

SUSAN D. RUDIN
Department of Anatomy, Laboratory of Developmental Genetics and Molecular Biology, Case Western Reserve University, Cleveland, OH 44106 [387]

H. EARL RULEY
Department of Microbiology and Immunology, Vanderbilt University School of Medicine, Nashville, TN 37232-2363 [89]

DARIO RUSCIANO
Friedrich Miescher Institute, Basel, Switzerland [357]

RANDI SCHEA
Department of Radiation Oncology, Division of Cancer Biology, University of Michigan Medical School, Ann Arbor, MI 48109 [163]

BERND W. SCHEITHAUER
Department of Laboratory Medicine and Pathology, The Mayo Clinic, Rochester, MN 55905 [1]

HENRY H. SCHMIDEK
Marion, MA 02738 [xv, 249,283]

BERND R. SEIZINGER
Oncology Drug Discovery, Bristol-Myers Squibb Pharmaceutical Research Institute, Princeton, NJ 08543-4000 [311]

M. PRISCILLA SHORT
Department of Neurology, Massachusetts General Hospital, Harvard Medical School, Charlestown, MA 02129 [303]

PETER A. STECK
Department of Tumor Biology, University of Texas, M.D. Anderson Cancer Center, Houston, TX 77030 [371]

JERZY TROJAN
Department of Anatomy, Laboratory of Developmental Genetics and Molecular Biology, Case Western Reserve University, Cleveland, OH 44106 [387]

MARK L. TYKOCINSKI
Department of Pathology, Case Western Reserve University, Cleveland, OH 44106 [387]

HIROYUKI UEDA
Laboratory of Virology, Jerome H. Holland Laboratory, American Red Cross, Rockville, MD 20855 [153]

SCOTT R. VANDENBERG
Department of Pathology, Neuropathology Laboratories, University of Virginia School of Medicine, Health Sciences Center, Charlottesville, VA 22908 [1]

TERRY A. VAN DYKE
Department of Biological Sciences, University of Pittsburgh, Pittsburgh, PA 15260 [287]

BELINDA J. WAGNER
Department of Biochemistry, Bowman-Gray School of Medicine, Winston-Salem, NC 27157 [101]

RAY WHITE
Department of Human Genetics, Eccles Institute of Human Genetics, Howard Hughes Medical Institute, University of Utah, Salt Lake City, UT 84112 [319]

Preface

This has been a remarkable time for fundamental research in oncology. The past twenty years have seen the discovery of those oncogenes and tumor suppressor genes that provide a new understanding of the origins and development of cancer. The diversity of mutations in these genes provides a new basis for studying the surprising diversity of different tumors, even of the same tissue types. It is now possible to make sense of the extraordinary karyotypic changes observed in tumor cells. The genes involved in the inherited predispositions to some cancers have been isolated, sequenced, and studied in great detail. For certain cancers, genetic testing and counseling are a reality. Having the persistence of cancer permits, for the first time, the development of a rational basis for pharmacologic or biologic treatment. The abnormal target proteins in cancer cells, at least in part, have been identified and are being used to develop drugs and other agents to selectively reverse or kill the tumorigenic cells. Further progress will require unraveling the functions of these oncogenes and tumor suppressor genes responsible for the origins of tumorigenesis and the progression of cancers within the nervous system.

The aim of this book is to bring together those threads which provide the intellectual underpinnings for a contemporary understanding of neural neoplasms. In order to accomplish this, material is presented from such diverse areas as neuropathology mapping and genetic engineering, which are the basis of the new concepts in oncology and which provide fresh insights into the pathogenesis, progression, and prognosis of neural tumors.

This volume results from the collaboration between a molecular biologist and a neurosurgeon and is specifically organized and written with the scientist and the clinician in mind. For molecular biologists the book provides an introduction to the categorization of these neoplasms and information about the pathogenesis and progression of some of these disorders. This material is presented to break down the barriers of specialized terminology and concepts which have evolved with the clinical interest in these tumors. Molecular biologists also will be exposed to the organism and its nervous system components, its development, and its responses to abnormal signals for cell proliferation that originate at the molecular level but express their phenotypes at the cellular, tissue, and organ levels.

For clinicians the last two decades has seen the convergence of major technological advances that have resulted in a quantum improvement in the diagnosis and treatment of patients with neoplasms affecting the nervous system. These advances, which have involved the development of sophisticated neurodiagnostic techniques and of microneurosurgery, allow for both the exquisite anatomic delineation of these tumors and their subsequent removal from almost anywhere within the human nervous system at an acceptable risk. However, although the outcomes in many benign tumors has improved dramatically, the treatment of malignant lesions of the nervous system remains grim. Further advances in diagnosis and treatment will be based on insights concerning the mechanisms involved in tumor initiation and progression.

This volume provides a path to understanding the ever-changing and informative world of the molecular basis of life. The results of studies reported in the chapters on the regulation of cell proliferation (oncogenes, cell cycle, tumor suppressor genes, DNA tumor viruses, etc.) are new, up-to-date, and organized in an informative way for the clinician and clinical investigator. Sections of the book also explore experimental neuro-oncogenesis and the impact of molecular genetic studies on a wide variety of specific types of primary and metastatic tumors of the nervous system.

The final section on the use of viral vectors or plasmids expressing anti-sense message brings together disparate fields of knowledge. With the use of engineered molecules to eradicate neural tumors we are entering the next major stage of diagnostic and therapeutic improvements in the treatment of neural tumors—the nascent field of "molecular neurosurgery."

We hope that the book will serve as a bridge between the scientist and the clinician and facilitate communication between these groups as well as inform the clinician concerning the latest advances in those areas of sciences impacting on our understanding of nervous system tumors.

Arnold J. Levine
Henry H. Schmidek

Chapter 1

The World Health Organization Classification of Nervous System Tumors in Experimental Neuro-Oncology

M. Beatriz S. Lopes, Scott R. VandenBerg, and Bernd W. Scheithauer

I. INTRODUCTION

During the past two decades, neuropathologists have acquired extensive information regarding the histogenesis of tumors of the nervous system, particularly their cytoskeletal and membrane proteins, growth factors, oncogenes, and growth kinetics. Increasing numbers of special morphologic procedures have contributed to our knowledge, especially immunohistochemistry and ultrastructural immunocytochemistry. Thus greater accuracy in the histopathologic diagnosis of tumors and the identification of a number of new tumor entities have changed the panorama of brain tumor pathology. In this context, a new classification of tumors of the nervous system was established in March 1990 by a consensus panel representing the World Health Organization (WHO) (Tables 1–7). The purpose of this chapter is to review current concepts of nervous system tumor classification in the context of experimental neuro-oncology.

II. GRADING MALIGNANT POTENTIAL OF NERVOUS SYSTEM TUMORS

The process of diagnosing brain tumors consists of imposing a static histopathological designation upon a variable and dynamic neoplastic process. A property of many brain tumors is transformation to more anaplastic forms. This process, although not unique to the central nervous system (CNS), is characteristic of gliomas. Another factor complicating the treatment of gliomas is their capacity to diffusely invade intact brain parenchyma, thus making complete surgical resection of all but ependymomas a virtual impossibility. Histopathologic criteria used to estimate the degree of anaplasia of various tumors of the nervous system are in large part similar to those employed for tumors at other sites. In grading neuroepithelial tumors, however, criteria are differently weighed, the same features not being of equal importance.

Diffuse astrocytic tumors (Table 1), the most common forms of glioma, comprise a spectrum from well-differentiated through anaplastic astrocytoma to glioblastoma. Dividing lines between these stages are not uniformly agreed upon and vary among laboratories. Clinical survival data, nonetheless, support the distinction of these three histopathologic groups. Features accompanying anaplastic transformation are generally agreed upon and include increasing cellularity, nuclear atypia, cellular pleomorphism, mitotic activity, endothelial or vascular proliferation, and necrosis (Davis, 1989). Daumas-Duport and co-workers (1988a) suggested the use of a four-tier numerical grading system for grading the diffuse infiltrative astrocytic tumors. Termed the St. Anne–Mayo System, it differs from previously published three-tiered systems in simply recording the presence of nuclear atypia, mitoses, endothelial proliferation, and necrosis without weighing the significance of particular combinations. According to this scheme, tumors with none of these features are grade 1, those with one are grade 2, those with two are grade 3, and those with three or four criteria are grade 4. The system is easily applied, reproducible, and shows close correlation with patient survival. In that the criteria make their appearance in a largely predictable order, grade 2 tumors exhibit nuclear atypia, grade 3 lesions show the addition of mitoses, and grade 4 tumors further show endothelial proliferation and/or necrosis. These three grades correspond roughly to astrocytoma, anaplastic or malignant astrocytoma, and glioblastoma of three-tier systems. The grade 1 astrocytoma of the St. Anne–Mayo scheme is so uncommon, representing but 1%, that for practical purposes,

Molecular Genetics of Nervous System Tumors, pages 1-36 ©1993 Wiley-Liss, Inc.

TABLE 1. Tumors of Neuroepithelial Tissue

Astrocytic tumors

1. Astrocytoma
   Variants: fibrillary, protoplasmic, gemistocytic, or mixed
2. Anaplastic (malignant) astrocytoma
3. Glioblastomas
   Variants: giant cell glioblastoma, gliosarcoma
4. Pilocytic astrocytoma
5. Pleomorphic xanthoastrocytoma
6. Subependymal giant cell astrocytoma (usually in association with tuberous sclerosis)

Oligodendroglial tumors

1. Oligodendroglioma
2. Anaplastic (malignant) oligodendroglioma

Ependymal tumors

1. Ependymoma
   Variants: cellular, papillary, epithelial, clear cell, mixed
2. Anaplastic (malignant) ependymoma
3. Myxopapillary ependymoma
4. Subependymoma

Mixed gliomas

1. Mixed oligoastrocytoma
2. Anaplastic (malignant) oligoastrocytoma
3. Others

Choroid plexus tumors

1. Choroid plexus papilloma
2. Choroid plexus carcinoma

Neuroepithelial tumors of uncertain origin

1. Astroblastoma
2. Polar spongioblastoma
3. Gliomatosis cerebri

it represents a three-tier system.

Criteria similar to those used in the grading of diffuse infiltrative astrocytomas are applied, albeit with less success, to oligodendrogliomas (Burger, 1989; Earnest et al., 1950; Mørk et al., 1986; Shaw et al., 1992; Smith et al., 1983). Prognostically precise "grading" on the basis of these features is problematic. Though univariate analysis shows nuclear atypia, a high mitotic rate, vascular proliferation, and necrosis to be statistically relevant, multivariate assessments demonstrate necrosis as the only independent prognostic feature.

For ependymal tumors, no precise combination of histopathologic features closely correlates with tumor recurrence or patient survival (Ross and Rubinstein, 1989). Nonetheless, from the histopathologic standpoint, anaplastic ependymomas are those tumors exhibiting high mitotic activity as well as endothelial proliferation. High cellularity and necrosis alone do not appear to have significant prognostic implications, since both may be present in otherwise obviously low-grade tumors.

The second most common group of CNS tumors, the meningiomas (Table 5), are divided into three histopathologic categories: meningioma, atypical meningioma, and anaplastic or malignant meningioma. Their distinction rests on cytologic and histologic features known to accompany aggressive or malignant behavior. In contrast to the gliomas, in which endothelial proliferation is an integral feature of anaplastic transformation, atypical meningiomas include cellular, cytologically atypical, and mitotically active tumors in which loss of meningothelial growth patterns with a shift to patternless "sheetlike" growth often accompanies foci of necrosis. Features of anaplastic transformation include frank cytologic malignancy, a high mitotic index, widespread necrosis, and the assumption of a papillary pattern. It is of note that brain invasion, often grossly apparent, is regarded as sufficient evidence for the designation *malignant meningioma* despite the occasional absence of other features of atypia or anaplasia.

In recent years the incorporation of tumor kinetics studies in the routine assessment of histopathological grade has demonstrated its utility in evaluating the proliferative potential of CNS tumors. The evaluation of cell cycle markers on routine specimens by histochemical methods, including the silver nucleolar organizer region or AgNOR technique, immunohistochemical methods, including Ki-67 and proliferating cell nuclear antigen (PCNA) and by *in vitro* and *in vivo* uptake of bromodeoxyuridine (BUdR), have been used both to grade neoplasms and to predict their biologic behavior. Although these techniques utilize different strategies, they apply the same principle—quantitative assessment of cell replication. In general, labeling indices based upon cell-cycle marker studies roughly correspond to histopathologic determinations of tumor anaplasia and to clinical assessments of aggressive tumor behavior (Burger et al., 1986; Giangaspero et al., 1987; Kajiwara et al., 1990; Louis et al., 1991; Plate et al., 1991). In combination with histopathologic grading, studies of tumor cell kinetics provide a valuable complementary parameter in the distinction of indolent from variably aggressive nervous system tumors. Nonetheless, a strict quantitative correlation between labeling indices and the biological behavior of individual tumors awaits further progress.

The most successful application of these techniques has been with the meningiomas, particularly for identifying those tumors that have a tendency to recur, the atypical meningioma group. Data derived from tumor cell proliferation studies, including the BUdR labeling index, Ki-67 immunoreactivity, and flow cytometric evaluation, have shown a clear association of these parameters with tumor recurrence (Cho et al., 1986; Lee et al., 1990; May et al., 1989). A labeling index (LI) of >1% appears to distinguish atypical from ordinary meningiomas (Jellinger, 1989; Lee et al., 1990). The AgNOR technique, in exhibiting somewhat of a correlation with labeling indices (Orita et al., 1990), also appears to distinguish tumors with a potential for recurrence, ones analogous to "atypical meningiomas" (Chin and Hinton, 1991).

Quantitative image processing, combined with either the *in situ* hybridization or immunohistochemical demonstration of abnormal growth factor receptors (Humphrey et al., 1990) or oncogene proteins (Salgaller et al., 1990), also promises to be an adjunct to routine histologic studies to identify and characterize anaplastic neuroepithelial tumors. In one report of cerebral anaplastic gliomas, metastasis beyond the nervous system was selectively accompanied by immunohistochemical detection of a ras protein epitope (Leifer et al., 1989).

III. TUMORS OF NEUROEPITHELIAL TISSUE

A. Astrocytic Tumors

The astrocytic tumors are divided into two major categories (Table 1). The first, the prognostically unfavorable diffuse tumors, comprises nearly 75% of the astrocytic tumors (Russell and Rubinstein, 1989). The principal groups are astrocytoma, anaplastic astrocytoma, and glioblastoma, all within a continuous spectrum exhibiting increasing anaplasia. Intratumoral heterogeneity, both morphologic and genetic, in diffuse infiltrative astrocytomas, increases with progressive "anaplastic transformation" (Bigner et al., 1981; Jenkins et al., 1989; Shapiro et al., 1981; Shapiro and Shapiro, 1984). This phenomenon makes histopathologic classification and grading of diffuse astrocytic tumors problematic, a process frustrated by tumor sampling. This is particularly the case in limited biopsies obtained by needle aspiration or stereotactic biopsy.

The second major group of astrocytic tumors, the prognostically favorable varieties, include pilocytic astrocytoma, pleomorphic xanthoastrocytoma, and subependymal giant cell astrocytoma. They exhibit distinctive clinicopathologic features that justify their separate classification. Despite significant histopathologic differences, all three are, in radiographic or gross terms, macroscopically circumscribed neoplasms when compared to astrocytic tumors of the diffuse type. In further distinction, these tumors exhibit slow growth and relatively indolent biologic behavior. Significant indicators of anaplasia in more ordinary gliomas, e.g., cytologic atypia, occasional mitoses, vascular proliferation, even necrosis, though not commonly present, do not necessarily indicate aggressive behavior or malignancy. As a result, the application of conventional grading schemes is inappropriate in that they grossly overestimate aggressive behavior. The clinical prognosis is usually favorable, even with surgical resection alone. Anaplastic progression is rare. To those with a bent toward "ontogenic" or "cell-of-origin" thinking, the distinctive histopathologic features of the favorable prognostic variants, as well as their proclivity for certain locations in the brain, suggest that these tumors may arise from transformation of astrocytes that may have regionally specific histogenesis.

A major goal of contemporary neuropathology is the identification of neurochemical properties that, perhaps in addition to morphology, might serve to identify tumors likely to behave aggressively or to undergo further anaplastic transformation. With reference to astrocytic tumors, particularly anaplastic variants, the abnormal expression of growth factor receptors provides interesting data complementary to that providing morphologic classification. For instance, the epidermal growth factor (EGF) receptor gene is found to be amplified only in high-grade tumors, mainly in primary glioblastomas, as well as in tumor lines (Bigner et al., 1987; Burgart et al., 1991; Humphrey et al., 1990; James et al., 1988; Jones et al., 1990; Sugawa et al., 1990). Furthermore, abnormal expression of growth factor ligands, including transforming growth factors (especially TGFβ), platelet-derived growth factors (PDGF), and basic fibroblast growth factor have also been shown to be associated with anaplastic astrocytic tumors (Betsholtz et al., 1989; Ekstrand et al., 1991; Mapstone et al., 1991; Takahashi et al., 1990). However, the functional implications of ectopic or aberrant expression in either low-grade or high-grade tumors, as related to prognostic factors, are yet to be clarified.

Current genetic studies also offer insight into the biology of gliomas. Allelotypic analysis by determination of restriction fragment length polymorphism may expand the histopathologic spectrum of glioblastoma and anaplastic astrocytoma (Fults et al., 1990). Whereas there is considerable loss of heterozygosity in many autosomes, loss of heterozygosity for loci on chromosomes 10 and 17 (17p) is relatively frequent and changes on chromosome 10 appear to be more selective for glioblastomas as compared to less anaplastic astrocytomas (Bigner et al., 1990; Fults et al., 1990; James et al., 1988; Wanatabe et al., 1990). Loss of heterozygosity for loci on chromosome 17 appears to occur in diffuse forms of astrocytoma regardless of their degree of anaplasia (Bigner et al., 1988; James et al., 1988).

1. Astrocytoma

Diffuse astrocytoma (Fig. 1) commonly appear in the third or fourth decade of a patient's life, anaplastic tumors in the fourth or fifth decades. In general, men are more often affected than women (Russell and Rubinstein, 1989).

The subdivision of astrocytomas into fibrillary, gemistocytic, and protoplasmic types is based upon their similarities to the morphology of normally occurring and reactive astrocytes. In reality, the majority of the tumors are fibrillary, but mixed lesions, particularly ones with gemistocytic elements, are common. Subclassifying astrocytomas on the basis of such cytologic characteristics has significant limitations in that these features are simply a reflection of morphologic heterogeneity and do little to

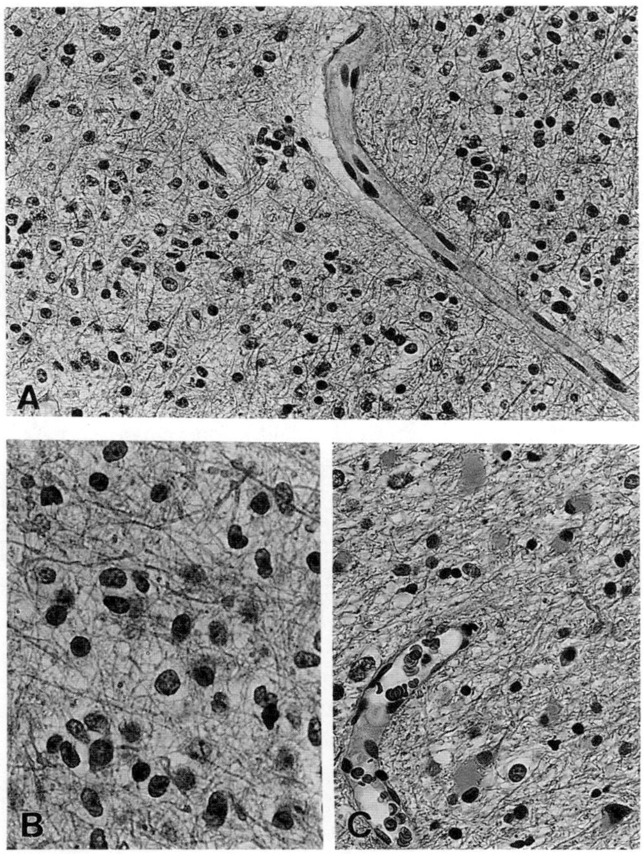

Fig. 1. Astrocytoma: Diffuse infiltrating low-grade astrocytomas demonstrate low cellularity and rather inconspicuous cellular pleomorphism in a fibrillary matrix (**A,B**). Delicate vessels are present (**A**). The majority of the tumors are fibrillary, but mixed population of astrocytes with gemistocytic elements are common. (H & E; A: ×200; B: ×400; C: ×200.)

forward our understanding of the pathobiology of neoplastic astrocytes (Denis-Donini et al., 1984; Hansson, 1988).

The *fibrillary astrocytoma* is the most frequently occurring variant of diffuse astrocytoma. Commonly, it involves the cerebral hemispheres of adults and the brain stem in children and adolescents. In macroscopic terms it is difficult to distinguish infiltrative astrocytoma from adjacent invaded brain parenchyma. Nevertheless, such tumors impart a firm texture and are often yellow-tinged. Classic features of astrocytoma cells include the presence of cytoplasmic processes, which are the basis of the fibrillary appearance seen in histologic preparations. Histochemical and immunohistochemical methods highlight the affinity of glial fibrils for phosphotungstic acid hematoxylin (PTAH) and confirm their composition of glial fibrillary acidic protein (GFAP), respectively (Cosgrove et al., 1989; Perentes and Rubinstein, 1987).

Gemistocytic astrocytomas are uncommon, representing no more than 20% of astrocytomas. As a pure histologic variant, arbitrarily defined as a tumor composed of greater than 60% gemistocytes, this lesion is largely confined to the cerebral hemispheres (Krouwer et al., 1991). Gemistocytic astrocytes are characterized by their round to slightly angulated shape, abundant, eosinophilic cytoplasm, eccentric, round nuclei, and an intercellular "matrix" not conspicuously fibrillated. The cells are immunoreactive for GFAP, in either a diffuse or peripheral cytoplasmic staining pattern, a feature consistent with the ultrastructural finding of cytoplasmic intermediate filaments dispersed either among organelles or beneath the plasmalemma (Russell and Rubinstein, 1989). Astrocytic tumors composed of greater than 20% gemistocytic cells appear to have a poor prognosis, clinically behaving more like anaplastic astrocytomas than fibrillary tumors of similar grade (Krouwer et al., 1991).

Protoplasmic astrocytomas in pure form are rare, comprising approximately 1% of diffuse astrocytomas (Russell and Rubinstein, 1989). Controversy exists regarding the nature of protoplasmic astrocytomas, in that virtually identical morphology may be seen in association with pilocytic astrocytoma and oligodendroglioma. Most occur in the cerebrum of children and young adults. The tumor is composed of a relatively homogeneous population of stellate cells lying within an abundant eosinophilic matrix characterized by microcystic degeneration. In that the delicate cytoplasmic processes lack PTAH positivity, GFAP immunoreactivity is correspondingly scant (Russell and Rubinstein, 1989).

2. *Anaplastic astrocytomas*

Like the diffuse astrocytoma noted above, anaplastic astrocytoma exhibits similar morphologic heterogeneity, usually in association with higher cellularity. Mitotic figures are a regular feature. All types of diffusely infiltrating astrocytic tumors appear capable of anaplastic transformation. Although the incidence of such progression may be as high as 80% (Russell and Rubinstein, 1989; Scherer, 1940), its latency and rapidity vary considerably. As a result, prognostication in individual cases of astrocytoma is not possible. Available data suggest that evidence of progression to higher grades is to be expected in the majority of recurrent astrocytomas (Russell and Rubinstein, 1989). In general, anaplastic tumors are less densely fibrillar and demonstrate increased cellularity, varying degrees of cytoplasmic/nuclear pleomorphism, and increased mitotic activity. Endothelial/pericytic proliferation is not well represented in these tumors; instead it accompanies necrosis in many glioblastomas.

3. *Glioblastoma*

Glioblastomas predominantly affect the cerebral hemispheres. They have a peak incidence in the fifth and sixth

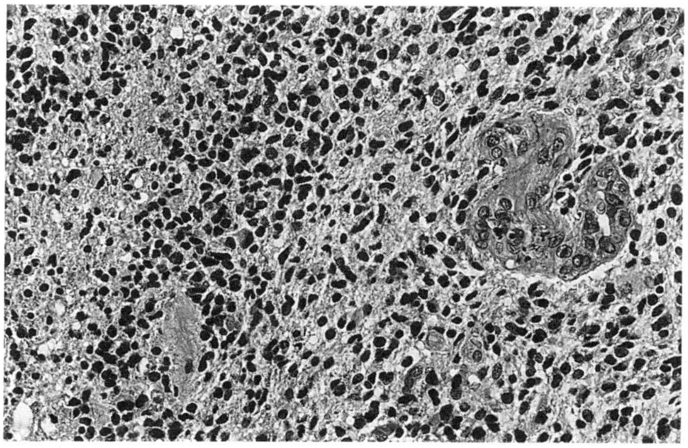

Fig. 2. Glioblastoma multiforme: Extensive cytoplasmic and nuclear pleomorphism, high mitotic activity, palisading necrosis, and exuberant endothelial proliferation constitute the histopathologic picture of glioblastomas. (H & E; ×200.)

decades, and are the most common intracranial tumor in this age group. Overall, this tumor type represents 15 to 20% of intracranial tumors and 50% of gliomas (Russell and Rubinstein, 1989). The majority have their origin in astrocytomas and anaplastic astrocytomas that undergo progression (Dropcho et al., 1987; Laws et al., 1984; Muller et al., 1977). Far less often, glioblastomas arise *de novo*. Pediatric cases, though infrequent, must be differentiated from embryonal tumors, particularly those of small cell type (primitive neuroectodermal tumors) (see below).

Glioblastomas are most often seen as expansive lesions characterized by "ring enhancement" on computed tomographic (CT) and magnetic resonance (MR) images. Enhancement correlates with either (or both) high cellularity or microvascular proliferation (Earnest et al., 1988), whereas the hypodense center corresponds to necrosis (Fig. 2). Despite their apparent radiographic and gross circumscription, glioblastomas invariably infiltrate adjacent brain parenchyma, often crossing commissures to infiltrate the opposite cerebral hemisphere ("butterfly" pattern). Their malignant nature is underscored by the fact that glioblastomas exhibit the highest incidence of systemic metastasis of any glial neoplasm (Liwnicz and Rubinstein, 1979). Histologically, they exhibit marked cytologic diversity, ranging from small-cell tumors with scant cytoplasm to ones composed of multinucleate giant cells. Mixed patterns are the rule. Mitoses, including atypical forms, are frequent but vary significantly in number within individual tumors. Exuberant vascular proliferation, often endothelial and pericytic, frequently accompanies necrosis. In the St. Anne–Mayo grading scheme, either feature confirms the diagnosis of glioblastoma.

Due to their unique features, two distinct subvariants of glioblastomas are recognized in the WHO classification. The first, *giant cell glioblastoma*, is characterized by a predominance of enormous, bizarre, multinucleated giant cells. Intercellular reticulin fibers, a not invariable feature, are often conspicuously related to blood vessels and areas of necrosis. Although the clinical features of giant cell glioblastoma are not unique, they are however remarkably circumscribed, both neuroimaging and at surgery. In that some are nearly totally resectable, the median survival of patients with this tumor exceeds that associated with glioblastomas (Russell and Rubinstein, 1989; Margetts and Kalyan-Raman, 1989).

The second, *gliosarcoma*, represents a variant of glioblastoma in which cellular elements of the vasculature are presumed to undergo sarcomatous transformation. The estimated frequency of sarcomatous change in glioblastomas ranges from 2 to 8% (Meis et al., 1991; Morantz et al., 1976). Interestingly, the proportions of glial and sarcomatous components are quite variable; in some instances the sarcomatous elements literally "overrun" the parent glioma. Sarcomatous areas are highlighted by reticulin or collagen stains, whereas S-100 protein and GFAP immunostains confirm the nature of the glial component. The sarcomatous portion varies in histologic appearance, though in most cases it resembles either malignant fibrous histiocytoma or fibrosarcoma. Immunohistochemical and ultrastructural studies suggest that the sarcoma probably arises from undifferentiated mesenchymal cells of the vascular adventitia, cells capable of differentiating along endothelial, smooth muscle, and pericytic cell lines (Grant et al., 1989; Haddad et al., 1991; Ho, 1990; Miller et al., 1991; Ng and Poon, 1990). Osseous and cartilaginous differentiation is uncommon. Rhabdomyoblastic components are rare (Barnard et al., 1986).

4. Pilocytic astrocytomas

Pilocytic astrocytomas usually occur in children and young adults. Most arise in the cerebellum, the region of the third ventricle, chiasm and optic nerves, thalamus, and, less frequently, the cerebrum (Burger et al., 1991; Russell and Rubinstein, 1989; Scheithauer et al., 1991). Many but not all appear circumscribed. Cystic lesions with a solid component or "mural nodule" are frequent.

Histologically, the archetypic pilocytic astrocytoma exhibits a biphasic growth pattern, one consisting of stellate cells with a sparsely fibrillary cytoarchitecture resembling "protoplasmic" astrocytes, the other of elongate, bipolar, highly fibrillated astrocytes (Fig. 3). The latter, termed *piloid* cells, are compactly arranged and often associated with Rosenthal fibers, whereas the stellate cells form a looseknit or "cobweb"-textured tissue in which eosinophilic protein droplets or "granular bodies" are numerous. Vessels are generally abundant, may show focal endothelial proliferation, and are often hyalinized. Degenerative atypia and multinucleated cells are common features, as is glomeruloid

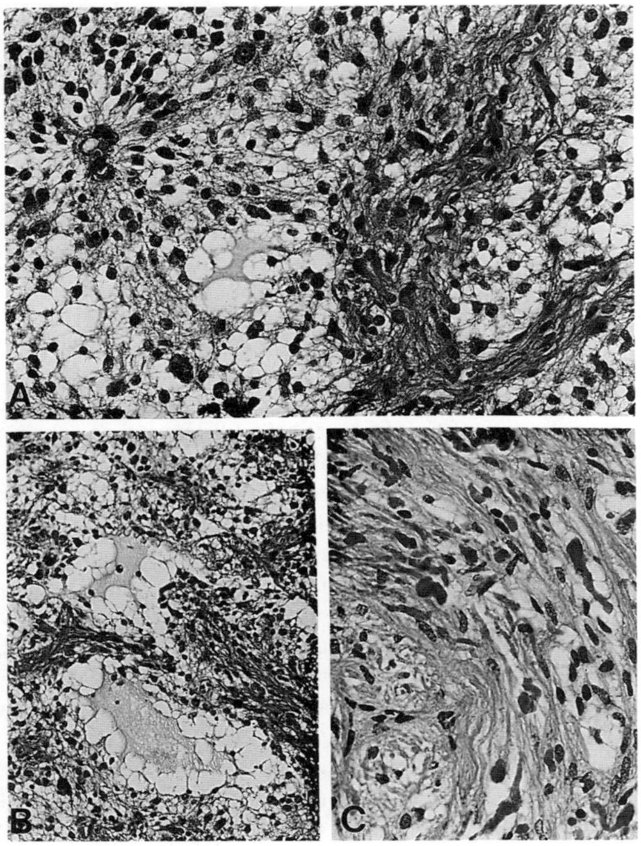

Fig. 3. Pilocytic astrocytoma: The biphasic growth arrangement of pilocytic astrocytomas is appreciated in this cerebellar example (**A–C**). Microcystic degeneration (**B**) and Rosenthal fibers (**C**) are commonly present in these tumors. (H & E; A: ×100; B,C: ×200.)

microvascular atypia and proliferation. Mitoses are absent or rare. None of these features indicates anaplasia or malignant clinical behavior. Brisk mitotic activity, extensive endothelial proliferation, and necrosis are not regular features of pilocytic astrocytoma and are associated with anaplastic progression, a rare event. Despite macroscopic circumscription, pilocytic astrocytomas not only infiltrate surrounding tissue, particularly white matter, but may extend into the leptomeninges. Such invasion does not indicate anaplastic or malignant progression, but may account for the rare observation of distant "seeding" of tumors with no histologic evidence of malignancy (Obana et al., 1991).

5. Pleomorphic xanthoastrocytoma

Relatively recently described (Kepes et al., 1979), pleomorphic xanthoastrocytoma (PXA) is a superficially situated cerebral tumor that most often involves the temporal lobe. Most affect adolescents and are seizure-associated (Kawano, 1991; Pasquier et al., 1985). In many instances, the solid portion of the tumor forms a nodule or mass involving the leptomeninges. The dura is typically uninvolved (Kawano, 1991; Kepes et al., 1979; Strom and Skullerud, 1983). Although a well-defined macroscopic border with the subjacent brain is identified, portions of the tumor do show parenchymal infiltration. In many cases the tumor is accompanied by an underlying cyst.

Cellular heterogeneity is the hallmark of PXA, a tumor composed of spindle, plump, and polygonal astrocytes arranged in interlacing bundles and with interspersed pleomorphic, often lipid-vacuolated, cells (Fig. 4). Bizarre hyperchromatic nuclei as well as multinucleated giant cells are invariably present. Cellularity is usually moderate but may be focally high. Foci of chronic inflammation are common. Eosinophilic protein droplets or "granular bodies" as well as occasional Rosenthal fibers may also be seen. Mitoses are absent or in low numbers. A very cellular low growth fraction has been confirmed by the cytofluorometric finding of an insignificant 0.25% of cells in S-phase analysis (Hosokawa et al., 1991). Cytoplasmic lipidization, usually

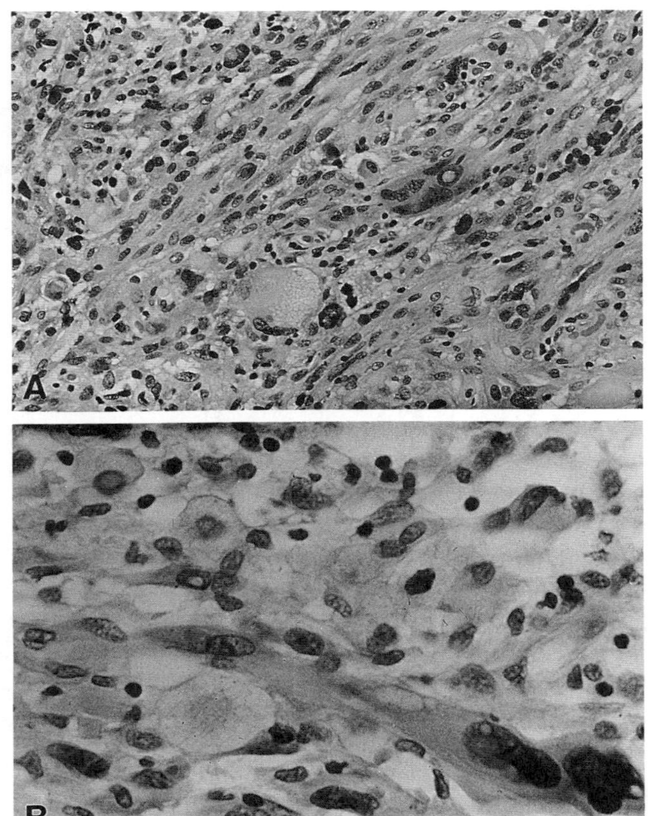

Fig. 4. Pleomorphic xanthoastrocytoma: A large variation of cytoarchitecture is a hallmark of the PXA. It is associated with moderate cellular pleomorphism and xanthomatous changes (**A,B**). The absence of necrosis and endothelial proliferation distinguish these tumors from glioblastomas with lipidized cells. (H & E; A:×200 B: ×400.)

of polygonal or giant cells, may be prominent but varies considerably among different tumors.

A remarkable feature of PXA tumors is the presence of intercellular reticulin staining in a pattern delineating either individual cells or fascicles of cells. Although this apparent stroma may be present in any portion of the tumor, it is most conspicuous in areas of leptomeningeal involvement and about the often prominent vasculature. Ultrastructural studies show little in the way of a collagenous stroma but rather the presence of basal lamina around cells (Kepes et al., 1989; Weldon-Linne et al., 1983). Since this is a feature shared with normal subpial astrocytes, a histogenic relationship between pleomorphic xanthoastrocytomas and this special astrocytic phenotype has been suggested (Russell and Rubinstein, 1989; Whittle et al., 1989).

6. Subependymal giant cell astrocytoma

The subependymal giant cell astrocytoma (SEGA) usually occurs in the first two decades of life, often (if not invariably) in association with the tuberous sclerosis complex (Shepherd et al., 1991). Symptoms referable to the tumor may be the first manifestations of the disease, but in many instances, seizures related to tubers are of long standing. Similar tumors may also occur without apparent signs of the phakomatosis. As a rule, SEGA are frequently nodular, solid tumors that are well delineated from underlying brain parenchyma (Russell and Rubinstein, 1989). They commonly arise from the wall of a lateral ventricle, often overlying the basal ganglia. Less commonly, they involve the third ventricle.

Histologically, SEGAs are composed of spindle and epithelial cells of astrocytic appearance that sweep from vessel walls in a manner reminiscent of ependymal pseudorosettes. However, the spindle cells commonly have broad cytoplasmic extensions, a feature not shared with typical ependymomas. Occasional large polygonal cells resemble ganglion cells (Fig. 5). Vesicular nuclei, prominent nucleoli, and amphophilic cytoplasm are common features. Some nuclear pleomorphism, occasional multinucleate cells, and some mitoses may be present but are not indicative of anaplasia (Shepherd et al., 1991). Despite a few examples of recurrent tumor, no cases of SEGA with malignant transformation have been described.

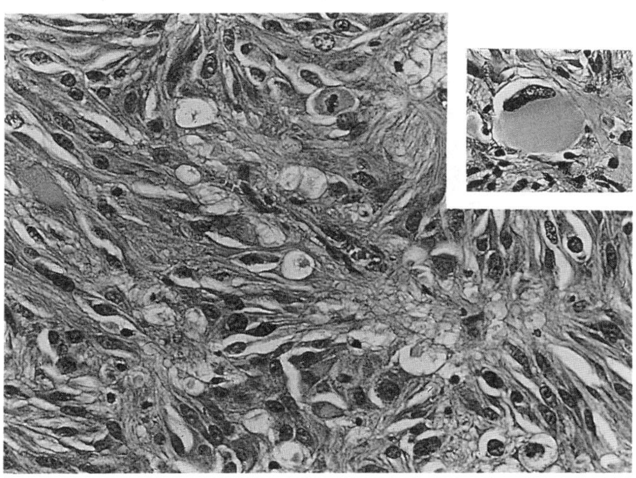

Fig. 5. Subependymal giant cell astrocytoma: Cellular heterogeneity, with extensive variation in cellular geometry, is also typical for SEGA. Ganglionic-like polygonal cells are commonly present (**inset**). (H & E; ×200.)

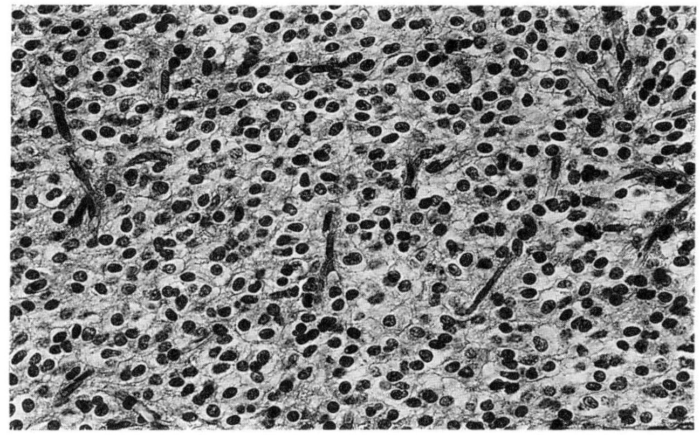

Fig. 6. Oligodendroglioma: Oligodendrogliomas are characterized by typically uniform cell population as demonstrated here. There are increased numbers of delicate blood vessels within the tumor. (H & E; ×200.)

The histogenesis and differentiation of SEGA are unclear. Astroglial, mixed glial–neuronal, and ependymal origins or differentiation have been suggested. The astrocytic nature of the tumor cells is demonstrated by both S-100 protein and GFAP immunoreactivity, as well as by ultrastructural studies (De Chadarévian and Hollenberg, 1979; Nakamura and Becker, 1983; Russell and Rubinstein, 1989; Sima and Robertson, 1979; Trombley and Mirra, 1981). Nevertheless, a significant population of tumor cells that are not immunoreactive for GFAP is present in many tumors, particularly in patients with tuberous sclerosis. Ultrastructural features suggesting neuronal differentiation have been reported in a small number of cases (Nakamura and Becker, 1983). Evidence for an ependymal origin includes single reports of tumors with 9 + 2 cilia associated with cytoplasmic lumina (Bancel et al., 1990) and of perivascular rosettes with blepharoplasts (Halmagyi et al., 1979).

B. Oligodendroglial Tumors

1. Oligodendroglioma

Tumors composed predominantly of oligodendroglia

represent 5 to 15% of intracranial gliomas and have a peak incidence in the fourth and fifth decades (Mørk et al., 1985). Oligodendrogliomas are more common in the frontotemporal region but arise throughout the neuraxis in relative proportion to the volume of white matter.

Oligodendrogliomas vary in cytologic appearance and exhibit a variety of histologic patterns. The most constant arrangement is the "pseudolobular" pattern, resulting from a conspicuous network of delicate capillaries encircling varying size aggregates of tumor cells. Reactive astrocytes are often found scattered throughout the tumor, particularly around blood vessels (Fig. 6). Minimal degrees of endothelial proliferation and occasional mitotic figures may be present.

Specific immunocytochemical markers of neoplastic oligodendrocytes are not presently available for routine use. The majority of oligodendrogliomas demonstrate immunoreactivity for Leu-7 antigen (Perentes and Rubinstein, 1986); however, this surface antigen is not specific for oligodendrogliomas, in that it is also expressed by astrocytomas, glioblastomas, and ependymomas (Perentes and Rubinstein, 1986). Myelin basic protein (MBP) and myelin-associated glycoprotein (MAG) immunoreactivity is present in the minority of oligodendrogliomas (Nakagawa et al., 1986) and do not discriminate these tumors from other gliomas (Figols et al., 1985). Neoplastic oligodendrocytes with a prominent eosinophilic skirt of eccentric cytoplasm, the so-called "gliofibrillary oligodendrocytes" (Herpers and Budka, 1984), demonstrate intense GFAP immunoreactivity. Such cells are reportedly present in about 50% of the oligodendrogliomas (Herpers and Budka, 1984; Nakagawa et al., 1986). Immunoreactivity for GFAP in neoplastic oligodendroglia has been said to recapitulate the ontogeny of normal oligodendrocytes. GFAP is transiently expressed in myelin-forming oligodendrocytes early in the development (Choi and Kim, 1985). Experimental studies in rodent suggest that type-2 astrocytes and oligodendroglia share a progenitor cell (O-2A) that can be immunohistochemically identified by an antiglycolipid antigen, the A2B5 (Raff et al., 1983). A study of 28 oligodendrogliomas has demonstrated that a high proportion contain A2B5 immunoreactive cells (de la Monte, 1989). Therefore, a proportion of oligodendroglial tumors may contain variable numbers of the "A2B5+" progenitor cells. The GFAP immunoreactive phenotypes in these tumors could originate from astrocytic type-2 differentiation. GFAP immunoreactivity in oligodendrogliomas increases as the tumors demonstrate more anaplastic features (Kros et al., 1990). However, the histogenesis of these GFAP immunoreactive cells remains unclear. They may arise from the neoplastic "gliofibrillary oligodendrocytic" phenotype or by astrocytic (type-2) evolution from the O-2A-like progenitor population (Kros et al., 1992). The distinction of this GFAP-immunoreactive oligodendroglia population from a true astrocytic component of a mixed oligoastrocytoma can be, however, problematic.

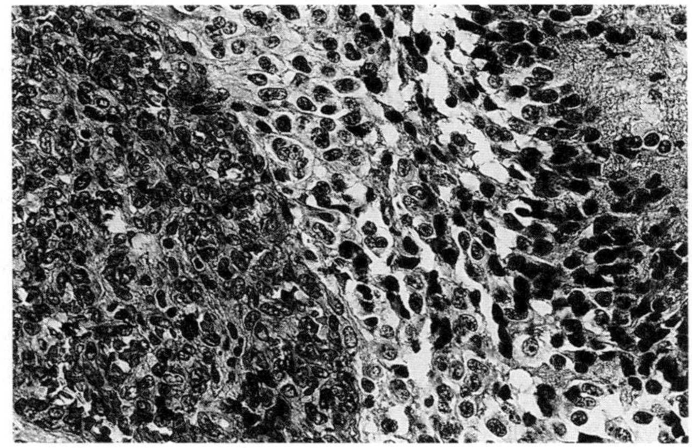

Fig. 7. Anaplastic oligodendroglioma: Palisading necrosis and endothelial/pericytic proliferation in conjunction with cellular pleomorphism and high mitotic activity are indicative features of anaplastic progression of oligodendrogliomas. The cellular elements, however, retain oligodendroglial cytologic features. (H & E; ×200.)

2. Anaplastic oligodendrogliomas

Oligodendrogliomas demonstrating increased high nuclear/cytoplasmic ratios, nuclear pleomorphism, markedly increased mitotic activity, and prominent endothelial proliferation are designated anaplastic. Palisading necrosis may also be seen (Fig. 7). These tumors are generally rapidly growing. As noted above, GFAP-immunoreactive oligodendrocytes are frequently present. Indeed, in one series such cells have been demonstrated in all anaplastic oligodendrogliomas (Wondrusch et al., 1991). Progression to glioblastoma is a recognized event, yet its distinction from particularly anaplastic oligodendrogliomas is not clear.

C. Ependymal Tumors

1. Ependymoma

Ependymomas represent approximately 6% of intracranial gliomas (Russell and Rubinstein, 1989). Although they occur over a wide range of patient age, most appear in childhood or adolescence, wherein they constitute almost 8% of intracranial gliomas. Ependymomas arise in relation to the ventricular system, the fourth ventricle being the most frequent site. In contrast, ependymomas in adults usually involve the spinal cord or filum terminale (Slooff et al., 1964). Multiple tumors are usually associated with von Recklinghausen's disease (Russell and Rubinstein, 1989).

In all locations, ependymomas are well defined with regard to surrounding parenchyma and jut into the adjacent ventricle. The tendency of ependymal cells to polarize and to form rosettes is evident even in culture (Russell and

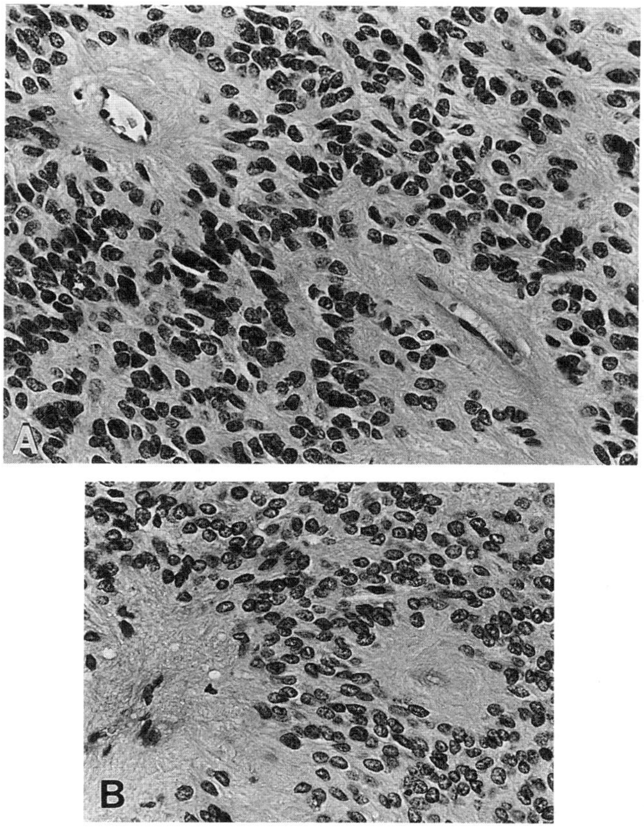

Fig. 8. Ependymoma: The tendency of the tumor cells to form pseudorosettes and ependymal rosettes, in a variably fibrillary and cellular background, is a hallmark of ependymomas. (H & E; A-B: ×200.)

Rubinstein, 1989). "True" ependymal rosettes encircle an extracellular space or lumen ranging from submicroscopic clefts to obvious tubules. Pseudorosettes, in contrast, form around blood vessels, and are numerically far more frequent (Fig. 8).

The current classification recognizes three histologic variants of ependymomas: *cellular*, *papillary*, and *clear cell*. Although these subtypes are of no prognostic importance, their recognition is essential when considering differential diagnoses. In the *cellular* ependymoma, the tumor cells form few rosetting structures and are disposed in rather patternless, homogeneous sheets. The *papillary* variant is more common in the cauda equina region, but when located in the ventricles, it may mimic choroid plexus neoplasia. The *clear cell* variant is of diagnostic importance in that it may be mistaken for oligodendroglioma. The correct diagnosis is suggested by the solid nature of the tumor, its circumscriptions, the presence, at least focally, of vague perivascular pseudorosettes, and its ultrastructural features of ependymoma. A small minority of ependymomas undergo craniospinal dissemination. Systemic metastases are rare (Russell and Rubinstein, 1989).

2. Anaplastic ependymoma

Anaplastic ependymomas are histologically recognized by their high mitotic activity index and, to some extent, by conspicuous endothelial proliferation. Nuclear atypia and necrosis, alone, are less reliable indicators of malignancy. The prognosis of anaplastic ependymomas is highly variable, and correlation of the clinical outcome and histopathological features are not well defined. In general, patients with high-grade tumors have a less favorable prognosis than those with low-grade tumors, the survivals being 30% and 70% at five years, respectively (Slooff et al., 1964).

A further differential to be made is between anaplastic ependymomas and ependymoblastoma. The latter constitutes a highly malignant embryonal tumor involving infants or children under the age of five years. The tumor is primarily composed of amorphous collections of poorly differentiated cells that are punctuated by remarkably well-formed true rosettes (see Embryonal Tumors, below).

3. Myxopapillary ependymoma

Myxopapillary ependymomas are a distinct variant of ependymomas, one nearly limited to the region of the cauda equina/filum terminale (Morantz et al., 1979; Pulitzer et al., 1988). They occur primarily in adults (third or fourth decades), and typically appear as sausage-shaped masses replacing the filum terminale and compressing or enveloping spinal nerve root of the cauda equina. Myxopapillary ependymomas are slow-growing. The clinical outcome depends on the extent of local involvement and of surgical removal. Although recurrence is seen in 10 to 20% of the cases, resection of intact tumors confined to the filum is curative (Sonneland et al., 1985).

Histologically, myxopapillary ependymomas are composed of cuboidal or columnar cells to spindle-shaped cells in papillary array surrounding vessels with myxoid or hyalinized stroma (Fig. 9). The hyalinized matrix stains positively for mucin. Other characteristic features are so-called "balloons," spherical pockets of periodic acid–Schiff (PAS)-positive extracellular matrix, and collagen surrounded by tumor cells. Ultrastructural findings confirm the ependymal character of these neoplasms, in that they possess a well-formed basal lamina. Depending upon whether the tumor consists primarily of epithelial or of glial-appearing cells, true lamina, cilia, microvilli, and cellular interdigitations vary in number (Rawlinson et al., 1973; Specht et al., 1986).

4. Subependymoma

Subependymomas are usually asymptomatic lesions located in the lateral and fourth ventricles, and not uncom-

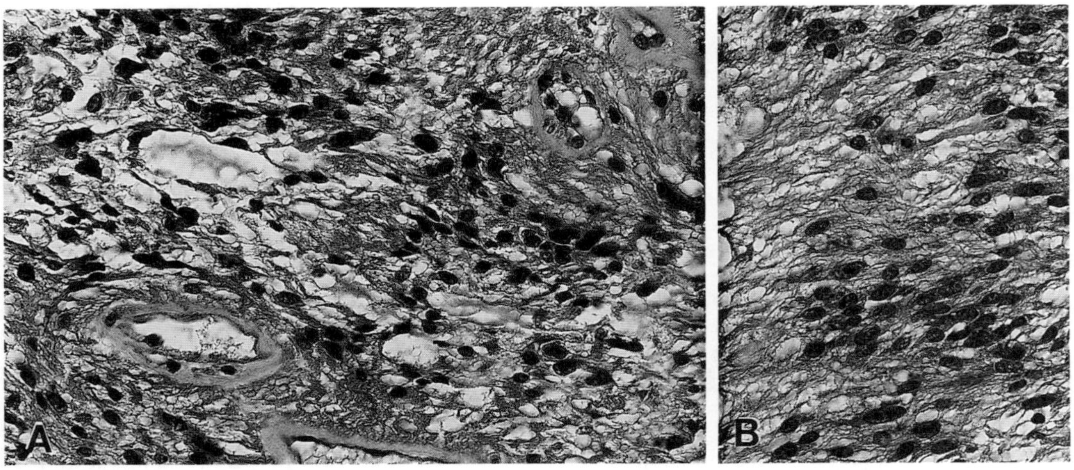

Fig. 9. Myxopapillary ependymoma: In this special variant of ependymomas, palisading arrangements of cells with long, fibrillated processes are arranged around hyalinized blood vessels. (H & E; A-B: ×200.)

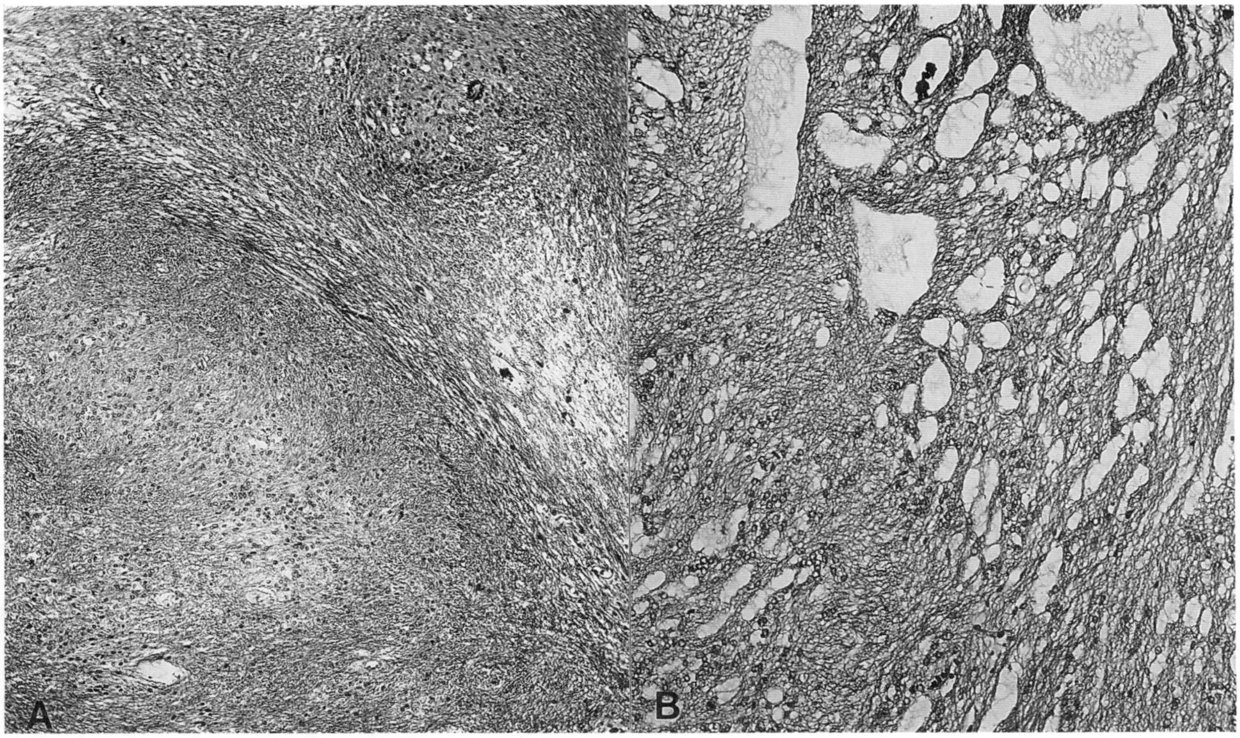

Fig. 10. Subependymoma: Subependymomas are nodular at low power and consist of aggregates of uniform ependymal and astrocytic cells surrounded by skeins of fibrillar processes (**A**). Microcystic formation is a common feature (**B**). (H & E; A: ×100; B: ×250.)

monly are incidental findings at autopsy. Grossly they are sharply demarcated at the ventricular surface and jut into the ventricles. Symptomatic subependymomas are far less common (Lombardi et al., 1991; Scheithauer, 1978). Extensive growth of these firm tumors may obstruct the flow of cerebrospinal fluid (CSF), with resultant increased intracranial pressure and hydrocephalus. Although the fourth ventricle is the most common site, accounting for 70% in one series (Scheithauer, 1978), other favored sites include the septum pellucidum, lateral ventricles, and foramen of Monro. Spinal cord examples are rare (Pagni et al., 1992). Resection is curative.

Subependymomas are hypocellular and are composed of a dense fibrillary glial stroma sweeping around small groups of uniform ependymal and/or astrocytic-appearing cells (Fig. 10). Such cells are usually arranged randomly but on occasion form ill-defined rosettes. Microcystic degeneration and calcification are common. The transitional astrocytic ependymal nature of subependymoma cells has been confirmed by electron microscopic and tissue culture studies (Azzarelli et al., 1977; Fu et al., 1974).

D. Mixed Gliomas

Tumors composed of more than one distinct glial cell type, e.g., astrocytes, oligodendroglia, and, less frequently, ependymal cells, are not rare. Recognition and proper classification of lesions are complicated by the fact that their constituents vary in proportion and in distribution. Only a minority are made up equally of distinct homogeneous cell populations. Ample sampling of mixed gliomas is therefore necessary. The apparent capacity of glial cells of one type to acquire phenotypic characteristics of another further complicates the matter, e.g., oligoastrocytomas (see below). Another consideration regarding mixed gliomas relates to the capabilities of different glial cell populations to undergo anaplastic transformation. For example, in oligoastrocytomas, criteria for anaplasia are not the same in both cell populations, i.e., as noted above, astrocytomas are more likely to undergo anaplastic transformation than oligodendrogliomas.

1. Mixed oligoastrocytoma

These mixed tumors are composed of oligodendrocytes and a significant, although not well-defined, population of neoplastic astrocytes that may be either geographically distinct or diffusely distributed (Fig. 11). Frequently, the prevalent cell component remains oligodendroglial (Rubinstein, 1972). The mixed tumors referred to are explicitly different from the approximately 60% of the oligodendrogliomas that contain conspicuous reactive astrocytes (Russell and Rubinstein, 1989). Furthermore, GFAP expression in the "gliofibrillary" oligodendrocytes previously referred to is also not to be interpreted as an astrocytic component of a mixed glioma. The origin and histogenesis of oligoastrocytomas are unclear, but a derivation from cells "transitional" between mature oligodendrocytes and astrocytes has been suggested (Herpers and Budka, 1984). The occurrence of bipotential cells associated with central gliogenesis in the rodent provides further, albeit indirect, evidence for the role of transitional cells in the genesis of these tumors (Miller et al., 1989; Lillien and Raff, 1990; Raff et al., 1983). Lastly, there is currently no data that would exclude the unlikely possibility that one glial component induces neoplastic transformation in a second cell population.

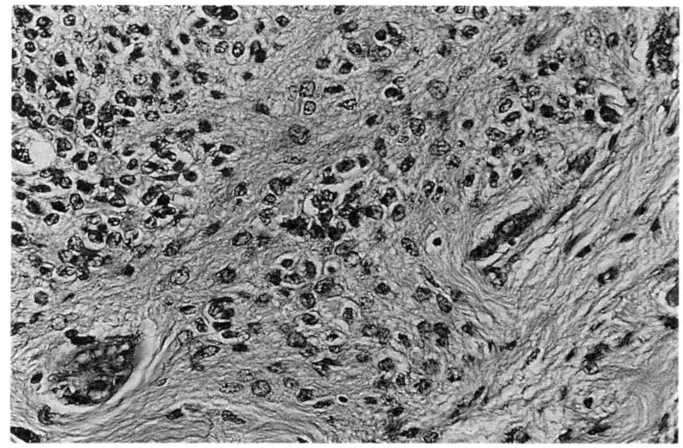

Fig. 11. Mixed oligoastrocytoma: Both components of mixed tumors can be either intimately admixed or juxtaposed, as illustrated here, with oligodendroglial elements (**left**) and a fibrillary astrocytic component (**right**). (H & E; ×200.)

a. Malignant oligoastrocytoma. The percentage of oligoastrocytomas that progress to anaplasia is unknown, although it appears that the astrocytic component of the mixed gliomas is more likely to do so (Muller et al., 1977; Russell and Rubinstein, 1989). The histopathologic features of such tumors are similar to those previously discussed under anaplastic astrocytoma and oligodendroglioma.

E. Choroid Plexus Tumors

1. Choroid plexus papilloma

Choroid plexus papillomas are rare tumors, representing only 2% of all intracranial gliomas (Rubinstein, 1972). Most occur in the pediatric population (Laurence, 1979), although adult cases are not uncommon (McGirr et al., 1988). These tumors arise throughout the ventricular system. The lateral ventricles are the preferential site of pediatric lesions (Russell and Rubinstein, 1989), whereas the fourth ventricle is more commonly affected in adults. Clinical symptoms are usually a reflection of increased intracranial pressure due either to obstruction of the CSF pathways or to increased production of CSF.

Plexus papillomas mimic the histopathologic features of normal choroid plexus in being composed of cuboidal to columnar epithelium surrounding delicate fibrovascular cores. In contrast to normal epithelium, papillomas are more cellular, exhibit greater cellular pleomorphism than normal choroid plexus, and often show deviation from the regimented cell orientation that characterizes the plexus. The immunophenotype of normal and neoplastic choroid plexus is of interest in that it includes reactivity for S-100 protein, keratin, vimentin, and, particularly in some papillomas, GFAP. Mitotic activity is scant, and limited microscopic invasion of underlying brain is not uncommon. The prognostic significance of these features is unclear. Some authors suggest both an increased potential for recurrence and anaplastic progression (Paulus and Jänisch, 1990; Russell and Rubinstein, 1989), whereas others find no such correlation (McGirr et al., 1988). Recurrence of histologically benign tumors has been reported as subarachnoid seeding (McGirr et al., 1988).

2. Choroid plexus carcinoma

Malignant transformation in choroid plexus papillomas is an uncommon phenomenon, occurring in less than 20% of cases (Laurence, 1979). That this is likely an overestimate is suggested by the frequent lack of an atypical or papillomatous component in association with carcinomas. Unlike the papillomas, choroid plexus carcinomas are frankly invasive lesions, which most often arise in the lateral ventricles of patients less than 10 years of age. Carcinomas demonstrate marked cytologic atypia, high mitotic indices, and sufficient architectural disorganization that many tumors lack a papillary pattern. The diagnosis of a metastatic adenocarcinoma must be rigorously excluded in the cases affecting adults. In children and adolescents, other neoplasms to be considered include embryonal and yolk sac carcinoma, as well as metastatic papillary carcinoma of the thyroid.

The value of prealbumin (transthyretin) immunohistochemistry in the differentiation of choroid plexus neoplasms from other epithelial tumors is limited (Albrecht et al., 1991). The absence of this marker may be associated with poor prognosis in choroid plexus carcinoma (Paulus and Jänisch, 1990).

F. Neuronal and Mixed Neuronal–Glial Tumors

This category of neuroepithelial tumors is comprised of rare lesions made up of mature and immature neurons and mixed populations of neuronal and glial cells (Table 2).

TABLE 2. Tumors of Neuroepithelial Tissue

Neuronal and mixed neuronal–glial tumors
1. Gangliocytoma
2. Dysplastic gangliocytoma of cerebellum (Lhermitte-Duclos)
3. Desmoplastic infantile ganglioglioma
4. Dysembryoplastic neuroepithelial tumor
5. Ganglioglioma
6. Anaplastic (malignant) ganglioglioma
7. Central neurocytoma
8. Olfactory neuroblastoma (Esthesioneuroblastoma)
Variant: olfactory neuroepithelioma
Pineal tumors
1. Pineocytoma
2. Pineoblastoma
3. Mixed pineocytoma/pineoblastoma

1. Neuronal tumors

a. Gangliocytoma. Among the tumors composed of mature neurons are the gangliocytomas. A special variant is the *dysplastic gangliocytoma of the cerebellum* (Lhermitte-Duclos disease), a slow-growing lesion wherein, despite bizarre cortical overgrowth of granule-like cells, the focal pattern is largely preserved. In contrast to gangliogliomas (see below), gangliocytomas lack neoplastic glial components. As a result, these tumors possess no potential for anaplastic progression (Faillot et al., 1990; Milbouw et al., 1988; Russell and Rubinstein, 1989; Yachnis et al., 1988).

b. Central neurocytoma. This rare but stereotypical intraventricular tumor most often occurs in young adults and, following surgical resection alone, is associated with an excellent prognosis (von Deimling et al., 1990). They are well-demarcated, partially calcified masses, typically arising in the septum pellucidum, which projects into the ventricles. Obstruction of the foramen of Monro is frequent. Massive tumors involving both the lateral and third ventricles have also been reported.

Neurocytomas demonstrate a distinctive histology, that of a homogeneous population of uniform cells with round nuclei, finely dispersed chromatin, ill-defined cytoplasm, and a delicate process forming a fibrillary matrix (Fig. 12). Anucleate areas composed of cellular processes often surround delicate blood vessels that form a network. Mitotic figures are rare, as are microfoci of necrosis. Such features are not indicative of anaplastic change.

Ultrastructural studies emphasize the high degree of neuronal maturation of the tumor cells, which at the light microscopic level do not resemble neurons. These include cellular processes with parallel arrays of microtubules, dense-core and clear vesicles, and various stages of synapse formation (Hassoun et al., 1982; Hessler et al., 1991; Nishio et al., 1988; Pearl et al., 1985; Townsend and Seaman, 1986; von Deimling et al., 1990; Wilson et al., 1985). The neurocytic nature of the tumor is also confirmed by immunohistochemical staining for synaptophysin and neuron-associated cytoskeletal proteins (Hessler et al., 1991; von Deimling et al., 1990). With rare exception (von Deimling et al., 1991), GFAP immunoreactivity is usually restricted to reactive stromal astrocytes (Hessler et al., 1991).

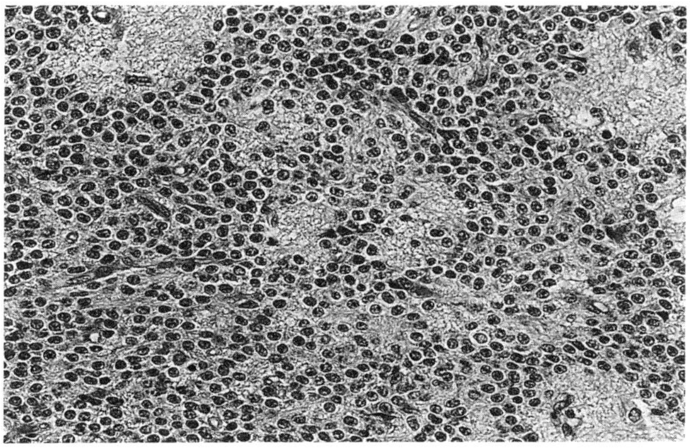

Fig. 12. Central neurocytoma: The appearance of central neurocytomas in tissue sections characteristically consists of homogeneous cell populations with ill-defined borders and a mild fibrillary matrix, particularly appreciated in the anuclear areas. (H & E; ×200.)

2. Mixed neuronal–glial tumors

a. Gangliogliomas. Gangliogliomas represent the most common of the mixed tumors. They consist of neoplastic neuronal and glial cells and account for only about 1% of brain tumors (Kalyan-Raman and Olivero, 1987). In the pediatric population, however, gangliogliomas are slightly more common, comprising 4 to 5% of CNS tumors (Sutton et al., 1983). The largest review, a report of 120 cases (Russell and Rubinstein, 1989), indicates that 80% of the tumors occur in patients under the age of 30, with 10% occurring in infants.

Although they may occur at any site in the neuraxis, gangliogliomas demonstrate a predilection for the temporal lobes. The majority are associated with seizures. Grossly, they appear demarcated and often show cystic change and foci of calcification. Histologically, gangliogliomas are benign-appearing, being composed of clusters of neoplastic, abnormal ganglion cells, and astrocytes, either pilocytic or fibrillary (Fig. 13). Oligodendroglial elements are far less common (Allegranza et al., 1990). Identification of the neuronal component is aided by immunohistochemistry for neuronal-associated cytoskeletal proteins, silver impregnation techniques, and staining for Nissl substance. The ultrastructural features of processes with microtubular arrays, synapses, and dense core vesicles are also diagnostic. A rich fibrocollagenous stroma, often containing a mild chronic inflammatory infiltrate, is most apparent within the ganglionic component of the tumor. A variable angiomatosis may also be present.

The glial component, usually astrocytic, confers the potential for recurrence, but anaplastic progression is rare. Though no primitive or embryonal elements are noted, the ganglion cells are thought to result from maturation of a neuroblastic cell population. In addition to its potential (albeit rare) role in anaplastic transformation, the glial component may also be involved in the regulation, maturation, and survival of the neuronal population by way of secretion of trophic substances such as nerve growth factor (NGF) (Cockram, 1990).

Although malignant transformation in gangliogliomas reportedly occurs in about 10% of gangliogliomas (Russell and Rubinstein, 1989), the figure is likely an overestimate. Clinically, the event should be suspected in patients with exacerbation of seizure activity and radiologic features suggesting progression of a glioma. The histopathologic picture of malignant gangliogliomas somewhat resembles that of anaplastic astrocytomas. The ganglionic elements do not contribute to the process.

b. Desmoplastic infantile ganglioglioma (DIG). This rare form of mixed neuronal–glial neoplasm affects only children under two years of age (VandenBerg, 1991). Typically supratentorial, the lesions show a slight predilection for the frontal and parietal lobes. Most are massive tumors with ill-defined borders and are partially cystic. The hallmark

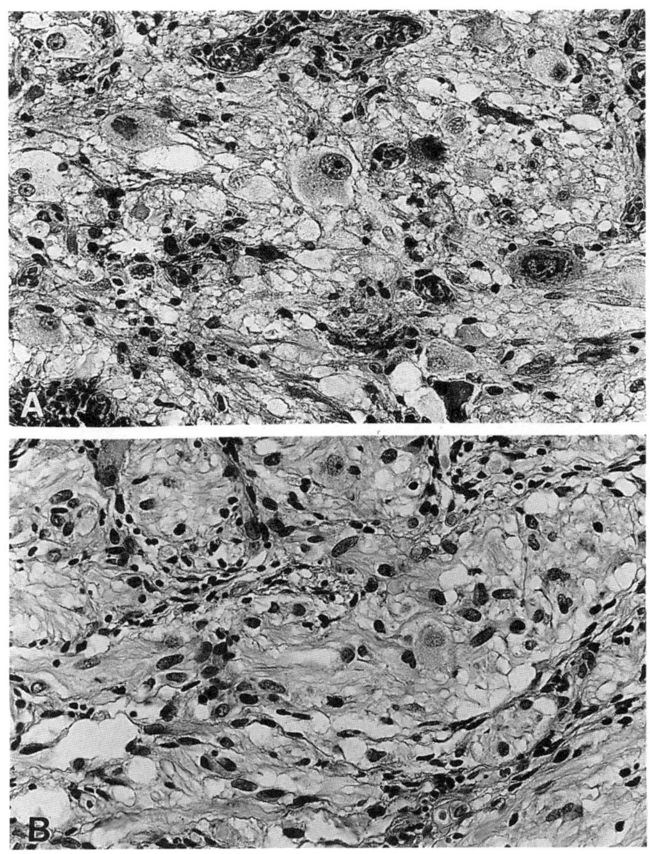

Fig. 13. Ganglioglioma: The mixed population of abnormal ganglionic cells and neoplastic astrocytes are typically intersected by a delicate fibrovascular stroma and variable lymphocytic infiltrates. (H & E; A-B: ×200.)

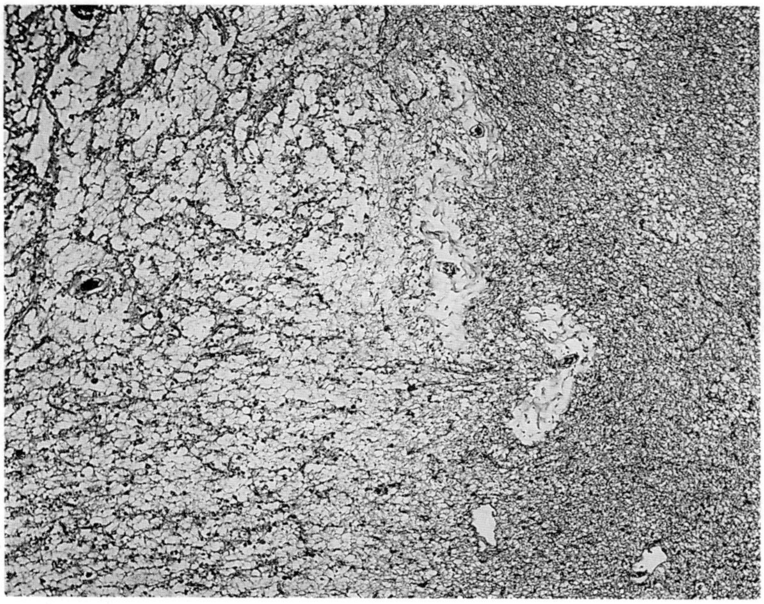

Fig. 14. Dysembryoplastic neuroepithelial tumor (DNT): A principal diagnostic feature is the finding of multiple, often "oligodendrocyte-rich," intracortical nodules. Their architecture is typically patterned and features pale intercellular mucosubstances. (H & E; ×64.)

feature of these neoplasms is their firm texture, a feature due to dense fibrosis, particularly in superficial portions, which often show dural attachment.

Histologically, DIG consists of an admixture of neoplastic neuroepithelial cells within a desmoplastic stroma. Astrocytes, globoid neurons, and in some cases a scant primitive or small cell population comprise the neoplastic elements. The cellularity of the tumor is variable. Frequently, particularly in areas more desmoplastic, cellular elements are less abundant. A typical gangliogliomatous component may also be present (Parisi et al., 1992). Mitoses and micronecrosis, if present, occur only in association with the primitive, small-cell component. The mixed nature of the neuroepithelial cells is confirmed on special stains and by immunohistochemistry. The nature of the primitive cell population has not yet been fully defined (VandenBerg, 1991). Interestingly, their presence, as well as that of mitoses and of foci of micronecrosis, is not related to poor prognosis. A favorable outcome follows gross or near-total resection (Gambarelli et al., 1982; Ng et al., 1990; VandenBerg et al., 1987a; VandenBerg, 1991). Death attributable to tumor growth is uncommon (Parisi et al., 1992).

As in the case of pleomorphic xanthoastrocytoma, the astrocytes comprised by DIG possess pericellular basal laminae (VandenBerg et al., 1987a). Their presence in both these lesions suggests that the glia play a role in the formation and modification of extracellular matrix. There is accumulating evidence that astrocytic cells are capable of elaborating extracellular matrix components, including complex chondromucoid matrices (Kepes et al., 1984) and specific forms of collagen (McComb et al., 1987; Paulus and Peiffer, 1988; Russell and Rubinstein, 1989; Sage and Iruela-Arispe, 1990). Surprisingly, neoplastic neuronal cells may also secrete collagens (DeClerck and Lee, 1985; Tsokos et al., 1985, 1987). The desmoplastic infantile ganglioglioma may, therefore, be a tumor in which the secretion, induction, and/or modification of a distinctive extracellular matrix accompanies differentiation. This property may, in fact, correlate with its limited capacity for brain invasion and characteristic lack of metastatic potential.

c. Dysembryoplastic neuroepithelial tumor. Dysembryoplastic neuroepithelial tumor (DNT) is an uncommon tumor, one that usually occurs in children and young adults with long-standing complex partial seizures. This entity was recently described in a detailed report of 39 cases (Daumas-Duport et al., 1988b). Its recognition is of importance in that long survivals follow even incomplete resection. The tumors are characteristically intracortical. Most are located in the temporal lobe.

The multinodular architecture of DNT is a hallmark feature that is best appreciated at low magnification (Fig. 14). The nodules exhibit concentric or "alveolar" patterns and are composed of oligodendroglial, astrocytic, and neuronal cells; neurons appear to float in the patterns in a mucoid matrix. Characteristically, all cases are associated with a more diffuse cortical dysplasia. The neuronal component may be inconspicuous, erroneously giving the impression of an oligoastrocytoma infiltrating cortex. This is particularly the case in small, nonrepresentative biopsies. Rare mitotic figures may be present, and atypia is minimal. Considering the varied aspect of these neoplasms, adequate sampling is important for the correct diagnosis.

G. Pineal Parenchymal Tumors

These rare lesions constitute less than 0.1% of intracranial neoplasms (Scheithauer, 1985) and about 11 to 28% of the pineal region tumors in children (D'Andrea et al., 1987; Edwards et al., 1988). The group of pineal parenchymal tumors includes three lesions. The pineocytoma represents the best differentiated member and is composed of cells displaying features similar to those of normal pineocytes. The pineoblastoma occupies the other end of the spectrum and is a highly cellular neoplasm composed of small, primitive-appearing cells not unlike those of medulloblastoma. The third group, the "mixed" pineocytoma/pineoblastoma, actually encompasses those tumors composed of both well- and poorly differentiated cells or ones intermediate in their cytoarchitectural features. This category includes tumors previously designated "malignant" pineocytomas or *pineocytomas without neuronal differentiation* (Herrick and Rubinstein, 1979).

1. Pineocytomas

Pineocytomas occur more commonly in adults than in children, the mean age being 42 ± 19 years (Borit et al., 1980; Herrick and Rubinstein, 1979; Vaquero et al., 1990). The extremes of age at initial diagnosis range from 11 to 78 years (Herrick and Rubinstein, 1979; Vaquero et al., 1990). Most are well demarcated from the adjacent parenchyma, showing little tendency to parenchymal infiltration. Although focal necrosis may be present, extensive necrosis is uncommon. Microscopically, pineocytomas are moderately cellular and show no histologic features of anaplasia. A hallmark feature is the arrangement of tumor cells in the so-called "pineocytomatous rosettes" (Borit et al., 1980) (Fig. 15), and their presence appear to be related to a more benign clinical course (see Russell and Rubinstein [1989] for review). Occasional giant cells and mitotic activity may be present but does not appear to directly correlate with clinical behavior. The cellular processes composing the pineocytomatous rosettes can be demonstrated by silver carbonate impregnation, which, in the normal pineal, identifies pineocytes (Herrick and Rubinstein, 1979). Although the pineocytomatous rosettes were previously described as evidence for "neuronal" differentiation (Herrick and Rubinstein, 1979), they should be more properly considered as an evidence of neoplastic pineocytic maturation.

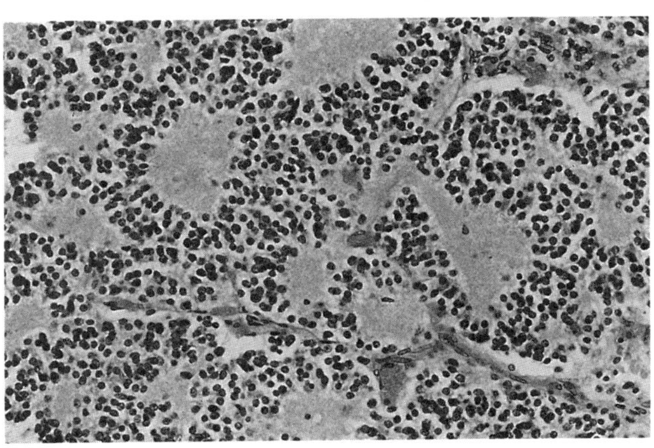

Fig. 15. Pineocytoma: The well-differentiated pineocytoma typically shows pineocytomatous rosettes. (H & E; ×200.)

The retinal S-antigen, a 48 kd protein involved in the phototransduction of vision, has been demonstrated in pineocytomas (Korf et al., 1986; Perentes et al., 1986). Pineal parenchymal cells may demonstrate a photosensory differentiation during ontogeny (Reiter, 1981) and a partial photosensory differentiation may analogously occur in the tumors (see pineoblastomas, below, for details).

Ultrastructural studies of pineocytomas have documented pineocytic differentiation with the presence of 9 + 0 neurosensory cilia and centriolar formations in addition to processes containing dense-core and clear vesicles, sheaths of microtubules, and bundles of intermediate filaments (Hassoun et al., 1983, 1984; Hassoun and Gambarelli, 1989; Kline et al., 1979; Markesbery et al., 1981). All of these features are suitable for a general neurosecretory differentiation. However, the collection of synaptic junctions observed in other CNS tumors with neoplastic neuronal differentiation have not been accurately documented in primary pineocytomas.

2. Pineoblastoma

The most primitive tumors of the pineal parenchyma, pineoblastomas comprise approximately 3 to 17% of pineal region tumors in children (D'Andrea et al., 1987; Edwards et al., 1988). The majority occur in the first decade, exhibiting a male-to-female ratio of 2:1 (Russell and Rubinstein, 1989). Macroscopically, pineoblastomas are poorly delineated, given a tendency to brain invasion and local invasion. Necrosis is common but calcification is less a feature than in pineocytoma. Cerebrospinal dissemination is frequent, but systemic metastases are rare (Banerjee and Kak, 1974). As a rule, the patient survival rate is less than two years (Borit et al., 1980; Herrick and Rubinstein, 1987). A single six-year survivor has been reported (Uematsu et al., 1988).

Pineoblastomas are highly cellular neoplasms, ones composed of small undifferentiated cells with round to oval nuclei containing dense, hyperchromatic chromatin. The resemblance to medulloblastoma may be striking. Cells are typically arranged in patternless sheets, but Homer Wright rosettes are frequently present. Special silver carbonate impregnations for pineocytes may demonstrate a small, unipolar, blunt process that forms a terminal, perivascular expansion. As in pineocytomas, multinucleated cells are occasionally found.

Exceptionally, retinoblastic differentiation occurs in pineoblastoma, a feature evidenced by the presence of specialized photosensory structures, including Flexner-Wintersteiner rosettes and fleurettes (Russell and Rubinstein, 1989; Sobel et al., 1981; Stefanko and Manschot, 1979). As previously discussed, human fetal pineal parenchymal cells exhibit morphologic degrees of photoreceptor differentiation not found in infant or adult glands (Kurumado and Mori, 1976, 1977). Interestingly, however, a biochemical link does exist between the adult pineal and photoreceptors in that both possess photoreceptor proteins. The normal adult pineal gland contains retinal S-antigen, but the substance is detected in greater quantity by immunohistochemistry in the perinatal pineal (Perentes et al., 1986). Such expression of photosensory substances early in development parallels that in primitive or embryonal tumors, e.g., retinoblastoma, pineoblastoma, and mixed pineocytoma/pineoblastomas, as opposed to mature lesions, e.g., pineocytoma. The phylogenetic relationship between pineoblastomas and retinoblastomas is underscored by the occurrence of so-called "trilateral retinoblastomas," the concurrence of bilateral retinoblastomas with pineoblastoma or a light-microscopically undifferentiated embryonal tumor (Bader et al., 1982; Johnson et al., 1985).

Ultrastructurally, pineoblastomas appear as primitive neuroepithelial neoplasms with variable degrees of photoreceptor differentiation with polarization of cytoplasmic organelles, 9 + 0 cilia, annulate lamellae, and well-developed smooth endoplasmic reticulum (SER) with whorl formation (Kline et al., 1979; Markesbery et al., 1981).

3. *Mixed pineocytoma/pineoblastoma*

Many previously reported "poorly differentiated" or "malignant" pineocytomas should, in fact, be incorporated into this category. The exact frequency of such tumors is difficult to evaluate. Mixed tumors appear to present a more aggressive clinical behavior with local infiltration and distant CSF dissemination (Fig. 16). Two principal histologic patterns of this tumor essentially bridge the features of pineocytoma and pineoblastoma. Some show a truly mixed pattern, clear-cut pineocytoma being associated with distinct zones of obvious pineoblastoma. The second, less well-defined form consists of a monomorphous lesion intermediate in terms of cytologic atypia, cellular pleomorphism, mitotic activity, cellularity, and other features of evolving anaplasia. Pineocytic rosettes are infrequent or absent. Limited necrosis may be observed.

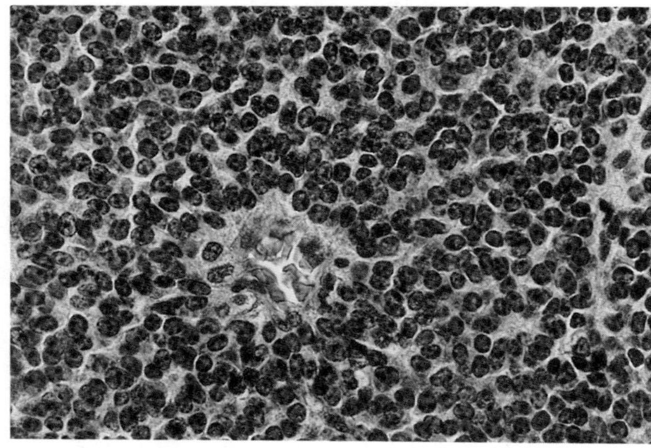

Fig. 16. Mixed pineocytoma/pineoblastoma: These intermediate tumors present a more aggressive clinical behavior than pineocytomas and may present with leptomeningeal dissemination. (H & E; ×400.)

H. Embryonal Tumors

The embryonal tumors constitute a group of primitive-appearing, clinically aggressive neoplasms, most of which occur in the first decade of life (Table 3). In the immature nervous system, not only are more diverse cell populations vulnerable to neoplastic transformation, but the populations demonstrate a greater capacity for differentiation than in adults (Kleihues et al., 1990). A large body of evidence indicates that embryonal neoplasms originate from different populations of transformed cells and that the molecular basis for malignant transformation differs from that of common anaplastic tumors in adults (Bigner et al., 1990; Bigner and Vogelstein, 1990; Thomas and Raffel, 1991; Wasson et al., 1990).

The designation *embryonal* does not imply that these tumors are biologically equivalent to the embryonal or immature cells of the normal, developing nervous system; however, it does indicate that they result from transformation of immature or undifferentiated neuroepithelial cells (Rorke, 1989; Rubinstein, 1989, 1991; VandenBerg et al., 1987b). Embryonal tumors share a number of common histologic features of high cellularity, including a small cell

TABLE 3. Tumors of Neuroepithelial Tissue

Embryonal tumors

1. Medulloepithelioma
2. Neuroblastoma
   Variant: ganglioneuroblastoma
3. Ependymoblastoma
4. Retinoblastoma
5. Primitive neuroectodermal tumors (PNETs) with multipotent differentiation: neuronal, astrocytic, ependymal, muscle, melanotic, etc.

a. Medulloblastoma
   Variants: medullomyoblastoma, melanocytic medulloblastoma

b. Cerebral or spinal PNETs

cytology and a usually high mitotic index. Single cell or focal necrosis is the rule. These features reflect the clinically aggressive behavior of embryonal tumors, lesions that exhibit a high tendency to leptomeningeal invasion as well as cerebrospinal metastasis.

The current WHO classification recognizes a number of embryonal tumors with restricted capacity for differentiation along neuronal, ependymal, and retinal cell lineages, e.g., neuroblastoma, ependymoblastoma, and retinoblastoma, respectively. The medulloepithelioma, a rare embryonal neoplasm with the greatest potential for divergent differentiation, occupies a unique niche.

1. Medulloepithelioma

This very rare tumor usually occurs early in the first decade of life, the cerebral hemispheres being the most frequently affected site (Caccamo et al., 1989; Russell and Rubinstein, 1989). Its histopathologic features recall the structure of the primitive epithelium of the embryonic neural tube (Fig. 17). These structures, often forming extensive ribbons or papillary structures, consist of pseudostratified columnar epithelium exhibiting a high mitotic rate. A PAS-positive basement membrane commonly defines the basal surface of the epithelium. The apical surface, where abundant mitotic activity may be present, is relatively simple, without evidence of cilia or blepharoplasts. Interposition of delicate stroma, often populated by differentiating neuroepithelial cells, may be present between the epithelial structures. Rare mesenchymal differentiation may include occasional skeletal muscle or bone formation, but care must be taken not to confuse immature teratomas with medulloepitheliomas. Ultrastructural studies in two tumors revealed a paucity of cytoplasmic organelles, moderately developed juxtaluminal zonulae adherents, and a conspicuous lack of apical differentiation (Pollak and Friede, 1977; Troost et al., 1990).

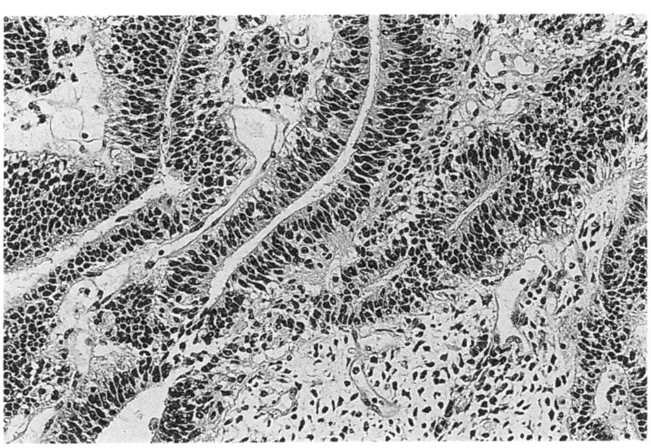

Fig. 17. Medulloepithelioma: The tubular structures in the medulloepitheliomas resemble the primitive epithelium of the neural tube. (H & E; ×200.)

Four cerebral tumors have been examined by immunohistochemistry (Caccamo et al., 1989); the primitive neuroepithelium showed vimentin immunoreactivity but only sparse expression of neuroblastic and glial cytoskeletal proteins, including class III ß-tubulin isotype and GFAP, respectively. Understandably, immunoreactivity for neuroblastic and glial cell-associated cytoskeletal proteins is uniformly expressed by the differentiating cell populations that compose the cellular "stroma" between the characteristic epithelial structures (Caccamo et al., 1989). Immunohistochemical studies have also demonstrated insulin-like growth factor I and basic fibroblastic growth factor immunoreactivity (Shiurba et al., 1991).

2. Neuroblastoma

Neuroblastomas are rare tumors that most often occur in children in the first decade. No sex predilection has been noted. The majority are supratentorial (Bennett and Rubinstein, 1984; Horten and Rubinstein, 1976) and show a predilection for the frontoparietal regions. In general, they are large, macroscopically discrete tumors. Varying degrees of stromal desmoplasia confer a firmer texture and, in some, a lobulated appearance. Focal necrosis and cyst formation may be present. The same is true of foci of hemorrhage. Neuroblastomas are highly cellular tumors consisting of monotonous sheets of small cells with ill-defined cytoplasm and round hyperchromatic nuclei. In some, intermingled large cells exhibit vesicular nuclei, some nucleolar prominence, more ample cytoplasm, and process formation. A delicately fibrillated matrix formed of these indistinct processes also makes up the centers of the characteristic Homer Wright rosettes. Rhythmic nuclear palisades, similar to ones observed in medulloblastomas, may also be present (Dehner et al., 1988; Ojeda et al., 1987).

Neuroblastic differentiation can be confirmed by silver impregnation methods, immunohistochemistry for neuron-associated cytoskeletal proteins, and by electron microscopy, wherein microtubule-containing processes and neurosecretory granules are seen. A minor ganglionic component may be seen but, by definition, is never predominant (Bennett and Rubinstein, 1984). Ultrastructural studies confirm neuronal maturation by showing increased organelle density and differentiation as well as the form synaptic junctions (Russell and Rubinstein, 1989). The presence of neurotrophic growth factor receptors, e.g., nerve growth factor, are also consistent with the histogenesis of these neuroblastic tumors. The identification of specific receptors to hormonal/growth factors may also have therapeutic implications (Baker et al., 1990) in addition to complementing current diagnostic features.

3. Ependymoblastoma

This rare ependymal tumor usually arises in children at a median age of two years (Mørk and Rubinstein, 1985). All reported examples have been intracranial, most supratentorial. A ventricular relationship may be difficult to verify since most tumors are enormous at presentation. They tend to be macroscopically defined but infiltration of adjacent brain and leptomeninges may occur.

Microscopically, ependymoblastomas are composed by small, poorly light-microscopically differentiated cells, sheets of which are interrupted by displaced pseudostratified, ependymoblastic rosettes and tubules (Fig. 18). Well-developed perivascular rosettes, of the type frequently seen in ependymomas, are relatively rare. The ultrastructural findings of luminal cytoplasmic specialization, including microvilli, 9 + 2 cilia, and juxtaluminal zonulae adherents, are indicative of ependymal differentiation (Langford, 1986). In clinical and histopathologic terms, ependymoblastomas can be distinguished from anaplastic ependymomas. Clinically, ependymoblastomas are more common in children below the age of five years, whereas anaplastic ependymomas are typically encountered in adults. The comparative morphologic features of ependymoblastoma include the absence of endothelial proliferation, a common feature of anaplastic ependymomas, a more restricted range of cellular pleomorphism, and the presence of multilayered, true rosettes with frequent mitoses. Necrosis may be found in either tumor, although diffuse, geographic, and pseudopalisading patterns of necrosis are largely limited to anaplastic ependymomas.

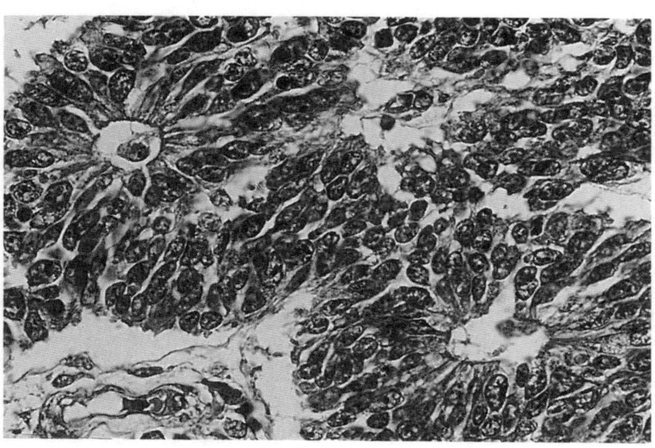

Fig. 18. Ependymoblastoma: Multilayered rosettes constitute the primary histopathologic feature of these embryonal tumors. (H & E; ×400.)

4. Retinoblastoma

Retinoblastoma is not only the most common intraocular tumor of childhood but is the only embryonal tumor for which a genetic basis of transformation is known. The event occurs in the immature retina, wherein both alleles of the rb tumor suppressor gene are inactivated within a single cell (Gallie, 1990; Gennett and Cavenee, 1990). Retinoblastomas are mainly composed of poorly differentiated small cells, but the majority exhibit Flexner-Wintersteiner and Homer Wright rosette formation. A small number of cases also display "fleurettes," remarkably well-developed photosensory structures. Although the rosettes frequently demonstrate immunoreactivity for neuronal-associated (Katsetos et al., 1991) and photosensory-related proteins (Lopes et al., 1991), no constant correlation between their presence and rosette or fleurette numbers has been established (Lopes et al., 1991).

The histogenetic nature of retinoblastoma is incompletely understood. Various theories have been proposed, including ones suggesting they arise from: (1) multipotential primitive neuroepithelial cells capable of differentiating along neuroretinal cells, retinal pigment epithelium, and astroglia (Campbell and Chader, 1988; Detrick et al., 1988; Kyritsis et al., 1986a,b; Tarlton and Easty, 1990; Terenghi et al., 1984); (2) neuroretinal progenitor cell restricted to neuronal differentiation (Katsetos et al., 1991; Perentes et al., 1987); and (3) photosensory cell-specific progenitor limited to the formation of photosensory cells (Popoff and Ellsworth, 1971; Tso et al., 1969, 1970). Recent data based upon a study of 22 *in situ* retinoblastomas suggests that they have a differentiating potential analogous to that of immature retinal neuroepithelium, including the formation of photosensory and Müller cell. These are evidenced by the presence within the same tumor of interphotoreceptor retinol-binding protein (IRBP) and cone/rod opsins, as well as of cellular retinaldehyde-binding protein (CRAlBP), respectively. In contrast, retinoblastomas do not appear to be pluripotential in the sense of demonstrating unrestricted retinal epithelial or neuroepithelial differentiation with the production of pigmented or astroglial cells (Gonzalez-Fernandez et al., 1992; Lopes et al., 1991).

5. Medulloblastoma

This poorly differentiated neoplasm represents the most common of the embryonal tumors and constitutes approximately 25% of all intracranial tumors in children. Its peak incidence occurs near the end of the first decade, there being a slight male predominance. The majority arise in the cerebellar vermis. Most tumors are, at least in part, macroscopically well demarcated from the adjacent brain. Small foci to more confluent zones of necrosis are commonly present, but massive necrosis is very unusual. Cyst formation is rare. Tumors arising laterally in the cerebellar hemi-

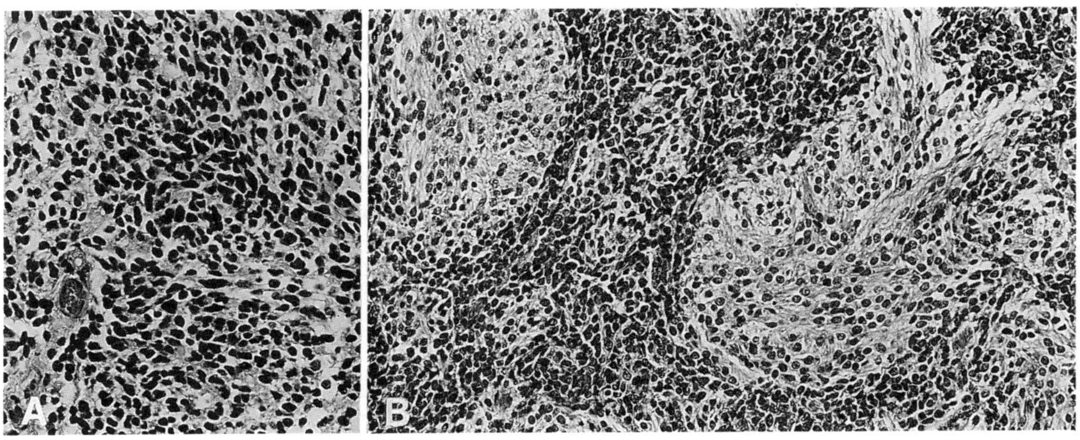

Fig. 19. Medulloblastoma: Medulloblastomas are highly cellular tumors composed of primitive cells in a delicately, and variably, fibrillated intercellular matrix (**A**). The desmoplastic variant demonstrate a lobular pattern with reticulin-free islands of more differentiated cells (**B**). (H & E; A-B: ×200.)

spheres tend to be more frequent in adults (Russell and Rubinstein, 1989).

Histopathologically, medulloblastomas are intensely cellular and composed by relatively small cells with hyperchromatic, ovoid to angular nuclei and scant cytoplasm with ill-defined cell borders (Fig. 19). Process formation is variable, as are mitoses. In some cases they are few, whereas in others they are abundant. Focal or geographic necrosis is common, but endothelial/pericytic proliferation is rare. Medulloblastomas are commonly considered as poorly or undifferentiated by routine light microscopy. Nonetheless, neuroblastic differentiation, demonstrated by the presence of Homer Wright rosettes, is found in approximately 40% of the cases (Kleihues et al., 1989). A neoplastic ganglionic component is also present in less than 10% of cases with obvious neuroblastic differentiation. Immunoreactivity for neuron-associated proteins such as class III ß-tubulin isotype, neurofilament epitopes, and synaptophysin is seen in neuroblastic areas. By comparison, astrocytic differentiation is less common. Definitive ependymal features are very rare.

The desmoplastic variant of medulloblastoma comprises 10 to 12% of cases (Burger et al., 1987; Garton et al., 1990; Hubbard et al., 1989; Kleihues et al., 1989) and is usually more prevalent laterally in the cerebellar hemispheres of older patients. They tend to be largely demarcated and exhibit a firm consistency. Microscopically, a biphasic pattern of highly cellular sheets and trabeculae of tumor cells alternates with hypocellular islands composed by cells with finely fibrillated processes. These architectural patterns are due, in large part, to variation in intercellular reticulin deposition, which is dense in highly cellular areas. The follicles are "reticulin-free islands" that demonstrate not only accentuation of synaptophysin reactivity but neuronal tubulin and neurofilament positivity (Katsetos et al., 1989).

Rare variants of medulloblastoma include tumors featuring pigmented, occasionally papillary epithelium, a tumor similar in histologic appearance to melanotic neuroectodermal tumor of infancy (melanotic progonoma) (Johnson et al., 1983), a lesion renewed by Russell and Rubinstein (1989). Tumors exhibiting rhabdomyosarcomatous or striated muscle differentiation are termed *medullomyoblastoma* (Rao et al., 1990). Myogenic differentiation has also been showed in other neuroectodermal neoplasms, including one example of subependymoma (Tomlinson et al., 1991). The relationship of these special variants to the more ordinary forms of medulloblastoma is unclear.

The differentiating capacity of medulloblastoma is broad and may reflect an innate heterogeneity of the tumors. Neuroblastic/neuronal differentiation in over 50% (Kleihues et al., 1989) suggests a close relationship of medulloblastoma to neuroblastoma. Unlike the latter, however, a significant number of medulloblastomas exhibit photoreceptor differentiation, as evidenced by opsin and S-Ag immunoreactivity (Bonnin and Perentes, 1988; Korf et al., 1987). Neoplastic astroglial differentiation accompanied by GFAP immunoreactivity is reported in approximately 10% of the medulloblastomas (Coffin et al., 1983; Kleihues et al., 1989; Schindler and Gullotta, 1983). A similar variety of differentiation has been documented in cell lines established from medulloblastomas (Bigner, 1989).

The term *primitive neuroectodermal tumor* (PNET) is a popular designation applied to tumors that exhibit similar histopathologic features and are assumed to arise from a hypothetical, common primitive neuroepithelial progenitor cell, one presumed to exist throughout the neuraxis. The implicit weakness of the PNET concept for histopathologic diagnosis is its implication that the rather orderly phenotypic differentiation shown by the embryonal tumors occurs after transformation of an undifferentiated neuroepithelial cell

(Gould, 1990). This premise does not particularly embrace, therefore, the well-accepted view that neoplastic differentiation reflects a more restricted genotype, one established preceding neoplastic transformation. Considering the generalizations implicit in the PNET concept to be somewhat too ambiguous for a histopathologic classification, the WHO designation limits the PNET category, reserving it for cerebellar medulloblastoma and other small-cell neoplasms indistinguishable from the medulloblastomas that arise at other sites in the neuraxis. The reluctant inclusion of the medulloblastoma in the PNET category reflects the uncertainty that surrounds its histogenesis as well as its varied differentiating potential. The pathobiology and cyto- as well as molecular genetics of medulloblastoma are the basis of a recent report (Tomlinson et al., 1992a) and a review by Tomlinson et al. (1992b).

IV. TUMORS OF CRANIAL AND SPINAL NERVES

In this category are included tumors derived from supportive or stromal cells of peripheral nerve, including Schwann and perineurial cells, as well as fibroblasts. Two principal variants comprise the majority of nerve sheath tumors; these include neurilemmoma (*schwannoma*) and *neurofibroma* (Table 4). The principal cell of the former is the Schwann cell, whereas the latter is composed of Schwann cells and fibroblasts. Both tumors exhibit distinct histopathologic patterns and stromal reactions. Neurilemmomas tend to be solitary and arise more often from cranial and spinal nerve roots than from peripheral nerves. Neurofibromas, on the other hand, are classically multiple and usually involve major peripheral nerves or their branches. Some are largely cutaneous.

Neurilemmomas and neurofibromas occur either sporadically or as part of the spectrum of neurofibromatosis. Both neurofibromatosis type 1 (von Recklinghausen's disease or peripheral neurofibromatosis) and neurofibromatosis type 2 (central neurofibromatosis) include the development of nerve sheath tumors. Bilateral "acoustic neuromas," in actuality neurilemmomas, are characteristic of neurofibromatosis type 2. In the peripheral form, or neurofibromatosis type 1, neurofibromas are the principal lesions.

TABLE 4. Tumors of Cranial and Spinal Nerves

1. Schwannoma (neurilemoma, neurinoma)
   Variants: cellular, plexiform, melanotic
2. Neurofibroma
   Variants: circumscribed, plexiform
3. Malignant peripheral nerve sheath tumor (MPNST) (neurogenic sarcoma, anaplastic neurofibroma, "malignant schwannoma")
   Variants: epithelioid, MPNST with divergent mesenchymal and/or epithelial differentiation, melanotic

These affect peripheral nerves and are far more common than neurilemmomas.

A. Schwannomas

Intracranial and intraspinal schwannomas are usually tumors developing in the adult life, with a female predominance reported in the intracranial examples (Russell and Rubinstein, 1989). Both intracranial and intraspinal tumors present a predilection for involving the sensory nerves. In the intracranial examples, the most frequently affected is the acoustic nerve, followed by the trigeminal but in a much lesser frequency. The thoracic and cervical sensory roots of the spinal cord are the most common sites of intraspinal schwannomas, although occasional tumors may affect the motor nerve roots. Intraspinal schwannomas are more frequently associated to neurofibromatosis type 2 and as a sporadic event than to neurofibromatosis type 1 (Halliday et al., 1991). Rare examples of intracranial schwannomas unrelated to a major cranial nerve but attached to the dura, intraparenchymal, or intraventricular may also occur. Small perivascular nerves and nerve terminations in the meninges are conjectured as origin for these tumors (for review, see Russell and Rubinstein, 1989).

Macroscopically, schwannomas are lobulated, well-circumscribed and encapsulated lesions on the surface of a stretched nerve. The acoustic tumors involve the cerebellopontine angle and expand the internal auditory meatus. The typical histopathological picture of schwannomas consists of two types of tissue, Antoni A and Antoni B, both of which, in all but acoustic tumors, are present in the majority of lesions. The Antoni A tissue is composed by compact bundles of fusiform cells with elongate nucleus, often regimented in a palisading pattern (Fig. 20). A particularly tight, focal, and often twisted arrangement of these palisades is termed a "Verocay body." Both nuclear palisading and Verocay bodies are a common feature of spinal neurilemmomas. A rich pattern of pericellular reticulin staining characterizes neurilemmomas and is due to basement membrane formation. Antoni B tissue is characterized by hypocellularity and loose texture verging upon a microcystic appearance. The cells are more stellate shaped, with vacuolated cytoplasm and occasionally pleomorphic nuclei. Unlike in Antoni A tissue, reticulin fibers are scant and irregularly distributed.

The vasculature of neurilemmomas varies, often showing sinusoidal dilatations that may either be collapsed or dilated to assume cavernous proportions. Mural thickening and hyalinization are frequently present. Spontaneous hemorrhages and thrombosis of the vessels with necrosis of the tumor are not uncommon, as is hemosiderin deposition.

Secondary changes such as enlargement and hyperchromasia of the nuclei are common. Even when accompanied by hypercellularity and mitotic activity, these

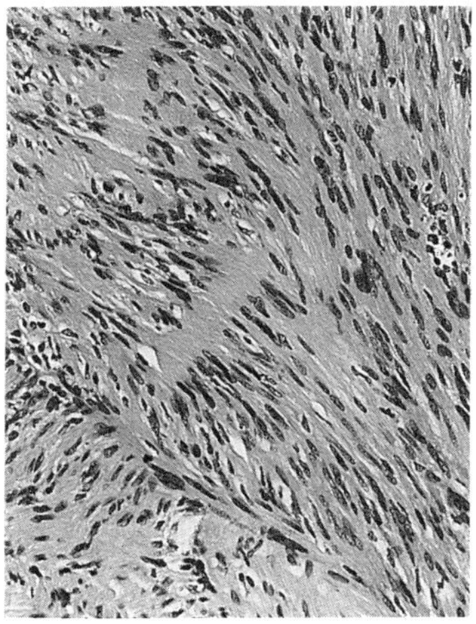

Fig. 20. Schwannoma: Cellular palisading arrangements within dense fascicles of cells, as demonstrated here, are features of the so-called Antoni A tissue. (H & E; ×100.)

are not to be considered as evidence of anaplastic change. Such tumors are termed *cellular schwannoma* and are associated with an excellent prognosis (White et al., 1990; Woodruff et al., 1981) (see below). In general these tumors also show classic features of neurilemmoma, including encapsulation, vascular hyalinization, schwannoma cytology, and, in some cases, cystic change.

The presence of reticulin staining in neurilemmomas corresponds to the ultrastructural level to pericellular basal lamina formation and to the presence of intercellular, granular basement membrane-like material (Erlandson, 1985). Tissue culture studies also demonstrate the ability of neoplastic Schwann cells to produce a basal lamina material (Cornbrooks et al., 1983), the most important constituent of which is laminin (Cornbrooks et al., 1983; Leivo et al., 1989).

1. Cellular schwannoma

This subtype of schwannoma is characterized by high cellularity, predominance of Antoni A tissue, and the presence of a variable mitotic activity (Woodruff et al., 1981). Although this tumor may demonstrate features suggestive of a malignant peripheral nerve sheath tumor (MPNST, see below), typical architectural features of an ordinary schwannoma, such as encapsulation and vascular hyalinization, and occasionally cystic change are also present. Focally, more ordinary Schwann cell cytology is usually also expressed. Like ordinary schwannomas, cellular schwannomas demonstrate strong generalized S-100 protein immunoreactivity and exhibit ultrastructural features of Schwann cell differentiation (White et al., 1990). Both of these features are rare in the MPNST.

B. Neurofibromas

Neurofibromas are divided in two principal groups, cutaneous/dermal and intraneural forms. The latter most often affect larger nerves and present as either solitary globular tumors or less frequently as ropy, multinodular lesions termed *plexiform* neurofibromas. The latter are pathognomonic of the neurofibromatosis type 1. Intraspinal neurofibromas are also frequently associated with neurofibromatosis type 1 (Halliday et al., 1991).

Macroscopically, solitary neurofibromas present as fusiform or cylindrical enlargements of a nerve. The plexiform pattern represents multinodular, less-defined enlargement of one or several nerves and their branches. Compared to schwannomas, neurofibromas tend to be softer, almost gelatinous in consistency. Both solitary and plexiform types demonstrate similar histopathologic features, including spindle-shaped, wavy Schwann cells disposed in loose bundles, separated by a polysaccharide-rich or fibrous matrix (Fig. 21). Nerve fibers, myelinated or nonmyelinated, may be separated by tumor cells. The presence of nerve fiber in the center of a tumor aids in the distinction of neurofibromas from schwannomas in that the latter tend to displace rather intact nerve. Focal concentrations of swirled Schwann cells may be seen and should not be mistaken as evidence of neurilemmoma. Such structures may be sufficiently distinctive as to resemble Meissner or Pacinian corpuscles.

Ultrastructural studies indicate that Schwann cells are the principal component of the neurofibromas, followed in frequency by the perineurial cells and fibroblasts (Enzinger and Weiss, 1988). All tend to be separated by an extensive extracellular space containing collagenous bundles.

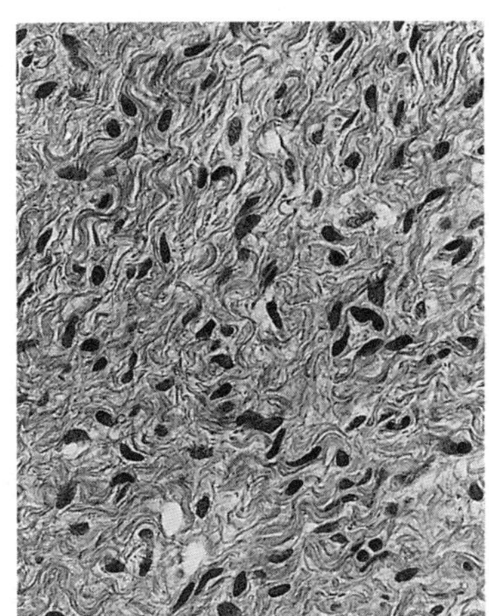

Fig. 21. Neurofibroma: A mixed population of Schwann cells, fibroblasts, and perineurial neoplastic cells compose a typically wavy cytoarchitectural pattern of neurofibromas. (H & E; ×200.)

In immunohistochemical terms, nerve sheath cell tumors share a number of features. Virtually all schwannomas are diffusely reactive for S-100 protein (Rubinstein, 1988). Leu-7 epitope (Perentes and Rubinstein, 1986) and myelin basic protein immunoreactivity have also been observed. In contrast, neurofibromas typically show uneven and less pronounced reactivity, a not unexpected finding given the mixed cell composition of these tumors. S-100 protein immunopositivity in neurofibromas confirms the presence of Schwann cells (Hirose et al., 1986). Epithelial membrane antigen (EMA)-reactive perineurial cells are far less frequent. The combination of S-100 protein and Leu-7 maximizes the likelihood of differentiating schwannomas and neurofibromas from other soft tissue tumors (Russell and Rubinstein, 1989). Vimentin is the predominant intermediate filament protein expressed in nerve sheath neoplasms (Gould et al., 1986; Gray et al., 1989), but its ubiquitous nature makes its identification of little diagnostic relevance. For reasons uncertain, immunopositivity for GFAP has been demonstrated in a number of both schwannomas and neurofibromas (Gould et al., 1986; Gray et al., 1989; Stanton et al., 1987).

Unlike schwannomas, neurofibromas, especially neurofibromas of the plexiform type, show a tendency to undergo malignant transformation. This is particularly true in tumors associated with neurofibromatosis type 1. Although the incidence of malignant transformation is low, approximately 5%, the event should be clinically suspected in the face of rapid enlargement of the tumoral mass, pain, and nerve deficits (Ducatman et al., 1986). From the histopathologic standpoint, cellular lesions with more than an occasional mitotic figure are suspect for anaplastic transformation. In that, transformation may be focal in the tumor, and extensive examination of the specimen is requested.

C. Malignant Peripheral Nerve Sheath Tumors (MPNST), Neurogenic Sarcoma, Neurofibrosarcoma

Malignant peripheral nerve sheath tumors are in almost all instances malignant counterparts of neurofibromas. Malignant transformation of schwannomas is an extremely rare event. In a large series from Mayo Clinic, the calculated incidence of these tumors is 0.001% in the general patient population and 4.6% among patients with von Recklinghausen's disease (Ducatman et al., 1986). MPNST may occur *de novo* within a normal nerve but more often arises by malignant transformation of a plexiform or solitary neurofibroma. The coexistence of a neurofibromatous component has been demonstrated in about 60% of all cases and in 81% of tumors arising in von Recklinghausen's disease (Ducatman et al., 1986). The association between prior radiation treatment and MPNST has been documented both in patients with and without von Recklinghausen's disease (Ducatman and Scheithauer, 1983; Ducatman et al., 1986). The distribution of the MPNST is similar to that of solitary and plexiform neurofibromas, e.g., head and neck, trunk and proximal limbs. Tumors in von Recklinghausen's disease are somewhat more often centrally located, a factor that may underlie their poorer prognosis. Intracranial examples are extremely rare; the few cases reported have involved the trigeminal or, more rarely, the acoustic nerve (see Russell and Rubinstein, 1989, for review).

Histologically, MPNST shows considerable cytologic and histologic variation. Most are composed by fascicle of spindle cells, sometimes exhibiting a herringbone or a storiform pattern. Arrangements resembling the palisading of ordinary neurilemmoma are infrequent. High cellularity, brisk mitotic activity, nuclear pleomorphism, microvasculature proliferation, and necrosis complete the picture. Necrosis is often prominent and may exhibit a geographic configuration accompanied by peripheral palisading. A small percentage of the cases demonstrate heterologous mesenchymal differentiation including skeletal muscle, bone, cartilage, and epithelial and neuroendocrine elements (Ducatman and Scheithauer, 1984; Enzinger and Weiss, 1988). Tumors presenting divergent differentiation, however, do not differ from ordinary MPNST regarding clinical behavior and prognosis (Ducatman and Scheithauer, 1984).

The clinical outcome is directly related to the presence or absence of von Recklinghausen's disease. Survival rates are significantly lower in patients with the disorder than patients free of the disease (16% versus 53% for five-year survival, respectively; Ducatman et al., 1986).

The epithelioid variant of MPNST or *malignant epithelioid schwannoma* represents a rare tumor composed of plump cells with round vesicular nuclei, prominent nucleoli, and moderate quantities of cytoplasm that are disposed in lobules or groups resembling the pattern of carcinomas or of amelanotic melanomas. Areas resembling more typical MPNST are frequently present. Strong S-100 protein immunoreactivity (Chu and Shmookler, 1988; Robey et al., 1987; Wick et al., 1987) and lack of reactivity for cytokeratins (Wick et al., 1987) differentiate these tumors from carcinoma. In our experience and that of others (Burger et al., 1991), malignant epithelioid MPNST may also demonstrate immunoreactivity for HMB-45, a propertied marker of melanoma cells.

V. TUMORS OF THE MENINGES

A. Tumors of Meningothelial Cells

1. Meningioma

Tumors of the meningothelial cells are termed *meningiomas*. They represent approximately 15% of intracranial neoplasms and presumably arise from the arachnoid cells residing in arachnoidal villi, perivascular pial extensions, and the choroid plexus (Kepes, 1982; Russell and Rubinstein, 1989). Meningiomas commonly occur in mid-life and show a distinct female predilection (3:1) (Nakasu et al., 1987). The less common childhood meningiomas show no sex predilection but nearly 25% are associated with neurofibromatosis type 2. Pediatric tumors show more of a tendency to aggressive behavior, no surprise in that a higher incidence of the papillary variant is noted in childhood (Deen et al., 1982). Multiple tumors occur in approximately 8% of cases (Nakasu et al., 1987). Although familial cases are typically associated with von Recklinghausen's disease (Russell and Rubinstein, 1989), other genetic abnormalities have been described in affected families (Smidt et al., 1990). In 40 to 80% of sporadic meningiomas, cytogenetic analyses show loss of heterozygosity for loci on chromosome 22 (Dumanski et al., 1990; Poulsgard et al., 1990; Zankle and Zang, 1972; Zang, 1982). The genetic defect has been detected on 22q12.3 qter (Lekanne Deprez et al., 1991).

Though meningiomas may arise at any location in the meninges, certain sites are preferred. Intracranial tumors affect the midline along the superior longitudinal sinus, the cerebral convexity, the tuberculum sellae and parasellar region, and the olfactory grooves and sphenoidal ridge. For reasons unknown, intraventricular tumors more often affect the left lateral ventricle (Criscuolo and Symon, 1986). In the posterior fossa, the cerebellopontine angle is a favored site. Spinal meningiomas most frequently occur in the thoracic region, followed by the cervical and lumbar levels. Extra calvarial and ectopic meningiomas have also been described (Burger et al., 1991; Russell and Rubinstein, 1989).

Meningiomas most frequently present as circumscribed, globular, or spherical dura-based masses that are readily separable from underlying brain. An uncommon diffuse form of meningioma termed *en plaque* characteristically occurs at the skull base and forms a carpetlike growth over the sphenoid ridge. Often acute, the tumors abut adjacent structures without infiltration. Nonetheless, involvement of dura/venous sinuses or bone is common, resulting in tumoral and thrombotic occlusion of vascular sinuses or of hyperostosis, respectively. Although this invasive tendency does not connote malignancy, complete surgical removal of such tumors is often impossible, the result being a high incidence of local recurrence.

Meningiomas are subdivided into three histopathologic categories in accord with their cytoarchitectural features and aggressive or malignant potential (Table 5). A spectrum of histologic variants has been described that reflects the mesenchymal and epithelial differentiating potential of arachnoidal cells (Scheithauer, 1990). By and large, the basic morphologic variants exhibit similar biologic behavior. Their importance resides in their recognition as meningiomas as well as in the clinical associations of some.

The meningotheliomatous (syncytial) and transitional variants are readily recognized by their characteristic histopathologic features, the formation of whorls frequently around blood vessels, and other stromal elements (Fig. 22A). The cells of the former are polygonal in shape, exhibiting indistinct cellular borders as well as nuclei with delicate, uniformly dispersed chromatin, and inconspicuous nucleoli. Nuclear cytoplasmic invaginations are frequently observed and appear as "intranuclear pseudoinclusions." Transitional tumors possess not only epithelial but spindle-shaped cells with a noticeable tendency to whorl formation. Psammoma bodies, laminated calcific concretions, are a common feature. At the ultrastructural level, the hallmark of meningothelial cells is their tendency to membrane interdigitations, the accumulation of cytoplasmic intermediate filaments, and their formation of desmosomes and hemidesmosomes (Kepes, 1982).

The principal mesenchymal-appearing variants include the fibrous, angiomatous, and metaplastic meningiomas that elaborate extracellular matrices, the light-microscopic appearance of which ranges from fibrous tissue through cartilage to bone. This capacity for connective tissue formation

TABLE 5. Tumors of the Meninges

Tumors of meningothelial cells

1. Meningioma
   Histologic types:
      a. meningothelial (syncytial)
      b. transitional/mixed
      c. fibrous (fibroblastic)
      d. psammomatous
      e. angiomatous
      f. microcystic
      g. secretory
      h. clear cell
      i. chordoid
      j. lymphoplasmacyte-rich
      k. metaplastic variants (xanthomatous, myxoid, osseous, cartilagenous, etc.)
2. Atypical meningioma
3. Anaplastic (malignant) meningioma
      a. variants of 1a–k (see above)
      b. papillary

Non-meningothelial tumors of the meninges

1. Mesenchymal tumors
   Benign neoplasms
      a. Osteocartilagenous tumors
      b. Lipoma
      c. Fibrous histiocytoma
   Malignant neoplasms
      a. Hemangiopericytoma
      b. Chondrosarcoma
      c. Mesenchymal chondrosarcoma
      d. Malignant fibrous histiocytoma
      e. Rhabdomyosarcoma
      f. Meningeal sarcomatosis
2. Tumors of uncertain origin
      a. Hemangioblastoma (capillary hemangioblastoma)

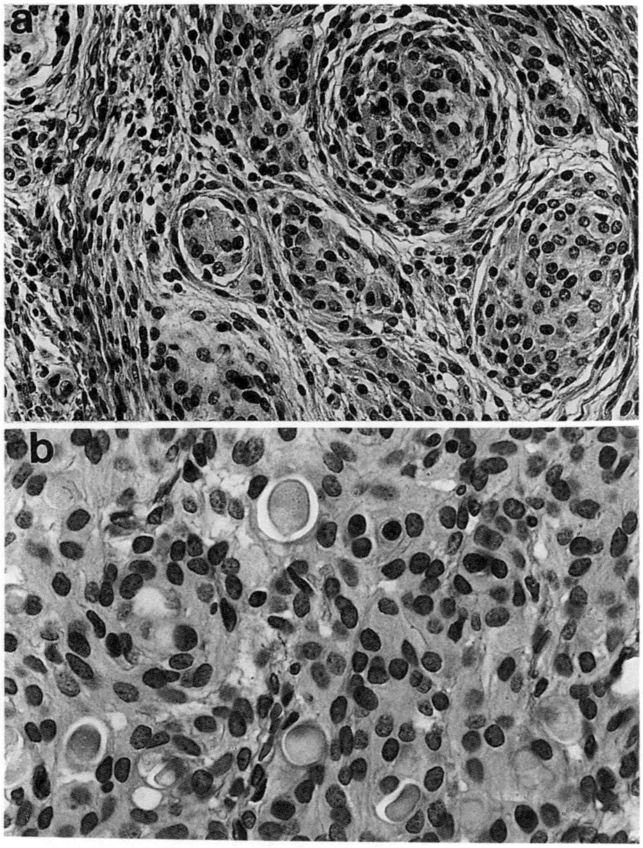

Fig. 22. Meningioma: The meningothelial variant of meningiomas displays the "syncytial" arrangement of cells. Whorl formations are most commonly present in this variant (**A**). Secretory meningiomas demonstrate eosinophilic cytoplasmic inclusions (**B**). (H & E; A: ×200; B: ×400.)

is most commonly exhibited by the fibrous variant, which demonstrates a reticulin and collagen-rich stroma. In biochemical terms these extracellular matrix proteins include laminin, fibronectin, and collagen types IV and V. Laminin expression is most prevalent in the fibrous meningiomas (Bellon et al., 1985; McComb and Bigner, 1985; Rutka et al., 1986).

The microcystic, secretory, clear-cell (glycogen-rich), chordoid, and papillary meningiomas are all examples of epithelial phenotypes. Eosinophilic, PAS-positive inclusion bodies are a common feature of the secretory subtype (Fig. 22B), which ultrastructurally demonstrates clear-cut epithelial features, including intracytoplasmic lumina with microvilli (Radley et al., 1989).

Immunohistochemical analyses support the concept that meningiomas express either or both mesenchymal and epithelial phenotypes. Vimentin is the major intermediate filament type demonstrated and appears to occur in almost 100% of meningiomas (Artlich and Schmidt, 1990; Russell and Rubinstein, 1989; Winek et al., 1989). Immunoreactivity for S-100 protein is variable, being present in 20 to 50% of cases, and is focally present with a predominance for fibroblastic over meningothelial areas (Artlich and Schmidt, 1990; Schnitt and Vogel, 1986; Winek et al., 1989). The epithelial nature of meningiomas is evidenced by immunoreactivity for EMA and, less often, by various cytokeratin epitopes. EMA immunoreactivity is variable, ranging from focal to diffuse reactivity, and with a incidence ranging from 50 to 100% of cases (Artlich and Schmidt, 1990; Meis et al., 1986; Schnitt and Vogel, 1986; Winek et al., 1989). Cytokeratin reactivity also tends to be focal and varies in frequency from an overall of less than 10% (Winek et al., 1989) to a peak of nearly 40% in some particularly epithelial variants (Artlich and Schmidt, 1990; Meis et al., 1986; Theaker et al., 1986). Cytokeratin immunoreactivity is rather common in meningothelial tumors but is a regular feature of secretory variants, wherein immunoreactive cells surround "pseudopsammomatous" inclusion bodies (Winek et al., 1989). Keratin reactivity may be coexpressed or absent in cells containing vimentin. Reactivity for GFAP is absent, an important differential diagnostic feature.

The results in the literature regarding the presence of estrogen and progesterone receptors in meningiomas are conflicting. In general it appears that estrogen receptors are far less frequent than progesterone receptors (Poisson et al., 1980; Russell and Rubinstein, 1989; Saitoh et al., 1989; Schrell et al., 1990; Tilzer et al., 1982). However, a recent report indicates that these hormone receptors are of an inactive form (Schrell et al., 1990). Further work is needed to more clearly define the role, if any, of various hormones in the growth and development of meningiomas, and the possible therapeutic implications of such a role.

2. Atypical meningioma

Atypical meningiomas are a somewhat ill-defined subpopulation of tumors that share an increased tendency to recurrence and aggressive behavior (Chen and Liu, 1990; de la Monte et al., 1986; Jääskelainen et al., 1986; Jellinger and Slowik, 1975; Jellinger, 1989; Scheithauer, 1990). Although they lack florid histopathologic anaplasia, significant histopathologic features serve to identify these tumors. Included are hypercellularity, nuclear pleomorphism, nucleolar prominence, increased mitotic activity, diffuse or sheet-like growth, and micronecrosis. In that these features are often not uniformly distributed throughout the tumor, careful sampling and the examination of multiple microsections are necessary for adequate pathologic evaluation. Although these histopathologic features can be encountered in most histologic variants, some types show a lesser tendency to atypia; these include microcystic, secretory, xanthomatous, and psammomatous tumors. An additional characteristic, the focal expression of a papillary pattern, is also indicative of atypia and likely recurrence (Chen and Liu, 1990).

As mentioned above in Section II, studies of tumor cell kinetics, based on the labeling indices as determined by immunohistochemical methods, e.g., Ki-67 and BUdR, flow cytometric evaluation, and the AgNOR technique, have been shown to be important adjuvants in the identification of atypical meningiomas and tumors likely to recur (Chin and Hinton, 1991; Cho et al., 1986; Jellinger, 1989; May et al., 1989; Lee et al., 1990; Orita et al., 1990).

3. Anaplastic (malignant) meningioma

The distinction of atypical from anaplastic meningiomas is, in histopathologic terms, a matter of degree. The same features that in a mild degree are synonymous with "atypia" are, when florid, indicative of anaplasia. These include a high mitotic index, marked cellular/nuclear pleomorphism, and necrosis. The evaluation of malignant potential cannot, in all instances, be based exclusively upon histopathological parameters. Frank brain invasion is considered a feature of malignancy despite the absence, in some cases, of histopathologic criteria of anaplasia or even significant atypia.

In addition to the above-mentioned indicators of proliferative activity (Section II), karyotypic analysis may be useful in the identification of tumors undergoing anaplastic or malignant transformation. For instance, recurrent and invasive meningiomas may undergo karyotypic evolution in addition to showing monosomy or deletions of chromosome 22 (Casartelli et al., 1989; Poulsgard et al., 1990).

The aberrant expression of growth factors or receptors as indicators of malignancy has not been fully explored. Indeed, the mechanisms of interaction of the growth factors and the meningiomas are poorly understood. Nonetheless, the presence of ß-PDGF receptors and PDGF-like growth factor has been documented in meningiomas as well as in primary meningioma cultures (Mauro et al., 1991). c-sis and c-myc expression have also been shown to occur in meningiomas (Kazumoto et al., 1990; Smidt et al., 1990). Subarachnoid spread and extraneural metastases are only rarely observed in the clinical course of malignant meningiomas (Kepes, 1982; Russell and Rubinstein, 1989). Subarachnoid dissemination is most frequently associated with tumors demonstrating histopathologic features of malignancy and postsurgical resections. However, leptomeningeal spread has also been reported in histologically benign meningiomas (Kepes, 1982). Extraneural metastases, although still uncommon, have been more frequently reported in cases that demonstrated leptomeningeal dissemination. Long clinical history, repeated local recurrences, and surgical procedures appear to contribute for this event (Russell and Rubinstein, 1989). About 25–20% of the reported cases were meningiomas of the hemangiopericytic variant (Kepes, 1982). However, this "variant" of meningiomas presents the same biologic, histopathologic, and immunohistochemical features as the hemangiopericytomas of the soft tissue and is now considered as a mesenchymal, non-meningothelial neoplasm of the meninges (see below, Mesenchymal, Non-meningothelial Tumors). The papillary variant is the second more frequent type of meningioma associated with extraneural metastases (Kepes, 1982) (Fig. 23). No apparent correlation between the other histological types of meningioma and the development of distant metastases have been noted (Kepes, 1982; Russell and Rubinstein, 1989).

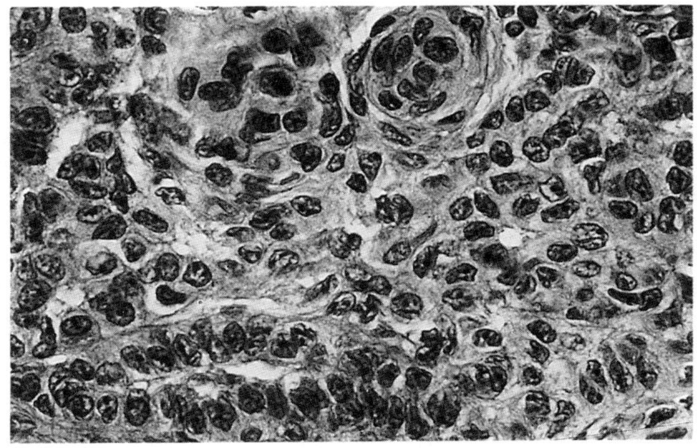

Fig. 23. Anaplastic meningioma: Focal papillary arrangement is one of histopathologic features suggestive of malignant transformation of meningiomas. Note a typical meningothelial whorl next to the papillary formation. (H & E; ×400.)

4. Papillary meningioma

This rare, often malignant variant of meningiomas occurs more frequently in children and young adults and demonstrates a significant tendency to recurrence and to local as well as brain invasion (Ludwin et al., 1975; Pasquier et al., 1986). Histologically the tumor is hypercellular and composed of epithelial-like cells disposed in papillary formations about blood vessels or arranged in "epithelial" lines resting upon a membrane (Kepes et al., 1983). The process of these tapered or columnar cells often contains fibrils that, like astrocytes or ependymal cells, show a remarkable affinity for PTAH. In contrast to the glial neoplasms, the papillary meningiomas do not manifest GFAP immunoreactivity. Ultrastructural studies demonstrate that these fibrils correspond to bundles of 10 nm filaments likely composed of vimentin (Piatt et al., 1986). Although the papillary pattern may predominate, in most instances classic features of meningiomas are readily recognized (Pasquier et al., 1986). High cellularity, mitotic activity, and brain invasion are often found.

B. Mesenchymal, Nonmeningothelial Tumors

A number of mesenchymal neoplasms may occasionally arise in the meninges. Generally, these tumors have similar features to their counterparts outside the CNS. The more common benign tumors in this category include chondroma, osteochondroma, osteoma, lipoma, and fibrous histiocytoma. The most commonly found sarcomas include hemangiopericytoma, chondrosarcoma, malignant fibrous histiocytoma, and rhabdomyosarcoma. It is particularly relevant to mention in this section the hemangiopericytomas, which have been removed from the category of tumors of meningothelial origin (in the context of angioblastic meningiomas) and are now classified as a nonmeningothelial, mesenchymal tumor. Therefore only this tumor is described in detail.

1. Hemangiopericytomas

Intense debate has surrounded the nature of this aggressive meningeal neoplasm. Many investigators consider them to be identical to hemangiopericytoma of the soft tissue (Burger et al., 1991; Guthrie et al., 1989; Jellinger and Paulus, 1991; Scheithauer, 1990), whereas others, applying the term *angioblastic meningioma*, feel that they share a common histogenesis with meningothelial neoplasms (Horten et al., 1977; Russell and Rubinstein, 1989). A review of the numerous studies that have shed light upon this matter is beyond the scope of this chapter (D'Amore et al., 1990; Holden et al., 1987; Nakamura et al., 1987; Theunissen et al., 1990; Winek et al., 1989). In the new WHO classification, hemangiopericytomas of the CNS have been placed in the category of malignant mesenchymal neoplasms.

Hemangiopericytomas of the meninges are rare, comprising 1 to 7% of all meningeal tumors. Most occur in adults, predominantly in men. Their aggressive behavior is confirmed by a high rate of recurrence (20 to 80%) and of metastases (10 to 55%) (Jellinger and Paulus, 1991; Mena et al., 1991). Overall survival is around 58 to 84 months (Guthrie et al., 1989; Jellinger and Paulus, 1991; Mena et al., 1991).

Histologically, the tumors exhibit similar features to those of hemangiopericytoma of the soft tissue, including high cellularity; a typical "staghorn" vascular pattern; round, oval, or slightly spindled cells; and a reticulin pattern that varies considerably from pericellular to sparse. The mitotic index varies considerably. Immunohistochemistry studies show vimentin positivity, but unlike meningiomas, EMA and cytokeratin epitopes are not demonstrable (D'Amore et al., 1990; Nakamura et al., 1987; Theunissen et al., 1990; Winek et al., 1989). A detailed ultrastructural study has shown a spectrum of differentiation varying from fibrous through myxoid to pericytic (Dardick et al., 1989).

C. Tumors of Uncertain Histogenesis

1. Capillary hemangioblastoma

Capillary hemangioblastomas represent approximately 2.5% of all intracranial tumors (Zülch, 1986) and primarily occur in adults (Russell and Rubinstein, 1989). Their typical location is infratentorial; though most involve the cerebellum, they also arise in the brain stem and spinal cord (Nakamura et al., 1985; Sanford and Smith, 1986). Cerebral examples are rare, the majority being associated with von Hippel-Lindau disease (Russell and Rubinstein, 1989).

Hemangioblastomas are often but not invariably cystic lesions in which the tumor forms a mural nodule. Most are cellular. Their histologic and cytologic features are variable in that they are minute clusters or large lobules of polygonal, often lipid-laden, "stromal" cells with hyperchromatic nuclei, in addition to a variably dense vascular stroma replete with pericytes. The nature of the stromal cells is unclear. Immunoreactivity for vimentin is a constant feature, but neuronal-specific enolase (NSE) and GFAP reactivity may also be noted. The general histopathologic architecture may closely resemble that of renal cell carcinomas, but the hemangioblastoma's lack of EMA reactivity serves to distinguish it from a metastasis (Hufnagel et al., 1989; Mills et al., 1990).

These tumors are usually discrete masses. Its "infiltration" of the adjacent parenchyma usually takes the form of creeping substitution associated with capillary outgrowth. Exuberant astrogliosis with Rosenthal fiber formation is frequently seen in the surrounding parenchyma, a feature that may prompt a mistaken diagnosis of pilocytic astrocytoma in small biopsies.

VI. HEMATOPOIETIC NEOPLASMS (TABLE 6)

A. Lymphomas

Primary CNS lymphomas are not uncommon tumors, representing approximately 1 to 5% of primary intracranial neoplasms. Recent epidemiologic studies suggest an increase in its incidence (Hochberg and Miller, 1988). Lymphomas occur sporadically but most are associated with inherited or acquired immunosuppression states, particularly the acquired immunodeficiency syndrome (AIDS) (Fig. 24). With inherited immunosuppression, children are

TABLE 6. Hematopoietic Tumors

1. Primary malignant lymphomas
2. Plasmocytoma
3. Granulocytic sarcoma
4. Others

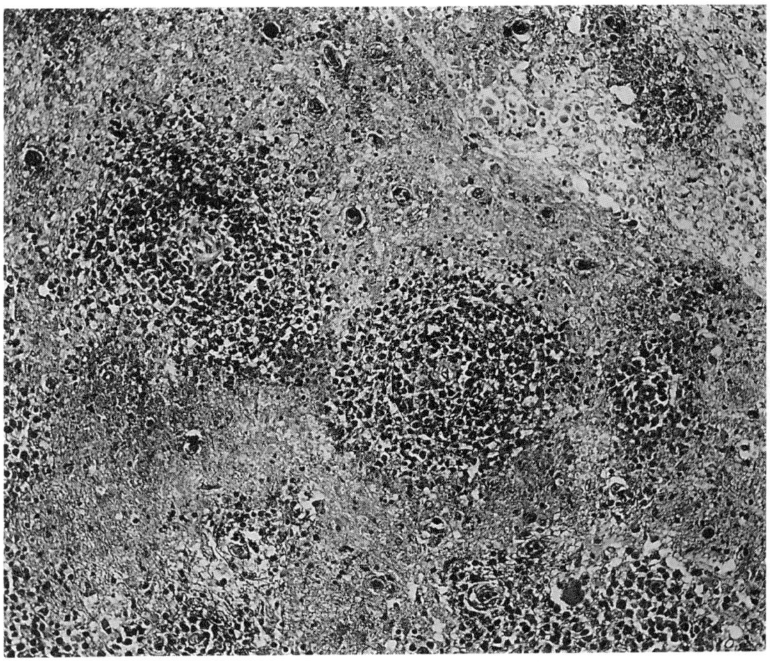

Fig. 24. Primary CNS lymphoma in acquired immunodeficiency syndrome (AIDS). Such tumors are typically large-cell or immunoblastic lymphomas that often feature necrosis. (H & E; ×160.)

most often affected. Transplant recipients and AIDS patients usually show signs of this disease in their fourth decade. In sporadic, immunosuppression-associated cases the peak incidence is after the fifth decade (Nakhleh et al., 1989). In all groups there is a slight predominance in men.

Primary CNS lymphomas are seen as solitary or multiple intracranial masses involving either or both the supra- or infratentorial regions. The frequency of multifocal lesions may be as high as 50% (Helle et al., 1984; Hochberg and Miller, 1988). Most lesions are deeply situated in the hemispheres. Perivascular, subependymal involvement is not uncommon. Intramedullary tumors, although rare, do occur and preferentially involve the upper thoracic or lower cervical levels (Slager et al., 1982). Preceding or concurrent ocular involvement, including vitreous, choroid, and/or retinal lesions, has been reported in approximately 10% of cases (Hochberg and Miller, 1988).

Although sporadic primary CNS lymphomas and those associated with immunosuppression display some differences with regard to clinical presentation and biological behavior (Hochberg and Miller, 1988), cytologically and immunohistochemically they are quite similar. Virtually all are non-Hodgkin's lymphomas with a diffuse architecture. Large-cell lymphomas are most common; among them, the immunoblastic variant accounts for approximately one-third (Hochberg and Miller, 1988). Nodular lymphomas are rare in the CNS. Independent of subtype, CNS lymphomas demonstrate a common and typical histopathologic feature—perivascular cuffing and infiltration of the vessel wall by the neoplastic cells. Hodgkin's disease is vanishingly rare, being limited to a few often-questioned cases (Ashby et al., 1988) or to instances of secondary involvement (Scheithauer, 1979).

Immunocytologic studies for cell surface immunoglobulins demonstrate that the great majority of CNS lymphomas are of B-cell type, exhibiting a monoclonal pattern of light chain production (Hamilton-Dutoit et al., 1991; Hochberg and Miller, 1988). T-cell lymphomas are relatively few.

Immunological studies have suggested a role for the Epstein-Barr virus (EBV) in the genesis of the primary CNS lymphomas. The presence of EBV genoma has been reported in primary CNS lymphomas of both sporadic- and immunosuppression-associated type (Bashir et al., 1989; Hamilton-Dutoit et al., 1991; Hochberg and Miller, 1988; MacMahon et al., 1991; Rouah et al., 1990). But EBV genoma is evident in only a few cases of primary CNS lymphomas in immunocompetent individuals (Bashir et al., 1989; Hochberg and Miller, 1988; MacMahon et al., 1991).

VII. GERM CELL TUMORS

Primary intracranial germ cell tumors are rare lesions that most commonly occur during childhood and adolescence. The histogenesis and differentiation of intracranial germ cell tumors are considered to be identical to that of gonadal and extragonadal tumors of the same type (Table 7). All variants of germ cell tumors have been described in the CNS; however, germinomas are the most common (Bjornsson et al., 1985; Hoffman et al., 1991; Russell and Rubinstein, 1989). Like their extragonadal counterparts, these tumors occur almost exclusively in the midline structures, particularly in the pineal and sellar regions, third ventricle and hypothalamus, and only seldom in the spinal cord.

A summary of the immunohistochemical characteristics of germ cell tumors appears in Table 8 and is the subject of a recent review (Niehans et al., 1988).

TABLE 7. Germ Cell Tumors

1. Germinoma
2. Embryonal carcinoma
3. Yolk sac tumor (endodermal sinus tumor)
4. Choriocarcinoma
5. Teratoma
   Variants: mature, immature, malignant
6. Mixed germ cell tumors

A. Germinoma

Germinomas represent the most common germ cell tumor of the neuraxis (Bjornsson et al., 1985; Hoffman et al., 1991; Russell and Rubinstein, 1989) and demonstrate a marked predominance in men. The vast majority arise in the pineal and sellar regions. They tend to be soft, infiltrative masses with ill-defined borders. Histologically, germinomas consist of pale polygonal cells with large, vesicular nuclei and prominent nucleoli. Lying singly or clustered, they incite a T-lymphocytic and often granulomatous reaction. Mitotic figures are frequent. Germinomas of the CNS show more of a tendency to epithelial differentiation than do those of the testis (Min and Scheithauer, 1990). Leptomeningeal dissemination is a frequent occurrence.

Compared to other malignant germ cell tumors, germinomas have, due to their radiosensitive nature, a more favorable clinical outcome. The presence of a less favorable element in the tumor, mainly embryonal carcinoma and choriocarcinoma, must be carefully sought, in that they have a very negative effect upon prognosis. Syncytiotrophoblastic differentiation evidenced by presence of multinucleated giant human chorionic gonadotropin(HCG)-immunoreactive cells can be found in these tumors and is of no prognostic significance.

B. Embryonal Carcinomas

After germinoma the embryonal carcinoma is the most "primitive" of germ cell tumors. Most present not in pure form but represent a component of mixed germ cell tumors in association with teratoma or yolk sac tumor. As pure tumors, embryonal carcinomas are rare, representing two of 70 tumors in the series of Bjornsson et al. (1985). Like germinomas, most occur in males during the second and third decades (Russell and Rubinstein, 1989). Histologically these tumors are epithelial in cytology and are composed of patternless sheets or cords or glands composed of cuboidal cells with large, vesicular nuclei with prominent nucleoli.

C. Yolk Sac Tumor

Yolk sac tumor (endodermal sinus tumor) is histologically characterized by cuboidal to columnar epithelial cells arranged in sheets, tubules, and papillary structures supported by a delicately vascularized connective stroma. The presence of special cellular arrangements termed "Shiller-Duval bodies" as well as of PAS-positive globules is typical of this lesion. Although yolk sac elements may be only one component of germ cell tumors, in rare circumstances they occur in pure form (four of 51 cases in one series; Hoffman et al., 1991). Immunoreactivity for α-fetoprotein is frequently present, as are elevated CSF and serum levels of this substance (Hoffman et al., 1991). Multinucleate HCG-immunoreactive syncytiotrophoblasts may occasionally be seen, but, as in germinoma, are neither indicative of choriocarcinoma nor of prognostic significance (Bjornsson et al., 1985; Nagamuna et al., 1984; Yamagami et al., 1987).

D. Choriocarcinoma

Pure choriocarcinomas are extremely rare (Bjornsson et al., 1986, case report and review) to occur in the neuraxis (two of 51 cases reported by Hoffman et al., 1991, and one of 70 in the series of Bjornsson et al., 1986). More often choriocarcinoma is but one element of a mixed germ cell tumor. Grossly, it is typified by hemorrhage. Microscopi-

TABLE 8. Immunocytology of Germ Cell Tumors

	PLAP	AFP	hCG	hPL	SP_1	CK	EMA
Germinoma*	+	-	±**	-	-	10%	-
Teratoma	+	±	-	-	-	±	±
Embryonal carcinoma	+	±	-	-	-	+	-
Yolk sac carcinoma	±	+	-	-	-	+	-
Choriocarcinoma	±	-	+	+	+	+	±

PLAP, placental alkaline phosphatase; AFP, α-fetoprotein; hCG, human chorionic gonadotropin; hPL, human placental lactogen; SP_1, pregnancy-specific protein; CK, cytokeratin; EMA, epithelial membrane antigen.

*Germinomas have been reported to show immunoreactivity for vimentin (15%) and neuron-specific enolase (90%).

**Isolated syncytiotrophoblastic cells may be present in 50% of germinomas.

(From Burger, P.C., B.W. Scheithauer, and F.S. Vogel [1991] Surgical Pathology of the Nervous System and Its Coverings, 3rd ed. New York: Churchill Livingstone.)

cally it is composed of syncytiotrophoblast and cytotrophoblast disposed in a bilayer about sinusoidal vessels. The syncytiotrophoblast cells are the source of HCG elevations.

E. Teratoma

Teratomas represent nearly 0.5% of all intracranial tumors (Zülch, 1986) and more frequently occur in men. The site most often affected is the pineal region (Russell and Rubinstein, 1989). As in gonadal and other extragonadal lesions, three principal variants are encountered. All are composed of tissues representing the three germ layers, i.e., ectoderm, endoderm, and mesoderm. As a rule, neuroectodermal elements are to be expected.

The *immature* form is most common and is composed of elements that resemble embryonal tissues. Not uncommonly, neuroepithelial elements may predominate with the formation of neuroepithelial tubules resembling primitive neural tubes to embryonal ependymal ribbons. Additional neural elements may consist of extensive fields of glial stroma and differentiating neuroblasts. Given the presumptive origin of these tumors, other elements showing differentiation consistent with all three "germ layers" are variably present in differing degrees of cellular maturation and tissue organization, including bone accompanying cartilage, skin with adnexa, and pigmented ocular tissue. The presence of these embryonal neuroectodermal tissues, unlike in embryonal tumors of the CNS or in immature teratomas of the gonads, do not in and of themselves indicate a less favorable prognosis (Hessler et al., 1992). Immature teratomas may be considerably more aggressive relative to tumor expansion than the mature teratomas, but the malignant growth potential of immature teratomas remains to be defined. Optimal treatment therefore is not well developed and the role of radiation/chemotherapy is unclear.

Mature teratomas are relatively uncommon but, like those in other locations, consist of fully mature tissue. Caution is in order, in that occasional mature teratomas are accompanied by malignant germ cell tumor, particularly germinoma. Sampling must be thorough.

The *malignant* variant of teratoma contains malignant component(s) resembling the common tumors occurring in normal adult organs and somatic tissues, e.g., adeno- and squamous carcinomas, sarcomas, etc.

VIII. TUMORS OF THE ANTERIOR PITUITARY

A. Pituitary Adenoma

Adenomas constitute the most common tumors of the sellar region and account for 8 to 10% of all intracranial tumors (Kovacs and Horvath, 1986; Zülch, 1986). They consist of hormone-producing cells of anterior pituitary type and, therefore, may demonstrate endocrine activity. They vary in size from microadenomas (≤1 cm) to macroadenomas (>1 cm), large tumors being associated with suprasellar extension, demineralization of the sellar floor, and an increased incidence of microscopic dural and gross, operatively apparent invasion (Scheithauer et al., 1986a; Selman et al., 1986). Nonencapsulated and soft in texture, adenomas displace and compress surrounding normal pituitary parenchyma. Approximately 40% show intraoperative evidence of invasion of dura, bone, cavernous sinus, etc. Histologically, pituitary adenomas may exhibit a wide morphologic appearance (trabecular, sinusoidal, diffuse, papillary arrangement, etc.). By and large, these histopathological patterns have no correlation to the hormone production or secretion. Multiple adenomas are not rare (Kontogeorgos et al., 1991). Pituitary adenomas are a key feature of multiple endocrine neoplasia, type I (Scheithauer et al., 1987). Although the etiology of pituitary adenomas is unclear, it is of interest that they occur in association with end-organ failure, e.g., Addison's disease (Scheithauer et al., 1983) and hypothyroidism (Scheithauer et al., 1985). The majority of pituitary adenomas are endocrinologically functioning tumors (Horvath and Kovacs, 1991) secreting prolactin (PRL), growth hormone (GH), adrenocorticotropic hormone (ACTH), or thyrotropin (TSH) (Horvath and Kovacs, 1991). Most of these are tumors producing gonadotropic hormones (LH, FSH), either singly or in combination, and the null cell adenomas, which are either chromophobic or oncocytic in appearance. Nearly 40% of the pituitary adenomas are clinically nonfunctioning. Null cell adenomas produce either no hormones or small quantities of glycoproteins, usually LH or FSH (Kovacs et al., 1980). Tumors producing multiple, often biochemically unrelated, pituitary hormones are termed *plurihormonal*; the most common example is the GH-PRL-TSH producing adenomas of acromegaly (Scheithauer et al., 1986b). For an updated review of the functional classification of the pituitary adenomas, see Horvath and Kovacs (1991), and Kovacs and Horvath (1986).

B. Pituitary Carcinoma

Carcinomas of the pituitary gland are extremely rare lesions (Pernicone and Scheithauer, 1992). Although histopathologically they may show cytologic anaplasia and mitotic activity, many do not. By definition, carcinomas are tumors that give rise to cerebrospinal or systemic metastases. The status of tumors exhibiting gross brain invasion as opposed to displacement or deep indentation remains to be clarified, but some tumors are also rare. Pituitary carcinomas may be either endocrinologically functional or silent.

REFERENCES

Albrecht, S., E. Rouah, L.E. Becker, et al. (1991) Immunoreactivity for transthyretin in choroid plexus neoplasms and brain metastases. J. Neuropathol. Exp. Neurol. 50:366.

Allegranza, A., S. Pileri, G. Frank, et al. (1990) Cerebral ganglioglioma with anaplastic oligodendroglial component. Histopathol. 17:439–441.

Artlich, A., and D. Schmidt (1990) Immunohistochemical profile of meningiomas and their histological subtypes. Hum. Pathol. 21:843–849.

Ashby, M.A., P.C. Barber, A.T. Holmes, et al. (1988) Primary intracranial Hodgkins' disease: A case report and discussion. Am. J. Surg. Pathol. 12:294–299.

Azzarelli, B., H.L. Rekate, and U. Roessman (1977) Subependymoma: A case report with ultrastructural study. Acta Neuropathol. 40:279–282.

Bader, J.L., A.T. Meadows, L.E. Zimmerman, et al. (1982) Bilateral retinoblastoma with ectopic intracranial retinoblastoma: Trilateral retinoblastoma. Cancer Genet. Cytogenet. 5:203–213.

Baker, D.L., U.R. Reddy, S. Pleasure, et al. (1990) Human central nervous system primitive neuroectodermal tumor expressing nerve growth factor receptors: CHP707m. Ann. Neurol. 28:136–145.

Bancel, B., M.F. Belin, A. Meiniel, et al. (1990) Contribution à l'étude de l'histogenèse des gliomes sous-épendymaires de la sclèrose tubéreuse de bourneville. Ann. Pathol. 10:109–116.

Banerjee, A.K., and V.K. Kak (1974) Pineoblastoma with spontaneous intra and extracranial metastasis. J. Pathol. 114:9–12.

Barnard, R.O., R. Bradford, T. Scott, et al. (1986) Gliomyosarcoma: Report of a case of rhabdomyosarcoma arising in a malignant glioma. Acta Neuropathol. 69:23–27.

Bashir, R.M., N.L. Harris, F.H. Hochberg, et al. (1989) Detection of Epstein-Barr virus in CNS lymphomas by in-situ hybridization. Neurology 39:813–817.

Bellon, G., T. Caulet, Y. Cam, et al. (1985) Immunohistochemical localization of macromolecules of the basement membrane and extracellular matrix of human gliomas and meningiomas. Acta Neuropathol. 66:245–252.

Bennett, J.P., Jr., and L.J. Rubinstein (1984) The biologic behavior of primary cerebral neuroblastoma: A reappraisal of the clinical course in a series of 70 cases. Ann. Neurol. 16:21–27.

Betsholtz, C., M. Nister, F. Rorsman, et al. (1989) Structural and functional aspects of platelet-derived growth factor and its role in the pathogenesis of glioblastoma. Mol. Chem. Neuropathol. 10:27–36.

Bigner, D.D., S.H. Bigner, J. Pontén, et al. (1981) Heterogeneity of genotypic and phenotypic characteristics of fifteen permanent cell lines derived from human gliomas. J. Neuropathol. Exp. Neurol. 40:201–229.

Bigner, D.D. (1989) Phenotypic analysis of medulloblastoma with monoclonal antibodies. In W.S. Fields (ed.). Primary Brain Tumors. A Review of Histologic Classification, chapter 5. New York: Springer Verlag, pp. 70–78.

Bigner, S.H., A.J. Wong, J. Mark, et al. (1987) Relationship between gene amplification and chromosomal deviations in malignant human gliomas. Cancer Genet. Cytogenet. 29:165–170.

Bigner, S.H., B. Vogelstein, and D.D. Bigner (1988) Chromosomal abnormalities and gene amplification in malignant gliomas. ISI Atlas Sci. Biochem. 1:333–336.

Bigner, S.H., J. Mark, and D.D. Bigner (1990) Cytogenetics of human brain tumors. Cancer Genet. Cytogenet. 47:141–154.

Bigner, S.H., and B. Vogelstein (1990) Cytogenetics and molecular genetics of malignant gliomas and medulloblastoma. Brain Pathol. 1:12–18.

Bjornsson J., B.W. Sheithauer, H. Okazaki, et al. (1985) Intracranial germ cell tumors: Pathobiological and immunohistochemical aspects of 70 cases. J. Neuropath. Exp. Neurol. 44:32–46.

Bjornsson, J., B.W. Scheithauer, and R.W. Leech (1986) Primary intracranial choriocarcinoma: A case report. Clin. Neuropathol. 5:242–245.

Bonnin, J.M., and E. Perentes (1988) Retinal S-antigen immunoreactivity in medulloblastomas. Acta Neuropathol. 76:204–207.

Borit, A., W. Blackwood, and W.G.P. Mair (1980) The separation of pineocytoma from pineoblastoma. Cancer 45:1408–1418.

Burgart, L.J., R.A. Robinson, S.F. Haddad, et al. (1991) Oncogene abnormalities in astrocytomas: EGF-R gene alone appears to be more frequently amplified and rearranged compared with other protooncogenes. Mod. Pathol. 4:183–186.

Burger, P.C. (1989) The grading of astrocytomas and oligodendrogliomas. In W.S. Fields (ed.). Primary Brain Tumors. A Review of Histologic Classification, chapter 13. New York: Springer Verlag, pp. 171–180.

Burger, P.C., F.C. Grahmann, A. Bliestle, et al. (1987) Differentiation in the medulloblastoma. A histological and immunohistochemical study. Acta Neuropathol. 73:115–123.

Burger, P.C., B.W. Scheithauer, and F.S. Vogel (1991) Surgical Pathology of the Nervous System and Its Coverings, 3rd ed. New York: Churchill Livingstone.

Burger, P.C., T. Shibata, and P. Kleihues (1986) The use of the monoclonal antibody Ki-67 in the identification of proliferating cells: Application to surgical neuropathology. Am. J. Surg. Pathol. 10:611–617.

Caccamo, D.V., M.M. Herman, and L.J. Rubinstein (1989) An immunohistochemical study of the primitive and maturing elements of human cerebral medulloepitheliomas. Acta Neuropathol. 79:248–254.

Campbell, M., and G.J. Chader (1988) Retinoblastoma cells in tissue culture. Ophthal. Paed. Genetics 9:171–199.

Casartelli, C., S.R. Rogatto, and J. Barbieri Neto (1989) Karyotypic evolution of human meningioma—progression through malignancy. Cancer Genet. Cytogenet. 40:33–45.

Chen, W.Y., and H.C. Liu (1990) Atypical (anaplastic) meningioma: Relationship between histologic features and recurrence—a clinicopathologic study. Clin. Neuropathol. 9:74–81.

Chin, L.S., and D.R. Hinton (1991) The standardized assessment of argyrophilic nucleolar organizer regions in meningeal tumors. J. Neurosurg. 74:590–596.

Cho, K.G., T. Hoshino, T. Nagashima, et al. (1986) Prediction of tumor-doubling time in recurrent meningiomas: Cell kinetic studies with bromodeoxyuridine labeling. J. Neurosurg. 65:790–794.

Choi, B.H., and R.C. Kim (1985) Expression of glial fibrillary acidic protein by immature oligodendroglia and its implications. J. Neuroimmunol. 8:215–235.

Chu T.-A., and B.M. Shmookler (1988) Malignant epithelioid Schwannoma: A light microscopic and immunohistochemical study. J. Surg. Oncol. 39:68–72.

Cockram, C.S. (1990) Growth factors, astrocytes and astrocytomas. Sem. Dev. Biol. 1:421–435.

Coffin, C.M., K. Mukai, and L.P. Dehner (1983) Glial differentiation in medulloblastomas. Histogenetic insight, glial reaction, or invasion of brain? Am. J. Surg. Pathol. 7:555–565.

Cornbrooks, C.J., D.J. Carey, J.A. McDonald, et al. (1983) *In vivo* and *in vitro* observations on laminin production by Schwann cells. Proc. Natl. Acad. Sci. USA 80:3850–3854.

Cosgrove, M., P.L. Fitzgibbons, A. Sherrod, *et al.* (1989) Intermediate filament expression in astrocytic neoplasms. Am. J. Surg. Pathol. 13:141–145.

Criscuolo, G.R., and L. Symon (1986) Intraventricular meningioma. A review of 10 cases of the National Hospital, Queen Square (1974–1985) with reference to the literature. Acta Neuropathol. 83:83–91.

D'Amore, E.S.G., J.C. Manivel, and J.H. Sung (1990) Soft-tissue and meningeal hemangiopericytomas: An immunohistochemical and ultrastructural study. Hum. Pathol. 21:414–423.

D'Andrea, A.D., R.J. Packer, L.B. Rorke, et al. (1987) Pineocytomas of childhood: A reappraisal of natural history and response to therapy.

Cancer 59:1353–1357.
Dardick, I., S. Hammar, B.W. Scheithauer, et al. (1989) Ultrastructural spectrum of hemangiopericytoma: A comparative study of fetal, adult, and neoplastic pericytes. Ultrastruct. Pathol. 13:111–154.
Daumas-Duport, C., B. Scheithauer, J. O'Fallon, et al. (1988a) Grading of astrocytomas. A simple and reproducible method. Cancer 62:2152–2165.
Daumas-Duport, C., B. Scheithauer, J.-P. Chodkiewicz, et al. (1988b) Dysembryoplastic neuroepithelial tumor: A surgically curable tumor of young patients with intractable partial seizures: Report of thirty-nine cases. Neurosurgery 23:545–556.
Davis, R.L. (1989) Grading of gliomas. In W.S. Fields (ed.). Primary Brain Tumors. A Review of Histologic Classification, chapter 11. New York: Springer Verlag, pp. 150–158.
De Chadarévian, J.-P., and R.D. Hollenberg (1979) Subependymal giant cell tumor of tuberose sclerosis. A light and ultrastructural study. J. Neuropathol. Exp. Neurol. 38:419–433.
de la Monte, S.M. (1989) Uniform lineage of oligodendrogliomas. Am. J. Pathol. 135:529–540.
de la Monte, S.M., J. Flickinger, and R.M. Linggood (1986) Histopathologic features predicting recurrence of meningiomas following subtotal resection. Am. J. Surg. Pathol. 10:836–843.
DeClerck, Y.A., and C. Lee (1985) Collagen synthesis by human neuroblastoma cells. In A.E. Evans, G.J. D'Angio, R.C. Seeger (eds). Advances in Neuroblastoma Research. New York: Alan R. Liss, pp. 239–247.
Deen, H.G., Jr., B.W. Scheithauer, and M.J. Ebersold (1982) Clinical and pathological study of meningiomas of the first two decades of life. J. Neurosurg. 56:317–322.
Dehner, L.P., P. Abenoza, and R.K. Sibley (1988) Primary cerebral neuroectodermal tumors: Neuroblastoma, differentiated neuroblastoma, and composite neuroectodermal tumor. Ultrastruct. Pathol. 12:479–494.
Denis-Donini, S., J. Glowinski, and A. Prochiantz (1984) Glial heterogeneity may define the three-dimensional shape of mouse mesencephalic dopaminergic neurones. Nature 307:641–643.
Detrick B., G.J. Chader, M. Rodrigues, et al. (1988) Coexpression of neuronal, glial, and major histocompatibility complex class II antigens on retinoblastoma cells. Cancer Res. 48:1633–1641.
Dropcho, E.J., J.H. Wisoff, R.W. Walker, et al. (1987) Supratentorial malignant gliomas in childhood: A review of fifty cases. Ann. Neurol. 22:355–364.
Ducatman, B.S., and B.W. Scheithauer (1983) Post-irradiation neurofibrosarcoma. Cancer 51:1028–1033.
Ducatman, B.S., and B.W. Scheithauer (1984) Malignant peripheral nerve sheath tumors with divergent differentiation. Cancer 54:1049–1057.
Ducatman, B.S., B.W. Scheithauer, D.G. Piepgras, et al. (1986) Malignant peripheral nerve sheath tumors. A clinicopathologic study of 120 cases. Cancer 57:2006–2021.
Dumanski, J.P., G.A. Rouleau, M. Nordenskjöld, et al. (1990) Molecular genetic analysis of chromosome 22 in 81 cases of meningioma. Cancer Res. 50:5863–5867.
Earnest, F. IV, P.J. Kelly, B.W. Scheithauer, et al. (1988) Cerebral astrocytomas: Histopathologic correlation of MR and CT contrast enhancement with stereotactic biopsy. Radiology 166:823–827.
Earnest, F. III, J.W. Kernohan, and W. McK. Craig (1950) Oligodendrogliomas. A review of 200 cases. Arch. Neurol. Psychiat. 63:964–976.
Edwards, M.S.B., R.J. Hudgins, C.B. Wilson, et al. (1988) Pineal region tumors in children. J. Neurosurg. 68:689–697.
Ekstrand, A.J., C.D. James, W.K. Cavenee, et al. (1991) Genes for epidermal growth factor receptor, transforming growth factor alpha, and epidermal growth factor and their expression in human gliomas in vivo. Cancer Res. 51:2164–2172.
Enzinger, F.M., and S.W. Weiss (1988) Soft Tissue Tumors, 2nd ed. St. Louis: CV Mosby, pp. 719–780.
Erlandson, R.A. (1985) Peripheral nerve sheath tumors. Ultrastruct. Pathol. 9:113–122.
Faillot, T., J.-P. Sichez, J.-L. Brault, et al. (1990) Lhermitte-Duclos disease (dysplastic gangliocytoma of the cerebellum). Report of a case and review of the literature. Acta Neurochir. 105:44–49.
Figols, J., J.R. Iglesias-Rozas, and E. Kazner (1985) Myelin basic protein (MBP) in human gliomas: A study of twenty-five cases. Clin. Neuropathol. 4:116–120.
Fu, Y-S., A.T.L. Chen, S. Kay, et al. (1974) Is subependymoma (subependymal glomerate astrocytoma) an astrocytoma or ependymoma? A comparative ultrastructural and tissue culture study. Cancer 34:1992–2008.
Fults, D., C.A. Pedone, G.A. Thomas, et al. (1990) Allelotype of human malignant astrocytoma. Cancer Res. 50:5784–5789.
Gallie, B.L., J.A. Squire, A. Goddard, et al. (1990) Biology of disease. Mechanism of oncogenesis in retinoblastoma. Lab. Invest. 62:394–408.
Gambarelli, D., J. Hassoun, M. Choux, et al. (1982) Complex cerebral tumor with evidence of neuronal, glial and Schwann cell differentiation: A histologic immunocytochemical and ultrastructural study. Cancer 49:1420–1428.
Garton, G.R., P.J. Schomberg, B.W. Scheithauer, et al. (1990) Medulloblastoma—prognostic factors and outcome of treatment: Review of the Mayo Clinic experience. Mayo Clin. Proc. 65:1077–1086.
Gennett, I.N., and W.K. Cavenee (1990) Molecular genetics in the pathology and diagnosis of retinoblastoma. Brain Pathol. 1:25–32.
Giangaspero, F., C. Doglioni, M.T. Rivano, et al. (1987) Growth factor in human brain tumors defined by the monoclonal antibody Ki-67. Acta Neuropathol. 74:179–182.
Gonzalez-Fernandez, F., M.B.S. Lopes, J.M. Garcia-Fernandez, et al. (1992) Expression of developmentally-defined retinal phenotypes in the histogenesis of retinoblastoma. Am. J. Pathol. 141:363-375.
Gould, V.E., R. Moll, I. Moll, et al. (1986) The intermediate filament complement of the spectrum of nerve sheath neoplasms. Lab. Invest. 55:463–474.
Gould, V.E., L.B. Rorke, D.S. Jansson, et al. (1990) Primitive Neuroectodermal tumors of the central nervous system express neuroendocrine markers and may express all classes of intermediate filaments. Hum. Pathol. 21:245–252.
Grant, J.W., P.V. Steart, A. Aguzzi, et al. (1989) Gliosarcoma: An immunohistochemical study. Acta Neuropathol. 79:305–309.
Gray, M.H., A.E. Rosenberg, G.R. Dickersin, et al. (1989) Glial Fibrillary acidic protein and keratin expression by benign and malignant nerve sheath tumors. Hum. Pathol. 20:1089–1096.
Guthrie, B.L., M.J. Ebersold, B.W. Scheithauer, et al. (1989) Meningeal hemangiopericytoma: Histopathological features, treatment, and long-term follow-up of 44 cases. Neurosurgery 25:514–522.
Haddad, S.F., S.A. Moore, R.L. Schelper, et al. (1991) Smooth muscle cells can comprise the sarcomatous component of gliosarcomas. J. Neuropathol. Exp. Neurol. 50:291.
Halliday, A.L., R.A. Sobel, and R.L. Martuza (1991) Benign spinal nerve sheath tumors: Their occurrence sporadically and in neurofibromatosis types 1 and 2. J. Neurosurg. 74:248–253.
Halmagyi, G.M., L.P. Bignold, and J.L. Allsop (1979) Recurrent subependymal giant-cell astrocytoma in the absence of tuberous sclerosis. J. Neurosurg. 50:106–109.
Hamilton-Dutoit, S.J., G. Pallesen, M.B. Franzmann, et al. (1991) AIDS-related lymphoma. Histopathology, immunophenotype, and association with Epstein-Barr virus as demonstrated by *in situ* nucleic acid hybridization. Am. J. Pathol. 138:149–163.
Hansson, E. (1988) Astroglia from defined brain regions as studied with primary cultures. Prog. Neurobiol. 30:369–397.

Hassoun, J., B. Devictor, D. Gambarelli, et al. (1984) Paired twisted filaments: A new ultrastructural marker of human pinealomas? Acta Neuropathol. 65:163–165.

Hassoun, J., and D. Gambarelli (1989) Pinealomas: Need for and ultrastructural diagnosis. In W.S. Fields (ed.). Primary Brain Tumors. A Review of Histologic Classification. New York: Springer Verlag, pp. 82–85.

Hassoun, J., D. Gambarelli, F. Grisoli, et al. (1982) Central neurocytoma: An electron-microscopic study of two cases. Acta Neuropathol. 56:151–156.

Hassoun, J., D. Gambarelli, J.C. Peragut, et al. (1983) Specific ultrastructural markers of human pinealomas: A study of four cases. Acta Neuropathol. 62:31–40.

Helle, T.L., R.H. Britt, and T.V. Colby (1984) Primary lymphoma of the central nervous system. Clinicopathological study of experience at Stanford. J. Neurosurg. 60:94–103.

Herpers, M.J.H.M., and H. Budka (1984) Glial fibrillary acidic protein (GFAP) in oligodendroglial tumors: Gliofibrillary oligodendroglioma and transitional oligoastrocytoma as subtypes of oligodendroglioma. Acta Neuropathol. 64:265–272.

Herrick, M.K., and L.J. Rubinstein (1979) The cytological differentiating potential of pineal parenchymal neoplasms (true pinealomas): A clinicopathologic study of 28 tumours. Brain 102:289–320.

Hessler, R.B., M.B.S. Lopes, A. Frankfurter, et al. (1991) Cytoskeletal immunohistochemistry of central neurocytomas. J. Neuropathol. Exp. Neurol. 50:363.

Hessler, R.B., M.E. Shaffrey, J.A. Jane, et al. (1992) Maturation in intracranial immature teratomas: A report of two cases. J. Neuropathol. Exp. Neurol. 51:368.

Hirose, T., T. Sano, and K. Kizawa (1986) Ultrastructural localization of S-100 protein in neurofibroma. Acta Neuropathol. 69:103–110.

Ho, K.-L. (1990) Histogenesis of sarcomatous component of the gliosarcoma: An ultrastructural study. Acta Neuropathol. 81:178–188.

Hochberg, F.H., and D.C. Miller (1988) Primary central nervous system lymphoma. J. Neurosurg. 68:835–853.

Hoffman, H.J., H. Otsubo, E.B. Hendrick, et al. (1991) Intracranial germ-cell tumors in children. J. Neurosurg. 74:545–551.

Holden, J., C.L. Dolman, and A. Churg (1987) Immunohistochemistry of meningiomas including the angioblastic type. J. Neuropathol. Exp. Neurol. 46:50–56.

Horten, B.C., and L.J. Rubinstein (1976) Primary cerebral neuroblastoma. A clinicopathologic study of 35 cases. Brain 99:735–756.

Horten, B.C., H. Urich, L.J. Rubinstein, et al. (1977) The angioblastic meningioma: A reappraisal of a nosological problem. J. Neurol. Sci. 31:387–410.

Horvath, E., and K. Kovacs (1991) The adenohypophysis. In: S.L. Asa and K. Kovacs (eds.). Functional Endocrine Pathology. Boston: Blackwell Scientific Publications, pp. 245-281.

Hosokawa, Y., Y. Tsuchihashi, H. Okabe, et al. (1991) Pleomorphic xanthastrocytoma. Ultrastructural, immunohistochemical, and DNA cytofluorometric study of a case. Cancer 68:853–859.

Hubbard, J.L., B.W. Scheithauer, D.B. Kispert, et al. (1989) Adult cerebellar medulloblastoma: The pathological, radiographic, and clinical disease spectrum. J. Neurosurg. 70:536-544.

Hufnagel, T.J., J.H. Kim, L.D. True, et al. (1989) Immunohistochemistry of capillary hemangioblastoma. Immunoperoxidase-labeled antibody staining resolves the differential diagnosis with metastalic renal cell carcinoma, but does not explain the histogenesis of capillary hemangioblastoma. Am. J. Surg. Pathol. 13:207–216.

Humphrey, P.A., A.J. Wong, B. Vogelstein, et al. (1990) Anti-synthetic peptide antibody reacting at the fusion junction of deletion-mutant epidermal growth factor receptors in human glioblastoma Proc. Natl. Acad. Sci. USA 87:4207–4211.

Jääskeläinen, J., M. Haltia, and A. Servo (1986) Atypical and anaplastic meningiomas: Radiology, surgery, radiotherapy, and outcome. Surg. Neurol. 25:233–242.

James, C.D., E. Carlbom, J.P. Dumanski, et al. (1988) Clonal genomic alterations in glioma malignancy stages. Cancer Res. 48:5546–5551.

Jellinger, K.L. (1989) Biologic behavior of meningiomas. In W.S. Fields (ed.). Primary Brain Tumors. A Review of Histologic Classification, chapter 17. New York: Springer Verlag, pp. 231–238.

Jellinger, K., and W. Paulus (1991) Mesenchymal, non-meningothelial tumors of the central nervous system. Brain Pathol. 1:79–87.

Jellinger, K., and F. Slowik (1975) Histological subtypes and prognostic problems in meningiomas. J. Neurol. 208:279–298.

Jenkins, R.B., D.W. Kimmel, C.A. Moertel, et al. (1989) A cytogenetic study of 53 human gliomas. Cancer Genet. Cytogenet. 39:253–279.

Johnson, D.L., R. Chandra, W.S. Fisher, et al. (1985) Trilateral retinoblastoma: Ocular and pineal retinoblastomas. J. Neurosurg. 63:367–370.

Johnson, R.E., B.W. Scheithauer, and D.C. Dahlin (1983) Melanotic neuroectodermal tumor of infancy. Cancer 52:661–666.

Jones, N.R., M.L. Rossi, M. Gregoriou, et al. (1990) Investigation of the expression of epidermal growth factor receptor and blood group A antigen in 110 human gliomas. Neuropathol. Appl. Neurobiol. 16:185–192.

Kajiwara, K., T. Nishizaki, T. Orita, et al. (1990) Silver colloid staining technique for analysis of glioma malignancy. J. Neurosurg. 73:113–117.

Kalyan-Raman, U.P., and W.C. Olivero (1987) Ganglioglioma: A correlative clinicopathological and radiological study of ten surgically treated cases with follow-up. Neurosurgery 20:428–433.

Katsetos, C.D., M.M. Herman, A. Frankfurter, et al. (1989) Cerebellar desmosplastic medulloblastomas. A further immunohistochemical characterization of the reticulin-free pale islands. Arch. Pathol. Lab. Med. 113:1019–1029.

Katsetos, C.D., M.M. Herman, A. Frankfurter, et al. (1991) Neuron-associated class III beta-tubulin isotype, microtubule-associated protein 2, and synaptophysin in human retinoblastomas *in situ*. Lab. Invest. 64:45–54.

Kawano, N. (1991) Pleomorphic xanthoastrocytoma (PXA) in Japan: Its clinico-pathologic features and diagnostic clues. Brain Tumor Pathol. 8:5–10.

Kazumoto, K., M. Tamura, H. Hoshino, et al. (1990) Enhanced expression of the sis and c-myc oncogenes in human meningiomas. J. Neurosurg. 72:786–791.

Kepes, J.J. (1982) Meningiomas. Biology, Pathology, and Differential Diagnosis. New York: Masson.

Kepes, J.J., S. Goldware, and R. Leoni (1983) Meningioma with pseudoglandular pattern. J. Neuropathol. Exp. Neurol. 42:61–68.

Kepes, J.J., L.J. Rubinstein, L. Ansbacher, et al. (1989) Histopathological features of recurrent pleomorphic xanthoastrocytomas: Further corroboration of the glial nature of this neoplasm. Acta Neuropathol. 78:585–593.

Kepes, J.J., L.J. Rubinstein, and H. Chiang (1984) The role of astrocytes in the formation of cartilage in gliomas. An immunohistochemical study of four cases. Am. J. Pathol. 117:471–483.

Kepes, J.J., L.J. Rubinstein, and L.F. Eng (1979) Pleomorphic xanthoastrocytoma: A distinctive meningocerebral glioma of young subjects with relatively favorable prognosis; a study of 12 cases. Cancer 44:1839–1852.

Kleihues, P., A. Aguzzi, T. Shibata, et al. (1989) Immunohistochemical assessment of differentiation and DNA replication in human brain tumors. In W.S. Fields (ed.). Primary Brain Tumors. A Review of Histologic Classification, chapter 8. New York: Springer Verlag, pp. 123–132.

Kleihues, P., A. Aguzzi, and O.D. Wiestler (1990) Cellular and molecular aspects of neurocarcinogenesis. Toxicol. Pathol. 18:193–203.

Kline, K.T., I. Damjanov, S.M. Katz, et al. (1979) Pineoblastoma: An electron microscopic study. Cancer 44:1692–1699.

Kontogeorgos, G., K. Kovacs, E. Horvath, et al. (1991) Multiple adenomas of the human pituitary. A retrospective autopsy study with clinical implications. J. Neurosurg. 74:243–247.

Korf, H.W., M. Czerwionka, J. Reiner, et al. (1987) Immunocytochemical evidence of molecular photoreceptor markers in cerebellar medulloblastomas. Cancer 60:1763–1766.

Korf, H.W., D.C. Klein, J.S. Zigler, et al. (1986) S-antigen-like immunoreactivity in a human pineocytoma. Acta Neuropathol. 69:165–167.

Kovacs, K., and E. Horvath (1986) Tumors of the pituitary gland. Atlas of Tumor Pathology, Fascicle 21. Washington, DC: Armed Forces Institute of Pathology.

Kovacs, K., E. Horvath, N. Ryan, et al. (1980) Null cell adenoma of the human pituitary. Virchows Arch. Pathol. Anat. 387:165–174.

Kros, J.M., A.A.W. de Jong, and T.H. van der Kwast (1992) Ultrastructural characterization of transitional cells in oligodendrogliomas. J. Neuropathol. Exp. Neurol. 51:186–193.

Kros, J.M., C.G. Van Eden, S.Z. Stefanko, et al. (1990) Prognostic implications of glial fibrillary acidic protein containing cell types in oligodendrogliomas. Cancer 66:1204–1212.

Krouwer, H.G.J., R.L. Davis, R. Silver, et al. (1991) Gemistocytic astrocytomas: A reappraisal. J. Neurosurg. 74:399–406.

Kurumado, K., and W. Mori (1976) Synaptic ribbon in the human pinealocyte. Acta Pathol. Jap. 26:381–384.

Kurumado, K., and W. Mori (1977) A morphological study on the pineal gland of human embryo. Acta Pathol. Jap. 27:527–531.

Kyritsis, A.P., M. Tsokos, T.J. Triche, et al. (1986a) Retinoblastoma: A primitive tumor with multipotential characteristics. Invest. Ophthal. Vis. Sci. 27:1760–1764.

Kyritsis, A.P., M. Tsokos, and G.J. Chader (1986b) Control of retinoblastoma cell growth by differentiating agents: Current work and future directions. Anticancer Res. 6:465–474.

Langford, L.A. (1986) The ultrastructure of the ependymoblastoma. Acta Neuropathol. 71:136–141.

Laurence, K.M. (1979) The biology of choroid plexus papilloma in infancy and childhood. Acta Neurochir. 50:79–90.

Laws, E.R., Jr., W.F. Taylor, M.B. Clifton, et al. (1984) Neurosurgical management of low-grade astrocytoma of the cerebral hemispheres. J. Neurosurg. 61:665–673.

Lee, K.S., T. Hoshino, L.A. Rodriguez, et al. (1990) Bromodeoxyuridine labeling study of intracranial meningiomas: Proliferative potential and recurrence. Acta Neuropathol. 80:311–317.

Leifer, D., T. Moore, T. Ukena, et al. (1989) Multifocal glioblastoma with liver metastases in the absence of surgery. J. Neurosurg. 71:772–776.

Leivo, I., E. Engvall, P. Laurila, et al. (1989) Distribution of merosin, a laminin-related tissue-specific basement membrane protein in human Schwann cell neoplasms. Lab. Invest. 61:426–432.

Lekanne Deprez, R.H., N.A. Groen, N.A. Van Biezen, et al. (1991) A t(4;22) in a meningioma points to the localization of a putative tumor-supressor gene. Am. J. Genet. 48:783–790.

Lillien, L.E., and M.C. Raff (1990) Differentiation signals in the CNS: Type-2 astrocyte development in vitro as a model system. Neuron 5:111–119.

Liwnicz, B.H., and L.J. Rubinstein (1979) The pathways of extraneural spread in metastasizing gliomas. A report of two cases and review of the literature. Hum. Pathol. 10:453–467.

Lombardi, D., B.W. Scheithauer, F.B. Meyer, et al. (1991) Symptomatic subependymoma: A clinicopathological and flow cytometric study. J. Neurosurg. 75:583–588.

Lopes, M.B.S., F. Gonzalez-Fernandez, S. Rosemberg, et al. (1991) Expression of temporally specific and divergent cell lineage in retinoblastoma. J. Neuropathol. Exp. Neurol. 50:292.

Louis, D.N., S. Edgerton, A.D. Thor, et al. (1991) Proliferating cell nuclear antigen and Ki-67 immunohistochemistry in brain tumors: A comparative study. Acta Neuropathol. 81:675–679.

Ludwin, S.K., L.J. Rubinstein, and D.S. Russell (1975) Papillary meningioma: A malignant variant of meningioma. Cancer 36:1363–1373.

MacMahon, E.M.E., J.D. Glass, S.D. Hayward, et al. (1991) Epstein-Barr virus in AIDS-related primary central nervous system lymphoma. Lancet 338:969–973.

Mapstone, T., M. McMichael, D. Goldthwait (1991) Expression of platelet-derived growth factors, transforming growth factors, and the ros gene in a variety of primary human brain tumors. Neurosurgery 28:216–222.

Margetts, J.C., and U.P. Kalyan-Raman (1989) Giant-celled glioblastoma of brain: A clinico-pathological and radiological study of ten cases (including immunohistochemistry and ultrastructure). Cancer 63:524–531.

Markesbery, W.R., R.M. Haugh, and A.B. Young (1981) Ultrastructure of pineal parenchymal neoplasms. Acta Neuropathol. 55:143–149.

Mauro A., A. Bulfone, A. Di Sapio, et al. (1991) Meningioma cells in culture secrete a PDGF-like growth factor and express functional PDGF β-receptors. J. Neuropathol. Exp. Neurol. 50:373.

May, P.L., J.C. Broome, J. Lawry, et al. (1989) The prediction of recurrence in meningiomas. A flow cytometric study of paraffin-embedded archival material. J. Neurosurg. 71:347–351.

McComb, R.D., and D.D. Bigner (1985) Immunolocalization of laminin in neoplasms of the central and peripheral nervous systems. J. Neuropathol. Exp. Neurol. 44:242.

McComb, R.D., J.M. Moul, and D.D. Bigner (1987) Distribution of type VI collagen in human gliomas: Comparison with fibronectin and glioma-mesenchymal matrix glycoprotein. J. Neuropathol. Exp. Neurol. 46:623–633.

McGirr, S.J., M.J. Ebersold, B.W. Scheithauer, et al. (1988) Choroid plexus papillomas: Long term follow-up of a surgically treated series. J. Neurosurg. 69:843–849.

Meis, J.M., K.L. Martz, and J.S. Nelson (1991) Mixed glioblastoma multiforme and sarcoma. A clinicopathologic study of 26 radiation therapy oncology group cases. Cancer 67:2342–2349.

Meis, J.M., N.G. Ordóñez, and J.M. Bruner (1986) Meningiomas: An immunohistochemical study of 50 cases. Arch. Pathol. Lab. Med. 110:934–937.

Mena, H., J.L. Ribas, G.H. Pezeshkpour, et al. (1991) Hemangiopericytoma of the central nervous system: A review of 94 cases. Hum. Pathol. 22:84–91.

Milbouw, G., J.D. Born, D. Martin, et al. (1988) Clinical and radiological aspects of dyplastic gangliocytoma (Lhermitte-Duclos disease): A report of two cases with review of the literature. Neurosurgery 22:124–128.

Miller, L.L., P.T. Ostrow, and R. Chau (1991) Characterization of gliosarcomas by image analysis of superimposed serial sections. J. Neuropathol. Exp. Neurol. 50:365.

Miller, R.H., C. French-Constant, and M.C. Raff (1989) The macroglial cells of the rat optic nerve. Ann. Rev. Neurosci. 12:517–534.

Mills, S.E., G.W. Ross, E. Perentes, et al. (1990) Cerebellar hemangioblastoma: Immunohistochemical distinction from metastatic renal cell carcinoma. Surg. Pathol. 3:121–132.

Min, K.W., and B.W. Scheithauer (1990) Pineal germinomas and testicular seminoma: A comparative ultrastructural study with special references to early carcinomatous transformation. Ultrastruct. Pathol. 14:483–496.

Morantz, R.A., I. Feigin, and J. Ransohoff III (1976) Clinical and pathological study of 24 cases of gliosarcoma. J. Neurosurg. 45:398–408.

Morantz, R.A., J.J. Kepes, S. Batnitzky, et al. (1979) Extraspinal ependymomas: Report of three cases. J. Neurosurg. 51:383–391.

Mørk, S.J., T.B. Halvorsen, K.F. Lindegaard, et al. (1986) Oligodendroglioma. Histologic evaluation and prognosis. J. Neuropathol. Exp. Neurol. 19:15–21.

Mørk, S.J., K.-F. Lindegaad, T.B. Halvorsen, et al. (1985) Oligodendroglioma: Incidence and biological behavior in a defined population. J. Neurosurg. 63:881–889.

Mørk, S.J., and L.J. Rubinstein (1985) Ependymoblastoma. A reappraisal of a rare embryonal tumor. Cancer 55:1536–1542.

Muller, W., D. Afra, and R. Schroder (1977) Supratentorial recurrences of gliomas: Morphological studies in relation to time intervals with 544 astrocytomas. Acta Neurochir. 37:75–91.

Nagamuna H., H. Inoue, S. Misumi, et al. (1984) Intracranial germ-cell tumors. Immunohistochemistry study of three autopsy cases. J. Neurosurg. 61:931–937.

Nakagawa, Y., E. Perentes, and L.J. Rubinstein (1986) Immunohistochemical characterization of oligodendrogliomas: An analysis of multiple markers. Acta Neuropathol. 72:15–22.

Nakamura, Y., and L.E. Becker (1983) Subependymal giant-cell tumor: Astrocytic or neuronal? Acta Neuropathol. 60:271–277.

Nakamura, M., H.K. Inoue, N. Ono, et al. (1987) Analysis of hemangiopericytic meningiomas by immunohistochemistry, electron microscopy and cell culture. J. Neuropathol. Exp. Neurol. 46:57–71.

Nakamura, N., H. Sekino, Y. Taguchi, et al. (1985) Successful total extirpation of hemangioblastoma originating in the medulla oblongata. Surg. Neurol. 24:87–94.

Nakasu, S., A. Hirano, T. Shimura, et al. (1987) Incidental meningiomas in autopsy study. Surg. Neurol. 27:319–322.

Nakhleh, R.E., J.C. Manivel, D. Hurd, et al. (1989) Central nervous system lymphomas. Immunohistochemical and clinicopathologic study of 26 autopsy cases. Arch. Pathol. Lab. Med. 113:1050–1056.

Ng, T.H.K., C.F. Furg, and L.T. Ma (1990) The pathological spectrum of desmoplastic infantile gangliogliomas. Histopathology 16:235–241.

Ng, T.H.K., and W.S. Poon (1990) Gliosarcoma of the posterior fossa with features of a malignant fibrous histiocytoma. Cancer 65:1161–1166.

Niehans, G.A., J.C. Manivel, G.T. Copland, et al. (1988) Immunohistochemistry of germ cell and trophoblastic neoplasms. Cancer 62:1113–1123.

Nishio, S., T. Tashima, I. Takeshita, et al. (1988) Intraventricular neurocytoma-clinicopathological features of six cases. J. Neurosurg. 68:665–670.

Obana, W.G., P.H. Cogen, R.L. Davis, et al. (1991) Metastatic juvenile pilocytic astrocytoma. J. Neurosurg. 75:972–975.

Ojeda, V.J., D.V. Spagnolo, and R.J. Vaughan (1987) Palisades in primary cerebral neuroblastoma simulating so-called polar spongioblastoma. A light and electron microscopy study of an adult case. Am. J. Surg. Pathol. 11:316–322.

Orita, T., K. Kajiwara, T. Nishizaki, et al. (1990) Nucleolar organizer regions in meningioma. Neurosurgery 26:43–46.

Pagni, C.A., S. Canavero, M.T. Giordana, et al. (1992) Spinal intramedullary subependymomas: Case report and review of the literature. Neurosurgery 30:115–117.

Parisi, J.E., B.W. Scheithauer, J.R. Priest, et al. (1992) Desmoplastic infantile ganglioglioma (DIG): A form of gangliogliomatosis? J. Neuropathol. Exp. Neurol. (in press)(abstr.).

Pasquier, B., F. Gasnier, D. Pasquier, et al. (1986) Papillary meningioma. Clinicopathologic study of seven cases and review of the literature. Cancer 58:299–305.

Pasquier, B., I. Kojder, F. Labat, et al. (1985) Le xanthoastrocytome du sujet jeune: Revue de la littérature à propos de deux observations d'évolution discordante. Ann. Pathol. 5:29–43.

Paulus, W., and W. Jänisch (1990) Clinicopathologic correlations in epithelial choroid plexus neoplasms: A study of 52 cases. Acta Neuropathol. 80:635–641.

Paulus, W., and J. Peiffer (1988) Does the pleomorphic xanthoastrocytoma exist? Problems in the application of immunological techniques to the classification of brain tumors. Acta Neuropathol. 76:245–252.

Pearl, G.S., Y. Takei, BaKay R.A.E., and Davis P. (1985) Intraventricular primary cerebral neuroblastoma in adults: Report of three cases. Neurosurg. 16:847-849.

Perentes, E., and L. Rubinstein (1986) Immunohistochemical recognition of human neuroepithelial tumors by anti-Leu 7 (HNK-1) monoclonal antibody. Acta Neuropathol. 69:227–233.

Perentes, E., and L.J. Rubinstein (1987) Recent applications of immunoperoxidase histochemistry in human neuro-oncology. Arch. Pathol. Lab. Med. 111:796–812.

Perentes, E., C.P. Herbort, L.J. Rubinstein, et al. (1987) Immunohistochemical characterization of human retinoblastomas in situ with multiple markers. Am. J. Ophthalmol. 103:647–658.

Perentes, E., L.J. Rubinstein, M.M. Herman, et al. (1986) S-antigen immunoreactivity in human pineal glands and pineal parenchymal tumors. A monoclonal antibody study. Acta Neuropathol. 71:224–227.

Pernicone, P., and B.W. Scheithauer (1992) Invasive pituitary adenoma and carcinoma. In: R. Lloyd (ed.). The Pituitary Gland. Major Problems in Pathology. Philadelphia: Saunders (in press)

Piatt, J.H., Jr., G.A. Campbell, and W.J. Oakes (1986) Papillary meningioma involving the oculomotor nerve in an infant: Case report. J. Neurosurg. 64:808–812.

Plate, K.H., J. Ruschoff, and H.D. Mennel (1991) Cell proliferation in intracranial tumours: Selective silver staining of nucleolar organizer regions (AgNORs). Application to surgical and experimental neuro-oncology. Neuropathol. Appl. Neurobiol. 17:121–132.

Poisson, M., H. Magdelenat, J.F. Foncin, et al. (1980) Récepteurs d'oestrogènes et de progestérone dans les méningiomes. Étude de 22 cas. Rev. Neurol. 136:193–203.

Pollak, A., and R.L. Friede (1977) Fine structure of medulloepithelioma. J. Neuropathol. Exp. Neurol. 36:7121–7125.

Popoff, N.A., and R.M. Ellsworth (1971) The fine structure of retinoblastoma. Lab. Invest. 25:238–240.

Poulsgard, L., H.D. Schrøder, and M. Rønne (1990) Cytogenetic studies of 11 meningiomas and their clinical significance. II. Anticancer Res. 10:535–538.

Pulitzer, D.R., P.C. Martin, P.C. Collins, et al. (1988) Subcutaneous sacrococcygeal ("myxopapillary") ependymal rests. Am. J. Surg. Pathol. 12:672–677.

Radley, M.G., P.A. Di Sant'Agnese, T.A. Eskin, et al. (1989) Epithelial differentiation in meningiomas. An immunohistochemical, histochemical and ultrastructural study—with review of literature. Am. J. Clin. Pathol. 92:266–272.

Raff, M.C., R.H. Miller, and M. Noble (1983) A glial progenitor cell that develops *in vitro* into an astrocyte or an oligodendrocyte depending on culture medium. Nature 303:390–396.

Rao, C., M.E. Friedlander, E. Klein, et al. (1990) Medullomyoblastoma in an adult. Cancer 65:157–163.

Rawlinson, D.G., M.M. Herman, and L.J. Rubinstein (1973) The fine structure of a myxopapillary ependymoma of the filum terminale. Acta Neuropathol. 25:1–13.

Reiter, R.J. (1981) The Pineal Gland. Anatomy and Biochemistry, vol. 1. Boca Raton, Fla.: CRC Press, pp. 121–154.

Robey, S.S., S.H. deMent, K.K. Eaton, et al. (1987) Malignant epithelioid peripheral nerve sheath tumor arising in a benign schwannoma. Surg. Neurol. 28:441–446.

Rorke, L.B. (1989) Primitive neuroectodermal tumor—A concept requiring an apologia? In: W.S. Fields (ed.). Primary Brain Tumors.

A Review of Histologic Classification, chapter 1. New York: Springer Verlag, pp. 5–15.
Ross, G.W., and L.J. Rubinstein (1989) Lack of histopathological correlation of malignant ependymomas with postoperative survival. J. Neurosurg. 70:31–36.
Rouah, E., B.B. Rogers, D.R. Wilson, et al. (1990) Demonstration of Epstein-Barr virus in primary central nervous system lymphomas by the polymerase chain reaction and in situ hybridization. Hum. Pathol. 21:545–550.
Rubinstein, L.J. (1972) Tumors of the Central Nervous System. Atlas of Tumor Pathology (Fascicle 6). Washington, DC: Armed Forces Institute of Pathology.
Rubinstein, L.J. (1988) Diagnostic markers in human neurooncology. A progress report. Ann. N.Y. Acad. Sci. 540:78–90.
Rubinstein, L.J. (1989) Justification for a cytogenetic scheme of embryonal central neuroepithelial tumors. In: W.S. Fields (ed.). Primary Brain Tumors. A Review of Histologic Classification, chapter 2. New York: Springer Verlag, pp. 16–27.
Rubinstein, L.J. (1991) Glioma cytology and differentiation viewed through the window of neoplastic vulnerability. In: M. Salcman (ed.). Neurobiology of Brain Tumors. Baltimore: William and Wilkins, pp. 35–51.
Russell, D.S., and L.J. Rubinstein (1989) Pathology of Tumours of the Nervous System, 5th ed. London: Edward Arnold, pp. 124–125.
Rutka, J.T., J. Giblin, D.V. Dougherty, et al. (1986) An ultrastructural and immunocytochemical analysis of leptomeningeal and meningioma cultures. J. Neuropathol. Exp. Neurol. 45:285–303.
Sage, H., and M.L. Iruela-Arispe (1990) Type VIII collagen in murine development. Association with capillary formation in vitro. Ann. N.Y. Acad. Sci. 580:17–31.
Saitoh, Y., Y. Oku, S. Izumoto, et al. (1989) Rapid growth of a meningioma during pregnancy: Relationship with estrogen and progesterone receptors—case report. Neurol. Medico-Chirurg. 29:440–443.
Salgaller, M., L. Agius, A. Yates, et al. (1990) Application of automated image analysis to demonstrate the correlation between *ras* p21 expression and severity of gliomas. Bioch. Biophys. Res. Comm. 169:482–491.
Sanford, R.A., and R.A. Smith (1986) Hemangioblastoma of the cervicomedullary junction: Report of three cases. J. Neurosurg. 64:317–321.
Scheithauer, B.W. (1978) Symptomatic subependymoma: Report of 21 cases with review of the literature. J. Neurosurg. 49:689–696.
Scheithauer, B.W. (1979) Cerebral metastasis in Hodgkin's disease. Arch. Pathol. Lab. Med. 103:284–287.
Scheithauer, B.W. (1985) Neuropathology of pineal region tumors. Clin. Neurosurg. 32:351–383.
Scheithauer, B.W. (1990) Tumors of the meninges: Proposed modifications of the World Health Organization classification. Acta Neuropathol. 80:343–354.
Scheithauer, B.W., C. Hayostek, E.G. Shaw, et al. (1991) Astrocytomas of the cerebellum: Parameters and prognosis in 132 cases. Pathol. Res. Prac. 187:763 (abstr.).
Scheithauer, B.W., K.T. Kovacs, E.R. Laws, et al. (1986a) Pathology of invasive pituitary tumors with special reference to functional classification. J. Neurosurg. 65:733–744.
Scheithauer, B.W., E. Kovacs, K. Kovacs, et al. (1986b) Plurihormonal pituitary adenomas. Semin. Diagn. Pathol. 3:69–82.
Scheithauer, B.W., K. Kovacs, and R.V. Randall (1983) The pituitary in untreated Addison's disease: A histologic and immunocytologic study of 18 adenohypophyses. Arch. Pathol. Lab. Med. 107:484–487.
Scheithauer, B.W., K. Kovacs, R.V. Randall, et al. (1985) Pituitary gland in hypothyroidism: Histologic and immunocytologic study. Arch. Pathol. Lab. Med. 109:499–504.
Scheithauer, B.W., E.R. Laws, K. Kovacs, et al. (1987) Pituitary adenomas of the multiple endocrine neoplasia type I syndrome. Semin. Diagn. Pathol. 4:205–211.
Scherer, H.J. (1940) Cerebral astrocytomas and their derivatives. Am. J. Cancer 40:159–198.
Schindler, E., and F. Gullotta (1983) Glial fibrillary acidic protein in medulloblastomas and other embryonic CNS tumours of children. Virchows Arch. [A] 398:263–275.
Schnitt, S.J., and H. Vogel (1986) Meningiomas: Diagnostic value of immunoperoxidase staining for epithelial membrane antigen. Am. J. Surg. Pathol. 10:640–649.
Schrell, U.M., E.F. Adams, R. Fahlbusch, et al. (1990) Hormonal dependency of cerebral meningiomas. I. Female sex steroid receptors and their significance as specific markers for adjuvant medical therapy. J. Neurosurg. 73:743–749.
Selman, W.R., E.R. Laws, B.W. Scheithauer, et al. (1986) The occurrence of dural invasion in pituitary adenomas. J. Neurosurg. 64:402–407.
Shapiro, J.R., and W.R. Shapiro (1984) Clonal tumor cell heterogeneity. Prog. Exp. Tumor Res. 27:49–66.
Shapiro, J.R., W.-K.A. Yung, and W.R. Shapiro (1981) Isolation, karyotype, and clonal growth of heterogeneous subpopulations of human malignant gliomas. Cancer Res. 41:2349–2359.
Shaw, E.G., B.W. Scheithauer, J.R. O'Fallon, et al. (1992) Oligodendrogliomas: The Mayo Clinic experience. J. Neurosurg. 76:428–434.
Shepherd, C.W., B.W. Scheithauer, M.R. Gomez, et al. (1991) Subependymal giant cell astrocytoma: A clinical, pathological, and flow cytometric study. Neurosurgery 28:864–868.
Shiurba, R.A., N.S. Buffinger, E.M. Spencer, et al. (1991) Basic Fibroblastic growth factor and somatomedin C in human medulloepithelioma. Cancer 68:798–808.
Sima, A.A.F., and D.M. Robertson (1979) Subependymal giant-cell astrocytoma. Case report with ultrastructural study. J. Neurosurg. 50:240–245.
Slager, U.T., R.L. Kaufman, K.L. Cohen, et al. (1982) Primary lymphoma of the spinal cord. J. Neuropathol. Exp. Neurol. 41:437–445.
Slooff, J.L., J.W. Kernohan, and C.S. MacCarty (1964) Primary Intramedullary Tumors of the Spinal Cord and Filum Terminale. Philadelphia: W.B. Saunders.
Smidt, M., I. Kirsch, and L. Ratner (1990) Deletion of Alu sequences in the fifth c-sis intron in individuals with meningiomas. J. Clin. Invest. 86:1151–1157.
Smith, M.T., C.L. Ludwig, A.D. Godfrey, et al. (1983) Grading of oligodendrogliomas. Cancer 52:2107–2114.
Sobel, R.A., J.E. Trice, S.L. Nielsen, et al. (1981) Pineoblastoma with ganglionic and glial differentiation: Report of two cases. Acta Neuropathol. 55:243–246.
Sonneland, P.R.L., B.W. Scheithauer, and B.M. Onofrio (1985) Myxopapillary ependymoma: A clinicopathologic and immunocytochemical study of 77 cases. Cancer 56:883-893.
Specht, C.S., T.W. Smith, U. DeGirolami, et al. (1986) Myxopapillary ependymoma of the filum terminale: A light and electron microscopic study. Cancer 58:310–317.
Stanton, C., E. Perentes, V.P. Collins, et al. (1987) GFA protein reactivity in nerve sheath tumors: A polyvalent and monoclonal antibody study. J. Neuropathol. Exp. Neurol. 46:634–643.
Stefanko, S.Z., and W.A. Manschot (1979) Pinealoblastoma with retinomatous differentiation. Brain 102:321–332.
Strom, E.H., and K. Skullerud (1983) Pleomorphic xanthoastrocytoma: Rreport of 5 cases. Clin. Neuropathol. 2:188–191.
Sugawa, N., A.J. Ekstrand, C.D. James, et al. (1990) Identical splicing of aberrant epidermal growth factor receptor transcripts from amplified rearranged genes in human glioblastomas. Proc. Natl. Acad. Sci. USA 87:8602–8606.
Sutton, L.N., R.J. Packer, L.B. Rorke, et al. (1983) Cerebral

gangliogliomas during childhood. Neurosurgery 13:124–128.
Takahashi, J.A., H.I. Mori, M. Fukumoto, et al. (1990) Gene expression of fibroblasts growth factors in human gliomas and meningiomas: Demonstration of cellular source of basic fibroblastic growth factor mRNA and peptide in tumor tissues. Proc. Natl. Acad. Sci. USA 87:5710–5714.
Tarlton, J.F., and D.L. Easty (1990) Immunohistochemical characterization of retinoblastoma and related ocular tissue. Br. J. Ophthalmol. 74:144–149.
Terenghi, G., J.M. Polak, J. Ballesta, et al. (1984) Immunocytochemistry of neuronal and glial markers in retinoblastoma. Virchows Arch. [A] 404:61–73.
Theaker, J.M., K.C. Gatter, M.M. Esiri, et al. (1986) Epithelial membrane antigen and cytokeratin expression by meningiomas: An immunohistological study. J. Clin. Pathol. 39:435–439.
Theunissen, P.H.M.H., M.D.-T. Baerts, and G. Blaauw (1990) Histogenesis of intracranial haemangiopericytoma and haemangioblastoma. An immunohistochemical study. Acta Neuropathol. 80:68–71.
Thomas, G.A., and C. Raffel (1991) Loss of heterozygosity on 6q, 16q and 17p in human central nervous system primitive neuroectodermal tumors. Cancer Res. 51:639–643.
Tilzer, L.L., F.V. Plapp, J.P. Evans, et al. (1982) Steroid receptor proteins in human meningiomas. Cancer 49:633–636.
Tomlinson, F.H., R. Jenkins, B.W. Scheithauer, et al. (1992a) Aggressive medulloblastoma with high level n-myc amplification. Mayo Clin. Proc. (in press).
Tomlinson, F.H., B.W. Scheithauer, and R.B. Jenkins (1992b) Medulloblastoma: A pathobiologic overview. J. Child. Neurol. (in press).
Tomlinson, F.H., B.W. Scheithauer, P.J. Kelly, et al. (1991) Subependymoma with rhabdomyosarcomatous differentiation: Report of a case and literature review. Neurosurgery 28:761–768.
Townsend, J.J., and J.P. Seaman (1986) Central neurocytoma—a rare benign intraventricular tumor. Acta Neuropathol. 71:167–170.
Trombley, I.K., and S.S. Mirra (1981) Ultrastructure of tuberous sclerosis: Cortical tuber and subependymal tumor. Ann. Neurol. 9:174–181.
Troost, D., G.H. Jansen, and K.P. Dingemans (1990) Cerebral medulloepithelioma—electron microscopy and immunohistochemistry. Acta Neuropathol. 80:103–107.
Tso, M.O., B.S. Fine, and L.E. Zimmerman (1969) The Flexner-Wintersteiner rosettes in retinoblastoma. Arch. Pathol. 88:664–671.
Tso, M.O., L.E. Zimmerman, and B.E. Fine (1970) The nature of retinolastoma. I. Photoreceptor differentiation. A clinical and histopathologic study. Am. J. Ophthalmol. 69:339–349.
Tsokos, M., R.A. Ross, and T.J. Triche (1985) Neuronal, Schwannian and melanocytic differentiation of human neuroblastoma cells *in vitro*. In: A.E. Evans, G.J. D'Angio, R.C. Seeger (eds.). Advances in Neuroblastoma Research. New York: Alan R. Liss, pp. 55–68.
Tsokos, M., S. Scarpa, R.A. Ross, and T.J. Triche (1987) Differentiation of human neuroblastoma recapitulates neural crest development. Study of morphology, neurotransmitter enzymes, and extracellular matrix proteins. Am. J. Pathol. 128:484–496.
Uematsu, Y., T. Itakura, S. Hayashi, et al. (1988) Pineoblastoma with an unusually long survival. Case report. J. Neurosurg. 69:287–291.
VandenBerg, S.R., E.E. May, L.J. Rubinstein, et al. (1987a) Desmoplastic supratentorial neuroepithelial tumors of infancy with divergent differentiation potential ("desmoplastic infantile gangliogliomas"). Report on 11 cases of a distinctive embryonal tumor with favorable prognosis. J. Neurosurg. 66:58–71.
VandenBerg, S.R. (1991) Desmoplastic infantile ganglioglioma: A clinicopathologic review of sixteen cases. Brain Tumor Pathol. 8:25–31.
VandenBerg, S.R., H.H. Herman, and L.J. Rubinstein, (1987b) Embryonal central neuroepithelial tumors: Current concepts and future challenges. Cancer Metast. Rev. 5:343–364.
Vaquero, J., J. Ramiro, R. Martínez, et al. (1990) Clinicopathological experience with pineocytomas: Report of five surgically treated cases. Neurosurgery 27:612–619.
von Deimling, A., R. Janzer, P. Kleihues, et al. (1990) Patterns of differentiation in central neurocytoma: An immunohistochemical study of eleven biopsies. Acta Neuropathol. 79:473–479.
von Deimling, A., P. Kleihues, P. Saremaslami, et al. (1991) Histogenesis and differentiation potential of central neurocytomas. Lab. Invest. 64:585–591.
Wasson, J.C., R.L. Saylors, P. Zeltzer, et al. (1990) Oncogene amplification in pediatric tumors. Cancer Res. 50:2987–2990.
Watanabe, K., M. Nagai, S. Wakai, T. Arai, and K. Kawashima (1990) Loss of constitutional heterozygosity in chromosome 10 in human glioblastoma. Acta Neuropathol. 80:251–254.
Weldon-Linne, G.M., T.A. Victor, D.R. Groothuis, et al. (1983) Pleomorphic xanthoastrocytoma: Ultrastructural and immunohistochemical study of a case with a rapidly fatal outcome following surgery. Cancer 52:2055–2063.
White, W., M.H. Shiu, M.K. Rosenblum, et al. (1990) Cellular Schwannoma. A clinicopathological study of 57 patients and 58 tumors. Cancer 66:1266–1275.
Whittle, I.R., A. Gordon, B.K. Misra, et al. (1989) Pleomorphic xanthoastrocytoma: Report of four cases. J. Neurosurg. 70:463–468.
Wick, M.R., P.E. Swanson, B.W. Scheithauer, et al. (1987) Malignant peripheral nerve sheath tumor. An immunohistochemical study of 62 cases. Am. J. Clin. Pathol. 87:425–433.
Wilson, A.J., D.H. Leaffer, and N.D. Kohout (1985) Differentiated cerebral neuroblastoma: A tumor in need of discovery. Hum. Pathol. 16:647–649.
Winek, R.R., B.W. Scheithauer, and M.R. Wick (1989) Meningioma, meningeal hemangiopericytoma (angioblastic meningioma), peripheral hemangiopericytoma and acoustic Schwannoma. A comparative immunohistochemical study. Am. J. Surg. Pathol. 13:251–261.
Wondrusch, E., M. Huemer, and H. Budka (1991) Production of glial fibrillary acidic protein (GFAP) by neoplastic oligodendrocytes: Gliofibrillary oligodendroglioma and transitional oligoastrocytoma revisited. Brain Tumor Pathol. 8:11–15.
Woodruff, J.M., T.A. Godwin, R.A. Erlandson, et al. (1981) Cellular schwannoma. A variety of schwannoma sometimes mistaken for a malignant tumor. Am. J. Surg. Pathol. 5:733–744.
Yachnis, A.T., J.Q. Trojanowski, M. Memmo, et al. (1988) Expression of neurofilament proteins in the hypertrophic granule cells of Lhermitte-Duclos disease: An explanation for the mass effect and the myelination of parallel fibers in the disease state. J. Neuropathol. Exp. Neurol. 47:206–216.
Yamagami T., H. Handa, J. Yamashita, et al. (1987) An immunohistochemical study of intracranial germ cell tumours. Acta Neurochir. (Wien) 86:33–41.
Zang, K.D. (1982) Cytological and cytogenetical studies on human meningioma. Cancer Genet. Cytogenet. 6:249–274.
Zankle, H., and K.D. Zang (1972) Cytological and cytogenical studies on brain tumors. IV. Identification of the missing G chromosome in human meningiomas as no. 22 by fluorescence technique. Humangenetik 14:167–169.
Zülch, K.J. (1986) Brain Tumors. Their Biology and Pathology, 3rd ed. Berlin: Springer-Verlag.

Chapter 2

Differentiation of Cell Types in the Central Nervous System

Timothy E. Hayes and Ronald D.G. McKay

INTRODUCTION

Classification systems for neural tumors are based on the idea that the properties of the tumor cells are a reflection of the normal cell type from which the tumor was generated. This notion has several corollaries. One corollary is that the frequency of occurrence of a given tumor is proportional to the normal occurrence of the precursor cell that gives rise to the tumor. Consequently, the most frequent tumors are those of the glia, which are the most numerous cells in the nervous system and are the last cells generated in the developing nervous system. Unlike neurons, glia retain their capacity to proliferate in response to stimuli such as injury. A second corollary is that the types of differentiated cells produced by the tumor or the expression of differentiation markers by tumor cells is a reflection of the developmental potential of the cell of origin; for example, a neuroblastoma would be considered a tumor of a committed neuronal precursor cell, while an epithelioependymoma, producing neuronal, astroglial, and oligodendroglial cells, would be considered to have originated from the most primitive precursor type in the nervous system.

This chapter discusses precursor cells in light of the determinants that cause them to follow specific developmental pathways. These determinants fall into two categories, intrinsic and extrinsic. Intrinsic determinants can be described in genetic terms as being cell autonomous—they are a consequence of a pattern of expresssion of genes in regulatory pathways which, once established, does not depend on interactions with other cells. The term *commitment*, used to describe the point at which a precursor cell becomes committed to a single fate, refers to the transition into a state dominated by intrinsic determinants. The committed cell often continues to divide for a short while but its final fate cannot be changed by altering the environment. In muscle cell precursors, these intrinsic determinants are thought to include a family of myogenic transcription factors, including myoD1, myogenin, myf5, and herculin; these transcription factors activate their own genes, setting up a self-reinforcing pattern of expression, as well as activating muscle-specific genes of the terminally differentiated phenotype, such as myosin light chain kinase. Extrinsic determinants are signals from outside the cell, such as growth factors, adhesion molecules, or cell recognition molecules, which can be recognized by receptors on the surface of the cell. Responses to such signals may be transient or may involve activation of a stable intrinsic change in the cell. This chapter reviews our current knowledge of the normal signals that control cell number and type in the mammalian central nervous system (CNS).

An Experimental Approach to the Role of Lineage in the CNS

Cell marking experiments and cultures of precursor cells define the origin of different cell types. These experiments provide data on the normal lineage of cells. It is tempting when presented with reproducible patterns of differentiation in vertebrates to conclude that lineage will play a dominant role in generating these patterns of development. However, the fact that precursor cell A gives rise to differentiated cell B does not necessarily mean that precursor cell A is committed or restricted to producing cell B; this relationship may be a result of the environment in which cell A is observed, rather than being a necessary consequence of the identity of cell A. The range of potentials of a precursor cell can only be tested by altering its environment.

Recent technological advances have made it possible to both extend the investigation of lineage by experimental intervention and to characterize external signals that play crucial roles in development. These experiments give us a more complete picture of the path traveled by neuronal

Molecular Genetics of Nervous System Tumors, pages 37-48

precursor cells in vertebrates as they generate the many types of cells found in the adult nervous system. In this laboratory we have approached the question of the origin of complexity in the central nervous system by using immortalizing oncogenes to make immortal cell lines of precursor cells from the CNS; these cell lines then serve as model systems for developmental studies. These cell lines have particular relevance to the study of neuro-oncology because their properties can yield information on what states of differentiation are represented in the immortalized population and how the oncogene affects the ability of the cell to terminally differentiate. Recent data pertaining to lineage and immortalization gathered using these cell lines is summarized in the last sections of this chapter.

INDUCTIVE INTERACTIONS IN FORMATION OF THE NEURAL TUBE

The nervous system is generated from an apparently homogeneous monolayer of cells called the neuroepithelium or neuroectoderm. These cells begin as part of the ectoderm, or outer layer of cells of the embryo. In amphibia it has been shown that large regions of the ectoderm have the potential to form neuroectoderm. The conversion of ectoderm into neuroectoderm occurs in response to signals produced by the underlying mesoderm. This induction begins a series of changes that result in the transformation of this simple epithelium into the vastly complex structures that are comprised in the central nervous system. This series of transformations begins with these cells becoming taller, forming a columnar epithelium. A mechanical process involving changes in cell shapes results in this columnar epithelial sheet rolling up into a tube (the neural tube), beginning at two points—the cervical–hindbrain junction and the midbrain–forebrain junction—and proceeding rostrally and caudally to the ends of the neuroepithelium. The neural tube closes by fusing not at the edges of the neuroepithelium but near the edges, leaving the tips outside the neural tube; these tip cells are called the neural crest and will detach from the neural tube and migrate away to form other structures. In the thorax and abdomen, they will form the peripheral nervous system, and in the head they will form not only the 12 cranial nerves but also contribute to the branchial arches, which are transient embryonic structures that give rise to many of the musculoskeletal structures of the face and neck. It is also at the time of closure of the neural tube that the cells within the tube begin to differentiate from one another.

Establishment of Dorsoventral Identity in the CNS

Underlying the neural tube is a second tube of cells, the notochord, which is derived from mesoderm. The notochord sends an inducing signal (which has been much studied but not yet defined) to the neural tube, causing the narrow strip of cells on the ventral side of the neural tube immediately overlying the notochord to become altered; this altered strip of cells is known as the floor plate and is set apart from the rest of the neuroepithelial cells by many special properties. One of these properties has to do with the guidance of developing axons. Developing commissural axons, but not association or motor axons, migrate through the early spinal cord to the floor plate and then turn toward their ultimate destination. It has been shown that floor plate cells secrete a diffusible factor that has a chemotropic effect on the commissural axons (Tessier-Lavigne et al., 1988). Another of the special properties of the floor plate is apparent when examining the segmentation of the CNS. During development, the central nervous system becomes subdivided rostrocaudally into forebrain, midbrain, hindbrain, and spinal cord. The spinal cord and hindbrain are further subdivided into segments or repeated units. In the hindbrain, cell-marking experiments show that cells in one segment or rhombomere do not cross the boundary into the adjacent rhombomeres. The notochord, in contrast, is not segmented at any level of the CNS and floor plate cells can cross the rhombomere boundaries (Fraser et al., 1990).

That the ability to respond to the notochord signal and form floor plate is a general property of the cells of the neural tube and not a property of a lineage-defined subset of the neural tube has been shown by transplantation of a second notochord into an ectopic position adjacent to the neural tube. In response to the transplanted notochord, a second set of floor plate cells is induced at the point in the neural tube closest to the transplant (Hirano et al., 1991). Dye marking of cells later in development has shown that while the progeny of cells outside of the floor plate may contribute to the floor plate, the progeny of floor plate cells always form floor plate cells (Fraser et al., 1990). Again, the fact that cells not originally in the floor plate can later contribute to the floor plate suggests that the special properties of the floor plate are not due to a unique lineage.

Following neural tube closure and induction of the floor plate, there is an ordered sequence of proliferation, increasing the number of cells in the neural tube, and differentiation, in which cells leave the proliferative cycle and express the attributes that will enable them to perform their mature functions. In the spinal cord there is a ventral-to-dorsal sequence of differentiation—cells in the ventral aspect of the neural tube that differentiate first will function as motor neurons. After the motor neurons have begun to diffferentiate, differentiation begins in the cells of the dorsal aspect of the spinal cord; these will form interneurons. In the experiment mentioned above, in which moving the notochord causes the floor plate to appear in a new position, the motor and sensory neurons also appear in new positions relative to the rest of the embryo, although they are in their normal positions with respect to the floor plate. As is the case for the floor plate cells, the identity of the motor and sensory neurons in the spinal cord appears to be determined not by their lineage but as a result of signals from other cells.

Molecular Mechanisms Involved in Fate Choice

The differentiation of cells in the just-closed neural tube occurs in response to several different signals and results in cells with very different phenotypes in response to each of them; for example, they can become floor plate in response to a signal from the notochord or they can become motor neurons in response to a different signal, which may come from the floor plate.

As the precursors in the nervous system are exposed to signals that alter their fate, several different types of responses may occur. The first is that in response to the signal, the cell stops proliferating and turns into a particular postmitotic cell type. This type of response has been shown to occur in the developing optic nerve in which the oligodendrocyte type 2 astrocyte (O2A) precursor chooses the astrocytic fate in response to an inducing signal, which is a combination of a soluble factor, ciliary neurotrophic factor (CNTF), and a component of the extracellular matrix (Lillien and Raff, 1990). The second type of response is the stable alteration of the pattern of gene expression in the proliferating cell so that its ability to respond to certain external signals is altered. For example, the cell could acquire the ability to respond to a new signal or lose the ability to respond to a signal due to up- or down-regulation of the receptor for that signal; or, the response apparatus downstream of the receptor could be altered to give a different response when the receptor is activated. In these ways the regulation of proliferation of the cell could be altered or the subsequent pathways of differentiation available to the precursor cell could be expanded or contracted at appropriate points in its lineage. An example of one signal inducing responsiveness to a second signal has been seen in the developing striatum (Cattaneo and McKay, 1990). Dispersed precursor cells from the developing striatum do not proliferate in response to nerve growth factor (NGF) but do proliferate in response to fibroblast growth factor (FGF). If the cells are treated with FGF for a short period, they then become stably responsive to NGF and will continue to proliferate until NGF is removed from the culture, at which point they differentiate into neurons. The simplest explanation of this phenomenon is that FGF induces the expression of a receptor for NGF. A similar mechanism has been inferred from the expression pattern of the low-affinity NGF receptor and one of the FGF receptors in the developing spinal cord as determined by *in situ* hybridization. The FGF receptor signal is seen in the proliferating cells occupying the ventricular zone and in some populations of differentiated neurons, while the NGF receptor signal is seen in the migrating cells of the mantle layer surrounding the ventricular layer, suggesting that the proliferating ventricular cells express the FGF receptor, switch to expression of the NGF receptor during migration, and switch back to the FGF receptor when migration is complete (Heuer et al., 1990).

It is also possible for a change in responsiveness to be transient rather than stable. In this case the response would decline with time after the inducing signal so that if the next signal did not come soon enough, the cell would not be able to respond to it. An example of this type of response has been observed in the response of quiescent cells to growth factors. In order to proliferate, quiescent cells need to be exposed to two types of growth factor, first a competence factor and then a progression factor. The competence factor induces the expression of proteins, including c-fos and c-myc, which have short half-lives. The progression factor must act before these proteins decay or the cells will return to quiescence, in which they will not proliferate in response to the progression factor. (A review of studies on competence and progression and the mechanisms that underlie these phenomena is found in Rollins and Stiles, 1989.)

Establishment of Rostrocaudal Identity in the CNS

In addition to the dorsoventral gradient of differentiation in the developing CNS, there are also rostrocaudal gradients of differentiation in the neural tube. Differentiation within the neural tube occurs initially at the parts of the neural tube that close first and a wave of differentiation proceeds from these points along the neural tube in both directions, following behind the moving points of tube closure. Along the rostrocaudal axis of the neural tube, regional identity again appears to be a result of position rather than of lineage. In the spinal cord, if not in the brain, rostrocaudal identity appears to involve the expression of members of the *Hox* clusters of homeobox-containing genes. These genes are expressed in a rostrocaudal gradient in the spinal cord and in stripes that correspond to the rhombomeres of the early hindbrain (Keynes and Lumsden, 1990; Lumsden, 1990; Wilkinson and Krumlauf, 1990). Inactivation of the gene *Hox 1.6* in the mouse causes late closure of the neural tube at the cervical–hindbrain junction and the absence of some cranial nerves and ganglia (Lufkin et al., 1991). This gene is also expressed in the lateral plate mesoderm, and it has not yet been shown whether these effects are the result of lack of expression in the neural tube itself or whether they are the result of a missing signal from the lateral plate mesoderm.

Transplant experiments demonstrate that overall regional identity within the neural tube—forebrain, midbrain, hindbrain, and spinal cord—is determined by the adjacent mesoderm (reviewed in Keynes and Lumsden, 1990). The periodic arrangement of spinal nerves in higher vertebrates is not the result of an intrinsic pattern but a result of the segmentation of the trunk mesoderm. The relative contribution of mesoderm to the segmentation of the hindbrain is less clear.

It is possible that the formation of rhombomeres in the hindbrain is a result of intrinsic rather than extrinsic patterning. It is interesting to note that the identity of the branchial neural crest, which comes from the rhombomeres of the hindbrain, is determined before migration. Transplant experiments in chick show that craniofacial mesoderm is patterned by the branchial neural crest and not vice versa.

Within the brain, moving pieces of developing tissue along the rostrocaudal axis tests whether the cells will maintain the rostrocaudal identity of the donor site or adopt the identity of the host site. For example, it has been shown that in the chick there is a gradient of expression of the homeobox-containing gene *engrailed-2 (En2)* at the metencephalic–mesencephalic junction. If a piece of tissue from this site is excised and replaced back-to-front (so that the gradient of *En2* expression in the transplant is reversed with respect to the host), then within 24 hours the transplant *En2* gradient changes to show the same orientation as that of the host (Martinez and Alvarado-Mallart, 1990). However, if the transplant is placed in the prosencephalon (part of the forebrain that does not express *En2*), then the transplant continues to express *En2* in its original pattern and induces the expression of *En2* in the surrounding host cells (Martinez et al., 1991).

This experiment demonstrates several points about the role of lineage in the vertebrate CNS. The first is that lineage can in fact play a role—when metencephalic/mesencephalic cells are moved to the forebrain, they act as a determined lineage, maintaining expression of the presumptive positional marker *En2*. Second, the role of lineage can be overridden by environmental factors. This point is apparent in both the forebrain, in which the transplant recruits forebrain cells to an *En2*-expressing state, and in the metencephalic/mesencephalic reversal experiment, in which the tissue is regulated to match the surrounding cells. Third, the dominance of either lineage or environment cannot be predicted *a priori*; it must be determined experimentally. In this transplant experiment, instances of both lineage dominance and environmental dominance were observed on the same phenotype, expression of *En2*.

Plasticity of Neuronal Fate in the Cortex

In addition to the experiments cited above, a number of other experiments have been done that shed light on the role of lineage in differentiation in the vertebrate nervous system. The common feature of these experiments is that they involve altering the environment of the cells in question. Systematic manipulation of the soluble and insoluble factors with which they interact is the only way to test the developmental plasticity of precursor cells and determine the contributions of lineage and environment to their fate.

One very informative set of experiments regarding the commitment of neuronal precursors to a particular laminar, and hence functional, fate in cortex involves the transplantation of cells in the visual cortex of the ferret from midembryonic development, when differentiating neurons are forming the deeper cortical layers 5 and 6, into newborn cortex, when differentiating neurons are forming the more superficial cortical layers 2 and 3 (McConnell, 1988; McConnell and Kaznowski, 1991). The neurons of layers 5 and 6 project to subcortical targets, including the thalamus. The neurons of layers 2 and 3 project to targets deeper within the cortex. Cells to be transplanted were labeled by injecting the donor with ^{3}H-thymidine, which labels cells that are going through the DNA synthesis phase of the cell cycle at the time of injection. When the transplant is made from newborn cortex into newborn cortex, the transplanted cells generate neurons in layers 2 and 3, as expected.

However, when the transplant is done placing cells from midembryonic cortex into newborn cortex, a more complex result is obtained. If the donor cells were labeled well before the transplant (4 to 24 hr), then the labeled transplanted cells are found in layers 5 and 6, to which they would have migrated had they remained in the donor. If the donor cells were labeled immediately prior to transplant, then the labeled transplanted cells are found in layers 2 and 3, to which the concurrently generated host neurons are migrating. In both conditions lightly labeled cells, which went through an additional mitosis in the host after transplantation, migrate to layers 2 and 3. These experiments suggest that in the cortex precursors remain uncommitted to a specific neuronal fate until the last minute and that the time window for commitment to laminar (and hence functional) neuronal fate occurs very late in the last cell cycle, in the G_2 or M phase immediately prior to the terminal mitosis and the onset of migration. These experiments also suggest that the critical factor in laminar fate choice is not intrinsic but is a result of reading signals from the environment during the critical period.

Specialized areas of the neocortex have unique architectural and functional properties. The question of whether these features are prespecified (i.e., a result of lineage) or are a result of local interactions has been investigated with interesting results. Primary visual cortex contains unique functional units called "blobs." Primary somatosensory cortex contains unique specializations called "barrels." Barrels are isolated aggregates of layer 4 neurons and have boundaries of glycoconjugated molecules not observed in other areas of cortex. They develop gradually from an apparently uniform cortical plate beginning in the first week after birth. The formation of barrels is dependent on sensory input during the critical period of development. When late embryonic visual cortex is transplanted into neonatal somatosensory cortex in the rat, the transplanted visual cortex forms barrels and barrel boundaries organized in an array similar to the pattern in the normal barrelfield (Schlaggar and O'Leary, 1991).

This experiment shows that the ability to form barrel-type processing structures is not unique to parietal cortex, which normally forms barrels *in situ*, but that embryonic occipital cortex also has this ability. Instead, the generation of this pattern must reflect the active response of the transplanted cortex to influences originating at the sensory periphery. A more general conclusion is that the neocortical epithelium is specified to form basic features such as a laminated structure composed of common cortical cell types with a common local circuitry, but that the cortex lacks a rigid specification of the functional and architectural specializations that distinguish cortical areas. These specialized organizations are generated in response to external cues to which any region of the developing neocortex can respond (Schlaggar and O'Leary, 1991).

CLONAL ANALYSIS

Cell Marking

One of the potential difficulties of interpreting the results of transplantation of large chunks of tissue is that changes in small numbers of cells may not be detected or may not be interpretable since the transplanted tissue contained many types of cells. This difficulty can be overcome through the use of clonal analysis, in which the progeny of a single precursor cell are followed. The most common method of clonal analysis is marking a precursor cell, such as by using a dye (which may last for several cell generations before it is diluted below the limit of detection) or by using a heritable marker such as a retrovirus expressing an easily assayable protein. As mentioned above, dye-marking experiments have been used to investigate the lineage relationship between the floor plate and the adjacent cells of the neural tube in the hindbrain. These experiments showed that cells in the floor plate generate only floor plate cells, while cells outside the floor plate can generate both non-floor plate and floor plate cells. These experiments also showed that once the boundaries between rhombomeres form, cells not in the floor plate do not cross these boundaries, so that each rhombomere is a polyclonal unit. Unlike the results obtained in the marking experiments described below, in the early hindbrain most clones produce a single cell type, suggesting that these cells may become determined early (Fraser et al., 1990).

Retroviral marking has provided a great deal of information about the lineage relationships in the maturing vertebrate nervous system (reviewed in McKay [1989]). However, the lineages produced vary depending on the region of the brain tested, the developmental stage at which the precursors are marked, and possibly on the species of animal examined. In the cortex and tectum, a common theme is that marking early in development shows clones containing both neurons and glia in which the cells are broadly distributed and the neurons are found in multiple laminae of the cortex, indicating that they will have different functions. Marking later in development often produces clones that contain only neurons or only glia with the cells narrowly distributed in a radial array from ventricle to pia. The results from the cortex and tectum have been interpreted to mean that there is an early common precursor to neurons and astrocytes that can later commit to either neurons only or astrocytes only. In the retina, results are obtained that are in some ways similar, in that early marking produces large clones that contain multiple types of cells. In contrast, late marking in the retina can produce two cell clones containing two cells of different types—either neurons of different types or a neuron and a glial cell. This result indicates that the sister cells produced by the terminal mitosis of a precursor can differentiate into cells of different types. This suggests that while lineage may restrict these precursors to neural fates or even to retinal fates, lineage is not the critical factor in determining what type of neural or retinal cell is produced.

Clonal Culture

The above marking experiments were done *in vivo*, so that the lineages produced were a reflection of those normally produced but do not indicate how plastic the precursors may be. Another type of clonal analysis experiment more amenable to experimental manipulation is clonal microculture, in which a single precursor cell is cultured *in vitro*. In one set of experiments of this type, a single cell from the septal region of embryonic day 14 rat forebrain was placed in a microwell that had embryonic day 14 rat striatal cells on the walls to provide soluble factors (Temple, 1989). The progeny of the single cells were then stained with antibodies to determine whether they were neuronal or glial. Most of the clones were entirely neuronal, some were a mixture of neuronal and glial cells, and a few were entirely glial. This result indicates that at this stage of development some of the septal precursor cells are not committed to a single fate. One of the potential advantages of this type of experiment is that it is possible to manipulate the culture conditions by the inclusion of growth factors or antibodies and determine how these interventions affect the differentiation of a single precursor cell. One of the drawbacks of this type of experiment is that tissues contain mixtures of precursor cells so that it is impossible to know ahead of time what class of precursor cell is being examined in a particular microculture experiment. Another difficulty is that genetic manipulation of single cells is difficult, which limits the range of experimental interventions to topical application of soluble agents or substratum.

IMMORTAL PRECURSOR CELL LINES

A strategy that combines the advantages of clonal analysis, the use of a defined cell type, ease of molecular biological manipulation, and analysis either *in vitro* or *in vivo* is that of using immortalized cell lines from the nervous system. Clonal cell lines can be generated from the desired region of the brain at the desired stage of development using an appropriate immortalization strategy. Most cell types require introduction of an immortalizing oncogene (or an activating mutation in one or more cellular oncogenes) for establishment of a permanent cell line. Some cell types may be kept in continuous culture by use of the appropriate growth factor(s), such as the O2A cell, precursor to oligodendrocytes and type 2 astrocytes in the optic nerve, which can be passaged for months if cultured using a combination of platelet-derived growth factor (PDGF) and FGF (Bogler et al., 1990). Once established, a stable clonal cell line can provide a large amount of genetically and behaviorally identical cells for characterization according to expression of antigens, expression of growth factors and receptors, substrate requirements, interactions with primary cells, and behavior after transplantation. It is also possible to use molecular biological techniques to introduce foreign genes or alter the patterns of expression of endogenous genes and generate subclones with these stable alterations. The subclones again provide a large supply of clonal cells for analysis and comparison to the parent cell line and to primary cells.

Identification of Precursors Using Nestin

One of the potential difficulties with making cell lines is determining whether a particular cell line represents the desired cell type. This problem is, of course, secondary to the larger question of whether the desired cell type is stable, i.e., whether the properties of the cell are self-maintaining. If the properties of the cell are not maintained when the tissue is dispersed and the immortalizing agent is active, then the cell lines produced will not include the desired population. Uncertainty about the existence and identification of the desired cell type is particularly great for neuronal precursor cells because these cells are not thoroughly characterized but are defined primarily in terms of the mature cell types they produce. This lack of definition is compounded by present ignorance of the signals required for neuronal precursors to acquire particular fates. This lab has addressed this difficulty by identifying a new marker that is expressed in neuronal precursor cells. This marker is the intermediate filament nestin (*ne*uroepithelial *st*em cell *in*termediate filament) (Frederiksen and McKay, 1988a; Lendahl et al., 1990), which was initially identified with the monoclonal antibody Rat401 (Hockfield and McKay, 1985). Nestin is expressed at some stage in most if not all of the neuroepithelial stem cells in the CNS. The use of this marker has allowed us to identify unequivocally cell lines that are derived from neuroglial precursor cells and not from other cell populations in the developing CNS.

Cell lines made from neuronal precursor cells have been used recently with success in the study of neuronal precursor cells (Bartlett et al., 1988; Birren and Anderson, 1990; Ryder et al., 1990). The remainder of this chapter describes work, primarily that done in this lab, using such immortal neuronal precursor cell lines to investigate the developmental pathway of neuronal precursor cells.

Making and Characterizing Cell Lines

As mentioned above, most primary cells will not grow continuously in culture; after a small number of mitoses they go through a crisis and most of the cells die. The cells that survive the crisis invariably carry mutations in one or more proto-oncogenes such as the gene for p53; these cells are then immortal and can be serially passaged indefinitely. This laboratory has chosen to make cell lines from the developing nervous system by the use of immortalizing oncogenes, obviating the need for activating mutations in the target cells. The oncogenes that we have tested are v-myc, neu, and the tsA58 allele of SV40 T antigen. v-myc is a viral analogue of the c-myc gene and was isolated from an avian retrovirus. Neu is an activated form of a receptor tyrosine kinase and is found in neuroblastomas and glioblastomas; the normal counterpart of this gene is expressed in the early nervous system, but the ligand that activates the normal receptor has not yet been identified. T antigen is the immortalizing gene from the simian DNA virus SV40; the tsA58 allele encodes a protein that is stable and active at the permissive temperature (33°C) but that is degraded at the nonpermissive temperature (39°C).

The coding sequences for these genes were introduced into the engineered proretroviral genome pZIPneoSVX (Cepko et al., 1984), which also contains the neomycin phosphotransferase gene, which confers resistance to antibiotics of the neomycin class, and infectious recombinant retroviral particles were produced. Infection of a cell by a retroviral particle results in the integration into the host cell genome of a copy of the retroviral genome that will express the introduced gene and the neomycin resistance gene, so that the cell becomes both immortal and neoresistant. These retroviruses have been engineered so that the integrated retrovirus cannot produce infectious viral particles; thus, an infected immortalized cell is unable to transmit the virus to other cells.

Retroviruses expressing v-myc, neu, or tsA58 were used to infect dispersed cultures of cells from the developing nervous system and immortalized neoresistant colonies were selected. We have used tissue from the developing cerebellum, hippocampus, optic nerve, striatum, retina, substantia nigra, and basal forebrain in our experiments. Some of this work has been described previously (Almazan and McKay, 1992; Frederiksen et al., 1988; McKay et al., 1990; Renfranz et al., 1991). We find that the different oncogenes tend to give cell lines with characteristic differences. Cells immortalized with neu grow very slowly; the four cell lines established from rat postnatal day 2 cerebellum were all positive for nestin (a precursor-specific intermediate filament), vimentin, and GFAP (an astrocyte-specific intermediate filament) and because of their slow growth have not been extensively characterized. Cell lines established with v-myc from rat postnatal day 2 cerebellum were also all positive for nestin, vimentin, and GFAP. Cell lines isolated from cerebellum by Ryder using v-myc also express some differentiated characteristics, such as neurofilament or GFAP (Ryder et al., 1990). It has been suggested that transient down-regulation of c-myc is necessary for terminal differentiation of lymphoid cells (Coppola and Cole, 1986). Thus, v-myc may immortalize precursor cells by preventing committed cells from leaving the cell cycle to become postmitotic. The coexpression of a precursor marker, nestin, with a differentiated marker, GFAP, suggests that the cells immortalized by neu and v-myc represent precursors from the point immediately prior to terminal differentiation.

We have further characterized one of our v-myc–immortalized lines, M15B. M15B does not express neurofilament or generate action potentials when passaged in serum; however, after treatment with retinoic acid and dibutyryl cyclic AMP, clusters of M15B cells are observed that have neuronal morphology and express neurofilament. One of the difficulties with using a dominant oncogene is that it is difficult for cells to override the proliferative effects of the oncogene *in vitro*; less than 1% of M15B cells respond to retinoic acid/cAMP treatment by expressing neurofilament. Of course, this makes such cell lines good candidates for the study of oncogenesis, but that is for another chapter.

Cell lines established with T antigen span a broader spectrum of properties than those generated with v-myc. For example, among the cell lines established from rat postnatal day 2 cerebellum were lines that strongly expressed nestin, lines with no detectable expression of nestin, and a series of lines expressing various intermediate levels of nestin. All of these lines express vimentin at 33°C but none of them expresses GFAP at the permissive temperature. One tentative conclusion that may be drawn from this data is that v-myc–immortalized cells represent very late stages of development, while T antigen-immortalized cells can represent a range of earlier stages of development. In this interpretation M15B represents a neuronal/astrocytic precursor, while ST15A represents a multipotential precursor that can generate neurons, oligodendrocytes, and multiple forms of astrocytes (see below).

There are other mutant alleles of T antigen besides tsA58 that may be useful in making cell lines. One such mutant is the allele U19, which is about two orders of magnitude more efficient for immortalizing cells than the wild type. A double mutant containing both the tsA58 and U19 mutations, designated tsU19, has been constructed in this lab and shown to retain temperature sensitivity while having an increased efficiency of immortalization (Almazan and McKay, 1992). This mutant has also been introduced into the retroviral backbone pZIPneoSVX and used to make cell lines from the optic nerve (Almazan and McKay, 1992), hippocampus (Renfranz et al., 1991), striatum, substantia nigra, and basal forebrain (E. Catteneo and R. McKay, unpublished results) and from mouse cerebellum and cortex (Redies et al., 1991). The cell lines produced with the tsU19 retrovirus, like those made using tsA58, include both nestin-positive and nestin-negative lines.

Work in this lab has focused on the nestin-positive lines, as nestin is a reliable marker for neuroepithelial stem cells and nestin-positive lines can differentiate into multiple fates when cocultured with primary cells or when transplanted into the developing brain (see below). The identity of the nestin-negative lines is unknown. One possibility is that these represent stages of differentiation that are either before or after the nestin- positive stage. Nestin is not expressed in the neural plate and first appears in both the spinal cord and cerebrum on embryonic day 10 in the rat, with about half of the cells positive on E10 and more than 95% positive on E11 (Frederiksen and McKay, 1988). In the cortex, the transition from the nestin-positive to the nestin-negative stage is quite abrupt and the disappearance of nestin mirrors the appearance of postmitotic neurons (Frederiksen and McKay, 1988), so that there may be a direct transition from the proliferating nestin-positive cell to the terminally differentiated neuron. In the cerebellum, there are several days between the time when nestin expression is down-regulated and the peak of neuronal differentiation and there is probably a proliferating nestin-negative precursor during this time.

A second possibility is that the nestin-negative cell lines represent cells that are not neuronal precursors, such as fibroblasts, microglia, or ependymal cells. A third possibility is that the nestin-positive state is not completely stable. Most of the proliferating cells in the primary cultures infected with the immortalizing retrovirus were nestin-positive. The fact that a significant fraction of the cell lines produced are nestin- positive indicates that the nestin-positive state is not inherently unstable. However, a significant representation of nestin-negative cell lines were isolated, indicating that either the nestin-negative cells are more efficiently immortalized, which we think unlikely, or that some nestin-positive cells are producing nestin-negative cell lines. One piece of data relevant to this issue was generated using the cell line ST15I, a nestin-negative cell line isolated from the same immortalization experiment on P2 cerebellum as ST15A (see below). When ST15I is cocultured with primary cells from postnatal cerebellum, it becomes nestin-positive and can produce cells with long neurofilament-positive processes (Valtz, 1989). This shows that a nestin-negative line is capable of expressing nestin and may represent a neuronal precursor cell but does not indicate whether this line is a pre-nestin state or whether nestin expression can be unstable.

TRANSPLANTATION OF IMMORTAL PRECURSORS INTO DEVELOPING BRAIN

In our experience, cell lines made from neuronal precursor cells do not spontaneously differentiate into neurons or glia; they require some as yet undefined signal or signals to acquire a terminal phenotype. As mentioned above, some immortalizing genes give cell lines that appear to be committed to a particular fate and that can differentiate (often not at high frequency) when treated with growth factors or chemical agents (Bartlett et al., 1988; Birren and Anderson, 1990; Frederiksen et al., 1988); these may be thought of either as an immortalized committed proliferating precursor or as a terminally differentiating cell in which the normal exit from the cell cycle has been overridden by the immortalizing oncogene. Precursor cells in primary cultures from the nervous system can differentiate in growth medium and differentiation can be encouraged by removal of serum or growth factors; nutrient starvation does not have this effect on most neuronal precursor cell lines. These differences allow cell lines to be used in the investigation of the signals that cause precursor cells to move to a later stage of differentiation or to acquire a particular terminal fate.

Cerebellum

Our lab has done a number of transplant experiments that are pertinent to the discussion of the role of lineage in the developing mammalian brain. One of the cell lines isolated from developing rat postnatal day 2 cerebellum using the tsA58 allele of T antigen, ST15A, has been studied by culture alone and with primary cells and by transplantation. In proliferating culture (33°C in the presence of serum), ST15A expresses the intermediate filament protein vimentin and the precursor-specific intermediate filament protein nestin. ST15A also expresses at low but detectable levels neuron-specific enolase (NSE) mRNA and neurofilament heavy chain protein (Frederiksen et al., 1988; Hayes et al., 1991). ST15A does not express detectable GFAP at 33°C. When ST15A is cultured at 39°C, the expression of nestin can be down-regulated and that of GFAP up-regulated; however, this change is dependent on the batch of serum used in the culture and does not occur in most batches. The expression of vimentin is unchanged at 39°C in culture, in contrast to the down-regulation that occurs upon terminal differentiation of normal cerebellar cells. In sum, the behavior of ST15A in culture, either at 33°C or at 39°C, is not dramatic and leads us to conclude that ST15A needs to be instructed through some external signal in order to execute a program of terminal differentiation.

Such instructions can be provided by coculturing ST15A with primary cells from developing cerebellum. In such a coculture at 39°C, ST15A cells lose the expression of vimentin and nestin and can gain the expression of GFAP and the flat morphology of astrocytes, can adopt the morphology of Bergmann glia, or can acquire the expression of neurofilament and the morphology of neurons, suggesting that in the coculture the ST15A cells underwent terminal differentiation into either neurons or astrocytes (Frederiksen et al., 1988; McKay et al., 1990).

Although changes in gene expression and morphology are indicative of terminal differentiation, concluding that ST15A can differentiate into normal cerebellar cells requires the more complex assay of behavior *in vivo*—migration to the appropriate location, proper orientation and connection of axons and dendrites, expression of appropriate transmitters, appropriate participation in local circuitry, etc. Upon transplantation into the postnatal rat cerebellum, most ST15A cells assume the location and morphology of either granule neurons, located in the granule cell layer and having a characteristic axonal morphology, or Bergmann glia (McKay et al., 1990). At a first approximation, then, ST15A cells appear to be able to respond to signals present in the developing cerebellum by terminally differentiating into normal cerebellar neurons and glia.

These transplantation experiments were done in the rat postnatal day 2 cerebellum. At this point the external granule layer of the cerebellum is undergoing massive proliferation and is generating large numbers of granule cells as well as small numbers of basket neurons and stellate neurons; the site of production of glia is unclear, although at least some glia appear to come from the ventricular zone (see Jacobson, 1991). It appears, then, that the fates ST15A adopts when transplanted into postnatal day 2 cerebellum are similar to or the same as those of the endogenous cells which terminally differentiate at that time.

Snyder et al. have done similar experiments using cerebellar cells immortalized with v-myc (Snyder et al., 1992). They find similar results—the transplanted cells can produce both neurons and glia. One of the interesting aspects of their experiments is that although their cell lines are immortalized with a dominant oncogene rather than a temperature-sensitive one, they do not find evidence of tumors or of unregulated proliferation of the transplanted cells. This suggests that the normal signal to differentiate can override the effects of a single oncogene, even though that oncogene is sufficient to immortalize the cell *in vitro*.

Hippocampus

The cell line HiB5 was isolated from rat embryonic day 16 hippocampus using the tsU19 double mutant allele of T antigen. In clonal culture it is nestin-positive at both 33°C and 39°C and expresses neither GFAP nor neurofilament. In several ways the properties of HiB5 are reminiscent of those of primary hippocampal precursor cells (Renfranz et al., 1991). However, like ST15A, HiB5 seems unable to initiate a program of terminal differentiation by itself. Also like ST15A, HiB5 cells placed in contact with developing brain are capable of differentiating into either neurons or glia. When transplanted into the postnatal hippocampus, the differentiation of HiB5 is quite dramatic, as the transplanted cell line extensively populates the granule cell layer of the dentate gyrus, as well as producing cells scattered through the hilar region and the occasional cell in the pyramidal layer. The cells scattered through the hilus are astrocytes as evidenced by their staining with antibodies to GFAP. The cells in the dentate gyrus are granule neurons, as demonstrated by several criteria: the location and size of their cell bodies; the direction and pattern of their dendrites and axons; the extension of their axons to the normal target site for granule neurons; and their expression of calbindin, glutamate, and NSE (Renfranz et al., 1991; M. Cunningham et al., in preparation).

Heterotopic Transplants

In contrast to its behavior in the developing hippocampus, HiB5 transplanted into the developing cerebellum produces what appear to be cerebellar cells—like ST15A, a majority of the transplanted cells appear to become astrocytes, including the Bergmann glia characteristic of the cerebellum, and a smaller population becomes cells similar in location, morphology, and antigenic profile (NSE-positive, calbindin-negative, glutamate-negative) to cerebellar granule neurons. Apart from the nomenclature "granule neuron" and the fact that both are late-born interneurons, cerebellar and hippocampal granule cells are quite different. They have different migratory patterns, different morphologies, different neurotransmitters, and express different antigens. Given two different sets of signals from different regions of the brain—those from postnatal hippocampus and those from postnatal cerebellum—HiB5 can differentiate into two different kinds of neurons, each appropriate for the local environment and each unique to its region of the brain.

Although not characterized as thoroughly, ST15A cells transplanted into developing hippocampus give results qualitatively similar to those of HiB5—both astrocytes and apparent dentate gyrus granule neurons are produced (M. Cunningham and R. McKay, unpublished). The apparent lack of regional restriction of the cell lines ST15A and HiB5 is surprising, as regional identity has been speculated to be the result of patterns of expression of genes, such as the homeobox-containing genes, which are fixed early in development. Were this true, the results obtained with the cell lines could be interpreted to be the result of the loss of such information during the cloning and establishment of the cell lines. We have tested this possibility by transplanting primary cells from cerebellum into postnatal hippocampus. Preliminary results indicate that the cerebellar cells produce both astrocytes and phenotypically normal dentate gyrus granule neurons (M. Cunningham and R. McKay, unpublished results).

These results are important for the understanding of the role of lineage in neural development. These results suggest that regional identity may be the result not of restriction, or establishment of an internal program, but of instruction, and that maintenance of regional identity during development requires instruction from the environment. These conclusions suggest that lineage plays a lesser role in neuronal development than previously thought but do not tell us what that role is. Other experiments, such as the transplant experiments showing both dominance and regulation of the *En2* phenotype (Martinez and Alvarado-Mallart, 1990; Martinez et al., 1991), show that the role of lineage must be experimentally determined on a case-by-case basis. All of these experiments in turn emphasize the significance of environmental signals in the determination of cell fate. This complicates the study of fate choice in two ways. First, how did the appropriate signals come to be expressed if not by an internal program, as a result of lineage? We are forced to consider the possibility that the vast complexity of the nervous system is the result of a series of contingent events, each successive signal dependent upon prior ones, a house of cards reproducibly and precisely built. Second, many of the signals that move the developmental processes along either have not yet been defined or may be the result of known signals used in combinations not yet defined. The biochemical characterization of the signaling agents and their receptors is a truly daunting task.

One simplifying result of this hypothesis is that fate choice may be studied at the level of the cell and the cell surface, emphasizing the role of growth factors and adhesion molecules and pushing back the necessity for genome-scale genetics of mammalian neural development. The connection between signal and response must still be traced through the nucleus and back to the cell surface, a task for which molecular methods are well suited. But the lesser the role played by lineage in the development of the CNS, the easier it will be to decipher the logic by which that development proceeds, especially if we can use the tools of molecular biology to isolate cells that represent defined precursor states and to grasp the signals and receptors by which these cells become the engines of our minds.

NEURO-ONCOLOGY AND DEVELOPMENT—THE MEDULLOBLAST

One of the central notions in molecular neuro-oncology is that tumors of the nervous system may arise from undifferentiated neuroepithelial cells. Medulloblastomas are a major class of childhood tumor associated with the cerebellum and brain stem. One of the characteristics of these tumors is that they often contain a mixture of undifferentiated and differentiated cells. Most often the differentiated cells of these tumors show characteristics of neuronal cells, glia, or striated muscle or a combination of these elements; there are also variants containing melanin-producing cells or ependymal-like cells.

In 1925, Bailey and Cushing proposed that medulloblastoma, rather than being multiple coincident tumors, was derived from a hypothetical multipotential stem cell, which they named the medulloblast (Bailey and Cushing, 1925). Because of the transient nature of the vulnerable cells and the lack of an animal model, little is known about the cells at risk and the "medulloblast hypothesis" has been difficult to confirm. In fact, a prominent text on neural tumors (Russell and Rubinstein, 1989) begins the discussion of medulloblastoma by stating that "the prevailing name 'medulloblastoma,' given by Bailey and Cushing (1925), is unfortunate because there is no embryonal cell that has been identified as a medulloblast." Our work is relevant at this point because the experimentally immortalized cell line ST15A described above has features expected of the medulloblast (Valtz et al., 1991). ST15A was isolated from developing cerebellum and has the potential to differentiate into apparently normal neurons or astrocytes, as described above. ST15A can also differentiate into muscle (Valtz et al., 1991), which is not a normal neuronal fate. The muscle cells produced by ST15A are apparently normal—they express troponin T, they express the myogenic genes myoD1 and myogenin, and they are electrically excitable and can spontaneously contract. These three cell types—neurons, glia, and muscle—are those most often found in medulloblastoma.

ST15A cells express the intermediate filament protein nestin, as do most proliferating precursors in the developing nervous system (Frederiksen and McKay, 1988). When we examined sections of medulloblastomas for the expression of nestin, each of the five tumors examined contained regions with large numbers of nestin-expressing cells. In contrast, cerebellar tissue from a normal 2.5-year-old contained no nestin-expressing cells. One of the sources of material for the study of tumors is cell lines generated by growing cells out of tumor samples. We examined a cell line generated by growing cells out of a medulloblastoma from an 11-year-old patient. All of the cells expressed nestin. The expression of this precursor-specific protein by the proliferating cells of a tumor, as well as the parallels with the developmental potentials of a cerebellar-derived precursor cell line, lend credence to the hypothesis that medulloblastoma is a tumor of a multipotential stem cell as suggested by Bailey and Cushing more than 65 years ago. ST15A, then, is presently the best model system in which to study the embryonic origin of medulloblastoma.

One of the conceptual hurdles facing the concept of the medulloblast is that muscle tissue is not normally produced in the cerebellum and that medulloblastomas contain a mixture of fates that are normally seen as deriving from distinct embryonic layers—neuronal and glial cells arise from ectoderm, while muscle is classified as having a mesodermal origin. In fact, many of the musculoskeletal structures of the face and neck are derived from neural crest tissue, a derivative of the neuroectoderm. Again, while the distinction between mesoderm and neuroectoderm is often phrased in terms of lineage, it may be possible for this boundary to be crossed as a result of instructive interactions overriding fate-restrictive mechanisms. The question in medulloblastoma is how the potential for the muscle fate becomes activated and how the tumorigenic cells are induced to leave the proliferative (and tumorigenic) state and differentiate.

The presence of combinations of differentiated cell types within a tumor type is a characteristic of primitive neuroectodermal tumors, a feature they share with medulloblastoma. The expression of nestin may also be a common feature of the renewing cells of these tumors (Tohyama et al., 1992). The discovery that an experimentally immortalized cell line retains the properties, including the surprising properties, hypothesized for the cell of origin of such a primitive tumor, suggests that the investigation of such cell lines may be a profitable avenue for the neuro-oncologist.

CONCLUSIONS

As we have considered the role of lineage in the development of the CNS, it has become apparent that early in development cells are not restricted by lineage—they are multipotential—and that late in development it is possible for proliferating cells to become committed to a particular fate, although it seems that some cells do not become committed until the time of their final mitosis. We have also examined data from experiments in which neural precursor cells were immortalized using oncogenes and the properties of the immortalized cells were examined both biochemically and developmentally. To what extent are the properties of proliferating precursors or experimentally immortalized precursors similar to the properties of the proliferating cells in neural tumors?

Neural tumors also show restrictions in the range of differentiated characteristics that they display. The most common neural tumors are those that display characteristics of the most abundant cells in the CNS, the glia. Proliferating cells are the most sensitive to mutagenic agents. Glia maintain the ability to proliferate throughout life and thus may be a target for mutational insult at any point in life. The events that lead to the neoplastic transformation of these cells seem to be independent of the mechanisms that determine their identity. In keeping with this, the cells immortalized from a population of precursor cells by an exogenously introduced oncogene have the properties expected of that population. These experimentally immortalized cells may then be appropriate models for the study of the properties of the proliferating cells of primitive tumors. A question that remains to be answered is the origin of adult neuronal tumors. Neurons are postmitotic and it is unclear whether neuronal tumors arise from the persistence of a small population of precursors into adulthood or whether a terminally differentiated neuron can be pushed back into the mitotic cycle as a result of or to become a target for the mutational sequence of events that result in neoplastic transformation. This sequence of events involves a number of independent mutations, the accumulation of which must take some time. In this context it is not difficult to ascribe childhood neuronal tumors to altered embryonic precursors.

Experimentally immortalized cell lines may be useful not only for identifying the cells from which particular tumors originate but also for studying the steps in their neoplastic transformation. The temperature-sensitive oncogene, the tsA58 allele of T antigen, we have used is inactivated at the normal body temperature of the rat, and we do not observe uncontrolled proliferation of tsA58-immortalized cell lines after transplantation as we have observed with some other cell types. We have not studied them over long times *in vivo*, but their conditional immortalization may make them appropriate hosts for examining the effects of other mutations on the proliferation and differentiation of precursor cells. As noted above, it may be possible that a single constitutively active oncogene such as v-myc can apparently be overridden by signals to differentiate. Snyder et al. report no occurrences of tumors in animals transplanted with v-myc–immortalized cell lines after periods of up to 22 months (Snyder et al., 1992). In contrast, embryonal carcinoma cells, which can be incorporated into a blastula and contribute to many normal tissues without producing a tumor, rapidly form tumors after transplantation into neonatal cerebellum (M. Cunningham and R. McKay, unpublished results). Many types of tumors produce cells with differentiated characteristics, which may be in part a reflection of the ability of differentiation signals to overcome the proliferative state even in transformed cells. The discovery that a soluble factor, leukemia inhibitory factor (LIF), can cause leukemic cells to differentiate has led to the investigation of LIF as a treatment for lymphoid tumors. A sophisticated understanding of the mechanisms that cause neural precursors to cease proliferation and terminally differentiate may allow a similar approach to be developed in neuro-oncology.

SUGGESTED READING LIST

For a detailed discussion of neural development and an up-to-date discussion of present thought on lineage and development, see Jacobson (1991).

A comparison of the development of many types of animals and a discussion of the mechanisms that might be involved in tissue specification is found in Slack, J.M.W. (1991) From egg to embryo: Regional specification in early development. Developmental and Cell Biology Series. Cambridge: Cambridge University Press.

An interesting discussion of the way in which an inductive signal induces a uniform rather than a patchy response in the responding tissue is Gurdon, J.B. (1988) A community effect in animal development. Nature 336:772–774.

The best text on the neural crest is LeDouarin, N.M. (1982) The Neural Crest. Cambridge: Cambridge University Press. The role of lineage in differentiation in the neural crest is reviewed in: (1) Anderson, D.J. (1989) The neural crest cell lineage problem: Neuropoiesis? Neuron 3:1–12 and (2) Patterson, P.H. (1990) Control of cell fate in a vertebrate neurogenic lineage. Cell 62:1035–1038.

The segmentation of the hindbrain is reviewed in Wilkinson, D.G., and R. Krumlauf (1990) Molecular approaches to the segmentation of the hindbrain. Trends in Neurosciences 13(8):335–339, and the involvement of expression of homeobox genes and tissue interactions in the establishment of regional identity in the nervous system is reviewed in Keynes, R., and A. Lumsden (1990) Segmentation and the origin of regional diversity in the vertebrate central nervous system. Neuron 2:1–9.

The signals that regulate proliferation and differentiation of the O2A precursor are reviewed in Lillien, L.E., and M.C. Raff (1990) Differentiation signals in the CNS: Type-2 astrocyte development in vitro as a model system. Neuron 5:111–119.

REFERENCES

Almazan, G., and R. McKay (1992) An oligodendrocyte precursor line from rat optic nerve. Brain Res. 579:234-245.

Bailey, P., and H. Cushing (1925) Medulloblastoma cerebelli: A common type of midcerebellar tumor of childhood. Arch. Neurol. Psychiatry 14:192–224.

Bartlett, P.F., H.H. Reid, K.A. Bailey, et al. (1988) Immortalization of mouse neural precursor cells by the c-myc oncogene. Proc. Natl. Acad. Sci. USA 85:3255–3259.

Birren, S.J., and D.J. Anderson (1990) A v-myc-immortalized sympathoadrenal progenitor cell line in which neuronal differentiation is initiated by FGF but not NGF. Neuron 4:189–201.

Bogler, O., D. Wren, S.C. Barnett, et al. (1990) Cooperation between two growth factors promotes extended self-renewal and inhibits differentiation of oligodendrocyte-type-2 astrocyte (O2A) progenitor cells. Proc. Natl. Acad. Sci. USA 87:6368–6372.

Cattaneo, E., and R.D.G. McKay (1990) Nerve growth factor regulates proliferation and differentiation of neuronal stem cells. Nature 347:762–765.

Cepko, C.L., B.E. Roberts, and R.C. Mulligan (1984) Construction and application of a highly transmissible murine retrovirus shuttle vector. Cell 37:1053–1058.

Coppola, J.A., and M.D. Cole (1986) Constitutive c-myc oncogene expression blocks mouse erythroleukemia cell differentiation but not commitment. Nature 320:760–763.

Fraser, S., R. Keynes, and A. Lumsden (1990) Segmentation in the chick embryo hindbrain is defined by cell lineage restrictions. Nature 344:431–435.

Frederiksen, K., P.S. Jat, N. Valtz, D. Levy, et al. (1988) Immortalization of precursor cells from the mammalian CNS. Neuron 1:439–448.

Frederiksen, K., and R. McKay (1988) Proliferation and differentiation of rat neuroepithelial precursor cells in vivo. J. Neurosci. 8:1144–1151.

Hayes, T.E., N.L.M. Valtz, and R.D.G. McKay (1991) Downregulation of CDC2 upon terminal differentiation of neurons. New Biologist 3:259–269.

Heuer, J.G., C.S. v. Bartheld, Y. Kinoshita, et al. (1990) Alternating phases of FGF receptor and NGF receptor expression in the developing chicken nervous system. Neuron 5:283–296.

Hirano, S., S. Fuse, and G.S. Sohal (1991) The effect of the floor plate on pattern and polarity in the developing central nervous system. Science 251:310–313.

Hockfield, S., and R. McKay (1985) Identification of major cell classes in the developing mammalian nervous system. J. Neurosci. 5:3310–3328.

Jacobson, M. (1991) Developmental Neurobiology. New York: Plenum Press.

Keynes, R., and A. Lumsden (1990) Segmentation and the origin of regional diversity in the vertebrate central nervous system. Neuron 2:1–9.

Lendahl, U., L. Zimmerman, and R.D.G. McKay (1990) CNS stem cells express a new intermediate filament protein. Cell 60:585–595.

Lillien, L.E., and M.C. Raff (1990) Differentiation signals in the CNS: Type-2 astrocyte development in vitro as a model system. Neuron 5:111–119.

Lufkin, T., A. Dierich, M. LeMeur, M. Mark, and P. Chambon (1991) Disruption of the *Hox-1.6* homeobox gene results in defects in a region corresponding to its rostral domain of expression. Cell 66:1105–1119.

Lumsden, A. (1990) The cellular basis of segmentation in the developing hindbrain. Trends in Neurosciences 13(8):329–335.

Martinez, S., and R.-M. Alvarado-Mallart (1990) Expression of the homeobox *chick-en* gene in chick/quail chimeras with inverted mes-metencephalic grafts. Dev. Biol. 139:432–436.

Martinez, S., M. Wassef, and R.-M. Alvarado-Mallart (1991) Induction of a mesencephalic phenotype in the 2-day-old chick prosencephalon is preceded by the early expression of the homeobox gene *en*. Neuron 6:971–981.

McConnell, S.K. (1988) Fates of visual cortical neurons in the ferret after isochronic and heterochronic transplantation. J. Neurosci. 8:945–974.

McConnell, S.K., and C.E. Kaznowski (1991) Cell cycle dependence of laminar determination in developing neocortex. Science 254:282–285.

McKay, R. (1989) The origins of cellular diversity in the mammalian central nervous system. Cell 58:815–821.

McKay, R.D.G., N. Valtz, M. Cunningham, et al. (1990) Mechanisms regulating cell number and type in the mammalian central nervous system. Cold Spring Harbor Symp. Quant. Biol. 50:291–301.

Redies, C., U. Lendahl, and R. McKay (1991) Differentiation and heterogeneity in T-antigen immortalized precursor cell lines from mouse cerebellum. J. Neurosci. Res. 30:601–615.

Renfranz, P.J., M.G. Cunningham, and R.D.G. McKay (1991) Region-specific differentiation of the hippocampal stem cell line HiB5 upon implantation into the developing mammalian brain. Cell 66:713–729.

Rollins, B.J., and C.D. Stiles (1989) Serum-inducible genes. Adv. Cancer Res. 53:1–32.

Russell, D., and L.J. Rubinstein (1989) Pathology of Tumors of the Nervous System. Baltimore: Williams and Wilkins.

Ryder, E.F., E.Y. Snyder, and C.L. Cepko (1990) Establishment and characterization of multipotent neural cell lines using retrovirus vector-mediated oncogene transfer. J. Neurobiol. 21(2):356–375.

Schlaggar, B.L., and D.D.M. O'Leary (1991) Potential of visual cortex to develop an array of functional units unique to somatosensory cortex. Science 252:1556–1560.

Snyder, E.Y., D.L. Deitcher, C. Walsh, et al. (1992) Multipotent neural cell lines can engraft and participate in development of mouse cerebellum. Cell 68:33–51.

Temple, S. (1989) Division and differentiation of isolated CNS blast cells in microculture. Nature 340:471–473.

Tessier-Lavigne, M., M. Placzek, A.G.S. Luimsden, et al. (1988) Chemotropic guidance of developing axons in the mammalian central nervous system. Nature 336:775–778.

Tohyama, T., V.M.-Y. Lee, L.B. Rorke, et al. (1992) Nestin expression in embryonic human neuroepithelium and in human neuroepithelial tumors. Lab. Invest. 66:303-313.

Valtz, N.L.M. (1989) Differentiation of immortal neural cell lines. Honors thesis, Harvard University, Cambridge, MA.

Valtz, N.L.M., T.E. Hayes, T. Norregaard, et al. (1991) An embryonic origin for medulloblastoma. New Biologist 3(4):364–371.

Wilkinson, D.G., and R. Krumlauf (1990) Molecular approaches to the segmentation of the hindbrain. Trends in Neurosciences 13(8):335–339.

Chapter 3

Nervous System Development in Drosophila melanogaster *and* Caenorhabditis elegans

Cornelia I. Bargmann

INTRODUCTION

In the past 10 years, the remarkable discovery has been made that molecular mechanisms of development are conserved among all animals, and that many of the same molecular components appear in signal transduction pathways of all eukaryotes from yeast to man. This mechanistic conservation means that molecules can be studied in the organism in which their properties are most transparent; general principles or specific predictions made from work in one organism can subsequently be explored in other organisms.

This chapter reviews aspects of nervous system development that have been studied using genetic approaches in two simple invertebrates, the fruit fly *Drosophila melanogaster* (herein referred to as *Drosophila*) and the soil nematode worm *Caenorhabditis elegans (C. elegans)*. The nervous systems of both of these organisms are extremely simple compared to those of mammals. The human brain consists of about a trillion neurons of many different types. By contrast, one fruit fly contains only a few hundred thousand neurons, and a single nematode only 302 neurons within its entire nervous system. This simplicity allows for identification of individual neurons or groups of neurons in many different animals with ease. Similarly, one generation time for a human is at least 15 years; for the fruit fly it is about two weeks and for the nematode, three days. Their short life cycles make the nematode and fruit fly well suited to genetic analysis of the nervous system, in which the effects and interactions of mutations that perturb nervous system function are examined. The basis of neurogenesis, specification of neural cell type, axon guidance and cell migration, and cell death are all being addressed in the fruit fly and the nematode. Each of these processes corresponds to an analogous process in vertebrate neural development.

Many of the oncogenes that trigger tumorigenesis are genes whose normal functions lie in developmental signaling or in neuronal or hormonal signaling. Understanding their normal mechanisms of action will shed light on their dysfunction in tumorigenesis. Strikingly, several genes identified as oncogenes in human tumors have close homologues that participate in well-defined aspects of *Drosophila* or *C. elegans* development. The implication of these results is that the same molecular mechanisms can regulate cell proliferation as well as normal development.

NEUROGENESIS

The Neurogenic Loci and Lateral Inhibition

In vertebrate and insect embryos, neurons and epidermal cell types are both generated from a single neurectodermal cell layer. In each case, inductive interactions lead to the separation of the two cell types (Fig. 1). In vertebrates, underlying mesoderm induces the formation of the neural tube. In insects, the relevant interactions appear to occur within the neurectoderm itself. In grasshoppers, if one neuroblast cell is killed with a laser beam, an adjacent cell in the neurepithelium will become a neuroblast in its place (Doe and Goodman, 1985). This experiment indicates that the nearby cells possess the developmental potential to become neuroblasts, but are inhibited by the existing neuroblast so that they form epidermoblasts. This sort of inhibitory interaction is called lateral inhibition.

The separation of neuronal and epidermal cell types is disrupted by mutations in any of seven genes of *Drosophila* called neurogenic genes (*N* = *Notch, Dl* = *Delta, bib* = *bigbrain, mam* = *mastermind, neu* = *neuralizing, amx* = *almondex, E(spl)* = *Enhancer of split*) (Campos-Ortega, 1988). The loss of function of any of these genes causes all of the cells of the neurectoderm to differentiate as neuroblasts. The embryo subsequently dies because of its lack of ectodermal cells. From the mutant phenotype, it is inferred that the

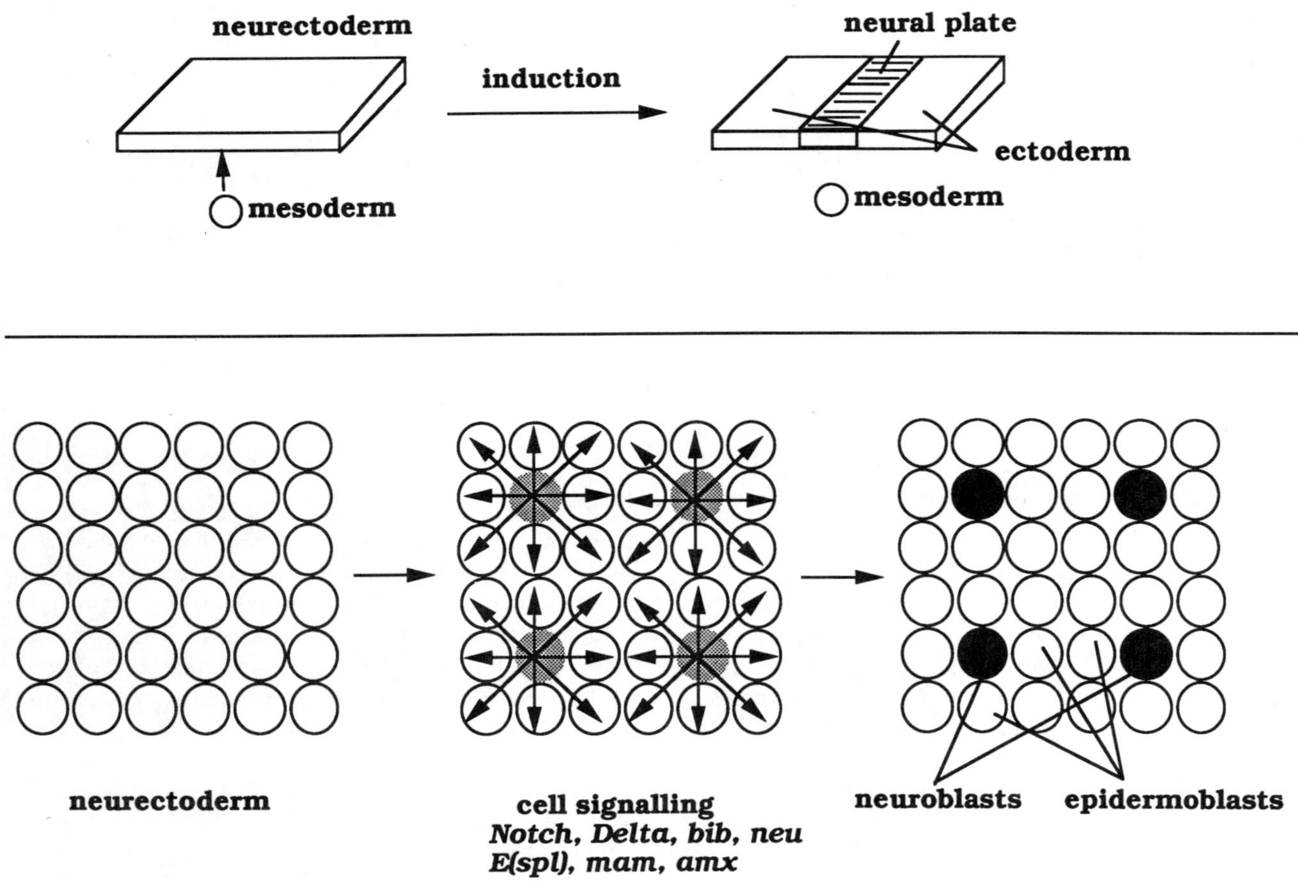

Fig. 1. Neurogenesis in vertebrate and *Drosophila* embryos. At **top** is shown neural induction in a vertebrate embryo. An undifferentiated ectoderm is induced by the underlying mesoderm to become neural plate. At **bottom** is shown *Drosophila* neurogenesis. Cell interactions among the cells of the neurectoderm lead some to become neuroblasts and some epidermoblasts. The neuroblasts inhibit the neighboring cells from becoming epidermoblasts by lateral signaling.

normal function of these genes prevents excessive neurogenesis.

The neurogenic genes are probably required for communication between neurectodermal cells. Cell communication in the form of lateral inhibition usually limits the number of neuroblasts; in the absence of cell communication, all of the cells can become neuroblasts, with disastrous consequences. The neurogenic genes could affect the production of the inhibitory signal, its receptor, or other responding mechanisms. Some genes could be required for epidermal differentiation, so that in their absence, neural differentiation occurs by default.

Several neurogenic genes have been cloned. *Notch* and *Delta* both encode large proteins with multiple cysteine-rich repeats similar to those observed in the low-density lipoprotein (LDL) receptor and in the epidermal growth factor (EGF), as well as membrane-spanning domains and intracellular regions. These structures suggest that *Notch* and *Delta* are directly involved in the cell interactions between neurectodermal cells, either as the signal or its receptor.

In order for epidermal development to occur, the signal molecule must be produced by the future neuroblast, while the receptor molecule would be required in the future epidermoblast. Considering the conceptual simplicity of the question, it has been surprisingly difficult to determine whether *Notch* and *Delta* act in the signaling or the responding cells during neurogenesis. Both clearly act over relatively short ranges, and both are expressed throughout the neurectoderm (in fact, *Notch* is expressed ubiquitously in the embryo). Two genes closely related to *Notch* have been identified in *C. elegans: lin-12* and *glp-1*. Both of these genes also participate in cell signaling, and both are required in the responding and not the signaling cell, suggesting that they are receptors and not the signal itself (Seydoux and

Greenwald, 1989). By analogy, it seems likely that *Notch* should act as a receptor within the future ectodermal cells. *Delta* expression is limited to the neuroblast (signaling) cell after induction takes place, consistent with the idea that it might be part of the lateral signal.

Notch and *Delta* mutations interact genetically, reflecting a close relationship between the two genes. If the two genes are each expressed on different *Drosophila* tissue culture cells, the two kinds of cells adhere to each other, suggesting a physical interaction between the two gene products (Fehon et al., 1990). *Notch* and *Delta* might be the receptor and signal in the lateral signaling pathway. Alternatively, they might be adhesive molecules required for cell interactions to take place, while some other genes encode the authentic signal and receptor.

Genetic and molecular data indicate that the target of the cell-signaling event mediated by *Notch* and *Delta* may be the *E(spl)* gene. *E(spl)* is a complex locus containing at least 11 transcripts. Several of these appear to encode possible transcriptional regulators of the helix–loop–helix class (Klaembt et al., 1989). More is said about these regulators below.

Another neurogenic gene, *bib*, also encodes a membrane-spanning protein. The *bib* product is related to the bovine lens major intrinsic protein (MIP) and several transport channels (Rao et al., 1990). *bib* may take part in a second form of cell communication during neurogenesis, perhaps acting as a specialized gap junction between neurectodermal cells.

The molecular data support the model that neurogenic genes are required for cell communication within the neurectoderm. The basic mechanism appears to be that cell communication over short distances takes place using transmembrane signaling proteins, and the activity of these proteins then sets the level of transcription factors that control cell-type-specific gene expression.

Comparison With Vertebrate Neurogenesis

In *Drosophila*, intrinsic interactions within the neurectoderm are important for neurogenesis, and if these interactions fail to take place, neuroblasts form by default. By contrast, in vertebrates the neurectoderm interacts with an inducing mesoderm, and in the absence of this interaction only ectodermal cells form. However, various lines of evidence hint that these two inductions may be more similar than they appear initially.

For example, cells of the vertebrate neurectoderm apparently give rise to neuroblasts in the absence of inducing mesoderm if they are separated into individual cells, rather than maintained in an intact monolayer (Godsave and Slack, 1991). This observation hints that in vertebrates, intercellular contacts within the neurectoderm may also inhibit neurogenesis; the inducing mesoderm might release cells from this inhibition. Though the relative contributions of the two components may differ, both neurectodermal interactions and regional information probably play roles in neurogenesis in both vertebrates and invertebrates.

Recently, vertebrate homologues of the *Drosophila Notch* gene have been cloned (Weinmaster et al., 1991). The rat *Notch*-like gene is expressed in the developing nervous system as well as areas outside the nervous system in which cell interactions between epithelial and mesenchymal cell types determine cell fates (the *C. elegans Notch*-like genes *lin-12* and *glp-1* also act outside the nervous system). The expression pattern of *Notch* homologues is consistent with the idea that *Notch*-like genes participate in similar cell interactions in flies, worms, and vertebrates.

DETERMINATION OF CELL LINEAGE AND CELL FATE BY TRANSCRIPTION FACTORS

Helix–Loop–Helix Proteins and Neurogenesis

Unlike the neurogenic genes, which normally act to prevent excessive neurogenesis, a locus called the *Achaete-scute* complex, or AS-C, promotes neurogenesis (Ghysen and Dambly-Chaudiere, 1988). If the AS-C genes are mutated, fewer neuroblasts than usual are formed; the cells that would have formed neuroblasts either form epidermoblasts or die. Both the AS-C genes and the neurogenic genes described above act at multiple times in nervous system development; the AS-C has been well characterized for its effects on easily visualized sensory neurons in the fly wing.

Four different transcripts from the AS-C each encode a protein in a class called helix–loop–helix (HLH) proteins (Table 1). These proteins contain a conserved helix–loop–helix motif that can bind DNA and form dimers; an individual HLH protein can dimerize with itself or with other members of this family (Murre et al., 1989). The genes in this family act directly or indirectly as regulators of transcription. The family includes the *myoD* genes, which direct muscle differentiation in mouse cells; the *myc* family of proto-oncogenes, whose alteration is one cause of cancer; E12/E47, proteins that regulate immunoglobulin expression in the mouse immune system; and numerous *Drosophila* genes, including the AS-C and the neurogenic locus *E(spl)*.

The ability of the AS-C gene products to dimerize with other family members provides a mechanism whereby molecules with different properties are generated through a combinatorial assortment of proteins. The active DNA-binding form of the AS-C gene product is dimerized with a helix–loop–helix protein called *da*, which is broadly distributed in many cell types (and may form a dimerization partner for other HLH proteins). Several other genes, including *hairy* and *extramacrochaetae (emc)*, antagonize the activity of the AS-C. This effect can be seen in the wing, where AS-

TABLE 1. Transcription Factors that Affect Neuronal Identity

Gene	Function (species)
Helix–loop–helix proteins	
Achaete-scute complex (AS-C)	Promotes neuronal differentiation (DM)
Enhancer of split (E(spl))	Promotes epidermal differentiation (DM)
daughterless (da)	Neuronal differentiation, also additional functions (DM)
hairy (h)	Inhibits AS-C neuronal differentiation, also additional functions (DM)
extra macrochaetae (emc)	Inhibits AS-C neuronal differentiation (DM)
Homeobox-containing proteins	
cut	Determines es sensory cell identity (DM)
mec-3	Determines mechanosensory identity (CE)
even-skipped (eve)	Determines subsets of neuronal fates, also additional functions (DM)
fushi tarazu (ftz)	Determines subsets of neuronal fates, also additional functions (DM)
unc-86	Mother–daughter lineage decision (CE)
rough	Photoreceptor cell inductions (DM)
Zinc fingers	
glass	Photoreceptor cell specification (DM)
seven-up (svp)	Photoreceptor identity, also additional functions (DM)
seven in absentia (sina)	Photoreceptor R7 identity (DM)
lin-29	Timing of developmental events (CE)
Other	
lin-14	Timing of developmental events (CE)

DM, *Drosophila melanogaster*; CE, *Caenorhabditis elegans*.

C mutants have missing sensory neuron bristles, while *hairy* or *emc* mutants have excessive numbers of bristles. *emc* encodes a member of the helix–loop–helix family that can dimerize but does not bind DNA. When *emc* forms dimers with the proteins of the AS-C, those dimers cannot bind DNA and are presumably incapable of carrying out the functions of the AS-C (Van Doren et al., 1991). Thus *emc* (and perhaps *hairy*, another HLH protein) represses the activity of the AS-C by soaking up and inactivating its helix–loop–helix proteins. The proper number of bristles is formed when a subset of the cells in the wing neurectoderm expresses the AS-C genes but not their repressors, *hairy* and *emc*.

Two genes homologous to the AS-C genes have been isolated from a rat sympathoadrenal cell line. One of these, called MASH1, is expressed only in the developing nervous system (Lo et al., 1991). Its expression is transient and precedes terminal neuronal differentiation. Similarly, the genes of the AS-C are required early in neurogenesis, but disappear during the later differentiation of the nervous system. By analogy to the homologous AS-C genes, MASH1 might promote neuronal development in the rat.

Homeobox Proteins and Cell Fate

Once the decision to become a neuron has been made, a cell must decide what sort of neuronal cell it will become. The expression of specific transcriptional regulators is key to the generation and differentiation of distinct cell types in the nervous system, as it is during other aspects of development (Table 1). In both *Drosophila* and *C. elegans*, many genes have been identified that are crucial for particular cell types to take on their appropriate fates. The most informative mutations have been those that result in cell transformations, whereby one cell takes on the properties of another cell type. Such a transformation implies that the mutated gene plays a role in distinguishing between the two cell types.

The *cut* gene of *Drosophila* is required for normal development of a class of larval sensory organs called es cells; in *cut* mutants, these cells are transformed into another kind of sensory structure called the chordotonal organ. *cut* expression is normally limited to the future es cells. If *cut* is inappropriately expressed in the future chordotonal cell, it differentiates as an es cell type instead (Blochlinger et al., 1991). Thus the expression of *cut* is necessary and sufficient for the es cells to be made instead of chordotonal cells. *cut* encodes a homeobox-containing protein. Homeoboxes are DNA-binding domains first identified in association with genes that determine segmental identity in *Drosophila*. Similar proteins regulate aspects of development and cell identity in *C. elegans* and vertebrates (Reid, 1990).

Some mutations lead to clear transformations among cell types; other mutations cause more ambiguous failures of specific neuronal identities. The *mec-3* gene of *C. elegans* is a homeobox gene that is required for the development of a class of sensory neurons (Way and Chalfie, 1988). When *mec-3* is absent, these neurons lack the characteristics of sensory neurons and have some of the properties of their lineal sisters instead. However, their identity appears to be confused rather than completely transformed. This confused identity can be explained if the difference between the sensory neurons and their sisters is determined by several genes acting together, so that no one gene controls the entire cellular differentiation program. The *mec-3* gene product is very similar to that of *lin-11*, another *C. elegans* gene involved in determining cell identity in sister cell lineages. Both of these genes encode homeoboxes along with a second domain called an LIM domain.

Homeobox genes identified for their patterns on the overall development of the embryo are also expressed in the nervous system. The *Drosophila eve* and *ftz* homeobox genes have general roles in development, where they act in large domains, and are also expressed in precise subsets of neurons in each segment of the fly body. Each of these neurons has a stereotypic pattern of axon outgrowth. If the *ftz* or *eve* genes are missing, axons grow out of the neurons but those axon patterns are abnormal, as though the unique identities of the neurons were conferred by expression of *eve* or *ftz* (Doe et al., 1988a,b).

Mouse homeobox-containing genes with spatially restricted patterns of expression have been identified and mutated by targeted gene disruption. Mice lacking the En-2 homeobox locus have abnormal cerebellar development (Joyner et al., 1991). Mice lacking two different spatially regulated homeoboxes in the *Hox* gene family have abnormal nervous system development along with abnormal development of other tissues in one "segment" of the developing mouse (Hunt and Krumlauf, 1991).

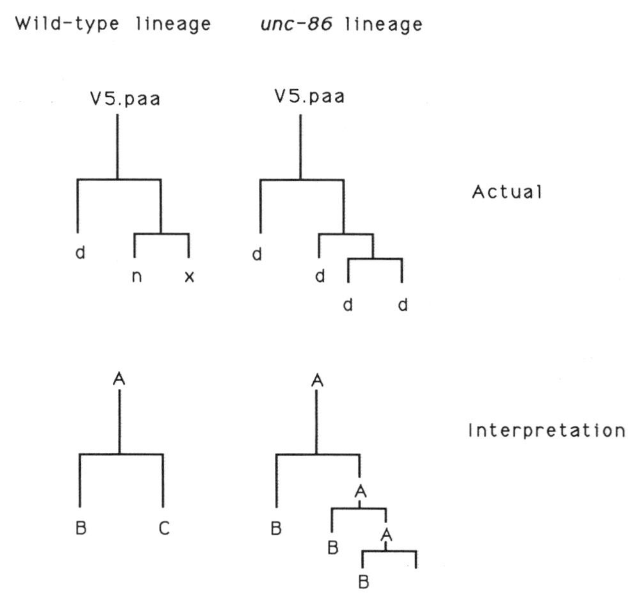

Fig. 2. Lineage abnormalities in *unc-86* mutants. Lineages of wild-type and *unc-86* V5.paa lineages are shown at **top**. The branching vertical lines below the cell name are cell divisions. V5.paa normally gives rise to d, a dopaminergic neuron; n, a nondopaminergic neuron; and x, a cell death. In *unc-86* mutants, extra dopaminergic neurons are generated at the expense of the other two cell types. The interpretation of these and other lineages is shown at **bottom**. If a cell A normally gives rise to daughters B and C, in *unc-86* mutants A gives rise to B and an additional A cell, which in turn can generate B and yet another A cell.

Developmental Timing and Transcription Factors in *C. elegans*

Many aspects of cell fate can be seen in a lineage-specific perspective: A gene determines the ability of a cell to know that it is different from its sister. However, the timing of developmental events can be as important as their occurrence. Several genes have been identified in *C. elegans* that are required for normal developmental timing during nervous system development and other aspects of development.

One such gene is the *unc-86* gene of *C. elegans*. Mutations in *unc-86* lead to reiterations in cell lineage in several neuronal lineages at different times in development (Chalfie et al., 1981). The normal function of *unc-86* appears to be required for certain daughter neuroblasts to know that they are different from their mothers (Fig. 2). *unc-86* encodes a protein containing a potential DNA-binding homeodomain in conjunction with a domain called the POU-specific domain that also participates in DNA binding, so *unc-86* is likely to be a transcriptional regulator. The *unc-86* gene product is expressed in daughter cells after cell division, presumably allowing them to express "daughter-specific" genes (Finney and Ruvkun, 1990). The question of how developmental timing is determined in these cells can thus be restated to ask how *unc-86* expression is induced in daughter cells but not their mothers.

POU-containing proteins include mouse immunoglobulin enhancer-binding proteins, general mammalian transcription factors, and a number of mammalian genes that are expressed in the nervous system (Ruvkun and Finney, 1991). It has been speculated that these mammalian genes may play roles in mammalian neurogenesis analogous to those that the *unc-86* gene plays in *C. elegans*.

Another set of genes that determines developmental timing in *C. elegans* is called heterochronic genes. Unlike *unc-86*, they are used throughout the body and not just in the nervous system. *unc-86* sets the developmental time within a lineage, and can be active at different times in different cells. By contrast, the heterochronic genes synchronize the general timing of development across the organism (Ambros and Horvitz, 1984). The best characterized heterochronic gene is called *lin-14*. Loss of *lin-14* function causes precocious development, in which cells act as adult cells early in larval life. Excessive *lin-14* function retards development, so that larval cells reiterate larval cell fates instead of maturing to their adult fates. Thus *lin-14* levels can "set" the developmental time in the animal. The *lin-14* gene encodes a novel nuclear protein that binds DNA. Consistent with its gene activity, it is normally expressed at high levels in young animals and low levels in adults. A second heterochronic locus that acts at later times called *lin-29* also encodes a DNA-binding protein of the zinc-finger type.

Sequential Inductions: Tyrosine Kinase and *ras* Genes

Along with lineage-specific information, interactions among groups of cells can be important in the determination of individual neuronal identities. One of the best characterized examples of interactions in determining neuronal cell fate is the development of the *Drosophila* eye (Banerjee and Zipursky, 1990).

The fly eye consists of about 800 repeating units called ommatidia, each of which contains eight neuronal photoreceptors and 12 nonneuronal cells. All cells of the ommatidia develop from an undifferentiated neurepithelium in a wave of morphogenesis; cell lineage plays no role in this development. Lineage-independent mechanisms appear in mammalian neuronal development as well; for example, the mouse retina develops by a largely lineage-independent route (Turner et al., 1990).

The eight photoreceptor cells in a single cluster of the fly eye are called R1–R8 and can be unambiguously identified by their position in the cluster (Fig. 3). During development, R8 cells are the first to differentiate. Additional neuronal cells are added to the ommatidial cluster in a stereotypic order: first R8, then R2 and R5, then R3 and R4, then R1 and R6, and finally R7. At each point, the cells that are added are in physical contact with the cells that are already contained within the cluster. Because of this proximity and the absence of lineage information, it is likely that cell interactions among the cells of the developing cluster determine the sequential developmental identities of R1–R8.

Mutations that disrupt eye development in *Drosophila* have been isolated by virtue of their visible defects in the adult eye. These mutants have been characterized to identify a genetic pathway that corresponds to the developmental pathway of the eye (Banerjee and Zipursky, 1990; Rubin, 1989).

At an early stage, the R8 precursors are generated in a regular pattern across the neurepithelium. The spacing of R8 precursors is disrupted by mutations in the neurogenic *Notch* gene. As in neurogenesis, *Notch* mutations cause excessive numbers of neuronal cells to form at the expense of nonneuronal cell types. *Ellipse*, an unusual dominant mutation in the *Drosophila* homologue of the epidermal growth factor receptor (a tyrosine kinase), also regulates cluster spacing. *Ellipse* mutations cause an increased spacing of ommatidial clusters with a corresponding reduction of the number of ommatidia in the eye. A model consistent with these data is that an epidermal growth factor-like activity inhibits ommatidial clusters from forming. The proposed activity could be expressed in a regulated pattern within the developing eye to allow proper cluster spacing.

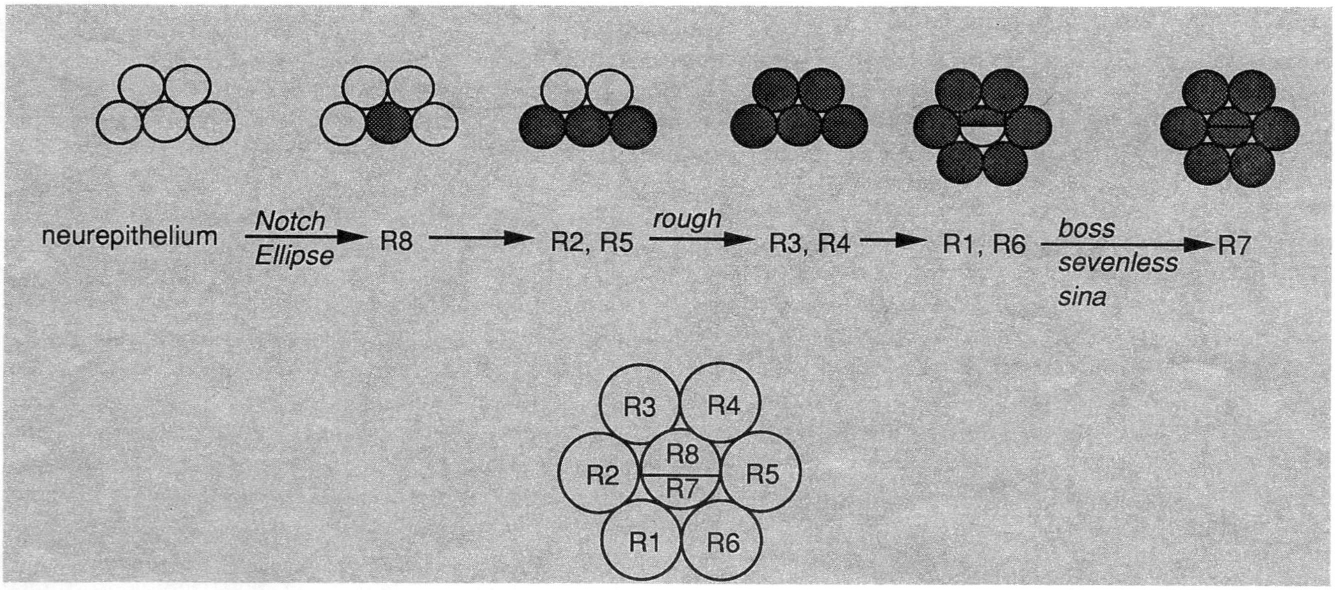

Fig. 3. Development of the *Drosophila* retina. The photoreceptor cells form in a neurepithelial layer. A five-cell precluster of photoreceptor cells contains one R8 neuron. Additional neurons differentiate in a precise series through interaction with the cells already in the cluster. R8 and R7 lie one on top of the other, so that both cell profiles would not be seen in a single section as diagrammed.

The R2 and R5 cells then join the R8 cell in the cluster, followed by R3 and R4. Mutations in the *rough* gene arrest the development of the cluster at the three-photoreceptor cell stage, before the addition of R3 and R4. The *rough* gene encodes a homeobox-containing protein, indicating that the determination of cell fate through inductive mechanisms involves this class of transcriptional regulatory proteins. Surprisingly, genetic mosaic experiments indicate that the *rough* gene is required within the R2 and R5 cells for the normal development of R3, R4, and subsequent cells in the cluster. Formally, *rough* acts as an element in a signaling process in the R2 and R5 cells that allows recruitment of R3 and R4 to the cluster.

The *rough* gene product (visualized by antibodies) is expressed within R2, R5, R3, and R4 photoreceptors. If it is aberrantly expressed in the R7 photoreceptor, that cell takes on properties more similar to those of R2–R5. Taken together with the effect of loss-of-function mutations in *rough*, it appears that *rough* (like *mec-3*, described above) controls some but not all of the properties of the R2 and R5 cells. These properties include the ability to induce R3 and R4 cells within the eye. Several other potential transcriptional regulators with DNA binding motifs of the zinc-finger family also appear to be required in eye development: The *sina* gene is required in and for R7; the *seven-up* gene represses R7 fates in non-R7 photoreceptors; and the *glass* gene is required for photoreceptor survival.

The last photoreceptor to be added to the developing eye is R7. Mutations in the *sevenless (sev)* gene prevent this cell from differentiating. The *sev* gene encodes a membrane-spanning tyrosine kinase protein. *sev* acts within the R7 cell, so it appears to be a receptor for a signal directing R7 development. Tyrosine kinases are widely known as receptors for growth factors and, in modified form, as active oncogenes in mammalian tumors. In *Drosophila* development, the tyrosine kinases *sev* and *Ellipse* appear to play roles in normal neuronal development.

The ligand for the *sev* receptor has also been identified through genetic analysis. The *boss* gene (for *Bride Of SevenlesS*) is required within the R8 (signaling) cell for normal R7 development. Most known ligands for tyrosine kinases are soluble molecules, but *boss* encodes a membrane-spanning protein with multiple transmembrane domains. The *boss* and *sev* genes appear to interact while both ligand and receptor are attached to their cell surfaces (Kraemer et al., 1991).

In summary, a cascade of genes is induced and regulated in fly eye development. Transcriptional regulators are likely to be important targets of the inductive events in the eye. At least some of the signals pass through tyrosine kinase receptors.

The importance of tyrosine protein kinases in oncogenesis and cell division has recently been extended, in this model and several others, to show important roles of tyrosine kinases in the normal development of *Drosophila*, *C. elegans*, and mice. An important question that remains only partly answered is the mechanism of signal transduction of these proteins. Insight into these mechanisms has recently been gained through genetics of *C. elegans* development. A gene called *let-23 (lethal gene 23)* participates in numerous cell-signaling events, including some that determine neuroblast identities, in *C. elegans* development. The function of *let-23* has been easiest to study in the development of a nonneuronal tissue, the developing vulva of *C. elegans*. The vulva is induced through the action of a gonadal cell called the anchor cell, which signals underlying epidermal cells to adopt vulval identities (Horvitz and Sternberg, 1991). In *let-23* mutants, this signaling process fails and no vulva is formed. *let-23* encodes the *C. elegans* epidermal growth factor receptor, so activity of this tyrosine kinase is crucial for normal *C. elegans* vulval development. The *lin-3* gene encodes an epidermal growth factor-like molecule that is probably the ligand for the *let-23* product (Hill and Sternberg, 1992).

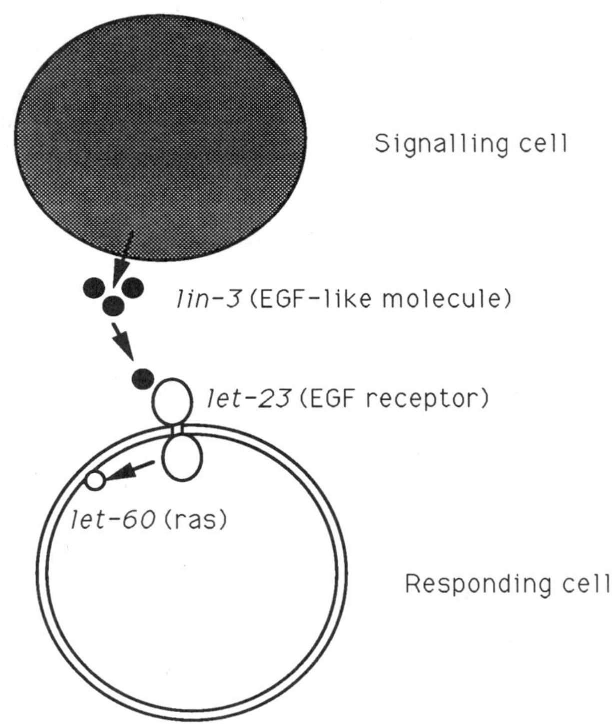

Fig. 4. A model for cell interactions in *C. elegans* development. A signaling cell induces the epithelial precursor cells to form vulval tissue. This signaling might occur through the release of an EGF-like molecule. An EGF receptor and a *ras* gene are required for vulval tissue to form. Here they are diagrammed in the responding epithelial cell; their actual site of action is unknown. Genetically, the *ras* gene acts downstream of the EGF receptor in this signaling pathway.

Loss-of-function mutations of the *let-60* gene lead to the absence of a vulva, like *let-23* mutations. Thus *let-60* acts in the same signaling pathway as *let-23*. Dominant mutations in *let-60* cause the opposite defect: Extra vulval tissue forms. In fact, extra vulval tissue forms in these mutants even in the presence of a *let-23* mutation. Thus the requirement for the *let-23* tyrosine kinase receptor is completely bypassed by overactivity in *let-60*, suggesting that *let-60* acts downstream of the *let-23* gene in a signaling pathway(Fig. 4).

let-60 encodes a member of the *ras* family. *ras* genes are oncogenes in many human and animal tumors, but their mode of action and normal function is unclear. The genetic data indicate that, in *C. elegans*, *ras* acts as part of the signaling pathway downstream of a tyrosine kinase receptor in normal development. It will be important to discover whether similar signaling pathways include *ras* genes in mammals. Many genes that are not yet cloned interact with *let-23* and *let-60* in vulval development; some of these genes may encode other regulators or targets of the *ras* proteins.

These studies emphasize the importance of the local environment in which a signal is released during inductive development. The EGF homologue in *C. elegans* development acts over only one or two cell diameters to exert its effects. The *boss* gene product is still attached to the membrane of the signaling cell when it contacts the *sevenless* tyrosine kinase receptor. These interactions take place over extremely small distances. Similarly, in mammals, platelets release growth factors precisely at a site of injury and hematopoetic factors operate over a limited distance in the bone marrow to regulate blood production.

Inductive interactions in vertebrate neuronal development have been demonstrated in contexts including developing spinal cord of the mouse and central and peripheral nervous system development of the chick. A tyrosine kinase receptor (the *W*, or Dominant White Spotted gene) and its ligand (the *Steel* gene product) are required for neural crest cell proliferation in the mouse (Reth and Bernstein, 1991). The *wnt-1* locus, a secreted ligand for an unknown receptor, is required for mouse midbrain and cerebellum development (McMahon and Bradley, 1990).

PROGRAMMED CELL DEATH

Regulated cell death is widespread in the developing nervous system of vertebrates and invertebrates. Although cell death can be induced by either deprivation of survival factors or exposure to killing factors, in many cases the death itself is the result of a cell-intrinsic suicide program. This form of death, which requires protein synthesis within the dying cell, is called apoptosis.

Cell death has been studied extensively in *C. elegans*, where 131 cells, many of them from neuronal lineages, die during normal development. The genes involved in programmed cell death fall into three general classes: cell-type-specific regulatory genes, suicide genes, and genes required to degrade the cell corpse.

The cell-type-specific regulatory genes are genes that act in only a few cell types to prevent or promote their deaths. Several genes that act generally to set up the developmental potential of cells in a lineage or a body region can also regulate cell deaths in that set of cells. In addition, a few genes have been identified that regulate only the onset of cell death in one or two types of cells (Ellis and Horvitz, 1991). These genes might link general developmental mechanisms to the cell death program.

Two genes, *ced-3* and *ced-4* (*ced* = *ce*ll *d*eath abnormal), are required for all cell deaths in the nematode. In animals mutant in either gene, all of the cells that normally die by programmed cell death survive. Genetic mosaic analysis revealed that *ced-3* and *ced-4* are both required within the dying cell for cell death to occur (Yuan and Horvitz, 1990). These two genes appear to be elements of a central cell suicide program that acts in all dying cells. This central mechanism can be triggered by the action of many different cell-specific genes.

Seven other genes, *ced-1,2,5,6,7,8,* and *10*, are required for the engulfment and degradation of cell debris after programmed cell death. In their absence, programmed cell death occurs but the cell corpses persist for abnormally long times. These genes might be induced by the dying cell to target it for destruction, or they might be required in the engulfing cell for its activity.

The sequences of the *C. elegans* genes in the programmed cell death pathway are not described, so it is not known if they are homologous to known mammalian genes. Candidate genes involved in mammalian apoptosis have been identified in nonneuronal cell types. Expression of the *bcl-2* gene can prevent programmed cell death of lymphocytes (Hockenbery et al., 1990). Persistent expression of *bcl-2* causes an excess of B lymphocytes to accumulate and ultimately results in lymphoma. Thus appropriate cell death can prevent oncogenesis.

AXON GUIDANCE AND CELL MIGRATION

In *C. elegans*, many neurological mutants have been isolated on the basis of their behavioral defects. The *unc* mutants are uncoordinated, the *egl* mutants are egg-laying defective, and so forth. These behavioral mutants have been screened using techniques that allow the visualization of axon processes to identify a subgroup of behavioral genes that are important in determining axon morphology.

The process of cell fate determination appears to involve successive narrowing of cell identities from neurectodermal to neuronal to a particular cell type. By contrast, the nematode genes required in axon guidance (like the pathway for programmed cell death) appear to act in many different cell types. For example, the genes *unc-5*, *unc-6*, and *unc-40* are required for normal dorsal–ventral guidance of numerous axon types as well as normal dorsal–ventral cell migration of mesodermal cells (Hedgecock et al., 1990). These genes influence the growth of neurons on the extracellular matrix, and could encode either neuronal or matrix molecules that regulate axon growth. Similarly, the *unc-34* gene has been implicated in the anterior–posterior axon growth of numerous different cell types (Desai et al., 1988). *unc-34* could mediate either neuron–substrate or neuron–neuron interactions.

In *C. elegans*, it appears that the specific axon morphology of a neuron is specified through the combinatorial action of genes with partly overlapping domains of action. Each broadly acting gene product would be responsible for a subset of the properties of the axon. Each neuron would be predicted to express or respond to some (but not all) of a group of gene products that are also expressed by other neurons. In general, the only genes that appear to have a cell-type-specific effect on axon guidance are those that alter cell identity. While the data do not eliminate the possibility that cell-type-specific axon guidance genes exist, they do indicate that any cell-type-specific gene would act through or together with the more generally acting genes like *unc-5* and *unc-6*.

In *Drosophila*, a different approach known as reverse genetics was taken to study axon guidance. Monoclonal antibodies that labeled surface proteins on insect axons were used to identify the genes that encoded those surface proteins. Subsequently, mutations in the genes were generated and their effect on axon growth was assessed.

The *fasciclin II*, or *fas II*, gene, encodes one antigen expressed on some *Drosophila* axons and on some ectodermal cells. *fas II* encodes a cell-surface protein with extracellular repeats of an immunoglobulinlike domain. Immunoglobulin domains are recognition domains that can be found in cell adhesion molecules. Interestingly, as in *C. elegans*, the guidance molecules are not restricted to an individual cell type or even to the nervous system.

Other cell-surface molecules in *Drosophila* have been analyzed using similar techniques (Grenningloh et al., 1990). Two additional cell-surface antigens called fasciclin III and neuroglian are also expressed in restricted axon pathways and are also members of the immunoglobulin gene family.

A novel, nonimmunoglobulinlike cell-surface protein called *fasciclin I*, or *fas I*, was also identified starting with a monoclonal antibody. *fas I* is expressed on subsets of axon pathways, but *fas I* mutations do not cause any defects in those pathways. This perplexing result was elucidated when it was found that *fas I* mutations interact with mutations in the *Drosophila abl* gene. *abl* encodes a cytoplasmic tyrosine kinase first identified in oncogenic retrovirus of the mouse. The *Drosophila* homologue of *abl* is expressed on developing axons but, like *fas I*, mutations in *abl* do not lead to overt defects in axon outgrowth. However, animals doubly mutant for *fas I* and *abl* have dramatic defects in axon morphology (Elkins et al., 1990). Thus *fas I* and *abl* act redundantly in directing axon guidance: Either alone can produce a normal pattern, but losing both genes leads to serious trouble.

Biochemical analysis of the *Drosophila* fasciclin and neuroglian molecules has revealed that each of these molecules can mediate homophilic adhesion. Expressing any one of these proteins on a nonadherent *Drosophila* cell line causes the cells to clump with one another. This property suggests that fasciclins direct axons to grow on preexisting pathways in the nervous system by self–self adhesion.

Surface proteins with restricted expression have been identified in the mammalian nervous system. Among the proteins expressed on subsets of neurons during mammalian development are N-CAM, TAG-1, and L1(NILE). Each of these proteins is a member of the immunoglobulin superfamily, indicating that the adhesion of this type of protein could be involved in mammalian nervous system development. Based on the genetics of *C. elegans* and *Drosophila*, one might predict that the subset of cell adhesion molecules expressed by a given cell would encode its axon morphology through a combination of their adhesive and signaling properties.

CONCLUSION

Genes that act in nervous system development are similar to (and may even be the same as) the genes that act in the development of other cell types. The problem of generating diverse cell types from a single precursor is the same whether the immune system or the nervous system is concerned. The *C. elegans* EGF-receptor-related molecule is used in both neuronal and nonneuronal tissues to mediate instructive signals for development. The *Drosophila eve* and *ftz* genes set up both segmental identity and neuronal identity. In vertebrates, leukemia inhibitory factor is a differentiation signal for lymphocytes and for cholinergic sympathetic neurons (Yamamori et al., 1989). Of course, there are likely to be some aspects of the nervous system that are more specific to its function, but not necessarily its developmental mechanisms.

The repeated isolation of certain kinds of molecules in *C. elegans* and *Drosophila*—homeobox-containing genes, other transcription factors, *Notch*-like genes, and tyrosine kinases—underscores their importance in nervous system development. The mammalian homologues of these genes are an excellent set of candidate genes for mediating similar processes in mammalian nervous system development.

Each of the developmental steps in neuronal differentiation represents a possible regulatory event that might be abnormal in nervous system tumors. By analogy to lymphoid and myeloid tumors, an early step in nervous system oncogenesis could be simply the accumulation of too many undifferentiated neuroblasts. Overproliferation of nervous system tissue could be due to faulty cell fate determination or interrupted cell interactions. *Notch*-like genes could act as recessive oncogenes in which mutations cause neuronal overproliferation. The tyrosine kinases that are involved in particular developmental events could be abnormal in tumor cells derived from the relevant cell types. Defects in kinases or transcription factors that determine cell identity could lead to arrested differentiation of neuroblasts. Mechanistic insights into the interacting heterodimers of the helix–loop–helix proteins may be relevant to understanding neuroblastomas that are associated with hyperactivity of the helix–loop–helix protein N-*myc*.

Some tumors might arise through a failure of developmental timing, in which a neuroblast continues to divide instead of differentiating. In studying such tumors, it might be useful to consider homologues of the *unc-86* or *lin-14* genes as possible tumor suppressor genes or oncogenes, respectively.

Preventing cell death in lymphocytes by expressing the *bcl-2* gene eventually leads to lymphoma. Blocking neuronal cell death could also lead to the accumulation of excessive numbers of potentially oncogenic cells. Finally, the genes that regulate axon outgrowth and cell migration are possible players in the regulation of adhesive properties that occurs during metastasis.

In principle, understanding the basis of the tumor cell defect could allow intervention to drive its differentiation to completion. Ultimately, understanding the developmental properties of tumor cells, and their relationship to the normal neuronal cells from which the tumors derive, could allow therapies to be designed on the basis of the intrinsic properties of the affected cells.

REFERENCES

Ambros, V., and H.R. Horvitz (1984) Heterochronic mutants of the nematode Caenorhabditis elegans. Science 226(4673):409–416.

Banerjee, U., and S.L. Zipursky (1990) The role of cell-cell interaction in the development of the *Drosophila* visual system. Neuron 4:177–187.

Blochlinger, K., L.Y. Jan, and Y.N. Jan (1991) Transformation of sensory organ identity by ectopic expression of Cut in *Drosophila*. Genes Dev. 5:1124–1135.

Campos-Ortega, J.A. (1988) Cellular interactions during early neurogenesis of *Drosophila melanogaster*. Trends Neurosci. 11:400–405.

Chalfie, M., H.R. Horvitz, and J.E. Sulston (1981) Mutations that lead to reiterations in the cell lineages of C. elegans. Cell 24(1):59–69.

Desai, C., G. Garriga, S.L. McIntire, et al. (1988) A genetic pathway for the development of the Caenorhabditis elegans HSN motor neurons. Nature 336(6200):638–646.

Doe, C.Q., and C.S. Goodman (1985) Early events in insect neurogenesis. II. The role of cell interactions and cell lineage in the determination of neuronal precursor cells. Dev. Biol. 111:206–219.

Doe, C.Q., Y. Hiromi, W.J. Gehring, et al. (1988a) Expression and function of the segmentation gene *fushi tarazu* during *Drosophila* neurogenesis. Science 239:170–175.

Doe, C.Q., D. Smouse, and S.C. Goodman (1988b) Control of neuronal fate by the *Drosophila* segementation gene *even-skipped*. Nature 333:376–378.

Elkins, T., K. Zinn, L. McAllister, et al. (1990) Genetic analysis of a Drosophila neural cell adhesion molecule: Interaction of fasciclin I and Abelson tyrosine kinase mutations. Cell 60:565–575.

Ellis, R.E., and H.R. Horvitz (1991) Two *C. elegans* genes control the programmed deaths of specific cells in the pharynx. Development 112:591–603.

Fehon, R.G., P.J. Kooh, I. Rebay, et al. (1990) Molecular interactions between the protein products of the neurogenic loci *Notch* and *Delta*, two EGF-homologous genes in Drosophila. Cell 61:523–534.

Finney, M., and G. Ruvkun (1990) The *unc-86* gene couples cell lineage and cell identity in *C. elegans*. Cell 63:895–905.

Ghysen, A., and C. Dambly-Chaudiere (1988) From DNA to form: The *achaete-scute* complex. Genes Dev. 2:495–501.

Godsave, S.F., and J.M.W. Slack (1991) Single cell analysis of mesoderm formation in the *Xenopus* embryo. Development 111:523–530.

Grenningloh, G., A. Bieber, J. Rehm, et al. (1990) Molecular genetics of neuronal recognition in *Drosophila*: Evolution and function of immunoglobulin superfamily cell adhesion molecules. Cold Spring Harbor Symp. Quant. Biol. 55:327–340.

Hill, R.J. and P.W. Sterling (1992) The gene*lin-3* encodes an inductive signal for vulval development in *C. elegans*. Natul`re 358:470-476.

Hedgecock, E.M., J.G. Culotti, and D.H. Hall (1990) The unc-5, unc-6, and unc-40 genes guide circumferential migrations of pioneer axons and mesodermal cells on the epidermis in *C. elegans*. Neuron 4(1):61–85.

Hockenbery, D., G. Nunez, C. Milliman, et al. (1990) Bcl-2 is an inner mitochondrial membrane protein that blocks programmed cell death. Nature 348:334–336.

Horvitz, H.R., and P.W. Sternberg (1991) Multiple intercellular signalling systems control the development of the *Caenorhabditis elegans* vulva. Nature 351:535–541.

Hunt, P., and R. Krumlauf (1991) Deciphering the Hox code: Clues to patterning branchial regions of the head. Cell 66:1075–1078.

Joyner, A.L., K. Herrup, B.A. Auerbach, et al. (1991) Subtle cerebellar phenotypes in mice homozygous for a targeted deletion of the En-2 homeobox. Science 251:1239–1243.

Klaembt, C., E. Knust, K. Tietze, et al. (1989) Closely related transcripts encoded by the neurogenic gene complex Enhancer of split of *Drosophila melanogaster*. EMBO J. 8:203–210.

Kraemer, H., R.L. Cagan, and S.L. Zipursky (1991) Interaction of bride of sevenless membrane-bound ligand and the sevenless tyrosine kinase receptor. Nature 352:207–212.

Lo, L.C., J.E. Johnson, C.W. Wuenschell, et al. (1991) Mammalian achaete-scute homolog 1 is transiently expressed by spatially restricted subsets of early neuroepithelial and neural crest cells. Genes Dev. 5:1524–1537.

McMahon, A.P., and A. Bradley (1990) The *Wnt-1 (int-1)* proto-oncogene is required for development of a large region of the mouse brain. Cell 62:1073–1085.

Murre, C., P.S. McGaw, H. Vaessin, et al. (1989) Interactions between heterologous helix-loop-helix proteins generate complexes that bind specifically to a common DNA sequence. Cell 58:537–544.

Rao, Y., L. Jan, and Y.N. Jan (1990) Similarity of the product of the Drosophila neurogenic gene *big brain* to transmembrane channel proteins. Nature 345:163–167.

Reid, L. (1990) From gradients to axes, from morphogenesis to differentiation. Cell 63:875–882.

Reth, A.D., and A. Bernstein (1991) Molecular basis of mouse developmental mutants. Genes Dev. 5:1115–1123.

Rubin, G.M. (1989) Development of the *Drosophila* retina: Inductive events studied at single cell resolution. Cell 57:519–520.

Ruvkun, G., and M. Finney (1991) Regulation of transcription and cell identity by POU domain proteins. Cell 64:475–478.

Seydoux, G., and I. Greenwald (1989) Cell autonomy of lin-12 function in a cell fate decision in C. elegans. Cell 57(7):1237–345.

Turner, D.L., E.Y. Snyder, and C.L. Cepko (1990) Lineage-independent determination of cell type in the embryonic mouse retina. Neuron 4:833–845.

Van Doren, M., H. Ellis, and J.W. Posakony (1991) The *Drosophila extramacrochaetae* protein antagonizes sequence-specific DNA binding by *daughterless/achaete-scute* protein complexes. Development 113:245–255.

Way, J.C., and M. Chalfie (1988) mec-3, a homeobox-containing gene that specifies differentiation of the touch receptor neurons in C. elegans. Cell 54(1):5–16.

Weinmaster, G., V. Roberts, and G. Lemke (1991) A homolog of *Drosophila Notch* expressed during mammalian development. Development 113:199–205.

Yamamori, T., Y. Fukada, R. Aebersold, et al. (1989) The cholinergic neuronal differentiation factor from heart cells is identical to leukemia inhibitory factor. Science 246:1412–1416.

Yuan, J., and H.R. Horvitz (1990) The *Caenorhabditis elegans* genes *ced-3* and *ced-4* act cell-autonomously to cause programmed cell death. Dev. Biol. 138:33–41.

Chapter
4

Cell Senescence and Apoptosis

J. Carl Barrett

INTRODUCTION: CELL DEATH AND CANCER

In normal tissues, homeostasis is maintained by a balance of cell birth and cell death, whereas tumor growth occurs when the birth rate of a population exceeds the death rate. This can be achieved either by increasing the proliferative rate of cells, decreasing the death rate, or both (Fig. 1). Considerable advances over the past 15 years have been made in our understanding of the genes controlling proliferation in normal and neoplastic cells. Positive growth signals can be transduced to cells by growth factors that bind to specific receptors, activate signal transduction pathways, and regulate the expression of specific genes. Many of these growth factors, receptors, signal-transducing proteins, and mitogen-activated genes are proto-oncogenes that are qualitatively or quantitatively altered in tumors and thereby act as oncogenes (Bishop, 1991). Tumors can also arise by inactivation of negative control mechanisms leading to increased growth and/or decreased cell death. A family of genes, commonly called tumor suppressor genes, acts as negative regulators of growth in normal cells. These genes are frequently lost, inactivated, or mutated in neoplastic cells. Although a half dozen or more of these genes are now cloned, the exact mechanisms by which they negatively control tumor cell growth are not clear (Boyd and Barrett, 1990).

Conceptually, there are two general mechanisms by which negative growth control can be exerted—growth arrest may be either reversible or irreversible (Table 1). Reversible controls on cell growth are important in the processes of normal development, tissue homeostasis, and in pathological conditions such as neoplasia. It is particularly challenging to explain how normal cells in a tissue maintain a balance between growth and growth arrest, yet respond to regenerative signals upon injury. Normally, the majority of stem cells or basal cells in a tissue are growth- arrested in a reversible state. Alterations in this growth-arrest mechanism may be common in neoplastic cells. A reversible block in cell proliferation may result from an extrinsic signal, such as a soluble growth inhibitory factor. A wide range of growth inhibitory factors, including the TGF- family of proteins, interferons and interleukins, tumor necrosis factors, and inhibitors of angiogenesis have been described and are under study (Boyd and Barrett, 1990).

TABLE 1. Mechanisms by Which Tumor Suppressor Genes Negatively Regulate Cell Proliferation

1. Reversible growth arrest
   - A. Extrinsic (soluble growth inhibitory substances)
   - B. Intrinsic (block in signal transduction or cell cycle progression)
2. Irreversible growth arrest
   - A. Cellular senescence
   - B. Terminal differentiation
   - C. Apoptosis
   - D. Necrosis

Normal Cells: Cell Renewal = Cell Death

Tumor Cells: Cell Renewal > Cell Death

Fig. 1. Cell proliferation versus cell death in normal and tumor cells.

Alternatively, reversible growth arrest may be regulated through intrinsic blocks to signal transduction or cell cycle progression. The underlying mechanisms for this important type of growth control remain unknown and comparatively few studies in this area have been performed. However, recent findings from our laboratory implicate a tumor suppressor gene product in the block of cellular growth responses to mitogens through an intrinsic block to growth factor signals in cells grown in agar. This is discussed in more detail elsewhere (Afshari and Barrett, 1992; Boyd and Barrett, 1990; Koi et al., 1989).

Molecular Genetics of Nervous System Tumors, pages 61-72 ©1993 Wiley-Liss, Inc.

Examples of irreversible growth arrest include terminal differentiation, cellular senescence, apoptosis, and necrosis. The role of terminal differentiation in carcinogenesis has been discussed by numerous authors and is not discussed here except to compare and contrast it with other forms of irreversible growth arrest. The major focus of this review is on cellular senescence and apoptosis as examples of negative growth arrest that are controlled by specific cellular genes.

CELLULAR SENESCENCE AND AGING

Both human and rodent cells in culture can be grown for only a limited number of cell divisions after which they exhibit morphological changes and cease proliferation, a process termed cellular senescence or cellular aging (Hayflick, 1976). Over 30 years ago, Hayflick and Moorhead (1961) reported that normal human fibroblasts are able to undergo a limited, fixed number of cell divisions after which they exhibit morphological changes and cease proliferation, a process termed cellular senescence (Hayflick, 1976). For example, embryonic human fibroblasts can be grown for 50–60 population doublings before senescence. The failure of the cells to grow beyond this limit is an inherent property of the cells that cannot be explained simply by inadequate media components (Hayflick, 1976). The key determinant in the life span of cells in culture is the number of cell doublings, not the length of time in culture (Hayflick, 1976). Normal cells transplanted serially *in vivo* also exhibit a finite life span, suggesting that cellular senescence is not a cell culture artifact (Daniel et al., 1968).

Three lines of evidence suggest that the aging of cells in culture may be related to the aging of the organism (Hayflick, 1976):

1. The doubling potential of cells in culture is inversely proportional to the age of the donor. Cells from embryonic tissue exhibit the longest life span (50–60 cell doublings). Cells from adult tissue can be grown for only 14 to 29 doublings, and there is a general decrease in the life span of cells in culture with increasing age of the donor tissue.
2. Cells derived from individuals who exhibit premature aging have a decreased life span in culture. Patients with progeria (Hutchinson-Gilford syndrome) manifest signs of aging at the end of their first decade of life that are typical of normal individuals in the seventh decade of life. Werner syndrome individuals also have accelerated aging but in later years—the mean age of death is 47. The main causes of death in these individuals are cancer and cardiovascular disease, similar to normal individuals. Fibroblasts derived from individuals with the premature aging characteristics senesce prematurely in culture, after only two to 10 population doublings.
3. The life span of cells in culture is correlated in general with the maximum life span of the species, although the variability in these experimental data is great. Cells from humans (maximum life span = 100–120 years) can be grown for 50–60 population doublings, whereas cells from rodents (3–5 year life span) can be grown for only 20–40 population doublings. Cells from the Galapagos tortoise, which has life span of 175–200 years, undergo a greater number of population doublings (90–125) in culture than human cells.

These lines of evidence, although not conclusive, provide provocative support for the hypothesis that aging of cells is related to the aging process of the organism.

CELL SENESCENCE AND CANCER

Escape from cellular senescence is an important step in neoplastic progression of human and rodent cancers (Barrett and Fletcher, 1987). Many, but not all, tumor cells can be grown indefinitely in culture and therefore have escaped senescence and are termed immortal. It is not clear whether the failure of some tumor cells to grow in culture is a technical artifact or an indication that escape from senescence is not required for these cancers. Many of these tumors cannot be maintained *in vivo* in nude mice, which indicates that only a small growth fraction of cells exists in the tumor and may explain the inability of some types of tumor cells to grow *in vitro*. Improvements in cell culture techniques have led to the establishment of cell lines from most tumor types, suggesting that it is possible to obtain immortal cell lines if the culture conditions are optimal. Since no property of cancer cells is universal, it is not necessary to demonstrate that escape from senescence has occurred in every cancer. However, in those cancers where this change is evident, it is probably a critical change based on the following additional lines of evidence (Barrett and Fletcher, 1987).

The observations that treatment of normal cells with diverse carcinogenic agents, including chemical carcinogens, radiation, viruses, and oncogenes, allows cells to escape senescence indicate that this change is important in cancer induction. While immortality is not sufficient for neoplastic transformation, most immortal cells have an increased propensity for spontaneous, carcinogen-induced or oncogene-induced neoplastic progression (Barrett and Fletcher, 1987). Therefore, escape from senescence is a preneoplastic change that predisposes a cell to neoplastic conversion. It is clear that immortal cells are further along the multistep pathway to neoplasia than normal cells. Since cellular senescence limits the growth of cells, it is reasonable that senescence might be one mechanism by which tumor suppressor genes operate (Barrett and Fletcher, 1987; Sager, 1986).

GENETIC BASIS FOR CELLULAR SENESCENCE

Two major theories of cellular senescence have been debated for many years (Macieira-Coelho, 1988). One is the error catastrophe or damage model, which proposes that the random accumulation of damage or mutations in DNA, RNA, or proteins leads to the loss of proliferative capacity. The experimental evidence supporting the error accumulation hypothesis has been criticized (Macieira-Coelho, 1988). A second hypothesis is that senescence is a genetically programmed process. Strong experimental support for a genetic basis for senescence is provided by studies of Pereira-Smith and Smith (1988) and by Sugawara et al. (1990), which are discussed below.

It is possible to fuse cells of different origins and to select for the hybrid cells using biochemical markers for drug sensitivity or resistance that differ in the parental cells. When cells with a finite life span are fused to immortal cells with an indefinite life span, the majority of these hybrids senesce (Koi and Barrett, 1986; Pereira-Smith and Smith, 1988). Even hybridization of two different immortal human cell lines with each other can result in senescence, indicating that different complementation groups exist for the senescence function lost in these cells. By fusing different immortal human cell lines with each other, Pereira-Smith and Smith (1988) have established four complementation groups, suggesting that loss or inactivation of one of multiple genes allows escape from senescence. If this hypothesis is correct, it should be possible to map the genes involved in cellular senescence. Recent findings with hamster and human interspecies hybrids and microcell-mediated chromosome transfer experiments have mapped putative senescence genes to specific human chromosomes (Klein et al., 1991; Ning et al., 1991; Sugawara et al., 1990).

When normal human lung fibroblasts (MRC-5), which have a finite life span, are fused to immortal hamster cells, the hybrids that form exhibit a finite life span (Fig. 2), which are characteristic of the normal human cells (Sugawara et al., 1990). At the end of this life span, the cells display signs of cellular senescence, which are indistinguishable from the parental human cells at the end of their life span. When MRC-5 cells, which have a life span of 60 population doublings, are used at a population doubling level of 40, the human–hamster hybrids grow for approximately 20 population doublings, i.e., the remaining life span of the parental human cells. Since the cell hybrids grow extensively before dying, the cessation of growth is not due to a toxic effect of the fusion protocol. Furthermore, when earlier passage MRC-5 cells are used (population doubling level 30), the hybrids grow longer, for up to 30 population doublings, again achieving the life span of the parental cells (Fig. 2). Therefore, the senescence of the hybrids is an active process dictated by the senescence program of the normal human cells. The limited life span of the hybrids indicates that cellular senescence is dominant in these hamster–human hybrids. A similar conclusion was drawn from studies of intraspecies hybrids, i.e., human–human (Pereira-Smith and Smith, 1988) and hamster–hamster (Koi and Barrett, 1986).

Although the majority of the hamster–human hybrids senesce, some of the hybrids ultimately escape senescence

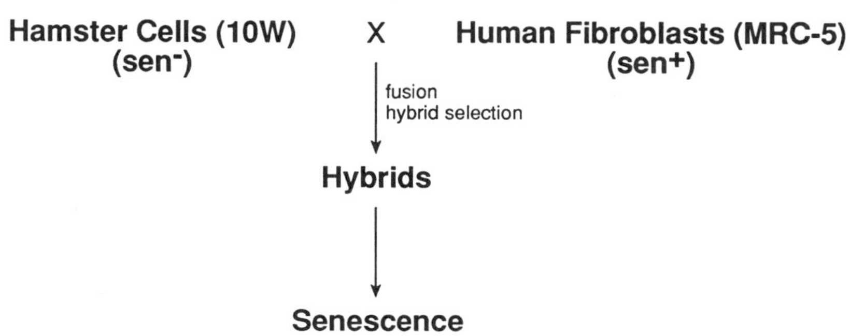

10W x MRC-5 (PDL30) hybrids
–senesce after approximately 30 population doublings

10W x MRC-5 (PDL30) hybrids
–senesce after approximately 20 population doublings

Fig. 2. Senescent phenotype is dominant in cell hybrids. When immortal cells are fused with normal cells, the resultant hybrid cells senesce. The life span of the hybrid is determined by the population doubling level (PDL) of the normal parental cell. The hybrids achieve approximately the life span remaining of the normal cell.

due to outgrowth of a few cells in the senescent population (Fig. 3). Because human chromosomes are usually lost in interspecies hybrids, the possibility exists that escape from senescence is due to loss of an essential chromosome or chromosomes, allowing the hybrids to escape senescence. The human–hamster hybrid clones that escaped senescence had lost both copies of human chromosome 1, whereas all other human chromosomes were present in one or two copies in at least one of the immortal hybrids (Sugawara et al., 1990). This suggested that a gene or genes on this chromosome may be necessary for cellular senescence.

This hypothesis was confirmed by two additional experimental approaches. One involved interspecies cell hybrids with human cells carrying a t(1;X) chromosome translocation. These experiments showed that applying selective pressure for the long arm of chromosome 1 increased the percentage of senescent hybrids. The second approach used microcell-mediated chromosome transfer (Figs. 4, 5) to introduce chromosome 1 into the immortal cell line (Sugawara et al., 1990). This chromosome, but not another control chromosome (chromosome 11), induces senescence in hamster 10W cells (Sugawara et al., 1990). Chromosome 1 induces senescence in some cell lines but not others (Yamada et al., 1990).

By the same technique of microcell-mediated chromosome transfer, Klein et al. (1991) have mapped another senescence gene to chromosome X, and Ning et al. (1991) have mapped a senescence gene for HeLa cells to chromosome 4. Thus, three senescence genes have now been mapped. In addition, several unpublished studies indicate the presence of senescent genes on other chromosomes (Table 2).

These results have led us to propose the following hypothesis: Cellular senescence is controlled by genes that are activated or whose functions become manifested at the end of the life span of the cell. Defects in the function of these gene products allow cells to escape the program of senescence and become immortal. Immortalization relieves one constraint on tumor cell growth, allowing malignant progression.

According to this hypothesis, a family of senescence genes exists. Immortalization occurs due to a defect in any one of these genes. This explains the complementation studies of Pereira-Smith and Smith (1988), which showed that different immortal cell lines could complement the defect in other immortal cells in hybrids that senesce after cell–cell fusion. This also explains our results that introduction of a specific human chromosome causes senescence in some cell lines but not others. It appears that senescence gene mutations are random. The senescence gene altered in a specific immortal cell line does not correlate with the tumor histology, activated oncogenes, or inactivated tumor suppressor genes in the cell. Current efforts are focused on the cloning of these genes and on understanding how they operate to arrest cell growth.

POSSIBLE ROLE OF CELL CYCLE CONTROLS IN CELLULAR SENESCENCE

The RB protein, which is involved in retinoblastoma and in several other malignancies, is differentially phosphorylated during the cell cycle (Buchkovich et al., 1989; Chen et al., 1989; DeCaprio et al., 1989; Furukawa et al., 1990;

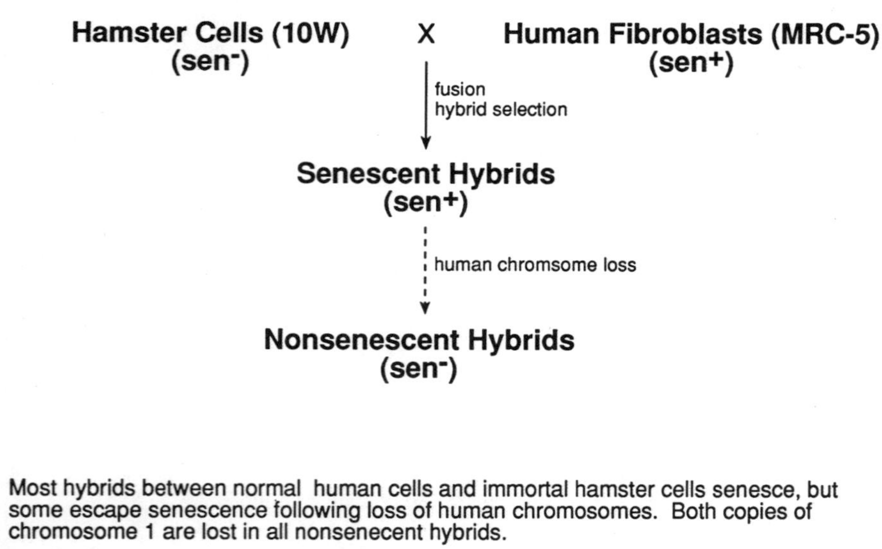

Fig. 3. Nonsenescent hybrids lose specific chromosomes. Examination of the nonrandom loss of human chromosomes in hamster–human hybrids allows the identification of chromosomes necessary for senescence (Sugawara et al., 1990).

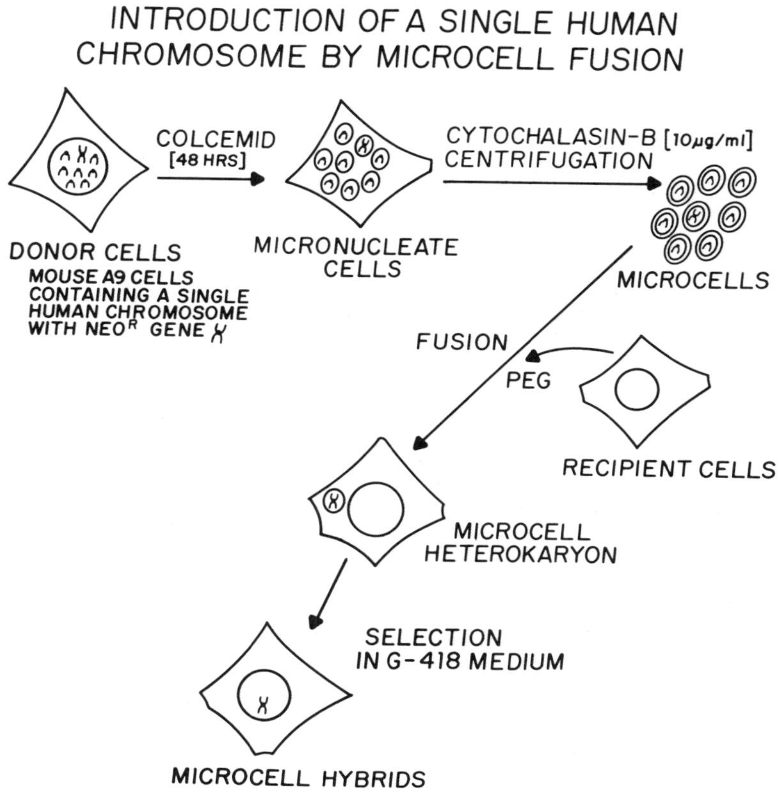

Fig. 4. Microcell-mediated chromosome transfer. Individual human chromosomes can be transferred into recipient cells of interest by this technique. Sugawara et al. (1990) confirmed that human chromosome 1 induces senescence of the hamster cell line 10W. Other authors have mapped senescence genes to other chromosomes by this approach.

Mihara et al., 1989). The RB protein is unphosphorylated in the G_1/G_0 compartment of the cell cycle and is phosphorylated as cells enter into S-phase. The protein is increasingly phosphorylated as the cells progress through G_2 and M and is again primarily in the unphosphorylated form as cells reenter G_1. Also, the RB protein is unphosphorylated in cells induced to differentiate (Buchkovich et al., 1989; Chen et al., 1989; Furukawa et al., 1990, Mihara et al., 1989). These data suggest that the unphosphorylated form of the protein is growth inhibitory and that the tumor suppressor function of the RB protein may be linked to cell cycle control and differentiation.

The product of the retinoblastoma tumor suppressor gene may be a key regulator of cellular senescence as well (Futreal and Barrett, 1991; Stein et al., 1990). Senescent cells express levels of RB protein comparable to young cells; however, only the unphosphorylated form of RB is observed. Quiescent cells, maintained in media containing low serum, exhibit only the unphosphorylated form of the RB protein. When the cells are stimulated with serum, phosphorylation of the RB protein is observed between 10 and 20 hours after stimulation, which corresponds to the time course for stimulation of DNA synthesis. However, when senescent cells are stimulated with serum, the RB protein remains unphosphorylated (Futreal and Barrett, 1991; Stein et al., 1990). This result indicates that senescent cells are blocked in their ability to phosphorylate the RB protein in response to normal growth stimuli (Fig. 6).

This finding implicates upstream modifiers of RB phosphorylation as possible crucial regulatory elements in mediating cellular senescence, with the end result being a block to proliferation caused by the presence of unphosphorylated RB protein acting on its own or through other effector molecules. Down-regulation of an RB kinase activity in senescent cells and/or up-regulation of an RB phosphatase are possible mechanisms for the alterations of RB phosphorylation in senescent cells. Recent studies indicate that the $p34^{cdc2}$ kinase, which is a candidate RB kinase, is down-regulated in senescent human and hamster cells (Richter et al., 1991; Stein et al., 1991). However, transfection of human $p34^{cdc2}$ under the control of a constitutive promoter into hamster cells fails to allow the cells to enter into S-phase even when the protein is synthesized (Richter et al., 1991). The $p34^{cdc2}$ kinase is the catalytic subunit of a serine/

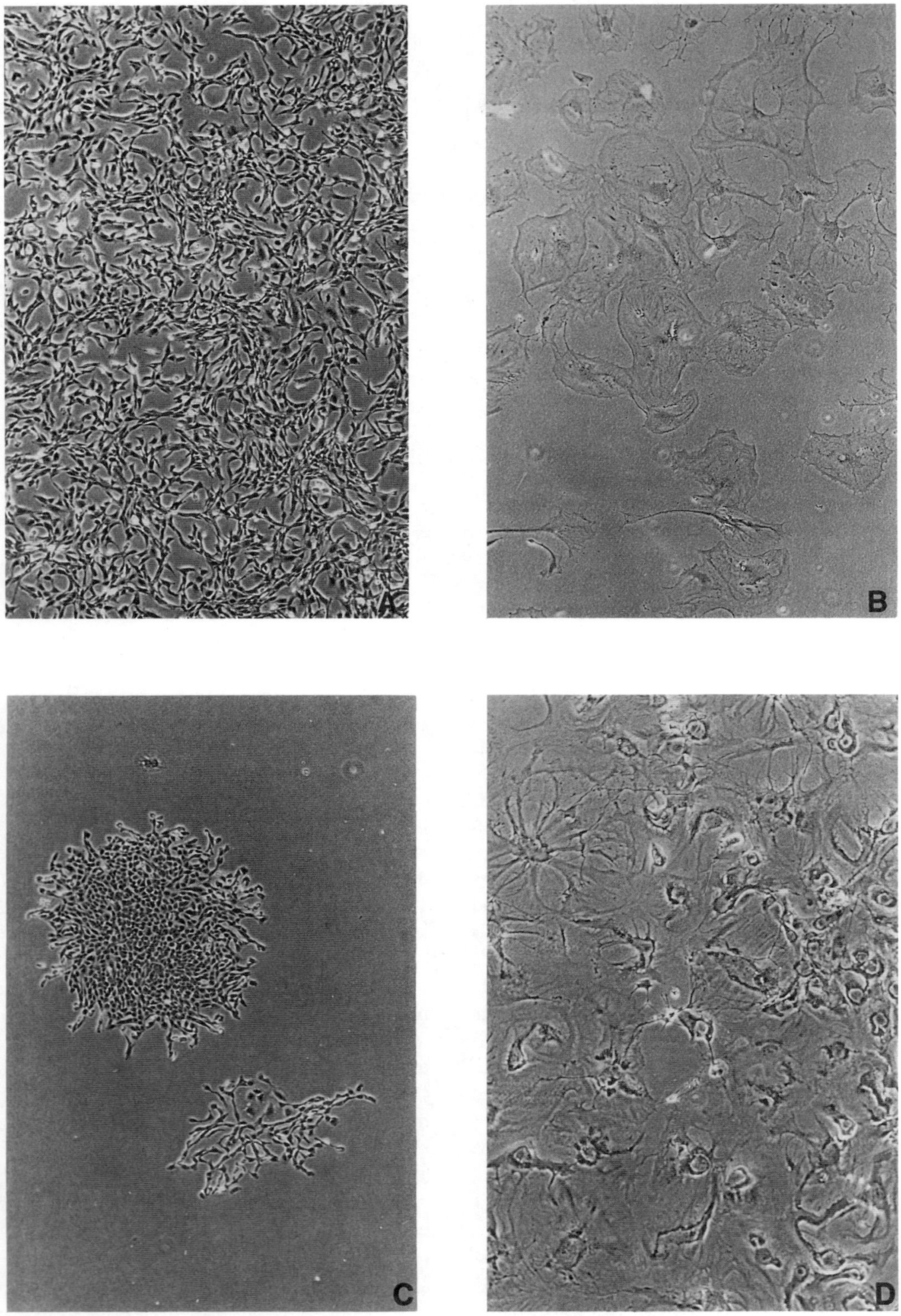

Fig. 5. Senescence following reintroduction of normal chromosome. The 10W cell line (**A**) is immortal. When normal human chromosome 1 is introduced into the cells by microcell-mediated chromosome transfer, the cells stop growing and display signs of senescence (**D**) that are typical of normal hamster cells at senescence (**B**). 10W cells with deleted portions of chromosome 1q continue to grow (**C**).

TABLE 2. Mapping of Putative Senescence Genes

Chromosome localization of Sen^+ gene	Cell line(s)	Reference
Chromosome 1	Syrian hamster 10W	Sugawara et al., 1990
	Human endometrial	Yamada et al., 1990
Chromosome 4	Cervical carcinoma (HeLa)	Ning et al., 1991
Chromosome X	Chinese hamster (Ni-2)	Klein et al., 1991
Chromosome 6	SV40 immortalized human fibroblasts	Ozer et al., unpublished
Chromosome 9	Human leukemia (K562)	Porterfield, Olopade, Rowley, and Diaz, unpublished
	Human melanoma (H3294T)	
Chromosome 11	Human bladder carcinoma (H-2)	Oshimura et al., unpublished

threonine kinase, which depends on interactions with members of a family of proteins called cyclins for activation (Murray and Kirschner, 1989). In growing cells, cyclins are synthesized in a periodic pattern and then rapidly degraded. Some cyclins function in G_1, while others are involved in cell cycle progression during S-phase, G_2 phase, and mitosis (Reed, 1991). Thus, the block to entry into DNA synthesis in senescent cells may also be due to the down-regulation of cyclin proteins and/or cdc2-like proteins (Afshari and Barrett, unpublished; Stein et al., 1991). Further studies of cell cycle control proteins should yield insights into negative growth control in cellular senescence.

APOPTOSIS IN DEVELOPMENT, TISSUE HOMEOSTASIS, AND CANCER

Cellular senescence is a form of cell death that occurs as cells age or reach the end of their division potential or life span. Cell death also occurs by other mechanisms throughout the life of a metazoan organism. Terminal differentiation, which is associated with the expression of a specialized function or product of a tissue, represents one form of cell death. Apoptosis, sometimes called programmed cell death, is a distinct type of cell death that occurs normally during development and continually in many tissues in the adult organism. Apoptosis was recognized as an important, distinct form of cell death by Kerr et al. (1972). They stressed the important distinction between apoptosis and necrosis. Necrosis is the classical, pathological form of cell death due to noxious injury or trauma. It often affects groups of cells and is associated with inflammatory responses. Necrosis is characterized by dilation of the endoplasmic reticulum and swelling of mitochondria, followed by breakdown of the plasma membrane, loss of ion transport, general cell swelling, and lysis. Flocculation of the chromatin in the nucleus and nonspecific degradation of DNA occurs, presumably because of release of cytoplasmic contents, including lysosomal enzymes. Release of cellular contents into the extracellular fluid also kills neighboring cells and elicits an inflammatory response (Kerr et al., 1972).

Apoptosis is a markedly different form of cell death. It generally affects single cells and is manifested by cell shrinkage (Kerr et al., 1972). Early reports even referred to this process as shrinkage necrosis. In apoptotic cells, lysosomes remain intact and cellular fragmentation rather than swelling occurs. Marked condensation and fragmentation of the chromatin is a key feature of apoptotic cells. The cytoplasm also condenses and the cell morphology is rapidly altered. Cellular blebbing and protuberances of the cell surface occur, resulting in the production of many membrane-bound cell fragments of varying sizes with well-preserved cellular organelles. (The word *apoptosis* was derived from the Greek word used to describe "dropping off" or "falling off"; for example, leaves from trees or petals from flowers [Kerr et al., 1972]. It was suggested by Kerr et al. [1972] that the penultimate syllable should be stressed and the second half of the word pronounced like *ptosis* with the *p* silent, which comes from the same root "to fall." This presumably emphasizes the falling apart of the cell in apoptosis.)

The formation of the apoptotic body of fragmented cells is the first stage of apoptosis, which is followed by phagocytosis of the fragments by surrounding parenchymal cells or macrophages. The entire process of apoptosis can occur within a few hours to a day. Apoptotic cells disappear following phagocytosis and degradation, which complicates quantitation of this form of cell death and is one reason this form of cell death is frequently overlooked.

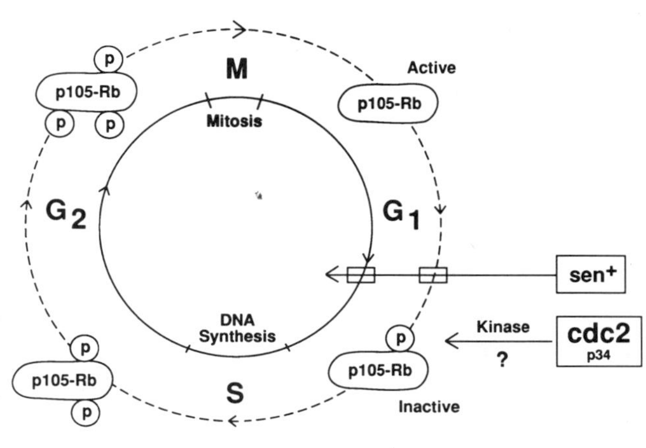

Fig. 6. RB and cellular senescence. The block to DNA synthesis in senescent cells correlates with the failure to phosphorylate the RB cell cycle control protein. The kinases and/or phosphatases controlling this phosphorylation are probably important in cellular senescence.

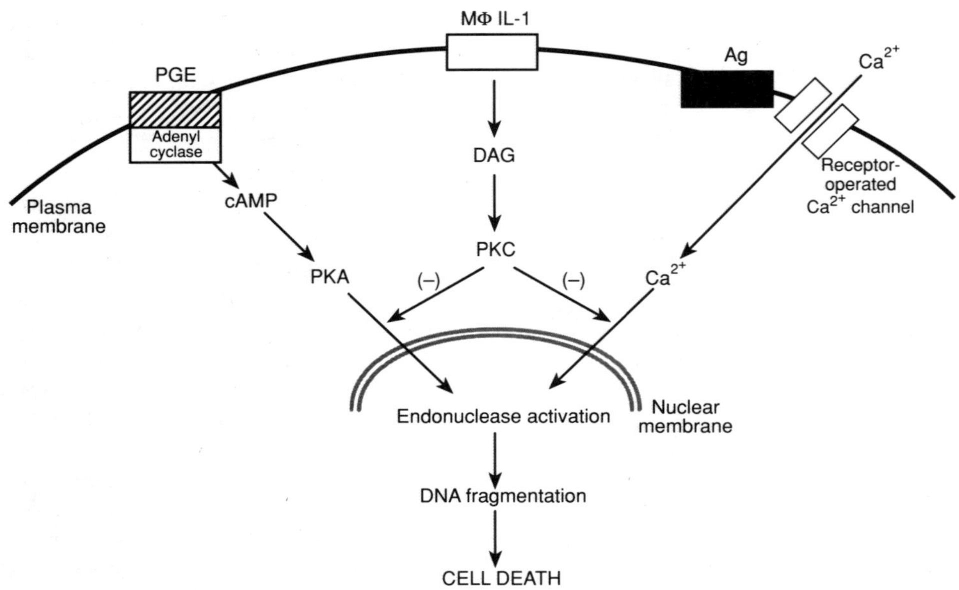

Fig. 7. Unbalanced signaling hypothesis for induction of apoptosis in thymocytes. A variety of signals induce apoptosis in thymocytes by causing sustained increases in cytosolic Ca^{2+} levels, activation of endonuclease, and DNA fragmentation. The same signals can be mitogenic if balanced by other signals that activate protein kinase C (PKC). (Adapted from McConkey et al., 1990, with permission.)

Apoptosis is an important form of cell death during development. The creation of life is highly dependent on cell death. For every ovum that is produced, three cells die and are extruded as polar bodies (Bowen and Bowen, 1990). Apoptosis also occurs during embryonic and fetal development, for example, during blastocyst inner cell mass remodeling and Mullerian duct regression. Removal of interdigital webs during avian wing development and amphibian tail regression during metamorphosis are further examples of apoptotic cell death in development (Arends and Wyllie, 1991; Walker et al., 1988). Programmed cell death in larvae of the nematode *Caenorhabditis elegans* is ultrastructurally similar to apoptosis but also has unique features (Walker et al., 1988). Apoptosis during development is regulated by intrinsic programs of certain cells and by cell–cell interactions.

In adult tissues, apoptosis is involved in tissue homeostasis, physiological responses to hormones, and pathological response to pharmacological and toxic substances. Apoptotic deletion of cells is observed in both rapidly and slowly dividing tissues, for example, in the intestinal crypts and the liver (Arends and Wyllie, 1991). Hormonal regulation of "cell growth" is often mediated by blocking apoptosis. For example, physiological lowering of trophic hormone levels results in apoptosis in the premenstrual endometrium and in the breast epithelium at the end of the menstrual cycle (Walker et al., 1988). In the prostate, testosterone inhibits apoptosis; removal of androgen stimulation by castration in rats results in regression and atrophy of the prostate due to cell death (Kyprianou and Isaacs, 1988). Apoptosis also plays a major role in immunological control. During education of thymocytes, cells recognizing self-antigens are clonally deleted by apoptosis to allow immunological tolerance (MacDonald and Lees, 1990). In the liver, numerous drugs, hormones, and environmental toxins induce liver enlargement and promote preferentially the growth of preneoplastic foci. Regression of these foci following removal of the stimuli is often accompanied by extensive apoptosis, suggesting that these agents affect growth by blocking apoptosis (Shulte-Hermann et al., 1990). Radiation, mild hyperthermia, cancer chemotherapeutic agents, and chemical carcinogens induce apoptosis in certain cells and tissues (Walker et al., 1988). These same agents often induce necrosis as well. The mode of cell death is determined by the cell type and by the dose of the chemical and presumably the extent of cellular injury. Less toxic insults often result in apoptotic cell death rather than necrosis.

Apoptosis is observed to occur spontaneously in a wide variety of neoplasms (Walker et al., 1988). This may explain discrepancies in actual versus calculated rates of tumor growth based solely on cell proliferation rates (Cotter et al., 1990). In contrast to necrosis, apoptosis is observed in small as well as large tumors and often affects scattered individual cells throughout the tumor rather than in necrotic centers. Estrogen-dependent tumors require the continued presence of estrogen for "growth." Withdrawal of estrogen, for example, by removal of the ovaries, results in tumor regression due to extensive apoptotic death of the tumor cells (Kyprianou et al., 1991). The incidence of apoptosis is higher in tumors that have extensive lymphocyte infiltration. Cytotoxic T cells and NK cells have been shown to kill tumor cells *in vitro* by inducing apoptosis (Cotter et al., 1990).

It is obvious, but not always appreciated, that tumor growth can be achieved by increasing the rate of cell proliferation or decreasing the rate of cell death (Fig. 1). Cell death, by apoptosis and necrosis, occurs in many tumors; modulation of cell death, particularly by apoptosis, may be an important feature of tumor promotion and progression. The genes involved in apoptosis are possibly altered during neoplastic development and may act as oncogenes or tumor suppressor genes.

BIOCHEMICAL MARKERS OF APOPTOSIS

Apoptosis was initially recognized on the basis of morphological changes at the cellular level, including cell separation in tissues, condensation of chromatin, cell fragmentation (apoptotic bodies), and phagocytosis of apoptotic bodies in the absence of an inflammation response. Biochemical and molecular changes involved in or associated with apoptosis are under investigation, but to date few definitive nonmorphological markers exist.

One of the earliest and most characteristic morphological features of apoptosis is the chromatin condensation, which results in sharply delineated masses of uniform density adjacent to the nuclear membrane (Arends and Wyllie, 1991). Wyllie and co-workers proposed that these nuclear changes are due to activation of an endogenous endonuclease during apoptosis (Arends and Wyllie, 1991). This endonuclease produces double strand breaks of the chromatin in the internucleosomal linker DNA, generating well-organized oligonucleotides with lengths in multiples of 180–200 base pairs (bp), the size of a DNA molecule wrapped around a single histone octamer. Upon isolation of DNA from cells undergoing apoptosis and separation by gel electrophoresis, a "DNA ladder" corresponding to DNA fragments of multiples of 200 bp is observed. Nucleosomal DNA ladder formation is a useful biochemical marker for apoptosis and is associated with nuclear chromatin changes with few exceptions (Arends and Wyllie, 1991). Identification of the endonuclease(s) involved and a further understanding of its mechanism of activation are key areas for future study in this field.

Few other biochemical markers of apoptosis are available. Fesus and co-workers have shown that induction and activation of a tissue transglutaminase (Ca^{2+}-dependent protein-glutamine-glutamyl transferase) occurs in cells undergoing apoptosis (Piacentini et al., 1991). This enzyme cross-links proteins and may be involved in the formation of detergent-insoluble structures found in apoptotic cells in the liver, which are similar to cornified envelopes in keratinocytes. The significance of this protein cross-linking in the apoptotic process and the generality of this marker for apoptosis remain to be established. Tenniswood and co-workers reported the isolation of an androgen-repressed gene in the rat ventral prostate, termed TRPM-2, which appears to be involved in the process of cell regression and programmed cell death in the prostate (Leger et al., 1988). The function of this gene and its role in apoptosis of other epithelial cells will be important to determine.

Apoptotic cells are recognized and phagocytized by neighboring cells and macrophages. Macrophages bind preferentially to apoptotic cells compared to nonapoptotic cells because of the expression of cell-surface ligands (Arends and Wyllie, 1991). The vitronectin receptor may play a role in this process (Savill et al., 1990).

CELLULAR SIGNALING IN APOPTOSIS

The cellular signals that initiate or inhibit apoptosis are partially understood. A divergent group of agents can induce apoptosis in many cell types. These include toxic agents such as irradiation, mild hyperthermia, and chemotherapeutic drugs (Arends and Wyllie, 1991). However, apoptosis can occur in the absence of toxic damage and many of the inducers of apoptosis are physiological mediators. Furthermore, apoptosis can occur spontaneously unless blocked by mitogenic or hormonal signals.

The mechanisms of apoptosis induction and the cellular signals involved in this process have been studied in detail in immature thymocytes (McConkey et al., 1990). Immature thymocytes undergo apoptosis in response to chemical toxins (e.g., dioxin), glucocorticoid hormones, calcium ionophores, and antibodies to the CD3 T-cell receptor complex. These diverse stimuli cause sustained increases in cytosolic Ca^{2+} levels, which may activate an endogenous endonuclease that is the trigger for the program of cell death.

The paradox is that sustained Ca^{2+} levels also occur in response to antigens and other mitogens that mediate cell proliferation as well as cell death. This paradox is explained by the "unbalanced signaling hypothesis" (Fig. 7) of McConkey et al. (1990). According to this hypothesis, the same signals can stimulate cell division or cell death, depending on the presence or absence of secondary signals. In the case of immature thymocytes, mitogenic signals such as antigens or lectins require secondary signals from accessory cells. These secondary signals can be provided also by phorbol esters that activate protein kinase C (PKC). PKC activation may block endonuclease activation in thymocytes caused by elevated calcium and allow a mitogenic response to these signals. Cell death occurs if the signals are unbalanced, whereas mitogenesis occurs when multiple signals are received by the cells. This allows a conservation of signaling pathways and also dual controls on cellular growth.

As a physiological example of this mechanism, thymic cortical epithelial cells, which lack full capacity for antigen presentation, require exogenous IL-1, which activates PKC, for growth. This hypothesis explains why many hematopoietic cells die in the absence of interleukins or growth factors. IL-2 withdrawal from thymocytes or IL-3 withdrawal from hematopoietic precursors activates apoptosis (Arends and Wyllie, 1991). Hematopoietic colony-stimulating factors (CSF), which are required for the survival, differentiation, and proliferation of precursor cells, promote cell survival by suppressing apoptosis without necessarily stimulating proliferation (Williams et al., 1990). Hormonal regulation of epithelial cell growth is also mediated in part by suppression of apoptosis. For example, estrogens suppress apoptosis of endometrial cells and androgens block apoptosis in the prostate (Arends and Wyllie, 1991). In the rat prostate, castration-induced apoptosis is associated with induction of the c-*fos* and c-*myc* immediate early genes, which further supports the hypothesis that the processes of cell death and proliferation share common signal pathways (Buttyan et al., 1988). Thus, what are classically viewed as mitogenic signals may actually provide signals that counterbalance intrinsic or extrinsic signals for cell death.

GENES CONTROLLING APOPTOSIS

An important area for future research is elucidation of the genes involved in the apoptotic program of cell death. This area is still in its initial phase, but a few interesting genes are currently under study.

Several investigators have explored the possibility of eliminating tumor cells by activating apoptosis via antibodies that trigger surface events. Trauth et al. (1989) identified a monoclonal antibody, called APO-1, that binds a 52-kd surface protein on lymphocytes and triggers apoptosis in these cells. This antibody could be activating a signal pathway for apoptosis or blocking the activity of a factor required for survival. Itoh et al. (1991) developed monoclonal antibodies to the FAS antigen expressed on the surface of various human cells including myeloid cells, T-lymphoblastoid cells, and diploid fibroblasts (FS7 fibroblasts were used as the initial immunogen for this antibody). The FAS antigen is a 36-kd transmembrane receptor with significant homology to the tumor necrosis factor (TNF) receptor (Itoh et al., 1991). TNFα and TNFβ also induce apoptosis, suggesting that a family of related receptor proteins may activate this process.

Cell death is important during animal development. In the nematode *C. elegans*, mutations in specific genes (*ced-3* and *ced-4*) prevent autonomous cell death (Yuan and Horvitz, 1990). The isolation of these genes should provide new insights into developmental control of cell death.

A potentially important mediator of apoptosis may be the *P53* gene. The *P53* gene is a tumor suppressor gene that is mutated or deleted in a wide variety of human malignancies. The wild-type gene inhibits cell division and the mutant gene may act as a dominant negative signal as well as providing positive growth signals for malignant cells (Levine and Momand, 1990). Yonish-Rouach et al. (1991) reported that when the wild-type p53 protein is transfected and expressed in murine myeloid leukemia cells, differentiation is not affected, but the cells lose viability and die by apoptosis. This effect is counteracted by interleukin-6. In other cell types, wild-type p53 causes growth arrest in a reversible manner without cell death (Levine and Momand, 1990). Therefore, whether apoptosis is mediated by p53 only in certain cell types or whether this is the fate of certain cells following growth arrest is not clear.

In addition to genes that induce apoptosis, there are genes that inhibit this program. An important example is the *bcl-2* gene, first isolated from the breakpoint of the translocation between chromosomes 14 and 18 in follicular B-cell lymphomas (Tsujimoto et al., 1984). Transgenic mice expressing the *bcl2*-immunoglobulin gene have an indolent follicular hyperplasia that progresses to a malignant diffuse large-cell lymphoma. The B cells in these transgenic mice have a prolonged life and progress to malignancy after a long latency and clonal selection progress presumably due to secondary changes, such as myc rearrangements (McDonnell and Korsmeyer, 1991). The prolongation of B-cell life by bcl-2 is due to its ability to block apoptosis without affecting cell proliferation (Hockenbery et al., 1990). Also, deregulated bcl-2 expression allows prolonged survival and blocks apoptosis of hematopoietic cell lines following growth factor deprivation. The *bcl-2* protein is an integral inner mitochondrial membrane protein of 25 kd (Hockenbery et al., 1990), but the mechanism of blocked programmed cell death by bcl-2 expression is not known. Certain mediators of apoptosis may function by modulation of bcl-2 expression. For example, Epstein-Barr virus (EBV) infection induces growth transformation and enhanced survival of B cells through the coordinate action of viral coded proteins. One of these proteins, latent membrane protein LMP-1, up-regulates bcl-2 expression and protects B cells from apoptosis (Henderson et al., 1991).

bcl-2 is expressed in nonhematopoietic cells, including glandular epithelial cells that are regulated by hormones (e.g., breast, thyroid, prostate), differentiating epithelial cells characterized by long-lived stem cells (e.g., skin and intestine), and long-lived postmitotic cells (e.g., neurons) (Hockenbery et al., 1991). These cells exhibit apoptotic regulation of growth and/or survival and bcl-2 is probably involved in this regulation. Future studies on the factors that control bcl-2 expression and its mechanism of action in blocking apoptosis will be extremely important.

CONCLUSION

Irreversible growth arrest can occur by different processes—terminal differentiation, cellular senescence, apoptosis, and necrosis. The signals controlling these forms of cell death and their mechanisms of activation and action are still under investigation. All of these processes are regulated by cellular genes and in that sense are forms of programmed cell death. Distinct features of each form of cell death exist. Differentiation involves expression of cell-type-specific (differentiation) markers. Cellular senescence is related to the age or cell division potential of the cell. Apoptosis differs from cellular senescence and terminal differentiation in terms of the time scale of the process. For example, apoptotic cells die within a matter of hours or days, whereas senescent cells remain metabolically active but nonproliferative for up to one year.

Despite these differences, common mechanisms of action may exist in these growth-arrest processes. Some terminally differentiated cells and senescent cells down-regulate the same cell cycle control proteins (Richter et al., 1991). Long-lived, postmitotic neurons express bcl-2, which blocks apoptosis. The same chemical or signal may induce apoptosis or necrosis, depending on the dose and the cell type. The possibility exists that common genetic and biochemical pathways exist in different forms of growth arrest. When the genes involved in the negative regulation of cell growth and death are identified, common and distinct mechanisms of these processes will be better understood.

REFERENCES

Afshari, C.A., and J.C. Barrett (1992) Negative regulation of mitogen stimulated anchorage-independent cell growth by a tumor suppressor gene function. Mol. Carc. (in press).

Arends, M.J., and A.H. Wyllie (1991) Apoptosis: Mechanisms and roles in pathology. In: International Review of Experimental Pathology, vol. 32. New York: Academic Press, pp. 233–254.

Barrett, J.C., and W.F. Fletcher (1987) Cellular and molecular mechanisms of multistep carcinogenesis in cell culture models. In: J.C. Barrett (ed.). Mechanisms of Environmental Carcinogenesis: Multistep Models of Carcinogenesis, vol. 2. Boca Raton: CRC Press, pp. 73–116.

Bishop, J.M. (1991) Molecular themes in oncogenesis. Cell 64:234–248.

Bowen, I.D., and S.M. Bowen (1990) Programmed Cell Death in Tumours and Tissues. New York: Chapman and Hill.

Boyd, J., and J.C. Barrett (1990) Tumor suppressor genes: Possible functions in the negative regulation of cell proliferation. Mol. Carc. 3:325–329.

Buchkovich, K., L.A. Duffy, and E. Harlow (1989) The retinoblastoma protein is phosphorylated during specific phases of the cell cycle. Cell 58:1097–1105.

Buttyan, R., Z. Zakeri, R. Lockshin, et al. (1988) Cascade induction of c-*fos*, c-*myc*, and heat shock 70K transcripts during regression of the rat ventral prostate gland. Mol. Endo. 2:650–657.

Chen, P.-L., P. Scully, J.-Y. Shew, et al. (1989) Phosphorylation of the retinoblastoma gene product is modulated during the cell cycle and cellular differentiation. Cell 58:1193–1198.

Cotter, T.G., S.V. Lennon, J.G. Glynn, et al. (1990) Cell death *via* apoptosis and its relationship to growth, development and differentiation of both tumour and normal cells. Anticancer Res. 10:1153–1160.

Daniel, C.W., K.B. DeOme, J.T. Young, et al. (1968). The in vivo span of normal and preneoplastic mouse mammary glands: A serial transplantation study. Proc. Natl. Acad. Sci. USA 61:53–60.

DeCaprio, J.A., J.W. Ludlow, D. Lynch, et al. (1989) The product of the retinoblastoma susceptibility gene has properties of a cell cycle regulatory element. Cell 58:1085–1095.

Furukawa, Y., J.A. DeCaprio, A. Freedman, et al. (1990) Expression and state of phosphorylation of the retinoblastoma susceptibility gene product in cycling and noncycling human hematopoietic cells. Proc. Natl. Acad. Sci. USA 87:2770–2774.

Futreal, P.A., and J.C. Barrett (1991) Failure of senescent cells to phosphorylate the RB protein. Oncogene 6:1109–1113.

Hayflick, L. (1976) The cell biology of human aging. N. Engl. J. Med. 295:1302–1308.

Hayflick, L., and P.S. Moorhead (1961) The serial cultivation of human diploid cell strains. Exp. Cell Res. 25:585–621.

Henderson, S., M. Rowe, C. Gregory, et al. (1991) Induction of *bcl*-2 expression by Epstein-Barr virus latent membrane protein 1 protects infected B cells from programmed cell death. Cell 65:1107–1115.

Hockenbery, D., G. Nunez, C. Milliman, et al. (1990) Bcl-2 is an inner mitochondrial membrane protein that blocks programmed cell death. Nature 348:334–336.

Hockenbery, D.M., M. Zutter, W. Hickey, et al. (1991) BCL2 protein is topographically restricted in tissues characterized by apoptotic cell death. Proc. Natl. Acad. Sci. USA 88:6961–6965.

Itoh, N., S. Yonehara, A. Ishii, et al. (1991) The polypeptide encoded by the cDNA for human cell surface antigen fas can mediate apoptosis. Cell 66:233–243.

Kerr, J.F.R., A.H. Wyllie, and A.R. Currie (1972) Apoptosis: A basic biological phenomenon with wide-ranging implications in tissue kinetics. Br. J. Cancer 26:239–257.

Klein, C.B., K. Conway, X.W. Wang, et al. (1991) Senescence of nickel-transformed cells by a mammalian X chromosome: Possible epigenetic control. Science 251:796–799.

Koi, M., C. Afshari, L.A. Annab, et al. (1989) Role of a tumor suppressor gene in the negative control of anchorage-independent growth of Syrian hamster cells. Proc. Natl. Acad. Sci. USA 86:8773–8777.

Koi, M., and J.C. Barrett (1986) Loss of tumor-suppressive function during chemically induced neoplastic progression of Syrian hamster embryo cells. Proc. Natl. Acad. Sci. USA 83:5992–5996.

Kyprianou, N., H.F. English, N.E. Davidson, et al. (1991) Programmed cell death during regression of the MCF-7 human breast cancer following estrogen ablation. Cancer Res. 51:162–166.

Kyprianou, N., and J.T. Isaacs (1988) Activation of programmed cell death in the rat ventral prostate after castration. Endocrinology 122:552–562.

Leger, J.G., R. Le Guellec, and M.P.R. Tenniswood (1988) Treatment with antiandrogens induces an androgen-repressed gene in the rat ventral prostate. Prostate 13:131–142.

Levine, A.J., and J. Momand (1990) Tumor suppressor genes: The p53 and retinoblastoma sensitivity genes and gene products. Biochim. Biophys. Acta 119–136.

MacDonald, R.R., and R.K. Lees (1990) Programmed death of autoreactive thymocytes. Nature 343:642–644.

Macieira-Coelho, A. (1988) Biology of normal proliferating cells in vitro. Relevance for in vivo aging. In: H.P. von Hang (ed.). Interdisciplimary Topics in Gerontology, vol. 23. Basel: S. Karger.

McConkey, D.J., S. Orrenius, and M. Jondal (1990) Cellular signalling in programmed cell death (apoptosis). Immunol. Today 11:120–121.

McDonnell, T.J., and S.J. Korsmeyer (1991) Progression from lymphoid hyperplasia to high-grade malignant lymphoma in mice transgenic for the t(14;18). Nature 349:254–256.

Mihara, K., X.-R. Cao, S. Chandler, et al. (1989) Cell cycle-dependent regulation of phosphorylation of the human retinoblastoma gene product. Science 246:2300–2303.

Murray, A.W., and M.W. Kirschner (1989) Dominoes and clocks: The union of two views of the cell cycle. Science 246:614–621.

Ning, Y., J.L. Weber, A.M. Killary, et al. (1991) Genetic analysis of indefinite division in human cells: Evidence for a cell senescence-related gene(s) on human chromosome 4. Proc. Natl. Acad. Sci. USA 88:5635–5639.

Pereira-Smith, O.M., and J.R. Smith (1988) Genetic analysis of indefinite division in humans cells: Identification of four complementation groups. Proc. Natl. Acad. Sci. USA 85:6043–6046.

Piacentini, M., F. Autuori, L. Dini, et al. (1991) "Tissue" transglutaminase is specifically expressed in neonatal rat liver cells undergoing apoptosis upon epidermal growth factor-stimulation. Cell Tissue Res. 263:227–235.

Reed, S.I. (1991) G1-specific cyclins: In search of an S-phase-promoting factor. Trends Genet. 7:95–99.

Richter, K.H., C.A. Afshari, L.A. Annab, et al. (1991) Downregulation of cdc2 in senescent human and hamster cells. Cancer Res. 51:6010–6013.

Sager, R. (1986) Genetic suppression of tumor formation: A new frontier in cancer research. Cancer Res. 46:1573–1580.

Savill, J., I. Dransfield, N. Hogg, et al. (1990) Vitronectin receptor-mediated phagocytosis of cells undergoing apoptosis. Nature 343:170–173.

Schulte-Hermann, R., I. Timmermann-Trosiener, G. Barthel, et al. (1990) DNA synthesis, apoptosis, and phenotypic expression as determinants of growth of altered foci in rat liver during phenobarbital promotion. Cancer Res. 50:5127–5135.

Stein, G.H., M. Beeson, and L. Gordon (1990) Failure to phosphorylate the retinoblastoma gene product in senescent human fibroblasts. Science 249:666–669.

Stein, G.H., L.F. Drullinger, R.S. Robetorye, et al. (1991) Senescent cells fail to express *cdc2, cdcA,* and *cycB* in response to mitogen stimulation. Proc. Natl. Acad. Sci. USA 88:11012–11016.

Sugawara, O., M. Oshimura, M. Koi, et al. (1990) Induction of cellular senescence in immortalized cells by human chromosome 1. Science 247:707–710.

Trauth, B.C., C. Klas, A.M.J. Peters, et al. (1989) Monoclonal antibody-mediated tumor regression by induction of apoptosis. Science 245:301–305.

Tsujimoto, Y., L.R. Finger, J. Yunis, et al. (1984) Cloning of the chromosome breakpoint of neoplastic B cells with the t(14;18) chromosome translocation. Science 226:1087–1099.

Walker, N.I., B.V. Harmon, G.C. Gobe, et al. (1988) Patterns of cell death. In: G. Jasmin (ed.). Methods and Achievements in Experimental Pathology. Basel: Karger, pp. 18–54.

Williams, G.T., C.A. Smith, E. Spooncer, et al. (1990) Haemopoietic colony stimulating factors promote cell survival by suppressing apoptosis. Nature 343:76–79.

Yamada, H., N. Wake, S. Fujimoto, et al. (1990) Multiple chromosomes carrying tumor suppressor activity for a uterine endometrial carcinoma cell line identified by microcell mediated chromosome transfer. Oncogene 5:1141–1147.

Yonish-Rouach, E., D. Resnitzky, J. Lotem, et al. (1991) Wild-type p53 induces apoptosis of myeloid leukaemic cells that is inhibited by interleukin-6. Nature 352:345–347.

Yuan, J., and H.R. Horvitz (1990) The *Caenorhabditis elegans* genes *ced-3* and *ced-4* act cell autonomously to cause programmed cell death. Dev. Biol. 138:33–41.

Chapter 5

Intercellular Interactions That Effect Nervous System Growth, Differentiation, and Regeneration

John Collins

INTRODUCTION

All organ systems develop through growth and differentiation of cells. Promoters within the genome extensively regulate these processes intracellularly; however, cells of the mammalian nervous system must also undergo an astounding sequence of events that results in their organization into an elaborate spatial array with myriads of specific interconnections. Maturing neurons migrate in association with glial cells to specific locations, then elaborate processes that extend over vast distances along specific pathways to precise target zones. Can intrinsic genetic events alone produce such complex cellular behavior? To what extent do these cells rely upon interaction with the surrounding microenvironment for prompts and cues that influence and direct their growth?

COMPONENTS OF THE MICROENVIRONMENT WITHIN THE NERVOUS SYSTEM

The extracellular microenvironment that neuroblasts encounter contains matrix material, diffusible factors, and other cells. Extracellular matrix, composed of collagen, noncollagenous glycoproteins (laminin, fibronectin, entactin, vitronectin), glycosaminoglycans (dermatan, chondroitin, keratan, heparan), and proteoglycans, functions as an adhesive that holds cells together in relation to one another (Rutka et al., 1988). Diffusible factors found in the nervous system include nerve growth factor (Johnson and Taniuchi, 1987; Whittemore and Seiger, 1987), brain-derived neurotrophic factor (Barde et al., 1982; Leibrock et al., 1989), neurotrophin-3 (Hohn et al., 1990; Jones and Reichardt, 1990; Maisonpierre et al., 1990), and ciliary neurotrophic factor (Lin et al., 1990; Stockli et al., 1989; Thoenen, 1991). Astrocytes and Schwann cells thus can influence other surrounding cells by synthesis of such diffusible neurotrophic factors (Assouline et al., 1987) or by production of extracellular matrix components (Chiu et al., 1991). Direct intercellular interactions, however, also occur, often mediated by membrane- bound receptors. The cellular milieu, therefore, can exert substantial influence over nervous system growth and development in a variety of different ways.

IMPORTANCE OF THE CELLULAR MILIEU WITHIN THE MICROENVIRONMENT OF THE NERVOUS SYSTEM

Cells that proliferate in conjunction with neurons differ, depending upon developmental age and location in the nervous system. Central nervous system (CNS) neurons encounter glial cells that later differentiate into astrocytes and oligodendroglia. Neurons whose processes extend into the peripheral nervous system (PNS) encounter Schwann cells and fibroblasts. Patterns of neuronal growth, differentiation, and regeneration vary depending upon the milieu of cells present in the microenvironment. In this chapter, we focus upon those intercellular interactions between neurons and their surrounding cells that appear to promote, inhibit, or regulate growth within the nervous system.

INTERCELLULAR INTERACTIONS BETWEEN MESODERM AND ECTODERM INDUCE THE DEVELOPMENT OF NEUROECTODERM

In the third week of development the embryo consists of a bilaminar disk with overlying columnar ectodermal cells exposed to amnion and underlying polyhedronal endodermal cells exposed to yolk sac. The process called gastrulation then begins with a rounding up of cells at the caudal end. As a result, mounds of cells collect on each side of midline, forming the primitive streak. This represents an invagination

Molecular Genetics of Nervous System Tumors, pages 73-82 Published 1993 by Wiley-Liss, Inc.

of cells migrating in between the two boundary layers to form a mesodermal layer. As ectodermal cells enter the mesodermal region and migrate forward and outward, they begin to change shape into bottlelike forms with their microfilaments aligned in a perpendicular fashion. Within this mesenchymal layer, the notochord develops in the midline and overlying ectoderm thickens and differentiates into neuroectoderm called the neural plate. From neural plate, both the neural tube and neural crest cells will develop and all elements of the central and peripheral nervous system will arise. By means of intercellular interactions, the developing notochord with its surrounding mesoderm induces determination of neuroectoderm and promotes its subsequent transformation into neural tube, folds, and neural crest (Cowan, 1979). Experiments with filters of different pore sizes placed between embryonic ectoderm and mesoderm reveal that neural induction does not require physical contact between these cell types (Toivonen et al., 1975). When filters obstruct the passage of macromolecules with molecular weights greater than 1,000, however, the induction fails to occur. These studies suggest a transmission of morphogenic signals from mesodermal to ectodermal cells in primary induction (Toivonen et al., 1976).

TARGET CELLS FOR NEURAL CREST INNERVATION INFLUENCE THE DIFFERENTIATION OF NEURAL CREST CELLS

Neural crest cells are vastly pluripotent. They migrate widely throughout the developing body and differentiate into different structures, depending upon the region that they come to occupy. They can give rise to sensory and autonomic neurons or ganglia, Schwann cells, melanocytes, or chromaffin cells of the adrenal medulla. By transplantation of embryonic quail neural crest tissue into disparate regions of chick embryo, Le Douarin revealed that the pattern of neural crest differentiation changed as a consequence of its location (Le Douarin, 1982). Other investigators have demonstrated that when immature sympathetic ganglion cells are raised in tissue culture in the presence of cells that ordinarily receive parasympathetic innervation, the sympathetic neurons will change their neurotransmitter expression and express features of cholinergic neurons (Furshpan et al., 1976; Patterson, 1978).

RADIAL GLIAL CELLS INFLUENCE NEURONAL MIGRATION AND DIFFERENTIATION BY CELL-TO-CELL INTERACTION

Neuroepithelial germinal cells in the cerebral periventricular germinal zone are believed to give rise both to glioblasts and neuroblasts. A subependymal population of glioblasts persists throughout life, providing a site for ongoing germination of new glial cells that subsequently migrate into overlying white matter. Neuroblasts in the germinal layer also undergo mitosis early in development. One daughter cell remains in the germinal layer, but the other daughter cell loses its basal attachment and begins a migration into the mantle layer (Jacobson, 1978; Sauer, 1935). That migration is believed to occur along the processes of the radially oriented glial cells that maintain one process in the periventricular region and extend one process out to the pial layer. Studies of the developing cerebellum and its particular type of radial glial cell, called a Bergman astrocyte, elucidate this interaction between glial and neuronal cell most comprehensively (Rakic, 1971). The cell bodies of these glia are located in the Purkinje cell layer and their processes extend to the pial surface. In the neurogenesis of the cerebellum, a population of neuronal cells called granule cells migrates to the external granular layer and there continues to proliferate for months after birth. As these cells finally become postmitotic, however, they migrate inwardly to reach their final location along the internal granule layer. That inward migration occurs in distinct association with the Bergman astrocytic process, which seems to form an extension of glial cell membrane along which the granule cell can move. Recent studies of this system have supported the presence of a neuronal surface glycoprotein (MW 100 kd) now referred to as astrotactin that is expressed transiently during cerebellar development in neurons undergoing glial-guided migration. Astrotactin antisera blocks the migration and disrupts subsequent neuronal positioning. Experiments suggest that astrotactin is a specific ligand that binds the cerebellar granule neuron to astroglia (Stitt and Hatten, 1990).

MEMBRANE CONTACT WITH NEURONAL CELLS MODIFIES GLIAL CELL PROLIFERATION AND DIFFERENTIATION

During brain development, astroglial cell proliferation and neuronal cell proliferation occur in a generally coordinated manner. Seeking clues to the basis of this coordination, M.E. Hatten developed an *in vitro* model system with which to study the neuron–glial relationship (Hatten, 1985). She created a cell suspension of dissociated cerebellar tissue harvested from mice at postnatal day 2–4 that she then fractionated both by size and propensity to bind onto polylysine-coated plates. This resulted in separate cultures of purified early postnatal mouse cerebellar neurons and astroglia. These cells types she allowed to grow either alone or in combination. When grown in tissue culture without other cells present, astroglial cells lost their highly elongated shape, became epithelioid in appearance, and proliferated rapidly. When she then added neurons to such a culture, though, they bound extensively to the glia and glial cell proliferation halted. The morphology of the glial cells also reverted to a complex shape with elaborate cell processes. Migration of neurons along those processes then began to occur. Morphologic evidence suggested that it was cell-to-cell contact with the neurons that promoted those responses in the astroglial cells (Mason et al., 1988). Furthermore, granule neuron membrane preparations alone could inhibit astroglial DNA synthesis, as shown by a two- to fivefold decrease in ^{3}H-thymidine incorporation depending upon the concentration of the membrane preparation added (Hatten, 1987).

Using similar methods, Hatten analyzed the influence of granular neurons on the growth of six different astrocytoma cell lines derived from primary tumors that occurred in different regions of human, rat, and mouse brains (Hatten, 1987). In four of the six astrocytoma cultures (three human and one mouse), the granular neurons made numerous cell-to-cell contacts with the malignant glia. In these cocultures the glioma cells underwent a significant reduction in numerical proliferation over the next five days as compared to similar tumor cell cultures not exposed to granular neurons. This was a reversible effect, however, in that vigorous shaking of the culture dishes promoted release of bound neurons and resumption of glioma growth. Glioma cell incorporation of ^{3}H-thymidine dropped off 2–5-fold within 12 hours after cell-to-cell contact with the neurons. Optimum inhibition occurred with a cell-to-cell ratio of four neurons to one glioma cell. Delayed labeling with ^{3}H-thymidine revealed that greater than 90% of the astrocytoma cells in coculture for 3–5 days experienced inhibition of DNA synthesis. Experiments using conditioned medium or coculture in separate chambers with intermixed culture medium resulted in no inhibition of glioma proliferation and thereby provided evidence against any mechanism mediated by a soluble factor. An enriched plasma membrane fraction from purified granule neurons, however, did arrest glial cell growth when a final concentration of added granule cell membrane protein material was 2–6 g/ml. As a control in these studies, she also applied PC12 neurons (derived originally from a pheochromocytoma) and membrane fraction preparations from PC12 cells to the glioma cultures. No growth inhibition occurred. These observations raise the possibility that differences in the biological behavior of low-grade versus high-grade astrocytomas may be partially explained by the degree to which the tumor cells retain this capacity to respond to intercellular regulation over their proliferation exerted by surrounding neurons.

OLIGODENDROCYTES INHIBIT NEURONAL GROWTH IN THE CENTRAL NERVOUS SYSTEM

Neurons appear to have the capacity to influence the growth of glial cells by intercellular contact, but the converse is also true. Certain types of glia permit neuronal growth, while other types clearly do not. Schwab (1988) has demonstrated this in a series of tissue culture experiments in which he plated fetal retinal cells, sympathetic or sensory ganglion, neuroblastoma cells, or 3T3 fibroblasts onto cultures of dissociated optic nerve glial cells from young rats. By features of cellular morphology and by using immunofluorescent labels, he could specifically identify the neurons, astrocytes, fibroblasts, and immature versus differentiated oligodendrocytes on the culture plates. Fibroblasts and astrocytes exhibited a growth pattern that distinctly avoided any contact with the highly branched mature oligodendrocytes. The retinal, ganglion, and neuroblastoma cells all sprouted dense networks of neuritic processes. Wherever these processes contacted astrocytes or immature oligodendrocytes, they made cellular contact, but then continued to elongate without inhibition across the surface of those cells. When neuritic processes encountered mature oligodendrocytes, however, further growth of those processes stopped, thus producing the appearance of "windows" on the plate where neurites were excluded. Time-lapse microscopic observation of the neuronal growth cones confirmed that their filopodia adhere to branches of the mature oligodendrocytes, but then grow no further. In contrast, contact with astrocytes clearly resulted in further extension of the neuronal cell processes (Schwab and Caroni, 1988).

Additional investigation revealed that CNS myelin produced this inhibitory effect on neuronal growth. More specifically, two protein fractions derived from CNS myelin, recovered from SDS-polyacrylamide gels at MW 35,000 and 250,000 Dalton, contained this inhibitory activity. These proteins he termed neurite growth inhibitors, NI-35 and NI-250 (Schwab, 1990). Monoclonal antibodies raised against these two proteins, when used as immunofluorescent labels, exclusively stain differentiated oligodendrocytes. The presence of these antibodies at low concentration in the coculture of neurons with rat retinal glia significantly diminishes the inhibitory effect of mature oligodendrocytes on neurite growth (Caroni and Schwab, 1988; Schwab, 1990).

CNS ASTROCYTES OR SCHWANN CELLS TRANSPLANTED INTO THE CNS CAN FACILITATE THE GROWTH OF NEURONAL PROCESSES

Schwab observed that although mature oligodendrocytes inhibited neurite outgrowth, mature astrocytes distinctly permitted such growth. This suggests that a naturally occurring cell in the CNS may have the capacity to facilitate growth of neuronal processes in a manner similar to the influence exerted by Schwann cells in the PNS. The PNS retains an ability to support the growth and regrowth of neuronal processes. Properties intrinsic to the Schwann cell itself account for this unique ability. Richard and Mary Bunge observed and described the interactions in tissue culture between dorsal root ganglion cells and Schwann cells. When these two cell types come into contact with one another, the Schwann cell begins production of extracellular matrix components such as laminin, collagen, and heparin sulfate proteoglycan. Schwann cell production of galactocerebroside also up-regulates. With sufficient vitamin C present in the medium, the Schwann cells generate a complete perineurium with its basal lamina. The growing axons then elongate between the Schwann cell and the basal lamina, within the environment of this extracellular matrix that the Schwann cell has produced, and axonal myelination occurs (Bunge, 1983; Bunge et al., 1982, 1983; Eldridge et al., 1987). Aligned corridors of Schwann cells with surrounding basal lamina tubes that result from this process produce the "bands of Bungner" that are noted to form in the distal segment of lesioned peripheral nerves. As early as 1928, Ramon y Cajal postulated the importance of these structures in peripheral axonal regeneration (Cajal, 1928).

More recently, Aguayo and Bray have demonstrated that a variety of intrinsic CNS neurons subjected to axotomy can extend regenerative processes a considerable distance into autologous peripheral nerve grafts implanted at the site of injury (Aguayo, 1985; Aguayo et al., 1983; Bray et al., 1987). Evidence from Smith further suggests that viable Schwann cells must be present within such peripheral nerve grafts to support CNS axonal regeneration (Smith and Stevenson, 1988). Kromer and Cornbrooks demonstrated *in vivo* the Schwann cell capacity to support CNS neuronal process regrowth. They transplanted Schwann cells raised in tissue culture to the lesion cavity of adult Sprague-Dawley rats subjected to bilateral fornix/fimbria transection. Control animals received transplantation of acellular collagen substrate. The Schwann cell transplants supported extensive regeneration of acetylcholine esterase-positive axons projecting through the transplanted tissue from septum into hippocampus. No growth of neuronal processes occurred through the acellular collagen substrate (Kromer and Cornbrooks, 1985).

To determine whether the Schwann cell itself or extracellular products produced by the Schwann cell facilitate CNS neurite outgrowth, Bunge compared retinal ganglion neurite growth on cultured Schwann cells with that on extracellular matrix organized by Schwann cells. These neuronal processes grew exclusively along Schwann cell surfaces rather than on the cell free matrix (Hopkins and Bunge, 1991; Kleitman et al., 1988).

Schwann cells are not the only cells, however, that can facilitate outgrowth of CNS neuronal processes. Fallon prepared >95% pure monolayers in tissue culture of neonatal rat astrocytes, sciatic nerve Schwann cells, and skin fibroblasts. He then tested various CNS explants on these monolayers to determine which promote neurite outgrowth preferentially. Although it was difficult in this model system to quantify the response, astrocytes strongly supported neurite growth to an extent that appeared similar to the growth response along the Schwann cells. Fibroblasts failed to support any growth of neuronal processes (Fallon, 1985). Tissue culture experiments performed by other investigators also reveal that Schwann cells and astrocytes offer similarly favorable substrates for growth of CNS neurons (Noble et al., 1984).

Taken together, these investigations suggest that in the developing nervous system, guidance of axonal growth may be facilitated in the CNS by immature astrocytes and in the PNS by Schwann cells. In the CNS, however, with the maturation of the oligodendrocytes, inhibitory cellular influences begin to restrict further growth of neuronal processes. In this way central axonal pathways become fixed tracts. This preserves the elaborate organization of the CNS, but at the expense of a loss of plasticity and regenerative capacity. In the PNS, inhibitory cellular influences do not arise. This preserves regenerative capacity. The specific organization of peripheral tracts, however, cannot be well maintained after injury. Aberrant reinnervation of injured peripheral nerves, therefore, commonly occurs.

MEMBRANE-BOUND PROTEINS PROPOSED AS MEDIATORS OF INTERCELLULAR INTERACTIONS IN THE NERVOUS SYSTEM

The increasing evidence that intercellular interactions play a crucial role in nervous system development now prompts efforts to identify cell surface molecules that may mediate these processes. One family of transmembrane glycoproteins called integrins participates in cell-to-matrix or cell-to-cell adhesion in several physiologic processes. Integrins consist of a noncovalent complex with an alpha and a beta subunit, both of which span the cell membrane. Their relatively small intracellular domains probably interconnect with actin filaments and actin-binding proteins of the cytoskeleton. The extracellular domain of these molecules interacts with a wide variety of ligands. Some possess a specificity for extracellular matrix proteins such as laminin fibronectin, vitronectin, and the collagens. Others have affinity for cell surface ligands (Hynes, 1987). Tissue culture experiments can probe the function of integrins in the migration of growth cones extended from sensory, retinal ganglion, and spinal cord neurons of chick embryos. These studies indicate that integrins in the nervous system can promote migration of growing neuronal processes across extracellular matrix substrates or across the surfaces of astrocytes. Antibody to the integrin complex, however, does not impede the migration of neuronal growth cones across glioma cells, suggesting that central nervous system tumors possess alternative mechanisms to promote intercellular interactions (Letourneau et al., 1988; Neugebauer et al., 1988; Tomaselli et al., 1988).

One particular member of the the integrin family of cell surface receptors, lymphocyte function-associated antigen-1 (LFA-1), plays an essential role in T lymphocyte-mediated cytotoxicity (Springer et al., 1987). By specific binding to a target cell surface glycoprotein called intercellular adhesion molecule-1 (ICAM-1), this integrin facilitates formation of antigen-independent conjugates, antigen-specific recognition, and cytolysis of particular target cells (Makgoba et al., 1988). Normal brain cells do not express significant levels of ICAM-1. This glycoprotein appears, however, on the cell surfaces of several CNS tumor lines, including glioblastoma, astrocytoma, ganglioglioma, medulloblastoma, and pineal tumor (Guarini et al., 1990; Kuppner et al., 1990). The highest levels of ICAM-1 occur in cells and intertumoral vessels of the glioblastoma and astrocytoma lines. In all of these tumor cultures, though, an increase in ICAM-1 expression occurs when the cells receive exposure to immune interferon-gamma or tumor necrosis factor-alpha. These observations suggest that interaction between LFA-1 and ICAM-1 may play a role in promoting lymphocytic infiltration into CNS tumors. Therapeutic interventions that enhance this intercellular interaction, therefore, might improve the clinical efficacy of adoptive immunotherapy trials that involve the direct infusion of lymphokine-activated killer (LAK) cells plus interleukin-2 (IL-2) into the tumors of patients with recurrent malignant gliomas (Barba et al., 1989).

Another family of membrane glycoproteins under investigation as possible mediators of intercellular interactions within the nervous system are the cell-adhesion molecules called CAMs. These fall into calcium-independent or calcium-dependent groups. The calcium-independent CAMs include two immunologically distinct but functionally related subtypes, N-CAM and L1 (Faissner et al., 1984). Both of these molecules serve as homophilic ligands in the formation of adhesions between cells (Rathjen and Schachner, 1984). During development, however, L1 appears only on postmitotic neurons, whereas N-CAM is detectable on all neuroepithelial cells already at the time of neural tube closure (Faissner et al., 1985; Hoffman et al., 1982). As a consequence of this difference, L1 appears to contribute only to neuron-to-neuron interaction, while N-CAM mediates neuron–neuron, neuron–astrocyte, and astrocyte–astrocyte adhesion (Keilhauer et al., 1983; Moos et al., 1988; Neugebauer et al., 1988; Rathjen et al., 1987). Calcium-dependent adhesion between neurons occurs by the action of a 130-kd cell-surface glycoprotein referred to as N-cadherin (Takeichi, 1988). In tissue culture, neurite outgrowth by both peripheral and central neurons onto astrocytes depends strongly on the function of N-cadherin (Neugebauer et al., 1988; Tomaselli et al., 1988).

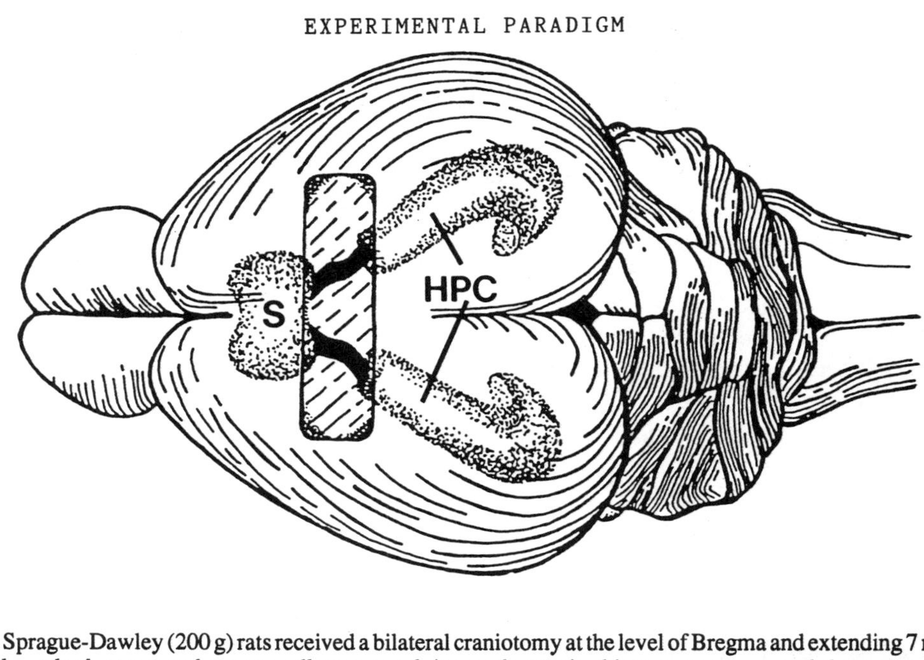

Fig. 1. Adult female Sprague-Dawley (200 g) rats received a bilateral craniotomy at the level of Bregma and extending 7 mm caudal. Aspiration of an area of parietal cerebral cortex and corpus callosum overlying each anterior hippocampus exposed these structures. Aspiration then continued to produce a complete bilateral lesion of the fimbria–fornix, thus axotomizing the dorsal septo-hippocampal pathways and leaving a lesion cavity. Implantation of the bridging microenvironments follows by inserting a whole-ARC rostrally into the septum and caudally into the dorsal hippocampus, then inserting a hemi-ARC similarly on the contralateral side of the brain. This schematic diagram illustrates the site of the bilateral aspiration lesion of the fimbria–fornex (rectangular grid) and the placement of whole and hemi-ARCs (solid tubes between the septum [S] and hippocampus [HPC]).

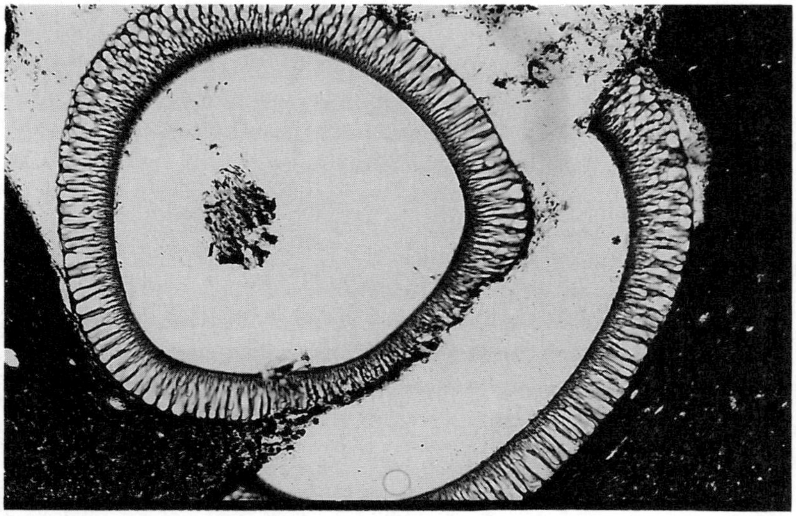

Fig. 2. This high-magnification coronal section taken at the insertion site of both the whole and the hemi-ARC into the medial septal nucleus reveals AChE + fibers within the matrix cable of the whole-ARC and coursing around the adjacent external surface of the whole-ARC in association with the open face of the hemi-ARC.

ENHANCING THE REGENERATIVE CAPACITY OF INJURED CNS NEURONS BY CONTROLLING THE CELLULAR MILIEU OF THEIR MICROENVIRONMENT

Observations in the past led to the conclusion that neuronal regeneration of severed axons with the reestablishment of function connections over long distances can occur only in the peripheral nervous system. Recent studies now reveal that CNS neurons can also regenerate severed processes provided they encounter a favorable cellular and molecular environment. Study of the complex events that direct embryologic growth and development within the nervous system can lead to concepts that will suggest new strategies to facilitate CNS axonal regeneration.

One such *in vivo* strategy currently under investigation by Kromer is to interpose a selectively permeable polymer tube filled with collagen or extracellular matrix with or without specific dispersed cell types to bridge the gap in a lesioned CNS axonal pathway. Adult Sprague-Dawley rats undergo a complete bilateral aspiration lesion of the dorsal septo-hippocampal pathways. Insertion of a 5 mm-long polymer tube with permeability of either 10, 30, or 100 kd MW (referred to as an axon regeneration channel or ARC) provides a bridge across the lesion cavity from the medial septum to one anterior hippocampus (Fig. 1). This ARC may contain entirely acellular purified collagen or commercially available extracellular matrix (ECM). In other experiments, though, the ARC contains cultured Schwann cells densely dispersed within the matrix. To provide an internal control, another acellular or cellular matrix-filled tube is sectioned longitudinally (refered to as a hemi-ARC) and interposed in the same lesion cavity between the medium septum and the contralateral anterior hippocampus. Postoperatively the animals survive for variable time periods up to six weeks before receiving paraformaldehyde perfusion under anesthesia. After removal, fixation, and freezing of the brains, thin transverse section preparation from septum to anterior hippocampus follows.

An immunohistochemical reaction with the acetylcholinesterase (AChE) that septal neurons express along their processes and immunofluorescent labels against GAP-43 (expressed in neurite growth cones) or L1 (expressed along neuronal surface membranes) reveal axons that migrate through the ARC or sprout in the lesion cavity. Cells in the matrix of the ARC or in the lesion cavity specifically fluoresce after immunostains for glial fibrillary acidic protein (specific for astrocytes), vimentin (nonspecific in many reactive cells), or nerve growth receptor (on Schwann cells and reactive astrocytes).

Preliminary observations demonstrate that within the restrictive microenvironment of a matrix-filled polymer tube, axotomized septal neurons will regrow and extend processes that reach their designated target in the hippocampus (Collins and Kromer, 1989; Massoudi et al., 1989). With initially acellular matrix in the ARC, regenerative axons do not begin to elongate into that matrix until the second postoperative week. Migration of reactive astrocytes (positive for both GFAP and vimentin) into the acellular ARC matrix occurs first, then axonal elongation into the ARC begins and is quite extensive at four weeks (Fig. 2). These axons elongate in a directional manner within the matrix cable of the whole-ARC (Fig. 3). Some bundles of these regrowing axons reach the hippocampus after 4 to 6 weeks (Fig. 3). A spatial association within the ARC matrix between axon bundles and astrocytes is seen. When the ARC matrix initially contains a dense population of cultured Schwann cells, no delay in the onset of axonal elongation into the ARC occurs. By the end of the first postoperative week, bundles of regenerative axons have

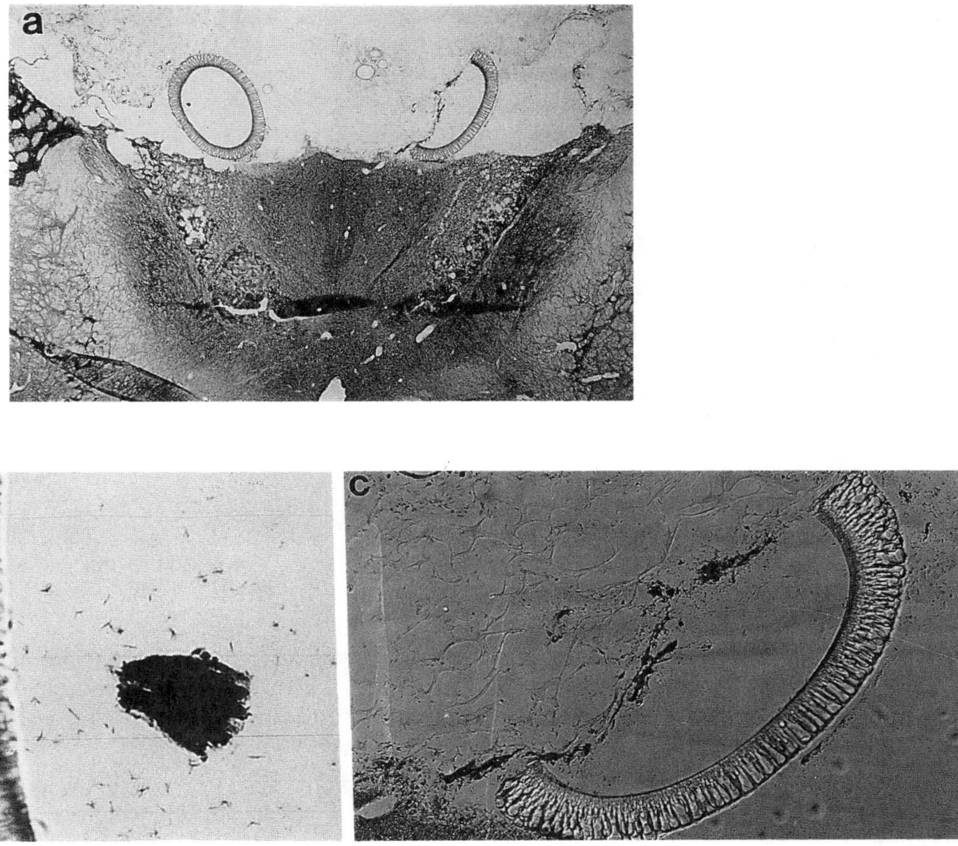

Fig. 3. **A**: Low-magnification coronal section showing a whole and hemi-ARC at a distance 1.5 mm caudal to the septal insertion (Fig. 2). **B**: High magnification of the matrix cable within the whole-ARC suspended in a fluid-filled space separated from cellular elements in the lesion cavity by the polymer tube. AChe + fibers extend densely through the whole-ARC matrix cable. **C**: High magnification of the matrix bridge attached to the hemi-ARC openly exposed to the contents of the lesion cavity. Throughout the lesion cavity, AChE + fibers freely associate with tissue strands extending from the hemi-ARC into the cavity and are not confined to the immediate vicinity of the hemi-ARC.

already completely traversed the entire distance and arrived at the hippocampus. Again within the ARC matrix a spatial association occurs between the axon bundles and the Schwann cells (Collins et al., 1992).

The control ARC undergoes longitudinal sectioning prior to implantation. The control situation for these studies, therefore, consists of a microenvironment without the confines of a closed polymer tube. This openly exposes the hemi-ARC matrix to the lesion cavity. When that matrix is initially acellular, directional migration of reactive astrocytes does not occur. Instead, the astrocytes migrate more randomly along tissue strands between the matrix and the lesion cavity (Fig. 3). As expected, axonal sprouts occur where the astrocytes have migrated, but no axons achieve directional elongation all the way across the hemi-ARC to reach the hippocampus (Fig. 4). With a dense population of Schwann cells in the hemi-ARC matrix, however, these cells retain a close association with one another and do not migrate widely; thus they create a direct cellular bridge between the septum and the hippocampus even though they are not confined by a complete polymer tube. Regenerative axons elongate directly along that cellular bridge without wandering into the cavity and achieve direct extension into the hippocampus.

These observations demonstrate *in vivo* in an adult mammalian system that CNS neurons subjected to axotomy can elongate regenerative processes with directionality toward an appropriate target if those processes are provided a suitable cellular milieu in the microenvironment. The findings support the concept that neuronal regeneration occurs as a consequence of intercellular interactions with specific cell types that the neuronal processes encounter.

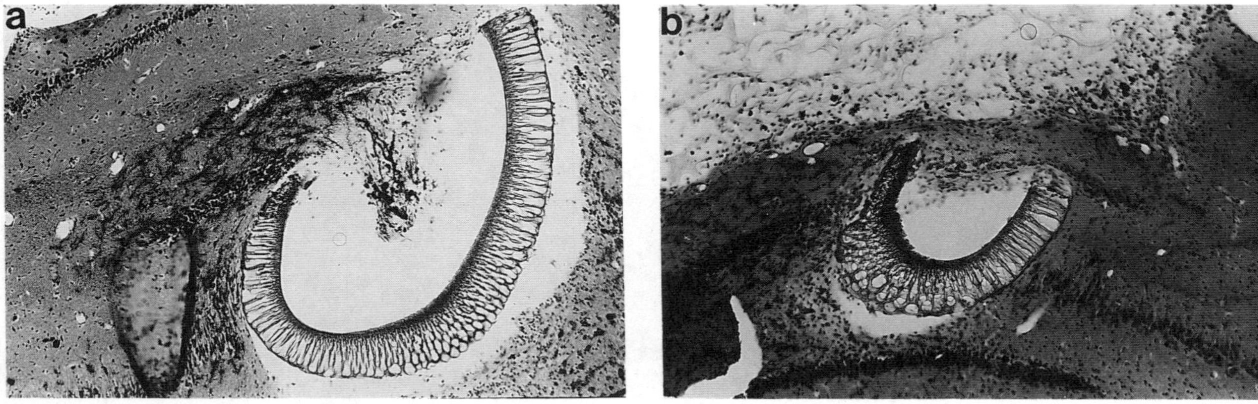

Fig. 4. Insertion of ARCs into the hippocampus. **A**: Whole-ARC stained for AChE revealing outgrowth of AChE + fibers from the matrix cable into the hippocampus. **B**: Hemi-ARC stained for AChE, revealing very few AChE + fibers entering the hippocampus directly from the matrix of the hemi-ARC. Adjacent areas of the hippocamus, however, contain some AChE + fibers that entered the hippocampus by traveling along tissue strands within the lesion cavity.

OVERVIEW

Intrinsic processes arising from the genomes of the various cells that comprise the mammalian nervous system do not direct growth, differentiation, and regeneration independently. The genome of one cell, either through the action of cell surface receptors or by means of soluble factors, must establish a kind of communication with the genomic expression of other differentiating cells in the nervous system to determine the pattern of growth that will occur. The highly elaborate complexity of cellular organization that develops in the nervous system occurs as a consequence of such extensive intercellular interactions. Neurons of the mature PNS and CNS subjected to axotomy possess significant regenerative capacity, provided they encounter a favorable microenvironment that includes the requisite intercellular interactions. As the study of neoplasia within the nervous system progresses, disruption in the mechanisms of these intercellular interactions will probably explain the lack of growth regulation that occurs in transformed cells. Furthermore, greater understanding of the molecules that mediate binding of immune cells to tumor cells may lead to new strategies for enhancing the immunologic susceptibility of CNS tumors.

REFERENCES

Aguayo, A.J. (1985) Axonal regeneration from injured neurons in the adult mammalian central nervous system. In: C.W. Cotman (ed.). Synaptic Plasticity. New York: Guilford Press.

Aguayo, A., M. Benfrey, and D. Samuel (1983) A potential for axonal regeneration in neurons of the adult mammalian nervous system. Birth Defects 19:327–346.

Assouline, J.G., P.V. Bosch, R. Lim, et al. (1987) Rat astrocytes and Schwann cells in culture synthesize nerve growth factor-like neurite-promoting factors. Dev. Brain Res. 31:103–118.

Barba, D., S. Saris, C. Holder, et al. (1989) Intratumoral LAK cell and interleukin-2 therapy of human gliomas. J. Neurosurg. 70:175–182.

Barde, Y.A., D. Edgar, and H. Thoenen (1982) Purification of a new neurotrophic factor from mammalian brain. EMBO J. 1:549–553.

Bray, G.M., M.P. Villegas-Perez, M. Vidal-Sanz, et al. (1987) The use of peripheral nerve grafts to enhance neuronal survival, promote growth and permit terminal reconnections in the central nervous system of adult rats. J. Exp. Biol. 132:5–19.

Bunge, M.B., R.P. Bunge, D.J. Carey, et al. (1983) Axonal and nonaxonal influences on Schwann cell development. In: Developing and Regenerating Vertebrate Nervous Systems. New York: Alan R. Liss, pp. 71–105.

Bunge, M.B., A.K. Williams, and P.M. Wood (1982) Neuron–Schwann cell interaction in basal lamina formation. Dev. Biol. 92:449–460.

Bunge, R. (1983) Recent observations on the control of Schwann cell function. In: Current Topics in Cellular Anatomy. New York: Alan R. Liss.

Cajal, S.R. (1928) In: R.M. May (ed.). Degeneration and Regeneration in the Nervous System. London: Oxford University Press.

Caroni, P., and M.E. Schwab (1988) Antibody against myelin-associated inhibitor of neurite growth neutralizes nonpermissive substrate properties of CNS white matter. Neuron 1:85–96.

Chiu, A.Y., A.E. De Los Monteros, R.A. Cole, et al. (1991) Laminin and s-laminin are produced and released by astrocytes, Schwann cells and schwannomas in culture. Glia 4:11–24.

Collins, J.J., L.F. Kromer (1989) Implanted polymer tubes facilitate axonal regeneration after fimbria/fornix lesion. J. Neurotrauma 6:200-201.

Collins, J.J., D.J. Messersmith, L.F. Kromer (1992) Effects of hemisectioned polymer channels on CNS axonal regeneration in transplants of extracellular matrix (ECM) terrains versus Schwann cell/ECM substrates. Res. Neurol. Neurosci. 4:152.

Cowan, W.M. (1979) The development of the brain. Sci. Am. 241:112–133.

Eldridge, C.F., M.B. Bunge, R.P. Bunge, et al. (1987) Differentiation of axon-related Schwann cells in vitro. I. Ascorbic acid regulates basal lamina assembly and myelin formation. J. Cell Biol. 105:1023–1074.

Faissner, A., J. Kruse, C. Goridis, et al. (1984) The neural cell adhesion molecule L1 is distinct from from the N-CAM related group of surface antigens BSP-2 and D3. EMBO J. 3:733–737.

Faissner, A., D.B. Teplow, D. Kubler, et al. (1985) Biosynthesis and membrane topography of L1 neural cell adhesion molecule. EMBO J. 4:3105–3113.

Fallon, J. (1985) Preferential outgrowth of central nervous system neurites on astrocytes and Schwann cells as compared with nonglial cells in vitro. J. Cell Biol. 100:198–207.

Furshpan, E.J., P.R. MacLeish, P.H. O'Lague, et al. (1976) Chemical transmission between rat sympathetic neurons and cardiac myocytes developing in microculture: Evidence for cholinergic, adrenergic and dual function neurons. Proc. Natl. Acad. Sci. USA 73:4225–4229.

Guarini, L., M. Temponi, J.N. Bruce, et al. (1990) Expression and modulation by cytokines of the intercellular adhesion molecule-1 (ICAM-1) in human central nervous system tumor cell cultures. Int. J. Cancer 46:1041–1047.

Hatten, M.E. (1985) Neuronal regulation of astroglial morphology and proliferation in vitro. J. Cell Biol. 100:384–396.

Hatten, M.E. (1987) Neuronal inhibition of astroglial cell proliferation is membrane mediated. J. Cell Biol. 104:1352–1380.

Hatten, M.E., and M. Shelanski (1988) Mouse cerebellar granule neurons arrest the proliferation of human and rodent astrocytoma cells in vitro. J. Neurosci. 8:1447–1453.

Hoffman, S., B.C. Sorkin, P.C. White, et al. (1982) Chemical characterization of a neural cell adhesion molecule (N-CAM) purified from embryonic brain molecules. J. Biol. Chem. 257:7720–7729.

Hohn, A., J. Leibrock, K. Bailey, et al. (1990) Identification and characterization of a novel member of the nerve growth factor/brain derived neurotrophic factor family. Nature 344:339–341.

Hopkins, J.M., and R.P. Bunge (1991) Regeneration of axons from adult rat retinal ganglion cells on cultured Schwann cells is not dependent on basal lamina. Glia 4:46–55.

Hynes, R.O. (1987) Integrins: A family of cell surface receptors. Cell 48:549–554.

Jacobson, M. (1978) Developmental Neurobiology, 2nd ed. New York: Plenum Press.

Johnson, E.M., and M. Taniuchi (1987) Nerve growth factor receptors in the central nervous system. Biochem. Pharm. 36:4189–4195.

Jones, K.R., and L.F. Reichardt (1990) Molecular cloning of a human gene that is a member of the nerve growth factor family. Proc. Natl. Acad. Sci. USA 87:8060–8064.

Keilhauer, G., A. Faissner, and M. Schachner (1983) Differential inhibition of neurone–neurone, neurone–astrocyte and astrocyte–astrocyte adhesion by L1, L2 and N-CAM antibodies. Nature 316:728–730.

Kleitman, N., P. Wood, M.I. Johnson, et al. (1988) Schwann cell surfaces but not extracellular matrix organized by Schwann cells support neurite outgrowth from embryonic rat retina. J. Neurosci. 8:653–663.

Kromer, L.F., and C.J. Cornbrooks (1985) Transplants of Schwann cell cultures promote axonal regeneration in the adult mammalian brain. Proc. Natl. Acad. Sci. USA 82:6330–6334.

Kuppner, M., C.M. Hamou, and N. de Tribolet (1990) Activation and adhesion molecule expression on lymphoid infiltrates in human glioblastomas. J. Neuroimmunol. 29:229–238.

Le Douarin, N. (1982) The Neural Crest. Cambridge: Cambridge University Press.

Leibrock, J., F. Lottspeich, A. Hohn, et al. (1989) Molecular cloning and expression of brain derived neurotrophic factor. Nature 341:149–152.

Letourneau, P.C., I.V. Pech, S.L. Rogers, et al. (1988) Growth cone migration across extracellular matrix components depends on integrin, but migration across glioma cells does not. J. Neurosci. Res. 21:286–297.

Lin, L.F., L.G. Armes, A. Sommer, et al. (1990) Isolation and characterization of ciliary neurotrophic factor from rabbit sciatic nerves. J. Biol. Chem. 265:8942–8947.

Maisonpierre, P.C., L. Belluscio, S. Squinto, et al. (1990) Neurotrophin-3: A neurotrophic factor related to NGF and BDNF. Science 247:1446–1451.

Makgoba, M.W., M.E. Sanders, G.E. Ginther Luce, et al. (1988) Functional evidence that intercellular adhesion molecule-1 (ICAM-1) is a ligand for LFA-1-dependent adhesion in T cell-mediated cytotoxicity. Eur. J. Immunol. 18:637–660.

Mason, C.A., J.C. Edmondson, and M.E. Hatten (1988) The extending astroglial process: Development of glial cell shape, the growing tip, and interactions with neurons. J. Neurosci. 8:3124–3134.

Massoudi, F., J.J. Collins, L.F. Kromer (1989) CNS axonal regeneration is influenced by transplants of matrix filled polymer tubes. Soc. Neurosci. Abstr. 15:1093.

Moos, M.R., R. Tache, H. Scherer, et al. (1988) Neural adhesion molecule L1 as a member of the immunoglobulin superfamily with binding domain similar to fibronectin. Nature 334:701–703.

Neugebauer, K.M., K.J. Tomaselli, J. Lilien, et al. (1988) N-Cadherin, N-CAM and integrins promote retinal neurite outgrowth on astrocytes in vitro. J. Cell Biol. 107:1177–1187.

Noble, M., J. Fok-Seang, and J. Cohan (1984) Glia are a unique substrate for the in vitro growth of central nervous system neurons. J. Neurosci. 4:1892–1903.

Patterson, P.H. (1978) Environmental determination of autonomic neurotransmitter functions. Annu. Rev. Neurosci. 1:1–17.

Rakic, P. (1971) Neuron–glia relationship during granule cell migration in developing cerebellar cortex. A golgi and electron microscopic study in macacus rhesus. J. Comp. Neurol. 141:283–312.

Rathjen, F.C., and M. Schachner (1984) Immunocytological and biochemical characterization of a new neuronal cell surface component (L1 antigen) which is involved in cell adhesion. EMBO J. 3:1–10.

Rathjen, F.C., J.M. Wolff, R. Framk, et al. (1987) Membrane glycoprotein involved in neurite fasciculation. J. Cell Biol. 104:343–353.

Rutka, J.T., G. Apodaca, R. Stern, et al. (1988) The extracellular matrix of the central and peripheral nervous systems: Structure and function. J. Neurosurg. 69:155–170.

Sauer, F.C. (1935) Mitosis in the neural tube. J. Comp. Neurol. 62:377–405.

Schwab, M. (1990) Myelin-associated inhibitors of neurite growth. Exp. Neurol. 109:2–5.

Schwab, M.E., and P. Caroni (1988) Oligodendrocytes and CNS myelin are nonpermissive substrates for neurite growth and fibroblast spreading in vitro. J. Neurosci. 8:2381–2393.

Smith, G.V., and J.A. Stevenson (1988) Peripheral nerve grafts lacking viable Schwann cells fail to support nervous system axonal regeneration. Exp. Brain Res. 69:299–306.

Springer, T.A., M.L. Dustin, T.K. Kishimoto, et al. (1987) The lymphocyte function-associated LFA-1, CD2, and LFA-3 molecules: Cell adhesion receptors of the immune system. Annu. Rev. Immunol. 5:223–252.

Stitt, T.N., and M.E. Hatten (1990) Antibodies that recognize astrotactin block granule neuron binding to astroglia. Neuron 5:639–649.

Stockli, K.A., F. Lottspeich, M. Senster, et al. (1989) Molecular cloning, expression and regional distribution of rat ciliary neurotrophic factor. Nature 342:920–923.

Takeichi, M. (1988) The Cadherins: Cell–cell molecules controlling animal morphogenesis. Development 102:639–655.

Thoenen, H. (1991) The changing scene of neurotrophic factors. Trends Neurosci. 14:165–170.

Toivonen, S., D. Tarin, L. Saxen, et al. (1975) Transfilter studies on neural induction in the newt. Differentiation 4:1–7.

Toivonen, S., D. Tarin, and L. Saxen (1976) The transmission of morphogenic signals from amphibian mesoderm to ectoderm in primary induction. Differentiation 5:49–55.

Tomaselli, K.J., K.M. Neugebauer, and J.L. Bixby (1988) N-Cadherin and integrins: Two receptor systems that mediate neuronal process outgrowth on astrocyte surfaces. Neurology 1:33–43.

Whittemore, S.R., and A. Seiger (1987) The expression, localization and functional significance of beta-nerve growth factor in the central nervous system. Brain Res. Rev. 12:439–464.

Chapter

6

The Cell Cycle

Mary Ellen Perry and Arnold J. Levine

The cell cycle is the term used to describe an orderly series of events that ensures the duplication of all cellular components (organelles, DNA) in their correct sequence and the faithful partitioning of these components into two parts. Traditionally, the cell cycle has been divided into four parts based upon measurable events in the cycle (Fig. 1). The S-phase, or synthetic phase of the cycle, results in the complete duplication of the nuclear DNA and does not permit any more or any less than that to occur. Mitosis, or M-phase, the condensation of chromosomes, attachment to the spindle, and segregation of these chromosomes into two equal parts, is easily visible in the light microscope. The M-phase and the S-phase are separated in time from each other by two gaps, termed G_1 between M and S and G_2 between S and M (Fig. 1). While these gap phases were originally named to indicate the degree of ignorance of the events that regulate entry into S-phase and entry into M-phase, this has changed dramatically (see Beach et al., 1991) thanks to studies from three related experimental systems: yeast genetics, biochemical analysis with *Xenopus* eggs, and the molecular biology of oncogenes and tumor suppressor genes.

It is now recognized that the cell cycle may be divided into two kinds of functions: functions that are obligatory to progress through the cell cycle, G_1 to S to G_2 to M, and checkpoints that monitor the efficacy and completion of events and stop the cycle if conditions to progress are not perfect (Hartwell and Weinert, 1989). An example of essential functions for the cell cycle is the synthesis or activation of enzymes, in late G_1, which are required to synthesize deoxyribonucleoside triphosphate precursors for DNA synthesis in S-phase. Without such precursors, no DNA could be synthesized, so the regulation of these enzyme activities is obligatory for progression in the cycle. Similarly, the synthesis of DNA polymerases, DNA ligases, primases, helicases, single strand DNA binding proteins, etc., are all required in S to progress through the cell cycle. Histone synthesis and DNA packaging is required in S-phase in an obligatory fashion to progress to G_2. Preparations for chromosome condensation occur in G_2 and condensation, spindle attachment, and segregation are all essential or obligatory functions in the cell cycle.

Checkpoint controls monitor and regulate these events. The cycle progresses unless stopped or restarted by a checkpoint control. Thus, checkpoint controls are not obligatory for the cycle to progress, but if they fail, the cycle may progress in an abnormal fashion, sometimes resulting in death, sometimes in cancer, or sometimes in the production of sterile offspring. There are several examples of G_1 checkpoint controls. One of these measures whether or not the nutrient levels in the environment are sufficient to duplicate the cell and, if not, the cycle is blocked at a specific place in G_1. Cells frequently measure their mass, in some fashion, and test whether or not they are large enough to progress through the cell cycle. Size control may be carried out either in G_1 as it is in *Saccharomyces cerevisiae* or in G_2 as it is in *Schizosaccharomyces pombe* (Nasmyth et al., 1992). Similarly, some mammalian cells rest in the cell cycle in early G_1, sometimes termed G_0, while other cells pause for long periods of time in G_2. Yet another G_1 checkpoint in the cell cycle measures whether or not the DNA template is normal or has been damaged by irradiation, chemical modification, or abnormal structural features. If these problems are recognized by a checkpoint control, normal cells are shut down in late G_1 and, where possible, repair processes correct these problems and permit cells to duplicate their DNA (Schimke et al., 1991). Mammalian cells frequently get signals from their neighbors that may be mediated by hormones (growth signals or inhibitory activities), adhesion molecules, integrins, or external events. Many of these signals act as checkpoints, most often in G_1 of the cell cycle.

One of the most critical checkpoints in G_1 is called start in the yeast cell cycle. Its likely homologue is the restriction point in mammalian cells. In yeast, several signals can act to stop progression through the cycle, such as the absence of glucose or other nutrients in the medium or even specific mating pheromones. The mating hormone alpha blocks cell cycle progression and prepares the cell for fusion and meiosis. Yeast cells kept in full nutrient medium through G_1

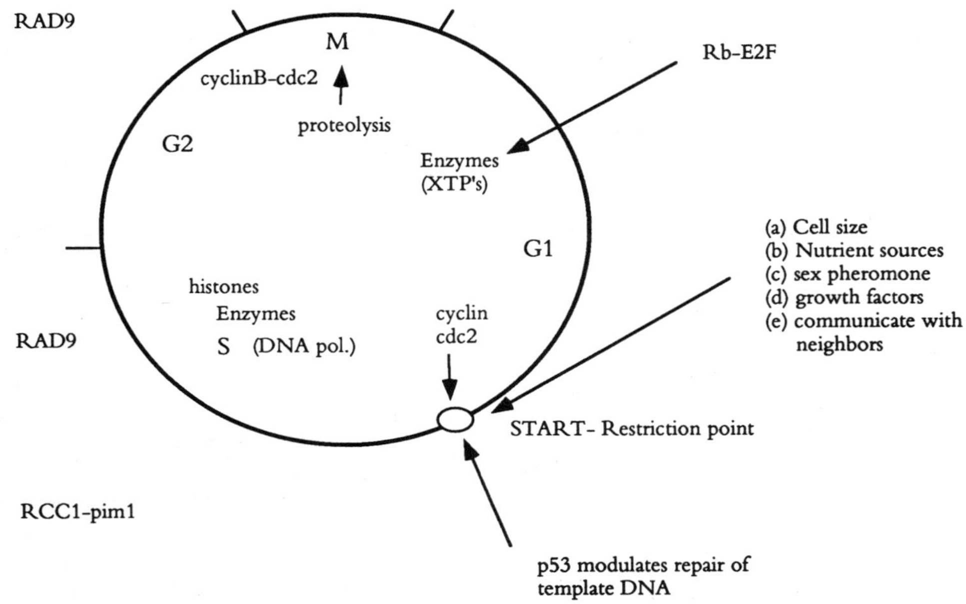

Fig. 1. The cell cycle is divided into four phases: G_1, S, G_2, and M. Inside the circle are listed the essential functions that carry out the progression through this cycle. In the absence of these functions, the cycle fails to progress and a mutation is lethal. Outside the circle are the nonessential functions that regulate the cell cycle, the checkpoints. Checkpoints negatively regulate the cell cycle, are not lethal to a cell if absent, but this often results in an inefficient cell cycle, leading to genomic instability, mutation, and a cancerous cell. In the absence of checkpoints, the default pathway of essential functions is programmed to promote cell division. The checkpoints often utilize signal transduction pathways to regulate the essential functions in the cell cycle. Some of the checkpoints are tumor suppressor genes and some of the activities in the signal transduction pathway are proto-oncogenes.

and past the start point in late G_1 can then be removed from these nutrients, but the cell now remains committed to complete the cell cycle. Thus, start is the point after which cells will progress through the cycle, even in the absence of sufficient nutrients or even in the presence of mating pheromone. Similarly, mammalian cells in culture respond to the application of growth hormones, such as platelet-derived growth factor or epidermal growth factor, by entering into the cell cycle and dividing. Removal of these hormones from the medium prior to the restriction point in late G_1 stops progression through the cell cycle. Removal of these hormones after the restriction point has been reached has no effect upon the progression through S, G_2, and M-phases. Thus, start (or the restriction point) is a critical checkpoint in the cycle where external influences act to shut down progression through the cell cycle. The start point in G_1 in *S. cerevisiae* is regulated by a set of biochemical events that have been defined by genetic analyses and these are reviewed below.

Checkpoints in S-phase include making sure that each DNA species is replicated once and only once. One S-phase checkpoint is regulated by the protein called RCC1 in humans or pim1 in yeast (Matsumoto and Beach, 1991; Seino et al., 1991). This is a protein that appears to examine the DNA to make sure that S-phase is complete before the cell enters G_2 and M-phase. RCC1 and pim1 mutants condense their chromosomes, form septa, and develop mitotic spindles even in the absence of DNA replication. pim1 acts, along with a GTPase, to signal that S-phase is complete and G_2 and M events can begin. As early as 1977, Rao and his colleagues (Rao et al., 1977) produced somatic cell hybrids between S-phase cells and M-phase cells. These hybrid cells underwent premature condensation of the S-phase chromosomes, implying that M-phase cells contained an activity that overrode the S-phase controls and induced mitotic chromosome condensation. It appears that RCC1 or pim1 prevent that M-phase condensation factor from being produced until all of the DNA is replicated in S-phase. The RCC1 or pim1 checkpoint acts in S-phase to block a late G_2/M-phase event.

One of the best studied checkpoints in G_2 is represented by the RAD9 pathway in yeast (Brown et al., 1991). Yeast cells treated with X-rays are forced to repair double-stranded breaks in their DNA. Since yeast cells are haploid, this can only be accomplished in G_2, where homologous chromosomes are required for repair. When yeast cells are treated with X-rays, they pause in G_2 to repair the damage. Mutations in any of six genes (RAD9, 17, 24, MEC1, 2, 3) eliminate this delay or block in the cell cycle and permit cells to enter mitosis directly. When they do so, these cells have very high error rates in chromosome segregation, producing aneuploid cells. The six genes in the RAD9 pathway define a checkpoint in the cell cycle in G_2 that prevents the cell from entering mitosis with damaged chromosomes. Interestingly, caffeine also allows a bypass of DNA-damage-induced G_2 arrest in mammalian cells (Lau and Pardee, 1982). RAD9 also acts to coordinate events in the S-phase with mitosis. Drugs that block cells in S-phase will induce premature entry into mitosis and cell death when in a cell mutant for any one of the six genes in the RAD9 pathway. Thus, the RAD9 pathway is a checkpoint required for G_2 arrest of mitosis in cells blocked in DNA replication or cells with damaged DNA. The former checkpoint is redundant with RCC1 or pim1 functions. Redundancy in checkpoint control of the cell cycle may be the rule.

Each of these checkpoints helps to ensure that a cell with damaged DNA does not propagate. Translocations, amplification, and aneuploidy are the result of aberrant cell cycles and are not found in normal cells. The frequency of amplification events is increased by agents that damage DNA and may reflect the number of unrepaired lesions. Efficient repair occurs after arrest of the cell cycle. Loss of a checkpoint control at any stage of the cycle could lead to chromosomal abnormalities and mutation. Loss of a G_1 control could allow replication of DNA under less than optimal conditions (e.g., nucleotide pool imbalances, which can lead directly to mutation). Since S-phase checkpoints ensure complete replication of the genome, it is possible that loss of an S-phase checkpoint would allow partially replicated chromosomes to segregate. Chromosomes with gaps are good substrates for the recombination events (including unequal sister chromatid exchange) that are thought to be important events in gene amplification (Perry et al., 1992). G_2 checkpoints serve to monitor and repair DNA before distribution of chromosomes to the daughter cells. Because of this redundancy, it appears likely then that more than one checkpoint may need to be lost in order for a cell to propagate a chromosome with an amplification. This may explain the instability of gene amplification events: Once the amplification event occurs, the cell must go through the cycle, passing each checkpoint before dividing. It is the balance between positive growth signals and the checkpoints that decides whether or not the amplification is propagated. In fact, overexpression of the myc oncogene enhances the frequency of amplification of the dihydrofolate reductase gene. This may occur by increasing the survival of such cells as a result of diminishing the remaining checkpoint controls.

It is clear from this discussion of checkpoint control over the cell cycle that the loss of a checkpoint control may not be lethal, but instead can lead to a loss of fidelity in the functioning of the cell cycle. This is precisely why the checkpoint controls may play a role in cancerous processes. Abnormal karyotypic alterations, gene amplifications, and segregation of chromosomes prior to repair processes provide a genomic instability for the selection of rare cancerous clones of cells to develop. A second process in the cell cycle can also result in genomic instability and the formation of abnormal cell types. When gene functions, which are essential to the orderly progression of cells through the cell cycle, are synthesized in abnormal levels it will commonly perturb the efficacy of the cell cycle. For example, histone synthesis is not considered a checkpoint event; rather it is essential for packaging of chromosomes in S, G_2, and M-phase and is clearly a structural element in this process. However, when the four essential core histones H2a, H2b, H3, and H4 are not synthesized in equimolar amounts during S-phase, there is a loss in the fidelity of mitotic chromosome transmission (Meeks-Wagner and Hartwell, 1986). Similarly, when the mitotic fidelity gene in yeast (MIF2) is synthesized in abnormally high levels, errors in chromosome segregation are increased and aneuploid cells are produced. These destabilizing events for the cell cycle are quite different from checkpoint functions. Checkpoint functions are dispensable (loss of a checkpoint is most often nonlethal to cells), act in a negative fashion to regulate downstream events, and enhance the fidelity of the progress of the cell cycle. Histones and MIF2 are essential genes (lethal if deleted), act as structural elements in essential cell cycle events (not as negative regulators of downstream events), but the result, disruption of the efficiency of cell cycle processes, is the same for histones, MIF2, and checkpoint functions. It is clear, however, that both the loss of checkpoint control or the abnormal expression levels of critical genes functioning in the cell cycle may contribute to cell cycle aberrations. One such critical gene is the myc oncogene, which is commonly expressed in abnormally high levels in tumor cells (as the result of chromosomal translocation promoter activation by integration of a viral long terminal repeat (LTR), or via gene amplification). Deregulated c-myc expression results in an inability to arrest the cell cycle even when growth factors or metabolites are depleted (Evan et al., 1992). Thus, activated oncogenes may override checkpoint controls.

How do these checkpoints act in the cell cycle to stop progression of the cells to the next stage of the cycle? While the complete answer to this is not yet available, it is clear that proteins called cyclins that activate catalytic subunits of protein kinases (p34 cdc2 kinases) play a critical role in providing essential functions in the cell cycle upon which several checkpoints act. Cyclins are proteins that were first detected (in sea urchins) and so named because they had the property of increasing in amount as the cell cycle progressed (from G_1 to M) and then were abruptly lost (degraded) during mitosis (Evans et al., 1983). This correlation between the increasing levels of cyclins and progress in the cell cycle was provocative and two different cyclin proteins were readily recognized and called cyclin A and B. At about the same time, Hartwell and his colleagues isolated a large number of mutants in yeast that were blocked in progression through the cell cycle. A central player in the G_1 block at start in *S. cerevisiae* (Lorincz and Reed, 1984) and the G_2 block before mitosis in *S. pombe* (Beach et al., 1982) was the same gene that encoded a 34 kd protein called cdc2 in *S. pombe* and cdc28 in *S. cerevisiae*. The cdc2/cdc28 (called cdc2 from now on) protein is the catalytic subunit of a protein kinase that phosphorylates serine residues in a serine–proline amino acid sequence, usually in the context of basic amino acid residues. The p34 cdc2 kinase subunit only functions when in a complex with an activating protein subunit, a cyclin. In *Xenopus*, cytoplasm from cells in mitosis contains an activity that will induce premature condensation of chromosomes in cells in S-phase (Masui and Market, 1971). This is most certainly a factor similar to that detected by Rao and Johnson (1970) while fusing S-phase to M-phase mammalian cells and obtaining premature condensation of chromosomes. When this factor was purified, it was composed of (at least in part) cyclin B and a p34 cdc2-like kinase (Hunt, 1992). Thus, from yeast through humans, cyclin B forms a tight protein complex with a p34 cdc2 kinase and when active sends a strong signal to enter M-phase and trigger mitosis. In *S. cerevisiae*, G_1 to S-phase transitions are regulated by three or four cyclin proteins that appear to be redundant and that bind to p34 cdc2 kinase and activate it. The function of this kinase is required for start in late G_1. Indeed, the functioning of the cyclin–p34 cdc2 kinase at late G_1 biochemically defines start, after which certain checkpoint controls no longer act to stop progress through the cell cycle. Both the mating pheromone alpha and the availability of nutrients (cell size) act as checkpoint signals by inhibiting the levels or activity of G_1 cyclins during this phase of the cell cycle. When these G_1 cyclins are not activated, no kinase activity is detected and start is not affected in the cycle. The process of regulating the level or activity of cyclins by environmental stimuli is complex, requiring a number of gene functions including (1) a receptor, (2) a G protein transducing the signal, (3) a series of protein kinases, (4) several transcription factors that regulate cyclin synthesis, and (5) some factors that regulate cyclins in a posttranslational mode. If a cell is too small to divide or a signal for sex attraction is received, there is no cyclin activation, no cdc2-like kinase activity, and no start of the cell cycle.

So, checkpoint controls act upon cyclins and cdc2 protein kinase activities, both in G_1 and G_2, where entry into S-phase and mitosis are regulated. Growth hormones, growth inhibitory signals, G proteins, protein kinases, phosphatases, proteases, and transcription factors all play a role in the signal transduction pathway between checkpoint pathways and essential events in the cell cycle. Several examples will illustrate this point. The cyclin–cdc2 complex is further regulated by phosphorylation of the cdc2 protein (and probably of the cyclin component as well). The cdc2 protein has two critical amino acid residues that regulate it by phosphorylation or dephosphorylation. Tyrosine-15 and threonine-161 in this protein only produce an active enzyme when tyr-15 is not phosphorylated and thr-161 is phosphorylated. The three other states, tyr-15 and thr-161 not phosphorylated, tyr-15 phosphorylated and thr-161 phosphorylated, or no phosphorylation at either site, all produce an inactive enzyme (Milarksi et al., 1991; Solomon et al., 1992). Several kinases (wee1, mik1) and phosphatases (cdc25, stf1) regulate the activity of the cyclin–cdc2 kinase via this posttranslational mechanism. The cdc25 phosphatase activates the cyclin B–p34 cdc2 kinase activity to trigger mitosis or premature condensation of chromosomes.

Late in mitosis, in anaphase–telephase, cyclin B is rapidly degraded by a ubiquitin-mediated pathway of proteolysis (Hunt, 1991; Solomon et al., 1991) returning cyclin levels to zero in G_0/G_1 and inactivating the kinase. Indeed, this event is required for entry into G_1 from M-phase so that the loss of cyclin B and/or kinase activity moves cells from M to G_0/G_1.

The genetics of the cell cycle in yeast and the biochemistry of the *Xenopus* egg have combined to provide an outline of events in the cell cycle that are both essential to progression and act as checkpoints of progression through the cell cycle. How does this apply to mammalian cells and the understanding of the origins of cancer? First, it should come as no surprise that the functions of proteins in the signal transduction pathways that link checkpoint modules with the cell cycle, i.e., growth hormones, inhibitors of growth, receptors, G proteins, protein kinases, phosphatases, proteases, and transcription factors, are the very same functions that oncogenes possess (see Chapters 7, 8, 9, this volume). Indeed, a novel cyclin (PRAD1) has been implicated to be an oncogene based upon its amplification and clonal rearrangement in several cancers: parathyroid tumors, B lymphoid, breast, and epidermoid tumors (Arnold et al., 1991). The cyclin A gene has been found once at an integration site of hepatitis B virus DNA in a hepatocellular carcinoma (Wang et al., 1990).

On the other hand, the tumor suppressor genes (Chapter 10, this volume) behave more like the checkpoint control mechanisms that supervise the various steps in the cell cycle. Checkpoint control gene functions have three properties in common: (1) They are dispensable; that is, a loss of function mutation is not lethal, (2) they act to negatively regulate progression through the cell cycle, and (3) when they fail to function, it results in a loss of fidelity in the cell cycle; that is, a higher mutation rate occurs and is imparted to the progeny cells. Both the retinoblastoma susceptibility gene (Rb) and p53 are examples of tumor suppressor genes (see Chapter 10, this volume). It is quite clear that loss of function for either Rb or p53 is not essential for the life of the cell. Both tumors and cell lines derived from these tumors can have deletion mutations or gene rearrangements that destroy all functions of these genes (see Levine, 1992). It is even clear that normal cells and normal cell divisions can proceed in the absence of Rb and p53. Mice that have insertional mutations that destroy both alleles of Rb or p53 (Donehower et al., 1992) have been produced and many generations of normal cell divisions have been documented in these mice. What is clear is that loss of the Rb gene in humans gives rise to retinoblastomas and predisposes to osteosarcomas at a very high rate (Weinberg, 1992). Loss of the p53 gene in mice (Donehower et al., 1992) and in humans (Malkin et al., 1990) also predisposes the host to cancers at a very high rate. Clearly, failure to express the Rb or p53 gene products is not lethal to the cell but leads to events that result in cancer. Indeed, p53 has yet another similarity to a checkpoint function, the RAD9 gene function. Gamma irradiation of cells with normal wild-type p53 protein results in a pause or block in G_1 prior to entry into S-phase. Cells with no p53 proteins fail to pause in G_1, enter S-phase, and die at much higher rates than cells with normal p53 proteins (Kastan et al., 1991). DNA damage by ultraviolet or gamma irradiation induces high levels of p53 by some posttranslational mechanism that appears to stabilize an otherwise unstable protein (Kastan et al., 1991; Maltzman and Czyzyk, 1984). Indeed, the overexpression of wild-type p53 blocks the cell cycle in late G_1 (Martinez et al., 1991; Michalowitz et al., 1990). Thus, it appears that p53 could act as a late G_1 checkpoint, making sure that the DNA template is intact and ready for normal DNA replication in S-phase. If the DNA is damaged, the cells are blocked in G_1 (at the restriction point) until proper repair occurs, and entry into S does not result in abnormal progeny cells that may contribute to clones of cancerous tissues.

Similarly, the Rb protein appears to negatively regulate a transcription factor E2F (see Chapter 10, this volume) that in turn plays a role in regulating enzyme activities that contribute to nucleoside triphosphate synthesis and DNA synthesis (Nevins et al., 1991). In the absence of Rb, gene functions that are essential to the late G_1 and S-phase cycle events are not in the proper stoichiometric balance and these enhanced levels of enzyme activities (and/or other gene functions) would contribute to the inefficient processing of cell cycle events.

It is unfortunately already clear that this outline of the events in the cell cycle and its relationship with oncogenes and tumor suppressor genes (see Fig. 1) is too simplistic and incomplete (see Hunt, 1991). Several diverse cyclins have now been identified from mammalian sources (cyclins A, B_1, B_2, C_1, D, and E) (Reed et al., 1991) and appear to function in G_1 (cyclin C at the restriction point) and G_1/S (cyclin E at the G_1/S border) (Reed et al., 1991) as well as cyclins that may be cell type or tissue specific. Cyclins that are stimulated in macrophages responding to colony-stimulating factor-1 (CSF-1) have expanded the list of G_1 cyclins in mammalian cells (Matsushime et al., 1991). Perhaps equally complex is a rapidly growing family of p34 cdc2-like kinase subunits, wherein more than 30 different gene products have now been characterized (Meyerson et al., 1992). This large number of cyclins and p34 cdc2-like kinases gives rise to the possibility of combinatorial regulation of enzyme activities using these two different subunits to create a large number of diverse enzyme activities. It is clear that the same cyclin can activate several different p34 cdc2-like kinase subunits and the same kinase subunit can be activated by different cyclins. In this way a small group of proteins can, via combinatorial interactions, regulate a large number of different targets in a specific fashion. Similarly, a single change in one cyclin molecule (such as a posttranslational modification) can now have a large and diverse effect when acting through several different p34 cdc2-like kinase subunits. Single events can result in immediate and diverse activities. It remains to be seen if different combinations of cyclins and p34 cdc2 kinases have different specificities or activities. To test this, we must understand and obtain the substrates for these molecules, and they are largely not yet available.

The subsequent chapters in this section deal in more detail with proto-oncogenes, tumor suppressor genes, and the oncogenes of DNA tumor viruses and bring the focus back to events closer to causing cancer or transforming cells (see Chapters 7–12, this volume). These events, however, must be played out in the cell cycle with its essential functions and checkpoints. Only by combining a fundamental understanding of the cell cycle with genes and products that lead to the pathogenesis and epidemiology of this disease can one hope to apply a rational basis for prevention or cures of cancer. It all starts with the cell cycle.

REFERENCES

Arnold, A., T. Motokur, T. Bloom, et al. (1991) The putative oncogene PRAD1 encodes a novel cyclin. In: D. Beach, B. Stillman, and J.D. Watson (eds.). The Cell Cycle. Cold Spring Harbor Symp. Quant. Biol. 56:93–97.

Beach, D., B. Durkacz, and P. Nurse (1982) Functionally homologous cell cycle control genes in budding and fission yeast. Nature 300:706–709.

Beach, D., B. Stillman, and J.D. Watson (eds.) (1991) The Cell Cycle. Cold Spring Harbor Symp. Quant. Biol., vol. 56. Cold Spring Harbor, N.Y.: Cold Spring Harbor Laboratory Press.

Brown, M., B. Garvik, L. Hartwell, et al. (1991) Fidelity of mitotic chromosome transmission. In: D. Beach, B. Stillman, and J.D. Watson (eds.). The Cell Cycle. Cold Spring Harbor Symp. Quant. Biol. 56:359–365.

Donehower, L.A., M. Harvey, B.L. Slagle, et al. (1992) Mice deficient for p53 are developmentally normal but susceptible to spontaneous tumours. Nature 356:215–221.

Evan, E.I., A.H. Wyllie, C.S. Gilbert, et al. (1992) Induction of apoptosis in fibroblasts by c-myc protein. Cell 69:119–128.

Evans, T., E.T. Rosenthal, J. Youngblom, et al. (1983) Cyclin: A protein specified by maternal mRNA in sea urchin eggs that is destroyed at each cleavage division. Cell 33:389–396.

Hartwell, L.H., and T.A. Weinert (1989) Checkpoints: Controls that ensure the order of cell cycle events. Science 246:629–634.

Hunt, T. (1991) Summary: Put out more flags. In: D. Beach, B. Stillman, and J.D. Watson (eds.). The Cell Cycle. Cold Spring Harbor Symp. Quant. Biol. 56:757–769.

Kastan, M.B., O. Onyekwere, D. Sidransky, et al. (1991) Participation of p53 protein in the cellular response to DNA damage. Cancer Res. 51:6304–6311.

Lau, C., and A. Pardee (1982) Mechanism by which caffeine potentiates lethality of nitrogen mustard. Proc. Natl. Acad. Sci. USA 79:2942–2946.

Levine, A.J. (1992) The p53 tumour suppressor gene and gene product. In: A.J. Levine (ed.). Cancer Surveys: Tumour Suppressor Genes, the Cell Cycle and Cancer, vol. 12. Cold Spring Harbor, N.Y.: Cold Spring Harbor Laboratory Press, pp. 59–79.

Lorincz, A.T., and S.I. Reed (1984) Primary structure homology between the product of yeast division control gene CDC28 and vertebrate oncogenes. Nature 307:183–185.

Malkin, D., F.P. Li, L.C. Strong, et al. (1990) Germ line p53 mutations in a familial syndrome of breast cancer, sarcomas, and other neoplasms. Science 250:1233–1238.

Maltzman, W., and L. Czyzyk (1984) UV irradiation stimulates levels of p53 cellular tumor antigen in nontransformed mouse cells. Mol. Cell. Biol. 4:1689–1694.

Martinez, J., I. Georgoff, J. Martinez, et al. (1991) Cellular localization and cell cycle regulation by a temperature sensitive p53 protein. Genes Dev. 5:151–159.

Masui, H., and C.L. Market (1971) Cytoplasmic control of nuclear behavior during meiotic maturation of frog oocytes. J. Exp. Zool. 117:129–146.

Matsumoto, T., and D. Beach (1991) The spil GTPase interacts with RCC1 in cell cycle dependency. In: D. Beach, B. Stillman, and J.D. Watson (eds.). The Cell Cycle. Cold Spring Harbor Symp. Quant. Biol. 56:385–398.

Matsushime, H., M.F. Roussel, and C.J. Sherr (1991) Novel mammalian cyclins (CYL genes) expressed during G_1. In: D. Beach, B. Stillman, and J.D. Watson (eds.). The Cell Cycle. Cold Spring Harbor Symp. Quant. Biol. 56:69–74.

Meeks-Wagner, D., and L.H. Hartwell (1986) Normal stoichiometry of histone dimer sets is necessary for high fidelity of mitotic chromosome transmission. Cell 44:43–52.

Meyerson, M., B. Faha, L.-K. Su, et al. (1991) The cyclin-dependent kinase family. In: D.Beach, B. Stillman, and J.D. Watson (eds.). The Cell Cycle. Cold Spring Harbor Symp. Quant. Biol. 56:177–186.

Michalovitz, D., O. Halevy, and M. Oren (1990) Conditional inhibition of transformation and of cell proliferation by a temperature-sensitive mutant of p53. Cell 62:671–680.

Milarski, K.L., W.G. Dunphy, P. Russell, et al. (1991) Cloning and characterization of wild-type *Xenopus* cdc2, a component of MPF. In: D. Beach, B. Stillman, and J.D. Watson (eds.). The Cell Cycle. Cold Spring Harbor Symp. Quant. Biol. 56:377–384.

Nasmyth, K., L. Dirick, U. Surana, et al. (1992) Some facts and thoughts on cell cycle control in yeast. In: D. Beach, B. Stillman, and J.D. Watson (eds.). The Cell Cycle. Cold Spring Harbor Symp. Quant. Biol. 56:9–20.

Nevins, J.R., S.P. Chellappan, M. Mudryj, et al. (1991) E2F transcription factor is a target for the Rb protein and the cyclin A protein. In: Cold Spring Harbor Symp. Quant. Biol. 56:157–162.

Perry, M.E., M. Commane, and G.R. Stark (1992) SV40 large T antigen alone or two cooperating oncogenes convert REF52 cells to a state permissive for gene amplification. Proc. Natl. Acad. Sci. USA 89:8112-8116.

Rao, P., and R. Johnson (1970) Mammalian cell fusion: Studies on the regulation of DNA synthesis and mitosis. Nature 225:159–164.

Rao, P.N., B. Wilson, and T.T. Puck (1977) Premature chromosome condensation and cell cycle analysis. J. Cell. Physiol. 91:131–142.

Reed, S.I., C. Wittenberg, D.J. Lew, et al. (1992) G_1 control in yeast and animal cells. In: D. Beach, B. Stillman, and J.D. Watson (eds.). The Cell Cycle. Cold Spring Harbor Symp. Quant. Biol. 56:61–67.

Schimke, R.T., A.L. Kung, D.R. Rush, et al. (1992) Differences in mitotic control among mammalian cells. In: D. Beach, B. Stillman, and J.D. Watson (eds.). The Cell Cycle. Cold Spring Harbor Symp. Quant. Biol. 56:417–425.

Seino, H., H. Nishitani, T. Seki, et al. (1992) RCC1 is a nuclear protein required for coupling activation of cdc2 kinase with DNA synthesis and for start of the cell cycle. In: D. Beach, B. Stillman, and J.D. Watson (eds.). The Cell Cycle. Cold Spring Harbor Symp. Quant. Biol. 56:367–375.

Solomon, M.J., J. Gautier, T. Lee, et al. (1992) Control of $p34^{cdc2}$ activation. In: D. Beach, B. Stillman, and J.D. Watson (eds.). The Cell Cycle. Cold Spring Harbor Symp. Quant. Biol. 56:427–435.

Wang, J., X. Chenivesse, B. Henglein, et al. (1990) Hepatitis B virus integration in a cyclin A gene in a hepatocellular carcinoma. Nature 343:555–557.

Weinberg, R.A. (1992) The retinoblastoma gene and gene product. Cancer Surv. 12:43–57.

Chapter 7

Oncogenes

H. Earl Ruley

INTRODUCTION

The proliferation of metazoan cells is tightly regulated so that cell division matches the need for additional cells during embryonic development, tissue renewal, or increased demand for cellular function. Neoplasia reflects the emergence of cells that replicate relatively autonomously in relation to the surrounding tissue. The majority of neoplasms are clonal in origin (Nowell, 1989) and arise largely as a result of somatic mutations that cause cells to disregard the controls that normally limit proliferation (Pitot, 1986). Genes contributing to neoplastic disease are classified according to the means by which they influence cell phenotypes. Genes that act in a dominant manner have been termed oncogenes, whereas recessive genes (that is, genes that contribute to malignant change when functionally lost from cells) are designated tumor suppressor genes or anti-oncogenes (Marshall, 1991; Weinberg, 1991).

Oncogenes have been identified by virtue of their associations with oncogenic retroviruses (Bishop and Varmus, 1985), DNA sequences that transform cultured cells to a malignant state (Weinberg, 1981), and as genes associated with specific chromosome abnormalities in tumor cells (Solomon et al., 1991). All oncogenes are derived from normal cellular genes (termed proto-oncogenes) by mutations that unmask, or activate, their oncogenic potential. Activating mutations may affect the expression of a proto-oncogene such that high levels of the normal protein are inappropriately expressed. Alternatively, activating mutations affect coding sequences, thereby deregulating intrinsic biochemical activities of the encoded proteins. In either event, activating mutations typically circumvent controls that regulate the expression or activity of the proto-oncogene in normal cells. In addition, some activating mutations have dominant-negative effects in which the oncogenic protein blocks the function of the normal proto-oncogene protein.

Oncogenes were first discovered associated with infectious retroviruses capable of transmitting malignant disease from one animal to another (Teich et al., 1985). The oncogenic potential of the viruses generally resulted from rare events in which portions of a cellular proto-oncogene recombined with the virus and were activated by mutations and/or by expression of the gene under viral controls (Bishop and Varmus, 1985). The pathogenic potential of the oncogene is also enhanced by the ability of the virus to express the gene in a large number of cells. Cellular proto-oncogenes may also be activated by integration of nearby retroviruses that may disrupt expression and truncate the gene (Peters, 1990). Given the low probability of these events, the carcinogenic process requires longer than when the virus already carries an activated oncogene (Teich et al., 1985).

A limited number of oncogenes have been identified in association with DNA sequences capable of transforming cultured cells. However, many more oncogenes have been discovered experimentally, following *in vitro* genetic manipulations. Indeed, genes that at one time were only indirectly linked to malignant transformation are capable of transforming cells under the appropriate circumstances. These include genes similar to known oncogenes by virtue of sequence or function, and genes associated with provirus integration sites and chromosome translocations. For the purpose of this chapter, an oncogene is defined as any gene that can participate in the transformation of cells to a malignant state.

Most proto-oncogenes encode components of signal transduction pathways (Fig. 1) by which extracellular factors regulate cell growth, differentiation, and survival (Cantly et al., 1991; Lewin, 1991; Ruley and Schmidek, 1990). Table 1 lists viral and cellular oncogenes, grouped according to their likely roles in a generalized signal transduction pathway. These include growth factors, growth factor receptors, transducer molecules that propagate or amplify signals from surface receptors to the interior of the cell, and factors that regulate transcription.

Molecular Genetics of Nervous System Tumors, pages 89-100 ©1993 Wiley-Liss, Inc.

TABLE 1. Viral/Cellular Oncogenes

Gene	Discovery	Function or activity	Activation in human tumors
Growth factors			
sis	RC	B-chain PDGF	Increased expression
PDGF-A	H, EA	A-chain PDGF	Amplification
EGF	EA	Epidermal growth factor	Increased expresssion
TGF-α	H, EA	EGF-like growth factor	Increased expression
GM-CSF	EA	Granulocyte-macrophage CSF	
M-CSF	EA	Macrophage-CSF	
IL-2	EA	T-cell growth factor	
IL-3	IA, EA	Multi-CSF	
aFGF	EA	Acidic FGF	
bFGF	EA	Basic FGF	
hst/KFGF	TX	FGF-like growth factor	Amplification
int-2	IA, EX	FGF-like growth factor	Amplification
FGF-5	TX	FGF-like growth factor	
FGF-6	H, EA	FGF-like growth factor	
Tyrosine kinase receptors			
*erb*B	RC, IA	EGF receptor	Amplification
fms	RC	CSF-1 receptor	
neu	TX	EGF receptor-like	Amplification
kit	RC	Receptor for steel prod.	
ros	RC, TX	Like insulin receptor	Amplification
trk	TX	NGF receptor	
met	TX	HGF receptor	
ret	TX	Ligand unknown	
sea	RC	Ligand unknown	
eph	H, EA	Ligand unknown	
axl	TX	Ligand unknown	
Other receptors			
mas	TX	Angiotensin receptor	
EpoR	EA	Erythropoietin receptor	
mpl	RC	? related to GH receptor	
Nonreceptor tyrosine kinases			
src	RC		Increased expression
yes	RC	src-like kinase	
fgr	RC	src-like kinase	
fyn	H, EA	src-like kinase	
lck	IA, EA	src-like kinase	
fps/fes	RC		
abl	RC		Translocation
lsk/tck	TX		
slk	H, EA		
hsc	TX		
Serine/threonine protein kinases			
raf-1(mil)	RC, TX		
A-*raf*	H, EA		
B-*raf*	H, EA		
mos	RC, IA		
akt	RC		
cot	TX		
Guanine nucleotide-binding proteins			
Ha-*ras*	RC, TX		Point mutations
Ki-*ras*	RC, TX		Point mutations Increased expression
N-*ras*	TX		Point mutations

(continued)

TABLE 1. Continued.

Gene	Discovery	Function or activity	Activation in human tumors
Nuclear location			
c-*myc*	RC, IA	?	Translocation, amplification
N-*myc*	H, EA	?	Amplification
L-*myc*	H, EA	?	Amplification
myb	RC	Binds DNA	Amplification
fos	RC	A component of transcription factor AP-1	
*fos*B	H, EA	*fos*-related gene	
fra 2	H, EA	*fos*-related gene	
ski	RC	?	
p53[a]	EA	? binds SV40 T antigen	Point mutations, gene loss[a]
jun	RC	A component of transcription factor AP-1	
*erb*A[a]	RC	T3/T4 hormone receptor	
evi-1	IA, EA		Putative Zn^{2+} finger
pbx	CT	Hybrid E2A transcription factor	Translocation
cbl	RC	Transcription factor motifs	
rel[a]	RC, TX	Transcription factor sim. NFκB, *dorsal*	
ets-1	RC	Transcription factor	
ets-2	H,EA	Transcription factor	
vav	TX	Zn finger protein	
Other			
crk	RC	Modulates tyrosine kinases	
dbl	TX		
lca	TX		
int-1	IA, EA	A secreted factor	
Bcl-2	CT, EA	G-protein?	Translocation

RC, retrovirus capture; IA, insertional activation; TX, DNA transfection; H, homology to other oncogenes; EA, experimental activation; CT, tumor-specific chromosome translocation.
[a]Dominant negative oncogene functionally equivalent to a tumor suppressor gene.

GROWTH FACTORS

Unicellular organisms, such as bacteria and yeast, will grow in simple media containing only metabolic nutrients. In contrast, mammalian cells normally cannot divide unless informational molecules are present in the extracellular environment. These include small diffusible factors, cell-associated proteins and insoluble components of the extracellular matrix. Specificity of the growth signal is provided by receptors on the cell surface that bind extracellular ligands, often with extremely high affinities (dissociation constants exceeding 10^{-9} M), and initiate the biochemical changes necessary for cell cycle progression.

Malignant cells typically require fewer growth factors than normal cells. Nevertheless, the discovery that oncogenes could encode polypeptide growth factors was made unexpectedly, when the protein sequence of the B-chain of platelet-derived growth factor (PDGF) proved similar to an oncogene, v-*sis*, carried by the simian sarcoma virus. Subsequently, a significant number of growth factor oncogenes have been identified, including members of the fibroblast growth factor (FGF) family (Goldfarb, 1990), and a number of hemopoietic colony-stimulating factors (CSFs) (Nicola, 1989).

Growth factor oncogenes are activated almost exclusively by mutations that allow the factor to be expressed continuously in a responsive cell. The secreted factor stimulates cell proliferation in the absence of environmental factors by an autocrine process. Consequently, the oncogenic potential of many growth factors has been determined experimentally, by simply forcing expression of the relevant genes in cells expressing the appropriate receptors. Mutations affecting the protein sequence do not activate growth factor oncogenes because the affinities of growth factors for receptors are extremely high, and therefore most mutations would only interfere with the ability of the protein to function as a ligand. However, in the case of basic FGF, experimental activation required the addition of a signal

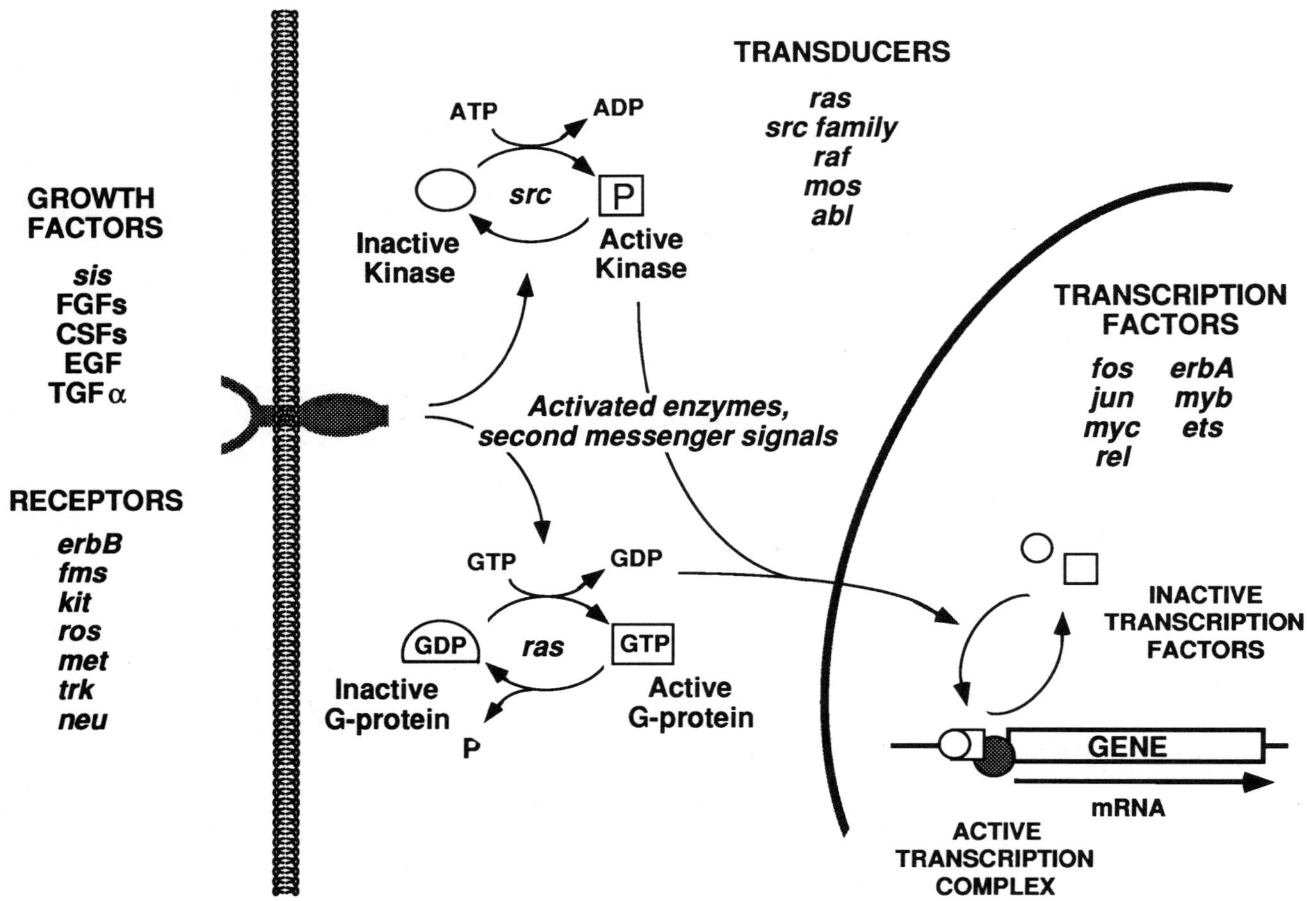

Fig. 1. Oncogenes and signal transduction. Most, and perhaps all, proto-oncogene proteins participate in the signal transduction pathways by which extracellular factors regulate cell growth and differentiation. Representative oncogenes are listed according to cellular location and function as growth factors, growth factor receptors, signal transducers, and transcription factors. (Adapted from Ruley and Schmidek, 1990.)

peptide in order for the protein to be secreted.

While many of the factors listed in Table 1 were discovered because of their ability to stimulate cell proliferation, most influence other phenotypes, including cell differentiation and survival (Burgess and Nicola, 1983; Cross and Dexter, 1991; Williams, 1991). However, for simplicity, our discussion of oncogenic factors and their receptors will emphasize cell growth phenotypes.

GROWTH FACTOR RECEPTORS

Growth factor receptors span the plasmid membrane such that binding of the extracellular factor elicits biochemical changes within cells that are required for cell proliferation (Aaronson, 1991). A large number of oncogenes are derived from genes that encode growth factor receptors (Table 1, Fig. 2). Many were initially discovered associated with oncogenic retroviruses and function enzymatically as protein tyrosine kinases (Ullrich and Schlessinger, 1990). The fact that many growth factor receptors and oncogenes are tyrosine kinases is significant since phosphotyrosine normally constitutes only a small fraction (0.1%) of the total protein phosphate. This suggests tyrosine phosphorylation is frequently reserved for pathways controlling cell growth and differentiation.

The receptors for a number of mitogenic cytokines lack intrinsic kinase activities. For example, several interleukins and hemopoietic CSFs bind proteins related to the growth hormone receptor (Arai et al., 1990), while vasoactive factors such as vasopressin and bombesin bind proteins related to the angiotensin receptor (Dohlman et al., 1991). Oncogenes derived from both types of receptor (the erythropoietin receptor and *mas*) have been identified (Fig. 2).

The first example of an oncogenic growth factor receptor was made unexpectedly when the protein sequence of the epidermal growth factor (EGF) receptor proved similar to an oncogene, *erb* B, carried by the avian erythroblastosis virus.

As with all receptor kinases, binding of the extracellular factor activates the tyrosine kinase domain located on the cytoplasmic side of the plasma membrane. Activating mutations uncouple the tyrosine kinase activity of the receptor from the binding of external ligand. Consequently, the mutated receptor stimulates cell proliferation in the absence of the appropriate environmental stimulus.

Cells expressing high levels of normal receptor proteins typically remain untransformed until exposed to the appropriate factor. However, even in the presence of factor, the cells are typically less transformed than cells expressing receptor oncogenes. One reason for this difference concerns mechanisms that normally down-regulate receptor functions. For example, once activated by factor binding, EGF and CSF-1 receptors are internalized and degraded; whereas mutationally activated v-ErbB and v-Fms persist on the cell surface.

Upon growth factor stimulation, several protein substrates become tightly associated with the PDGF and EGF receptors. These include proteins thought to participate in signal transduction: (1) an isoform of phospholipase C (PLCγ1), (2) a phosphatidylinositol 3'kinase (PI3K), (3) the product of the *Raf*-1 proto-oncogene, a serine threonine kinase related to protein kinase C, and (4) a protein (GAP) that activates the *ras* p21 GTPase.

Physical interactions between receptors and PLCγ1 and PI3K illustrate the role of inositol phospholipids in cell signaling (Rana and Hokin, 1990). PLCγ1 catalyzes the breakdown of phosphatidylinositol 4,5-bisphosphate to generate two second messenger metabolites, inositol trisphophate and diacylglycerol, which mobilize calcium and activate protein kinase C, respectively. PI3K is involved in the synthesis of inositol lipid precursors.

The interactions with Raf-1 and GAP illustrate how growth factor receptors may regulate proto-oncogenes involved in intracellular signaling. Raf-1 was initially identified as a retroviral oncogene. Oncogenic activation of *Raf*-1 (and two related genes, A-*Raf* and B-*Raf*) results from the loss of a negative regulatory domain located in the amino-terminus. The Raf-1 kinase can be metabolically activated by either tyrosine or serine phosphorylations, both as a direct and indirect result of growth factor stimulation (Rapp,

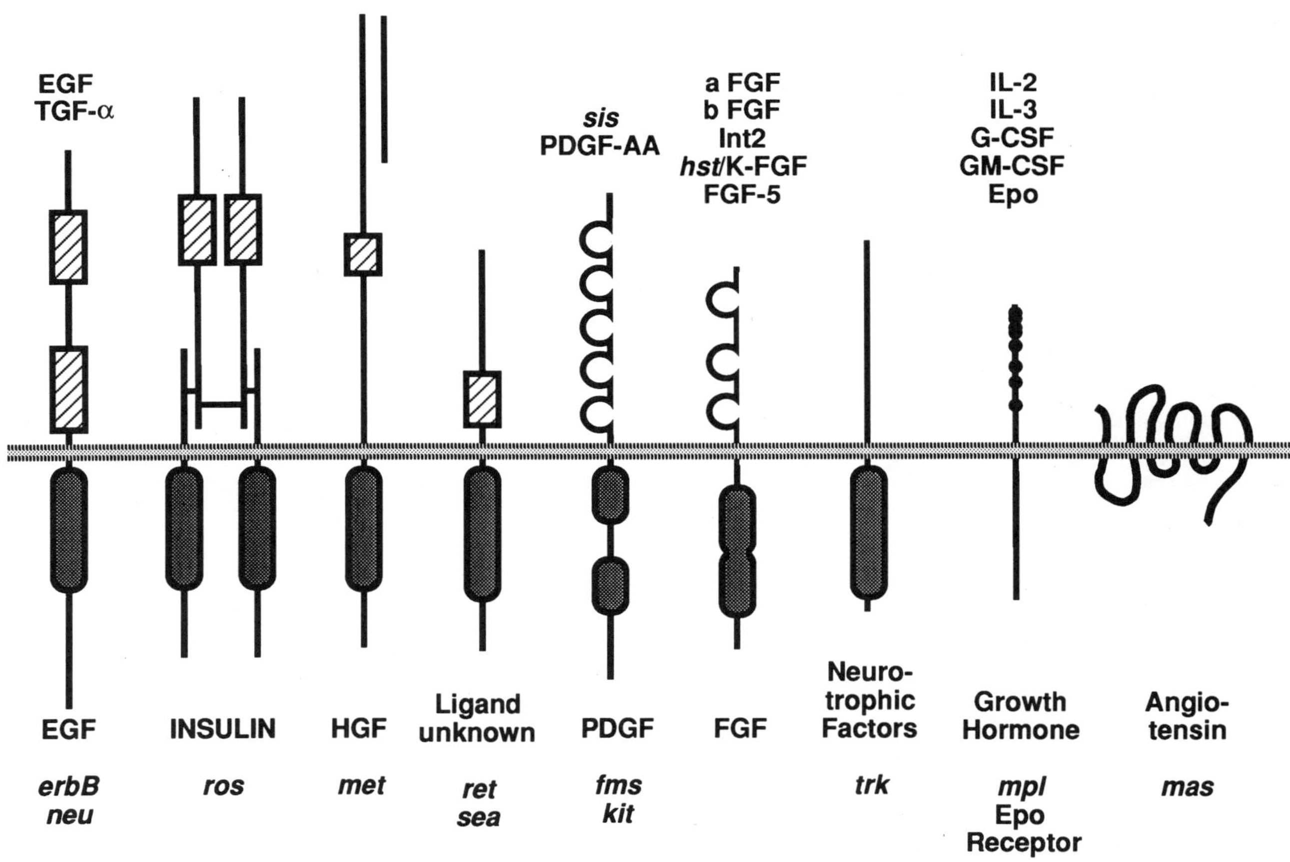

Fig. 2. Growth factor and growth factor receptor oncogenes. Examples of oncogenes derived from most of the known families of growth factors and their receptors have been identified, as shown. Structural features characteristic of different receptor types are shown as follows: tyrosine kinase domains, shaded ovals; cysteine-rich domains, hatched boxes; immunoglobinlike domains, open circles.

1991). As discussed later, GAP down-regulates proteins encoded by the *ras* family of proto-oncogenes (Bourne et al., 1991). Tyrosine phosphorylation apparently enables the GAP protein to complex with the receptor and with several other tyrosine phosphoproteins. This provides the best biochemical evidence linking Ras with activated tyrosine kinase receptors.

A number of mitogenic factors bind to membrane-spanning glycoproteins belonging to the growth hormone receptor superfamily (Aaronson, 1991). These include interleukins (IL-2, IL-4, IL-6, and IL-7) and hematopoietic colony-stimulating factors (G-CSF, GM-CSF, IL-3, and erythropoietin). Since these receptors lack tyrosine kinase or other enzymatic domains (Bazan, 1990; Nicola, 1989), they may interact with accessory kinases within the cell. For example, IL-2 activates Lck, an Src-related tyrosine kinase. This raises the possibility that other members of the growth hormone family of receptors may associate with different *src* family tyrosine kinases. The formation of high-affinity CSF receptors appears to require interactions between the ligand-binding subunit and other cell surface proteins (Nicola and Metcalf, 1991). Competition among receptors that bind different ligands for the same accessory protein may account for the negative interference observed among hemopoietic colony-stimulating factors. Similarly, the transforming activity of receptor oncogenes such as *mpl* could reflect the ability of the oncogenic protein to interfere with the function of related receptors.

Several circumstances illustrate the oncogenic potential of the growth hormone receptor family. Forced expression of IL-2, IL-3, G-CSF, and GM-CSF in the appropriate context can promote oncogenic transformation (Nicola, 1989). The ability of the spleen focus-forming virus to induce erythroleukemia requires the gratuitous activation of the erythropoietin receptor by the viral *env* gene product. The *mpl* retroviral oncogene is a truncated growth hormone-type receptor. Finally, experimentally selected mutations in the erythropoietin receptor promote the transformation of certain hematopoietic cell types (Longmore and Lodish, 1991).

MEMBRANE TRANSDUCERS: *SRC*-RELATED TYROSINE KINASES

A number of oncogene tyrosine kinases do not span the plasma membrane and therefore probably do not encode growth factor receptors (Cantly et al., 1991; Koch et al., 1991). These include members of the *src* gene family and *abl*. These enzymes are frequently located on the inner surface of the plasma membrane and interact with a variety of growth factor receptors. Notably, $p56^{lck}$ interacts with CD4 and CD8, two receptors in T cells that lack kinase domains (Eiseman and Bolen, 1990). In addition, the catalytic activity of these enzymes is sensitive to phosphorylation on both tyrosine and serine/threonine, suggesting they may participate in cascades of protein kinases involved in intracellular signaling.

Src is the best characterized of the nonreceptor tyrosine kinases. The metabolic regulation of the c-*src* protein ($pp60^{c\text{-}src}$) is complex, but illustrates molecular themes that are likely to be shared by other nonreceptor kinases. $pp60^{c\text{-}src}$ itself is substrate for both tyrosine and serine kinases. The tyrosine kinase activity of Src is regulated by phosphorylation and probably by interactions with a number of accessory proteins (Fig. 3). For example, phosphorylation of tyrosine 527 suppresses the activity of the enzyme, and as a result, mutations that eliminate tyrosine 527 frequently activate the oncogenic potential of $pp60^{src}$.

The amino-terminus contains two protein domains designated SH2 and SH3 (Src Homology 2 and 3). The SH2 sequence occurs in nonreceptor, cytoplasmic tyrosine kinases and in a number of other proteins, including PLCγ, PI3K, and GAP (Koch et al., 1991), while the SH3 domain is a common feature of proteins that interact with the cytoskeleton. SH2 domains bind with broad specificity to peptides containing phosphotyrosine. Consequently, a wide variety of protein–protein interactions (i.e., between kinase substrates and proteins containing SH2 domains) are regulated by tyrosine phosphorylation. These include the complexes discussed earlier, involving PLCγ, PI3K, GAP, and activated growth factor receptors.

The catalytic activities of tyrosine kinases are also regulated by SH2–phosphotyrosine interactions. For example, $pp60^{c\text{-}src}$ is oncogenically activated by mutations in the SH2 region of the protein. Such mutations presumably allow Src to escape negative controls exerted by other proteins and/or by carboxy-terminal sequences of Src containing Tyr527 (Fig. 3).

The regulatory potential of the SH2 domain is further illustrated by the protein encoded by the *crk* retroviral oncogene ($p47^{gag\text{-}crk}$). $p47^{gag\text{-}crk}$ consists of little more than viral *gag* sequences appended to a single SH2 and SH3 domain. Several observations suggest that Crk activates cellular tyrosine kinases by binding to regions containing phosphotyrosine. First, although Crk does not function enzymatically as a tyrosine kinase, *crk*-transformed cells express higher levels of tyrosine phosphorylated proteins. Second, Crk complexes with a number of tyrosine phosphorylated proteins *in vivo* and with $pp60^{v\text{-}src}$ *in vitro*. Finally, the SH2 region of the Crk protein is required for interactions between Crk and other proteins, activation of cellular tyrosine kinases, and cell transformation.

MEMBRANE TRANSDUCERS: *RAS*

The first oncogenes isolated from human tumors were related to the *ras* oncogenes captured by murine retroviruses (Barbacid, 1987). In mammalian cells, three *ras* genes

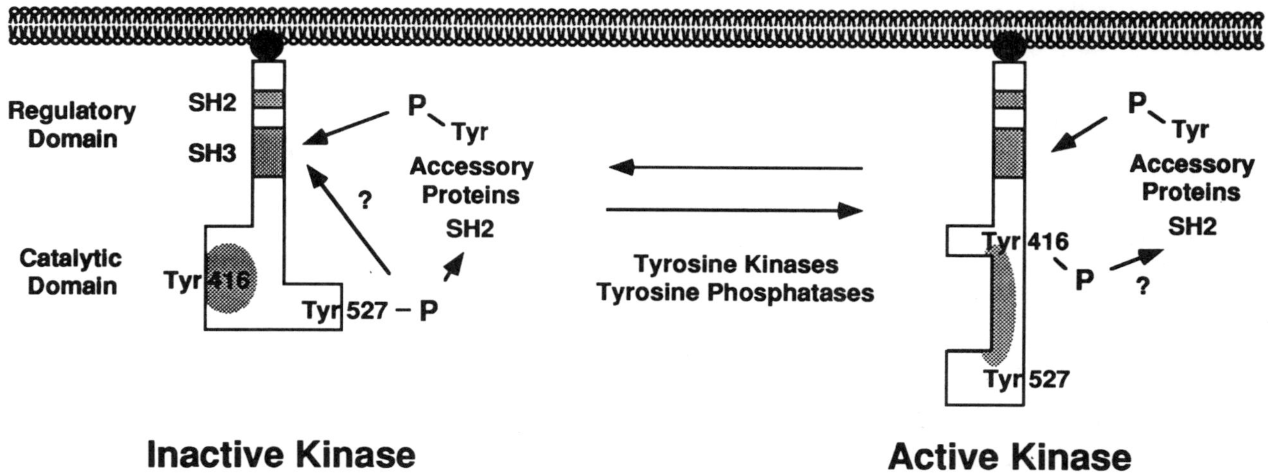

Fig. 3. Metabolic control of pp60^{src}. Multiple factors modulate the tyrosine kinase activity of pp60^{src}. These include other kinases, phosphatases, and accessory proteins. Phosphorylation of tyrosines (Tyr) 527 and 416 are associated with inactive and active kinases, respectively. The catalytic activities of Src are also regulated by interactions between SH2 domains and peptides containing phosphotyrosine.

encode four closely related proteins of 21,000 Daltons in size, designated p21s. The p21s are expressed in a large variety of cell types and are located on the inner surface of the plasma membrane. Several lines of evidence suggest that p21s transduce or amplify growth-promoting signals from cell surface receptors. First, oncogenic p21s stimulate cell proliferation even in the absence of external growth factors. Second, microinjected Ras antibodies block the ability of growth factors to stimulate DNA replication in quiescent cells. Finally, the association of GAP with activated growth factor receptors provides a direct biochemical connection to Ras.

While the biochemical functions of the p21s are unknown, the proteins appear to be regulated by guanine nucleotides, as shown in Figure 4. Thus, p21s are inactive when complexed with GDP, activated by the exchange of GTP for GDP and inactivated when the bound GTP is hydrolyzed. Such regulation is reminiscent of the G proteins that couple adenyl cyclase to adrenergic receptors.

The intrinsic rates of GTP/GDP exchange and hydrolysis associated with purified p21 are very low, suggesting that accessory factors may stimulate both activities in cells. Efforts to identify proteins that regulate *ras* p21 have assayed fractionated cell extracts for factors capable of enhancing the rates of GTP exchange or hydrolysis. One protein, designated GAP (GTPase activating protein), increases the rate of GTP hydrolysis of normal p21 by several hundred fold (Bourne et al., 1991). Significantly, most mutations that activate *ras* interfere with the ability of GAP to stimulate GTP hydrolysis. Consequently, the oncogenic proteins escape controls that otherwise limit Ras functions and stimulate cell proliferation in the absence of external signals.

In principle, exchange factors could act upstream of p21 and couple the protein to specific cell surface receptors. Unfortunately, while potential exchange factors have been purified, the biochemical and biological significance of these activities has not yet been established. Some activating mutations decrease the affinity of p21 for GDP, thereby shifting the equilibrium in favor of p21:GTP.

The gene encoding GAP has been cloned and characterized (Bourne et al., 1991). Nevertheless, the role of GAP in regulating Ras functions is not clear. High levels of GAP activity are found in a variety of tissues, and expression is not linked to proliferation. The protein is normally cytoplasmic, but as mentioned earlier, stimulation by growth factors such as PDGF results in the translocation of GAP to the plasma membrane. The translocated GAP is phosphorylated on tyrosine and forms a tight complex with receptors and other proteins involved in signal transduction. Since p21 is also located on the plasma membrane, translocation may allow GAP to switch off p21 activated by nucleotide exchange.

However, several studies argue against such a scheme, at least in all cell types, and have conflicting implications concerning the relative positions of Ras and GAP in intracellular signaling. Conversion of p21:GDT to p21:GTP in activated T cells does not result from any measurable increase in the rate of nucleotide exchange; rather, it is the inhibition of GAP activity that leads to the accumulation of p21 in an active, GTP-bound form. This suggests that GAP

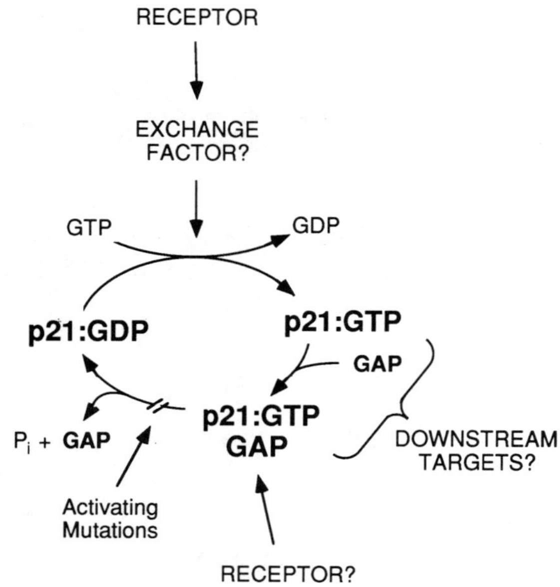

Fig. 4. Metabolic regulation of p21^{ras}. p21^{ras} is regulated by the binding and hydrolysis of guanine nucleotides. Exchange of GTP for GDP activates p21 by allowing the protein to interact with an as yet unknown downstream target(s). The rate of GTP hydrolysis is regulated by the GTPase activating protein (GAP). Activating mutations interfere with the ability of GAP to stimulate GTP hydrolysis. Receptors may activate p21^{ras} functions (i.e., promote the formation of Ras:GTP) by stimulating GTP exchange or by suppressing GAP. Activating mutations interfere with the ability of GAP to stimulate GTP hydrolysis.

may function upstream of p21. GAP also binds to the effector domain of p21, a region of the protein required for transformation. Thus, GAP may function as a downstream target, or may compete with downstream targets for binding to p21.

Clearly it will not be possible to understand how Ras functions until the biochemical targets regulated by Ras have been identified. The problem is complicated by the fact that Ras and GAP interact with related proteins expressed by members of their respective gene families (Bourne et al., 1991; Chardin, 1991; Hall, 1990). For example, Krev (Rap1A), a Ras-like G protein, suppresses phenotypic transformation induced by v-Ki-*ras*. Krev and v-Ki-Ras may have opposing effects on the same targets or compete with one another for an upstream activator. Ras also interacts with NF-1, a GAP-related tumor suppressor involved in von Recklinghaus neurofibromatosis. Ras has a greater affinity for NF-1 than for GAP, although the Ras GTPase is less sensitive to NF-1. Whether loss of NF-1 activates Ras is presently unknown.

Several of the Ras-related G proteins (and associated GAP activities) appear to be involved in secretion and endocytosis, in which GTP binding and hydrolysis provide directional specificity to vesicular transport. This raises the possibility that the function of p21^{ras} and/or GAP/receptor interactions concern problems of membrane trafficking (e.g., receptor-mediated endocytosis) rather than the production of second messenger signals. While it is not clear how regulatory changes affecting membrane traffic or organization could promote oncogenic transformation, processes such as membrane biogenesis and cycling are tightly linked to cell growth (van Meer, 1989).

TRANSCRIPTION FACTORS

Upon binding to cell surface receptors, growth factors affect many biochemical processes within cells; however, stable changes in cell phenotype associated with growth and differentiation ultimately depend on nuclear processes that regulate cellular gene expression. This suggests that the distal targets of signal transduction pathways will include transcription factors, proteins that regulate the transcription of genes by RNA polymerase.

Significantly, a number of proto-oncogene products are localized to the cell nucleus (Table 1) and appear to function as distal targets of signal transduction pathways (Lewin, 1991). This occurs in at least two different ways. First, transcription of several proto-oncogenes (for example, *fos*, c-*myc*, and *jun*) is induced in cells stimulated by growth factors (Pardee, 1989). The transcriptional regulation is important in growth control, since mutations that allow the genes to be expressed continuously in the absence of growth factors are often sufficient for oncogenic activation. Second, the proteins encoded by a variety of proto-oncogenes (e.g., *fos*, *jun*) function as transcription factors. These proteins bind DNA and control the transcription of growth factor-regulated genes.

That oncogenes such as *myc* and *myb* might encode transcription factors had been suspected given their nuclear localization and ability to bind DNA (Dang, 1991; DePinho et al., 1991; Luscher and Eisenman, 1990a,b). Furthermore, the Jun sequence revealed a distant relationship to the yeast transcription factor GCN4. However, conclusive evidence resulted from a biochemical analysis of AP-1, a DNA binding activity involved in the transcription of genes induced by tumor-promoting phorbol esters (Angel and Karin, 1991; Ransone and Verma, 1990).

As with many transcription factors, AP-1 was characterized by a series of experiments that included the identification of a DNA sequence (TGACTCA) that allow genes to respond to phorbol esters, and the purification of proteins that bound to the sequence. Subsequent experiments demonstrated that AP-1 preparations cross-reacted with v-Jun antibodies and contained the c-Jun protein. AP-1 preparations purified by DNA affinity chromatography contain second proto-oncogene protein, c-Fos. A sequence in both Fos and Jun, known as the leucine zipper, allows the proteins to form homodimers and heterodimers. Fos/Jun heterodimers and Jun homodimers bind the AP-1 site with much higher affinities than Fos homodimers.

How *fos* and *jun* regulate transcription and transform cells is presently unknown. Both genes are transcriptionally induced by growth factors. High levels of Fos and Jun accumulate, but only transiently, because the proteins and their mRNAs are rapidly degraded. Mutations that increase the expression of the normal proteins are sufficient for oncogenic activation, although loss of N-terminal sequences significantly enhances the transforming activities of v-Jun. Mutations in the zipper region that preclude dimer formation also abolish functions required for transcription activation and transformation. This suggests that v-Fos and v-Jun transform by constitutively activating the expression of genes normally induced by growth factors (Fig. 1).

c-Fos and c-Jun are members of larger gene families that include FosB, Fra1, Fra2, JunA, JunB, and JunD. All of the corresponding proteins pair with one another to varying extents via their leucine zipper regions, and the heterodimers bind DNA. Forced expression of FosB and Fra2 also transforms susceptible cells (Table 1); however, JunB prevents high levels of c-Jun from transforming.

AP-1 also binds DNA sites recognized by transcription factors responsive to cyclic AMP (CREB/ATF), and interacts with steroid hormone receptors and proteins encoded by the c-*ets* family of proto-oncogenes. These interactions are complex, and can have both positive and negative effects on gene expression. Indeed, given the enormous combinatorial complexity of the Fos- and Jun-related subunits, it is not clear how specific multimeric transcription complexes are assembled on different genes. Consequently, the targets of Jun and Fos involved in transformation may be difficult to identify.

c-*myc* belongs to a family of genes, including N-*myc* and L-*myc*, each of which is expressed in a tissue-specific manner (Dang, 1991; DePinho et al., 1991; Luscher and Eisenman, 1990a). *myc* levels frequently increase and decline during proliferation and differentiation, respectively, and the oncogenes are activated by genetic events (either natural or experimental) that permit constitutive gene expression. For example, Myc genes are frequently activated in human tumors by gene amplification and chromosome translocations.

Evidence that Myc proteins function as transcription factors has recently improved but is less compelling than in the case of Fos or Jun. The carboxy-terminus of the three proteins contain leucine-zipper and helix–loop–helix (HLH) sequence motifs. The HLH sequence is present in a large number of transcription factors, including E12/E47 and *MyoD*1, factors controlling immunoglobin gene expression and myocyte differentiation, respectively (Garrell and Campuzano, 1991; Murre et al., 1989; Weintraub et al., 1991). Like the leucine zipper, the HLH motif enables transcription factors to form homo- and heterodimeric DNA-binding proteins. The Myc leucine-zipper/HLH region is required for both oligomerization and transformation.

Drawing on the analogy of Fos and Jun, efforts to understand Myc functions have sought to identify partners that promote DNA binding. One such protein, Max, was isolated by screening cDNA expression libraries for proteins capable of binding the Myc HLH sequence. Max complexes with c-Myc, N-Myc, and L-Myc via a HLH domain. Max does not transform by itself but does augment transformation by Myc.

Cells transformed by *myc* frequently express *myc* RNA at levels exceeding those found in normal, proliferating cells. Similarly, with few exceptions (e.g., B-cell tumors), participation of *myc* in oncogenesis is associated with higher than physiological levels of expression. This implies that genetic events that activate *myc* do not simply circumvent controls that link expression of the proto-oncogene to growth-promoting stimuli. High levels of expression may allow *myc* to interact with lower-affinity DNA or protein targets, either to activate or to antagonize their functions. Unfortunately, if this reasoning is correct, methods to identify proteins or DNA sequences with the highest affinities for *myc* proteins may not uncover the actual cellular targets involved in transformation.

DOMINANT NEGATIVE ONCOGENES

Fos, Jun, and Myc illustrate a common characteristic of the nuclear oncoproteins, namely, each functions as members of multimeric DNA binding complexes. Such proteins are susceptible to dominant-negative mutations in which the mutant protein antagonizes the function of the wild-type protein (Herskowitz, 1987). For example, mutant and normal proteins may assemble into nonfunctional heteromeric complexes or compete for binding to DNA. Examples of dominant-negative oncogenes include v-*erb*A, p53, and v-*rel*. As with p53, such genes may also function as tumor suppressors, and contribute to malignant transformation when functionally lost from cells.

erbA is carried, along with *erbB*, by the avian erythroblastosis virus. While sequencing identified the oncogene as a member of the steroid hormone receptor family, biochemical studies identified c-ErbA as the thyroid hormone receptor (Guernsey and Fisher, 1990; Rosen et al., 1991). Both c-ErbA and v-ErbA bind DNA in the absence of hormone (triiodothyroxine). However, v-ErbA blocks the ability of c-ErbA to activate transcription in the presence of thyroid hormone. Expression of v-ErbA alone is not known to transform cells. However, v-ErbA collaborates with v-ErbB to induce erythroleukemia. Presumably, v-ErbA blocks the transcriptional activities of thyroid hormone that promote erythrocyte differentiation and preclude transformation by v-ErbB alone.

v-*rel* was discovered as an oncogene associated with the avian reticuloendotheliosis virus (Hannink and Temin, 1991). The c-*rel* gene is related to the DNA-binding subunit of NF-κB, a transcription factor initially characterized as an element controlling immunoglobin κ gene expression. However, NF-κB appears to be involved in the transcription of genes induced in response to a variety of external stimuli (Baeuerle, 1991). The DNA binding subunit of NF-κB is normally maintained in the cytoplasm by an inhibitor, IκB. Phosphorylation of IκB by cytoplasmic kinases (e.g., protein kinase C) releases the factor and allows transport into the nucleus. c-Rel and NF-κB belong to a larger gene family, members of which are transcriptionally induced by growth factors. Like AP-1, heteromeric complexes with NF-kB DNA binding activity are formed by different combinations of Rel-related proteins. Oncogenic activation of v-Rel can involve the loss of sequences that regulate translocation between the cytoplasm and nucleus. v-Rel blocks transcriptional activation by NF-κB, suggesting that the protein acts in a dominant-negative manner.

p53 was first identified as a protein associated with the transforming protein (the large T antigen) of simian virus 40, and later with transforming proteins expressed by human adenoviruses and papillomaviruses (Lane and Benchimol, 1989; Levine et al., 1991). The gene was initially classified as an oncogene because early molecular clones possessed *in vitro* transforming activities. However, with the discovery that the early clones inadvertently expressed mutant p53 proteins, this view has been revised. Subsequent studies suggest the transforming forms of p53 act in a dominant-negative manner, suppressing p53 functions in cells expressing both normal and mutant proteins. First, many tumors either lose p53 or express dominant transforming proteins. This suggests that loss of gene function mimics the effects of activating mutations. Second, the wild-type and mutant proteins form mixed (and presumably inactive) oligomers *in vivo*. Finally, wild-type p53 suppresses transformation mediated by a number of oncogenes, including mutationally activated p53. The biochemical functions of wild-type p53 are not known, but may involve cell cycle control and selective cell death (apoptosis) (Levine et al., 1991; Yonish-Rouach et al., 1991).

NATURAL HISTORY OF PROTO-ONCOGENE ACTIVATION

Several oncogenes have been independently activated in different circumstances. For example c-*myc* is transduced by the avian myelocytomatosis virus, associated with provirus integration sites and activated in human tumors by gene amplification and translocation. Related genes (N-myc and L-myc) have also been experimentally activated. This provides some assurance that oncogenes associated with experimental tumors in animals are relevant to our understanding of cancer in humans.

Nevertheless, many of the oncogenes listed in Table 1 are not frequently activated human tumors. The reasons for this concern differences in the circumstances that allow oncogene activation, some of which are listed in Table 2. For example, association of a proto-oncogene with a transmissible retrovirus allows the gene to accumulate multiple activating mutations, to be expressed at high levels under viral transcriptional controls and to be expressed in a large number of cells, including those predisposed to transformation. In

TABLE 2. Natural History of Proto-oncogene Activation

	Mechanism of activation		
Properties of activated oncogene	Retrovirus capture	Experimental activation	Somatic mutations
Number of cells expressing oncogene	Large, gene transduced by virus	Unlimited, e.g., transgenic animals	Limited to descendants of single cell
Opportunity for multiple mutations in the same gene	Yes, high mutation rates; many cycles of virus replication	Unlimited, site directed mutagenesis	None unless cell with mutation proliferates extensively
Level of oncogene expression	High, under control of virus LTR	High, wide variety of expression vectors	Natural promoter or translocation targets
Interactions with other oncogenes	Multiple genes carried by virus or activated in target cells	Multiple genes in target transferred or activated in target cells	Limited, additional genes activated during tumor progression

The table compares certain biological properties of oncogenes identified as (1) genes associated with oncogenic retroviruses, (2) genes whose oncogenic potential can be experimentally activated by *in vitro* genetic manipulations, and (3) genes altered by somatic mutations in natural tumors. As a result of such differences, many oncogenes identified using experimental animal models are not frequently activated in human tumors in naturally occurring tumors.

contrast, genes activated by somatic mutations are present in a single cell initially and subsequently in the descendants of that cell. This greatly reduces the probability that the gene will accumulate multiple mutations or will be expressed in cell along with other oncogenes. Furthermore, with the exception of translocated alleles, the mutated gene will be expressed from the gene's natural promoter.

It should be noted that individual oncogenes are seldom able to convert normal cells to a malignant state, particularly when expressed in a context different from that in which the oncogene was originally discovered (Hunter, 1991; Ruley, 1990; Weinberg, 1989). In some cases, conversion of normal cells to a malignant state requires two or more oncogenes acting in concert. For example, *ras* oncogenes fail to transform normal cells explanted from animal tissues unless cotransfected with a second oncogene such as *myc*. Similarly, erythroleukemia induced by AEV requires both v-ErbA and v-ErbB. Interactions between oncogenes are potentially significant with regard to the metabolic pathways discussed above and mechanisms important in multistep carcinogenesis.

REFERENCES

Aaronson, S.A. (1991) Growth factors and cancer. Science 254:1146–1153.

Angel, P., and M. Karin (1991) The role of Jun, Fos and the AP-1 complex in cell-proliferation and transformation. Biochim. Biophys. Acta 1072:129–157.

Arai, K.-I., F. Lee, A. Miyajima, et al. (1990) Cytokines: Coordinators of immune and inflammatory responses. Annu. Rev. Biochem. 59:783–836.

Baeuerle, P.A. (1991) The inducible transcription activator NF-kappa B: Regulation by distinct protein subunits. Biochim. Biophys. Acta 1072:63–80.

Barbacid, M. (1987) *ras* genes. Annu. Rev. Biochem. 56:779–827.

Bazan, J.F. (1990) Structural design and molecular evolution of a cytokine receptor superfamily. Proc. Natl. Acad. Sci. USA 87:6934–6938.

Bishop, J.M., and H. Varmus (1985) Functions and origins of retrovirual transforming genes. In: R. Weiss, N. Teich, H. Varmus, and J. Coffin (eds.). RNA Tumor Viruses, Molecular Biology of Tumor Viruses. Cold Spring Harbor, N.Y.: Cold Spring Harbor Laboratory, pp. 249–356.

Bourne, H.R., D.A. Sanders, and F. McCormick (1991) The GTPase superfamily: Conserved structure and molecular mechanism. Nature 349:117–127.

Burgess, A., and N. Nicola (1983) Growth Factors and Stem Cells. New York: Academic Press.

Cantly, L.C., K.R. Auger, C. Carpenter, et al. (1991) Oncogenes and signal transduction. Cell 64:281–302.

Chardin, P. (1991) Small GTP-binding proteins of the Ras family: a Conserved functional mechanism? Cancer Cells 3:117–126.

Cross, M., and T.M. Dexter (1991) Growth factors in development, transformation and tumorigenesis. Cell 64:271–280.

Dang, C.V. (1991) *c-myc* oncoprotein function. Biochim Biophys Acta 1072:103–113.

DePinho, R.A., A.N. Schreiber, and F.W. Alt (1991) *myc* family oncogenes in the development of normal and neoplastic cells. Adv. Cancer Res. 57:1–46.

Dohlman, H.G., J. Thorner, M.G. Caron, et al. (1991) Model systems for the study of seven transmembrane segment receptors. Annu. Rev. Biochem. 60:653–688.

Eiseman, E., and J. Bolen (1990) *src*-related tyrosine protein kinases as signaling components in hematopoietic cells. Cancer Cells 2:303–310.

Garrell, J., and S. Campuzano (1991) The helix-loop-helix domain: A common motif for bristles, muscles and sex. Bioessays 13:493–498.

Goldfarb, M. (1990) The fibroblast growth factor family. Cell Growth Diff. 1:439–445.

Guernsey, D.L., and P.B. Fisher (1990) Thyroid hormone and neoplastic transformation. Crit. Rev. Oncog. 1:389–408.

Hall, A. (1990) The cellular functions of small GTP-binding proteins. Science 249:635–640.

Hannink, M., and H.M. Temin (1991) Molecular mechanisms of transformation by the v-rel oncogene. Crit. Rev. Oncog. 2:293–309.

Herskowitz, I. (1987) Functional inactivation of genes by dominant negative mutations. Nature 329:219–222.

Hunter, T. (1991) Cooperation between oncogenes. Cell 64:249–270.

Koch, C.A., D. Anderson, M.F. Moran, et al. (1991) SH2 and SH3 domains: Elements that control interactions of cytoplasmic signaling proteins. Science 252:668–674.

Lane, D.P., and S. Benchimol (1989) p53 Oncogene or anti-oncogene? Genes Dev. 4:1–8.

Levine, A.J., J. Momand, and C.A. Finlay (1991) The p53 tumour suppressor gene. Nature 351:453–456.

Lewin, B. (1991) Oncogenic conversion by regulatory changes in transcription factors. Cell 64:303–312.

Longmore, G.D., and H.F. Lodish (1991) An activating mutation in the murine erythropoietin receptor induces erythroleukemia in mice: A cytokine receptor superfamily oncogene. Cell 67:1089–1102.

Luscher, B., and R.N. Eisenman (1990a) New light on Myc and Myb. I. Myc. Genes Dev. 4:2025–2035.

Luscher, B., and R.N. Eisenman (1990b) New light on Myc and Myb. II. Myb. Genes Dev. 4:2235–2241.

Marshall, C.J. (1991) Tumor suppressor genes. Cell 64:313–326. Murre, C., S.P. McCaw, and D. Baltimore (1989) A new DNA binding and dimerization motif in immunoglobin enhancer binding, *daughterless*, *Myo*D and *myc* proteins. Cell 56:777–783.

Nicola, N.A. (1989) Hemopoietic cell growth factors and their receptors. Annu. Rev. Biochem. 58:45–77.

Nicola, N., and D. Metcalf (1991) Subunit promiscuity among hemopoietic growth factor receptors. Cell 67:1–4.

Nowell, P.C. (1989) Chromosomal and molecular clues to tumor progression. Semin. Oncol. 2:116–127.

Pardee, A.B. (1989) G_1 events and regulation of cell proliferation. Science 246:603–608.

Peters, G. (1990) Oncogenes at viral integration sites. Cell Growth Diff. 1:503–510.

Pitot, H.C. (1986) Fundamentals of Oncology. New York: Marcel Dekker.

Rana, R.S., and L.E. Hokin (1990) Role of phosphoinositides in transmembrane signaling. Physiol. Rev. 70:115–162.

Ransone, L.J., and I.M. Verma (1990) Nuclear proto-oncogenes *FOS* and *JUN*. Annu. Rev. Cell Biol. 6:539–557.

Rapp, U.R. (1991) Role of Raf-1 serine/threonine protein kinase in growth factor signal transduction. Oncogene 6:495–500.

Rosen, E.D., A.L. O'Donnell, and R.J. Koenig (1991) Protein–protein interactions involving erbA superfamily receptors: Through the TRAPdoor. Mol. Cell Endocrinol. 78:C83–88.

Ruley, H.E. (1990) Transforming collaborations between *ras* and nuclear oncogenes. Cancer Cells 2:258–268.

Ruley, H.E., and H.H. Schmidek (1990) Oncogenes and nervous system tumors. In: R.H. Wilkins and S.S. Rengachary (eds.). Neurosurgery Update. New York: McGraw-Hill, pp. 226–232.

Solomon, E., J. Borrow, and A. Goddard (1991) Chromosome aberrations and cancer. Science 254:1153–1160.

Teich, N., J. Wyke, and P. Kaplan (1985) Pathogenesis of retrovirus-induced disease. In: R. Weiss, N. Teich, H. Varmus, and J. Coffin (eds.). RNA Tumor Viruses, Molecular Biology of Tumor Viruses. Cold Spring Harbor, N.Y.: Cold Spring Harbor Laboratory, pp. 187–248.

Ullrich, A., and J. Schlessinger (1990) Signal transduction by receptors with tyrosine kinase activity. Cell 61:203–212.

van Meer, G. (1989) Lipid traffic in animal cells. Annu. Rev. Cell Biol. 5:247–275.

Weinberg, R.A. (1981) Use of transfection to analyze genetic information and malignant transformation. Biochem. Biophys. Acta 651:25–35.

Weinberg, R. (1989) Oncogenes, antioncogenes, and the molecular bases of multistep carcinogenesis. Cancer Res. 49:3713–3721.

Weinberg, R.A. (1991) Tumor suppressor genes. Science 254:1138–1146.

Weintraub, H., R. Davis, S. Tapscott, et al. (1991) The *myo*D gene family: Nodal point during specification of the muscle cell lineage. Science 251:761–766.

Williams, G.T. (1991) Programmed cell deaths: Apoptosis and oncogenesis. Cell 65:1097–1098.

Yonish-Rouach, E., D. Resnitzky, J. Lotem, et al. (1991) Wild-type p53 induces apoptosis of myeloid leukaemic cells that is inhibited by interleukin-6. Nature 352:345–347.

Chapter 8

Growth Factors: The PDGF Paradigm

Belinda J. Wagner and Brent H. Cochran

INTRODUCTION

Growth factors were originally identified as proteins that induce cellular proliferation by specifically binding to high-affinity receptors expressed on the surface of cells. These proteins exert their effects on cellular proliferation by activating intracellular signaling mechanisms via their cognate receptors. More recently, it has been found that growth factors play a wide variety of roles in the growth, development, and maintenance of multicellular organisms. In early vertebrate development, growth factors are the key informational molecules that mediate early inductive events in the embryo. Although our understanding is incomplete, growth factors undoubtedly play key roles in coordinating the balance between the proliferation and differentiation throughout development.

In addition to their effects on proliferation and differentiation, growth factors are necessary for the survival of certain cell types and often serve as chemotractants and angiogenic factors. Moreover, growth factors play critical roles in wound healing and cellular activation of nondividing cells such as macrophages. To perform such diverse tasks, mammals have evolved a large and complex array of growth factors and receptors. The combination of lineage and spatially restricted expression of growth factors and their cognate receptors gives rise to the complex orchestration of developmental events.

Because of the multitude of important roles played by growth factors, understanding their function has become a key endeavor in modern biology. The understanding of growth factor action has been made all the more urgent by the findings that oncogenic mutations in tumor cells are most often found in genes that are in growth factor-regulated pathways. This first became apparent with the study of the retroviral oncogenes. Many highly oncogenic retroviruses can carry genes responsible for their tumorigenic potential. These oncogenes are usually mutated derivatives of normal cellular genes. The discovery that the oncogene of the simian sarcoma virus, v-*sis*, encodes a molecule that is essentially identical to the platelet-derived growth factor (PDGF) solidified the molecular link between growth factors and cancer (Doolittle et al., 1983; Waterfield et al., 1983). Other retroviral oncogenes have been shown to contain mutated forms of growth factor receptors, growth factor-regulated signal transduction molecules, and growth factor-regulated genes. The vast majority of known dominantly active retroviral oncogenes fall into one of the above categories. The probable role of many oncogenes in mediating signal transduction is diagrammed in Figure 1. Thus, a complete understanding of growth factor action is essential for understanding of cellular and developmental biology as well as cancer biology.

It is important to keep in mind that the term *growth factor* is often a misnomer and that many so-called growth factors have a variety of biological properties. Thus, for instance, the nerve growth factor—one of the earliest growth factors discovered—usually does not stimulate cellular proliferation at all, but rather promotes cell survival and neurite outgrowth. Moreover, a given growth factor may well have more than one function. The fibroblast growth factor (FGF) can be angiogenic as well as stimulate cellular proliferation (Folkman and Klagsbrun, 1987). A given growth factor can have different effects on different tissue types. The growth factor LIF can induce a differentiation of some myeloid cells, but prevents the differentiation of embryonic stem cells (Williams et al., 1988b). Also, a given growth factor can be expressed at different times and places during the life of an organism. Thus, the platelet-derived growth factor was originally discovered as a component of platelet α-granules that are released upon the wounding of the endothelium (Ross et al., 1986). This means of delivery ensures that the growth factor is delivered to the right place at the right time. More recent studies have shown that forms of the platelet-derived growth factor play important roles in the growth and differentiation of neuronal cells (Yeh et al., 1991).

Is it possible to understand such a complexity of response? The cellular response to growth factors is the product of the signals generated by the growth factor–

Molecular Genetics of Nervous System Tumors, pages 101–121

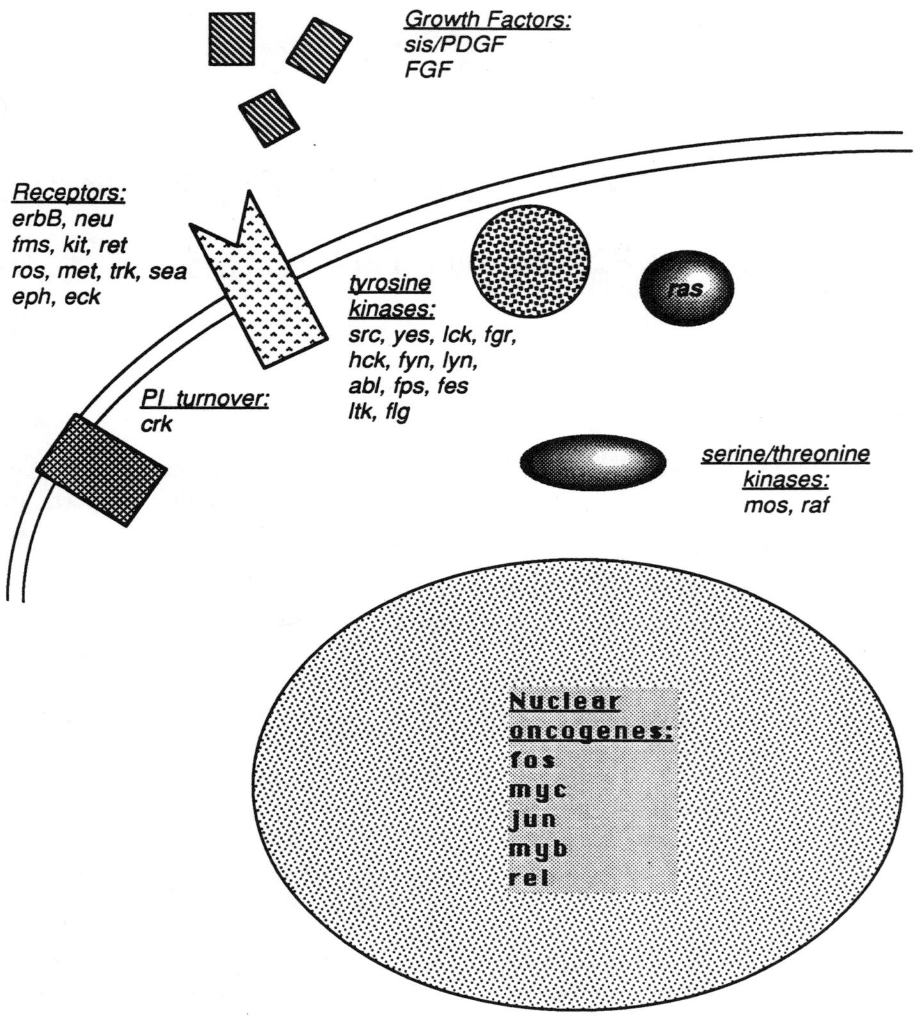

Fig. 1. Various oncogenes that show homology to cellular components of signal transduction mechanisms are shown. Growth factors and receptors can act at the cell surface or intracellularly within the secretory pathway. The phospholipase C-like oncogene *crk* is a membrane-bound protein. *ras* also acts at the cytoplasmic face of the cell membrane. Tyrosine kinases and serine/threonine kinases are cytoplasmic, but can undergo translocation to the cell membrane or the perinuclear region. Nuclear oncoproteins are thought to transmodulate gene expression (Bishop, 1987). This list is not comprehensive. See text for further explanation.

receptor interaction and the capability of a cell to respond to those signals. Fortunately, the major second signal pathways activated by growth factors appear to be fairly similar between different cell types. For instance, the *trk* gene family encodes receptors for several neurotropic factors, including NGF, BDNF, and NT-3, that are usually only expressed in the nervous system. However, when these genes are introduced into fibroblasts, the fibroblasts become responsive to neurotropic factors. Moreover, mutated forms of *trk* genes can oncogenically transform fibroblasts in culture. Thus, the tumorigenic phenotype need not be caused by genes that are usually expressed in the tissue of origin. Though the signaling pathways are often similar in different cell types, a single growth factor receptor does not necessarily activate all the pathways. Thus, the PDGF receptor stimulates phosphatidyl inositol turnover, whereas the CSF-1 receptor does not. Moreover, the cellular readout from a given signal can vary dramatically from one cell type to another. NGF will cause the pheocytomachroma cell line PC12 to stop growing and then differentiate in culture, yet will cause proliferation in fibroblasts that express the NGF receptor. By the same token, an activated *ras* oncogene can cause a proliferative response in fibroblasts, but drive differentiation in PC12 cells (Bar-Sagi and Feramisco, 1985). This difference in response must be due to the fact that the neuronal PC12 cells interpret the signal generated by the *ras* gene in a matter very differently than fibroblasts. It is important to note that most cells can respond to multiple growth factors and can do so in a synergistic manner. PDGF and the insulinlike growth factors have a synergistic proliferative effect on fibroblastic cells in culture (Stiles et al., 1979). This is also the case in neuronal cells, where it has been shown that NGF and FGF induce vastly different responses in neuronal stem cells in culture when given either singly or in combination (Cattaneo and McKay, 1990). The synergistic effects between two or more growth factors is commonly seen in many different cell types. This synergy is likely due to the fact that different receptors activate different signaling pathways, so that growth factor combinations activate more pathways simultaneously than does a given growth factor individually.

The remainder of this chapter reviews our current understanding of the mechanisms of action of the platelet-derived growth factor. The platelet-derived growth factor is perhaps the best understood growth factor in terms of its mechanism of action and serves as a paradigm for a variety of other growth factors, many of which work through related receptors. The platelet-derived growth factor is expressed in the developing brain and is inappropriately expressed in several tumors of the nervous system. The simian sarcoma virus encodes the v-*sis* oncogene, which makes a molecule nearly identical to the platelet-derived growth factor and causes gliomas when injected into the brains of newborn baboons (Doolittle et al., 1983). Thus, an understanding of the biology of PDGF serves as a solid foundation to the understanding of the mechanisms of actions of other growth factors and of the tumorigenic process.

PLATELET-DERIVED GROWTH FACTOR (PDGF)

Platelet-derived growth factor was originally identified as a human serum component that can stimulate the growth of glial cells, smooth muscle cells, and fibroblasts in the presence of plasma factors (see Ross et al., 1986, for review). PDGF is an important component in the repair of vascular injury. However, inappropriate expression of PDGF has been proposed to contribute to disease states such as atherosclerosis and cancer (Cochran, 1985; Ross et al., 1986). Normal human plasma contains undetectable amounts of PDGF and when exogenous PDGF is injected intravenously into baboons, it is rapidly cleared within minutes. This suggests that local concentrations of PDGF are more important than circulating levels. Therefore, the normal functions of PDGF are primarily determined by the relative anatomic location of cells capable of synthesizing and secreting PDGF and responder cells (Ross et al., 1986).

In addition to the mitogenic effects of PDGF on responder cells, it is likely that PDGF plays a role in development. PDGF is chemotactic for mesenchymal cells and may contribute to cell migration during embryogenesis as well as during wound repair (Martinet et al., 1986; Ross et al., 1986). PDGF-like activity is detected in embryonal carcinoma cells induced to differentiate (Gudas et al., 1983; Rizzino and Bowen-Pope, 1985). High levels of PDGF are found in cytotrophoblasts of the placenta and may contribute to autocrine stimulation of placental growth (Goustin et al., 1985).

Recently, a role for PDGF in oligodendrocyte development has been elucidated. In the developing central nervous system, two glial cell lineages are involved in cell–cell interactions that regulate the normal timing of oligodendrocyte differentiation. Type-1 astrocytes secrete a factor that drives proliferation of a second population of oligodendrocyte/type-2 astrocytes (O-2A) progenitor cells. In the absence of type-1 astrocytes or conditioned medium from these cells, O-2A cells stop dividing and undergo premature differentiation. The secreted factor important in regulating the normal timing of O-2A differentiation has recently been identified as PDGF (Noble et al., 1988; Raff et al., 1988).

PDGF from human platelets is a heterodimer composed of two related polypeptide chains, A and B (PDGF-AB) (Antoniades et al., 1979). Homodimeric forms PDGF-AA and PDGF-BB are also found in sera from other species or conditioned medium from certain cell lines (Bowen-Pope et al., 1989). The precursor proteins are about 40% homologous, whereas the mature chains are about 60% homologous (Betsholtz et al., 1986).

PDGF-B gene is the cellular homologue of the v-*sis* oncogene, the transforming protein of the simian sarcoma virus (SSV) (Doolittle et al., 1983). Constitutive expression of PDGF proteins in cells that express PDGF receptors is associated with transformation (Bowen-Pope et al., 1989; Owen et al., 1984). The v-*sis* gene may transform cells via an autostimulatory loop (autocrine) whereby a cell constitutively proliferates by producing a growth factor for which it also has a receptor. An intracellular, autocrine loop has been proposed as mediating transformation because of several criteria: (1) Both receptor and growth factor are synthesized and processed through the endoplasmic reticulum and Golgi apparatus; therefore the potential for intracellular activation exists, (2) intracellular, immature forms of v-*sis* are mitogenic (Lokeshwar et al., 1990), (3) both partially and fully processed forms of the receptor are phosphorylated on tyrosine in v-*sis*-transformed cells, whereas in normal cells tyrosine phosphorylation only occurs on the cell surface form of the receptor after ligand stimulation (Fleming et al., 1989; Keating and Williams, 1988), and (4) cells constitutively expressing v- or c-*sis* exhibit accelerated turnover of PDGF receptor, which cannot be mimicked by continual exposure to exogenous protein (Huang and Huang, 1988). However, the possibility exists that transient cell surface expression of activated receptor is required for functional coupling to signal transduction pathways (Hannink and Donoghue, 1988).

PDGF-AA, PDGF-AB, and PDGF-BB are proposed to have different biological activities (Nister et al., 1988). PDGF-AA and PDGF-BB differ in their transforming potentials. PDGF-A-expressing cells grow to somewhat higher saturation densities than control cells but do not form foci in monolayer culture or colonies in soft agar (Beckmann et al., 1988; Bywater et al., 1988). In contrast, cells transfected with a PDGF-B expression vector formed foci and grew to very high saturation densities in monolayer culture, formed large colonies in soft agar (Beckmann et al., 1988; Bywater et al., 1988), and were associated with anchorage-independent growth and transient, regressing tumor nodules when injected into nude mice (Stevens et al., 1988). Actin reorganization and chemotaxis in response to PDGF in human fibroblasts has been demonstrated with PDGF-BB, but not PDGF-AA (Kazlauskas et al., 1988; Nister et al., 1988). However, this result may be a consequence of PDGF receptor isoform expression in human fibroblasts and not a reflection of different biological activities of the PDGF isoforms (discussed below).

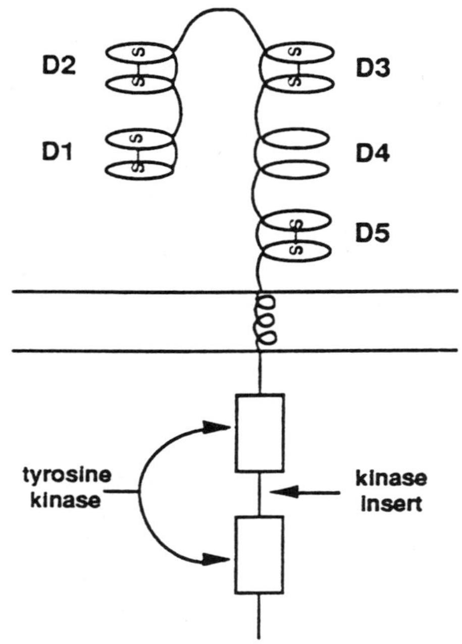

Fig. 2. The extracellular portion of the receptor consists of five immunoglobulinlike domains. Disulfide bonds are represented by S–S. The single transmembrane segment is probably an α-helix. The cytoplasmic portion contains sequences homologous to other tyrosine kinases (boxes). The kinase domain is interrupted by the kinase insert region. This same basic structure is shared by all the members of the PDGF receptor family.

PLATELET-DERIVED GROWTH FACTOR RECEPTOR

The PDGF receptor is a 180 to 190 kd membrane glycoprotein expressed on fibroblasts, smooth muscle cells, and glial cells. Its extracellular portion consists of five immunoglobulinlike domains (Yarden et al., 1986). The receptor has a single membrane-spanning segment. Like the insulin, epidermal growth factor (EGF), colony-stimulating factor-1 (CSF-1), and insulinlike growth factor-1 (IGF-1) receptors, the PDGF receptor has an intrinsic tyrosine kinase activity (Williams, 1989). The tyrosine kinase domain is homologous to the kinase domains of other transmembrane and intracellular tyrosine kinases. However, the coding sequence of the PDGF receptor tyrosine kinase domain is interrupted by a large region of about 100 amino acids. The presence of this kinase insert region defines a family of related proteins that include the PDGF receptor, the CSF-1 receptor, and the c-*kit* protein (Yarden et al., 1986, 1987). A schematic representation of the receptor is shown in Figure 2.

The binding of PDGF to its cognate receptor triggers numerous cellular responses culminating in DNA synthesis. Cellular responses can be categorized on the basis of their requirements for protein synthesis and transcription into immediate, early, and late events (Stiles, 1983). Immediate events require neither protein synthesis nor transcription. Early events do not require protein synthesis, but do require transcription. Late events require both. Full mitogenesis requires plasma growth factors in addition to PDGF. These responses are summarized in Table 1 and are discussed below.

There are two forms of the PDGF receptor. Binding studies revealed two populations of receptor based on their relative affinities for the three isoforms of PDGF (Hart et al., 1988; Heldin et al., 1988). One form of receptor exhibits equal affinity for all three isoforms of growth factor (Claesson-Welsh et al., 1989b; Hart et al., 1988; Heldin et al., 1988). The other form of receptor binds PDGF-BB with high affinity, PDGF-AB with lower affinity, and PDGF-AA only at very high concentrations (Escobedo et al., 1988c; Hart et al., 1988; Heldin et al., 1988). Subsequent cloning of genes encoding PDGF receptor proteins confirmed the existence of two receptor genes (Claesson-Welsh et al., 1988, 1989a; Matsui et al., 1989a). The α-PDGF receptor gene resides on chromosome 4 adjacent to the gene for c-*kit*. The gene encoding the α-PDGF receptor is found on chromosome 5 adjacent to that for the CSF-1 receptor. It is postulated that the adjacent genes arose by duplication, followed by a chromosomal duplication event that resulted in four copies of an ancestral gene (Matsui et al., 1989a).

Binding of PDGF to its receptor activates the intrinsic tyrosine kinase activity, causing receptor autophosphorylation (Bishayee et al., 1986) and phosphorylation of other cellular proteins (Kaplan et al., 1987; Roberts et al., 1988). Autophosphorylation results in a conformational change that may be important in signal transduction and the phosphorylation of exogenic substrates (Bishayee et al., 1988; Keating et al., 1988).

Studies using cloned expression vectors of the β receptor in Chinese hamster ovary cells have demonstrated that the intrinsic receptor tyrosine kinase activity is required for mitogenesis but not ligand binding. Therefore, although necessary for DNA synthesis, tyrosine kinase activity is not required for receptor down-regulation (Escobedo et al., 1988a; Escobedo and Williams, 1988).

The role of the kinase insert region in receptor-mediated signal transduction is less clear. Two groups report mostly similar results using receptor clones containing a deletion of the kinase insert region (Escobedo and Williams, 1988; Fantl et al., 1989; Severinsson et al., 1990). Both studies indicate that the kinase insert region is required to transmit a mitogenic response. Both deletion mutants retained high-affinity ligand binding and down- regulation activities. It is likely that this domain mediates binding of the receptor to the PI-3 kinase substrate (see below).

Recent evidence suggests that the functional form of the PDGF receptor is a dimer (Bishayee et al., 1989; Heldin et al., 1989). Ligand binding induces dimerization and stabilizes a receptor dimer complex (Heldin et al., 1989). Although transient dimerization of receptor molecules in the absence of ligand may occur, they are unstable. Overexpression of a PDGF-β receptor that lacks most of the cytoplasmic domain inhibits wild-type receptor functions (Veno et al., 1991), including autophosphorylation. This result strongly suggests that dimerization of wild-type receptors is required for signal transduction. Dimer formation appears to occur prior to receptor tyrosine kinase activation (Bishayee et al., 1989). Using purified membrane components in the absence of adenosine triphosphate (ATP), receptor dimerization occurs even though autophosphorylation was inhibited. In the presence of ^{32}P-ATP, binding and subsequent removal of ligand results in dissociation of receptor dimer complexes and the appearance of ^{32}P-labeled monomeric receptors.

Receptor dimerization or aggregate formation may be a general mechanism for transduction of an extracellular signal to the intracellular domain. Ligand-induced dimerization has been demonstrated for the EGF and insulin receptors. The formation of dimeric EGF and insulin receptor complexes has also been implicated in receptor kinase activation (Boni-Schnetzler et al., 1988; Cochet et al., 1988). Ligand binding could induce an intramolecular conformational change that alone activates receptor tyrosine kinase. Alternatively, ligand-induced changes in receptor conformation could alter the affinity of the receptor kinase for substrate proteins, including other receptor molecules. In this case, "autophosphorylation" would actually be the result of one receptor molecule intermolecularly phosphorylating another. The finding that PDGF-induced dimerization can occur independently of tyrosine kinase activity supports an intermolecular model for receptor tyrosine kinase activation. The different binding specificities between the α and β PDGF receptors and the three isoforms of PDGF growth factor have also been exploited to examine the role of dimerization in signal transduction. PDGF-AA can only bind to α receptors, therefore, it can be used to "clear" a cell surface of α receptors. In the absence of β receptors, responses to PDGF-AB are inhibited, while responses to PDGF-BB are only minimally affected (Hammacher et al., 1989; Seifert et al., 1989). Similarly, high concentrations of PDGF-AB can block responses to limiting amounts of PDGF-BB (Hammacher et al., 1989).

The dual agonist–antagonist role of PDGF supports the idea that the functional form of the receptor is a dimeric or aggregate complex. It also has led to a model of isoform-specific receptor complexes, as illustrated in Figure 3 (Hammacher et al., 1989; Seifert et al., 1989; Williams,

TABLE 1. Metabolic Responses to PDGF*

Immediate events: 1 to 10 min and transcription independent
Dimerization and autophosphorylation of PDGF receptor
Internalization and down-regulation of PDGF receptor
Activation of phospholipase C—increased phosphotidylinositol turnover
Tyrosine phosphorylation of cellular proteins: 85, 75, and 34 kd and others
Posttranslational modification of DNA binding proteins—SIF and others (?)
Inhibition of EGF binding
Phosphorylation of ribosomal protein S6
Elevation of cellular cyclic AMP (transient in fibroblasts)
Increase in diacylglycerol
Arachidonic acid and prostaglandin release
Reorganization of actin filaments
Stimulation of Na^+/H^+ exchange
Increased Na^+/K^+ ATPase activity
Release of Ca^{2+} from intracellular stores
Stimulation of polysome formation
Early events: 5 to 180 min and transcription dependent
Induction of immediate early gene transcription: c-*fos*, c-*myc*, c-*jun*, and others
Late events: 30 min to 6 hr and protein synthesis dependent
Down-regulation of immediate early gene transcription
Acquisition of intracellular mitogenic signal: competence
Appearance of new proteins in cytoplasm
Stimulation of somatomedin C binding
Increase in LDL receptors
Chemotaxis
DNA synthesis: 12 to 24 hr and requires progression factors

[8]Adapted from Cochran, 1985.

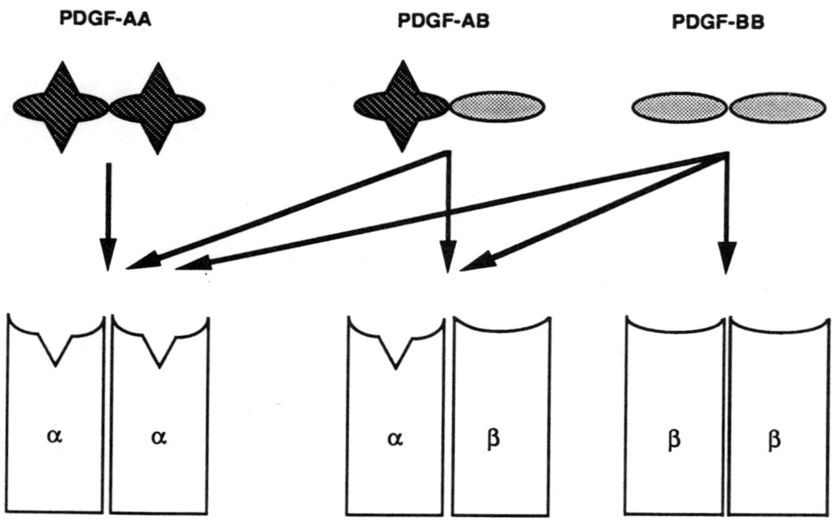

Fig. 3. PDGF receptor dimer model. PDGF A can only bind to the form of the receptor. PDGF B can bind to either isoform. The ability of each PDGF dimer to induce receptor dimerization is shown by arrows.

1989). In this model, PDGF-AA binds only to an α-α receptor dimer. PDGF-AB can bind to either an α-β or α-α receptor dimer. PDGF-BB is capable of binding to all three receptor isoform combinations. Thus, the ability of PDGF-AA or PDGF-AB isoforms to elicit a response would be dependent on the relative abundance of α and β receptor isoforms expressed on the cell surface. If α receptors were expressed at a low level relative to β receptors, then the probability that PDGF-AA could bind two receptors and form a functional dimer would be low. This is indeed the case in human fibroblasts, which express many more than receptors (Kazlauskas et al., 1988). This may explain the inability of PDGF-AA and the different abilities of PDGF-AB and PDGF-BB to mediate actin reorganization in these cells (Hammacher et al., 1989; Kazlauskas et al., 1988).

Alternatively, each receptor isoform could mediate distinct intracellular signals (Hammacher et al., 1989). Both receptor isoforms, when expressed in naive cells lacking either receptor, can couple to mitogenic pathways (Claesson-Welsh et al., 1988, 1989a; Escobedo et al., 1988b; Matsui et al., 1989a,b). It remains to be determined whether the α receptor couples to exactly the same constellation of second messenger pathways. The kinase insert regions of the α and β receptors are not highly conserved (Claesson-Welsh et al., 1989a). Therefore any potential substrate specificity mediated by the kinase insert region could differ between the two receptor isoforms.

Several growth factor receptors have been introduced into cells that do not normally express them and have been shown to functionally couple with endogenous signaling pathways (Roussel and Sherr, 1989). These findings reinforce the idea that cellular responses to growth factors are mediated by near-ubiquitous signal transduction mechanisms. The ability of a cell to respond to a particular growth factor is primarily dependent only upon expression of the appropriate receptor. A given signaling pathway activated by a particular receptor would be expected to be similarly activated by other growth factor receptors. However, ligand binding activates multiple pathways and a particular receptor is likely to activate a distinct combination of second messenger pathways. Identifying components of pathways involved in the transmission of extracellular signals to the nucleus remains an active area of inquiry.

SECOND MESSENGER PATHWAYS

From Table 1 it is clear that the number of metabolic events stimulated by PDGF is quite large. Sorting out the mechanisms by which these events are regulated and in what order they occur is clearly a formidable task. It is readily apparent that multiple second messenger pathways must be activated in order to generate such a pleiotropic response. To begin to understand these events, it would be helpful to know which events are regulated as a direct consequence of phosphorylation by the PDGF receptor itself. Fortunately, we have been helped in this endeavor by the fortuitous finding that several key receptor substrates actually bind to the autophosphorylated receptor with high enough affinity to form a stable complex. These molecules are almost certainly direct receptor substrates and from what is known of their function, they initiate key signaling pathways. These substrates and their pathways are diagrammed in Figure 4 and discussed below.

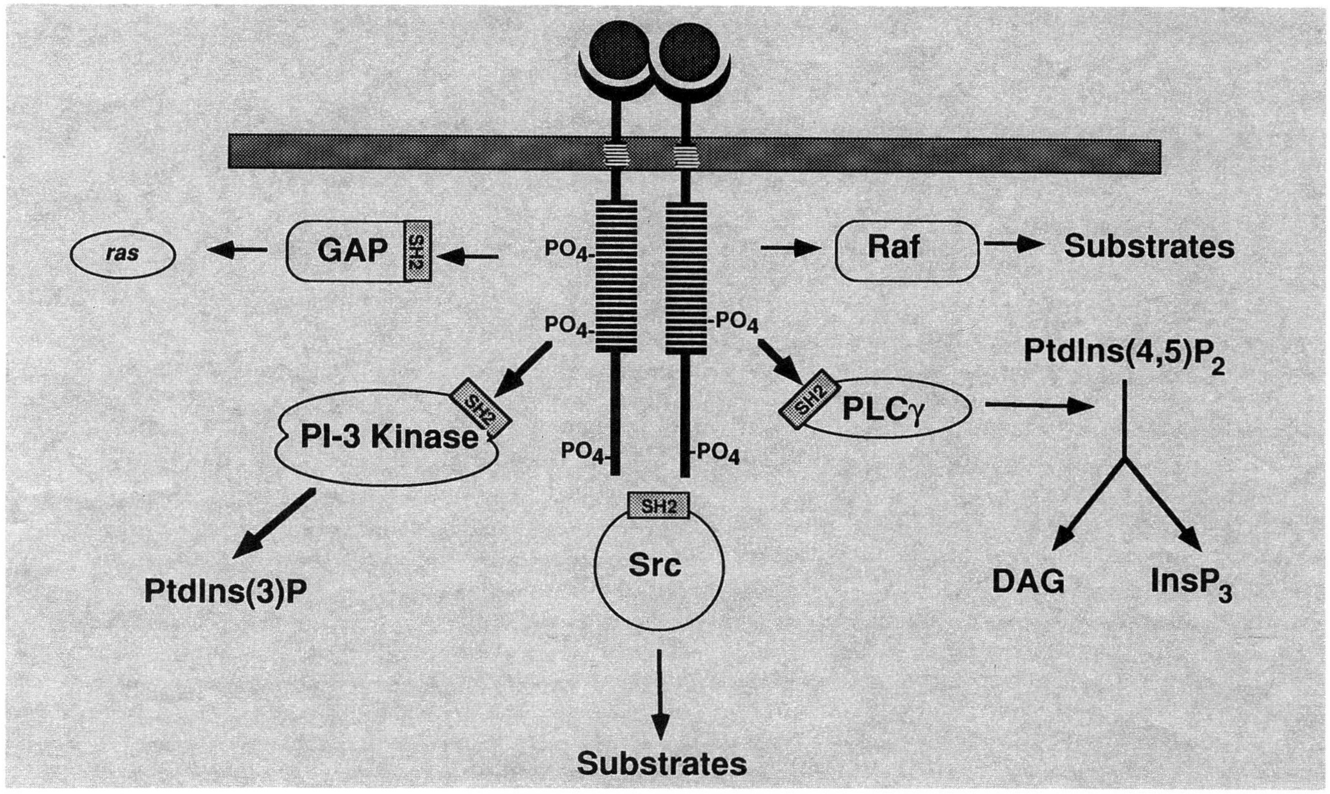

Fig. 4. PDGF receptor substrates.

Substrates of PDGF Receptor Protein Tyrosine Kinase

Phospholipase C-γ

Phosphoinositides reside in the cell membrane and constitute storage forms of second messenger molecules that transmit extracellular signals across the membrane. The breakdown of phosphoinositides to diacylglycerol (DAG) and various inositol phosphates is catalyzed by the phospholipase C family of enzymes. Phospholipase C enzymes are largely cytosolic, but are regulated by membrane lipids and are inactive unless bound to a lipid bilayer that contains appropriate substrate molecules (Majerus et al., 1986).

PDGF rapidly stimulates tyrosine phosphorylation of phospholipase C-γ in BALB/c 3T3 cells. Tyrosine phosphorylation of PLC-γ correlates with increased enzymatic activity (>50-fold) (Wahl et al., 1989). PLC-γ is likely to be a substrate for PDGF receptor protein kinase activity because antibodies to PDGF receptor coprecipitate PLC-γ (Kumjian et al., 1989). Ligand occupancy of the receptor is required for coprecipitation. PDGF receptor-mediated autophosphorylation occurs in response to ligand binding and is accompanied by a conformational change in the receptor (Bishayee et al., 1988). Removal of ligand by acid wash at 4°C results in tyrosine dephosphorylation within 10 minutes and a concomitant dissociation of PDGF receptor and PLC-γ. Therefore, interaction between PDGF receptor and PLC-γ may be dependent upon the conformation of the PDGF receptor (Kumjian et al., 1989).

Turnover of phosphatidylinositol 4,5-bisphosphate (PIP_2) likely plays a direct role in cell proliferation. Microinjection of antibody to PIP_2 abolishes thymidine uptake induced by PDGF and bombesin in NIH 3T3 fibroblasts (Matuoka et al., 1988). This same antibody blocks cell growth in yeast. Mutants of yeast that are more sensitive to the antibody than wild-type cells have alterations in phosphoinositide metabolism (Uno et al., 1988).

The cleavage of PIP_2 by phospholipase C generates two well-characterized second messenger molecules. One is inositol-(1,4,5)-trisphosphate, which releases calcium from intracellular stores. Calcium, in turn, regulates a variety of enzymatic pathways. The other product of phospholipase C action is diacylglycerol, which is the endogenous form of tumor-promoting phorbol esters. Diacylglycerol in turn activates protein kinase C.

Protein kinase C

Protein kinase C (PKC) is a calcium- and lipid-regulated family of serine/threonine protein kinases that are the endogenous receptors of tumor promoter phorbol esters. At least seven members of the family have been identified and cloned. Two members (βI and βII) appear to originate from a single gene by alternative splicing resulting in a difference at their carboxy-terminal ends. PKC α is expressed ubiquitously in all tissues. Most tissues have either βI or βII. PKC γ is found only in brain tissues. Several other members of the PKC family appear to have some tissue specificity. Each protein also exhibits subtle differences in regulation by diacylglycerol, Ca^{2+}, and phorbol esters (Kikkawa et al., 1988; Nishizuka, 1988).

PKC proteins are segregated into two subgroups by their structures. α, βI, βII, and γ have four conserved domains, C_1–C_4. C_1 contains a sequence similar to the zinc finger regions of many metalloproteins and DNA-binding proteins (Berg, 1986). C_3 and C_4 comprise regions with homology to other protein kinases and C_3 contains the ATP-binding site. PKC δ, ε, and ζ are very similar to the other four, but lack a C_2 region.

Diacylglycerol (DAG) activates PKC by lowering the Ca^{2+} requirement for enzymatic activity. Phorbol esters can mimic the action of DAG and directly activate PKC. When activated, PKC associates with the inner surface of the plasma membrane (Hunter et al., 1988). PKC can also be activated via limited proteolysis by the Ca^{2+}-dependent protease, calpain. Proteolysis leads to rapid removal of PKC from the cell and may be the basis for PKC down-regulation by prolonged treatment with phorbol esters (Nishizuka, 1988).

Phorbol esters have been useful in demonstrating the role of protein kinase C in altering gene expression. The tumor promoter 12-*O*-tetradecanoyl-phorbol-13-acetate (TPA) is an analogue of the 1,2-*sn* configuration of diacylglycerol, the physiological activator of PKC. This compound activates PKC without Ca^{2+} mobilization (Nishizuka, 1986). A number of genes are induced by TPA treatment, including c-*fos*, c-*myc*, and c-*jun* (Greenberg and Ziff, 1984; Kelly et al., 1983; Kruijer et al., 1984; Lamph et al., 1988; Muller et al., 1984a; Ryder et al., 1988).

In addition to the positive effects of PKC activation on gene expression, PKC participates in feedback regulation of other cell-signaling pathways, such as decreasing the inositol-(1,4,5)-trisphosphate-induced elevation of intracellular Ca^{2+} level. The epidermal growth factor (EGF) receptor and the catalytic subunit of adenylate cyclase are substrates for PKC. Phosphorylation of the EGF receptor results in decreased high-affinity EGF binding. Thus, PKC activation leads to functional down-regulation of the EGF receptor (Nishizuka, 1988). The effect of PKC activity on adenylate cyclase seems to be more cell-type specific. In some cells, TPA treatment leads to a decrease in receptor-mediated stimulation of adenylate cyclase. In others, such as frog erythrocytes, TPA treatment increases agonist-stimulated adenylate cyclase activity (Yoshimasa et al., 1987).

Phosphatidylinositol-3-kinase

Another protein that associates with the PDGF receptor is phosphatidylinositol (PtdIns)-3-kinase (Kaplan et al., 1987; Roberts et al., 1988). This kinase activity produces 1,3-PtdInsP rather than the classical 1,4-PtdInsP. PtdIns-3-kinase also phosphorylates 1,4-PtdInsP to produce 1,3,4-$PtdInsP_2$. These phosphoinositides are not cleavable by phospholipase C and could constitute novel second messengers (Kaplan et al., 1987; Majerus et al., 1986; Roberts et al., 1988; Whitman et al., 1988).

Studies using mutants of polyoma middle T that are defective in transforming activity suggest that PtdIns-3-kinase plays a role in growth regulation. PtdIns-3-kinase fails to associate with most nontransforming mutants of a middle T/pp$60^{c\text{-}src}$ complex (Fukui and Hanafusa, 1989; Whitman et al., 1985). However, some transformation-defective mutants of both middle T protein (Morgan et al., 1988) and pp$60^{v\text{-}src}$ (Fukui and Hanafusa, 1989) retain wild-type levels of PtdIns-3-kinase association and activity, suggesting that PtdIns-3-kinase activation alone is not sufficient for transformation (Varticovski et al., 1989).

Consistent with the hypothesis that PtdIns-3-kinase activity is important in transduction of the mitogenic response to PDGF, activated PDGF receptor directly associates with such an activity (Coughlin et al., 1989). PtdIns-3-kinase activity coprecipitates with PDGF receptor using antibodies to either extracellular or intracellular portions of the receptor. Immunoprecipitates from PDGF-stimulated 3T3 cell cultures contained approximately 10-fold more PtdIns-3-kinase activity than those from quiescent cultures. Additionally, CHO cells expressing mutant forms of the cloned β-PDGF receptor that are defective in stimulating cell division are unable to associate with PtdIns-3-kinase activity (Coughlin et al., 1989; Williams, 1989; Williams et al., 1988a). CSF-1 receptor also activates and associates with PtdIns-3-kinase (Varticovski et al., 1989), although this association may not be essential for mitogenesis by CSF-1 (Roussel et al., 1991). PtdIns-3 kinase is composed of at least two subunits, one with an 85 kd molecular weight and another with a 110 kd molecular weight. The 85 kd subunit has been cloned and found to have an SH2 (*src* homology) domain. SH2 domains appear to mediate specific binding of proteins to short peptides that contain phosphorylated tyrosine residues. Short peptides of defined sequence containing phosphorylated tyrosine residues can compete with PtdIns-3 kinase for binding to the receptor.

c-raf

The c-*raf* proto-oncogene is one member of a family of serine/threonine kinases. In humans, there are two other forms of *raf*: A-*raf* and B-*raf*. They share some homology to the protein kinase C family of serine/threonine kinases. The proteins can be divided into a zinc finger domain, an ATP binding site, and a kinase domain. Since *raf* proteins can be rendered transforming by truncation at the amino-terminal end, thus eliminating the zinc-finger region, it is postulated that activation of *raf* occurs by an unfolding mechanism that unmasks the ATP binding site and substrate binding site. Alternatively, protein cleavage could occur, leaving a labile but active enzymatic domain. Protein kinase C can be activated directly by diacylglycerols or phorbol esters. Similar "ligands" may exist for the *raf* proteins, but have not been found (Rapp et al., 1988a).

The c-*raf* protein is likely a substrate of the tyrosine kinase activity of the PDGF receptor (Morrison et al., 1989). PDGF-AB increases the phosphorylation of c-*raf* as detected by antiphosphotyrosine antibodies. Increased serine and threonine phosphorylation of c-*raf* is also observed in PDGF-AB stimulated cells when compared to quiescent BALB/c 3T3 cells (Roberts et al., 1988). c-*raf* phosphorylation is rapid and transient (maximal at 5 min with a half-life of approximately 2 hr). PDGF-induced phosphorylation of c-*raf* is independent of PKC activity as it is unaffected by TPA down-regulation. Exposure of cells to dibutyryl cAMP or insulin also fails to elicit c-*raf* phosphorylation (Roberts et al., 1988).

raf proteins appear to function as direct mediators of signal transduction by several criteria. v-*raf* (or v-*mos*) oncogenes can overcome conditional and permanent inactivation of *ras* (Rapp et al., 1988a,b). In primary macrophages, a combination of v-*raf* and v-*myc* abrogates CSF-1 dependence, suggesting that the CSF-1 receptor triggers at least two pathways that are both required for proliferation: a *myc* pathway and a *raf* pathway (Blasi et al., 1985).

Treatment of cells with either PDGF or TPA alters both the kinase activity and subcellular distribution of *raf*. In quiescent cells, *raf* exhibits a diffuse cytoplasmic distribution by immunofluorescence. Upon treatment with PDGF (maximal at 5 min) or TPA (maximal at 20 min), the distribution shifts to a perinuclear location (Rapp et al., 1988b).

In cells that express membrane oncogenes such as v-*src*, v-*sis*, v-*fms*, activated c-Ha-*ras*, or polyomavirus middle T antigen, c-*raf* displays altered electrophoretic mobility and enhanced phosphorylation. In contrast, in cells expressing nuclear oncogenes such as v-*fos* and SV40 large T antigen, or in cells overexpressing c-*myc* proteins, no change in c-*raf* is observed (Roberts et al., 1988). These results suggest that *raf* proteins may function as signal transmitters and probably act downstream of membrane oncogenes (Rapp et al., 1988b).

ras *GTPase activating protein, GAP*

ras was the first viral oncogene found to be mutated in human tumors (Shih and Weinberg, 1982). It is membrane bound and capable of binding the guanine nucleotides GDP and GTP. Normal *ras* also has a GTPase activity. Because of these properties, *ras* is thought to be analogous to the well-characterized GTP-binding or G proteins, which regulate cyclic nucleotide metabolism. G proteins, when bound to GTP, activate important second messenger enzymes, including adenylate cyclase and cyclic GMP phosphodiesterase (Gilman, 1984). The hydrolysis of GTP to GDP then returns the G protein to the inactive state. *ras* likely functions in a similar manner, but it is not known what enzyme it may regulate. GAP was originally characterized as a cytosolic factor whose activity increased the GTPase activity of normal *ras* proteins by >200-fold (Gibbs et al., 1988; Trahey and McCormick, 1987). GAP has no effect on GTP turnover for oncogenic forms of *ras* (Trahey and McCormick, 1987). Purified normal *ras* proteins and many oncogenic forms have very low rates of GTP hydrolysis. Yet, normal *ras* proteins exist mainly in an inactive GDP-bound state despite the fact that intracellular concentrations of GTP are much higher than GDP. Oncogenic forms of *ras* are primarily in a GTP-bound, active state (Trahey and McCormick, 1987). Therefore, it has been postulated that the function of GAP in normal cells is to regulate the activity of *ras* proteins by maintaining them in an inactive, GDP-bound state (Trahey and McCormick, 1987).

The site of GAP interaction with *ras* proteins maps to a region previously defined as important in mediating downstream effector functions (Adari et al., 1988). In addition, GAP interacts preferentially with GTP-bound *ras* proteins, both normal and mutant (Vogel et al., 1988). In the case of mutant *ras* proteins, this association is constitutive, although it has no effect on GTP hydrolysis. These results imply that GAP could function as a downstream effector of *ras*-mediated signal transduction, although they do not rule out an upstream regulatory function (McCormick, 1989).

Cloning of GAP revealed the presence of two *src* homology region 2 (SH2) domains (Trahey et al., 1988; Vogel et al., 1988). SH2 domains mediate binding to phosphotyrosine residues in a sequence-specific manner (Anderson et al., 1990). The SH2 domain was originally identified as a noncatalytic region of 80 to 100 amino acids that is highly conserved among all nonreceptor protein-tyrosine kinases whether of cellular or viral origin (Pawson, 1988). SH2 domains have also been found in phospholipase C-γ and the related v-*crk* oncoprotein (Mayer et al., 1988; Stahl et al., 1988). Both PLC-γ and GAP contain two SH2 domains. The SH2 domain in v-*fps* has been shown to modulate tyrosine kinase activity. The SH2 region functions in *cis* as a positive effector of tyrosine kinase activity, possibly as a result of interactions in *trans* with cellular components (DeClue et al., 1987; Koch et al., 1989; Sadowski et al., 1986). It is currently unknown whether the SH2 domain in GAP performs a similar regulatory function to modulate activity. The SH2 domains likely mediate binding between phosphorylated tyrosine residues on the PDGF receptor and GAP and PLC-γ.

Recent studies suggest that GAP is a substrate for receptor tyrosine kinase activity. GAP is phosphorylated on tyrosine in response to EGF and PDGF and in cells expressing v-*fms* (Ellis et al., 1990; Molloy et al., 1989). It is not known what effect tyrosine phosphorylation has on GAP activity, but it does suggest a mechanism of connecting growth factor stimulation to *ras*-mediated signal transduction. If GAP acts as a downstream effector of *ras*, then perhaps tyrosine phosphorylation decreases GAP's ability to stimulate *ras* GTPase activity. This would prolong the life of a *ras*–GAP complex and the duration of the putative signal output (McCormick, 1989). Termination of the signal would be dependent on dephosphorylation of GAP to restore GTPase activation or some other mechanism to promote dissociation of GAP from *ras*.

Alternatively, GAP may function upstream of *ras*. Phosphorylation of GAP by receptor and nonreceptor tyrosine kinases could decrease the affinity of GAP for GTP-bound *ras*. This would serve to increase the half-life of activated *ras* in the cell. Evidence for this model of GAP function comes from studies in yeast. The yeast genes IRA1 and IRA2 are probably functional equivalents of mammalian GAP and function as a negative regulators of activated *ras*–GTP (Ballester et al., 1989; Tanaka et al., 1990).

Ligand binding to receptor elicits pleiotropic effects, many of which are not correlated with transcription-dependent events required for cell division. Some substrates for the protein kinase A (PKA), PKC, and PLC pathways contribute to regulatory feedback mechanisms. Cross-talk between pathways can result in synergism of positive forward effects or negative regulation, depending on cell type (Nishizuka, 1986, 1988; Yoshimasa et al., 1987).

Other important second messenger pathways activated by PDGF include c-^{src}, pp90^{rsk}, and cyclic AMP. The PDGF receptor may directly phosphorylate c-^{src} or be associated with the PtdIns-3 kinase complex (Kypta et al., 1990). pp90^{rsk} is a serine/threonine kinase that is activated and phosphorylated after growth factor stimulation (Chen et al., 1991). One of its known substrates is the ribosomal protein S6. PDGF also stimulates a transient increase in cyclic AMP levels, presumably mediated by the release of prostagladins (Rozengurt et al., 1983). Effects of cAMP are largely mediated through protein kinase A.

cAMP-dependent protein kinase

cAMP-dependent protein kinase, or protein kinase A (PKA), was one of the first protein kinases characterized in eukaryotic cells (Walsh et al., 1968). Although the structure of the catalytic and regulatory subunits of PKA have served as a prototype for the study of many other protein kinases, its regulation is unusual and perhaps unique. In the absence of cAMP, the enzyme is a tetrameric protein, consisting of two regulatory (R) subunits and two catalytic (C) subunits. R subunits contain cAMP binding sites; C subunits are the catalytic proteins. When intracellular cAMP levels are increased, cAMP binds to R subunits and lowers their affinity for C subunits. The tetrameric complex dissociates and frees active catalytic subunits (Taylor et al., 1988). The C subunits catalyze phosphorylation of substrate proteins which in turn alters their activity (Bramson et al., 1984; Krebs and Beavo, 1979).

The affinity of the R subunits for the C subunits is determined by the intracellular level of cAMP, which is regulated by the combined activities of adenylate cyclase and cyclic nucleotide phosphodiesterase. Another molecule, protein kinase inhibitor (PKI), can also regulate the activity of C subunits via a mechanism similar to that of the R subunit proteins. However, this protein lacks the cAMP-binding sites of the R subunit proteins; therefore its ability to inhibit PKA activity is not dependent on intracellular cAMP levels (Taylor et al., 1988; Walsh et al., 1971).

The PKA catalytic subunits transduce extracellular signals by their ability to phosphorylate cellular proteins. Two recent studies provide evidence supporting a direct role for the catalytic subunit of PKA in altering gene expression. In a C6 glioma cell line harboring a stably integrated fusion gene consisting of the human vasoactive intestinal protein (VIP) gene promoter/enhancer region linked to the *E. coli lacZ* gene, expression of *lacZ* is responsive to elevated cAMP levels. Microinjection of purified C subunit of PKA resulted in the induction of *lacZ* expression, bypassing the requirement for increased intracellular cAMP. Injection of either R_I or R_{II} alone or of equal amounts of C and R_{II} subunits did not result in elevated *lacZ* expression (Riabowol et al., 1988). Similarly, in cotransfection assays using reporter constructs and expression vectors containing wild-type and mutant subunits of PKA, overexpression of catalytic subunit induced expression from three cAMP-responsive promoters without increasing intracellular cAMP levels (Mellon et al., 1989). These studies establish that the C subunit alone is sufficient to induce the expression of genes known to be transcriptionally activated by elevated levels of cAMP.

IMMEDIATE EARLY GENES

Given the complexity of signal transduction via growth factor receptors, it is important to relate the activation of a second messenger pathway to an event essential to the mitogenic response. The expression of immediate early genes provides a means of correlating extracellular stimuli with nuclear events leading to cellular proliferation.

PDGF can function synergistically with other growth factors present in plasma. PDGF or platelet-poor plasma alone are poor mitogens, but when combined, they can induce mitosis in >95% of quiescent cells (Cochran, 1985). In addition, PDGF is required only transiently to produce this effect. Exposure to PDGF establishes a relatively stable state in which the cells can mitogenically respond to platelet-poor plasma for up to 24 hours. The primed state established by transient exposure to PDGF has been termed *competence* (Pledger et al., 1977). The competent state is stable and its establishment requires RNA synthesis (Smith and Stiles, 1981). Competent cells do not emerge from the G_0–G_1 boundary until the addition of the growth factors in platelet-poor plasma. These factors have therefore been termed "progression factors" (Pledger et al., 1977). Growth factors can be categorized based on their ability to complement known competence or progression factors.

The requirement for RNA synthesis to attain a competent state has led to the discovery of genes whose expression is induced by growth factors. PDGF activates the expression of between 30 and 100 genes (Cochran et al., 1983; Lau and Nathans, 1985, 1987). These genes are termed "immediate early" or "competence" genes (Jones et al., 1988; Lau and Nathans, 1987). Induction of immediate early genes is rapid and transient and does not depend upon protein synthesis. In fact, in the presence of protein synthesis inhibitors, induction is potentiated (Cochran et al., 1983; Greenberg et al., 1986a; Kelly et al., 1983; Kruijer et al., 1984; Lau and Nathans, 1985). The immediate early genes vary in the time course of their induction. However, they are typically expressed only transiently and have messenger RNAs with short half-lives.

Several immediate early gene products encode transcription factors. The proto-oncogenes c-*fos*, c-*myc* (Cochran et al., 1984; Greenberg and Ziff, 1984; Kelly et al., 1983; Kruijer et al., 1984; Lau and Nathans, 1985; Muller et al., 1984a), and c-*jun* (Ryder et al., 1988) are among the immediate early genes. These genes encode nuclear proteins that have DNA binding activity. c-*fos* and c-*jun* proteins combine to form a complex, AP-1, that is important in the transcriptional regulation of many genes including some that are responsive to TPA (discussed below) (Rauscher et al., 1988b; Schonthal et al., 1988). Other immediate early genes encode cytokines and zinc finger family transcription factors.

Both c-*fos* and c-*myc* have functional roles in cell division. When a c-*myc* expression vector is placed under the control of the inducible mouse mammary tumor virus (MMTV) promoter, hydrocortisone will function as a competence factor (Armelin et al., 1984). Similarly, microinjection of c-*myc* protein will support cell division in plasma-containing medium without added competence factors (Kaczmarek et al., 1985). Antisense RNA experiments also establish a role for c-*myc* and c-*fos* in cellular proliferation. When synthesis of c-*myc* or c-*fos* protein is inhibited by antisense RNA, a dramatic decrease in cell proliferation ensues (Holt et al., 1986; Mercola et al., 1987; Wickstrom et al., 1988).

Immediate early genes are thought to be direct targets of signal transduction mechanisms because they are induced in response to growth factors in the absence of protein synthesis. Yet, a single pathway does not entirely account for their regulation. Different combinations of second messenger pathways converge upon each of the immediate early genes. c-*fos* and c-*myc* expression are induced upon treatment of quiescent BALB/c 3T3 cells with either PDGF or TPA. In contrast, JE and KC, two other PDGF-inducible genes, are induced with PDGF treatment, but not TPA (Hall and Stiles, 1987). JE and KC are induced by interleukin-1 (IL-1) in BALB/c 3T3 cells, whereas neither c-*myc* nor c-*fos* respond (Jones et al., 1988; Rollins et al., 1988). Additionally, c-*myc* and JE are expressed differently depending on the proliferation state of the cells. In growing cells, the steady-state level of c-*myc* expression is high. When the cells reach confluence, the level of c-*myc* expression diminishes greatly. JE gene expression is undetectable in subconfluent cultures but is activated as the cells reach confluence (Rollins et al., 1987).

EUKARYOTIC TRANSCRIPTION

Biochemical and genetic studies in prokaryotes have led to a detailed understanding of transcriptional regulation (Ptashne, 1986). However, attaining similar knowledge about eukaryotic transcription has been hampered by the fact that purified RNA polymerase II is unable to accurately initiate transcription in cell-free systems. Transcription from minimal promoters requires the addition of various factors, including the TATA box binding factor TFIID, which interact with core promoter elements and RNA polymerase II (RNA pol II) to form a competent transcription complex (Nakajima et al., 1988).

Minimal promoters of most eukaryotic genes transcribed by RNA polymerase II are relatively inactive in transcription initiation (Maniatis et al., 1987; Mitchell and Tjian, 1989). Within a region upstream and downstream of the start site of transcription, numerous target sites for sequence-specific DNA binding proteins can be found. Many of these *cis*-acting sequences are important for regulation of signal-dependent transcriptional initiation.

Proteins that bind to *cis*-acting regulatory sequences have been characterized biochemically and the domains responsible for their DNA binding and transcriptional activation (or repression) activities have been defined (Maniatis et al., 1987; Mitchell and Tjian, 1989). In general, DNA binding and transcriptional activation are mediated by separate domains (Maniatis et al., 1987; Mitchell and Tjian, 1989; Ptashne, 1988). Chimeric proteins that fuse a DNA binding domain of one protein with the transcriptional activation domain of another will function to stimulate transcription in *in vitro* assays from a reporter construct based on the sequence specificity of the DNA binding domain (Ptashne,

1988).

However, exactly how upstream sequences and their binding proteins interact with core promoter sequences to facilitate the formation of preinitiation complexes is unknown.

C-*FOS* GENE EXPRESSION

The regulation of expression of the c-*fos* proto-oncogene is the most well studied of the immediate early genes. A diagram of the human c-*fos* promoter is presented in Figure 5. c-*fos* is the cellular homologue of v-*fos*, the transforming gene found in two murine retroviruses that cause osteosarcomas with short latency and transform fibroblasts in culture (Kruijer et al., 1984). The hallmarks of c-*fos* proto-oncogene expression are the variety of extracellular stimuli that can activate its transcription and the rapid and transient nature of its induction. The c-*fos* gene is highly conserved among vertebrates and is differentially expressed during development. Prenatal expression of c-*fos* is primarily within extraembryonic tissues and midgestational fetal liver (Muller et al., 1982, 1983a,b, 1984b). In postnatal tissues, c-*fos* expression is correlated with either the induction or maintenance of differentiation in several cell types, including F9 teratocarcinoma cells, PC12 cells (Curran and Morgan, 1985; Greenberg et al., 1985; Kruijer et al., 1985; Muller et al., 1982, 1983a,b, 1984b; Muller and Wagner, 1984), and adipocytes (Distel et al., 1987; Herrera et al., 1989a).

c-*fos* can be induced in many cells by exposure to serum or purified growth factors such as EGF, FGF, and PDGF (Bravo et al., 1985; Cochran et al., 1984; Greenberg and Ziff, 1984; Kruijer et al., 1984; Muller et al., 1984a). c-*fos* transcription is activated within minutes of treatment. Transcription reaches a maximum level at 10 minutes and decreases to quiescent levels within 60 minutes (Greenberg and Ziff, 1984). Direct activation of second messenger pathways by elevated cAMP, Ca^{2+} ionophores, and phorbol esters can also induce c-*fos* expression in a variety of cell types (Greenberg et al., 1985, 1986; Greenberg and Ziff, 1984; Moore et al., 1986; Morgan and Curran, 1986). Transcriptional activation is independent of protein synthesis and in the presence of protein synthesis inhibitors, the level of c-*fos* mRNA is "superinduced"; that is, the increase in c-*fos* mRNA is enhanced relative to treatment with growth factors alone (Cochran et al., 1984; Muller et al., 1984a).

The c-*fos* promoter sequence between -64 and -404 bp has been shown to act as a transcriptional enhancer element (Deschamps et al., 1985). There are two highly conserved regions within this sequence. One is centered around the TATA box and CAP site, the other extends between -240 and -475 bp. These two regions are also sites of DNAse I hypersensitivity (Deschamps et al., 1985).

Serum Response Element (SRE)

Using gel retardation assays, several groups have shown that the region between -60 and -370 contains target sites for sequence-specific DNA binding proteins (Verma and Sassone-Corsi, 1987). Three of these sites are completely conserved between the human, mouse, and chicken c-*fos* promoters. The first sequence is the dyad symmetry element (DSE) located between -299 and -319 relative to the start of c-*fos* transcription (Gilman et al., 1986; Treisman, 1986). Its position is within the region previously defined as the serum response element (SRE). The DSE can mediate c-*fos* induction by serum, TPA, NGF, and EGF in several cell lines (Fisch et al., 1987; Gilman, 1988; Sheng et al., 1988; Siegfreid and Ziff, 1989; Treisman, 1986). When placed upstream of a heterologous promoter, it can confer inducibility by serum (Mohun et al., 1987; Ray et al., 1989; Subramaniam et al., 1989). The DSE appears to function as a convergence point for a number of growth factors and activators of the protein kinase C pathway.

Several proteins that bind to the c-*fos* DSE have been characterized. The first to be identified was the serum response factor (SRF), which binds specifically to the DSE (Prywes and Roeder, 1986; Treisman, 1986). SRF has been purified and cloned (Norman et al., 1988; Prywes and Roeder, 1987; Schroter et al., 1987; Treisman, 1987). Its apparent molecular mass is 64–67 kd. Phosphorylation is

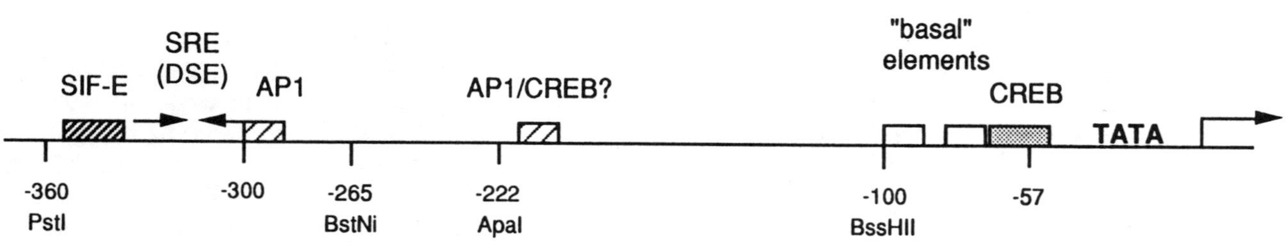

Fig. 5. Diagram of c-*fos* promoter sequences between +1 and -400 bp (Verma and Sassone-Corsi, 1987; see text). The TATA box is indicated by bold type. The known CREB site centered at -60 is shown by the lightly shaded box, binding sites for AP1 as thinly hatched boxes, and basal elements as open boxes. The dyad symmetry element (DSE) is indicated by arrows and the sequence originally defined as the serum response element (SRE) is delineated. The binding site for the *sis*-inducible factor (SIF), designated SIF-E, at -346 is shown by the heavily hatched box. Distances from the start site of transcription are shown along the bottom with restriction enzyme sites.

required for maximal DNA binding activity (Prywes et al., 1988b). SRF binding activity is constitutively expressed in a variety of cell types. Induction by growth factors does not alter the DNA binding activity or cause any change in the phosphorylation state of the protein (Fisch et al., 1987; Gilman et al., 1986; Hayes et al., 1987b; Prywes et al., 1988a,b; Prywes and Roeder, 1986, 1987; Sheng et al., 1988; Treisman, 1986). Genomic footprinting experiments have revealed no difference in the pattern of protein–DNA contacts between quiescent and induced nuclei (Herrera et al., 1989b).

Since no modification of SRF protein or binding activity can be detected that correlates with induction, activation of c-*fos* expression via the SRE could be mediated by the action of other proteins. Induction could be via a protein–protein interaction with SRF or involve protein–DNA contacts that are facilitated by SRF bound to the DSE. Alternatively, another protein could displace the SRF.

The former possibility is suggested from studies in yeast. The yeast homologue of the SRF is the product of the MCM1 locus (Hayes et al., 1988; Norman et al., 1988). MCM1 can bind to c-*fos* SRE and SRF can bind to core sequences present in mating-type-specific genes (Hayes et al., 1988). MCM1 can mediate both positive and negative regulation of gene expression, depending on flanking sequences. In cells, two products of the MAT locus can interact with MCM1 to regulate the expression of **a**- and α-specific genes. MATα1 and MCM1 combine to activate α-specific genes (Bender and Sprague, 1987). MATα2 and MCM1 repress α-specific genes (Keleher et al., 1988). Therefore, depending on the sequence and cellular context, MCM1 can function either as an activator or repressor (Hayes et al., 1988). Both MATα1 and MATα2 contact DNA sequences adjacent to the MCM1 binding site (Bender and Sprague, 1987; Sauer et al., 1988). However, neither 1 or 2 can bind DNA independently; they require MCM1 (Ammerer, 1990).

In mammalian systems, several proteins have been identified as interacting with either the SRE or SRF. A 62 kd protein has been found to bind to sequences 5' to the SRE (Shaw et al., 1989b). It forms a ternary complex with the SRF. It cannot bind DNA in the absence of SRF. This protein has been implicated in the serum response of c-*fos*, in that mutations in the binding site for the 62 kd protein that do not abolish SRF binding reduce serum-induced transcription (Shaw et al., 1989b). Graham and Gilman (1991) have shown that mutations in the 5' flank of the DSE specifically abolish the response of the element to TPA. These properties are analogous to that of MATα1 in yeast.

MAPF1 and MAPF2 are proteins involved in the expression of muscle-specific actin in developing myocytes (Walsh and Schimmel, 1987, 1988a,b). Chicken skeletal actin gene promoter contains a 28 base pair muscle-response element (MRE) that is sufficient for muscle-specific expression (Walsh and Schimmel, 1988a). The MRE contains a partial dyad symmetry element that has homology to the DSE. Both MAPF1 and MAPF2 can bind to the DSE (Walsh, 1989). MAPF1 shows a related, but distinct methylation interference pattern over the MRE and DSE. The DSE can functionally substitute for the MRE dyad symmetry element when placed within the context of the MRE. Although this construct does display increased activity in muscle versus nonmuscle cells, the expression level in nonmuscle cells is higher than that of the native MRE. MAPF1, MAPF2, and SRF are present in both muscle and nonmuscle cells and there is potential for cross-binding activity (Walsh, 1989).

Using DNA-affinity precipitation, a 62 kd protein was isolated with binding affinity to the DSE (Ryan et al., 1989). This protein is distinct from the 62 kd protein identified by Shaw et al. (1989), but may be related to MAPF1. It has a similar pattern of methylation interference with MAPF1. The 62 kd protein is a phosphoprotein, but whether phosphorylation is required for binding has not been determined. It has distinct binding specificities from the SRF (Ryan et al., 1989). Its function in regulating expression via the DSE is unknown.

These results make it difficult to assign a functional role to the SRF in c-*fos* induction. Complex protein–protein interactions are likely to be important in coupling the factors that interact with the DSE to the core transcriptional apparatus.

CaRE Element

A second sequence element located at -60 bp has been shown to mediate c-*fos* induction by cAMP and Ca^{2+} influx. It is believed to be the target of the cAMP-dependent protein kinase (PKA) and calmodulin kinase (CaM) pathways (Berkowitz et al., 1989; Fisch et al., 1989; Sheng et al., 1988). Multiple sequence elements that conform to a consensus sequence are found within the c-*fos* enhancer region (Berkowitz et al., 1989; Fisch et al., 1989). Although the sequence at -60 is the primary mediator of cAMP response (CRE), some of the other sequence elements appear to contribute to its magnitude (Berkowitz et al., 1989).

CREB/ATF represents a family of proteins that can bind to cAMP-responsive elements (CRE) *in vitro* (Hai et al., 1988, 1989; Lin and Green, 1988). Six different genes that produce proteins that can bind to the element have been cloned (Gonzalez et al., 1989; Hai et al., 1989; Hoeffler et al., 1988; Maekawa et al., 1989). However, by gene fusion it has been shown that the first CREB cloned can mediate responses to both cAMP and calcium (Sheng et al., 1991).

CRE sites have also been implicated in regulating basal transcription of c-*fos* (Fisch et al., 1987; Gilman et al., 1986). Perhaps certain forms of CREB/ATF function to couple proteins that bind to inducible promoter elements with the core transcriptional apparatus.

CREB/ATF *in vitro* DNA binding activity does not

correlate with the growth state of the cell. It is constitutive, like SRF. Most likely, phosphorylation of CREB increases its transcriptional activity in response to elevated cAMP and Ca levels (Montminy and Bilezsikjian, 1987; Montminy et al., 1986; Yamamoto et al., 1988).

SIF (SCM)

A third sequence, located at -346 bp, is the target of a protein whose binding activity is induced by treatment with purified v-*sis* or conditioned medium from v-*sis*-transformed cells. Induction is specific for *sis*-PDGF, as treatment of BALB/c-3T3 cells with insulin, FGF, EGF, or TPA fails to stimulate binding activity (Hayes et al., 1987a,b). In contrast to SRF and CREB/ATF proteins, the DNA binding activity of the *sis*-inducible factor (SIF) is negligible in extracts from quiescent BALB/c-3T3 fibroblasts. Upon treatment with conditioned medium from v-*sis*-transformed NRK cells, or purified v- or c-*sis*, SIF binding activity is markedly increased. SIF activity cannot be induced by treatment with phorbol esters, Ca^{2+} ionophores or elevated levels of cAMP. The time course of SIF induction is rapid and is unaffected by protein synthesis inhibition (Hayes et al., 1987a,b). Therefore, SIF is present in unstimulated cells in an inactive or non-DNA binding form.

Treatment with *sis*-PDGF proteins results in activation of SIF binding activity, presumably as a direct consequence of signal transduction through the PDGF receptor. Activators of the PKA and PKC pathways fail to induce SIF activity. Therefore SIF activation is likely to occur via a distinct second messenger pathway.

The SIF element can confer PDGF responsiveness onto the c-*fos* promoter in the absence of the SRE (Wagner et al., 1990). This element specifically responds to PDGF and not serum or phorbol esters. However, it does require cooperation with an element nearer the promoter for activity.

Autoregulation

Rapid and transient induction of c-*fos* expression is achieved by the coordinate action of positive and negative regulation at transcriptional and posttranscriptional levels. One mechanism of posttranscriptional negative regulation is mRNA stability. c-*fos* mRNA is highly unstable, with a half-life of approximately 15 minutes. The c-*fos* message has multiple sequence determinants that affect mRNA half-life (Bonnieu et al., 1989; Shyu et al., 1989; Treisman, 1985; Wilson and Treisman, 1988b). At least two AU-rich sequences in the 3' untranslated region of the c-*fos* message can confer instability onto normally stable heterologous mRNAs (Bonnieu et al., 1989; Treisman, 1985). At least one other sequence determinant resides within the coding sequence (Shyu et al., 1989).

Down-regulation of c-*fos* following stimulation is also mediated by transcriptional repression. Transcriptional repression requires protein synthesis (Greenberg et al., 1986a) and it was originally postulated that the FOS protein itself functions in an autoregulatory fashion. It has since been shown that the FOS protein does indeed participate in negative regulation of c-*fos* gene transcription. Overexpression of c-*fos* protein by the use of constitutive or inducible heterologous promoters diminishes c-*fos* promoter-driven induction by serum growth factors (Konig et al., 1989; Sassone-Corsi et al., 1988b; Schonthal et al., 1988, 1989). Mutations in the C-terminal region of FOS also abolish transcriptional repression (Wilson and Treisman, 1988a). Expression vectors replacing the c-*fos* coding region with that from v-*fos* are much less efficient in down-regulating c-*fos* transcription (Sassone-Corsi et al., 1988b). One mechanism by which the viral form of FOS causes transformation could be its reduced autorepression activity.

Another component that seems to be involved in transcriptional repression of c-*fos* is c-*jun*. Overexpression of c-*jun* can diminish c-*fos* induction in response to serum growth factors (Konig et al., 1989; Schonthal et al., 1989). Cotransfection of both c-*fos* and c-*jun* expression vectors results in a dramatic repression of serum-induced c-*fos* transcription (Konig et al., 1989).

The DSE is a target for repression of serum-induced c-*fos* transcription (Konig et al., 1989; Rivera et al., 1990; Sassone-Corsi et al., 1988b; Shaw et al., 1989a). FOS and JUN proteins are postulated to act through the DSE to mediate this repression (Konig et al., 1989; Sassone-Corsi et al., 1988b; Shaw et al., 1989a). Autoregulation of c-*fos* transcription does not appear to be mediated by direct interaction of the FOS protein with the c-*fos* promoter, but may occur indirectly (Rivera et al., 1990). However, other elements involved in positive regulation may also be targets of repression (Schonthal et al., 1989). The exact role of FOS/JUN complexes and *cis*-acting sequences in transcriptional repression of c-*fos* following growth factor induction remains to be elucidated.

c-*jun* is identical to the *fos*-associated protein p39 (Rauscher et al., 1988a) and the related yeast protein GCN4 (Struhl, 1987; Vogt et al., 1987). FOS and JUN proteins form heterodimeric complexes that have DNA-binding activity (Franza et al., 1988; Herrera et al., 1989a; Rauscher et al., 1988b). FOS/JUN complexes have been implicated in both positive and negative regulation of various genes. The transcription factor AP-1 is a FOS/JUN complex (Chiu et al., 1988; Curran and Franza, 1988; Rauscher et al., 1988b). AP-1 appears to be important in the activation of some genes responsive to TPA (Angel et al., 1987; Lee et al., 1987).

Multiple forms of both FOS and JUN proteins exist (Cochran et al., 1984; Cohen et al., 1989; Franza et al., 1987, 1988; Hirai et al., 1989; Ryder et al., 1988, 1989; Zerial et al., 1989). These proteins can dimerize in a variety of heteroduplex combinations (Cohen et al., 1989; Nakabeppu et al.,

1988; Zerial et al., 1989). JUN proteins, but not FOS proteins, can also form homodimers (Halazonetis et al., 1988; Nakabeppu et al., 1988; O'Shea et al., 1989; Rauscher et al., 1988c; Sassone-Corsi et al., 1988a; Sellers and Struhl, 1989; Smeal et al., 1989; Turner and Tjian, 1989). Dimer formation is achieved via a "leucine zipper" domain found in several transcription factors (Landschulz et al., 1988).

CONCLUSION

In this chapter we have used PDGF as a paradigm for growth factors. Other growth factors will of course not function identically to PDGF, but our current indications are that the second signaling pathways utilized by most growth factors are fairly ubiquitous. Many of the differences in the functions of growth factors may be due to the particular constellation of second signaling pathways activated by a given factor. PDGF may be unique among growth factors in the sense that it may activate more of these second signaling pathways than do other factors. As such, lessons learned from the biology of PDGF can be readily extended to the biology of other growth factors. In any case, PDGF likely plays a direct role in human oncogenesis and in the etiology of human nervous system tumors such as gliomas. Thus, understanding of the mechanism of action of PDGF is necessary for the understanding of these tumor types. Moreover, insofar as the signaling pathways activated by PDGF are general, then understanding these pathways will have importance for understanding the etiology of tumors that do not involve a PDGF autocrine stimulation.

Our understanding of growth factor action is far from complete. The general outlines are becoming apparent, however. The largest gaps in our knowledge at the present time are (1) how are the signals transmitted from the cell surface to the nucleus? and (2) what do the nuclear oncogenes do? Many of the receptor substrates such as phospholipase C-γ and GAP have been identified. However, our knowledge of the function of these substrates themselves is often woefully inadequate. For instance, the *ras* gene, which is one of the most commonly mutated genes in human tumors, has an almost completely unknown mechanism of action. The same is true for the PtdIns-3 kinase, which although correlated with oncogenic transformation, produces a phosphoinositide product that has a completely unknown function. Moreover, most of these molecules are located at the internal face of the cytoplasmic membrane and how these molecules send signals that ultimately result in activation of gene expression of proto-oncogenes such as *myc* and *fos* is not understood. We've only begun to answer these questions.

Moreover, although it seems clear that the nuclear proto-oncogenes that are regulated by growth factors are required and perhaps even sufficient for mitogenesis, they have functions that are far from being identified. Note that these molecules are located in the nucleus and can specifically bind to DNA sequences. In the case of *fos* and *jun*, it is known that these molecules can regulate transcription in a positive and negative fashion. In the case of c-*myc*, it is less clear what the function of this gene is, or what genes it may regulate. It is also possible that these molecules may have a direct involvement in the initiation of the replication of DNA synthesis in the cell, perhaps at origins of replication. It is still unclear how these nuclear oncogenes may potentially interact with cell cycle proteins such as *cdc2* and cyclins. These, in turn, may be regulated by recessive or anti-oncogenes. However, many of the tools needed to study these problems are at our disposal and answers to these questions should be forthcoming, paving the way for new therapeutic treatment of human tumors.

ACKNOWLEDGMENTS

The authors thank Margaret Wiegel for help with preparation of the manuscript. This work was supported in part by NIH grant P01 CA42063.

REFERENCES

Adari, H., D.R. Lowy, B.M. Willumsen, et al. (1988) Guanosine triphosphatase activating protein GAP interacts with the p21 *ras* effector binding domain. Science 240:519–521.

Ammerer, G. (1990) Identification, purification, and cloning of a polypeptide (PRTF/GRM) that binds to mating-specific promoter elements in yeast. Genes Dev. 4:299–312.

Anderson, D., C.A. Koch, L. Grey, et al. (1990) Binding of SH2 domains of phospholipase C gamma 1, GAP, and Src to activated growth factor receptors. Science 250:979–982.

Angel, P., M. Imagawa, R. Chiu, et al. (1987) Phorbol ester-inducible genes contain a common *cis* element recognized by a TPA-modulated *trans*-acting factor. Cell 49:729–739.

Antoniades, H.N., C.D. Scher, and C.D. Stiles (1979) Purification of the human platelet-derived growth factor. Proc. Natl. Acad. Sci. USA 76:1809–1813.

Armelin, H.A., M.C.S. Armelin, K. Kelly, et al. (1984) Functional role for c-*myc* in mitogenic response to platelet-derived growth factors. Nature 310:655–660.

Ballester, R., T. Michaeli, K. Ferguson, et al. (1989) Genetic analysis of mammalian GAP expressed in yeast. Cell 59:681–686.

Bar-Sagi, D., and R.J. Feramisco (1985) Microinjection of the *ras* oncogene protein into PC12 cells induces morphological differentiation. Cell 42:841–848.

Beckmann, M.P., C. Betsholtz, C.-H. Heldin, et al. (1988) Comparison of the biological properties and transforming potential of human PDGF-A and PDGF-B chains. Science 241:1346–1349.

Bender, A., and G.F. Sprague (1987) Matα1 protein, a yeast transcription activator, binds synergistically with a second protein to a set of cell-type-specific genes. Cell 50:681–691.

Berg, J. (1986) Potential metal-binding domains in nucleic acid binding proteins. Science 232:485–487.

Berkowitz, L.A., K.T. Riabowol, and M.Z. Gilman (1989) Multiple sequence elements of a single functional class are required for cyclic AMP responsiveness of the mouse c-*fos* promoter. Mol.

Cell. Biol. 9:4272–4281.

Betsholtz, C., A. Johnsson, C.-H. Heldin, et al. (1986) cDNA sequence and chromosomal localization of human platelet-derived growth factor A-chain and its expression in tumour cell lines. Nature 320:695–699.

Bishayee, S., S. Majumdar, C.D. Scher, et al. (1988) Characterization of a novel anti-peptide antibody that recognizes a specific conformation of the platelet-derived growth factor receptor. Mol. Cell. Biol. 8:3696–3702.

Bishayee, S., S. Majumdar, J. Khire, et al. (1989) Ligand-induced dimerization of the platelet-derived growth factor receptor. J. Biol. Chem. 264:11699–11705.

Bishayee, S., A. Ross, R. Womer, et al. (1986) Purified human platelet-derived growth factor receptor has ligand-stimulated tyrosine kinase activity. Proc. Natl. Acad. Sci. USA 83:6756–6760.

Blasi, E., B. Mathieson, L. Varesio, et al. (1985) Selective immortalization of murine macrophages from fresh bone marrow by a *raf/myc* recombinant murine retrovirus. Nature 318:667–670.

Boni-Schnetzler, M., A. Kaligian, R. DelVecchio, et al. (1988) Ligand-dependent intersubunit association within the insulin receptor complex activates its intrinsic kinase activity. J. Biol. Chem. 263:6822–6828.

Bonnieu, A., J. Rech, P. Jeanteur, et al. (1989) Requirements for c-*fos* mRNA down regulation in growth stimulated murine cells. Oncogene 4:881–888.

Bowen-Pope, D.F., C.E. Hart, and R.A. Seifert (1989) Sera and conditioned media contain different isoforms of platelet-derived growth factor (PDGF) which bind to different classes of PDGF receptor. J. Biol. Chem. 264:2502–2508.

Bramson, H.N., A.S. Mildvan, and E.T. Kaiser (1984) Mechanistic studies of cAMP-dependent protein kinase action. CRC Crit. Rev. 15:93–124.

Bravo, R., J. Burckhardt, T. Curran, et al. (1985) Stimulation and inhibition of growth by EGF in different A431 cell clones is accompanied by the rapid induction of c-*fos* and c-*myc* proto-oncogenes. EMBO J. 4:1193–1198.

Bywater, M., F. Rorsman, E. Bongcam-Rudloff, et al. (1988) Expression of recombinant platelet-derived growth factor A- and B-chain homodimers in Rat-1 cells and human fibroblasts reveals differences in protein processing and autocrine effects. Mol. Cell. Biol. 8:2753–2762.

Cattaneo, E., and R. McKay (1990) Proliferation and differentiation of neuronal stem cells regulated by nerve growth factor. Nature 347:762–765.

Chen, R.H., J. Chung, and J. Blenis (1991) Regulation of pp90rsk phosphorylation and S6 phosphotransferase activity in Swiss 3T3 cells by growth factor-, phorbol ester-, and cyclic AMP-mediated signal transduction. Mol. Cell. Biol. 11:1861–1867.

Chiu, R., W.J. Boyle, J. Meek, et al. (1988) The c-*fos* protein interacts with c-*jun*/AP-1 to stimulate transcription of AP-1 responsive genes. Cell 54:541–552.

Claesson-Welsh, L., A. Eriksson, A. Moren, et al. (1988) cDNA cloning and expression of a human platelet-derived growth factor (PDGF) receptor specific for B-chain containing PDGF molecules. Mol. Cell. Biol. 8:3476–3486.

Claesson-Welsh, L., A. Eriksson, B. Westermark, et al. (1989a) cDNA cloning and expression of the human A-type platelet-derived growth factor receptor establishes structural similarity to the B-type receptor. Proc. Natl. Acad. Sci. USA 86:4917–4921.

Claesson-Welsh, L., A. Hammacher, B. Westermark, et al. (1989b) Identification and structural analysis of the A type receptor for platelet-derived growth factor. J. Biol. Chem. 264:1742–1747.

Cochet, C., O. Kashles, E.M. Chembaz, et al. (1988) Demonstration of epidermal growth factor-induced receptor dimerization in living cells using a chemical covalent cross-linking agent. J. Biol. Chem. 263:3290–3295.

Cochran, B.H. (1985) The molecular action of platelet-derived growth factor. Adv. Cancer Res. 45:183–216.

Cochran, B.H., A.C. Reffel, and C.D. Stiles (1983) Molecular cloning of gene sequences regulated by platelet-derived growth factor. Cell 33:939–947.

Cochran, B.H., J. Zullo, I.M. Verma, et al. (1984) Expression of the c-*fos* gene and of an *fos*-related gene is stimulated by platelet-derived growth factor. Science 226:1080–1082.

Cohen, D.R., P.C.P. Ferreira, R. Gentz, et al. (1989) The product of a *fos*-related gene, *fra-1*, binds cooperatively to the AP-1 site with Jun: Transcription factor AP-1 is comprised of multiple protein complexes. Genes Dev. 3:173–184.

Coughlin, S.R., J.A. Escobedo, and L.T. Williams (1989) Role of phosphatidylinositol kinase in PDGF signal transduction. Science 243:1191–1193.

Curran, T., and B.R. Franza, Jr. (1988) Fos and Jun: The AP-1 connection. Cell 55:393–397.

Curran, T., and J.I. Morgan (1985) Superinduction of c-*fos* by nerve growth factor in the presence of peripherally active benzodiazepines. Science 229:1265–1268.

DeClue, J., I. Sadowski, G.S. Martin, et al. (1987) A conserved domain regulates interactions of the v-*fps* protein-tyrosine kinase with the host cell. Proc. Natl. Acad. Sci. USA 84:9064–9068.

Deschamps, J., R. Meijlink, and I.M. Verma (1985) Identification of a transcriptional enhancer element upstream from the proto-oncogene *fos*. Science 230:1174–1177.

Distel, R.J., H.S. Ro, B.S. Rosen, et al. (1987) Nucleoprotein complexes that regulate gene expression in adipocyte differentiation: Direct participation of c-*fos*. Cell 49:835–844.

Doolittle, R.F., M.W. Hunkapiller, L.E. Hood, et al. (1983) Simian sarcoma virus *onc* gene, v-*sis*, is derived from the gene (or genes) encoding a platelet-derived growth factor. Science 221:275–277.

Ellis, C., M. Moran, F. McCormick, et al. (1990) Phosphorylation of GAP and GAP-associated proteins by transforming and mitogenic tyrosine kinases. Nature 343:377–381.

Escobedo, J.A., P.J. Barr, and L.T. Williams (1988a) Role of tyrosine kinase and membrane-spanning domains in signal transduction by the platelet-derived growth factor receptor. Mol. Cell. Biol. 8:5126–5131.

Escobedo, J.A., M.T. Keating, H.E. Ives, et al. (1988b) Platelet-derived growth factor receptors expressed by cDNA transfection couple to a diverse group of cellular responses associated with cell proliferation. J. Biol. Chem. 263:1482–1487.

Escobedo, J.A., S. Navankasatussas, L.S. Cousens, et al. (1988c) A common PDGF receptor is activated by homodimeric A and B forms of PDGF. Science 240:1532–1534.

Escobedo, J.A., and L.T. Williams (1988) A PDGF receptor domain essential for mitogenesis but not for many other responses to PDGF. Nature 335:85–87.

Fantl, W.J., J.A. Escobedo, and L.T. Williams (1989) Mutations of the platelet-derived growth factor receptor that cause a loss of ligand-induced conformational change, subtle changes in kinase activity, and impaired ability to stimulate DNA synthesis. Mol. Cell. Biol. 9:4473–4478.

Fisch, T.M., R. Prywes, and R.G. Roeder (1987) c-*fos* sequences necessary for basal expression and induction by epidermal growth factor, 12-*O*-tetradecanoyl phorbol-13-acetate and the calcium ionophore. Mol. Cell. Biol. 7:3490–3502.

Fisch, T.M., R. Prywes, M.C. Simon, et al. (1989) Multiple sequence elements in the c-*fos* promoter mediate induction by cAMP. Genes Dev. 3:198–211.

Fleming, T.P., T. Matsui, C.J. Molloy, et al. (1989) Autocrine mechanism for v-*sis* transformation requires cell surface localization of internally activated growth factor receptors. Proc. Natl. Acad. Sci.

USA 86:8063–8067.

Folkman, J., and M. Klagsbrun (1987) A family of angiogenic peptides. Nature 329:671–672.

Franza, B.R., Jr., F.J. Rauscher III, S.F. Josephs, et al. (1988) The Fos complex and Fos-related antigens recognize sequence elements that contain AP-1 binding sites. Science 239:1150–1153.

Franza, B.R., Jr., L.C. Sambucetti, D.R. Cohen, et al. (1987) Analysis of Fos protein complexes and Fos-related antigens by high-resolution two-dimensional gel electrophoresis. Oncogene 1:213–221.

Fukui, Y., and H. Hanafusa (1989) Phosphatidylinositol kinase activity associates with viral $p60^{src}$ protein. Mol. Cell. Biol. 9:1651–1658.

Gibbs, J.B., M.D. Shaber, W.J. Allard, et al. (1988) Purification of *ras* GTPase activating protein from bovine brain. Proc. Natl. Acad. Sci. USA 85:5026–5030.

Gilman, A.G. (1984) G proteins and dual control of adenylate cyclase. Cell 36:577–579.

Gilman, M.Z. (1988) The c-*fos* serum response element responds to protein kinase C-dependent and -independent signals but not to cyclic AMP. Genes Dev. 2:394–402.

Gilman, M.Z., R.N. Wilson, and R.A. Weinberg (1986) Multiple protein-binding sites in the 5'-flanking region regulate c-*fos* expression. Mol. Cell. Biol. 6:4305–4316.

Gonzalez, G.A., K.K. Yamamoto, W.H. Fischer, et al. (1989) A cluster of phosphorylation sites on the cyclic AMP-regulated nuclear factor CREB predicted by its sequence. Nature 337:749–752.

Goustin, A.S., C. Betsholtz, S. Pfeifer-Ohlsson, et al. (1985) Coexpression of the *sis* and *myc* proto-oncogenes in developing human placenta suggests autocrine control of trophoblast growth. Cell 41:301–312.

Graham, R. and M. Gilman (1991) Distinct protein targets for signals acting at the c-*fos* serum response element. Science 251:189-192.

Greenberg, M.E., L.A. Greene, and E.B. Ziff (1985) Nerve growth factor and epidermal growth factor induce rapid transient changes in proto-oncogene transcription in PC12 cells. J. Biol. Chem. 260:14101–14110.

Greenberg, M.E., A.L. Hermanowski, and E.B. Ziff (1986a) Effect of protein synthesis inhibitors on growth factor activation of c-*fos*, c-*myc* and actin gene transcription. Mol. Cell. Biol. 6:1050–1057.

Greenberg, M.E., and E.B. Ziff (1984) Stimulation of 3T3 cells induces transcription of the c-*fos* proto-oncogene. Nature 311:433–438.

Greenberg, M.E., E.B. Ziff, and L.A. Greene (1986b) Stimulation of neuronal acetylcholine receptors induces rapid gene transcription. Science 234:80–83.

Gudas, L.J., J.P. Singh, and C.D. Stiles (1983) Secretion of growth regulatory molecules by teratocarcinoma stem cells. In: L.M. Silver, G.R. Martin, and S. Strickland (eds.). Teratocarcinoma Stem Cells. Cold Spring Harbor, N.Y.: Cold Spring Harbor Laboratory, pp. 229–236.

Hai, T., F. Liu, E.A. Allegretto, M. Karin, et al. (1988) A family of immunologically related transcription factors that includes multiple forms of ATF and AP-1. Genes Dev. 2:1216–1226.

Hai, T., F. Liu, W.J. Coukos, et al. (1989) Transcription factor ATF cDNA clones: An extensive family of leucine zipper proteins able to selectively form DNA-binding heterodimers. Genes Dev. 3:2083–2090.

Halazonetis, T.D., K. Georgopoulos, M.E. Greenberg, et al. (1988) c-*jun* dimerizes with itself and with c-*fos*, forming complexes of different DNA binding affinities. Cell 55:917–924.

Hall, D.J., and C.D. Stiles (1987) Platelet-derived growth factor-inducible genes respond differentially to at least two distinct intracellular second messengers. J. Biol. Chem. 262:15302–15308.

Hammacher, A., K. Mellstrom, C.-H. Heldin, et al. (1989) Isoform-specific induction of actin reorganization by platelet-derived growth factor suggests that the functionally active receptor is a dimer. EMBO J. 8:2489–2495.

Hannink, D., and D.J. Donoghue (1988) Autocrine stimulation by the v-*sis* gene product requires a ligand-receptor interaction at the cell surface. J. Cell Biol. 107:287–298.

Hart, C.E., J.W. Forstrom, J.D. Kelly, et al. (1988) Two classes of PDGF receptor recognize different isoforms of PDGF. Science 240:1529–1531.

Hayes, T.E., A.M. Kitchen, and B.H. Cochran (1987a) Inducible binding of a factor to the c-*fos* regulatory region. Proc. Natl. Acad. Sci. USA 84:1272–1276.

Hayes, T.E., A.M. Kitchen, and B.H. Cochran (1987b) A rapidly inducible DNA-binding activity which binds upstream of the c-*fos* proto-oncogene. J. Cell. Physiol. (Suppl.) 5:63–68.

Hayes, T.E., P. Sengupta, and B.H. Cochran (1988) The human c-*fos* serum response factor and the yeast factors GRM/PRTF have related DNA-binding specificities. Genes Dev. 2:1713–1722.

Heldin, C.-H., G. Backstrom, A. Ostman, et al. (1988) Binding of different dimeric forms of PDGF to human fibroblasts: Evidence for two separate receptor types. EMBO J. 7:1387–1393.

Heldin, C.-H., A. Ernlund, C. Rorsman, et al. (1989) Dimerization of B-type platelet-derived growth factor receptors occurs after ligand binding and is closely associated with receptor kinase activation. J. Biol. Chem. 264:8905–8912.

Herrera, R., H.S. Ro, G.S. Robinson, et al. (1989a) A direct role for C/EBP and the AP-1-binding site in gene expression linked to adipocyte differentiation. Mol. Cell. Biol. 9:5331–5339.

Herrera, R.E., P.E. Shaw, and A. Nordheim (1989b) Occupation of the c-*fos* serum response element *in vivo* by a multi-protein complex is unaltered by growth factor induction. Nature 340:68–70.

Hirai, S.-I., R.-P. Ryseck, F. Mechta, et al. (1989) Characterization of *junD*: A new member of the *jun* proto-oncogene family. EMBO J. 8:1433–1439.

Hoeffler, J.P., T.E. Meyer, Y. Yun, et al. (1988) Cyclic AMP-responsive DNA-binding protein: Structure based on a cloned placental cDNA. Science 242:1430–1433.

Holt, J.T., T.V. Gopal, A.D. Moulton, et al. (1986) Inducible production of c-*fos* antisense RNA inhibits 3T3 cell proliferation. Proc. Natl. Acad. Sci. USA 83:4794–4798.

Huang, S.S., and J.S. Huang (1988) Rapid turnover of the platelet-derived growth factor receptor in *sis*-transformed cells and reversal by suramin. J. Biol. Chem. 263:12608–12618.

Hunter, T., P. Angel, W.J. Boyle, et al. (1988) Targets for signal-transducing protein kinases. CSH Symp. Quant. Biol. 53:131–142.

Jones, S.D., D.J. Hall, B.J. Rollins, et al. (1988) Platelet-derived growth factor generates at least two distinct intracellular signals that modulate gene expression. CSH Symp. Quant. Biol. 53:531–536.

Kaczmarek, L., J.K. Hyland, R. Watt, et al. (1985) Microinjected c-*myc* as a competance factor. Science 228:1313–1315.

Kaplan, D.R., M. Whitman, B. Schaffhausen, et al. (1987) Common elements in growth factor stimulation and oncogenic transformation: 85 kd phosphoprotein and phosphatidylinositol kinase activity. Cell 50:1021–1029.

Kazlauskas, A., D. Bowen-Pope, R. Seifert, et al. (1988) Different effects of homo- and heterodimers of platelet-derived growth factor A and B on human and mouse fibroblasts. EMBO J. 7:3727–3735.

Keating, M.T., J.A. Escobedo, and L.T. Williams (1988) Ligand activation causes a phosphorylation-dependent change in platelet-derived growth factor receptor conformation. J. Biol. Chem. 263:12805–12808.

Keating, M.T., and L.T. Williams (1988) Autocrine stimulation of intracellular PDGF receptors in v-*sis*-transformed cells. Science 239:914–916.

Keleher, C.A., C. Goutte, and A.D. Johnson (1988) The yeast cell-type-specific repressor a2 acts cooperatively with a non-cell-type-specific protein. Cell 53:927–936.

Kelly, K., B.H. Cochran, C.D. Stiles, et al. (1983) Cell specific regulation of c-*myc* gene by lymphocyte mitogens and platelet-derived growth factor. Cell 35:603–610.

Kikkawa, U., K. Ogita, M.S. Shearman, et al. (1988) The family of protein kinase C: Its molecular heterogeneity and differential expression. CSH Symp. Quant. Biol. 53:97–102.

Koch, C.A., M. Moran, I. Sadowski, et al. (1989) The common *src* homology region 2 domain of cytoplasmic signaling proteins is a positive effector of v-*fps* tyrosine kinase function. Mol. Cell. Biol. 9:4131–4140.

Konig, H., H. Ponta, U. Rahmsdorf, et al. (1989) Autoregulation of *fos*: The dyad symmetry element as the major target of repression. EMBO J. 8:2559–2566.

Krebs, E.G., and J.A. Beavo (1979) Phosphorylation-dephosphorylation of enzymes. Annu. Rev. Biochem. 48:923–959.

Kruijer, W., J.A. Cooper, T. Hunter, et al. (1984) Platelet-derived growth factor induces rapid but transient expression of the c-*fos* gene and protein. Nature 312:711–716.

Kruijer, W., D. Schubert, and I.M. Verma (1985) Induction of the proto-oncogene *fos* by nerve growth factor. Proc. Natl. Acad. Sci. USA 82:7330–7334.

Kumjian, D.A., M.I. Wahl, S.G. Rhee, et al. (1989) Platelet-derived growth factor (PDGF) binding promotes physical association of PDGF receptor with phospholipase C. Proc. Natl. Acad. Sci. USA 86:8232–8236.

Kypta, R.M., Y. Goldberg, E.T. Ulug, et al. (1990) Association between the PDGF receptor and members of the src family of tyrosine kinases. Cell 62:481–492.

Lamph, W.W., P. Wamsley, P. Sassone-Corsi, et al. (1988) Induction of proto-oncogene *JUN*/AP-1 by serum and TPA. Nature 334:629–631.

Landschulz, W.H., P.F. Johnson, and S.L. McKnight (1988) The leucine zipper: A hypothetical structure common to a new class of DNA binding proteins. Science 240:1759–1764.

Lau, L.F., and D. Nathans (1985) Identification of a set of genes expressed during the G0/G1 transition of cultured mouse cells. EMBO J. 4:3145–3151.

Lau, L.F., and D. Nathans (1987) Expression of a set of growth related immediate early genes in Balb/c-3T3 cells: Coordinated regulation with c-*fos* or c-*myc*. Proc. Natl. Acad. Sci. USA 84:1182–1186.

Lee, W., P. Mitchell, and R. Tjian (1987) Purified transcription factor AP-1 interacts with TPA-inducible enhancer elements. Cell 49:741–752.

Lin, Y.-S., and M.R. Green (1988) Interaction of a common cellular transcription factor, ATF, with regulatory elements in both E1a- and cyclic AMP-inducible promoters. Proc. Natl. Acad. Sci. USA 85:3396–3400.

Lokeshwar, V.B., S.S. Huang, and J.S. Huang (1990) Intracellular turnover, novel secretion, and mitogenically active intracellular forms of v-*sis* gene product in simian sarcoma virus-transformed cells. J. Biol. Chem. 265:1665–1675.

Maekawa, T., H. Sakura, C. Kanei-Ihsii, et al. (1989) Leucine zipper structure of the protein CRE-BP1 binding to the cyclic AMP response element in brain. EMBO J. 8:2023–2028.

Majerus, P.W., T.M. Connolly, H. Deckmyn, et al. (1986) The metabolism of phosphoinositide-derived messenger molecules. Science 234:1519–1526.

Maniatis, T., S. Goodbourn, and J.A. Fischer (1987) Regulation of inducible and tissue-specific expression. Science 236:1237–1245.

Martinet, Y., P.B. Bitterman, J.-F. Mornex, et al. (1986) Activated human monocytes express the c-*sis* proto-oncogene and release a mediator showing PDGF-like activity. Nature 319:158–160.

Matsui, T., M. Heidaran, T. Miki, et al. (1989a) Isolation of a novel receptor cDNA establishes the existence of two PDGF receptor genes. Science 243:800–804.

Matsui, T., J.H. Pierce, T.P. Fleming, et al. (1989b) Independent expression of human a or b platelet-derived growth factor receptor cDNAs in a naive hematopoietic cell leads to functional coupling with mitogenic and chemotactic signaling pathways. Proc. Natl. Acad. Sci. USA 86:8314–8318.

Matuoka, K., K. Fukami, O. Nakanishi, et al. (1988) Mitogenesis in response to PDGF and bombesin abolished by microinjection of antibody to PIP_2. Science 239:640–643.

Mayer, B.J., M. Hamaguchi, and H. Hanafusa (1988) A novel viral oncogene with structural similarity to phospholipase C. Nature 332:272–275.

McCormick, F. (1989) *ras* GTPase activating protein: Signal transmitter and signal terminator. Cell 56:5–8.

Mellon, P.L., C.H. Clegg, L.A. Correll, et al. (1989) Regulation of transcription by cyclic AMP-dependent protein kinase. Proc. Natl. Acad. Sci. USA 86:4887–4891.

Mercola, D., A. Rundell, J. Westwick, et al. (1987) Antisense RNA to the c-*fos* gene: Restoration of density-dependent growth arrest in a transformed cell line. Biochem. Biophys. Res. Commun. 147:288–294.

Mitchell, P.J., and R. Tjian (1989) Transcriptional regulation in mammalian cells by sequence-specific DNA binding proteins. Science 245:371–378.

Mohun, T., N. Garrett, and R. Treisman (1987) *Xenopus* cytoskeletal actin and human c-*fos* gene promoters share a conserved protein-binding site. EMBO J. 6:667–673.

Molloy, C.J., D.P. Bottaro, T.P. Fleming, et al. (1989) PDGF induction of tyrosine phosphorylation of GTPase activating protein. Nature 342:711–714.

Montminy, M.R., and L.M. Bilezsikjian (1987) Binding of a nuclear protein to the cyclic AMP response element of the somatostatin gene. Nature 328:175–178.

Montminy, M.R., K.A. Sevarino, J.A. Wagner, et al. (1986) Identification of a cyclic-AMP-responsive element within the rat somatostatin gene. Proc. Natl. Acad. Sci. USA 83:6682–6686.

Moore, J.P., J.A. Todd, T.R. Hesketh, et al. (1986) c-*fos* and c-*myc* gene activation, ionic signals, and DNA synthesis in thymocytes. J. Biol. Chem. 261:8158–8162.

Morgan, J.I., and T. Curran (1986) Role of ion flux in the control of c-*fos* expression. Nature 322:552–555.

Morgan, W.C., D.R. Kaplan, D.C. Pallas, et al. (1988) Recombinant retroviruses that transduce middle T antigen cDNAs derived from polyoma virus mutants. J. Virol. 62:3407–3411.

Morrison, D.K., D.R. Kaplan, J.A. Escobedo, et al. (1989) Direct activation of the serine/threonine kinase activity of Raf-1 through tyrosine phosphorylation by the PDGF beta-receptor. Cell 58:649–657.

Muller, R., R. Bravo, J. Burckhardt, et al. (1984a) Induction of c-*fos* gene and protein by growth factors precedes activation of c-*myc*. Nature 312:716–720.

Muller, R., D. Muller, and L. Guilbert (1984b) Differential expression of c-*fos* in hematopoietic cells: Correlation with differentiation of monomyelocytic cells *in vitro*. EMBO J. 3:1887–1890.

Muller, R., D.J. Slamon, J.M. Tremblay, et al. (1982) Differential expression of cellular oncogenes during pre- and postnatal development of the mouse. Nature 299:640–644.

Muller, R., J.M. Tremblay, E.D. Adamson, et al. (1983a) Tissue and cell type-specific expression of two human c-*onc* genes. Nature 304:454–456.

Muller, R., I.M. Verma, and E.D. Adamson (1983b) Expression of c-*onc* genes: c-*fos* transcripts accumulate to high levels during development of mouse placenta, yolk sac, and amnion. EMBO J. 2:679–684.

Muller, R., and E.F. Wagner (1984) Differentiation of F9 teratocarcinoma stem cells after transfer of c-*fos* proto-oncogene. Nature 311:438–442.

Nakabeppu, Y., K. Ryder, and D. Nathans (1988) DNA binding activities of three murine *jun* proteins: Stimulation by *fos*. Cell 55:907–915.

Nakajima, N., M. Horikoshi, and R.G. Roeder (1988) Factors involved in specific transcription by mammalian RNA polymerase II: Purification, genetic specificity and TATA box-promoter interactions of TFIID. Mol. Cell. Biol. 8:4028–4040.

Nishizuka, Y. (1986) Studies and perspectives of protein kinase C. Science 233:305–312.

Nishizuka, Y. (1988) The molecular heterogeneity of protein kinase C and its implication for cellular regulation. Nature 334:661–665.

Nister, M., A. Hammacher, K. Mellstrom, et al. (1988) A glioma-derived PDGF A chain homodimer has different functional activities from a PDGF AB heterodimer purified from human platelets. Cell 52:791–799.

Noble, M., K. Murray, P. Stroobant, et al. (1988) Platelet-derived growth factor promotes division and motility and inhibits premature differentiation of the oligodendrocyte/type-2 astrocyte progenitor cell. Nature 333:560–562.

Norman, C., M. Runswick, R. Pollock, et al. (1988) Isolation and properties a cDNA clones encoding SRF, a transcription factor that binds to the c-*fos* serum response element. Cell 55:989–1003.

O'Shea, E.K., R. Rutkowski, W.F. Stafford III, et al. (1989) Preferential heterodimer formation by isolated leucine zippers from Fos and Jun. Science 245:646–648.

Owen, A.J., P. Pantazis, and H.N. Antoniades (1984) Simian sarcoma virus-transformed cells secrete a mitogen identical to platelet-derived growth factor. Science 225:54–56.

Pawson, T. (1988) Non-catalytic domains of cytoplasmic protein-tyrosine kinases: Regulatory elements in signal transduction. Oncogene 3:491–495.

Pledger, W.J., C.D. Stiles, H.N. Antoniades, et al. (1977) Induction of DNA synthesis in BALB/c 3T3 cells by serum components: Reevaluation of the commitment process. Proc. Natl. Acad. Sci. USA 74:4481–4485.

Prywes, R., A. Dutta, J.A. Cromlish, et al. (1988a) Phosphorylation of serum response factor, a factor that binds to the serum response element of the c-*fos* enhancer. Proc. Natl. Acad. Sci. USA 85:7206–7210.

Prywes, R., T.M. Fisch, and R.G. Roeder (1988b) Transcriptional regulation of c-*fos*. CSH Symp. Quant. Biol. 53:739–748.

Prywes, R., and R.G. Roeder (1986) Inducible binding of a factor to the c-*fos* enhancer. Cell 47:777–784.

Prywes, R., and R.G. Roeder (1987) Purification of the c-*fos* enhancer-binding protein. Mol. Cell. Biol. 7:3482–3489.

Ptashne, M. (1986) A Genetic Switch, Gene Control and Phage 1. Cambridge: Cell Press.

Ptashne, M. (1988) How eukaryotic transcriptional activators work. Nature 335:683–689.

Raff, M.C., L.E. Lillien, W.D. Richardson, et al. (1988) Platelet-derived growth factor from astrocytes drives the clock that times oligodendrocyte development in culture. Nature 333:562–565.

Rapp, U.R., J.L. Cleveland, T.I. Bonner, et al. (1988a) The *raf* oncogenes. In: P. Reddy et al. (eds.). Handbook of Oncogenes. Amsterdam: Elsevier, 213 pp.

Rapp, U.R., G. Heidecker, M. Huleihel, et al. (1988b) *raf* family serine/threonine protein kinases in mitogen signal transduction. CSH Symp. Quant. Biol. 53:173–184.

Rauscher, F.J. III, D.R. Cohen, T. Curran, et al. (1988a) Fos-associated protein p39 is the product of the *jun* proto-oncogene. Science 240:1010–1016.

Rauscher, F.J. III, L.C. Sambucetti, T. Curran, et al. (1988b) Common DNA binding site for Fos protein complexes and transcription factor AP-1. Cell 52:471–480.

Rauscher, F.J., P.J. Voulalas, B.R. Franza, et al. (1988c) Fos and Jun bind cooperatively to the AP-1 site: Reconstruction in vitro. Genes Dev. 2:1687–1699.

Ray, A., P. Sassone-Corsi, and P.B. Sehgal (1989) A multiple cytokine- and second messenger responsive element in the enhancer of the human interleukin-6 gene: Similarities with c-fos gene regulation. Mol. Cell. Biol. 9:5537–5547.

Riabowol, K.T., J.S. Fink, M.Z. Gilman, et al. (1988) The catalytic subunit of cAMP-dependent protein kinase induces expression of genes containing cAMP-responsive elements. Nature 336:83–86.

Rivera, V.M., M. Sheng, and M.E. Greenberg (1990) The inner core of the serum response element mediates both the rapid induction and subsequent repression of c-fos transcription following serum induction. Genes Dev. 4:255–268.

Rizzino, A., and D.F. Bowen-Pope (1985) Production of PDGF-like factors by embryonal carcinoma cells and response to PDGF by endoderm-like cells. Dev. Biol. 110:15–22.

Roberts, T.M., D. Kaplan, W. Morgan, et al. (1988) Tyrosine phosphorylation in signal transduction. CSH Symp. Quant. Biol. 53:161–171.

Rollins, B.J., E.D. Morrison, and C.D. Stiles (1987) A cell cycle constraint on the regulation of gene expression by platelet-derived growth factor. Science 238:1269–1271.

Rollins, B.J., E.D. Morrison, and C.D. Stiles (1988) Cloning and expression of JE a PDGF-inducible gene with cytokine-like properties. Proc. Natl. Acad. Sci. USA 85:3738–3742.

Ross, R., E.W. Raines, and D.F. Bowen-Pope (1986) The biology of platelet-derived growth factor. Cell 46:155–169.

Roussel, M.F., J.L. Cleveland, S.A. Shurtleff, et al. (1991) Myc rescue of a mutant CSF-1 receptor impaired in mitogenic signaling. Nature 353:361–363.

Roussel, M.F., and C.J. Sherr (1989) Mouse NIH 3T3 cells expressing human colony-stimulating factor 1 (CSF-1) receptors overgrow in serum-free medium containing human CSF-1 as their only growth factor. Proc. Natl. Acad. Sci. USA 86:7924–7927.

Rozengurt, E., P. Stroobant, M.D. Waterfield, et al. (1983) Platelet-derived growth factor elicits cyclic AMP accumulation in Swiss 3T3 cells: Role of prostaglandin production. Cell 34:265–272.

Ryan, W.A., Jr., B.R. Franza, Jr., and M.Z. Gilman (1989) Two distinct cellular phosphoproteins bind to the c-*fos* serum response element. EMBO J. 6:1785–1792.

Ryder, K., A. Lanahan, E. Perez-Albuerne, et al. (1989) *Jun*-D: A third member of the *Jun* gene family. Proc. Natl. Acad. Sci. USA 86:1500–1503.

Ryder, K., L.F. Lau, and D. Nathans (1988) A gene activated by growth factors is related to the oncogene v-*jun*. Proc. Natl. Acad. Sci. USA 85:1487–1491.

Sadowski, I., J.C. Stone, and T. Pawson (1986) A noncatalytic domain conserved among cytoplasmic protein-tyrosine kinases modifies the kinase function and transforming activity of Fujinami sarcoma virus $p130^{gag\text{-}fps}$. Mol. Cell. Biol. 6:4396–4408.

Sassone-Corsi, P., L.J. Ransone, W.W. Lamph, et al. (1988a) Direct interaction between *fos* and *jun* nuclear onco-proteins: Role of the 'leucine zipper' domain. Nature 336:692–695.

Sassone-Corsi, P., J.C. Sisson, and I.M. Verma (1988b) Transcriptional autoregulation of the proto-oncogene *fos*. Nature 334:314–319.

Sauer, R.T., D.L. Smith, and A.D. Johnson (1988) Flexibility of the yeast $\alpha 2$ repressor enables it to occupy the ends of its operator leaving the center free. Genes Dev. 2:807–816.

Schonthal, A., M. Buscher, P. Angel, et al. (1989) The Fos and Jun/AP-1 proteins are involved in the downregulation of Fos transcription. Oncogene 4:629–636.

Schonthal, A., P. Herrlich, H.J. Rahmsdorf, et al. (1988) Requirement for *fos* gene expression in the transcriptional activation of collagenase by other oncogenes and phorbol esters. Cell 54:325–334.

Schroter, H., P.E. Shaw, and A. Nordheim (1987) Purification of intercalator-released p67, a polypeptide that interacts specifically with the c-*fos* serum response element. Nucleic Acids Res. 15:10145–10158.

Seifert, R.A., C.E. Hart, P.E. Phillips, et al. (1989) Two different subunits associate to create isoform-specific platelet-derived growth factor receptors. J. Biol. Chem. 264:8771–8778.

Sellers, J.W., and K. Struhl (1989) Changing Fos oncoprotein to a Jun-independent DNA-binding protein with GCN4 dimerization specificity by swapping 'leucine zippers'. Nature 341:74–76.

Severinsson, L., B. Ek, K. Mellstrom, et al. (1990) Deletion of the kinase insert sequence of the platelet-derived growth factor β-receptor affects receptor kinase activity and signal transduction. Mol. Cell. Biol. 10:801–809.

Shaw, P.E., S. Frasch, and A. Nordheim (1989a) Repression of c-*fos* transcription is mediated through $p67^{SRF}$ bound to the SRE. EMBO J. 8:2567–2574.

Shaw, P.E., H. Schroter, and A. Nordheim (1989b) The ability of a ternary complex to form over the serum response element correlates with serum inducibility of the human c-*fos* promoter. Cell 56:563–572.

Sheng, M., S.T. Dougan, G. McFadden, et al. (1988) Calcium and growth factor pathways of c-*fos* transcriptional activation require distinct upstream regulatory sequences. Mol. Cell. Biol. 8:2787–2796.

Sheng, M., M.A. Thompson, and M.E. Greenberg (1991) CREB: A Ca(2+)-regulated transcription factor phosphorylated by calmodulin-dependent kinases. Science 252:1427–1430.

Shih, C., and R.A. Weinberg (1982) Isolation of a transforming sequence from a human bladder carcinoma cell line. 29:161-169.

Shyu, A.B., M.E. Greenberg, and J.G. Belasco (1989) The c-*fos* transcript is targeted for rapid decay by two distinct mRNA degradation pathways. Genes Dev. 3:60–72.

Siegfreid, Z., and E.B. Ziff (1989) Transcription activation by serum, PDGF, and TPA through the c-*fos* DSE: Cell type specific requirements for induction. Oncogene 4:3–11.

Smeal, T., P. Angel, J. Meek, et al. (1989) Different requirements for formation of Jun:Jun and Jun:Fos complexes. Genes Dev. 3:2091–2100.

Smith, J.C., and C.D. Stiles (1981) Cytoplasmic transfer of the mitogenic response to platelet-derived growth factor. Proc. Natl. Acad. Sci. USA 78:4363–4367.

Stahl, M.L., C.R. Ferenz, K.L. Kelleher, et al. (1988) Sequence similarity of phospholipase C with the non-catalytic region of *src*. Nature 332:269–272.

Stevens, C.W., W.H. Brondyk, J.A. Burgess, et al. (1988) Partially transformed, anchorage-independent human diploid fibroblasts result from overexpression of the c-*sis* oncogene: Mitogenic activity of an apparent monomeric platelet-derived growth factor 2 species. Mol. Cell. Biol. 8:2089–2096.

Stiles, C.D. (1983) The molecular biology of platelet-derived growth factor. Cell 33:653–655.

Stiles, C.D., G.T. Capone, C.D. Scher, et al. (1979) Dual control of cell growth by somatomedins and platelet-derived growth factor. Proc. Natl. Acad. Sci. USA 76:1279–1283.

Struhl, K. (1987) The DNA-binding domains of the *jun* onco-protein and the yeast GCN4 transcriptional activator are functionally homologous. Cell 50:841–846.

Subramaniam, M., L.J. Schmidt, C.E. Crutchfield III, et al. (1989) Negative regulation of serum-responsive enhancer elements. Nature 340:64–66.

Tanaka, K., M. Nakafuku, T. Satoh, et al. (1990) S. cerevisae genes *IRA1* and *IRA2* encode proteins that may be functionally equivalent to mammalian *ras* GTPase activating protein. Cell 60:803–807.

Taylor, S.S., J.A. Buechler, L.W. Slice, et al. (1988) cAMP-dependent protein kinase: A framework for a diverse family of enzymes. CSH Symp. Quant. Biol. 53:121–130.

Trahey, M., and F. McCormick (1987) A cytoplasmic protein stimulates normal N-*ras* p21 GTPase, but does not affect oncogenic mutants. Science 238:542–545.

Trahey, M., G. Wong, R. Halenbeck, et al. (1988) Molecular cloning of two types of GAP complementary DNA from human placenta. Science 242:1697–1700.

Treisman, R. (1985) Transient accumulation of c-*fos* RNA following serum stimulation requires a conserved 5' element and c-*fos* 3' sequences. Cell 42:889–902.

Treisman, R. (1986) Identification of a protein-binding site that mediates transcriptional response of the c-*fos* gene to serum factors. Cell 46:567–574.

Treisman, R. (1987) Identification and purification of a polypeptide that binds to the c-*fos* serum response element. EMBO J. 6:2711–2717.

Turner, R., and R. Tjian (1989) Leucine repeats and an adjacent DNA binding domain mediate the formation of functional cFos-cJun heterodimers. Science 243:1689–1694.

Uno, I., K. Fukami, H. Kato, et al. (1988) Essential role for phosphatidylinositol 4,5-bisphosphate in yeast cell proliferation. Nature 333:188–190.

Varticovski, L., B. Druker, D. Morrison, et al. (1989) The colony stimulating factor-1 receptor associates with and activates phophatidylinositol-3-kinase. Nature 342:699–702.

Veno, H., H. Colbert, J.A. Escobedo, et al. (1991) Inhibition of PDGF beta receptor signal transduction by coexpression of a truncated receptor. Science 252:844–848.

Verma, I.M., and P. Sassone-Corsi (1987) Proto-oncogene *fos*: Complex but versatile regulation. Cell 51:513–514.

Vogel, U.S., R.A.F. Dixon, M.D. Schaber, et al. (1988) Cloning of bovine GAP and its interaction with oncogenic *ras* p21. Nature 335:90–93.

Vogt, P.K., T.J. Bos, and R.F. Doolittle (1987) Homology between the DNA-binding domain of the GCN4 regulatory protein of yeast and the carboxy-terminal region of a protein coded for by the *onc* gene *jun*. Proc. Natl. Acad. Sci. USA 84:3316–3319.

Wagner, B.J., T.H. Hayes, C.J. Hoban, et al. (1990) The SIF binding element confers *sis*/PDGF inducibility onto the c-*fos* promoter. EMBO J. 9:4477–4484.

Wahl, M.I., N.E. Olashaw, S. Nishibe, et al. (1989) Platelet-derived growth factor induces rapid and sustained tyrosine phosphorylation of phospholipase C-γ in quiescent BALB/c 3T3 cells. Mol. Cell. Biol. 9:2934–2943.

Walsh, D.A., C.D. Ashby, C. Gonzalez, et al. (1971) Purification and characterization of a protein inhibitor of adenosine 3',5'-monophosphate-dependent kinases. J. Biol. Chem. 246:1977–1985.

Walsh, D.A., J.P. Perkins, and E.G. Krebs (1968) An adenosine 3',5'-monophosphate dependent protein kinase from rabbit skeletal muscle. J. Biol. Chem. 243:3763–3765.

Walsh, K. (1989) Cross-binding of factors to functionally different promoter elements in c-*fos* and skeletal actin genes. Mol. Cell. Biol. 9:2191–2201.

Walsh, K., and P. Schimmel (1987) Two nuclear factors compete for the skeletal actin promoter. J. Biol. Chem. 262:9429–9432.

Walsh, K., and P. Schimmel (1988a) DNA-binding site for two skeletal actin promoter factors is important for expression in muscle cells. Mol. Cell. Biol. 8:1800–1802.

Walsh, K., and P. Schimmel (1988b) Preliminary characterization of muscle actin promoter factors 1 and 2 and their interaction with the skeletal actin promoter. UCLA Symp. Mol. Cell. Biol. New Ser.

93:669–676.

Waterfield, M.D., G.T. Scrace, N. Whittle, et al. (1983) Platelet-derived growth factor is structurally related to the putative transforming protein p28*sis* of simian sarcoma virus. Nature 304:35–39.

Whitman, M., C.P. Downes, M. Keeler, et al. (1988) Type I PI kinase makes a novel inositol phospholipid, phosphatidylinositol-3-phosphate. Nature 322:664-666.

Whitman, M., D.R. Kaplan, B. Schaffhausen, et al. (1985) Association of PI kinase activity with polyoma middle T competent for transformation. Nature 315:239–242.

Wickstrom, E.L., T.A. Bacon, A. Gonzalez, et al. (1988) Human promyelocytic leukemia HL-60 cell proliferation and c-*myc* protein expression are inhibited by an antisense pentadecadeoxynucleotide targeted against c-*myc* mRNA. Proc. Natl. Acad. Sci. USA 85:1028–1032.

Williams, L.T. (1989) Signal transduction by the platelet-derived growth factor receptor. Science 243:1564–1570.

Williams, L.T., J.A. Escobedo, M.T. Keating, et al. (1988a) Signal transduction by the platelet-derived growth factor receptor. CSH Symp. Quant. Biol. 53:455–465.

Williams, R.L., D.J. Hilton, S. Pease, et al. (1988b) Myeloid leukaemia inhibitory factor maintains the developmental potential of embryonic stem cells. Nature 336:684–687.

Wilson, T., and R. Treisman (1988a) Fos C-terminal mutations block down-regulation of c-*fos* transcription following serum stimulation. EMBO J. 7:4193–4202.

Wilson, T., and R. Treisman (1988b) Removal of poly(A) and consequent degradation of c-*fos* mRNA facilitated by 3' AU-rich sequences. Nature 336:396–399.

Yamamoto, K.K., G.A. Gonzalez, W.H. Biggs III, et al. (1988) Phosphorylation-induced binding and transcriptional efficacy of nuclear factor CREB. Nature 334:494–498.

Yarden, Y., J.A. Escobedo, W.J. Kuang, et al. (1986) Structure of the receptor for platelet-derived growth factor helps to define a family of closely related growth factor receptors. Nature 323:226–232.

Yarden, Y., W.-J. Kuang, T. Yang-Feng, et al. (1987) Human proto-oncogene c-*kit*: A new cell surface receptor tyrosine kinase for an unidentified ligand. EMBO J. 6:3341–3351.

Yeh, H.J., K.G. Ruit, Y.X. Wang, et al. (1991) PDGF A-chain gene is expressed by mammalian neurons during development and in maturity. Cell 64:209–216.

Yoshimasa, T., D.R. Sibley, M. Bouvier, et al. (1987) Cross-talk between cellular signaling pathways suggested by phorbol-ester-induced adenylate cyclase phosphorylation. Nature 327:67–70.

Zerial, M., L. Toschi, R.-P. Ryseck, et al. (1989) The product of a novel growth factor activated gene, *fos B*, interacts with JUN proteins enhancing their DNA binding activity. EMBO J. 8:805–813.

Chapter 9

The Trk Family of Neurotrophin Receptors: Molecular Characterization and Oncogenic Activation in Human Tumors

Mariano Barbacid

INTRODUCTION

To date, oncogenes have been identified in more than one-third of all human malignancies. Most of these oncogenes have been extensively characterized at the molecular level and their mechanism of activation has been elucidated. Recently, the development of molecular diagnostic tecnhiques has made it possible to utilize these oncogenes as probes for a variety of clinical applications, including cancer diagnosis, prognosis, and quantitation of residual disease in patients in remission. However, the ultimate goal of molecular oncology is to utilize oncogenes as targets for cancer therapy by generating oncogene inhibitors that will not interfere with the physiological function of their normal alleles. The development of such specific inhibitors will require a more profound knowledge of the mechanisms by which oncogenes contribute to neoplasia and, even more importantly, a precise understanding of the role that proto-oncogenes play in normal cells.

During the last decade, the physiological role of various proto-oncogene products has been elucidated. First, the v-*sis* oncogene was found to be allelic to the B chain of platelet-derived growth factor (PDGF) (Deuel et al., 1983; Doolittle et al., 1983). Subsequently, the v-*erb*B and v-*fms* retroviral oncogenes were found to be derived from cellular proto-oncogenes encoding the EGF and CSF-1 receptors, respectively (Donward et al., 1984; Sherr et al., 1985). More recently, the c-ErbA protein was shown to be a high-affinity receptor for thyroid hormones (Sap et al., 1986), the c-Kit proto-oncogene product was identified as the receptor for steel factor (SF) (Huang et al., 1990; Zsebo et al., 1990), and the Met protein was found to be the receptor for the hepatocyte growth factor (HGF) (Bottaro et al., 1991; Naldini et al., 1991). Other examples of recent assignments of defined biological functions to the products of proto-oncogenes include the identification of the c-Mos protein as a component of the cytostatic factor responsible for preventing oocytes from entering the mitotic cycle (Sagata et al., 1989), and the identification of the c-Fos, c-Jun, and c-Rel proteins as members of the AP-1 (c-Fos and c-Jun) (reviewed in Curran and Franza, 1988) and NF-kB (c-Rel) (Ghosh et al., 1990; Kieran et al., 1990) transcription factors.

Last year, the normal function of another proto-oncogene, *trk*, was elucidated. The *trk* locus was first identified as a transforming gene in a human colon carcinoma biopsy (Martin-Zanca et al., 1986; Pulciani et al., 1982). Molecular analysis of *trk* gene sequences indicated that this locus encodes a novel cell surface receptor of the tyrosine protein kinase family (Martin-Zanca et al., 1989). Subsequent studies revealed that the *trk* proto-oncogene product serves as prototype for a subfamily of tyrosine kinase receptors that includes the products of the related *trk*B (Klein et al., 1989; Middlemas et al., 1991) and *trk*C (Lamballe et al., 1991) genes. In this chapter I review recent studies indicating that the *trk* family of tyrosine protein kinases are the functional receptors for the NGF family of neurotrophins, which include nerve growth factor (NGF), brain-derived neurotrophinic factor (BDNF), neurotrophin-3 (NT-3), neurotrophin-4 (NT-4), and neurotrophin-5 (NT-5) (Hefti, 1992). In addition, I summarize our current knowledge about the incidence of *trk* oncogenes in human tumors and their various mechanisms of activation.

II. THE *TRK* RECEPTOR

IIA. Molecular Structure

The human *trk* proto-oncogene encodes a protein of 790 amino acid residues that has features characteristic of tyrosine protein kinase cell surface receptors (Fig. 1). They include a 32 amino acid-long signal peptide, a putative ligand-binding domain rich in consensus N-glycosylation sites, a single transmembrane domain, a catalytic region highly related to that of known tyrosine protein kinases, and a short carboxy-terminal tail of 15 amino acid residues (Martin-Zanca et al., 1989).

Analysis of the deduced amino acid sequence of the *trk* proto-oncogene product indicates that this gene does not belong to any of the well-characterized subfamilies of tyrosine kinase cell surface receptors (Hanks et al., 1988). For instance, its extracellular domain, along with those of the related *trk*B and *trk*C receptors (see below), displays a unique array of structural motifs indicative of a distinct subfamily of tyrosine protein kinase receptors (Fig. 1). The amino-terminal moiety of the *trk* extracellular domain contains three tandem leucine-rich motifs (LRM) of 24 amino acid residues flanked by two cysteine clusters (CC) that include eight of the 12 cysteine residues conserved among all known *trk* receptors (Schneider and Schweiger, 1991). This structural feature is characteristic of the LRM superfamily of proteins, which includes proteins of diverse function such as the human platelet von Willebrand factor receptor, the ribonuclease/angiogenin inhibitor, cell adhe-

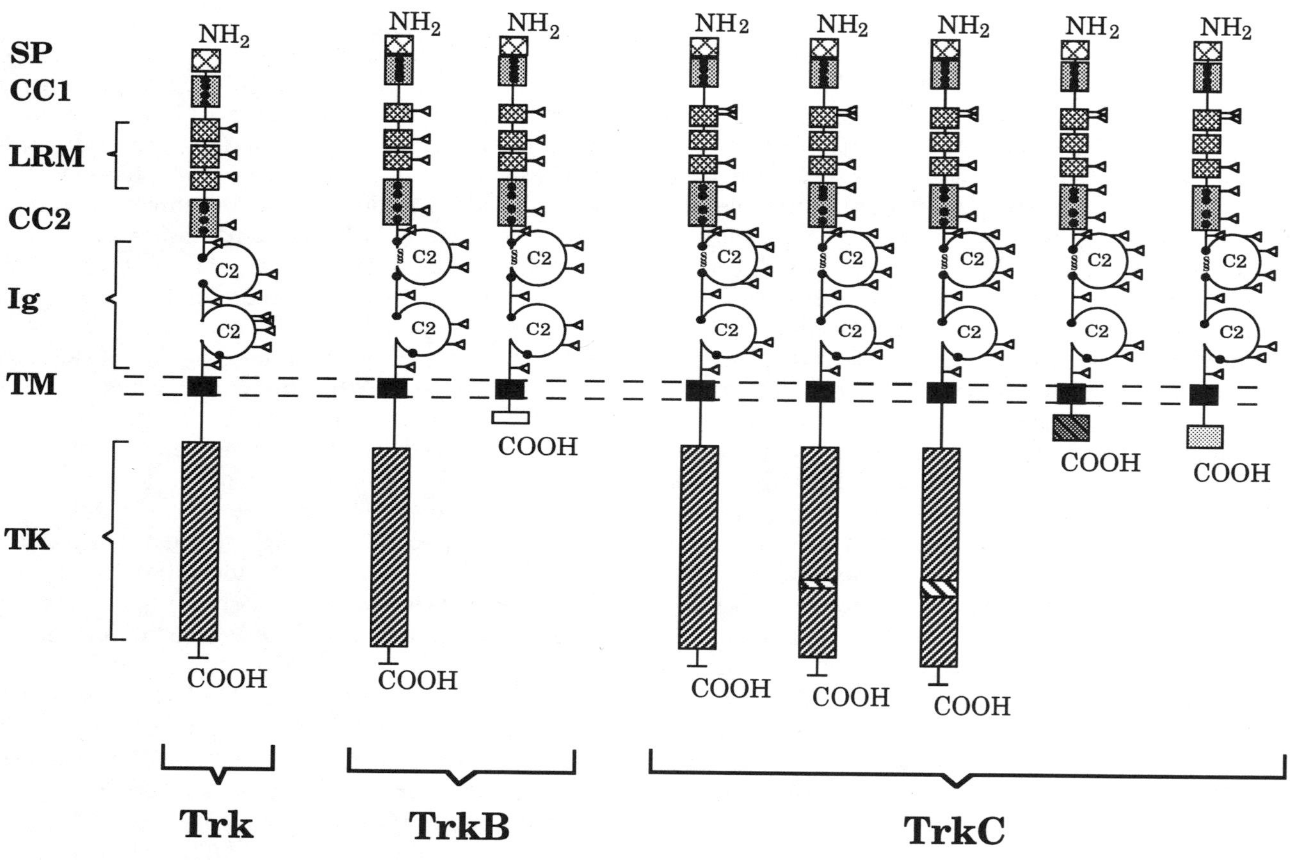

Fig. 1. Schematic representation of the structures of the various members of the *trk* family of neurotrophin receptors. The depicted extracellular structural motifs include a signal peptide (dotted box), three leucine-rich motifs (LRM), two cysteine clusters (CC) and two Ig-like domains (IgC2), conserved cysteine residues (black dots), and N-glycosylation sites (inverted triangles) are also indicated. The transmembrane domains are represented by black boxes. The tyrosine protein kinase domain is indicated by hatched boxes. Unique amino acid sequences present at the carboxy-terminal end (COOH) of the three noncatalytic receptors are indicated by open (*trk*B), thick hatched, and dotted (*trk*C) boxes. Finally, the additional amino acid residues present in the tyrosine protein kinase domain of the gp145^{trkC} isoform receptors are indicated by boxes hatched in the opposite direction as the kinase domain.

sion proteins, and extracellular matrix components (Schneider and Schweiger, 1991). These motifs are supposed to be involved in mediating tight and specific protein–protein interactions.

The carboxy moiety of the *trk* extracellular domain contains two immunoglobulin (Ig)-like domains of the C2 type (Schneider and Schweiger, 1991) (Fig. 1). The first Ig-like domain contains two conserved cysteines presumably involved in generating the canonical disulphide bond. The second Ig-like domain has retained one of the two cysteine residues, whereas the other cysteine has been replaced by a leucine. Ig-like domains have been previously identified in other cell surface tyrosine protein kinases, such as the receptors for PDGF, CSF-1, SF, and FGFs. However, the *trk* Ig-like domains appear to be more closely related to those of cell adhesion molecules such as N-CAM. Interestingly, the *Drosophila* homologue of the mammalian *trk* receptor, D*trk*, also contains two C2 Ig-like domains very similar to those present in their mammalian counterparts. The D*trk* protein possesses four additional Ig-like motifs (two of the C2 type and two of the V type) in its amino-terminal moiety instead of the LRMs (Pulido et al., 1992). Recent studies have shown that the D*trk* protein mediates cell aggregation *in vitro*, a process that leads to the activation of its tyrosine protein kinase domain (Pulido et al., 1992). Understanding the precise biological role of the D*trk* receptor in *Drosophila* may explain the nature of the evolutionary pressure required to maintain the two carboxy-terminal C2-type Ig-like domains in the mammalian *trk* receptors while allowing the divergent evolution of their corresponding amino-terminal moieties.

Close examination of the amino acid sequences of the second C2 Ig-like domain in each of the known members of the *trk* family of receptors indicates a highly conserved region (Fig. 2). These sequences are likely to play a significant role in modulating the catalytic activity of the kinase domain (Barbacid et al., 1991), since a 51 amino acid deletion encompassing this region results in the oncogenic activation of the *trk* receptor (Coulier et al., 1990). Similar results were obtained by replacing the conserved cysteine residue present in the middle of this domain by serine (Coulier et al., 1990). Whether this C2 Ig-like domain participates in ligand interaction is currently under investigation.

The predicted amino acid sequence of the *trk* gene product (Martin-Zanca et al., 1989) indicates the presence of a single transmembrane domain and a 76 amino acid-long juxtamembrane region that contains a putative ATP binding site (GKGSGLQGX_{17}K). Whether this domain binds ATP or is required for biological activity remains to be determined. The region of homology with the catalytic domain of other tyrosine protein kinases extends for 266 amino acid residues, of which 99 (37%) are homologous to those of other tyrosine kinase cell surface receptors (Martin-Zanca et al., 1989). However, only 66 of them (25%) are conserved in the *src* family of tyrosine kinases. The extent of the homology of the *trk* protein to other tyrosine kinase receptors is similar to that observed among the different members of this class of proteins and did not reveal any striking relatedness with any particular receptor.

The kinase domain of the *trk* proto-oncogene exhibits several unique features, including a single-amino acid gap between residues 542 and 543, a threonine instead of alanine at position 647, a tryptophan instead of the conserved tyrosine at position 722, and a histidine instead of the helix-breaker proline at position 766 (Fig. 2). In addition, the *trk* protein contained several additional residues (five with respect to the insulin and IGF-1 receptors and 14 with respect to the EGF receptor, *erb*B2, *met*, and *eph* proteins) in the region corresponding to the long intervening domain characteristic of the c-*kit* protein and the PDGF and CSF-1 receptors. Finally, the *trk* proto-oncogene product contains a very short carboxy-terminal tail of just 15 amino acid residues (Martin-Zanca et al., 1989). One of these residues (Tyr-785) is a tyrosine. Phosphorylation of carboxy-terminal tyrosine residues in the EGF and CSF-1 receptors as well as in $pp60^{src}$ has a profound effect on their respective biological activities. Whether phosphorylation of Tyr-785 also plays a role in the regulation of the enzymatic activity of the *trk* proto-oncogene product remains to be determined.

Biochemical analysis of NIH3T3 cells expressing *trk* proto-oncogene cDNA clones indicates that the primary translational product of this gene is a 110 kd glycoprotein. This molecule becomes further glycosylated, presumably during its transport to the plasma membrane, to yield a 140 kd glycoprotein. Both $gp110^{trk}$ and $gp140^{trk}$ proteins exhibit *in vitro* kinase activity specific for tyrosine residues. ^{125}I-surface labeling of NIH3T3 cells expressing a complete cDNA clone of the *trk* proto-oncogene indicated that only the heavily glycosylated 140 kd isoform is located on the cell surface. These observations indicate that $gp140^{trk}$ is the mature human *trk* gene product (Martin-Zanca et al., 1989).

IIB. Biological Function

Recent studies have indicated that the product of the *trk* proto-oncogene, $gp140^{trk}$, is the functional receptor for nerve growth factor (NGF) (Fig. 3). Experimental evidence supporting this conclusion includes:

1. NGF binds to cells expressing the *trk* proto-oncogene and can be chemically cross-linked to $gp140^{trk}$ receptors (Hempstead et al., 1991; Kaplan et al., 1991b; Klein et al., 1991a).
2. Addition of NGF to rat PC12 pheochromocytoma cells or to NIH3T3 cells ectopically expressing the human *trk* proto-oncogene results in the rapid phosphorylation of $gp140^{trk}$ on tyrosine residues (Kaplan et al., 1991a,b; Klein et al., 1991a).

3. Addition of NGF to NIH3T3 cells ectopically expressing gp140*trk* receptors induced c-*fos* expression, DNA synthesis, and ultimately morphologic transformation (Cordon-Cardo et al., 1991).
4. NGF can induce the meiotic maturation of *Xenopus* oocytes expressing gp140*trk* (Nebreda et al., 1991).
5. Transfection of the *trk* proto-oncogene into NGF-unresponsive PC12 mutant cells restores their full response to NGF, including their ability to differentiate into neuron-like cells (Loeb et al., 1991).

Previous studies have indicated that NGF can interact with two classes of receptors that differ in their respective affinity for this neurotrophin (Bothwell, 1991). The low-affinity receptor (K_d in the nM range) has been identified as a 75 kd glycoprotein, gp75^{LNGFR}, which shares certain structural homologies with other cytokine receptors such as the receptor for TNF. gp75^{LNGFR} has a cysteine-rich extracellular domain and a short cytoplasmic region (Johnson et al., 1986; Radeke et al., 1987). gp75^{LNGFR} also binds each of the other known members of the NGF family of neurotrophins with similar low affinity (Rodriguez-Tebar et al., 1990; Squinto et al., 1991). It is generally accepted that cells that only express low-affinity gp75^{LNGFR} receptors do not respond to NGF nor to the other NGF-related neurotrophins. However, there have been several reports indicating that transfection of mutant PC12 cells or neuroblastoma cell lines with the LNGFR gene restores some of the biological responses attributed to NGF (Bothwell, 1991). The mechanism by which gp75^{LNGFR} may mediate signal transduction remains to be elucidated.

As indicated above, gp140*trk* receptors appear to be responsible for mediating the neurotrophic properties of NGF. However, the molecular nature of the high-affinity NGF receptors remains to be resolved. Hempstead et al. (1991) have proposed that such receptors consist of heterodimers of gp140*trk* and gp75^{LNGFR}. This model is based on binding studies that indicate that NGF binds with low

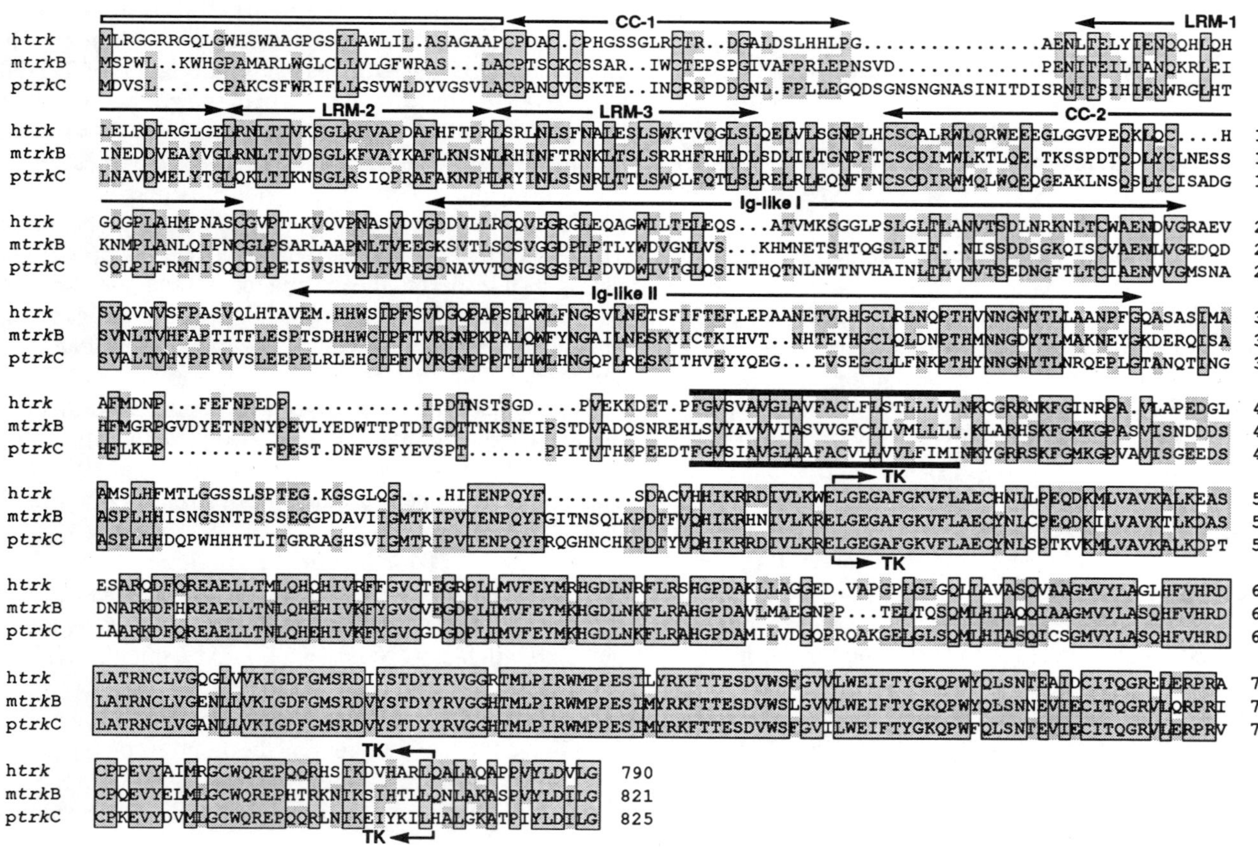

Fig. 2. Comparison of the deduced amino acid sequences of the human gp140*trk*, mouse gp145*trkB*, and porcine gp145*trkC* neurotrophin receptors. Sequences corresponding to the putative signal peptide (open box), cysteine clusters (CC-1 and CC-2), leucine-rich motifs (LRM-1, LRM-2 and LRM-3), Ig-like motifs (Ig-I and Ig-II) transmembrane domain (dark box), and tyrosine kinase catalytic region (TK) are indicated. Amino acid residues shared by two of these receptors are shaded. Residues shared by the three receptors are shaded and boxed. Dots indicate gaps introduced to maximize alignment.

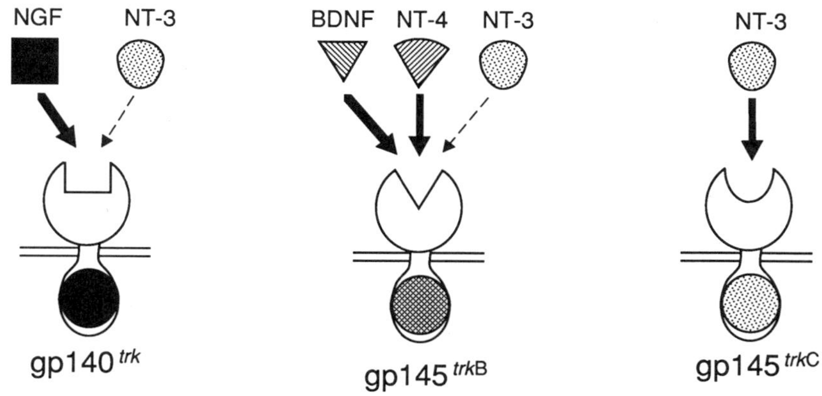

Fig. 3. Schematic representation of the known interactions between the NGF family of neurotrophins and the *trk* family of tyrosine protein kinase receptors. Dotted lines indicate limited biological activity.

affinity (nanomolar range) to either $gp140^{trk}$ or $gp75^{LNGFR}$ alone, but with high affinity (picomolar range) to cells coexpressing both NGF receptors. Similar experiments conducted in our laboratory have yielded different results (Klein et al., 1991a). NIH3T3 cells ectopically expressing $gp140^{trk}$ display high-affinity receptors capable of binding NGF with an affinity close to 10^{-11} M (Klein et al., 1991a). This high-affinity binding is solely mediated by $gp140^{trk}$ since NIH3T3 cells do not express $gp75^{LNGFR}$ receptors. The observed high-affinity NGF receptors represent 3 to 5% of the $gp140^{trk}$ molecules present in these cells. These findings are reminiscent of those found previously reported for the EGF and PDGF receptors and suggest that the generation of functional high-affinity NGF receptors may require the oligomerization of $gp140^{trk}$ molecules (Ullrich and Schlessinger, 1990).

Coexpression of a large number (>100,000) of $gp75^{LNGFR}$ receptors in NIH3T3 cells expressing $gp140^{trk}$ does not alter either the K_d or the percentage of the high-affinity NGF binding sites (Jing et al., in press). Moreover, the presence of $gp75^{LNGFR}$ receptors has no effect on any of the biochemical (NGF-induced tyrosine phosphorylation of $gp140^{trk}$) or biological (DNA synthesis and morphologic transformation) effects elicited by NGF in these $gp140^{trk}$-containing NIH3T3 cells (Jing et al., in press). Perhaps the most compelling evidence against the proposed $gp140^{trk}/gp75^{LNGFR}$ heterodimer model is the limited coexpression of $gp140^{trk}$ and $gp75^{LNGFR}$ receptors in the CNS. Whereas certain cholinergic neurons of the basal forebrain express both receptors, those present in the striatiatum only contain $gp140^{trk}$ (Steininger et al., submitted). The recent availability of mice lacking the LNGFR locus (Lee et al., 1992) should resolve whether the low-affinity $gp75^{LNGFR}$ receptors are required to mediate the neurotrophic properties of NGF through $gp140^{trk}$ receptors.

In addition to NGF, $gp140^{trk}$ may also serve as a receptor for the related neurotrophin-3 (NT-3) (Cordon-Cardo et al., 1991) (Fig. 3). NT-3 partially competes for NGF binding to cells expressing $gp140^{trk}$. Moreover, NT-3 induces the phosphorylation of $gp140^{trk}$ on tyrosine residues and induces the malignant transformation of NIH3T3 cells when coexpressed with $gp140^{trk}$. However, quantitative analysis of each of these parameters indicates that the biological effects of NT-3 mediated through $gp140^{trk}$ receptors are about two orders of magnitude less intense than those induced by NGF (Cordon-Cardo et al., 1991). Moreover, addition of NT-3 to PC12 cells fails to induce their differentiation into neuronal-like cells. Therefore, it remains to be established whether NT-3 can mediate any of its neurotrophic properties through $gp140^{trk}$ receptors *in vivo*.

IIC. Expression Pattern

To date, the only published information regarding the *in vivo* expression of the *trk* proto-oncogene has been obtained from studies using mouse embryos (Martin-Zanca et al., 1990). Northern blot analysis of RNAs isolated from mouse embryos at various developmental stages indicates that the earliest detection of *trk* gene transcripts coincides with the onset of neurogenesis (E9.5 embryos). Similar results were obtained by *in situ* hybridization where *trk* gene expression could be first observed in E9.5 embryos at the site of condensing dorsal root ganglia (DRG) (Martin-Zanca et al., 1990). In E13.5 embryos, where most of the major organs can be identified, *trk* expression appears to be confined to very defined structures of the nervous system, including DRG and visceral sensory ganglia such as the fifth (trigeminal), superior ninth (superior), and tenth (jugular) cranial ganglia (Table 1). This pattern of expression is strikingly specific, since other neighboring cranial ganglia such as the eight (acoustic), inferior ninth (petrosal), and tenth (nodose) ganglia do not exhibit detectable *trk* gene transcripts (Martin-Zanca et al., 1990).

TABLE 1. Expression of the *trk* Gene Family of Neurotrophin Receptors in the Murine Nervous System

Structure	*trk*	*trk*B	*trk*C
Central nervous system			
Cerebral cortex	-	++*	++
Corpus callosum	-	-	ND
Thalamus	-	++*	+
Hippocampus	-	++*	++
Fimbria hippocampus	-	++**	ND
Choroid plexus	-	++**	-
Ependymal layer of ventricles	-	++*	-
Cerebellum: Purkinje cells	-		-
Cerebellum: Granular cell layer	-	+*	++
Brain stem	-	++*	++
Spinal cord	-		++
Peripheral nervous system			
Cranial ganglia: V	++	+	
VII	-	++	++
VIII	-	++	++
IX/X superior	++	++	++
IX/X inferior	-	++	++
Dorsal root ganglia	++	+	++
Autonomic nervous system	-	++	++
Eye: retina, ophthalmic nerve	-	++	ND
Vestibular system: inner ear	–	+	ND
Nasal chamber	-	+	+
Snout	-	+	+
Submaxillary gland	-	+	++
Palate, tongue, tooth buds	-	+	+
Gut	-	-	++
Artery walls	-	-	++

*Probe specific for transcripts containing tyrosine protein kinase sequences.
**Probe specific for transcripts encoding noncatalytic *trk*B receptors.

trk transcripts are also absent in other related structures of the brain, such as the trigeminal nerve that arrives into the *trk*-expressing trigeminal ganglion from the base of the myelencephalon. Analysis of late embryos (E17.5) indicates that *trk* gene expression continues to be limited to visceral sensory ganglia of neural crest origin (dorsal root, trigeminal, and superior ninth and tenth ganglia) (Martin-Zanca et al., 1990). No *trk* expression could be detected in other sensory ganglia of nonneural crest origin (acoustic ganglion) or in other cranial ganglia that are not strictly sensory (Table 1). These studies, albeit informative, only provide a limited window of the pattern of *trk* expression in the nervous system. A systematic analysis of the expression of the *trk* proto-oncogene in mouse embryos as well as in adult tissues should provide a more comprehensive understanding of the involvement of this receptor in the development and function of the nervous system. Moreover, these studies will also serve to establish a functional correlation between the presence of *trk* receptors and the biological effects attributed to NGF.

III. *TRKB* RECEPTORS

IIIA. Molecular Structure

The *trk*B locus was first identified in our laboratory by screening a mouse brain cDNA library with a human *trk* proto-oncogene probe at low stringency (Klein et al., 1989, 1990a). Rat *trk*B cDNA clones have been subsequently isolated from a rat cerebellar cDNA library using a probe specific for the conserved HRDLAAR kinase domain sequence (Middlemas et al., 1991). Unlike the *trk* proto-oncogene, the *trk*B locus exhibits a complex transcriptional pattern (at least eight different transcripts ranging in size from 0.7 kb to 9.0 kb) and encodes at least two different classes of receptors.

One of the products of the *trk*B gene is $gp145^{trkB}$, a heavily glycosylated molecule of 821 amino acid residues that contains all of the classical motifs of cell surface receptors of the tyrosine protein kinase family (Klein et al., 1989) (Fig. 1). The predicted 31 amino acid-long signal peptide is followed by an extracellular domain that contains the same basic structural motifs as $gp140^{trk}$, a three-subunit LRM flanked by two cysteine-rich regions followed by two Ig-like domains of the C2 type (Schneider and Schweiger, 1991) (Fig. 1). The overall homology between the extracellular domains of the human $gp140^{trk}$ and the mouse $gp145^{trkB}$ proteins is 57% (38% identity). This homology is not randomly distributed. As shown in Figure 2, most of the homologous residues map within the second and third LRM motifs and particularly within the second Ig-like domain. In addition, each of the 12 extracellular cysteines present in $gp140^{trk}$ are conserved in $gp145^{trkB}$.

As expected, the highest homology (88%) between $gp145^{trkB}$ and $gp140^{trk}$ was found in their catalytic domains (Klein et al., 1989) (Fig. 2). Indeed, the *trk*B protein contains 92 of the 99 amino acid residues conserved between the *trk* proto-oncogene product and the other growth factor receptors of the tyrosine protein kinase gene family. These conserved residues include Tyr^{705}, the tyrosine presumably involved in autophosphorylation reactions. In *trk*B, Tyr^{705} is followed by another tyrosine residue, a feature previously observed in the *trk* proto-oncogene and in the insulin receptor subfamily (insulin, IGF-I receptors, and c-*ros*). In addition, the *trk*B kinase contains each of four distinctive features present in the human *trk* proto-oncogene receptor. They include (1) a single amino acid gap between residues 575 and 576 (residues 542 and 543 in human *trk*), (2) a threonine in position 678 (Thr^{647} in human *trk*), (3) a tryptophan in position 753 (Trp^{772} in human *trk*), and (4) the absence of the helix-breaking proline in position 797 (position 766 in human *trk*) (Fig. 2). gp145^{trkB} also resembles the human gp140^{trk} receptor in its carboxy-terminal domain. Both molecules exhibit a very short, highly conserved tail of 15 amino acids that includes a conserved tyrosine located six residues from their respective carboxy-terminus (Klein et al., 1989).

The *trk*B locus also codes for a second class of receptors that lack a catalytic kinase domain (Klein et al., 1990a) (Fig. 1). One of these receptors is gp95^{trkB}, a glycoprotein identical to gp145^{trkB} from amino acid residues 1 to 465, a region that encompasses the signal peptide, the extracellular domain, and the transmembrane region. However, gp95^{trkB} has a very short cytoplasmic domain of 23 residues, of which the last 11 bear no resemblance with any of the sequences present in gp145^{trkB}. Nucleotide sequence analysis of rat *trk*B cDNA clones suggests the existence of a second noncatalytic *trk*B protein that differs from gp95^{trkB} in its carboxy-terminus (Middlemas et al., 1991). Since the point of sequence divergence from the catalytic receptor occurs at the same residue, it is likely that these noncatalytic receptors are generated by differential splicing of alternative exons.

Northern blot analysis of mouse brain transcripts with various cDNA probes suggests that two abundant mRNAs (8.0 kb and 2.0 kb) may code for an additional noncatalytic receptor with a cytoplasmic domain identical to that of gp95^{trkB} but with a different amino-terminus (Klein et al., 1990a). Rat brain contains three additional *trk*B transcripts (0.7 to 1.6 kb) that do not have sufficient information to code for any of the known *trk*B products (Middlemas et al., 1991). These observations raise the possibility that the rat *trk*B locus may code for an additional cytoplasmic kinase (1.6 kb mRNA) and/or secreted extracellular domains (0.9 kb and 0.7 kb mRNAs) in a fashion reminiscent of some of the recently described EGF and FGF receptor isoforms. However, this possibility must await isolation of the corresponding cDNA clones and identification of their putative gene products.

IIIB. Biological Function

Recent studies have indicated that the gp145^{trkB} tyrosine protein kinase is a functional receptor for at least two different members of the NGF family of neurotrophins, brain-derived neurotrophic factor (BDNF) and the recently isolated *Xenopus* NT-4 (*X*NT-4) (Fig. 3). Experimental evidence supporting this claim includes:

1. BDNF (Klein et al., 1991b; Soppet et al., 1991; Squinto et al., 1991) and *X*NT-4 (Klein et al., 1992) efficiently bind to NIH3T3 cells expressing gp145^{trkB} but not gp140^{trk} or gp145^{trkC} receptors. In addition, BDNF has been shown to bind to mouse hippocampal cells known to express gp145^{trkB} (Soppet et al., 1991).
2. BDNF can be chemically cross-linked to gp145^{trkB} (Soppet et al., 1991; Squinto et al., 1991).
3. BDNF (Klein et al., 1991b; Soppet et al., 1991; Squinto et al., 1991) and *X*NT-4 (Klein et al., 1992) induce the rapid phosphorylation of gp145^{trkB} on tyrosine residues. No such effect can be observed with either gp140^{trk} or gp145^{trkC} receptors.
4. Coexpression of gp145^{trkB} with either BDNF (Klein et al., 1991b) or *X*NT-4 (Klein et al., 1992) in NIH3T3 cells leads to their morphologic transformation.
5. BDNF induces the survival of NIH3T3 cells in serum free media only when these cells express gp145^{trkB} (Glass et al., 1991).
6. Addition of BDNF or *X*NT-4 to PC12 cells ectopically expressing gp145^{trkB} receptors results in their differentiation into sympatheticlike neurons (Klein et al., 1992; Squinto et al., 1991).

Since NIH3T3 cells do not express gp75^{LNGFR} receptors (Cordon-Cardo et al., 1991), the above observations indicate that gp145^{trkB} must be sufficient to mediate signal transduction processes initiated by BDNF and *X*NT-4. Whether gp75^{LNGFR} receptors are required for other biological processes such as the observed differentiation of PC12 cells remains to be demonstrated. However, it is unlikely that gp75^{LNGFR} receptors are needed for the *in vivo* activities of BDNF and NT-4 through *trk*B receptors since this gene is expressed in a large variety of CNS and PNS structures (see below) in which there is no detectable expression of the LNGFR gene.

gp145^{trkB} can also serve as a receptor for NT-3 (Klein et al., 1991b; Soppet et al., 1991; Squinto et al., 1991) (Fig. 3). It has been shown that NT-3 can bind and be cross-linked to gp145^{trkB}. Moreover, NT-3 can induce the phosphorylation of gp145^{trkB} in tyrosine residues with an efficiency comparable to that of BDNF. However, the biological responses induced by NT-3 through gp145^{trkB} receptors are rather limited. For instance, NT-3 is at least 10-fold less efficient than BDNF in inducing the survival of NIH3T3 and the differentiation of PC12 cells expressing gp145^{trkB} receptors (Glass et al., 1991; Squinto et al., 1991). The differential effect of NT-3 and BDNF (or *X*NT-4) through gp145^{trkB} is even more dramatic in the NIH3T3 cell transformation assay, where NT-3 is at least two orders of magnitude less potent than BDNF or *X*NT-4 (Klein et al., 1991b, 1992). These observations raise the possibility that the interaction of NT-3 with gp145^{trkB} receptors may not have a major biological significance *in vivo*.

Berkemeier et al. (1991) have recently reported that gp145^{trkB} may also serve as a receptor for a novel mammalian neurotrophin designated NT-5. In addition, NT-5 appears to activate gp140^{trk} receptors and induce the differentiation of PC12 cells (Berkemeier et al., 1991). Interestingly, NT-5 is most closely related to XNT-4, a neurotrophin for which its mammalian homologue has not been yet identified. However, XNT-4 and NT-5 appear to differ in their ability to interact with gp140^{trk} receptors (Berkemeier et al., 1991; Klein et al., 1992). Additional work will be required to determine the functional and evolutionary relationships between these two neurotrophins.

IIIC. Expression Pattern

The *trk*B locus is widely expressed in the central nervous system (CNS) and peripheral nervous system (PNS) (Klein et al., 1990a,b) (Table 1). During mouse development, *trk*B transcripts can be first detected in E8.5 embryos. At this time, *trk*B is expressed in a wide variety of rudimental structures of the developing CNS and PNS, including the forebrain, caudal midbrain, hindbrain, spinal cord, the trigeminal ganglion, and differentiating neural crest cells, which will form the dorsal root ganglia (Klein et al., 1990b). In 13.5-day-old embryos, the pattern of *trk*B expression appears to reflect the entire network of the developing CNS and PNS (Table 1). *trk*B expression is maintained throughout this network during late fetal development, once the scaffolding for the entire nervous system is fully constructed (Klein et al., 1990b).

In the adult mouse, most *trk*B expression is found in the brain and the spinal cord (Klein et al., 1990a). However, low levels of certain *trk*B transcripts could also be observed in lung and muscle as well as in ovaries. Traces of two distinct mRNA species (2.8 kb and 2.5 kb) were found in testis. Whether these mRNAs correspond to *trk*B transcripts that use different polyadenylation signals or correspond to transcripts of a related locus remains to be determined. Adult rats also have the highest levels of *trk*B transcripts in the brain (Middlemas et al., 1991). Other tissues exhibiting low, albeit detectable, *trk*B expression include spleen, testes, and submaxillary gland.

In situ hybridization analysis of mouse brain tissue revealed that *trk*B transcripts can be identified in most structures of the cerebrum, including the cortical layers, thalamus, hippocampus, and choroid plexus (Klein et al., 1990a,b) (Table 1). Either low levels or absence of *trk*B transcripts occur in the corpus callosum, the parafascicular and reticular thalamic nuclei, and in the fimbria hippocampus. In the cerebellum, *trk*B expression appears to be more restricted, with high levels in the Purkinje cell layer and the caudal penduncle, but low levels or no expression in other neurons and glial and granule cells (Table 1).

Not all *trk*B transcripts are equally distributed throughout mouse neural tissues. Indeed, *in situ* hybridization studies using probes specific for transcripts encoding the catalytic (gp145^{trkB}) and noncatalytic (gp95^{trkB}) *trk*B gene products revealed a highly specialized pattern of expression (Klein et al., 1990a). Transcripts coding for gp145^{trkB} are present in the cerebral cortex and the pyramidal cell layer of the hippocampus, but not in the choroid plexus or the ependymal lining of the cerebral ventricles. In contrast, transcripts encoding the noncatalytic receptorlike gp95^{trkB} protein were found in the choroid plexus and the ependymal lining of the ventricles but not in the cerebral cortex or the hippocampus (Table 1).

This pattern of expression has provided some insights regarding the possible function of the noncatalytic *trk*B gene product(s) (Klein et al., 1990a). The ependyma is a ciliated epithelial cell layer that participates in the circulation of the cerebrospinal fluid from the ventricles into the interstitial space between neurons and neuroglia. The choroid plexus is known to serve as an active barrier between the blood supply and the cerebrospinal fluid. Therefore, it is possible that gp95^{trkB} may have a role in the active transport of any of the *trk*B ligands within the brain or even across the blood–brain barrier. Alternatively, gp95^{trkB} may serve as a clearance receptor. Interestingly, the receptor implicated in clearance of the atrial natriuretic peptide lacks the cytoplasmic guanyl cyclase domain present in a related atrial natriuretic peptide receptor involved in signal transduction. Further expression studies involving probes specific for each of the *trk*B transcripts products will be needed to unveil the functional complexity of the mammalian *trk*B locus.

IV. *TRK*C RECEPTORS

IVA. Molecular Structure

Recent studies aimed at identifying additional members of the *trk* gene family of neurotrophin receptors have led us to the isolation of a novel gene, *trk*C (Lamballe et al., 1991). The *trk*C locus encodes a wide repertoire of receptor isoforms including some without catalytic tyrosine kinase domains (Fig. 1). So far, we have identified five putative *Trk*C receptors based on molecular analysis of various mouse and porcine *trk*C cDNA clones. Below is a brief description of their structural features.

One of the predicted *trk*C products is a 825 amino acid polypeptide that exhibits the same structural features as the gp140^{trk} and gp145^{trkB} receptors (Lamballe et al., 1991; Schneider and Schweiger, 1991). They include a 31 amino acid-long signal peptide, a long extracellular domain with LRM, cysteine-rich and Ig-like C2 motifs, a single transmembrane domain, a 90 amino acid-long juxtamembrane region, a kinase catalytic domain, and a short (15 residues) carboxy-terminal tail with a single tyrosine residue (Fig. 1). The overall homology of this porcine *Trk*C protein to the products of the human *trk* and mouse *trk*B genes is 67% and 68%, respectively (Lamballe et al., 1991) (Fig. 2). Their external domains exhibit 54% (*trk*C and *trk*) and 53% (*trk*C and *trk*B) similarity. As expected, this homology is significantly higher in the kinase catalytic domain. The 267 amino acid residues of this region are 76% identical (87% homologous) to those of gp140^{trk} and 83% identical (88% homologous) to those of the mouse gp145^{trkB} receptor. More importantly, each of the 12 cysteine residues present in the extracellular domain of gp145^{trkB} and gp140^{trk} is also present in this *trk*C product. Similarly, its catalytic kinase domain exhibits the characteristic features of the *trk* family of receptors, including a threonine residue instead of the alanine present in the HRDLAARN motif of all other tyrosine protein kinases, a single amino acid gap (between residues 576 and 577), a tryptophan in position 757, and the absence of the helix-breaking proline in position 801 (Lamballe et al., 1991) (Fig. 2). Expression of this cDNA clone in NIH3T3 cells revealed a 145 kd glycoprotein that has been designated gp145^{trkC} (Lamballe et al., 1991).

The mammalian *trk*C locus encodes two additional tyrosine protein kinase receptor (Lamballe et al., unpublished observations). Molecular analysis of cDNA clones isolated from porcine and mouse cDNA libraries revealed the existence of two receptor isoforms with a sequence identical to that of gp145^{trkC} except for the presence of a stretch of 14 (TrkCK2) and 25 (TrkCK3) additional amino acid residues located two residues downstream from YY^{703}, the putative autophosphorylation site. Since these additional sequences are identical in the murine and porcine isolates, it is likely that they correspond to true receptor isoforms rather than to cloning artifacts. Moreover, biochemical studies have revealed that these receptors, have tyrosine protein kinase activity (see below).

In addition to these catalytic receptors, the mammalian *trk*C locus may encode noncatalytic isoforms similar to the gp95^{trkB} product of the related *trk*B gene (Lamballe et al., unpublished observations). Molecular analysis of *trk*C cDNA clones has revealed the existence of two putative noncatalytic receptors, each of which contains signal peptides, extracellular domains, and transmembrane regions identical to those of gp145^{trkC} (Fig. 1). However, these receptors have short cytoplasmic domains of 49 and 51 amino acid residues that include different amino acid sequences at the carboxy-terminus. In one case, the point of divergence with gp145^{trkC} occurs at the same position previously observed for the gp145^{trkB} and gp95^{trkB} receptors. In the other *trk*C noncatalytic receptor, the point of divergence occurs 30 amino acids downstream from the predicted cytoplasmic end of the transmembrane region (Fig. 1). Pending identification of these receptors *in vivo*, our findings suggest that the mammalian *trk*C locus may encode multiple cell surface receptors with distinct biological properties.

IVB. Biological Function

Recent studies have demonstrated that the gp145^{trkC} tyrosine protein kinase is a primary receptor for NT-3 (Lamballe et al., 1991) (Fig. 1). Experimental evidence supporting this claim includes:

1. NT-3 but not NGF, BDNF, or NT-4 bind to NIH3T3 cells expressing gp145^{trkC} receptors (Lamballe et al., 1991; Klein et al., 1992). As previously observed with gp140^{trk} receptors, NT-3 binds to gp145^{trkC} with high affinity (26 pM) without requiring the contribution of low-affinity gp75^{LNGFR} receptors (Lamballe et al., 1991).
2. NT-3 can be chemically cross-linked to gp145^{trkC}.
3. Addition of NT-3 but not of any of the other known members of the NGF neurotrophin family induces the phosphorylation of gp145^{trkC} receptors on tyrosine residues (Lamballe et al., 1991).
4. NT-3 induces DNA synthesis in NIH3T3 cells only when they express gp145^{trkC} receptors. Moreover, constitutive coexpression of NT-3 and gp145^{trkC} led to the efficient morphologic transformation of NIH3T3 cells (Lamballe et al., 1991).
5. NT-3 can induce the differentiation of PC12 cells into neuronlike cells upon ectopic expression of gp145^{trkC} receptors (Lamballe et al., unpublished observations).

As indicated above, NT-3 can also bind and mediate some of its biological properties through the related gp140^{trk} and gp145^{trkB} receptors. However, quantitative studies have indicated that gp145^{trkC} is 10- to 100-fold more efficient in mediating the neurotrophic properties of NT-3 and is likely to serve as its physiological receptor (Lamballe et al., 1991).

To date, we do not have sufficient information to postulate a biological role for the TrkC receptor isoforms. However, preliminary studies suggest that this receptor may play a physiological role *in vivo*. For instance, expression of cDNA clones encoding gp145^{trkC} K2 and K3 in NIH3T3 cells revealed a 150-155 kd glycoprotein with tyrosine protein kinase activity. Moreover, addition of NT-3 to these cells induces the rapid and efficient phosphorylation of TrkC isoforms on tyrosine residues (Lamballe et al., unpublished observations). Ongoing studies aimed at determining their pattern of expression in the CNS and PNS should provide relevant clues regarding the physiological significance of these TrkC receptors.

IVC. Expression Pattern

*trk*C is abundantly expressed in the adult mouse brain (Lamballe et al., 1991). *In situ* hybridization studies have revealed *trk*C expression in distinct brain structures, including the pyramidal cell layer of the hippocampus, the dentate gyrus, the cerebral cortex, and possibly the thalamus (Table 1). Moreover, we observed *trk*C transcripts in the granular cell layer of the cerebellum but not in the adjacent white matter or in the Purkinje cell layers (Lamballe et al., 1991).

TABLE 2. Involvement of *trk* Oncogenes in Human Neoplasia

Human tumor	Incidence	Mechanism of activation
Colon carcinoma	Sporadic	Replacement of ligand-binding domain by tropomyosin
Thyroid papillary carcinoma	17% (7/42)	Replacement of ligand-binding domain by heterologous sequences including tropomyosin and *tpr*

This pattern of expression in the cerebellum is somewhat complementary to that of *trk*B, which is expressed in Purkinje cells but not in granular cell layer (Table 1).

More recent studies conducted in developing mouse embryos have revealed that the *trk*C locus is as widely expressed throughout the CNS and PNS as *trk*B (Lamballe et al., unpublished observations). As summarized in Table 1, we have detected *trk*C transcripts in the developing mouse brain, spinal cord, dorsal root ganglia, cranial ganglia, snout, submaxillary gland, and various innervating tissues. In addition, we have observed intense hybridization in the artery walls and in the lining of the gut, two structures that so far do not appear to contain *trk*B transcripts (Lamballe et al., unpublished observations). These findings suggest that some *trk*C receptors may play a role in the development and/or maintenance of the vascular and enteric nervous systems. Ongoing studies utilizing probes specific for the various *trk*C receptor isoforms shall provide important information regarding their specific roles within the mammalian nervous system.

V. *TRK* GENES AND HUMAN NEOPLASIA

The *trk* oncogene (initially designated as *onc*D) was one of the first transforming genes identified in human malignancies (Martin-Zanca et al., 1986; Pulciani et al., 1982). This oncogene was detected during the course of gene transfer assays using DNA isolated from a colon carcinoma biopsy. However, recent studies indicate that *trk* oncogenes are more frequently activated in thyroid papillary carcinomas (Bongarzone et al., 1989) (Table 2). Below is a description of the molecular structures of the various human *trk* oncogenes characterized to date.

VA. *trk* Oncogenes: Molecular Structure

Molecular characterization of the colon carcinoma *trk* oncogene revealed a hybrid molecule consisting of seven of the eight exons of a nonmuscle tropomyosin gene followed by sequences coding for the transmembrane and cytoplasmic domains of the *trk* proto-oncogene (Martin-Zanca et al., 1986) (Fig. 4). Comparison of the deduced amino acid sequences of the *trk* oncogene and proto-oncogene products indicated that the transforming protein lacks the signal peptide and most of the putative extracellular ligand-binding domain (360 out of a total of 375 residues) of gp140^{trk}. However, no differences were observed within the remaining *trk* sequences, including the transmembrane and kinase catalytic domain. These observations indicate that the fusion of tropomyosin sequences is sufficient to confer transforming properties to the *trk* tyrosine kinase (Martin-Zanca et al., 1989).

The molecular structure of some of the *trk* oncogenes identified in thyroid tumors has been recently unveiled (Greco et al., 1992; M. Pierotti, personal communication). These *trk* oncogenes exhibit an overall structure similar to the original colon carcinoma isolate, coding for hybrid molecules containing at least the kinase catalytic domain and the short carboxy-terminal tail of the *trk* gene product (Fig. 4). However, the *trk* oncogenes isolated from thyroid tumors differ in the nature of the activating sequences that replaced the *trk* extracellular domain. Three of seven thyroid *trk*

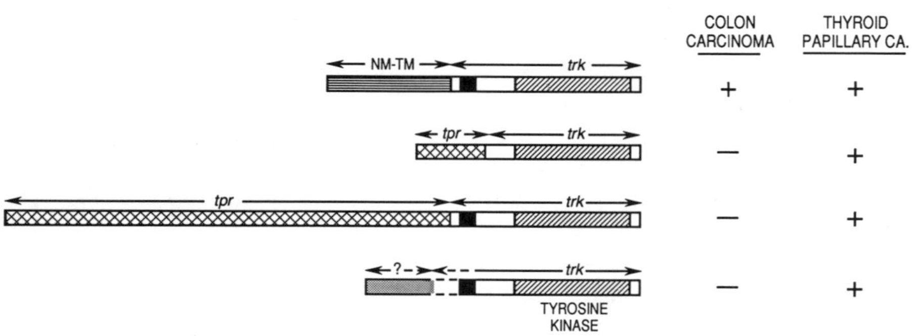

Fig. 4. Schematic representation of *trk* oncogenes isolated from human colon and thyroid papillary carcinomas. Sequences derived from the *trk* proto-oncogene product (*trk*), nonmuscle tropomyosin (NM-TM), and the *tpr* gene product (*tpr*) are indicated by arrows. Sequences of unknown origin are indicated by a question mark. Dotted lines indicate that the precise point of recombination has not been determined. The tyrosine protein kinase domain present in each of these *trk* oncogenes is indicated by a hatched box. The black box represents the transmembrane domain of the *trk* proto-oncogene product.

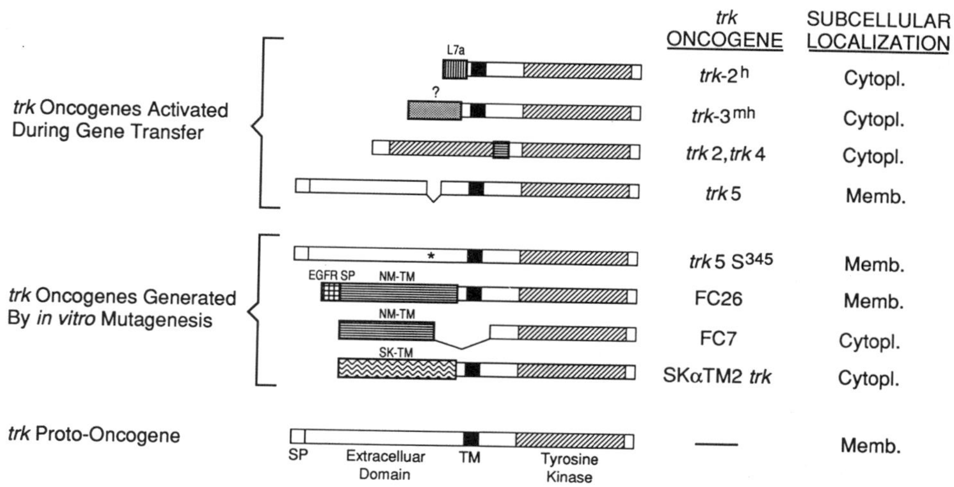

Fig. 5. Schematic representation of the various *trk* oncogenes activated *in vitro*. Sequences derived from the *trk* proto-oncogene are represented by the smaller boxes. Thicker boxes correspond to sequences derived from the ribosomal protein L7a (*trk*-2^h oncogene), nonmuscle tropomyosin (NM-TM) (*trk*2, *trk*4, FC26 and FC7 oncogenes), EGF receptor signal peptide (EGFRSP) (FC26 oncogene), and skeletal tropomyosin (SK-TM) (SKTM2*trk* oncogene). The unknown sequences responsible for the activation of the *trk*-3^{mh} oncogene are indicated by a question mark. The lines angling down indicate deleted sequences (*trk*5 and FC7 oncogenes). The asterisk present in the *trk*5 S^{345} oncogene represents a point mutation replacing the cysteine residue located at position 345 of the *trk* proto-oncogene by a serine. The subcellular localization of each of these *trk* oncoproteins is also indicated.

oncogenes so far characterized contain the same activating nonmuscle tropomyosin sequences that participated in the generation of the colon carcinoma oncogene (M. Pierotti, personal communication). In contrast, three of these *trk* oncogene isolates contain sequences derived from the *tpr* locus (Greco et al., 1992), a gene first identified as part of the human *met* oncogene. The breakpoint that led to the generation of these *tpr-trk* oncogenes appears to be different in each tumor. One of the *tpr-trk* oncogenes, *trk*T1, encodes a 55 kd protein that only contains cytoplasmic *trk* sequences (Greco et al., 1992) (Fig. 4). In contrast, the second *tpr-trk* isolate, *trk*T2, encodes a large molecule of about 150 kd that contains additional *tpr* sequences linked to the same subset of *trk* sequences present in the colon carcinoma *trk* oncogene (M. Pierotti, personal communication). Finally, the nature of the sequences present in the seventh thyroid *trk* oncogene isolate remains to be identified.

The three loci so far implicated in the generation of *trk* oncogenes, *trk*, nonmuscle tropomyosin, and *tpr*, map in the long arm of chromosome 1. Preliminary molecular studies have indicated that tumor cells contain the rearranged extracellular *trk* sequences displaced by the activating tropomyosin or *tpr* genes (M. Pierotti, personal communication). Therefore, it is likely that *trk* oncogenes become activated in human neoplasia by chromosomal abnormalities involving either inversions or unequal crossovers between the long arm of chromosome 1.

trk oncogenes can also be generated by *in vitro* recombinational events. Transfection of rodent fibroblasts with either human genomic DNA or a recombinant phase containing the transmembrane and cytoplasmic domains of the human *trk* proto-oncogene resulted in the generation of two additional *trk* oncogenes designated as *trk*-2^h and *trk*-3^{mh}, respectively (Kozma et al., 1988). These oncogenes depict the same breakpoint within *trk* gene sequences as the colon carcinoma isolate. However, they differ in the amino-terminal sequences, one of which (*trk*-2^h) has been shown to be derived from the human ribosomal protein L7a (Ziemiecki et al., 1990) (Fig. 5).

Transfection of rodent fibroblasts with cDNA clones encompassing either the entire *trk* proto-oncogene coding sequences or its tyrosine kinase catalytic domain result in the generation of *trk* oncogenes with an unusually high frequency (one *trk* oncogene per 10^3–10^4 transfected NIH3T3 cells) (Oskam et al., 1988). Forty-one different *trk* oncogenes have been identified by this experimental approach. Immunoprecipitation analysis of each of these *trk* oncogene NIH3T3-transformed cells revealed *trk*-related proteins of abnormal sizes. These observations suggest that these *trk* oncogenes were generated by either rearrangements within the transfected *trk* sequences or by their recombination with the host genome.

Three of these *in vitro*-generated *trk* oncogenes, *trk*2, *trk*4, and *trk*5, have been characterized at the molecular level (Coulier et al., 1990) (Fig. 5). *trk*2 and *trk*4 consist of similar head-to-tail arrangements of two tyrosine kinase catalytic domains separated by short hydrophilic regions. In contrast, *trk*5 encodes a cell surface glycoprotein identical to $gp140^{trk}$

except for a 51 amino acid residue deletion within the second extracellular Ig-like domain (Fig. 5). This deletion encompasses one of the 12 cysteine residues conserved among members of the mammalian *trk* gene family. Replacement of this conserved cysteine residue by serine using site-directed mutagenesis resulted in the generation of a novel *trk* transforming gene (Coulier et al., 1990). These results illustrate that small mutations within the *trk* locus (without intervention of foreign sequences) can also activate the neoplastic potential of the human *trk* proto-oncogene (Fig. 5).

VB. Mechanism of Activation of *trk* Oncogenes

Comparison of the various amino-terminal sequences of *trk* oncogenes activated in human tumors failed to reveal any common structural motif that might be considered responsible for their transforming properties. A common feature of these *trk* oncogenes is that they have lost their signal peptide sequences, therefore their products are preferentially localized in the cytoplasm. However, several lines of evidence argue against the possibility that a change in the subcellular localization of the *trk* kinase might be responsible for its neoplastic properties (Barbacid et al., 1991).

Biochemical analysis of the various *trk* oncogene products revealed that the levels of the *in vitro* tyrosine protein kinase activity of these oncoproteins was similar to that observed with the normal $gp140^{trk}$ receptor. Similar results have been obtained *in vivo* when we have examined the levels of tyrosine phosphorylation of these proteins by Western blot analysis using antiphosphotyrosine antibodies. However, under these conditions, tyrosine phosphorylation of $gp140^{trk}$ was absolutely dependent upon the presence of its cognate ligand, NGF. These observations suggest that *trk* oncogenes owe their transforming activity to the constitutive activation of their respective catalytic domains. Further support for this hypothesis is provided by our recent observations that constitutive expression of each of the *trk* receptors with their respective ligands in NIH3T3 cells induces their morphologic transformation with an efficiency comparable to that observed with the colon carcinoma *trk* oncogene (Cordon-Cardo et al., 1991).

Physiological activation of tyrosine protein kinase receptors by their cognate ligands is likely to be mediated by an oligomerization step that results in *trans*-phosphorylation of specific tyrosine residues (Ullrich and Schlessinger, 1990). Preliminary results obtained in our laboratory suggest that $gp140^{trk}$ receptors also become activated by an NGF-induced dimerization step (Jing et al., unpublished observations). Therefore, it is possible that *trk* oncoproteins may owe their transforming properties to the formation of stable dimeric structures that result in their constitutive activation. For instance, the activating deletions and miscoding mutations in the extracellular domain of certain *trk* oncogenes may confer to their gene products a structure that mimics the conformational changes induced by NGF in the normal *trk* receptor. Dimerization may also take place in the products of the tropomyosin-containing *trk* oncogenes, since tropomyosins are known to exist as dimeric α-helical coiled-coil structures. In support of this hypothesis, replacement of the nonmuscle tropomyosin sequences present in certain *trk* oncogenes by those encoding the skeletal muscle tropomyosin isoforms results in novel *trk* chimeric genes with efficient transforming activity (Barnes et al., 1989) (Fig. 5).

The *in vitro*-generated *trk*2 and *trk*4 oncogenes may become activated by reciprocal phosphorylation in *cis* of their two kinase catalytic domains. The *tpr*/*trk* oncoproteins may also dimerize since the *tpr*-derived *tpr* sequences are rich in leucine residues, which may form leucine zipper-like structures. Moreover, preliminary results indicate that *tpr* proteins may also dimerize through the formation of disulfur bridges (M. Park, personal communication). These findings, taken together, strongly suggest that dimerization of the *trk* tyrosine protein kinase domain plays a major role in the malignant activation of *trk* oncogenes. However, the possibility that *trk* oncoproteins may owe their neoplastic properties to mechanisms that do not involve protein kinase dimerization cannot be ruled out at the present time.

To date, there is no evidence for the involvement of the *trk*B and *trk*C genes in human neoplasia. However, studies conducted in our laboratory indicate that both of these receptors can acquire transforming properties by recombinational events similar to those described above for the *trk* locus. Moreover, the demonstration that constitutive coexpression of each of the members of the *trk* family of receptors with their cognate ligands leads to morphologic transformation of NIH3T3 cells raises the possibility that these receptors may participate in human neoplasia by mechanisms that do not require structural rearrangements. For instance, ectopic expression of normal *trk* receptors in cells either physiologically or abnormally exposed to their cognate neurotrophins may alter their normal proliferative and/or developmental programs and contribute to the development of certain human neoplasias.

REFERENCES

Barbacid, M., F. Lamballe, D. Pulido, et al. (1991) The *trk* family of tyrosine protein kinase receptors. BBA Reviews on Cancer 1072:115–127.

Barnes, D., L. Clayton, G. Chumbley, et al. (1989) Activation of the *trk* oncogene by alternatively spliced muscle and non-muscle tropomyosin sequences. Oncogene 4:259–262.

Berkemeier, L.R., J.W. Winslow, D.R. Kaplan, et al. (1991) Neurotrophin-5: A novel neurotrophic factor that activates *trk* and *trk*B. Neuron 7:857–866.

Bongarzone, I., M.A. Pierotti, N. Monzini, et al. (1989) High frequency of activation of tyrosine kinase oncogenes in human papillary thyroid carcinoma. Oncogene 4:1457–1462.

Bothwell, M. (1991) Keeping track of neurotrophin receptors. Cell 65:915–918.

Bottaro, D.P., J.S. Rubin, D.L. Faletto, et al. (1991) Identification of the hepatocyte growth factor receptor as the c-met proto-oncogene product. Science 251:802–804.

Cordon-Cardo, C., P. Tapley, S.Q. Jing, et al. (1991) The *trk* tyrosine protein kinase mediates the mitogenic properties of nerve growth factor and neurotrophin-3. Cell 66:173–183.

Coulier, F., R. Kumar, M. Ernst, et al. (1990) Human *trk* oncogenes activated by point mutation, in-frame deletion, and duplication of the tyrosine kinase domain. Mol. Cell. Biol. 10:4202–4210.

Curran, T, and B.R. Franza, Jr. (1988) Fos and Jun: The Ap-1 connection. Cell 55:395–397.

Deuel, T.F., J.S. Huang, S.S. Huang, et al. (1983) Expression of a platelet-derived growth factor-like protein in simian sarcoma virus transformed cells. Science 221:1348–1350.

Doolittle, R.F., M.W. Hunkapiller, L.E. Hood, et al. (1983) Simian sarcoma virus onc gene, v-sis, is derived from the gene (or genes) encoding a platelet-derived growth factor. Science 221:275–277.

Downward, J., Y. Yarden, E. Mayes, et al. (1984) Close similarity of epidermal growth factor receptor and v-erb-B oncogene protein sequences. Nature 307:521–527.

Ghosh, S., A.M. Gifford, L.R. Riviere, et al. (1990) Cloning of the p50 DNA binding subunit of NF-kappa B: Homology to rel and dorsal. Cell 62:1019–1029.

Glass, D.J., S.H. Nye, P. Hantzopoulos, et al. (1991) *Trk*B mediates BDNF/NT-3-dependent survival and proliferation in fibroblasts lacking the low affinity NGF receptor. Cell 66:405–413.

Greco, A., M.A. Pierotti, I. Bongarzone, et al. (1992) *TRK*-T1 is a novel oncogene formed by the fusion of *TPR* and *TRK* genes in human papillary thyroid carcinomas. Oncogene (in press).

Hanks, S.K., A.M. Quinn, and T. Hunter (1988) The protein kinase family: Conserved features and deduced phylogeny of the catalytic domains. Science 241:42–52.

Hefti, F. (1992) Growth factors and neurodegeneration. Neurodegenerative Diseases (in press).

Hempstead, B.L., D. Martin-Zanca, D.R. Kaplan, et al. (1991) High-affinity NGF binding requires coexpression of the *trk* proto-oncogene and the low-affinity NGF receptor. Nature 350:678–683.

Huang, E., K. Nocka, D.R. Beier, et al. (1990) The hematopoietic growth factor KL is encoded by the S1 locus and is the ligand of the c-kit receptor, the gene product of the W locus. Cell 63:225–233.

Jing, S., P. Tapley, and H. Barbacid (1992) Nerve growth factor mediates signal transduction through TrK homodimer receptors. Neuron, in press.

Johnson, D., A. Lanahan, C.R. Buck, et al. (1986) Expression and structure of the human NGF receptor. Cell 47:545–554.

Kaplan, D.R., D. Martin-Zanca, and L.F. Parada (1991a) Tyrosine phosphorylation and tyrosine kinase activity of the *trk* proto-oncogene product induced by NGF. Nature 350:158–160.

Kaplan, D.R., B.L. Hempstead, D. Martin-Zanca, et al. (1991b) The *trk* proto-oncogene product: A signal transducing receptor for nerve growth factor. Science 252:554–558.

Kiernan, M., V. Blank, F. Logeat, et al. (1990) The DNA binding subunit of NF-kappa B is identical to factor KBF1 and homologous to the rel oncogene product. Cell 62:1007–1018.

Klein, R., D. Conway, L.F. Parada, et al. (1990a) The *trk*B tyrosine protein kinase gene codes for a second neurogenic receptor that lacks the catalytic kinase domain. Cell 61:647–656.

Klein, R., S.Q. Jing, V. Nanduri, et al. (1991a) The *trk* proto-oncogene encodes a receptor for nerve growth factor. Cell 65:189–197.

Klein, R., F. Lamballe, S. Bryant, and M. Barbacid (1992) The *trk*B tyrosine protein kinase is a receptor for neurotrophin-4. Neuron 8:947-956.

Klein, R., D. Martin-Zanca, M. Barbacid, et al. (1990b) Expression of the tyrosine kinase receptor gene trkB is confined to the murine embryonic and adult nervous system. Development 109:845–850.

Klein, R., V. Nanduri, S.A. Jing, et al. (1991b) The *trk*B tyrosine protein kinase is a receptor for brain-derived neurotrophic factor and neurotrophin-3. Cell 66:395–403.

Klein, R., L.F. Parada, F. Coulier, et al. (1989) *trk*B, a novel tyrosine protein kinase receptor expressed during mouse neural development. EMBO J. 8:3701–3709.

Kozma, S.C., S.M. Redmond, X.C. Fu, et al. (1988) Activation of the receptor kinase domain of the *trk* oncogene by recombination with two different cellular sequences. EMBO J. 7:147–154.

Lamballe, F., R. Klein, and M. Barbacid (1991) *trk*C, a new member of the *trk* family of tyrosine protein kinases, is a receptor for neurotrophin-3. Cell 66:967–979.

Lee, K.F., L.J. Huber, S.C. Landis, A.H. Sharpe, M.V. Chao, and R. Jaenisch (1992) Targeted mutation of the gene encoding the low affinity NGF receptor p75 leads to deficits in the peripheral sensory nervous system. Cell 69:737-749.

Loeb, D.M., J. Maragos, D. Martin-Zanca, et al. (1991) The *trk* proto-oncogene rescues NGF responsiveness in mutant NGF-nonresponsive PC12 cell lines. Cell 66:961–966.

Martin-Zanca, D., M. Barbacid, and L.F. Parada (1990) Expression of the *trk* proto-oncogene is restricted to the sensory cranial and spinal ganglia of neural crest origin in mouse development. Genes Dev. 4:683–694.

Martin-Zanca, D., S.H. Hughes, and M. Barbacid (1986) A human oncogene formed by the fusion of truncated tropomyosin and protein tyrosine kinase sequences. Nature 319:743–748.

Martin-Zanca, D., R. Oskam, G. Mitra, et al. (1989) Molecular and biochemical characterization of the human *trk* proto-oncogene. Mol. Cell. Biol. 9:24–33.

Middlemas, D.S., R.A. Lindberg, and T. Hunter (1991) *trk*B, a neural receptor protein-tryrosine kinase: Evidence for a full-length and two truncated receptors. Mol. Cell. Biol. 11:143–153.

Naldini, L., E. Vigna, R.P. Narsimhan, et al. (1991) Hepatocyte growth factor (HGF) stimulates the tyrosine kinase activity of the receptor encoded by the proto-oncogene c-MET. Oncogene 6:501–504.

Nebreda, A.R., D. Martin-Zanca, D.R. Kaplan, et al. (1991) Induction by NGF of meiotic maturation of *Xenopus* oocytes expressing the *trk* proto-oncogene product. Science 252:558–561.

Oskam, R., F. Coulier, M. Ernst, et al. (1988) Frequent generation of oncogenes by *in vitro* recombination of TRK proto-oncogene sequences. Proc. Natl. Acad. Sci. USA 85:2964–2968.

Pulciani, S., E. Santos, A.V. Lauver, et al. (1982) Oncogenes in solid human tumours. Nature 300:539–542.

Pulido, D., S. Campuzano, T. Koda, et al. (1992) D*trk*, a *Drosophila* gene related to the *trk* family of neurotrophin receptors, encodes a novel class of neural cell adhesion molecule. EMBO J. 11:391–404.

Radeke, M.J., T.P. Misko, C. Hsu, et al. (1987) Gene transfer and molecular cloning of the rat nerve growth factor receptor. Nature 325:593–597.

Rodriguez-Tebar, A., G. Dechant, and Y.A. Barde (1990) Binding of brain-derived neurotrophic factor to the nerve growth factor receptor. Neuron 4:487–492.

Sagata, N., N. Watanabe, G.F. Vande Woude, et al. (1989) The c-mos proto-oncogene product is a cytostatic factor responsible for meiotic arrest in vertebrate eggs. Nature 342:512–518.

Sap, J., A. Munoz, K. Damm, et al. (1986) The c-erb-A protein is a high-affinity receptor for thyroid hormone. Nature 324:635–640.

Schneider, R., and M. Schweiger (1991) A novel modular mosaic of cell adhesion motifs in the extracellular domains of the neurogenic *trk* and *trk*B tyrosine kinase receptors. Oncogene 6:1807–1811.

Sheer, C.J., C.W. Rettenmier, R. Sacca, et al. (1985) The c-fms proto-oncogene product is related to the receptor for the mononuclear phagocyte growth factor, CSF-1. Cell 41:665–676.

Soppet, D., E. Escandon, J. Maragos, et al. (1991) The neurotrophic factors brain-derived neurotrophic factor and neurotrophin-3 are ligands for the *trk*B tyrosine kinase receptor. Cell 65:895–903.

Squinto, S.P., T.N. Stitt, T.H. Aldrich, et al. (1991) trkB encodes a functional receptor for brain-derived neurotrophic factor and neurotrophin-3 but not nerve growth factor. Cell 65:886–893.

Ullrich, A., and J. Schlessinger (1990) Signal transduction by receptors with tyrosine kinase activity. Cell 61:203–212.

Ziemiecki, A., R.G. Muller, X.C. Fu, et al. (1990) Oncogenic activation of the human *trk* proto-oncogene by recombination with the ribosomal large subunit protein L7a. EMBO J. 9:191–196.

Zsebo, K.M., D.A. Williams, E.N. Geissler, et al. (1990) Stem cell factor is encoded at the Sl locus of the mouse and is the ligand for the c-*kit* tyrosine kinase receptor. Cell 63:213–224.

Chapter 10

Tumor Suppressor Genes

Arnold J. Levine

INTRODUCTION: HISTORICAL PERSPECTIVES

The idea that one or more gene products could suppress or inhibit the division of cells in a tumor derives from experiments first carried out using somatic cell genetics (Harris et al., 1969). In these experiments a normal cell in culture was fused with a cancer cell that was able to produce tumors in animals. The hybrid cell most often grew well in cell culture and even retained many of the properties of the cancerous cell, but it no longer produced tumors in animals. This result was obtained independently of the etiology of the tumorigenic cell line, which could be derived from a viral or chemical induction of tumors or even from spontaneous tumors. The result was also independent of the tumor type or lineage of the tumor tissue, so that tumorigenic epithelial cells, fibroblasts, or lymphocytes, when fused with a normal cell, usually failed to produce tumors when injected into animals. As these experiments progressed, occasionally a hybrid cell derived from one normal parent cell and a tumorigenic cell would produce a tumor in an animal. When the cells from these tumors were analyzed it became clear that the hybrid cell had lost one or more chromosomes supplied from the normal parent. This suggested that the normal cell could contribute a gene(s) whose product suppressed the ability to form tumors. The analysis of which chromosomes carried these putative tumor suppressor genes indicated that several distinct chromosomes could suppress tumor formation, depending upon the origin of the tumor-producing cell in the hybrid. The notion that the cancerous cells had a mutant form of this tumor suppressor gene that was not functioning and that different tumors carried different mutant tumor suppressor genes was made more plausible by experiments in which two different tumorigenic cell lines were fused together to produce a hybrid cell. This hybrid cell no longer produced tumors in animals, suggesting that genetic complementation could explain this observation and that at least two distinct genes in different cell lines were involved.

This idea received support and even confirmation from epidemiological studies of childhood cancers. In 1971 Knudson was reviewing the possible role of inherited alterations in retinoblastoma. He noted that 40% of the cases of retinoblastoma occurred in very young infants (mean age of 14 months). These tumors were commonly bilateral (occurred in both eyes) and these patients had an average of three independent tumors in both eyes (Knudson, 1971). These patients were sometimes found in families with a history of retinoblastoma, indicating an inherited predisposition. In such families there is a 95% chance to develop retinoblastoma if the mutant allele is passed to the offspring. Knudson explained these observations by suggesting that the susceptible offspring inherited one mutant allele at a locus called the retinoblastoma susceptibility gene (Rb gene) and one normal allele at that locus. A subsequent spontaneous mutation in the wild-type gene would then destroy both alleles and a cancer would develop. Knudson went on to note that 60% of the retinoblastomas did not fit this pattern. In these cases there was no family history of retinoblastomas. These tumors were first detected at an older age in children (mean age of 30 months) and were always unilateral with only one tumor per patient. In fact, these tumors were quite rare, occurring in one in 30,000 people. Knudson postulated that these tumors must arise from two independent mutational events, each one at the Rb locus in the two different chromosomes and in the same cell, the retinoblast. When both alleles were inactivated by mutation, a tumor would arise. In this way the same hypothesis was used to explain the two observed different patterns of retinoblastomas in the population: inherited and spontaneous.

Inherent in this hypothesis was the concept of a gene and product that normally prevents a cancer or tumorous growth and must be lost or mutated to initiate a cancer. That is a tumor suppressor gene. The Rb gene was mapped to chromosome 13 band 14q and was isolated by DNA cloning at this locus (Friend et al., 1987). The availability of these DNA

Molecular Genetics of Nervous System Tumors, pages 137-143 ©1993 Wiley-Liss, Inc.

clones permitted a test of the Knudson hypothesis, which proved to be correct. Retinoblastomas have both alleles altered by mutation or, more commonly, one allele is mutant while the second allele is lost (via deletion or chromosome loss) or reduced to homozygosity by a gene conversion or a recombination event (Cavenee et al., 1983). The individuals who have bilateral retinoblastoma and multiple independent tumors inherited one mutant allele. The late-onset unilateral retinoblastoma was associated with two mutational events at the Rb locus in the retinoblast tumor and normal alleles in the remainder of the somatic tissue.

A third line of evidence that suggested the existence of tumor suppressor genes came from an examination of the karyotypes of cancer cells. Tumor cells can reproducibly lose selected chromosomes or specific portions of chromosomes (deletions) (Rowley, 1984). During the past 10 years, specific DNA probes detecting restriction site polymorphisms have allowed the documentation and mapping of these lost or altered DNA species. An example of these types of alterations and the gene involved (when known) is presented in Table 1. These chromosomal alterations permitted the mapping and DNA cloning of several tumor suppressor genes. This in turn has led to a clear test of the concept that some genes are inactivated by mutation in both alleles and that this loss of function correlates with the cancerous cell growth in a tumor. The availability of these genes has permitted a test of the idea that returning these genes to a tumorigenic cell would restore the cell to a normal phenotype, i.e., the cells would no longer produce tumors in animals. A test of this has shown that this is the case for Rb (Huang et al., 1988) and p53 (Chen et al., 1990). That of course provided us with the original definition of a tumor suppressor gene in the somatic cell genetic experiments.

TABLE 1. Human Tumor Suppressor Loci

Tumor	Chromosomes involved (lost or altered)	Known tumor suppressor genes
1. Retinoblastoma	13q14	Rb
2. Wilms' tumor	11p13	Wt-1
3. Glioblastoma multiforme (astrocytoma)	17p13	p53
	9p, 10	Not known
4. Breast cancer	13p14	Rb
	17p13	p53
	1p, 1q, 11p, 18q	Not known
5. Small-cell lung cancer	13q14	Rb
	17p13	p53
	3p	Not known
6. Colorectal cancer	5q21	APC, MCC
	17p13	p53
	18q21	DCC
7. Neurofibroma	17q	NF-1
8. Neuroblastoma	1p, 14q, 17	Not known
9. Meningioma	22	Not known

TABLE 2. Properties of Oncogenes and Tumor Suppressor Genes

Property of the mutant	Proto-oncogenes	Tumor suppressor genes
1. Number of events required for cancer	One	Two
2. Function of the mutant allele	Gain of function, acts in a dominant fashion	Loss of function, acts in a recessive fashion
3. Mutant allele inherited through the germline	Never	Frequently has an inherited form
4. Somatic mutation in cancerous tissue	Yes	Yes
5. Tissue specificity of mutational event	Some, but can act in many tissues	The inherited form commonly shows a tissue preference

There are a number of distinctions between the oncogenes (see Chapter 7, this volume) and the tumor suppressor genes, and these are reviewed in Table 2. It is clear that the inherited predispositions of cancers are best explained by the existence of tumor suppressor genes. In addition, the tissue-specific cancers (Rb; retinoblastoma, osteogenic sarcoma) or tissue preferences (p53; adrenocortical carcinoma, osteosarcoma, breast) (Malkin et al., 1990) associated with genetic predispositions in families are best understood in the context of the tumor suppressor genes. Both oncogene and tumor suppressor gene mutations can occur in somatic cells. Together they contribute to cancerous cell growth and division. In this chapter we review in more detail each of the known tumor suppressor genes that recently have been isolated.

THE RETINOBLASTOMA GENE AND PROTEIN

The Rb gene is localized on chromosome 13q14 and encompasses about 200 kb of DNA. The gene has 27 exons from which the primary transcript is spliced together to produce an mRNA that is synthesized in many different cell or tissue types. The Rb protein is 105–110 kd and it is located in the nucleus of the cell. In addition to retinoblastomas, which carry mutant Rb alleles, all osteogenic sarcomas and all small-cell lung carcinomas also suffer mutations in both Rb alleles. About one-third of the bladder cancers and mammary carcinomas are found to have Rb mutations in both copies of this gene (Horowitz et al., 1990). Deletion mutations in this gene are known (true loss-of-function mutations) but many point mutations, which result in polypeptide chain termination (nonsense or stop codons) or errors in splicing the mRNA, functionally deleting an exon, are also known to occur. Point mutations producing missense proteins have been observed and tend to cluster in two regions of the protein, located between amino acids 393–572 and 646–772. These are the very same portions of the protein that are thought to interact with other cellular proteins and even viral oncogene products (see Chapter 11, this volume) (Hu et al., 1990). Thus, point mutations in these critical regions of the protein appear to inactivate Rb function as a negative regulator of cell division.

Many cells that divide in culture have normal levels of the wild-type Rb protein throughout the cell cycle. During the G_0 and G_1 phase of the cell cycle, the Rb protein is not phosphorylated or is underphosphorylated. In late G_1 and S-phase, however, the Rb protein becomes heavily phosphorylated at multiple sites (Buchkovich et al., 1989). This is carried out by one of the cell cycle kinases, cdc-2-like kinase, at serine–proline sites in the protein (Lees-Miller and Anderson, 1991). Thus, a cyclin-dependent kinase activity, active in late G_1–S-phase, alters the Rb protein as cells enter the S-phase. During mitosis, a phosphatase activity removes these phosphate groups from Rb protein in the cell. During G_0 and G_1, the unphosphorylated Rb protein is found to be associated with a number of other cellular proteins. Two closely related transcription factors called E2F (Chellappan et al., 1991) and DRTF1 (Bandara et al., 1991) are known to bind to Rb, but when Rb becomes phosphorylated in late G_1–S-phase the Rb protein no longer is complexed with E2F/DRTF1 proteins (Chellappan et al., 1991). The release of E2F/DRTF1 from this Rb complex is correlated with an increased affinity or ability of these transcription factors to bind to specific DNA sequences that regulate genes and an increased level of transcription of these genes (Raychaudhuri et al., 1990). E2F binding sites have been identified in the region of a number of genes that produce enzymes that synthesize nucleotides and polymerize DNA. Recently the Rb protein was shown to bind *in vitro* with members of the myc family of oncogenes, which are another class of transcription factors (Rustgi et al., 1991) involved in cell cycle control. Thus, it appears that the function of Rb is to regulate the level of transcriptional activity of several key gene products that are critical for entry into the S-phase of the cell cycle. In G_0–G_1 the Rb-E2F/DRTF1 complex is inactive for transcription. In late G_1 the cdc-2-like kinase phosphorylates Rb, which releases free E2F/DRTF1, which in turn positively regulates or activates critical genes for entry into S-phase. Dephosphorylation of Rb in M-phase restores the Rb-E2F/DRTF1 complex in G_0. The cyclin-activated kinase (cdc-2) regulates the timing of these events in the cell cycle and Rb is then a negative regulator of key transcription factors. Loss of Rb in a cancerous cell should result in constitutive activity of E2F/DRTF1. As will become clear in the next chapter (Chapter 11, this volume), the DNA tumor viruses act upon Rb in much the same manner as the cdc-2 kinase activity.

THE P53 GENE AND PROTEIN

The p53 gene in humans encompasses about 20 kilobases located on chromosome 17p13.1 (Benchimol et al., 1985). The gene is composed of 11 exons and the first exon (213 base pairs) is a noncoding mRNA leader sequence. The gene produces an mRNA of 2.2–2.5 kb in size and the p53 protein is expressed in all cells and tissues of the body. Little is known about the regulation of this gene in different tissues.

Individuals with the Li-Fraumeni syndrome come from families where many closely related members acquire cancer, frequently at a young age (25–35 years old). These same people are at high risk for acquiring a second independent tumor. Many Li-Fraumeni families carry and pass on germline mutations in one p53 allele (Li, 1988; Malkin et al., 1990; Srivastava et al., 1990). The cancers that are observed in these families are varied but certain tissue types tend to be more common (breast cancer, osteogenic sarcomas). These inherited p53 mutations are all missense mutations producing an altered or faulty protein and in all cases the individual is heterozygous for these mutations. Mutations of this type occur in several different codons but they are not randomly distributed in the gene. There is no indication that the presence of one mutant allele has disturbed any developmental process or otherwise normal functions. Interestingly, the introduction of a missense mutant p53 gene in transgenic mice produces a similar phenotype (Lavigueur et al., 1989). In families of mice expressing a mutant p53 gene, 20% of the mice develop cancer at six to nine months of age. As with Li-Fraumeni patients, the time lag to develop cancer indicates the need for additional events to occur, presumably mutations at other loci and reduction to homozygosity at the p53 locus. The transgenic mouse is born with one mutant allele and two wild-type genes. The fact that cancer is much more common in that mouse than a normal mouse (two wild-type alleles) suggests either that gene conversion plays a role here or the altered missense p53 protein promotes a higher probability of cancer.

Like Rb, p53 mutations are common in a number of cancers. Mutations in the p53 gene have been documented in cancers of the anus, bone, bladder, brain, breast, colon, esophagus, stomach, liver, lung, lymphoid system, ovary, and prostate. Unlike Rb, the great majority of these mutations are point mutations that produce a missense or altered product. The position of this missense mutation in the p53 gene is not random. Most mutations are found between amino acid residues 118–309 (Levine et al., 1991). Some mutations at specific codons occur many times in independently produced cancers. These hot spots are located at codons 175, 248, and 273 (out of 393 codons). There is a tendency for specific hot-spot mutations to differ between different tissue types. Liver cancers tend to have a hot spot at codon 249, lung at 154 and 273, and colon at 175, 248, and 273 (Levine et al., 1991). This unusual tissue-dependent distribution could result from the different mutagens that differing tissues are exposed to in the body or alternatively could result from selection by a specific cell type for a mutation that produces a protein that optimally promotes cell division in that cell type.

The strong selection for missense mutations of a specific type (between codons 118–309) and the high expression of these altered proteins in tumors suggests that the altered p53 protein (unlike Rb) has a function to promote cell division. In support of this idea, the cDNA clones of p53 mutant genes from murine or human cancers have the ability to immortalize cells in culture (Jenkins et al., 1984) and cooperate with the ras oncogene to transform cells in culture (Eliyahu et al., 1984; Parada et al., 1984). These of course are assays for oncogenes, not tumor suppressor genes. The fact that a p53 missense mutant gene can act as an oncogene is consistent with a new function to promote cell growth with an altered p53 protein. Perhaps most convincing is the observation that a mutant p53 cDNA can enhance the tumorigenic potential of a clone of cells that have no endogenous p53 gene expression (Wolf et al., 1984). Here the p53 mutant protein confers a gain of function, i.e., enhanced tumorigenic potential. Clearly, then, the altered mutant p53 protein has a new function and helps promote cell division in tumors. It is also clear that the presence of a wild-type p53 allele in a tumor is selected against because it negatively regulates cell division. In colon cancers that express only one p53 allele (those cells that have undergone a reduction to homozygosity), 91% of the time the mutant allele is expressed. Thus, there is a strong preference for no wild-type alleles, and that is the definition of a tumor suppressor gene.

The p53 gene is unusual in that the wild-type gene is a negative regulator of cell division and a tumor suppressor gene but a missense mutation produces a faulty protein product that appears to promote cell division under certain circumstances (acts like an oncogene). The ratio of mutant protein to wild-type protein in a cell could be an important variable in regulating cell division.

When wild-type p53 is expressed at high levels in a cell, cell division stops and most cells are blocked in the late G_1 phase of the cell cycle (Martinez et al., 1991; Michalovitz et al., 1990). The available evidence suggests that p53 may act as a transcriptional regulator protein (Fields and Jang, 1990; Raycroft et al., 1990). The wild-type p53 protein binds to a specific DNA sequence (Bargonetti et al., 1991; Kern et al., 1991). In addition, the wild-type protein can activate the expression of some genes (Weintraub et al., 1991) or repress the expression of other genes (Santhanam et al., 1991). It appears then that p53, either in combination with other proteins or alone, can regulate gene expression, probably at the level of transcription. It is not yet clear what the mutant p53 protein does to promote cell division. It is also not clear how wild-type p53 in a cell is inactivated (like Rb with phosphorylation) so as to allow the cell to divide normally.

WILMS' TUMOR, WT-1 GENE, AND PROTEIN

Wilms' tumor is a pediatric kidney cancer that, in some cases, has a genetic predisposition. This nephroblastoma occurs in about 1 in 10,000 children. The tumor can be bilateral (10% of the cases) and associated with congenital malformations. The first clue to mapping the putative tumor suppressor gene for Wilms' tumor came from an individual with a deletion on the short arm of chromosome 11 at band 13p (Riccardi et al., 1978). Individuals with this mutation have several abnormalities including aniridia, genitourinary abnormalities, mental retardation, and Wilms' tumor. Using this deletion to identify DNA clones that were missing in this Wilms' tumor and were altered via a different mutation in other Wilms' tumors, Housman and colleagues isolated the WT-1 gene (Call et al., 1990). This is the tumor suppressor gene whose loss appears to be responsible for these tumors.

The WT-1 gene contains 10 exons that span 50 kb of DNA. The mRNA is about 3 kb in size and several alternative spliced mRNAs are known. The protein product is 46–49 kd in size and contains four zinc-finger motifs. Such structures predict a DNA binding property, and a specific DNA sequence recognized by this protein has already been identified (Rauscher et al., 1990). The other major feature of this protein is a proline–glutamine-rich domain that is also a property of transcription factors. WT-1 shares some amino acid homology with known transcription factors in yeast and mammalian cells. The WT-1 gene is expressed in a limited number of tissues: kidney, spleen, gonads, and uterus (Call et al., 1990). In the mouse the WT-1 gene is first expressed during the eighth day of gestation (the mouse has a 21-day gestation period). These mRNA levels peak at day 17 and then decline dramatically to the low level seen in the adult by three days after birth. WT-1 is expressed in specific cell types in the developing kidney and gonads, suggesting an important role in development of these tissues.

The mutations in this gene are occasionally large deletions (loss of function) but it is more common to observe small deletions internal to the gene (Haber et al., 1990). In many cases the normal (wild-type) allele is lost in the Wilms' tumor, but in at least one case a wild-type allele was present in the tumor tissue. More examples will be needed before a clear picture emerges here.

There is some evidence to suggest a second Wilms' tumor locus at a chromosome 11p15 location. Patients with Beckwith-Wiedemann syndrome develop a Wilms' tumor and show chromosomal abnormalities at 11p15 and not at 11p13. Patients with Wilms' tumors and aniridia have chromosomal abnormality at 11p13 and not 11p15. Furthermore, in rare familial Wilms' tumors unassociated with either Beckwith-Wiedemann or aniridia, a third genetic locus, not 11p13 or 11p15, appears to be involved. Clearly there is more than one way (gene) to obtain a Wilms' tumor.

NEUROFIBROMATOSIS, NF-1 GENE, AND ITS PROTEIN

Von Recklinghausen's neurofibromatosis is a result of an abnormal proliferation of cells of neural crest origin. The mild form of the disease presents with *café-au-lait* spots thought to be caused by abnormal growth of melanocytes. More severe forms of this disorder feature disfiguring neurofibromas composed mostly of Schwann cells. The disease affects about one in every 3,500 people. About half of these cases are due to transmission of an inherited defective allele. This inheritance pattern follows those seen with Rb, p53, and WT-1, suggesting the need for both alleles to be inactivated.

The gene for NF-1 has now been cloned (Cawthon et al., 1990) and mutations in this gene detected in neurofibromas (Viskochil et al., 1990; Wallace et al., 1990). The amino acid sequence of the NF-1 protein contains a domain with good homology to the GTPase activating proteins (GAP) that regulate ras (Xu et al., 1990). The NF-1 protein has closest homology to two yeast proteins termed IRAs, which stimulate the GTPase activity of ras in yeast, resulting in the ras-GDP form, which no longer signals for growth of the cells. Mutations in IRA act in a recessive fashion only and leave ras in the GTP form for a longer time, thus stimulating cyclic AMP levels and cell division. NF-1 protein binds to the p21 ras from human cells and stimulates GTPase activity (Martin et al., 1990). This binding is inhibited by low levels of lipids, suggesting a regulatory role here for some lipids. NF-1 is expressed in all cells and tissues but mutations in this gene are not observed in other types of tumors. These observations then lead to a simple model for the role of NF-1 as a tumor suppressor gene. In the absence of NF-1, ras remains in the GTP-bound state and signals for growth in an uncontrolled fashion. There are, however, several observations that do not fit this hypothesis. First, the mutations in NF-1 all lie outside the GTPase activating domain of NF-1 and may not alter NF-1/ras interactions. Second, in Schwann cells the ras GTP form results in growth arrest, not abnormal proliferation. Third, in NF-1 patients, Schwann cells from neurofibromas often contain a wild-type allele (are not reduced to homozygosity). Clearly there are some parts of this puzzle still missing. Perhaps the simplest model is to suggest that NF-1 mutations result in hyperactive or unregulated GTPase activity, placing ras in a GDP form that permits abnormal proliferation.

ADENOMATOUS POLYPOSIS COLI: THE APC GENE AND PROTEIN

Familial adenomatous polyposis occurs in one in 10,000 people and results in the growth of hundreds to thousands of visible adenomatous polyps in the colon during the second or third decade of life. A small fraction of these benign tumors develop into carcinomas. This is correlated with the appearance of p53 mutations and a reduction to homozygosity at the p53 locus (Baker et al., 1989; Hollstein et al., 1991). The gene responsible for familial adenomatous polyposis was mapped to chromosome number 5 (Bodmer et al., 1987) and a gene identified (adenomatous polyposis coli or APC) when deletions at 5q21 were found in two unrelated patients (100–200 kb deletions). YAK vectors derived from this region indicated the deletion spanned at least three genes and finally point mutations in one of these genes (APC) were found in the germline of nine patients with familial adenomatous polyposis (Groden et al., 1991; Joslyn et al., 1991; Nishisho et al., 1991). Frameshift mutations, nonsense mutations (stop codons), and missense mutations are all represented in this group. This same gene is also a target for somatic mutations in colorectal cancers. The APC gene is large and encodes the information for a protein of 2,843 amino acids. Scanning the primary sequence of this protein provides no clues as to a possible function or mode of action. The APC gene appears to be expressed in many diverse adult tissues. Once again, this is an example of a ubiquitous gene product that only produces tumors in a specific tissue type when it is lost or inactivated. This tissue preference is a common tumor suppressor gene property.

In colon carcinomas, two other genetic loci have been implicated as possible tumor suppressor genes. On chromosome 5, near APC, is a gene termed mutated in colorectal cancer (MCC), which was found to commonly contain somatic mutations in both alleles. On chromosome 18 at location 18q21 a gene termed DCC, or deleted in colorectal carcinoma, has been identified as a putative tumor suppressor gene. DCC covers about 370 kb of DNA and encodes a protein with homology to the neural adhesion molecules called NCAM (Fearon et al., 1990). The DCC protein appears to be a transmembrane protein with a large extracellular domain that could be important in the extracellular matrix functions and cell–cell communication. Mutations and allelic loss or reduction to homozygosity is common (70%) at the DCC locus. The DCC gene is also expressed in many diverse adult tissue types. A colorectal cancer syndrome that is inherited but does not involve polyposis has been mapped to chromosome 18 in the region of 18q11.1-q21.1 (Lynch et al., 1985). More studies are required to confirm that MCC and DCC represent clear examples of tumor suppressor genes.

TABLE 3. Review of the Tumor Suppressor Genes

Name	Chromosome	Location	Inherited form	Somatic mutations functions proposed
1. Rb	13q4	Retinoblastoma	Breast, osteogenic sarcoma, small-cell lung	Regulates transcription factors E2F/DRTF1
2. p53	17p13.1	Li-Fraumeni syndrome	Colon, lung, breast, etc.	Transcription factor (?)
3. WT-1	11p13	Wilms' tumor with aniridia	None detected	Transcription factor
4. NF-1	17q	Neurofibromatosis	None detected	GTPase activating protein for *ras*
5. APC	5q21	Familial adenomatous polyposis	Colon	Unknown
6. DCC	18q21	Not clear	Colon	NCAM-like protein
7. MCC	5q21	Not clear	Colon	Unknown

CONCLUSION

From the very first experiments in somatic cell genetics (Harris et al., 1969) it became clear that there were multiple tumor suppressor genes with distinct sets of functions (see Table 3). The best example of a pure tumor suppressor gene is Rb, with a true loss-of-function mutation giving rise to the tumor. p53 is a complicated example, with a true loss of function for negative growth regulation but a gain of some function due to the altered or mutated protein that is expressed in most cancers. For Rb, p53, NF-1, and APC, all of which are expressed in most of the adult tissues of the body, it remains unclear why each of these genes shows a marked tissue preference for the tumors that arise in their absence. The simplest explanation is that in some cells the absence of Rb is sufficient for a tumor to arise (retinoblastoma), while in other cells (lung, breast) it is maybe necessary but not sufficient. This would mean that cancers of different tissues arise via genetic alterations in a set of genes that regulate division in a cell in tissue-specific fashion.

These tumor suppressor genes have, for the first time, put the inherited basis of cancer in a clear perspective. They all appear to be inherited as autosomal dominant traits because of the high probability of mutation or reduction to homozygosity in the wild-type allele. In some cases, such as Rb in retinoblastomas, no second event appears to be necessary for tumor formation, while in other cases (p53) loss of both alleles is not sufficient for cancer to occur. Similarly, the type of mutations in the Rb gene clearly result in a loss of function for this protein, while the types of mutations in the p53 gene (missense, altered protein produced) seem to provide a loss of negative regulation of cell division but a gain of a new function promoting cell division in a tumor. In spite of this, there is still continued selection in a cell heterozygous for a p53 mutation for a second mutation in the p53 wild-type gene or a loss of the wild-type allele. Wild-type p53 in cancer cells behaves as a tumor suppressor gene. The evidence that WT-1, NF-1, APC, DCC, and MCC clearly function as tumor suppressor genes is still incomplete but each of these genes has some features that suggest that category of function. Clearly this list will grow with time, and we will know more about the functions of these critical genes and their products.

ACKNOWLEDGMENTS

The author thanks K. James for secretarial help and M.E. Perry and C. Finlay for excellent advice and criticism of the manuscript. This chapter was written while the author's laboratory was supported by grants from the National Institutes of Health, CA49271, CA38757, and CA41086.

REFERENCES

Baker, S.J., E.R. Fearon, J.M. Nigro, et al. (1989) Chromosome 17 deletions and p53 gene mutations in colorectal carcinoma. Science 244:217–221.

Bandara, L., J. Adamczewski, T. Hunt, et al. (1991) Cyclin A and the retinoblastoma gene product complex with a common transcription factor. Nature 352:249–251.

Bargonetti, J., P.N. Friedman, S.E. Kern, et al. (1991) Wild-type but not mutant p53 immunopurified proteins bind to sequences adjacent to the SV40 origin of replication. Cell 65:1083–1091.

Benchimol, S., P. Lamb, L.V. Crawford, et al. (1985) Transformation associated p53 protein is encoded by a gene on human chromosome 17. Som. Cell. Mol. Gen. 11:505–509.

Bodmer, W.F., C. Bailey, J. Bodmer, et al. (1987) Localization of the gene for familial polyposis on chromosome 5. Nature 328:614–616.

Buchkovich, K., L.A. Duffy, and E. Harlow (1989) The retinoblastoma protein is phosphorylated during specific phases of the cell cycle. Cell 58:1097–1105.

Call, K.M., T. Glaser, C.Y. Ito, et al. (1990) Isolation and characterization of a zinc finger polypeptide gene at the human 11 Wilms' tumor locus. Cell 60:409–520.

Cavenee, W.K., T.P. Dryja, R.A. Phillips, et al. (1983) Expression of recessive alleles by chromosomal mechanisms in retinoblastoma. Nature 305:779–784.

Cawthon, R.M., R. Weiss, G. Xu, et al. (1990) A major segment of the neurofibromatosis type 1 gene: cDNA sequence, genomic structure and point mutations. Cell 62:192–201.

Chellappan, S.P., S. Hiebert, M. Mudryj, et al. (1991) The E2F transcription factor is a cellular target for the TB protein. Cell 65:1053–1061.

Chen, P.-L., Y. Chen, R. Bookstein, et al. (1990) Genetic mechanisms of tumor suppression by the human p53 gene. Science 250:1576–1579.

Eliyahu, D., A. Raz, P. Gruss, et al. (1984) Participation of p53 cellular tumor antigen in transformation of normal embryonic cells. Nature 312:646–649.

Fearon, E.R., K.R. Cho, J.M. Nigro, et al. (1990) Identification of a chromosome 18q gene that is altered in colorectal cancer. Science 247:49–56.

Fields, S., and S.K. Jang (1990) Presence of a potent transcription activating sequence in the p53 protein. Science 249:1046–1049.

Friend, S.H., J.M. Horowitz, M.R. Gerber, et al. (1987) Deletions of a DNA sequence in retinoblastomas and mesenchymal tumors: Organization of the sequence and its encoded protein. Proc. Natl. Acad. Sci. USA 84:9059–9063.

Groden, J., A. Thlweres, W. Samowitz, et al. et al. (1991) Identification and characterization of the familial adenomatous polyposis coli gene. Cell 66:589–600.

Haber, D.A., A.J. Buckler, T. Glaser, et al. (1990) An internal deletion within an 11p13 zinc finger gene contributes to the development of Wilms' tumor. Cell 61:1257–1269.

Harris, H., O.J. Miller, G. Klein, et al. (1969) Suppression of malignancy by cell fusion. Nature 223:363–368.

Hollstein, M., D. Sidransky, B. Vogelstein, et al. (1991) p5 mutations in human cancers. Science 253:49–53.

Horowitz, J.M., S.H. Park, E. Boegnmann, et al. (1990) Frequent inactivation of the retinoblastoma antioncogene is restricted to a subset of human tumors. Proc. Natl. Acad. Sci. USA 87:2775–2779.

Hu, Q., N. Dyson, and E. Harlow (1990) Regions of the retinoblastoma protein needed for binding to adenovirus E1A or SV40 large T-antigen are common sites for mutations. EMBO J. 9:1147–1155.

Huang, H.J.S., J.K. Yee, J.Y. Shew, et al. (1988) Suppression of the neoplastic phenotype by replacement of the Rb gene in human cancer cells. Science 242:1563–1566.

Jenkins, J.R., K. Rudge, and G.A. Currie (1984) Cellular immortalization by a cDNA clone encoding the transformation-associated phosphoprotein p53. Nature 312:651–654.

Joslyn, G., M. Carlson, A. Thlweres, et al. (1991) Identification of deletion mutations and three new genes at the familial polyposis locus. Cell 66:601–613.

Kern, S.E., K.W. Kinzler, A. Bruskin, et al. (1991) Identification of p53 as a sequence-specific DNA-binding protein. Science 252:1708–1711.

Knudson, A.G., Jr. (1971) Mutation and cancer: Statistical study of retinoblastoma. Proc. Natl. Acad. Sci. USA 68:820–823.

Lavigueur, A., V. Maltby, D. Mock, et al. (1989) High incidence of lung, bone, and lymphoid tumors in transgenic mice overexpressing mutant alleles of the p53 oncogene. Mol. Cell Biol. 9:3982–3991.

Lees-Miller, S.P., and C.W. Anderson (1991) The DNA activated protein kinase, DNA-PK: A potential coordinator of nuclear events. In Cancer Cells, vol. 3, no. 9. Cold Spring Harbor, N.Y.: Cold Spring Harbor Laboratory Press, pp. 341–346.

Levine, A.J., J. Momand, and C.A. Finlay (1991) The p53 tumor suppressor gene. Nature 351:453–456.

Li, F.P. (1988) Cancer families: Human models of susceptibility to neoplasia—The Richard & Hinda Rosenthal Foundation Award Lecture. Cancer Res. 48:5381–5386.

Lynch, H.T., G.S. Schuelke, W.J. Kimberling, et al. (1985) Hereditary nonpolyposis colorectal cnacer (Lynch syndromes I and II). Biomarker studies. Cancer 56:939–951.

Malkin, D., F.P. Li, L.C. Strong, et al. (1990) Germ line p53 mutations in a familial syndrome of breast cancer, sarcomas, and other neoplasms. Science 250:1233–1238.

Martin, G.A., D. Viskochil, G. Bollag, et al. (1990) The GAP-related domain of the neurofibromatosis type 1 gene product interacts with ras p21. Cell 63:843–849.

Martinez, J., I. Georgoff, J. Martinez, et al. (1991) Cellular localization and cell cycle regulation by a temperature sensitive p53 protein. Genes Dev. 5:151–159.

Michalovitz, D., O. Halevy, and M. Oren (1990) Conditional inhibition of transformation and of cell proliferation by a temperature-sensitive mutant of p53. Cell 62:671–680.

Nishisho, I., Y. Nakamura, Y. Miyoshi, et al. (1991) Mutations of chromosome 5q21 genes in FAP and colorectal cancer patients. Science 253:665–669.

Parada, L.F., H. Land, R.A. Weinberg, et al. (1984) Cooperation between gene encoding p53 tumour antigen and ras in cellular transformation. Nature 312:649–651.

Rauscher, F.J., J.F. Morris, O.E. Tournay, et al. (1990) Binding of the Wilms' tumor locus zinc finger protein to the EGR-1 consensus sequence. Science 250:1259–1262.

Raychaudhuri, P., S. Bagchi, S. Neill, et al. (1990) Activation of the E2F transcription factor in adenovirus infected cells involves an E1A-dependent stimulation of DNA binding activity and induction of cooperative binding mediated by an E4 gene product. J. Virol. 64:2702–2710.

Raycroft, L., H. Wu, and G. Lozano (1990) Transcriptional activation by wild-type but not transforming mutants of the p53 anti-oncogene. Science 249:1049–1051.

Riccardi, V.M., E. Sujansky, A.C. Smith, et al. (1978) Chromosomal imbalance in the aniridia-Wilms' tumor association: 11p deletion. Pediatrics 61:604–610.

Rowley, J.D. (1984) Consistent chromosomal rearrangements in human malignant disease and oncogene location. In: G.F. Vande Woude, A.J. Levine, W.C. Topp, and J.D. Watson (eds.). Cancer Cells II: Oncogenes and Viral Genes. Cold Spring Harbor, N.Y.: Cold Spring Harbor Laboratory Press, pp. 221–226.

Rustgi, A.K., N.J. Dyson, and R. Bernards (1991) Amino terminal domains of c-myc and N-myc proteins mediate binding to the retinoblastoma gene product. Nature 352:541–544.

Santhanam, U., A. Ray, and P.B. Sehgal (1991) Repression of the interleukin 6 gene promoter by p53 and the retinoblastoma susceptibility gene product. Proc. Natl. Acad. Sci. USA 88:7605–7609.

Srivastava, S., Z. Zou, K. Pirollo, et al. (1990) Germ-line transmission of a mutated p53 gene in a cancer-prone family with Li-Fraumeni syndrome. Nature 348:747–749.

Viskochil, D., A.M. Buchberg, G. Xu, et al. (1990) Deletions and a translocation interrupt a cloned gene at the neurofibromatosis type 1 locus. Cell 62:187–192.

Wallace, M.R., D.A. Marchuk, L.B. Anderson, et al. (1990) Type 1 neurofibromatosis gene: Identification of a large transcript disrupted in three NF1 patients. Nature 249:181–186.

Weintraub, H., S. Hauschka, and S.J. Tapscott (1991) The MCK enhancer contains a p53 responsive element. Proc. Natl. Acad. Sci. USA 88:4570–4571.

Wolf, D., N. Harris, and V. Rotter (1984) Reconstitution of p53 expression in a nonproducer Ab-MuLV-transformed cell line by transfection of a functional p53 gene. Cell 38:119–126.

Xu, G.F., B. Lin, K. Tanaka, et al. (1990) The catalytic domain of the neurofibromatosis type 1 gene product stimulates ras GTPase and complements IRA mutants of S. Cerevisiae. Cell 63:835–841.

Chapter 11

The Oncogenes of the DNA Tumor Viruses

Arnold J. Levine

INTRODUCTION

The DNA-containing viruses that can cause or initiate cancers in animals and humans encompass a very heterogeneous group of agents: simian virus 40 (SV40) and polyoma virus, human and animal papilloma viruses, the adenoviruses, some human (Epstein-Barr virus) and animal herpesviruses, and some poxviruses (myxoma-fibroma group). This chapter focuses upon the smaller and simpler viruses, SV40, papilloma viruses, and adenoviruses, for three reasons: (1) Their simple genome and small number of genes and products have made these viruses more dependent upon cellular functions to replicate the virus. Because of this we can learn a great deal about normal and abnormal functioning of cell division processes when we study these viruses. (2) Each of these viruses has small, simple genomes and so one or only a few viral genes and proteins are responsible for tumor formation. This has permitted a detailed and intensive study of these mechanisms. (3) All three virus groups share a common set of mechanisms to transform cells in culture or to produce tumors in animals. These three virus groups will thus provide us with general and central principles of abnormal cell growth and division.

Viruses have evolved with their host cells over millions of years. These parasitic agents have developed ways to use normal cellular control mechanisms for the benefit of the virus, i.e., to duplicate the virus efficiently. For the small DNA tumor viruses like SV40, papilloma viruses, or the adenoviruses, optimal virus replication occurs in actively growing cells that supply high levels of nucleotides and amino acid precursors, enzyme activities that duplicate DNA, and histones that package some of the viral DNA genomes (SV40 and papilloma viruses). It is most often the case, however, that these viruses encounter and infect resting cells in the adult body of an animal or human. Because of this, each of these viral agents has developed mechanisms to force the cell out of the resting state and into a growth or division cycle. In each case the viruses encode one or more genes that produce proteins designed to force a cell into the replication cycle. As such, these viruses and the oncogenes they encode have a great deal to tell us about how a cell keeps its machinery for division in a resting state. The viral oncogene products must either inactivate the negative regulators of cell division and/or promote the positive regulators of cell division. Remarkably, all three virus groups have evolved oncogenes that function similarly to commit their host cells to a division cycle. The study of these viral oncogenes has told us a great deal about how normal cells remain at rest or how cancer cells are released from these negative controls and replicate.

THE DNA TUMOR VIRUS PARADIGM

An injection of SV40 or some of the human adenoviruses into a newborn hamster frequently results in the formation of a tumor at the site of injection after a long latent period (five to 12 months). These tumors never contain infectious viruses. However, each of the cancer cells contains a portion of the viral genome integrated into a chromosome of the host. In any one tumor the site of integration into the host genome is the same in all cells of the tumor. This demonstrates that the cancer arose from a single cell and is clonal. Different independently produced tumors have different integration sites, indicating that the site of insertion of viral DNA is not critical to tumor formation. In the tumor, a subset of viral genes are expressed and the mRNA produces a subset of viral proteins in each of the tumor cells. As the tumor grows, these viral proteins are presented to the host's immune system and they are recognized as foreign. The tumor-bearing animal responds by producing antibodies against the viral gene products, and for this reason the products of viral oncogenes have come to be called tumor antigens or T antigens. In some cases these T antigens (from SV40, for example) also elicit a killer, CD-8, T-cell response, and, under the appropriate circumstances (with prior

Molecular Genetics of Nervous System Tumors, pages 145-151 ©1993 Wiley-Liss, Inc.

immunization), the T-antigen is also a tumor-specific transplantation rejection antigen, or TSTA (Tevethia, 1990). The antisera from tumor-bearing animals have been very useful in studying the tumor antigens of the DNA tumor viruses.

The first line of evidence demonstrating that the tumor antigens of the DNA tumor viruses are important for the initiation and maintenance of the tumor derived from the observation that all tumors retained the viral gene for these proteins and all tumors expressed these proteins (Tooze, 1973). Second, transgenic mice that contained and expressed the SV40 T-antigen gene developed tumors (Brinster et al., 1984) while viral genes with mutations in them failed to produce tumors (Palmiter et al., 1985). Similarly, a virus with a mutation in the T-antigen gene fails to initiate tumors. In one interesting case, a polyoma virus with a temperature-sensitive mutation in the T-antigen gene produced tumors in mice in locations on the body of lower temperature (Fried, 1965) but not at higher temperatures. Clearly this genetic analysis has led to the conclusion that the viral genes encoding T antigens are indeed oncogenes or cancer-causing genes. The expression of these viral gene products is needed to initiate the tumor and to maintain tumor growth.

The papilloma viruses of vertebrates, including humans, induce benign skin lesions, including squamous warts and papillomas. There are more than 65 different human papilloma viruses. About 20 of these have been associated with anogenital tract lesions. These human papilloma viruses (HPV) have been divided into two groups termed (1) the low-risk viruses, such as HPV6 or 11, which are associated with lesions that usually remain benign, and (2) the high-risk viruses, such as HPV16 or 18, which tend to be found with lesions that progress to cancer (zur Hausen and Schneider, 1987). About 85% of cervical carcinomas contain HPV DNA sequences integrated into the chromosomes of each tumor cell. A subset of viral genes are expressed in these tumors and two HPV-associated oncogenes (termed E6 and E7) have been identified (Münger et al., 1992). Thus, the paradigm of the DNA tumor viruses carries over to human neoplasia, in this case, cervical carcinomas.

TRANSFORMATION OF CELLS IN CULTURE

Tumor formation in an animal is a cumbersome experimental procedure and does not permit a detailed genetic analysis of the viral oncogenes. For this reason, a system to measure the phenotypes of viral oncogenes in cell culture was developed. Basically, five phenotypes of cells in culture have been shown to respond by changing in the presence of the oncogenes of DNA tumor viruses: (1) the ability to immortalize cells in culture so that they grow indefinitely; (2) the acquired ability of cells to grow on top of each other and form a dense focus of cells in a colony on the surface of a culture dish; (3) the ability of cells to grow in agar, where no substratum is available for cellular attachment to a surface; (4) the ability to form tumors in animals when the cells are inoculated into these hosts; and (5) the alteration in the morphology of these cells (Tooze, 1973).

Primary cells directly explanted from a fetus or young animal into cell culture can divide a limited number of times and then cease cell division. The cell culture slows down in its ability to duplicate itself and a "crisis period" ensues in which most cells never divide again. With cells from many species, rare variants can then appear and replicate, taking over the culture. It is common with mouse cells that these rare variants are observed to contain mutations in selected genes such as the p53 gene (see Chapter 10, tumor suppressor Genes, Levine) (Harvey and Levine, 1991). In some species such as humans, no rare cell variants arise and the culture terminates. These cells have a limited life span. Infection of these cells with SV40 or adenoviruses can confer upon them the ability to replicate themselves and form a permanent cell line. The cells have gained an immortality and this immortalization is dependent upon the expression of the T antigen or oncogenes of the virus. Just as in the case with tumor formation, an immortalized cell line contains viral DNA integrated into the chromosomes of the host cell (it is a clonal event also) and this genome expresses the T antigens of the virus. Mutations in the T-antigen genes eliminate the ability of the virus to immortalize cells in culture. A temperature-sensitive T-antigen mutation produces a cell line that proliferates at 32°C but not at 39°C, showing that T antigen is required to maintain the transformed or immortalized phenotype.

A subset of these viral transformed and immortalized cells will express additional phenotypes. Some clones of cells will grow in a culture dish in an irregular pattern so that cells will grow on top of each other, forming a dense focus of cells. The number of cells per square centimeter of surface of a dish is therefore much higher with viral transformed cells than with normal cells. This is often termed the saturation density of cells, which is high for transformed cells and low for normal cells. A further subset of the transformed cell clones will divide and form colonies when plated into agar suspension cultures. Normal cells or nontransformed cells require a surface to attach to and duplicate upon. These cells are said to be anchorage dependent for growth, while transformed cells replicate in agar. Finally, a subset of these cells will be able to produce tumors in animals, while normal cells do not.

These phenotypes of transformed cells are all controlled by the viral oncogenes and their proteins. They are, however, very complex in their patterns and expression. Some transformed cells may express all five phenotypes (altered morphology, immortalization, high saturation density, growth in agar, and tumorigenic potential), while other cells transformed by the same viral oncogene express only one or two of these phenotypes. The transformed phenotype is not

TABLE 1. The Structural and Functional Domains of T Antigen

Amino acid residues	Replicative functions			
1–82, 390–610	DNA polymerase alpha binds			
Serine 106, 111, 112, 123	Phosphorylation sites			
Threonine 124	cdc-2-like kinase phosphorylates, required for optimal DNA binding and replication			
126–132	Nuclear localization signal			
130–260	DNA binding domain, origin specific			
260–390	Zinc-finger domain			
390–610	ATPase activity			
130–620	Helicase activity			
Serine 639, 676, 677, 679, Threonine 701	Phosphorylation sites			
		Transforming functions		
	Immortalization assay	C3H 10T½	REF-52	Cellular protein binding
1–82 (Δ-27)	Reduced cooperation with *ras* in rat cells, not required for mouse cells	Essential for Tx	Essential for Tx	Putative 300 Kd protein
105–114	Not required for mouse cells	Not required for Tx	Essential for Tx	Rb protein
272–517 region				
(point mutant 402)	Human cells, essential			
(insertion mutation at 409, 424)	Mouse cells, essential	Not required for Tx	Essential for Tx	p53

coordinately controlled by the viral oncogene. Furthermore, almost all of the phenotypes have a quantitative variation in their expression. For some transformed cell lines it will take 100 cells to form a tumor in five days and for others 1 million cells are needed to form a tumor that takes three weeks to develop. Cell lines may have all gradations of saturation densities or plating efficiencies in agar. The tremendous heterogeneity of the transformed cell phenotype is a true reflection of the heterogeneity of cancer cells in a tumor. The cell culture model for tumorigenesis may well be a good one.

THE TRANSFORMING FUNCTIONS OF SV40

Virtually all of the transforming functions and tumorigenic potential of SV40 resides in its T antigen. This is a multifunctional protein that has two related tasks during productive infection of a cell with this virus. First, the T antigen must force the cell into a replicative mode, increasing nucleotide and amino acid pools, synthesizing cellular DNA with cellular enzyme activities involved in this process kept at maximum levels, producing histones to package viral and cellular DNA, and maximizing replicative activities. The virus uses the transforming activities of T antigen to accomplish this. T antigen also acts to replicate the viral DNA. The protein is synthesized as a monomer and rapidly oligomerizes into a hexamer. A cdc-2-like kinase (whose activity is increased as a signal of cell division) then phosphorylates T antigen at the threonine-145 residue (T antigen is 708 amino acids long) (McVey et al., 1989), which then permits three phosphorylated hexamers of T antigen to bind to the origin of DNA replication (via DNA sequence recognition) on the viral DNA. The DNA binding domain of T antigen resides between amino acid residues 130–260 (see Table 1) and is apparently aided by a zinc-finger domain (residues 260–390). T antigen is an ATP-dependent (ATPase activity at residues 390–610) helicase (at residues 130–620) that helps unwind the DNA helix at the origin of replication. A cellular primase activity associated with the cellular DNA polymerase alpha binds to T antigen (residues 1–82 and 390–610) and synthesizes RNA chains that act as primers for DNA synthesis by DNA polymerase alpha. T antigen continues to act at the replication fork with a bidirectional synthesis of viral DNA that is concomitantly packaged by cellular histones. At the end of the replication of viral DNA, additional phosphorylation sites on T antigen located at serines 106, 111, 112, 123, and serines 639, 656, 677, and 679 and threonine 701 are phosphorylated (they are not phosphorylated by a cdc-2-like kinase activity) (Chen et al., 1991; Lees-Miller and Anderson, 1991) and this form of T antigen can no longer initiate DNA replication. This form of T antigen, however, does bind to cellular proteins such as Rb and p53 (see Table 1).

Three different assays for the transformation by T antigen have been useful in demonstrating that at least three regions of this protein are utilized to stimulate cellular division processes. Immortalization of human, rat, or mouse cells has been used as a test for T-antigen transformation. In addition, the transformation (foci formation, growth in agar) of two permanent mouse (C3H 10T½) or rat (REF–52) cell lines have been helpful in dissecting out different T-antigen functional domains involved in transformation. In practice, the wild-type T antigen transforms all of these cell lines or cell types. Different T-antigen mutants, however, fail to transform different combinations of these and the ability of a mutant to transform a cell line indicates the missing portion of the T antigen (mutant) is not essential for that function. On the other hand, when mutant T antigens fail to transform one cell type but do transform another cell type, that indicates that the lost function is indeed essential for the specific steps required to transform that particular cell. Employing this approach (Lin and Simmons, 1991; Peden et al., 1989; Srinivasan et al., 1989; Zhu et al., 1991), three regions of the SV40 T antigen, at residues 1–82, 105–114, and 272–517 (especially mutations at residues 402, 409, 424) (see Table 1) have been identified as distinct transforming domains of this protein. Two of these three transforming domains correspond to regions of T antigen that bind to known tumor suppressor gene products. The amino acid residues 105–114 bind to the cellular Rb protein (DeCaprio et al., 1988) and the amino acid residues 272–517 bind to p53 (Lane and Crawford, 1979; Linzer and Levine, 1979; Schmieg and Simmons, 1988). Since these cellular proteins are negative regulators of cell division, it is presumed that T-antigen binding to these proteins (Rb and p53) inactivates their functions. But it remains possible that the T-Rb and T-p53 complexes gain a new function and promote cell division. The third region of T antigen at residues 1–82 that is also involved in transformation has not been shown to be associated with a cellular tumor suppressor protein. However, one of the oncogene products of adenoviruses, called the E1A protein, is known to associate with a cellular 300 kd nuclear phosphoprotein in a domain of the E1A protein that is essential for it to transform cells in culture. Mutants in this adenovirus E1A domain fail to transform cells in culture. A protein containing amino acid residues 1–121 from the SV40 large T antigen can complement this defect in transformation by the E1A mutants and T antigen with a deletion of amino acid residues 7–27 fails to complement the E1A defective mutant. Thus, this genetic analysis suggests that the 1–82 domain of SV40 T antigen is equivalent to the E1A domain binding to the cellular 300 kd protein in transforming cells in culture (see Table 1) (Chen et al., 1991).

Thus, there is good evidence that three regions of the SV40 T antigen have evolved to interact with cellular proteins and stimulate the cell into a growth cycle. Two of the proteins, Rb and p53, have been clearly identified as major negative regulators of cell division and their loss or inactivation is common in some human tumors (see Chapter 10, Tumor Suppressor Genes, Levine).

THE TRANSFORMING FUNCTIONS OF ADENOVIRUSES

In adenovirus-induced tumors or transformed cells, three viral genes are expressed and produce proteins or tumor antigens that are required to initiate and, in some cases, maintain the transformed phenotypes of these cells. These genes encode proteins that have been called the E1A, E1B-55 kd, and E1B-19 kd proteins (see Table 2). The E1A proteins of adenoviruses are synthesized immediately after virus infection of a cell. One of their major functions is to promote the transcription of additional adenovirus genes in the viral chromosome. There are five E1A transcripts and proteins that result from complex splicing patterns across this gene (see Dyson and Harlow, 1992, and Flint and Shenk, 1989, for a review). For the purposes of this chapter, we can focus upon one major E1A gene product synthesized from a 13S mRNA to produce a protein of 289 amino acids. When the amino acid sequences of E1A proteins from many different serotypes of this virus are compared, three regions of this E1A protein have been strongly conserved in all

TABLE 2. Transforming Functions of Adenovirus

E1A	Immortalization	
E1B-19 kd, E1B-55 kd	Focus formation, agar growth and tumorigenic potential	
E1A + *Ras*	Primary rat embryo fibroblasts—fully transformed	
E1A, 13S mRNA, 289 amino acids		
Conserved domains in all adenovirus serotypes		
Site 1	Residues 40–80	Transformation
Site 2	Residues 121–139	Transformation
Site 3	Residues 140–188	Transcription
E1A binding proteins		
300 kd protein	Residues 1–80 (1–49)	
Rb protein	Residues 1–80 and 121–127	
p107 protein	Residues 1–80 and 121–127	
cyclin A-cdk-2 kinase	Residues 1–80 and 121–127	
p130	Residues 1–80 and 121–127	
E1B-55 kd: binds to p53		
Mutants that do not bind do not transform		
Mutants that bind to p53 inactivate p53-mediated transcription		
E1B-19 kd: nuclear membrane location, some mutants result in DNA degradation		

groups of viruses. These regions map at amino acid residues 40–80 (termed site 1), 121–139 (site 2), and 140–188 (site 3) (see Table 2). Site 3 is all that is required to promote the transcription of the other adenovirus genes (E1B, E2A, E2B, E3, and E4 genes). The E2A and E2B transcripts in turn will produce proteins that begin the synthesis of adenovirus DNA. Sites 1 and 2, however, are essential for transformation of cells in culture and for stimulating the virus-infected cell into a replicative cycle. The E1A gene by itself is sufficient to immortalize rat cells in culture, while mutations in site 1 or 2 inhibit this phenotype (van der Eb, 1979). These cells, however, are not fully transformed and do not produce foci, grow in agar, or produce tumors in animals. Several cellular proteins have been shown to bind to E1A sites 1 and 2 (Harlow et al., 1986). A 300 kd nuclear phosphoprotein binds to amino acid sequences 1–80 overlapping site 1 (it binds to a protein fragment containing residues 1–49). The Rb protein binds to the E1A polypeptide between residues 30–60 and 120–127 (Whyte et al., 1989). An Rb-related protein termed p107 also binds to E1A in the same regions of the protein. A complex of cyclin A-cdk-2 protein kinase activity binds to E1A, possibly by binding directly to Rb or p107, which it can then phosphorylate. Phosphorylation of Rb releases it from SV40 T-antigen or adenovirus E1A protein interactions. Finally, a cellular protein of 130 kd (p130) is also bound to E1A protein in this same region. Mutant proteins that fail to bind these cellular proteins also fail to immortalize rat cells in culture. There is an excellent genetic correlation between E1A–Rb interactions and the ability to transform cells in culture. Thus, E1A appears to act upon these cellular proteins in the process of immortalizing cells in culture.

In order to obtain fully transformed cells in culture that produce foci, grow in agar, and are able to form tumors in animals, the two E1B genes, E1B-55 Kd and E1B-19 Kd, are required. The E1B-55 Kd protein from type 5 adenovirus binds to the cellular p53 protein (Sarnow et al., 1982). Mutants in this E1B-55 Kd protein that fail to bind to p53 do not transform cells in culture and binding of E1B-55 kd to p53 inactivates p53-mediated transcription of a test gene (Yew and Berk, 1992) (Table 2). Thus, it appears that the E1B-55 kd protein functions to transform cells via its interactions with p53. The function of E1B-19 kd is less clear at this time.

THE HUMAN PAPILLOMA VIRUSES

The human papilloma viruses HPV16 and 18 are commonly (85% of the cases) associated with cervical carcinomas in humans. In these tumors the viral DNA has integrated into the host cell DNA and two viral genes, termed E6 and E7, are expressed. As such, these genes are good candidates for the viral oncogenes producing the E6 and E7 tumor antigens. In fact, the E6 and E7 proteins have a number of properties that demonstrate their oncogenic potential (Table 3). The E6 protein is a DNA binding protein with two zinc-finger domains. It is required, along with the E7 protein, to immortalize primary human epithelial cells in culture and as such behaves as an oncogene. The E6 protein can form a complex with p53 *in vitro* and when this occurs it promotes the ATP-dependent addition of ubiquitin to p53, followed by the proteolytic breakdown of this protein (Scheffner et al., 1990; Werness et al., 1990). In cells expressing E6, p53 levels are consequently kept very low. The high-risk viruses such as HPV16 or 18 which are associated with cervical cancers have E6 proteins that bind to p53, while the low-risk viruses, HPV6 or 11, produce E6 proteins that do not detectably bind to p53 *in vitro* (Werness et al., 1990). There is thus an interesting correlation between molecular and epidemiological evidence. The E6-mediated proteolytic degradation of p53 is an example of a true loss of function for p53. This observation reinforces the role of p53 in negatively regulating cell division.

The E7 protein is also a zinc-finger protein of 98 amino acid residues (Table 3). The amino-terminal 38 residues of this protein has striking amino acid sequence homology to the transforming regions of the adenovirus E1A protein (sites 1 and 2) and two of the transforming regions of the SV40 T antigen (residues 6–18 and 99–117). The E7 gene is sufficient to fully transform immortalized mouse 3T3 cells so that they produce foci, grow in suspension or agar cultures, and make tumors in animals. The E7 gene also can cooperate with the ras oncogene to transform primary rat embryo fibroblasts and it can cooperate with the E6 gene to

TABLE 3. Properties of HPV E6 and E7 Oncogenes

E6 protein	151 amino acids, zinc finger with four copies of Cys-X-X-Cys sequence motifs, binds to DNA
Transformation	Cooperates with E7 for efficient immortalization of primary human epithelial cells
Protein interactions	Forms a complex with p53 and promotes its degradation *in vitro* via an ATP-dependent ubiquitin reaction
E7 protein	98 amino acids, zinc finger with two Cys-X-X-Cys sequence motifs, the amino-terminal 1–38 residues are homologous to E1A sites 1 and 2 and T-antigen residues 6–18 and 99–117
Transformation	Fully transforms mouse 3T3 cells, cooperates with *ras* to transform primary rat embryo fibroblasts, cooperates with E6 to immortalize human epithelial cells
Protein interactions	Complexes with the cellular Rb protein, can activate E2F activity

immortalize primary human epithelial cells in culture. These cells are not transformed for other criteria such as focus formation or tumorigenic potential. The E7 protein appears to act, in part, by complexing with the cellular Rb protein in transformed cells, presumably inactivating its function. Consistent with this is the observation that E7 can activate the E2F transcription factor to function on an adenovirus template (E2 gene, see Chapter 10). When Rb is complexed to E2F, it blocks its function as a transcription factor and the E7–Rb complex appears to release active E2F.

About 85% of the cervical carcinomas are associated with HPV and express E6 and E7 gene products. If these viral proteins function so as to inactivate (or in the case of E6, degrade) p53 and Rb, then there would be no selection pressure to find p53 or Rb mutations in such tumors. On the other hand, the 15% of the cervical carcinomas with no associated HPV DNA would still have to deal with Rb and p53 function in a cell and therefore might well select for p53 and Rb mutations. When this was examined in a number of cell lines derived from cervical carcinomas with HPV present, all contained the wild-type or normal Rb and p53 genes. However, in cell lines derived from HPV-negative cervical carcinomas, both the Rb gene and the p53 gene sustained mutations (Crook et al., 1991; Scheffner et al., 1990). As discussed previously (Chapter 10), some p53 mutations appear to confer a gain of function (enhanced tumorigenic potential) upon a cell. In this case a cell with E6 promoting the proteolytic degradation of p53 would provide a null phenotype, i.e., the absence of the protein. It could be better for the patient to have a null phenotype than a gain-of-function (growth potential) phenotype. It is of some interest then that a recent study of HPV-negative cervical carcinomas shows that they have a poorer clinical prognosis than HPV-positive carcinomas in that they progress more rapidly and have a higher metastatic potential (Riou et al., 1990).

TABLE 4. DNA Tumor Virus Oncogenes

Oncogene Product	Cellular Tumor Suppressor
SV40 T antigen	
Residues 105–114	Rb
Residues 272–517	p53
Adenovirus	
E1A, site 1, 2	Rb
E1B-55 kd	p53
Papilloma viruses 16, 18	
E7	Rb
E6	p53

DNA TUMOR VIRUSES AND CELL DIVISION

Over the past 12 years of studying the mechanisms of action of the oncogene products from the DNA tumor viruses, it has become quite clear that these agents have targeted common cellular gene products that are central tumor suppressor genes (Table 4). The simple hypothesis that results from these studies, i.e., that the inactivation of Rb and p53 results from these viral oncogene–cellular anti-oncogene protein complexes may not be totally correct. It certainly remains possible that the T–p53, T–Rb, E1A–Rb, E1B–55 kd–p53, E6–p53, and E7–Rb complexes gain a new function that promotes cell growth. It is also clear that the binding of the viral oncogene proteins to cellular Rb and p53 proteins cannot fully explain transformation of cells in culture. The adenovirus E1A protein which binds Rb is sufficient to immortalize rat or mouse cells in culture. The T-antigen fragment 1–121 which binds Rb is not sufficient to immortalize mouse cells in culture, which requires the p53 binding region also. Either E1A associates with additional cellular proteins (not p53) resulting in immortalization of cells or E1A–Rb does something that T(1–121 residues)–Rb does not do in a cell. These kinds of comparisons point out the limitations in our present knowledge.

One might think our newfound knowledge about the role of HPV E6 and E7 in cervical carcinoma cells and their interactions with tumor suppressor gene products (p53 and Rb, respectively) would now permit a clear interpretation of the early experiments with tumor suppressor genes using somatic cell genetics (see Chapter 10). For example, in 1982 Stanbridge and colleagues demonstrated that somatic cell hybrids made between two HPV-positive cell lines HeLa and SiHa and normal human cells had a nontumorigenic phenotype. When the tumor suppressor genes involved in this process were mapped to a chromosome, one might in retrospect (Table 4) have thought that the overproduction of Rb protein (on chromosome 13) or p53 proteins (on chromosome 17) in these cells might well now explain how larger levels of a tumor suppressor gene product could overwhelm the viral oncogenes E6 or E7. The prediction is clear; such tumor suppressor genes could map to chromosome 13, 17, or both. That is not the case, however. Studies with tumorigenic revertants of these hybrid cells or with microcell transfer of single chromosomes have mapped the tumor suppressor gene for HeLa or SiHa cells to chromosome 11 (Koi et al., 1989; Saxon et al., 1986). Perhaps the gene on chromosome 11 results in a reduced level of E6 or E7 proteins in the cell. In either case, these experimental results once again point out that we have more to understand and the DNA tumor viruses continue to provide important and useful lessons in our studies of tumorigenesis and transformation.

ACKNOWLEDGMENTS

The author thanks K. James for secretarial help and M.E. Perry and C. Finlay for excellent advice and criticism of the manuscript. This chapter was written while the author's laboratory was supported by grants from the National Institutes of Health, CA49271, CA38757, and CA41086.

REFERENCES

Brinster, R.L., H.Y. Chen, A. Messing, et al. (1984) Transgenic mice harboring SV40 T antigen genes develop characteristic brain tumors. Cell 37:367–379.

Chen, Y.R., S.P. Lees-Miller, P. Tegtmeyer, et al. (1991) The human DNA-dependent protein kinase phosphorylates SV40 T-antigen amino and carboxy-terminal sites. J. Virol. 65:5131–5140.

Crook, T., D. Wrede, and K.H. Vousden (1991) p53 point mutation in HPV negative cervical carcinoma cell lines. Oncogene 6:873–875.

DeCaprio, J.A., J.W. Ludlow, J. Figge, et al. (1988) SV40 large tumor antigen forms a specific complex with the product of the retinoblastoma susceptibility gene. Cell 54:275–283.

Dyson, N., and E. Harlow (1992) Adenovirus E1A targets key regulators of cell proliferation. Cancer Surv. (in press).

Flint, S.J., and T. Shenk (1989) Adenovirus E1A protein paradigm viral transactivator. Ann. Rev. Genet. 23:141–161.

Fried, M. (1965) Cell transforming ability of a temperature sensitive mutant of polyoma virus. Proc. Natl. Acad. Sci. USA 53:486–497.

Harlow, E., P. Whyte, B.J. Franza, et al. (1986) Association of adenovirus early region 1A proteins with cellular polypeptides. Mol. Cell. Biol. 6:1579–1589.

Harvey, D., and A.J. Levine (1991) p53 alteration is a common event in the spontaneous immortalization of primary Balb/c murine embryo fibroblasts. Genes Dev. 5:2375–2385.

Koi, M., H. Morita, H. Yamada, et al. (1989) Normal human chromosome 11 suppresses tumorigenicity of human cervical tumor cell line SiHa. Mol. Carcin. 2:12–21.

Lane, D.P., and L.V. Crawford (1979) T antigen is bound to a host protein in SV40-transformed cells. Nature 278:261–263.

Lees-Miller, S.P., and C.W. Anderson (1991) The DNA activated protein kinase, DNA-PK: A potential coordinator of nuclear events. In: Cancer Cells, vol. 3, no. 9. Cold Spring Harbor, N.Y.: Cold Spring Harbor Laboratory Press, pp. 341–346.

Lin, J.-Y., and D.T. Simmons (1991) The ability of large T-antigen to complex with p53 is necessary for the increased life span and partial transformation of human cells by SV40. J. Virol. 65:6447–6453.

Linzer, D.I.H., and A.J. Levine (1979) Characterization of a 54K dalton cellular SV40 tumor antigen in SV40 transformed cells. Cell 17:43–52.

McVey, D., L. Brizuela, I. Mohr, et al. (1989) Phosphorylation of large T-antigen by cdc-2 stimulates SV40 DNA replication. Nature 341:503–507.

Münger, K., M. Scheffner, J.M. Huibregtse, et al. (1992) The interactions of the HPV E6 and E7 oncoproteins with tumor suppressor gene products. Cancer Surv. (in press).

Palmiter, R., H. Chen, A. Messing, et al. (1985) SV40 enhancer and large T antigen are instrumental in development of choroid plexus tumors in transgenic mice. Nature 316:457–460.

Peden, K.W.C., A. Srinivasan, J.M. Farber, et al. (1989) Mutants with changes in or near the hydrophobic region of SV40 large T-antigen are defective for binding cellular protein p53. Virology 168:13–21.

Riou, G., M. Favre, D.J. Jeannel, et al. (1990) Association between poor prognosis in early stage invasive cervical carcinomas and nondetection of HPV DNA. Lancet 335:1171–1174.

Sarnow, P., Y.S. Ho, J. Williams, et al. (1982) Adenovirus E1B-58Kd tumor antigen and SV40 large tumor antigen are physically associated with the same 54Kd cellular protein in transformed cells. Cell 28:387–394.

Saxon, P.J., E.S. Srivatsan, and E.J. Stanbridge (1986) Introduction of chromosome 11 via microcell transfer controls tumorigenic expression of HeLa Cells. EMBO J. 5:3461–3466.

Scheffner, M., B.A. Werness, J.M. Huibregtse, et al. (1990) The E6 oncoprotein encoded by human papillomavirus 16 or 18 promotes the degradation of p53. Cell 63:1129–1136.

Schmieg, F.I., and D.T. Simmons (1988) Characterization of the in vitro interaction between SV40 T-antigen and p53: Mapping the p53 binding site. Virology 164:132–140.

Srinivasan, A., K.W.C. Peden, and J.M. Pipas (1989) The large tumor antigen of Simian Virus 40 encodes at least two distinct transforming functions. J. Virol. 63(12):5459–5463.

Stanbridge, E., C.J. Dez, D.J. Dorsen, et al. (1982) Human cell hybrids: Analysis of transformation and tumorigenicity. Science 215:252.

Tevethia, S.S. (1990) Recognition of SV40 T-antigen by cytotoxic T-lymphocytes. Mol. Biol. Med. 7:83–96.

Tooze, J. (ed.) (1973) Molecular Biology of the Tumor Viruses. Cold Spring Harbor, N.Y.: Cold Spring Harbor Laboratory Press, pp. 350–403.

van der Eb, A., H. van Ormondt, P.I. Schrier, et al. (1979) Structure and function of the transforming genes of human adenoviruses and SV40. In CSH Symp. Quant. Biol., vol. 44, Cold Spring Harbor, N.Y.: Cold Spring Harbor Laboratory Press, pp. 383–399.

Werness, B.A., A.J. Levine, and P.M. Howley (1990) Association of human papillomavirus types 16 and 18 E6 proteins with p53. Science 248:76–79.

Whyte, P., N.M. Williamson, and E. Harlow (1989) Cellular targets for transformation by the adenovirus E1A proteins. Cell 56:67–75.

Yew, P.R,. and A.J. Berk (1992) Inhibition of p53 transactivation required for transformation by adenovirus E1B 55 Kd protein. Nature (in press).

Zhu, M.A., R.P.W. Rice, and C.N. Cole (1991) The ability of Simian Virus 40 large T-antigen to immortalize primary mouse embryo fibroblasts cosegregates with its ability to bind to p53. J. Virol. 65:6872–6880.

zur Hausen, H., and A. Schneider (1987) The role of papillomaviruses in human anogenital cancer. In: P.M. Howley and N.P. Salzman (eds.). Papillomaviruses. The Papovaviridae, vol. 2. New York: Plenum Press, pp. 245–263.

Chapter 12

Viral Etiology of Neurological Malignancies in Transgenic Mice: Olfactory Neuroblastoma, Adrenal and Myenteric Plexus Tumors, and Nonfamilial von Recklinghausen Neurofibromatosis

Lionel Feigenbaum, Hiroyuki Ueda, and Gilbert Jay

INTRODUCTION

Based upon both epidemiologic and experimental data, it is becoming increasingly clear that viruses are a major risk factor for the development of human cancers. Perhaps the most convincing are the correlations between human papillomavirus (HPV) and cervical carcinoma, hepatitis B virus (HBV) and hepatocellular carcinoma, Epstein-Barr virus (EBV) and both Burkitt's lymphoma and nasopharyngeal carcinoma, and human T-lymphotropic virus and adult T-cell leukemia. It has been estimated that these four viruses alone account for approximately one-sixth of the worldwide cancer incidence (Zur Hausen, 1991).

So far, there has been no direct evidence for an association between viruses and the development of nervous system tumors in humans. However, there is no reason to exclude the possibility of such an involvement. Indeed, a variety of human viruses have been found in the brain and some of them have also been shown to possess transforming activities on appropriate target cells in culture. Furthermore, there are increasing examples of viruses causing neurological malignancies in experimental animals. It is not unreasonable, therefore, to suspect that viruses may ultimately be found to play a similar etiological role in human cancers of the nervous system.

Traditionally, the study of human viruses has been hampered by a variety of experimental factors, including their inability to infect laboratory animals because of their restricted host range, their route of entry in determining disease outcome, and the status of the host immune system both at the time of infection and subsequent to infection. Together, these factors have made it difficult to establish appropriate animal models for the study of viral carcinogenesis. The use of tissue culture cells, while useful in many instances, has not proven particularly successful with regard to defining the mechanism of pathogenesis. The manifestation of disease is invariably a consequence of perturbed cell–cell interactions, and attempts to recreate such cell–cell communications *ex vivo* have been difficult at best.

To overcome some of these difficulties, we have attempted to apply the transgenic technology to understand how viruses play a role in the development of human cancers. By placing viral genes directly into the germline of animals, we avoid many of the problems inherent with viral infections. Since virus attachment, penetration, uncoating, and replication are entirely circumvented, we can study viral gene expression and function in animals that are normally refractory to infection. Since the progeny of a transgenic founder animal almost invariably express the transgene in a characteristic subset of tissues, the disease phenotype that develops will be independent of virus multiplicity and route of infection. Since specific viral genes and not necessarily the entire viral genome may be introduced into the fertilized egg, one can potentially dissect an otherwise complex array of disease manifestations that is associated with infection by a single virus. However, a transgenic animal model is not without flaws, one of which being the argument that no laboratory animal can fully mimic the human host. Suffice it to say, with proper perspective and appropriate controls, transgenic animals can prove particularly rewarding in our understanding of human cancers.

Molecular Genetics of Nervous System Tumors, pages 153-161

In this chapter, we discuss three potential mechanisms whereby viruses may be responsible for causing nervous system tumors in transgenic mice. First, we discuss the possibility of endogenous viral sequences present within our genome contributing to the development of olfactory neuroblastoma. Second, we give an example of how evolutionary changes in the regulatory region of a virus can alter its tissue or host specificity, and lead to the development of tumors in the adrenal medulla and the myenteric plexus. Finally, we provide a suggestion of how the product of a regulatory gene from a virus may alter the expression of host genes and cause the development of neurofibromas. These examples illustrate not only the value of the transgenic approach but also how endogenous and exogenous viruses may, directly or indirectly, induce host cells to acquire a malignant state of growth.

POSSIBLE ROLE OF ENDOGENOUS RETROVIRUSES IN THE DEVELOPMENT OF OLFACTORY NEUROBLASTOMA

The early (E1) region of the human adenovirus type 12 (Ad12) encodes two gene products, designated E1A and E1B, which are necessary for the initiation and maintenance of transformation of a variety of cells (Byrd et al., 1988). At least in tissue culture cells, E1A and E1B complement each other in function; it appears that E1A is required for immortalization of cells that then become susceptible to transformation by E1B (Kelekar and Cole, 1987; Ruley, 1983).

We were curious whether E1A and E1B can also cooperate with each other *in vivo* to induce tumors in targeted tissues of transgenic mice (Koike et al., 1989). When the E1A and E1B genes were separately placed under the control of the mouse mammary tumor virus (MMTV) long terminal repeat (LTR) and introduced into the germline of mice, neither gene by itself was sufficient to induce a phenotype. However, when the entire E1 region, containing both the E1A and E1B genes, was put under the control of the MMTV LTR, the resulting transgenic mice almost invariably developed gastric carcinoma and died prematurely. This observation supports the hypothesis of functional cooperativity between genes that induce immortalization and transformation in inducing malignant tumors *in vivo*.

In our study, one of the transgenic mice, designated G5, failed to develop gastric carcinoma and was able to be bred into a continuous line. Unlike the others, which all showed restricted expression of both the E1A and E1B genes in the stomach, the G5 mouse and its progeny did not (Koike et al., 1990). Instead, expression of E1A and E1B genes was detected only in the brain. Since this discrepancy was observed in only one transgenic line, it is fair to assume that the site of integration of the transgene resulted in the altered expression. It is likely that the transgene has been inserted next to a cis-regulatory element which restricts the expression of E1A and E1B to the brain.

At about six to nine months of age, the G5 transgenic offspring became progressively lethargic and developed bulgings anteriorly of the cranium (Fig. 1a). Upon dissection, a large tumor mass that occupied the frontal lobe of the brain was observed in each mouse. The tumors were found to originate from the nasal sinuses and extended through the cribriform plate to involve the olfactory bulb. Penetrance of disease was 100% among the homozygous offspring. Given the young age when these mice develop the brain tumors, the high penetrance of disease is inconsistent with the requirement of a second genetic event other than integration of the transgene.

Histologic examination of the nasal masses revealed that they were composed of neoplastic cells. These cells, round to moderately elongate in appearance, extended through the sensory nasal mucosa and connective tissues in solid sheets. They occasionally formed neuroepithelial rosettes with basally oriented nuclei and elongate apical cytoplasm (Fig. 1b). The site of these tumors and the morphological characteristics of the tumor cells are consistent with a diagnosis of olfactory neuroblastoma.

While olfactory neuroblastomas are tumors that originate from primitive neural precursor cells in the nasal sinuses and cranium, the exact cell of origin is uncertain. In order to obtain more insight into the derivation of these tumors, electron microscopy was performed on tumor tissues from several animals. The tumor cells were poorly differentiated and contained both neurosecretory granules and neurofilaments. Immunoperoxidase staining for both neuron-specific enolase and S-100 was negative. Surprisingly, large numbers of mature retrovirus particles were observed within the lumens of the neuroepithelial rosettes (Fig. 1c). The morphology of these viruses was reminiscent of type C retroviral particles. Some were found budding from the membranes of the tumor cells. Analysis of tumor extracts by the XC plaque assay has consistently demonstrated high titers of mouse ecotropic retroviruses that were not detected in control littermates.

Given the high penetrance of the disease and the tight correlation with transgene expression, it is likely that E1A and E1B were responsible for the pathological changes observed. However, it is not clear whether this effect was direct or indirect. Several possible mechanisms could have led to these observations. First, the site of integration may have altered the tissue specificity of the transgenic promoter and the brain-specific expression of the transgene may have led to the development of the phenotype, either directly as a result of the activities of E1A and E1B, or indirectly through the activation and function of an endogenous retrovirus sequence. Second, integration of the transgene may have disrupted the expression of an essential cellular gene, whose altered expression may have been responsible for either

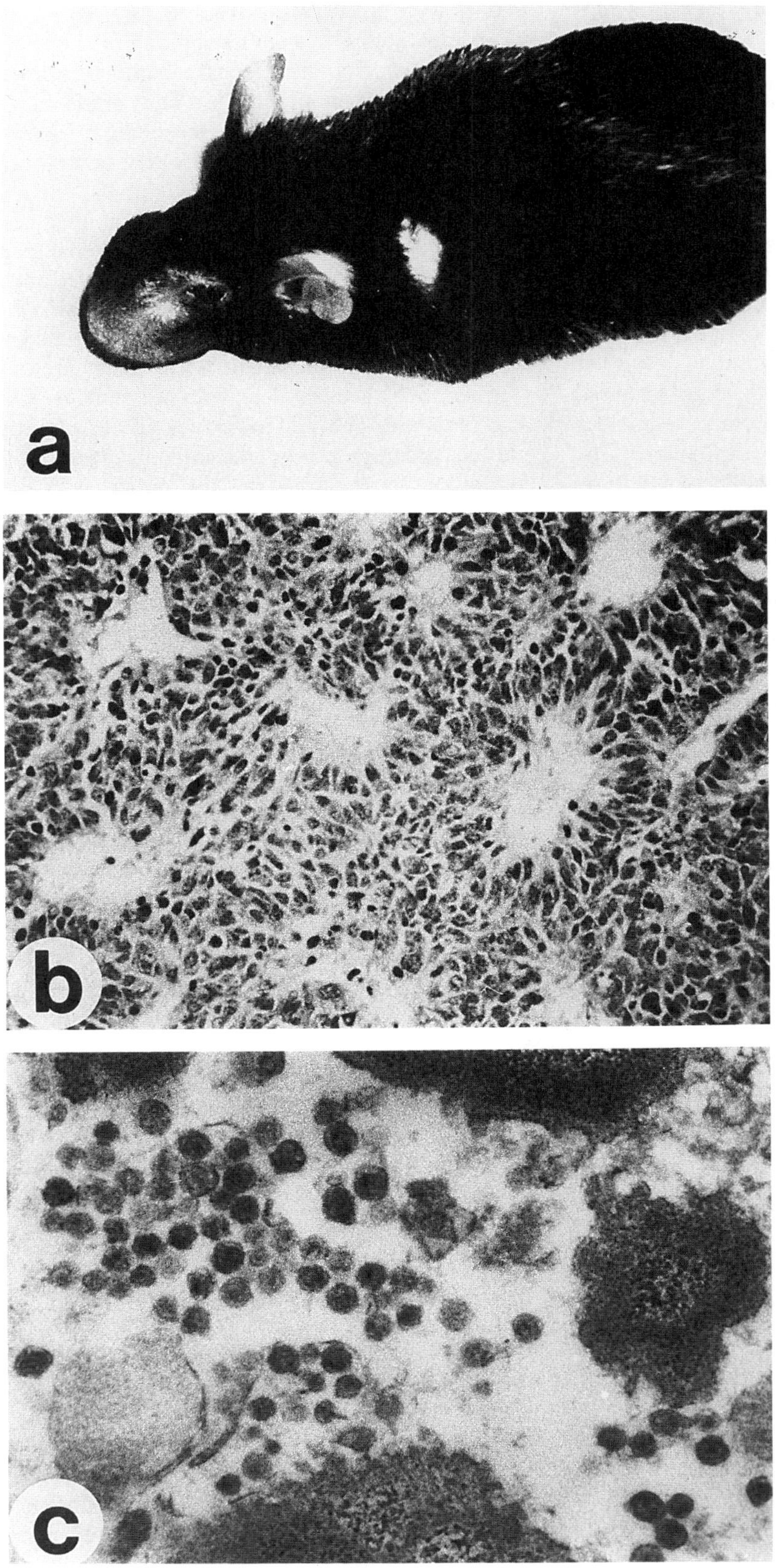

Fig. 1. Gross and microscopic analysis of transgenic mice with olfactory neuroblastoma. **a**: Mouse with tumor bulging anteriorly of the cranium. **b**: Hematoxylin and eosin (H & E) staining of an olfactory bulb tumor, with neuroepithelial cells arranged in rosettes. **c**: Electron microscopic analysis depicting type C retroviral particles located within the lumen of a neuroepithelial rosette.

directly inducing the phenotype or indirectly activating an endogenous retroviral sequence, which led to the development of disease.

In an attempt to further define whether the presence of the retroviral particles is the cause or the consequence of the tumor growth, we have analyzed domestic cats that have been diagnosed with olfactory neuroblastoma. Of the three animals available for examination, all had primitive neoplastic cells extending from the olfactory mucosa and forming characteristic neuroepithelial rosettes. Interestingly, each of the animals also had large numbers of retroviral particles with type C morphology within both the lumens of the tumor rosettes and the intercellular spaces (Koike et al., 1990).

While these observations do not by themselves prove that viruses are the etiologic agents of olfactory neuroblastoma, they are suggestive that endogenous retroviral sequences may play an important role in the development of this disease, and they beg for a careful examination of appropriate human tumor tissues for a similar finding. Endogenous type C retroviral sequences have been detected in many animal species; they are present in 50–100 copies per haploid human genome (Bonner et al., 1982; Rabson et al., 1985). At least in mice these endogenous sequences may be activated by different physical and chemical agents, and their activation has been correlated with the development of disease (Coffin, 1984). It is likely that the mode of activation of these viruses was quite different in our transgenic mice and in the domestic cats, although each mechanism subserved the same hypothetical endpoint.

ROLE OF A VIRAL REGULATORY ELEMENT IN CONFERRING NEUROTROPISM AND IN INDUCING NERVOUS SYSTEM TUMORS

The human DNA virus, JCV, is associated with the neurodegenerative disease termed progressive multifocal leukoencephalopathy (PML) (Padgett et al., 1971). The early JCV protein, T antigen, plays a critical role not only in the transformation activity of this virus but also in viral DNA replication and late gene expression (Frisque et al., 1984). Expression of JCV T antigen is controlled by the viral enhancer/promoter region, which demonstrates highly restricted tissue specificity (Feigenbaum et al., 1987; Kenney et al., 1984; Khalili et al., 1988). The neurotropism of JCV is, therefore, controlled at the transcriptional level (Feigenbaum et al., 1987; Lynch and Frisque, 1991).

Transgenic mice were derived that contained the entire early region of JCV, including the enhancer/promoter element and the complete coding sequence of T antigen (Feigenbaum et al., 1992; Small et al., 1986a,b). Several mice transgenic for the JCV sequence died between two and four months of age. Upon autopsy, greatly enlarged adrenal glands, reaching half the size of a kidney, were observed. In each case, the involvement was bilateral. Intestinal masses located in the ileum were also observed. Some of these masses were extensive in size.

Histopathological analysis of the adrenal glands revealed tumors arising from the adrenal medulla. A normal adrenal gland is shown in Figure 2a. This organ contains two different types of endocrine tissues that have distinctly different embryological derivation. The outer layer, or adrenal cortex, is of mesodermal origin; the inner layer, or adrenal medulla, is derived from neural crest cells. Compared to the normal adrenal gland, the adrenal gland from a transgenic mouse showed a greatly enlarged medulla that was darkly stained by hematoxylin and eosin (Fig. 2b). A high magnification of the same region revealed cells with well-rounded hyperchromatic nuclei and numerous mitoses (Fig. 2c, arrowheads). Cells with large bizarre nuclei were also observed. The vast expansion of the adrenal medulla may cause compression of the adrenal cortex. It is important to note that these adrenal medulla tumors were well encapsulated. These histologic changes are characteristic of adrenal medulla neuroblastoma.

Histological analysis of the large masses located in the intestine revealed that the lesions were derived from the muscularis layer. The muscularis externa consists of two substantial layers of smooth muscles that undergo peristaltic contractions. The peristaltic waves are coordinated by efferent impulses from a plexus known as the myenteric plexus, situated between the circular and longitudinal layers of the muscularis externa (Fig. 3a, arrowheads). In early lesions, tumor cells were seen arising from the nerve ganglia that made up the myenteric plexus (Fig. 3b, arrows). A normal ganglion is composed of neural cell bodies with nuclei that are round and lightly stained by hematoxylin and eosin. In early lesions, the ganglion showed complete loss of cell morphology. Nuclei were hyperchromatic and cell bodies were highly proliferative (Fig. 3b). Given time, these aggressive tumors displaced the muscularis externa, erroneously suggesting that they arose from this structure. Unlike transgenic mice with the olfactory neuroblastoma, in which tissue specificity is dictated by the site of integration of the transgene, the neural pathology observed in the JCV transgenic mice was the result of specific expression of the transgene in neural tissues. When the enhancer/promoter region of a highly related virus, simian virus 40 (SV40), was employed to direct expression of the JCV T antigen, tumors were observed selectively in epithelially derived cells of the choroid plexus, and hyperplasia is seen in the thyroid and the kidney (Feigenbaum et al., 1992). These results demonstrate that JCV T antigen has cellular targets not only in neural tissues but also in cells of other derivation, and would argue for the tissue specificity of this virus being determined by its transcriptional enhancer/promoter region. In support of this argument is the finding that JCV variants with fine mutations within the enhancer/promoter element were not expressed in neural tissues (Loeber and Dorries, 1988).

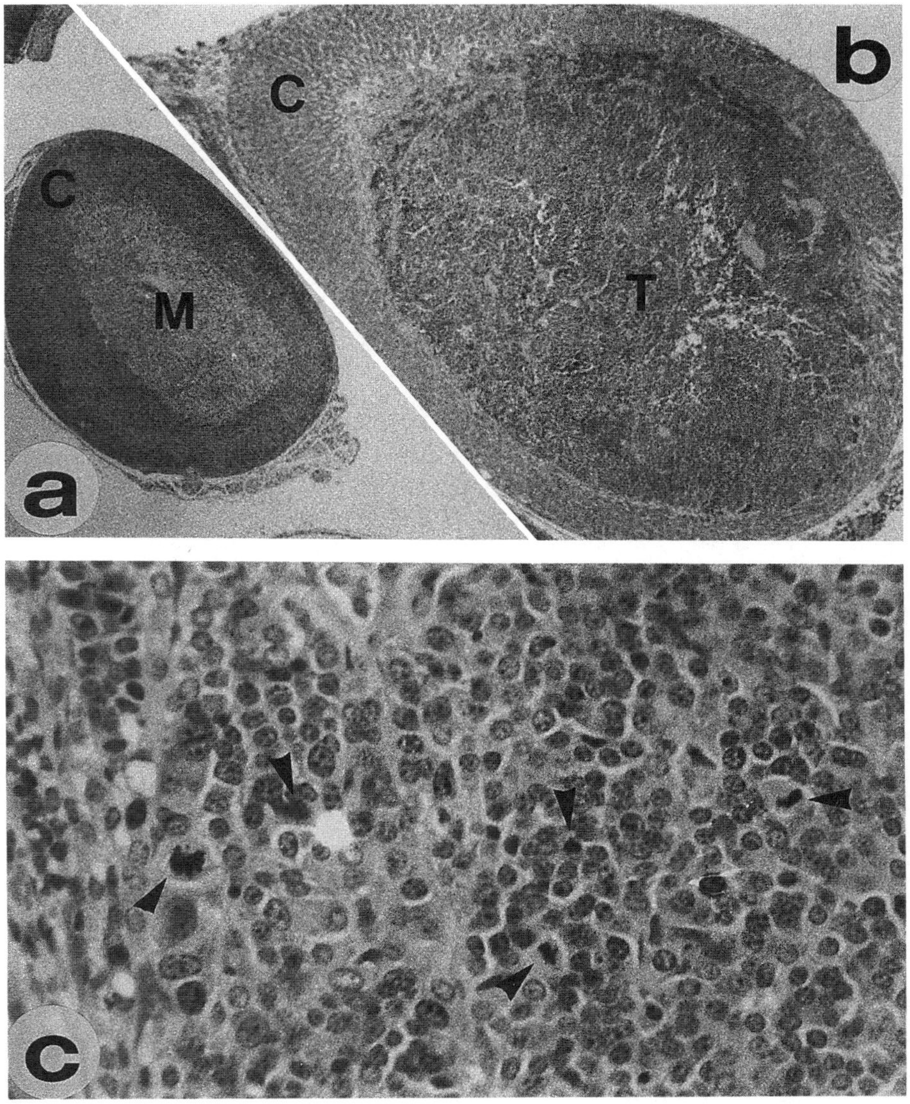

Fig. 2. Microscopic analysis of the adrenal glands of the JCV transgenic mice. **a**: H & E stain of a cross section from the adrenal gland of a normal mouse. The darkly stained adrenal cortex (C) and adrenal medulla (M) are represented. **b**: Enlarged adrenal gland from a JCV transgenic mouse. The adrenal medulla has been replaced by a large tumor (T), and the adrenal cortex was somewhat compressed. **c**: High magnification of the adrenal medulla neuroblastoma with numerous mitoses (arrowheads).

The type of neural tumors that developed in our transgenic mice has not been documented in humans infected with JCV. However, this does not imply that JCV is incapable of inducing these neoplasms in man, but may suggest finer levels of specificity of the virus *in vivo*. It is possible that access to ganglia of the myenteric plexus or to the cells of the adrenal medulla is difficult during a normal viral infection. That JCV can cause neural tumors in humans is suggested by the occurrence of gliomas in several patients with PML (Castaigne et al., 1974; Sima et al., 1983). The somewhat rare occurrence of this neoplasm among the JCV-infected individuals is most likely due to the poor prognosis and short life span following the onset of PML. Similarly, the shortened life span of our transgenic mice that succumbed to the aggressive myenteric plexus tumors may not allow other types of neural tumors to develop. Indeed, intracranial inoculations of JCV in hamsters and owl monkeys have led, after a long latency, to the development of various brain tumors (Major et al., 1987; Nagashima et al., 1984; Padgett et al., 1977). The site of injection of the virus, together with the long incubation period, apparently facilitated the outcome of these studies. It is likely that JCV has multiple neural

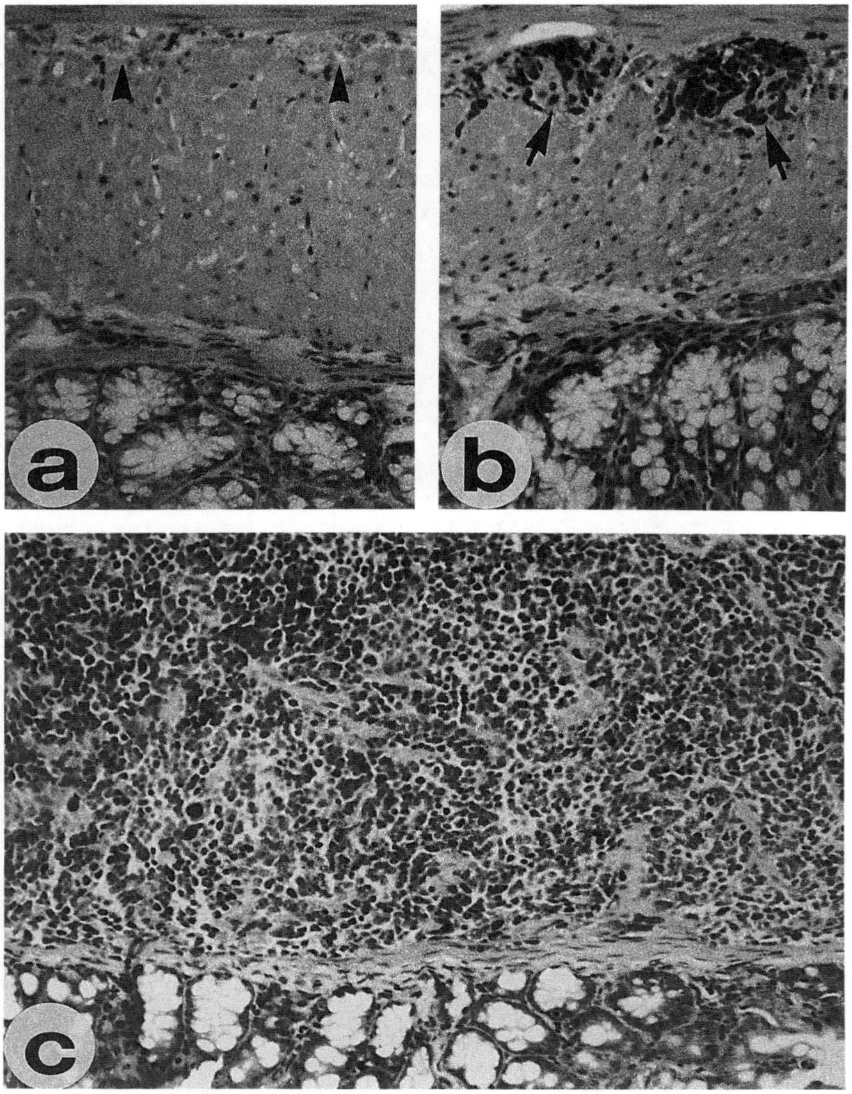

Fig. 3. Microscopic analysis of the myenteric plexus tumors from the JCV transgenic mice. **a**: H & E stain of a cross section of the intestine of a normal mouse. The nerve ganglia that make up the myenteric plexus (arrowheads) are surrounded by two muscle layers that border the intestinal mucosa. **b**: The greatly enlarged and proliferative nerve ganglia of the myenteric plexus (arrows) from a transgenic mouse are composed of cell bodies with hyperchromatic nuclei (×200). **c**: The myenteric plexus tumor has displaced the muscle layer and is now juxtaposed next to the intestinal mucosa (×50).

targets in infected individuals, and that the route of uptake and the period of incubation dictate the neurologic malignancy that develops.

ROLE OF A VIRAL TRANSACTIVATOR PROTEIN IN INDUCING NEUROFIBROMAS

Human T-cell leukemia virus type I (HTLV-I) has been associated with adult T-cell leukemia (ATL), a malignancy characterized by monoclonal proliferation of CD4-positive T cells (Poiesz et al., 1980; Yoshika et al., 1982), and tropical spastic paraparesis (TSP), a degenerative disorder involving spastic paralysis as well as facial nerve paralysis and sensory deficits (Gessain et al., 1985; Rogers-Johnson et al., 1985). Evidence linking HTLV-I to these two diseases is derived from epidemiological studies that in themselves do not prove direct involvement of the virus.

To determine the role HTLV-I plays in these lymphoid and neurological disorders, transgenic mice were derived

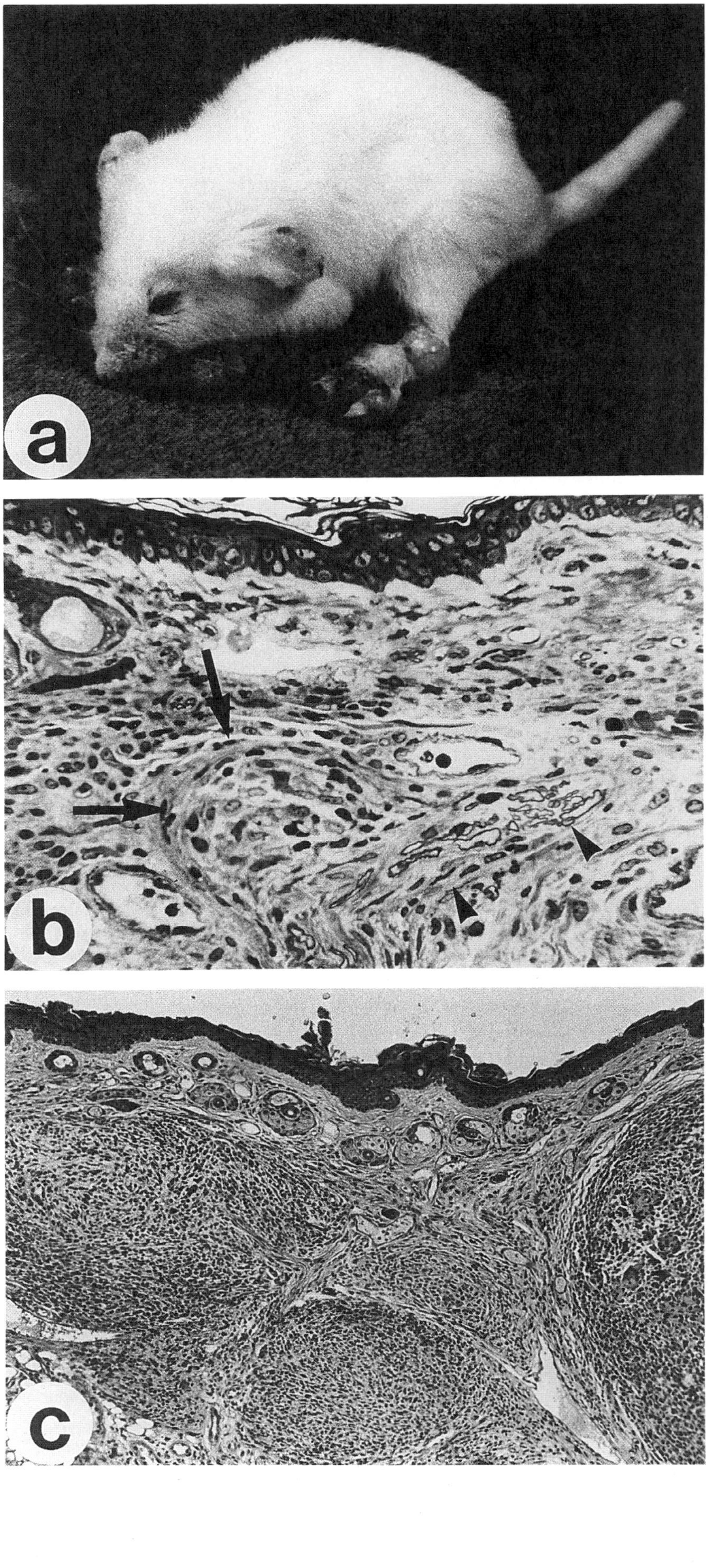

Fig. 4. Gross and microscopic analysis of transgenic mice with the HTLV *tax* gene. **a**: Mouse with tumors located on the front and hind legs, ears, and tail. **b**: Methylene Blue Azur II stain of a section of an ear tumor with spindle-shaped cells (arrows). The small nodule extends from the perineurium, characterized by the presence of myelinated nerve axons (arrowheads) (×100). **c**: H & E stain through a neurofibroma located in the tail. This plexiform tumor shows distinctive nodules.

that expressed the viral regulatory gene *tax*. The Tax protein acts as a transregulator of both viral genes by acting on sequences located within the viral LTR (Sodroski et al., 1984) and cellular genes involved in cell growth such as those that encode the interleukin 2 receptor (Inoue et al., 1986), granulocyte-macrophage colony-stimulating factor (GM-CSF) (Chan et al., 1986), and c-*fos* (Fujii et al., 1988). Transgenic mice carrying the HTLV *tax* gene under the control of the viral LTR developed soft tissue tumors between the ages of 13 and 17 weeks (Nerenberg et al., 1987). These tumors were localized primarily on the ear, nose, mouth, foot, and tail (Fig. 4a). The occurrence of similar tumors in three independent founder lines clearly demonstrates the role of Tax in the development of this phenotype. Histological analysis of an early lesion, shown in Figure 4b, shows that the tumor is composed of spindle-shaped cells (arrows), characteristic of tumors of mesenchymal derivation. These spindle-shaped cells were associated with peripheral nerves (arrowheads). As the tumors progressed, they usually retained their benign appearance, but occasionally extended into neighboring connective tissues. Regardless of their location, these multicentric tumors formed discrete nodules (Fig. 4c).

Further electron microscopic analysis to pinpoint the cell of origin of these tumors revealed that they were of nerve sheath origin (Hinrichs et al., 1987). On occasion, some of the females did not develop any apparent tumors on their extremities, but still died prematurely with progressive wasting. Analysis of these animals showed tumors of the cranial nerves. Involvement was seen especially with cranial nerve V and VII (Hinrichs et al., 1987). Tumors arising from the nerve sheath and located at extremities had great resemblance to neurofibromas in patients with von Recklinghausen neurofibromatosis (VRNF). This human genetic disease, which affects one in 3,000 individuals, manifests itself by the presence of *café au lait* spots, as well as dermal neurofibromas. It has also been shown that cranial nerve V is often the site of plexiform tumors in humans, a pathology also observed in the *tax* transgenic mice.

The gene (NF1) that is thought to cause VRNF has been identified, and is localized on chromosome 17 (Cawthon et al., 1990). For the 50% of individuals who have inherited the disease in an autosomal dominant fashion, mutations ranging from single point mutations to entire deletions of the NF gene have been observed (Viskochil et al., 1990; Wallace et al., 1990). The rest of the VRNF cases have no familial history and presumably arose spontaneously. This high frequency of spontaneous cases has suggested the involvement of infectious agents, such as HTLV-I, as demonstrated by the striking similarities between the tumors in our HTLV *tax* transgenic mice and those seen in VRNF patients. Since only the Tax protein from HTLV-I is expressed in these mice and at high levels in the tumors, a mechanism by which it could directly or indirectly down-regulate the expression of the gene could conceivably occur, and suffice the same endpoint as structurally altering the gene by mutation. Given the role of the Tax protein as a transregulator of different cellular genes, this proposed mechanism of etiology of NF in our transgenic mice and in spontaneous cases of VRNF is currently being tested.

In contrast with JCV T antigen, which has the ability to transform cells in a variety of tissues but is restricted in expression to cells of the nervous system by the viral transcriptional regulatory element, the HTLV-I Tax protein while expressed in different tissues under the control of the viral LTR seems to transform selectively only neurofibroblasts. It seems that JCV restricts its promiscuous transforming protein with a highly tissue-specific control element, while HTLV allows its highly targeted transforming protein to be expressed in apparently irrelevant tissues without functional consequences. This illustrates how these two viruses differ in their modes of action in inducing pathogenesis in the host.

CONCLUSIONS

The studies presented show the important role viruses play in the development of nervous system tumors in transgenic mice, and have suggested three potential mechanisms through which neural tumors may arise. The conclusions drawn by the various experiments are provocative. We have suggested that olfactory neuroblastoma may be caused by the activation of endogenous retroviruses, that humans who are infected with JCV may potentially develop adrenal medulla neuroblastomas and myenteric plexus tumors, and that individuals infected with HTLV-I are at risk of developing the nonfamilial form of VRNF. Our transgenic mouse models provide a molecular handle on the study of the various forms of nervous system tumors and offer testable hypotheses as to their possible mechanisms of etiology. Isolation of the endogenous retrovirus suspected of causing olfactory neuroblastoma in our transgenic mice and its inoculation into normal mice to test its ability to cause disease is technically feasible. Testing the effect of the HTLV Tax protein on the expression of the NF gene in mice, as well as its ability to down-regulate the activity of the NF promoter in transfection studies, are experiments of paramount importance. Finally, identification of subcellular targets for the JCV T antigen in neural tissues to determine events leading to tumor formation is of equal interest. Valuable information has been gained through the study of our transgenic mice. Many more questions are awaiting to be explored.

REFERENCES

Bonner, T.I., C. O'Connell, and M. Cohen (1982) Cloned endogenous retroviral sequences from human DNA. Proc. Natl. Acad. Sci. USA 79:4709–4713.

Byrd, P.J., R.J. Grand, D. Breiding, et al. (1988) Host range mutants of adenovirus type 12 El defective for lytic infection, transformation and oncogenicity. Virology 163:155–165.

Castaigne, P., P. Rondot, R. Escourolle, et al. (1974) Leucoencéphalopathie multifocale progressive et 'gliomes' multiples. Rev. Neurol. 130:379–392.

Cawthon, R.M., R. Weiss, G. Xu, et al. (1990) A major segment of the neurofibromatosis type 1 gene: cDNA sequence, genomic structure, and point mutations. Cell 62:193–201.

Chan, J.Y., D.J. Slamon, S.D. Nimer, et al. (1986) Regulation of expression of human granulocyte/macrophage colony-stimulating factor. Proc. Natl. Acad. Sci. USA 83:8669–8673.

Coffin, J. (1984) Endogenous viruses. In: R. Weiss, N. Teich, H. Varmus, and J. Coffin (eds.). RNA Tumor Viruses. Cold Spring Harbor, N.Y.: Cold Spring Harbor Laboratory Press, pp. 1109–1203.

Feigenbaum, L., S.H. Hinrichs, and G. Jay (1992) JCV and SV40 enhancers and transforming proteins: Role in determining tissue specificity and pathogenicity in transgenic mice. J. Virol. 66:1176–1182.

Feigenbaum, L., K. Khalili, E. Major, et al. (1987) Regulation of the host range of human papovavirus JCV. Proc. Natl. Acad. Sci. USA 84:3695–3698.

Frisque, R.J., G.L. Bream, and M.T. Canella (1984) Human polyomavirus JC virus genome. J. Virol. 51:458–469.

Fujii, M., P. Sassone-Corsi, and I.M. Verma (1988) c-fos promoter trans-activation by the tax_1 protein of human T-cell leukemia virus type I. Proc. Natl. Acad. Sci. USA 85:8526–8530.

Gessian, A., J.C. Vernant, L. Maurs, et al. (1985) Antibodies to human T-lymphotropic virus type-I in patients with tropical spastic paraparesis. Lancet 2:407–410.

Hinrichs, S.H., M. Nerenberg, R.K. Reynolds, et al. (1987) A transgenic mouse model for human neurofibromatosis. Science 237:1340–1343.

Inoue, J., M. Seiki, T. Taniguchi, et al. (1986) Induction of interleukin 2 receptor gene expression by $p40^x$ encoded by human T-cell leukemia virus type 1. EMBO J. 5:2883–2888.

Kelekar, A.C., and M.D. Cole (1987) Immortalization by c-myc, H-ras, and E1A oncogenes induces differential cellular gene expression and growth factor responses. Mol. Cell. Biol. 7:3899–3907.

Kenney, S., B. Natarajan, D. Strike, et al. (1984) JC virus enhancer-promoter active in human brain cells. Science 236:1337–1339.

Khalili, K., J. Rappaport, and G. Khoury (1988) Nuclear factors in human brain cells bind specifically to the JCV regulatory region. EMBO J. 7:1205–1210.

Koike, K., S.H. Hinrichs, K.J. Isselbacher, et al. (1989) Transgenic mouse model for human gastric carcinoma. Proc. Natl. Acad. Sci. USA 86:5615–5619.

Koike, K., G. Jay, J.W. Hartley, et al. (1990) Activation of retrovirus in transgenic mice: Association with development of olfactory neuroblastoma. J. Virol. 64:3988–3991.

Loeber, G., and K. Dorries (1988) DNA rearrangements in organ-specific variants of polyomavirus JC strain GS. J. Virol. 62:1730–1735.

Lynch, K.J., and R.J. Frisque (1991) Factors contributing to the restricted DNA replicating activity of JC virus. Virology 180:306–317.

Major, E.O., D.A. Vacante, R.G. Traub, et al. (1987) Owl monkey astrocytoma cells in culture spontaneously produce infectious JC virus which demonstrates altered biological properties. J. Virol. 61:1435–1441.

Nagashima, K., K. Yasui, J. Kimura, et al. (1984) Induction of brain tumors by a newly isolated JC virus (Tokyo-1 strain). Am. J. Pathol. 116:455–463.

Nerenberg, M., S.H. Hinrichs, R.K. Reynolds, et al. (1987) The *tat* gene of human T-lymphotropic virus type 1 induces mesenchymal tumors in transgenic mice. Science 237:1324–1329.

Padgett, B.L., D.L. Walker, G.M. ZuRhein, et al. (1971) Cultivation of a papova-like virus from human brain with progressive multifocal leukoencephalopathy. Lancet 1:1257–1260.

Padgett, B.L., D.L. Walker, G.M. ZuRhein, et al. (1977) Differential neurooncogenicity of strains of JC virus, a human polyoma virus, in newborn Syrian hamsters. Cancer Res. 37:718–720.

Poiesz, B.J., F.W. Ruscetti, A.F. Gazdar, et al. (1980) Detection and isolation of type C retrovirus particles from fresh and cultured lymphocytes of a patient with cutaneous T-cell lymphoma. Proc. Natl. Acad. Sci. USA 77:7415–7419.

Rabson, A.B., Y. Hamagishi, P.E. Steele, et al. (1985) Characterization of human endogenous retroviral envelope RNA transcripts. J. Virol. 56:176–182.

Rogers-Johnson, P., D.C. Gajdusek, O.S. Morgan, et al. (1985) HTLV-I and HTLV-III antibodies and tropical spastic paraparesis. Lancet 2:1247–1248.

Ruley, H.E. (1983) Adenovirus early region 1A enables viral and cellular transforming genes to transform primary cells in culture. Nature 304:602–606.

Sima, A.A.F., S.D. Finkelstein, and D.R. McLachlan (1983) Multiple malignant astrocytomas in a patient with spontaneous progressive multifocal leukoencephalopathy. Ann. Neurol. 14:183–188.

Small, J.A., G. Khoury, G. Jay, et al. (1986a) Early regions of JC virus and BK virus induce distinct and tissue-specific tumors in transgenic mice. Proc. Natl. Acad. Sci. USA 83:8288–8292.

Small, J.A., G.A. Scangos, L. Cork, et al. (1986b) The early region of human papovavirus JC induces dysmyelination in transgenic mice. Cell 46:13–18.

Sodroski, J.G., C.A. Rosen, and W.A. Haseltine (1984) Trans-acting transcriptional activation of the long terminal repeat of human T-lymphotropic viruses in infected cells. Science 225:381–385.

Viskochil, D., A. Buchberg, G. Xu, et al. (1990) Deletions and a translocation interrupt a cloned gene at the neurofibromatosis type 1 locus. Cell 62:187–192.

Wallace, M., D. Marchuk, L. Andersen, et al. (1990) Type 1 neurofibromatosis gene: Identification of a large transcript disrupted in three NF1 patients. Science 249:181–186.

Yoshika, M., I. Miyoshi, and Y. Hinuma (1982) Isolation and characterization of retrovirus from cell lines of human adult T-cell leukemia and its implication in the disease. Proc. Natl. Acad. Sci. USA 79:2031–2035.

Zur Hausen, H. (1991) Viruses in human cancers. Science 254:1167–1173.

Chapter 13

Radiobiological Effects and Changes in Gene Expression in the Central Nervous System in Response to Ionizing Radiation

Patrick W. McLaughlin, Randi Schea, Paul E. McKeever, and David A. Boothman

INTRODUCTION

For decades, therapeutic radiation has held a proven role in the treatment of malignant central nervous system (CNS) tumors. Due to improved localization of tumors by computed tomographic (CT) scan and magnetic resonance imaging (MRI), and improved focal radiation delivery based on three-dimensional (3-D) treatment planning, it is now possible to selectively deliver high-dose therapy to tumor and immediately adjacent brain tissues. The tolerance of this tissue is a critical dose-limiting factor in the treatment of brain tumors. Although injury and necrosis of normal brain tissue are recognized complications of radiation treatment, the actual mechanism of necrosis initiation is poorly understood. Improved therapy and outcome for patients afflicted with brain tumors may be achieved by better understanding cellular changes at the molecular level after ionizing radiation.

CENTRAL NERVOUS SYSTEM INJURY FROM RADIATION

Three clinically distinct syndromes of CNS injury following ionizing radiation are recognized: acute, early-delayed, and late-delayed responses (Leibel and Sheline, 1987). The syndromes differ in pathogenesis and prognosis. The risk of development of these syndromes is influenced by radiation fractionation (Marks et al., 1981), total radiation dose, volume irradiated, concurrent chemotherapy, and host factors such as hypertension and age.

The acute reaction rarely occurs during the course of fractionated radiation treatment and is characterized by symptoms of increased intracranial pressure or progression of initial neurologic symptoms. Acute reactions are commonly reported after treatment schedules employing large fraction size (Young et al., 1974). Although acute reactions have been attributed to radiation-induced edema, CT scans obtained during the course of radiotherapy fail to demonstrate a radiographically measurable increase in edema (Deck, 1980).

Early-delayed injury appears within weeks to months after radiation (Rider, 1963). Clinically, patients have somnolence or worsening preexisting signs and symptoms. Radiographic changes suggest tumor progression, but the usual clinical course is spontaneous resolution in six to eight weeks. Occasionally, the reaction is fulminant and fatal. Autopsy specimens demonstrate extensive demyelination with central necrosis, gliosis, and a decrease in oligodendroglial cells (Lampert and Davis, 1964). In contrast to late-delayed injury, there is little evidence for vascular injury, and neurons and axons are preserved. The pathogenesis is ascribed to transient interruption in myelin synthesis secondary to injury to oligodendroglial cells or actual loss of oligodendroglial cells. In addition, increased capillary permeability has been associated with demyelination in animal models and may contribute to the early-delayed reaction (Delattre et al., 1988).

Late-delayed injury occurs months to years following radiation (Fike et al., 1984; Leibel and Sheline, 1987; Marks et al., 1981). Clinically, patients have progressive neurologic symptoms and radiographic changes indistinguishable from tumor progression (contrast-enhancing mass lesion). Positron emission tomography (PET) is the only noninvasive study that may allow the distinction of tumor from radiation-induced necrotic areas (Patronas et al., 1982). In contrast to early-delayed injury, the pathological changes observed with late-delayed injury are predominantly associated with the fine vasculature. There is hyalinization, fibrinoid necrosis, and endothelial hyperplasia. Radiation-induced paren-

Molecular Genetics of Nervous System Tumors, pages 163-177 ©1993 Wiley-Liss, Inc.

chymal changes may be distinguished from other etiologies by the multiplicity quality of the necrotic foci and the presence of fibrin deposition along the border of gray and white matter within the brain. Fibrin deposition may extend into the deep layers of the cortex and may not be associated with other conditions (Burger and Boyko, 1991; Burger et al., 1979).

Due to difficulties in distinguishing tumor progression from delayed radiation reactions, surgical intervention is often necessary for the diagnosis and treatment of these lesions. Four patterns have been identified in pathology specimens at the time of suspected recurrence: (1) pleomorphic and hypocellular "inactive" neoplasia; (2) radiation necrosis; (3) parenchymalgliosis; and (4) hypercellular recurrent neoplasm (Burger and Boyko, 1991). In individual specimens, one or all patterns may be present. Before discussing the pathophysiology of delayed CNS injury from radiation, factors known to affect the sensitivity of cells and tissues to radiation are addressed below (Table 1).

Factors Affecting Radiation Sensitivity

The most common form of radiation in clinical use is megavoltage radiation generated by a linear accelerator. Radiation at this energy primarily interacts with tissue by transferring the energy of a photon to loosely bound electrons. These electrons then stream through tissue, causing further ionization and the formation of associated free radicals. Most free radicals are formed by the interaction of electrons with water. Hydrated electrons and hydrogen free radicals combine with oxygen to form toxic hydroperoxy radicals and hydrogen peroxide. Hypoxic areas of tissues can actually be protected from radiation damage, since the formation of these reactive oxygen intermediates is prevented. Reactive oxygen intermediates that are formed interact with macromolecules, causing a variety of lesions. Intracellular free radical scavengers (both bound and unbound to DNA) limit radiation damage to DNA (Biaglow et al., 1988; Ljungman et al., 1991), which is considered the critical target for lethal injury.

There are a variety of other factors which determine the radiation responsiveness of a given tissue. The configuration of chromatin may provide protection from DNA damage following ionizing radiation. When cellular chromatin is decondensed by hypotonic treatment (Warters and Lyons, 1990), acetylation (Nackerdien et al., 1989), or protein depletion (Ljungman, 1991), there is a marked increase in DNA damage as measured by neutral elution. DNA damage that occurs in spite of these natural defenses can be reversed through an elaborate and yet poorly understood DNA repair system. Differences in DNA repair capacity are thought to be major factors determining relative radiation sensitivity of normal and tumor tissues. Mutant cells that lack the capacity to repair radiation damage (e.g., cancer-prone human cells) have contributed to our understanding of these repair processes.

TABLE 1. Summary of Radiation Effects on the Nervous System

Proposed target cells	Possible pathogenesis	Altered physiology	Clinical syndromes
Acute Damage (0-2 months)			
Endothelia and Type 1 astrocytes	Loss of nonviable cells	Increased intracranial edema	Headache, nausea, vomiting Herniation Worsening of CNS symptions
"Early" late damage (4-6 months)			
Oligodendroglia			
Moderate damage	Glial damage Nodal widening	Transient demyelination (self limited)	Lhermitte's syndrome (PNS) Somnolence syndrome (CNS)
Severe damage	Segemental demyelination Glial depletion Extensive demyelination Radionecrosis	Irreversible demyelination (global)	Global CNS dysfunction
Endothelial	Altered vascular permeability	Transient demyelination	
"Delayed" late damage (>6 months)			
Endothelia	Endothelial hyperplasia Fibrinoid necrosis Hyaline thickening Thrombosis	Luminal narrowing/ischemia Infarction	Paralysis (PNS) Radionecrosis (cerebral/CNS) Global CNS dysfunction
Oligodendroglia	Demyelination (diffuse and extensive)	Permanent disruption of neuroconduction	

The determinants of radiation lethality at the cellular level are complex and poorly understood. A vast spectrum of DNA lesions is created by radiation. Many factors determine whether or not DNA damaged sites are repaired, misrepaired, or left unrepaired. Resting cells (i.e., G_0 cells), whether arrested due to nutritional depletion, confluency (Iliakis, 1988), or terminal differentiation, are less vulnerable to radiation damage than proliferating cells. A growth-arrested state is thought to allow full expression of all repair mechanisms, which mend lethal lesions before DNA replication.

The response of a whole tissue or population of cells to a radiation insult is determined, in part, by the fraction of surviving cells remaining after treatment. One possible adaptive survival response of damaged cells is accelerated proliferation. The phenomenon of accelerated repopulation in response to ionizing radiation has been documented in normal tissues (Withers et al., 1988). During daily radiation treatments to the oral cavity, for example, the mucosal epithelial cells rapidly proliferate and replace cells that were initially lost to radiation damage. This phenomenon may occur in human tumors as well. Accelerated repopulation has been proposed as a mechanism for failure of radiotherapy (Withers et al., 1988). Although cellular proliferation as a mechanism of resistance seems to contradict resistance due to the G_0 state discussed above, extremely rapid proliferation between radiation fractions may allow replacement of dying cells. The molecular signal initiating accelerated repopulation in normal tissues and tumors is unknown.

The discussion to this point assumes that radiation lethality occurs due to accumulation of lethal damage to DNA. At the tissue level, however, another mechanism of radiation-induced killing is the destruction of the vascular infrastructure (Reinhold et al., 1974). If radiation destroys the blood vessels to a dependent population of cells, the entire population may die due to infarction, apart from individual cell damage. In both tumors and normal tissues, vascular damage and infarction contribute to radiation lethality.

PATHOPHYSIOLOGY OF CNS RADIATION INJURY

The tolerance of whole organs such as lung, liver, and kidney to ionizing radiation is literally half the tolerance of CNS tissue. For this reason early literature viewed the brain as a radioresistant organ. Although this review has emphasized mechanisms of radiation injury to the CNS, it is important to consider reasons for its radioresistance as well as its vulnerability.

The primary controversy over pathogenesis of radiation injury concerns which target cell is responsible for the observed CNS injury syndromes (van der Kogel, 1986). There are three major cell types present: (1) neurons; (2) glia (including oligodendrocytes); and (3) vascular cells. Interestingly, the neuron or brain tissue proper is not implicated as a primary target for radiation injury in any model. Fully differentiated, nonproliferating cells such as neurons are extremely resistant to radiation. As discussed above, resting cells have unlimited time for repair. The controversy at present is whether vascular elements or glial elements are the critical target(s) for radiation damage. Both cell types proliferate, accounting for their vulnerability to radiation damage. In addition, both proliferate at an extremely slow rate, accounting for the protracted time course before damage is expressed. Results of radiation-induced injury in immature animals shows the critical role of cellular proliferation rate in the expression of damage. During early stages of development, the proliferation rate of glial cell elements is more rapid than in adults. Although the same total dose is required to induce comparable damage in adult and immature brain tissue, the onset and expression of lethality is much earlier in immature irradiated animals (van der Kogel, 1986).

The argument for a primary role for glia in CNS injury is based on analyses of specimens taken at various time intervals following radiation exposure. Glial cell populations include both astrocytes, which are thought to have a protective and supportive function, and oligodendrocytes, which are responsible for myelin synthesis. Myelin synthesis is critical for axon insulation and conduction. Initially, serial sections reveal segmental demyelination secondary to oligodendrocyte injury. Oligodendrocyte repopulation accounts for reversal of injury and remyelination. With increasing X-ray doses, segmental demyelination is followed by progressive depletion of glial elements and finally necrosis. The depletion of glial elements prior to necrosis has led some investigators to implicate glia as the primary target cell (van der Kogel, 1986).

Other investigators argue that glial damage alone is not sufficient to account for large areas of necrosis, and emphasize the importance of vascular damage. Luminal narrowing from endothelial hyperplasia in response to damaged endothelium is implicated as a critical lesion. Hopewell et al. (1974) reported endothelial hyperplasia and associated luminal narrowing in rat brains that preceded white matter necrosis. In a recent study, dogs were sacrificed three to 30 months following a single dose of 15 Gy to the brain. Coincident with the appearance of necrotic foci, small vessel and capillary changes including endothelial cell hyperplasia with resultant luminal narrowing were evident (Yamaguchi et al., 1991). The greater incidence of white matter necrosis was explained by differences in vascular density, which was lower in white matter compared to gray matter (Yamaguchi et al., 1991).

Finally, a potential late effect of radiation to the CNS is the formation of secondary tumors, which may occur years after the initial exposure to ionizing radiation. Although most adult patients with malignant brain tumors succumb to their disease, patients with curable benign tumors have an increased rate of tumor induction, including the formation of malignant gliomas and meningiomas (Kumar et al., 1987;

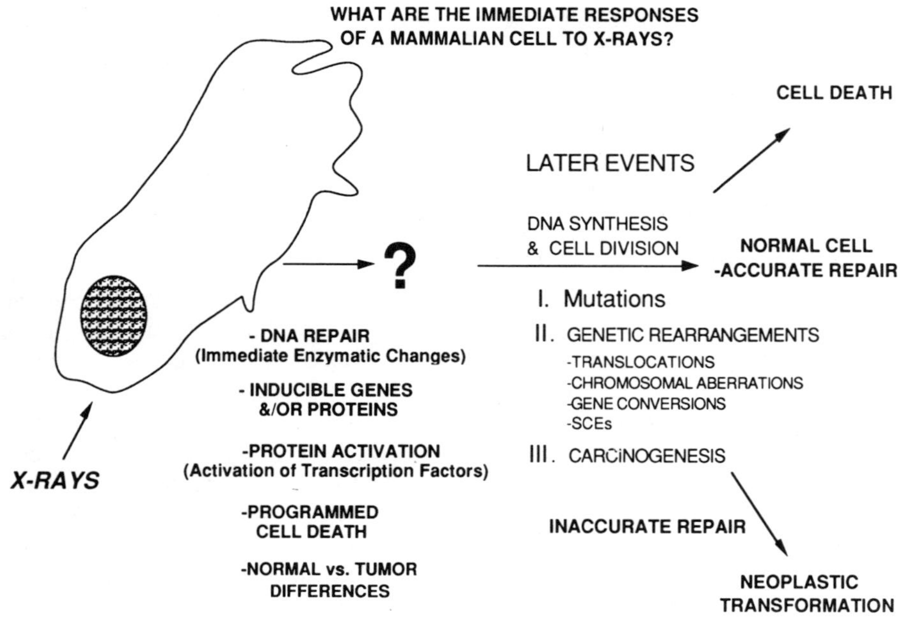

Fig. 1. Could immediate responses of mammalian cells have long-term consequences?

Ron et al., 1988). Glioblastomas indistinguishable from spontaneously formed tumors have also been induced by ionizing radiation in monkeys (Haymaker et al., 1979).

DNA DAMAGE-INDUCIBLE PROTEINS AND GENES IN HUMAN CELLS

Very little is known about the immediate genetic and enzymatic changes that occur in mammalian cells following the stress of DNA damage, including that which follows ionizing radiation exposure (Fig. 1). The existence of DNA damage-inducible processes in mammalian cells (including those that may be involved in DNA repair) could have a significant impact upon our approach in evaluating the mutagenic, carcinogenic, and therapeutic potential of radiation and chemicals in our environment. Naturally occurring and carcinogen-induced (including X-ray-induced) tumors display not only point mutations in cellular oncogenes but also more complex changes in other genes. Therefore, DNA damage in mammalian cells might induce alterations in gene expression with both short- and long-term consequences.

Although much of the evidence is indirect, several experiments using mammalian cells have indicated that physiologic responses affecting mutagenesis, carcinogenesis, and DNA/chromosomal repair may be induced by DNA damage. DNA damage can cause amplification and mutation-associated rescue of various viral particles (DasGupta and Summers, 1978; Sarasin and Hanawalt, 1978), much like that observed in the bacterial SOS responses. Ultraviolet irradiation of human fibroblasts enhanced the coordinate production of new mRNAs and the synthesis of several proteins, identified mostly as general stress-response genes (Kartasova et al., 1988; Schorpp et al., 1984). The damage-inducible response(s) of eukaryotic cells to ultraviolet (UV) irradiation appears to be similar to the SOS response in bacteria (Walker, 1985), since it was not specific for UV irradiation but was also induced by heat shock and other carcinogenic insults. Various signal transduction pathways may be induced with UV irradiation (Schorpp et al., 1984). The significance of induction of these and other genes and their role(s) in mutagenesis or carcinogenesis remains to be determined.

Could the induction of genes following DNA damage be directly or indirectly involved in DNA repair and, as a result, affect cell survival, mutagenesis, and/or carcinogenesis? If so, are there differences in gene expression and regulation in normal vs. tumor cells that could be exploited for enhanced tumor therapy, anticarcinogenesis, and/or normal cell tolerance?

Evidence for X-ray-inducible DNA repair genes in mammalian cells has been limited. In the past five years several laboratories have reported the induction of various gene products in mammalian cells by a wide range of DNA-damaging agents (Ronai et al., 1990). The regulation of new transcripts and gene products following various DNA insults include: (1) DNA binding transcription factors, such as the UV-responsive element (URE) binding factors and

$XIP_{bp}175$ (Boothman et al., 1990b); (2) proto-oncogenes, such as c-fos, c-jun, c-myc, and c-fes/fps; (3) several growth-related genes, such as protein kinase C, IL-1, p53, and thymidine kinase; and (4) a variety of other genes, including proteases, RNA polymerase-ß, tumor necrosis factor-alpha, metallothioneins, DT diaphorase, and several as yet unidentified gene transcripts (Table 3). Only a certain subset of genes are induced by any one particular DNA-damaging agent, although there are common genes that are induced to different levels by various agents. To complicate matters even more, mammalian cells may respond differently in both level and type of individually induced genes following a particular DNA-damaging stress. These induced gene products are synthesized immediately following DNA damage. Their influence upon later events occurring within mammalian cells has yet to be determined (Fig. 1).

X-Ray-Induced Genes/Proteins and DNA Repair

DNA repair in mammalian cells following X-irradiation exhibits biphasic kinetics in terms of survival recovery over time. This type of repair has been termed potentially lethal damage repair (PLDR), since delayed plating of damaged cells results in increased survival in very short periods of time (i.e., 2 to 8 hr). New gene transcripts may be required for, or regulate, at least a portion of these DNA repair processes, especially those required to reconstitute gross chromosomal damage, which occurs during the second slow phase of repair (Boothman et al., 1989). The fast component of repair of ionizing radiation damage may be due to constitutively synthesized enzymes (e.g., DNA ligases, topoisomerases, or polymerases), which act immediately to repair X-ray-damaged DNA (Boothman et al., 1989). In contrast, the second slower phase of repair observed in human cells may require the induction of specific genes and gene products that regulate or are directly involved in the repair of potentially lethal or carcinogenic DNA lesions, including more complex chromosomal damage (Boothman et al., 1989).

Recognizing certain types of DNA damage, and the stimulation of DNA repair processes that would then act to mend them, may require the activation (possibly not involving *de novo* protein/mRNA synthesis) of existing DNA binding proteins in the cell (Boothman et al., 1990b). UV-induced DNA binding proteins have already been described (Singh and Lavin, 1990). Activation of specific proteins may stimulate DNA binding directly, or cause the stimulation of cooperative binding of numerous proteins to DNA at specific promoter regions. Such binding would then block or induce the expression of other genes, including those responsible for the slow phase of repair. DNA binding proteins have recently been visualized by DNA gel mobility shift analyses using the human TK promoter region (Boothman et al., unpublished data), which contains many common DNA binding sites. A similar activation of DNA binding proteins following X-irradiation has been observed using the SV40 promoter region (Singh and Lavin, 1990). Therefore, both the initial fast phase and slower second phase(s) of repair may be stimulated by X-ray activation of proteins that bind to DNA.

Synthesis of X-Ray-Induced Proteins in Human Melanoma Cells

To test our hypothesis, we asked if new proteins were synthesized by an extremely radioresistant human malignant melanoma (U1-Mel) cell line immediately following X-irradiation under conditions that would promote repair (Boothman et al., 1989). U1-Mel cells were chosen since they have one of the highest capacities for survival recovery (repair) known. We identified and partially characterized the pleiotropic expression of several X-Ray-Induced proteins (XIPs) in U1-Mel cells and in a variety of other human cells using two-dimensional (2-D) gel electrophoresis (Boothman et al., 1989). Eight major polypeptides of $M_r = 126, 138, 141, 145, 147, 269,$ and 275 kDa (XIP126 through XIP275) were coordinately induced five- to 10-fold following X-irradiation. The expression of these particular XIPs was delayed, since they were not evident until 4 to 5 hours following X-irradiation.

Three classes of proteins were affected by X-irradiation. The expression of class I proteins XIP269 and XIP145 increased linearly with dose. The expression of six class II proteins, XIP-126, -135, -138,-141, -147, and -275 increased linearly with low X-ray levels, but reached plateau levels at higher doses, ranging from 50 cGy for XIP147 to 300 cGy for XIP275. In contrast, the expression of two class III proteins, ß-actin and a protein of $M_r = 254$ kd (p254), decreased with increasing doses of X-rays. All of the proteins induced or repressed following X-irradiation were present at very low levels and, except for XIP269, were not visualized by silver staining. The specific syntheses of these XIPs were also induced in unirradiated U1-Mel cells transfected (or infected) with X-irradiated plasmid or X-irradiated HSV-1 DNA (exposed to 10–100 Gy) (Boothman et al., unpublished data).

Unirradiated log-phase U1-Mel cells synthesized polypeptides with identical apparent molecular weights and isoelectric points (pIs) as the XIPs described above. X-irradiation of these log-phase cells resulted in an additional increase in their syntheses. Thus, XIPs were normally synthesized in proliferating U1-Mel cells, switched off upon growth arrest at confluency, and were once more synthesized in response to X-irradiation (Boothman et al., 1989). These proteins thus appear to be under very tight cell cycle regulatory control, which is altered in some manner following X-ray-induced DNA damage. Identification and characterization of these XIPs may also reveal critical cell cycle regulatory roles for each gene.

XIPs were not induced by other environmental stresses. They were not synthesized following exposure of confluence-arrested U1-Mel cells to melphalan (a bifunctional alkylating agent), heat shock, or hypoxic treatments (Boothman et al., 1989). XIPs observed in U1-Mel cells were different from those proteins previously induced by UV irradiation in other mammalian cells (Kartasova et al., 1988), and from those proteins produced by U1-Mel cells following UV damage. Cycloheximide treatment inhibited the synthesis of XIPs and partially prevented repair in terms of survival recovery (Boothman et al., 1989; Hughes and Boothman, 1991).

Evidence for a Role of XIP269 in DNA Repair

Of the XIPs synthesized in U1-Mel cells, only XIP269 was observed in nearly all of the human cells tested. Induction levels of XIP269 correlated well with the repair capacities of various cell types; the higher the induction of XIP269, the greater the capacity of the cell for repair (Fig. 2). XIP269 was not observed in any of the unirradiated, quiescent human cells tested.

Cells derived from patients with cancer-prone diseases, and with suspected DNA repair deficiencies, synthesized XIPs at extremely low levels and induced only very low molecular weight proteins (Boothman et al., 1989). Cells derived from patients with Bloom's syndrome and ataxia telangiectasia (AT) showed no induction of XIP269 and no repair capacity following X-ray treatment. X-irradiation of cells from Fanconi's anemia and xeroderma pigmentosum variant resulted in the appearance of low levels of XIP269, and showed only a slight capacity to repair (Boothman et al., 1989). Cancer-prone cells also demonstrated the new synthesis of a variety of low molecular weight proteins, which may signify the abnormal synthesis of proteases within these cells. For example, since no clear DNA break repair defect has yet been found in AT cells, the induction of intracellular proteases may be responsible for destroying inducible or constitutively synthesized regulatory proteins that control DNA repair and chromosomal repair processes. Interestingly enough, higher levels of tissue-type plasminogen activator (t-PA) have recently been noted following ionizing radiation exposure in nearly all of the human cancer-prone cells tested to date (Fukunaga et al., in press). Higher, or unregulated, intracellular tissue-type plasminogen activator (t-PA) levels may account for the low molecular weight proteins detected within these cells after ionizing radiation (Boothman et al., 1989).

Exposure of U1-Mel cells to caffeine, a known inhibitor of repair (survival recovery) following X-irradiation, not only reduced the level of XIP269 but concomitantly inhibited survival recovery to a similar extent. The expression of the other XIPs remained unchanged following caffeine

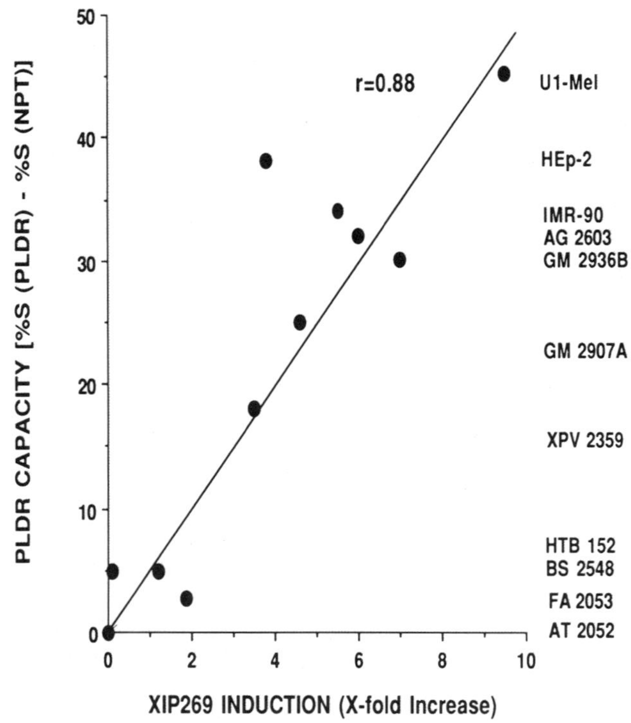

Fig. 2. Correlation of XIP269 induction with increased PLDR capacity in normal, cancer-prone, and neoplastic confluence-arrested human cells. (Reprinted from Boothman et al., 1989, by permission from Cancer Research.)

treatment (Hughes and Boothman, 1991). These data support the hypothesis that XIP269 is a key regulatory protein required for repair and may play a role in cell cycle regulatory events, such as in maintaining an arrested condition to allow repair (e.g., maintenance of a G_0 condition or pausing in G_1 following DNA damage in cycling cells) (Hughes and Boothman, 1991).

ISOLATION AND cDNA CLONING OF X-RAY-INDUCED GENE TRANSCRIPTS IN HUMAN CELLS

Differential hybridization techniques were used to isolate and clone 12 X-ray-induced gene products (i.e., xips 1 to 12; Table 2) from U1-Mel cells under identical conditions as those used to demonstrate xip synthesis as described above (Boothman et al., 1989). It was our hope that isolated cDNA clones would correspond to a few or all of the XIPs above.

TABLE 2. X-ray-induced Genes (xips) Isolated From U1-Mel Cells

cDNA clone name	Identity (DNAsequence homology)
xip1	High homology cytochrome P450, ETOH-inducible
xip2	Unknown
xip3	DT Diaphorase
xip4	Unknown
xip5	Some homology to Human Growth Hormone
xip6	Tissue-type plasminogen activator
xip7	Homology to RAG-1
xip8	Unknown
xip9	Some homology to c-fes/fps protooncogene
xip10	Unknown
xip11	Thymidine kinase
xip12	High homology to human Angiogenin mRNA

Partial DNA sequence data was obtained for all 12 cDNA clones isolated using the T3 and T7 bacterial promoters on the pBluescript plasmid. Roughly 300 base pairs from both ends of the cDNA inserts were sequenced for each of the 12 cDNA clones. DNA sequences for each clone were then separately compared to those of known genes in GenBank (a DNA sequence data bank).

Of the 12 cDNA clones sequenced, only three were known genes. Thymidine kinase (TK), t-PA, and DT diaphorase have been confirmed at both the mRNA (using authentic mRNA probes) and enzymatic levels; expression increased 5-100, and 20-fold, respectively (Boothman et al., 1991, unpublished data). The other cDNA clones showed significant homologies to certain regions of the: (1) ETOH-inducible cytochrome P450 gene (67%), (2) c-fes/fps proto-oncogene (67%), (3) Human Growth Hormone (47%), (4) human angiogenin mRNA (63%), and (5) RAG-1 gene (62%). Four of the 12 cDNA clones (xip2, xip4, xip8, and xip10) contained less than 20% DNA sequence homology to any known gene in GenBank (Table 2, Boothman et al., unpublished data).

Based upon DNA sequence homologies, it is possible that these cDNA copies of X-ray-induced transcripts are functionally related to their respective counterpart genes, but they may or may not be identical to them. Assuming that DNA sequence homology indicates similar function, these clones can be classified into five separate but related groups. The first of which, thymidine kinase, the sole known member, could constitute DNA metabolism genes used to replenish needed nucleotides for DNA repair. The induction of redox cycling genes, DT diaphorase, and a gene with similar function to the ethanol-inducible cytochrome P450 could allow the cell to detoxify potential DNA-damaging free radicals and thus provide resistance to radiation-induced DNA damage. t-PA and a gene homologous to human angiogenin mRNA are genes implicated in metastasis and could allow tumor or normal tissues to remodel following damaging insults. Genes homologous to the c-fes/fos proto-oncogene, and RAG-1 (recombinant activating-1 gene) may be involved in recombination, oncogenesis, and DNA binding and gene expression.

The final set of X-ray-induced gene transcripts (xip2, xip4, xip8, and xip10) that we isolated in U1-Mel cells constitute a group of currently unknown genes whose DNA sequences were not homologous to any gene(s) registered in GenBank.

As a group, therefore, the repertoire of X-ray-induced gene transcripts (xips) may act in the cell to: (1) prevent mutagenesis via induced DNA repair processes, (2) suppress tumor formation via induced "tumor suppressor" genes, (3) build up an adaptive survival advantage through delayed accumulation of inducible DNA repair enzymes or genes involved in other mechanisms of radioresistance, (4) shut down oncogenes that might otherwise lead to carcinogenesis, and (5) induce oncogenes, shut off (or mutate) tumor suppressor genes, or induce tissue remodeling genes (t-PA and human angiogenin), which may increase tumor spreading and thereby reduce the clinical utility of ionizing radiation. As witnessed by the variety of select gene products induced by ionizing radiation damage, mammalian cells react by synthesizing a variety of new gene transcripts.

Unfortunately, very little is understood about the function of these genes. Following damage, mammalian cells may synthesize these gene products to produce a unified response to ionizing radiation. Alternatively, there may be independent competing processes that are controlled by as yet unidentified signaling events, such as the posttranslational activation of existing transcription factors. Other factors, such as the growth state of the cell, the initial extent and type of DNA damage created by ionizing radiation, cell type, microenvironment, and capacity for repair, may determine which gene transcript(s) govern the radiobiological response observed. For example, if initial DNA damage to confluence-arrested U1-Mel cells is minimal, survival recovery would be the observed phenomenon in the majority of cells. Yet, if the damage is increased or if the cell is in S-phase, cell death in terms of apoptotic events or divisional delays may be the observed radiobiological response. In both instances the same repertoire of genes could be produced, but at different levels. For example, a few of the inducible genes above may interact with different growth regulatory genes, which are synthesized at different points and at different times in cycling as compared to resting cells.

Induction of Tissue-Type Plasminogen Activator (t-PA)

Once t-PA was observed to be induced by ionizing radiation in human U1-Mel cells (xip6, Table 2), we further investigated its mode of regulation. Two differently timed

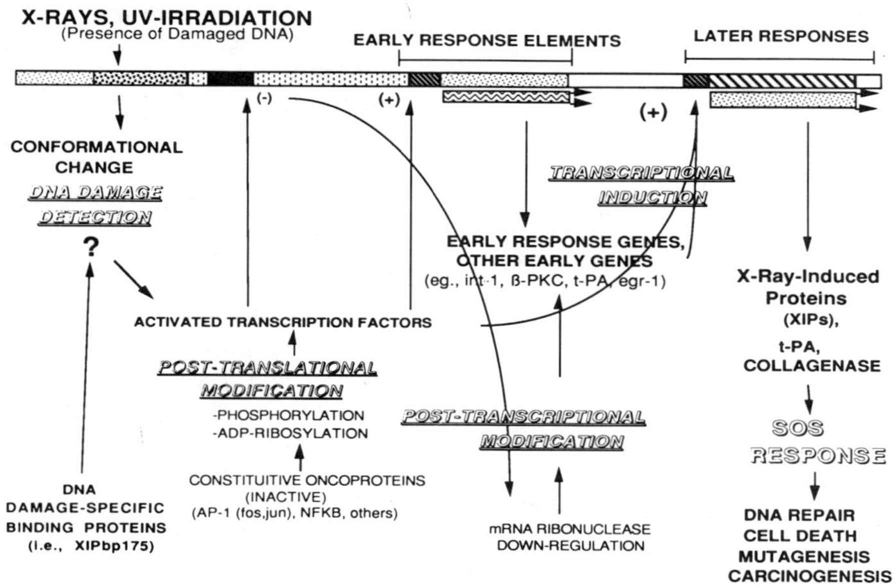

Fig. 3. Temporal model of proposed pathways of gene regulation initiated following ionizing radiation.

extracellular and intracellular enzymatic and mRNA peaks of t-PA were induced following ionizing radiation; one early minor t-PA peak and another later, major enzymatic peak (Boothman et al., 1991). The first t-PA enzymatic peak was shown to be the result of stabilization of t-PA mRNA transcripts, since the appearance of new mRNA transcripts within the first 30 min was: (1) not observed by nuclear run-on analyses (which measures the level of new synthesis of a particular gene); (2) not prevented by actinomycin D, an inhibitor of RNA transcription; and (3) not blocked by cycloheximide, a protein synthesis inhibitor.

Nearly 12 hours following X-irradiation, a second major enzymatic peak of t-PA was observed. t-PA mRNA levels were induced greater than 100-fold in 4 hours following ionizing radiation, as assayed by Northern or nuclear run-on analyses. Unirradiated U1-Mel cells showed only basal levels of t-PA mRNA and enzymatic activity. Similar induction responses were found following UV irradiation or 12-*O*-tetradecanoyl-phorbol-13-acetate (PMA) treatments.

Normal human fibroblast (GM 2936B, GM 2907A, and IMR-90) cells also showed induction of t-PA (Boothman et al., 1991). The induction of t-PA mRNA levels and intracellular and extracellular enzymatic activities for these cells were 50-fold lower than with U1-Mel cells given equitoxic doses of X-rays. In contrast, cells from cancer-prone patients with Bloom's syndrome, Fanconi's anemia, and ataxia telangiectasia (these cells have suspected DNA repair deficiencies) induced five- to 20-fold higher levels of t-PA when compared to normal cells following ionizing radiation (Fukunaga et al., 1992). As with U1-Mel cells, induced t-PA enzymatic activities following ionizing radiation were blocked by cycloheximide or actinomycin D treatments. Kinetic induction of mRNA levels for normal and cancer-prone cells were consistent with the observed enzymatic activities. We are currently examining possible relationships between a known genetic cancer-prone predispositions and possible DNA repair deficiencies of these cells, and the overabundant synthesis of t-PA compared to normal cells.

The induction of intracellular and extracellular t-PA enzymatic activities may play a number of roles within human cells after a cellular insult, including cell death, survival, carcinogenesis, and metastasis. The induction of intracellular t-PA may activate, or be induced by, an enzyme involved in a rec A-like, error-prone SOS response of damaged mammalian cells (Fukunaga et al., 1992). This response may act in a similar fashion to SOS processes occurring in bacteria (Walker, 1985). Such a mutagenic response may explain the high rates of transformation observed in rodent cells during early periods of repair following ionizing radiation treatment (Boothman et al., 1990a).

The role of t-PA in neoplastic processes has been the subject of intense research over the past several years. Increased secretion of proteases may contribute to the malignant behavior of cells by decreasing their adhesion to the substratum and allowing invasion into surrounding tissue and, therefore, promoting metastatic potential. In addition, protease inhibitors are known to prevent both inflammation and carcinogenesis. Thus, both inflammation and carcinogenesis may be mediated, at least in part, by the intracellular and extracellular production of t-PA from X-irradiated human cells.

REGULATORY CONTROLS OF GENE EXPRESSION INDUCED BY IONIZING RADIATION

Although it is far too early to describe all the major pathways induced in mammalian cells following ionizing radiation treatment, a few major genetic regulatory mechanisms have been described (Boothman and Lee, 1991). These include: (1) posttranslational modification, (2) mRNA stabilization, (3) inhibition of rRNA processing and/or decreased stability of mature rRNA species, (4) the induction of new genetic transcripts, and (5) X-ray-induced transcriptional down-regulation. These mechanisms can be combined into one overall picture of how mammalian cells might control their genetic responses to ionizing radiation (Fig. 3). Since the data is incomplete, the model contains several ambiguous elements. Such a model may, however, be used to construct experiments to confirm and extend what is already known. Once clarified, we may better understand the mechanisms by which these regulatory controls become faulty and thus lead to the formation of cancer, the induction of programmed cell death, and the emergence of tumor cells with increased metastatic potential and radioresistance.

Five major regulatory mechanisms by which mammalian cells control the expressions of gene products following ionizing radiation have been discovered using DNA band shift assays, cDNA cloning, Northern RNA blotting, nuclear run-on, and enzymatic assay techniques. The exact functional role that each of these controlling mechanisms plays within the cell are not entirely clear. Below is a discussion of each regulatory mechanism known to date and a schematic (Fig. 3) that attempts to fit the regulatory mechanisms into one cohesive model.

Posttranslational Modification

Evidence for posttranslational modification as a regulatory mechanism in mammalian cells has come through the use of DNA gel shift (or retardation) assays and specific promoters, such as the SV40 (Singh and Lavin, 1990) or human thymidine kinase (Boothman et al., unpublished data) DNA regions. The appearance of a transcription factor that bound to the SV40 promoter was present in the cytoplasm of unirradiated human cells, but appeared in the nucleus only after X-irradiation. The activation of this protein did not require *de novo* protein synthesis, since DNA binding activity was noted in cycloheximide-treated cells (Singh and Lavin, 1990). Southwestern analyses showed that both the nuclear and cytoplasmic proteins had equivalent molecular weights of 43 kDa.

We have identified a similar factor in X-irradiated human cells that binds to the human thymidine kinase promoter. Like the DNA binding activity in SV40, this DNA–protein interaction did not require the synthesis of new protein since cycloheximide-treated, X-irradiated human cells also revealed the activation of this transcription factor (Boothman et al., unpublished data). Unlike the SV40 DNA–protein complex, binding to the human thymidine kinase promoter involves more than one protein, and is not apparent in the cytoplasm of unirradiated human cells. Although we have not narrowed the DNA binding activity to a specific DNA sequence, the binding activity has been reduced to a 67 bp DNA region. The multiprotein nature of DNA binding to the human thymidine kinase promoter appears to be a very complex DNA–protein interaction that is difficult to elucidate. Binding to this promoter is believed to be responsible for the induction of new thymidine kinase mRNA transcripts that were discovered by differential hybridization and cDNA cloning (Table 2).

We have also noted the activation and induction of a protein (XIPbp175) derived from radioresistant human melanoma cells that binds specifically to ionizing radiation-damaged genomic DNA, but not to undamaged genomic DNA (Boothman et al., 1990b). The activation/induction of this protein (and others) within the cell may be an early DNA damage recognition mechanism by which the cell initiates the repair of its damaged DNA. Similar damage-specific DNA binding activities have been described for apurinic sites and UV-damaged DNA (Ronai et al., 1990).

Posttranscriptional Modification (mRNA Stabilization)

The first example of mRNA stabilization following ionizing radiation exposure was with the early induction of human t-PA, as previously discussed (Boothman et al., 1991). It remains to be determined if mRNA stabilization is unique to U1-Mel cells or occurs in other human tumor cells. Stabilization of the c-jun proto-oncogene mRNA has recently been observed in other mammalian cells following UV-irradiation or X-irradiation.

RNA Processing Inhibition/Degradation

Several laboratories have described major changes in the distribution of rRNA species following ionizing radiation. After ionizing radiation, the relative amount of 18S rRNA dramatically decreases, presumably due to increased degradation. In addition, processing of the 45S precursor form of rRNA was inhibited following X-irradiation. The inhibition of rRNA processing and increased degradation of 18S rRNA is accompanied by a relative excess of 28S rRNA. This excess may trigger some sort of induced cell-killing phenomenon (Fuchs et al., 1990).

In our laboratory we have also noticed similar, albeit not as dramatic, changes in rRNA processing as those noted above. Four hours following X-irradiation, an accumulation of the 45S and 32S species of rRNA was noted (Boothman and Lee, 1991). In terms of percentage of total RNA loaded, the 45S rRNA species increased 1.8-fold as compared to RNA isolated from unirradiated U1-Mel cells. Similarly, the relative amount of 32S rRNA found in X-irradiated U1-Mel cells decreased 2.4-fold over the total percentage found in unirradiated U1-Mel cells. Although these effects on rRNA processing are not as dramatic as those previously described (Fuchs et al., 1990), a lower dosage of ionizing radiation was delivered (450 cGy) to radioresistant U1-Mel cells as compared to the doses delivered to HeLa cells (750 cGy); at these doses, U1-Mel cells are 50% viable as compared to less than 5% for HeLa cells under similar irradiation conditions (Fuchs et al., 1990).

What significant role does processing inhibition of rRNA play within the cell? It is possible that a similar mechanism controls the expression of other gene products and thereby down-regulates their transcription. Similar inhibition of rRNA processing has also been described following other cellular stresses, such as with heat shock and antimetabolite treatments. Studying these processes should help explain fundamental rRNA processing mechanisms that occur within the cell under normal conditions, as well as their role within the cell following genotoxic stresses.

INDUCTION OF NEW GENE TRANSCRIPTS

As previously discussed, a number of new gene transcripts and associated gene products (i.e., proteins) have been found to be induced following various types of ionizing radiation. As with UV-irradiation, exposure to ionizing radiation seems to stimulate certain signal transduction responses in mammalian cells, most notably the induction of ß-protein kinase C (ß-PKC). For example, the agents that have been shown to elicit the enzymatic and mRNA induction of t-PA (PMA, UV irradiation, and ionizing radiation) indicate an induction pathway possibly involving ß-PKC (Burger et al., 1979). In the next few years it will be interesting to discover whether or not certain X-ray-Responsive Elements (XREs) exist in the promoter regions of these inducible genes that ultimately regulate their transcription (Table 3). Another mode of induction of gene transcription following ionizing radiation may be through the production of the AP-1 protein complex. The AP-1 protein complex is a potent DNA binding complex that is the dimeric product of both c-fos and c-jun proto-oncogenes, both inducible by ionizing radiation. Simultaneous induction of both of these genes has not been noted, however, and some doubts as to the relevance of this pathway still exist. The appearance of AP-1 could influence (both positively and negatively) the transcriptional expression of a variety of genes. Further investigation is required to substantiate this AP-1-mediated mode of gene regulation and to determine if cell-type specificity exists.

In general, however, it is far too early to describe all the major pathways induced in mammalian cells following ionizing radiation since the complete repertoire of X-ray-inducible genes has not been described. Further research will be required not only to identify the cell type and tissue-specific induction of X-ray-induced genes, but also to determine the functional role(s) of their induced gene products

TABLE 3. Genes/Proteins Influenced by Ionizing Radiation[a]

Affected gene/ protein	Type of ionizing radiation used	Cell system
A. Induced genes		
p42, p60	Alpha particles	Chinese hamster ovary
c-fos	250 KVp	Human tumor/ fibroblasts
c-jun	250 KVp	Human tumor/ fibroblasts
Protein kinase C	50 KVp, ^{60}Co	Syrian hamster fibroblasts
XIP126 to XIP275 (8 total)[b]	250 KVp	Human tumor/ fibroblasts
Interleukin-1	50 KVp	Human tumor lines
Platelet-derived growth factor	50 KVp, ^{60}Co, 0.85 MeV neutrons	Human/rodent fibroblasts
MHC class II invariant chain	^{60}Co	Murine myeloma
Collagenase	^{60}Co	Human fibroblasts
Tumor necrosis factor alpha	250 KVp	Human tumor lines
Tissue-Plasminogen activator[b]	250 KVp	Human tumor/ fibroblasts
DT diaphorase[b]	250 KVp	Human tumor/ fibroblasts
Thymidine kinase[b]	250 KVp	Human tumor/ fibroblasts
egr-1	250 KVp	Human tumor cells
B. Repressed genes		
β-actin	50 KVp, 250 KVp, ^{60}Co	
	0.85 MeV neutrons	Human/Rodent
	250 KVp	Tumors/fibroblasts
Ornithine decarboxylase	50 KVp, ^{60}Co, 0.85 MeV neutrons	Syrian hamster fibroblasts
PCNA	^{137}Cs	Rat fibroblasts

[a]Updated version of a table first published by Boothman and Lee, 1991.
[b]Genes/proteins identified by Boothman et al., 1989, 1991 unpublished data.
MHC, Major histocompatibility complex; PCNA Proliferating cell nuclear antigen; XIP, X-ray-Induced Protein.

following the genotoxic insult created by exposure to ionizing radiation.

DOWN-REGULATION OF EXISTING GENE TRANSCRIPTS

Even less is known about the number and extent of gene transcripts that are down-regulated due to X-ray exposure. A number of gene products, including ß-actin, PCNA, ornithine decarboxylase, and a protein of 254 kDa molecular weight, were dramatically down-regulated following X-irradiation (Table 3). Unfortunately, no significant functional relevance has been attributed to their down-regulation. It is possible, for example, that a ribonuclease that degrades t-PA mRNA transcripts is transiently down-regulated following X-irradiation, thus allowing increased levels (or stability) of t-PA mRNA to be translated into active enzymatic protein (Boothman et al., 1991). Functional significance of genetic transcriptional repression due to the stress of ionizing radiation-induced damage remains a mystery.

Model of Altered Gene Expression Caused by Ionizing Radiation

Using the fragmented information concerning regulatory controls, a picture of the complex manner by which mammalian cells control gene expression in response to ionizing radiation is emerging. Cellular stress induced by ionizing radiation somehow triggers complex responses in mammalian cells. Damage to DNA appears to be the triggering event in mammalian cells, but the exact mechanism(s) by which gene expression is altered by ionizing radiation exposure is very poorly understood. Transfection/infection of X-ray-damaged plasmid DNA or HSV-1 viral DNA, respectively, into mammalian cells does lead to the induction of XIPs previously described (Boothman et al., 1989, and unpublished data). Once DNA damage has been created by ionizing radiation, conformational changes and/or DNA–protein interactions that specifically detect certain types of DNA lesions (such as the DNA binding activity of $XIP_{bp}175$) probably cause the activation of preexisting transcription factors, which then bind to specific DNA promoter regions. The number and extent of transcription factors induced probably depend upon a number of internal and external factors, such as the dose of ionizing radiation used, the cell type, microenvironment, (e.g., pH changes) and the growth state of the cell. Posttranslational activation of transcription factors may occur by phosphorylation and ADP-ribosylation reactions (Fig. 3).

The activation of transcription factors may then result in the simultaneous induction of early response genes and repression of other normally expressed genes. The down-regulation of ribonuclease genes, which act intracellularly to destroy mRNAs through binding of activated transcription factors to certain promoter regions, may explain the post-transcriptional mRNA stabilization of gene transcripts, such as with t-PA mRNA. In addition, DNA–protein interactions could result in the stimulation of gene expression and the appearance of additional "early response" genes (such as c-jun and egr-1) (Weichselbaum et al., 1991), whose mRNA transcripts accumulate almost immediately following DNA damage. These gene products may play a role in DNA damage detection or by stimulating DNA repair. Thus, through mRNA stabilization and early gene induction, a number of early response gene products appear within the damaged cell.

A second set of gene transcripts then appears at later times (1–14 hr) following ionizing radiation exposure. The delayed induction of a number of proteins (XIPs) and a number of genes (Table 2) have been demonstrated (Boothman et al., 1989, 1990b, 1991; Hughes and Boothman, 1991). It is unlikely that these later response proteins (with the exception of XIP269) or genes play any significant role in mending single- or double-stranded DNA lesions created by X-ray damage. They may, however, regulate or be involved in: (1) suppressing tumor formation; (2) restructuring and repairing gross chromosomal DNA damage; (3) the error-correcting phase of repair, which proofreads DNA lesions and mutations missed in fast repair, but possibly corrected during the slowly repaired, nontranscribed regions of the genome (Bohr et al., 1987); (4) programmed cell death (apoptosis); (5) adaptive survival responses, which may lead to increased radioresistance; and (6) enhancing the metastatic potential of tumor cells (the elevated levels of t-PA and human angiogenin may lead to increases in tumor spreading and the ability of metastasized tumor cells to establish a blood supply).

Correlation of Gene Expression Changes with CNS Pathology

Our understanding of the changes in gene expression occurring in mammalian normal and tumor cells following ionizing radiation has increased over the past five years due to the application of molecular biology techniques. Due to difficulties in maintaining normal neural cells in culture, most studies have used established human and rodent tumor cell lines. Improved techniques in isolating neural cells may make it possible to evaluate their altered gene expressions following exposure to ionizing radiation. These advances will also allow us to compare changes in gene expression in normal as compared to neoplastic neural cells. Assays using state-of-the-art protein chemistry technology and molecular biology techniques (differential/subtractive hybridizations, cDNA cloning, antisense constructs, the polymerase chain reaction [PCR]) do not require established cell lines and may be easily applied to study normal and tumor tissues from experimental animals or human patients. For example, Narayan et al. (1984) have analyzed 2-D gels of human brain, peritumoral irradiated brain, and normal brain tissues in an attempt to define changes in protein patterns following radiation exposure.

How might the X-ray-induced genes that have been described above (Tables 2, 3) be related to the response(s) of various types of neural cells? The speculative discussion below relies upon the assumption that many, if not all, of the genes induced in confluence-arrested human melanoma cells (a neural crest derivative), or in other mammalian cells, may be induced in various neural cell populations.

Based upon numerous investigations, we have broken down the responses of normal neural cells within the CNS after ionizing radiation therapy into four major groups (Table 4): (1) neuron radioresistance (repair); (2) demyelination; (3) endothelial hyperplasia; and (4) induced tumor formation. Each response group is discussed below in terms of X-ray-induced gene products that may play a role in the mechanism of the cellular reaction observed. Since ionizing radiation causes a broad spectrum of damage that cells must react to, the discussion below does not presume that induced gene products are the sole determinants of the response observed. Gene products induced by ionizing radiation may control, or be integral members of, a concerted response occurring in some (but not all) neural cells following ionizing radiation.

Neuron Radioresistance: Repair

Neurons are remarkably resistant to the lethal effects of ionizing radiation due to: (1) their resting state at the time of radiation; (2) their failure to proliferate following injury (possibly due to their normal induction of gene products that prevent hyperplasia); and (3) the possible induction of free radical scavenging gene products and DNA repair gene products. The degree of resistance of neural cells is most evident from studies in sharks following X-irradiation (Brightman et al., 1970). Due to a unique vascular architecture, the shark nervous system is resistant to the development of edema. Radiation doses as high as 300 Gy have been administered to shark brain without causing acute necrosis. These data reinforce the hypothesis that the vulnerability of neurons is a function of damage to critical supporting cells, and not to neural cells directly.

Although clinical necrosis of neurons is due to damage to supporting cells, neurons are still vulnerable to free radical attack on macromolecules. Studies on the effects of ionizing radiation on neuronal DNA and chromatin structure using cells in culture have begun to define temporal changes that may determine survival. Since neurons are not proliferating, they provide a unique opportunity to study changes in chromatin structure following X-irradiation (changes due to the cell cycle are not present). Confluence-arrested U1-Mel cells were chosen for the gene expression studies above for similar reasons. Jaberaboansari et al. (1988, 1989) studied alterations in neuronal chromatin of rats following radiation. Immediately following radiation, bulk chromatin was more accessible to endonucleolytic digestion, suggesting an alteration of superhelical structure. There was also a marked decrease in the accessibility of the nuclear-matrix-associated DNA to digestion by DNase I. Two mechanisms were proposed to account for this change. DNA may be cross-linked to nuclear matrix proteins as a consequence of radiation, or proteins may bind the nuclear matrix and mask DNA. These apparent DNA binding proteins may be involved in DNA repair, much like the earlier proposals for XIP269 and/or $XIP_{bp}175$ (Boothman et al., 1989, 1990b; Hughes and Boothman, 1991). The restoration of normal chromatin originally damaged by ionizing radiation in neural cells follows a biphasic pattern, exhibiting a rapid early phase lasting minutes and a later slow phase lasting several hours. The observed biphasic restoration of X-ray-damaged neuronal chromatin mimics PLDR responses. As previously discussed, this biphasic pattern of repair may be dependent in

TABLE 10.4. Immediate Cellular Responses of Normal Neural Cells to Ionizing Radiation and Possible Function(s) of X-ray Induced Gene Products

Normal neural cellular response	Gene products involved	Examples
Neuron radioresistence (repair)	DNA repair-related genes; free radical scavenging gene products; genes involved in growth-arrested	XIP269; $XIP175_{bp}$; DT diaphorase; cytochrome P450 (ETOH-induced); down regulation of PCNA
Demyelination	Induction of intracellular proteases; stimulation of of negative transcription	Tissue-type plasminogen activator; collagenase; tumor necrosis factor-alpha; down regulation of ornithine decarboxylase
Endothelia hyperplasia	Growth factor induction; induction of regulatory genes (e.g., transcription factors); induced proto-oncogenes	c-fes/fps; egr-1; thymidine kinase; c-fos; c-jun; protein kinase C; platelet-derived growth factor (PDGF); XIP126-XIP275 proteins; human growth hormone (GH-1, GH-2); p53, cyclin A

*Examples of X-ray-induced gene products which may be involved in a given radiobiological response.

part upon X-ray-induced gene products appearing at various times following damage.

A few of the X-ray-induced gene products listed in Tables 3 and 4 may play a role in the radioresistance of neurons. Radioresistance may be proportional to repair processes. The induction of DNA repair or DNA binding proteins may act to stabilize broken chromatin strands or naked DNA. Radioresistance could also be conferred to neurons through the induction of free radical scavenging gene products, such as DT diaphorase or cytochrome P450 (ETOH-induced). Finally, the cell must have enough time to reseal and restore DNA/chromosomal damage. Thus, damaged neural cells must be able to control hyperplasia but at the same time induce repair enzymes, which are probably also needed during normal DNA synthesis during S-phase. For example, a damaged cell might perform both functions at once following X-irradiation by rapidly down-regulating growth-related genes (e.g., PCNA) and at the same time inducing a series of proteins that are found in log phase, but not in confluence-arrested cells (Boothman et al., 1989; Haimovitz-Friedman et al., 1991). These induced genes may have dual roles in DNA repair and replication.

Demyelination: Oligodendroglial Response to Ionizing Radiation

Demyelination following X-irradiation is the major primary change that occurs during the initial delayed reaction to the CNS following ionizing radiation damage. Demyelination may result from injury to oligodendroglial cells, which results in impaired myelin synthesis, and oligodendroglial cell death. It is not known, however, what mechanism results in the down-regulation of myelin synthesis and ultimate death of oligodendroglial cells following ionizing radiation. Active proteolytic degradation by intracellularly "trapped" proteases, such as collagenase or tissue-type plasminogen activator (which are normally released outside the cell), may lead to myelin degradation or synthesis shutdown by the degradation of transcription factors. Myelin synthesis may also be controlled by the synthesis or down-regulation of regulatory gene products. The induction of tumor necrosis factor and the down-regulation of ornithine decarboxylase may affect the production of myelin.

Hyperplasia: An Endothelial Cell Response to Ionizing Radiation

Endothelial cells of the CNS have been implicated in major dose-limiting complications following X-irradiation treatment, including complications caused by radiation necrosis. Although it is unlikely that vascular damage caused by ionizing radiation is the sole cause of CNS necrosis, endothelial hyperplasia with luminal obstruction has been shown to precede necrosis in animal studies. The delayed appearance of necrosis of the CNS has been a perplexing problem for decades, but it is probably due to the slow proliferation of endothelial cells which delays the expression of lethal damage. Ironically, it is not the loss of endothelial cells but the hyperplasia of repopulating endothelial cells that appears to contribute most to the pathogenesis of CNS necrosis. Endothelial repopulation may be analogous to accelerated repopulation in normal tissues and in some tumors following radiotherapy.

The mechanism by which damaged endothelial cells of the CNS initiate entry into the cell cycle and ultimately repopulate (and possibly overpopulate) damaged areas remains unclear. Are specific growth factors induced by radiation? Does radiation disrupt the normal control of repair processes, resulting in extreme hyperplasia, obstruction, and necrosis? Yamaguchi and colleagues (1991) studied sequential changes in the CNS of dog brains following radiation and isolated endothelial cells to determine qualitative changes in transcriptional activity by determining the rate constant of apurinic acid production. Transcriptional activity decreased to a minimum at three months and reached maximum activity at nine months following ionizing radiation. This increase in transcription preceded necrosis.

Of the X-ray-induced gene products listed in Tables 3 and 4, several of them may be involved in initiating, controlling, or maintaining CNS endothelial hyperplasia (Table 4). The proto-oncogenes egr-1, c-fos, c-jun, and c-fes/fps are early-response genes involved in the regulation of the entry of cells into their divisional cycle. The stimulation and binding of platelet-derived growth factor (PDGF) (Haimovitz-Friedman et al., 1991) and the up-regulation of protein kinase C may also be involved in the induction of hyperplasia. Both gene products initiate a cascade of pathways through phosphorylation, which signals the cell to proceed through its cycle. Recently, a PDGF-like protein was found to be produced by endothelial cells. The vascular endothelial cell proliferation of gliomas contained large quantities of mRNA for the ß-chain of platelet-derived growth factor (c-sis) and its receptor (Di Corleto and Bowen-Pope, 1983; Hermansson et al., 1988). Although the functions of the XIPs observed in human melanoma cells (i.e., XIP 126-275; Boothman et al., 1989, 1990b) are not known, XIPs were present in log phase cells and are probably involved in processes common DNA repair, DNA synthesis and cell division. The most interesting question in the future will be whether it is possible to block hyperplasia through the down-regulation of these or other genes via antisense experiments and whether this prevents X-ray-induced necrosis.

Secondary Tumor Induction

The induction of secondary tumors following ionizing radiation treatments is a well-documented complication in the CNS (Haymaker et al., 1979; Ron et al., 1988). In instances where carcinogenesis does occur, several previously described X-ray-inducible gene products may play a role. Either immediate or delayed (through genetic translocations) alterations in the control of proto-oncogenes and ß-PKC have been shown to be related to the induction of neoplasms. Since gross genetic rearrangements are common events following ionizing radiation treatment of mammalian cells, it is possible that these oncogenes may be up-regulated due to alterations in regulatory promoter regions. Ionizing radiation may also induce gene transcripts that result in DNA repair processes. These may have functions analogous to SOS repair processes in bacteria and are extremely error-prone and mutagenic. Very little is known about gene transcripts that may be responsible or induced in response to SOS repair processes. By analogy to rec A, an intracellular protease such as tissue-plasminogen activator may initiate or be the result of SOS repair. This enzyme is dramatically induced in cancer-prone radiation-hypersensitive human cells and its intracellular substrates (if any) are not known.

CONCLUSION

Therapeutic radiation has held a proven role in the treatment of malignant CNS tumors for decades. Normal tissue tolerance has, however, been a limiting factor in the treatment of neoplasms of the CNS. The reaction of normal neural cells within the CNS after ionizing radiation treatment can be separated into four major responses: (1) neuron radioresistance (repair); (2) demyelination; (3) vascular damage; and (4) induced formation of secondary tumors. These responses collectively determine CNS toxicity.

Our understanding of gene expression changes occurring in normal and tumor cells following X-irradiation has dramatically increased due to the application of molecular biology techniques. Isolation of gene transcripts induced by X-irradiation have identified both known and as yet unidentified gene products that may play a role in CNS toxicity. Ionizing radiation may induce gene products that: (1) suppress the formation of secondary tumors (i.e., X-ray-induced tumor suppressor genes, such as p53 (Kastan et al., 1992); in (2) restructure and repair gross chromosomal damage; (3) correct errors created during initial repair periods following ionizing radiation; (4) induce programmed cell death (i.e., apoptosis); (5) act in adaptive survival responses, which may lead to increased radioresistance; and (6) enhance the metastatic potential of tumor cells. Future studies must address the possibility of altering gene expression to reduce CNS cytotoxicity and enhance antitumor efficacy.

ACKNOWLEDGMENTS

The authors are grateful to Drs. Allen S. Lichter, Jeffrey M. Trent, Nina Fukunaga, and Theodore S. Lawrence, to Heather Burrows, Mark Meyers, Christine Leichtman, Janet K. Owens, and Dr. Thomas E. Carey for their critical reading of this work. This research was supported by grant PDT-412 to David A. Boothmann, Ph.D., from the American Cancer Society.

REFERENCES

Biaglow, J.C., J.E. Biaglow, M.E. Varnes, et al. (1988) The role of thiols in cellular response to radiation and drugs. Radiat. Res. 95:437–455.

Bohr, V.A., D.H. Phillips, and P.C. Hanawalt (1987) Heterogeneous DNA damage and repair in the mammalian genome. Cancer Res. 47:6426–6436.

Boothman, D.A. and Lee, S.W. (1991) Regulation of gene expression in mammalian cells following ionizing radiation. Yokohama Med. Bull. 42:137-149.

Boothman, D.A., I. Bouvard, and E.N. Hughes (1989) Identification and characterization of X-ray-induced proteins in human cells. Cancer Res. 49:2871–2878.

Boothman, D.A., E.N. Hughes, and A.B. Pardee (1990a) The role of X-ray-induced DNA repair processes in mutagenesis and carcinogenesis. In: M. Mendelsohn (ed.). Mutation and the Environment. Prog. Clin. Biol. Res., vol. 340E, pp. 307–318.

Boothman, D.A., S.W. Lee, D.K. Trask, et al. (1990b) X-ray-inducible proteins and genes in human cells. In: S.S. Wallace and R.B. Painter (eds.). Ionizing Radiation Damage to DNA: Molecular Aspects. UCLA symposia on Molecular and Cellular Biology, New Series. New York: Wiley-Liss, vol. 86, pp. 309–317.

Boothman, D.A., M. Wang, and S.W. Lee (1991) Induction of tissue-type plasminogen activator by ionizing radiation in human malignant melanoma cells. Cancer Res. 51:5587–5595.

Brightman, M.W., I. Klatzo, Y. Olsson, et al. (1970) The blood-brain barrier to proteins under normal and pathological condition. J. Neurol. Sci. 10:215–239.

Burger, P.C., and O.B. Boyko (1991) The pathology of central nervous system radiation injury. In: R.H. Gutin, S.A. Leibel, and G.E. Sheline (eds.). Radiation Injury to the Nervous System. New York: Raven Press.

Burger, P.C., M.S. Mahaley, L. Dudka, et al. (1979) The morphologic effects of radiation administered therapeutically for intracranial gliomas. Cancer 44:1256–1272.

DasGupta, U.B., and W.C. Summers (1978) Ultraviolet reactivation of herpes simplex virus is mutagenic and inducible in mammalian cells. Proc. Natl. Acad. Sci. USA 75:2377–2381.

Deck, M.D.F. (1980) Imaging techniques in the diagnosis of radiation damage to the nervous system. In: H.A. Gilbert and R.A. Kogan (eds.). Radiation Damage to the Nervous System. A Delayed Therapeutic Hazard. New York: Raven Press, pp. 107–127.

Delattre, J.Y., M.K. Rosenblum, H.T. Thaler, et al. (1988) A model of radiation myelopathy in the rat. Pathology, regional capillary permeability changes, and treatment with dexamethasone. Brain 111:1319–1336.

Di Corleto, P.E., and D.F. Bowen-Pope (1983) Cultured endothelial cells produce a platelet-derived growth factor-like protein. Proc. Natl. Acad. Sci. USA 80:1919–1923.

Fike, J.R., G.E. Sheline, C.E. Cann, et al. (1984) Radiation necrosis. Prog. Exp. Tumor Res. 28:136–151.

Fuchs, P., J.M. Krolak, D. McClain, et al. (1990) 18S rRNA degradation is not accompanied by altered rRNA transport at early times following irradiation of HeLA cells. Radiat. Res. 121:67–70.

Fukunaga, N., R.A. Schea, H.L. burrows, M. Meyers, and D.A. Boothman. Enhanced induction of tissue-type plasimnogen activation in normal human cells compared to cancer-prone cells following ionizing radiation. Int. J. Radiat. Oncol. Biol. Phys. (in press).

Haimovitz-Friedman, A., I. Vlodavsky, A. Chaudhuri, et al. (1991)

Autocrine effects of fibroblast growth factor in repair of radiation damage in endothelial cells. Cancer Res. 51:2552–2558.

Haymaker, W., L.J. Ruibinstein, and J. Mique (1979) Brain tumors in irradiated monkeys. Acta Neuropathol. 20:267–277.

Hermansson, M., M. Mistei, C. Betsholtz, et al. (1988) Edothelial cell hyperplasia in human glioblastoma. Coexpression of mRNA for platelet-derived growth factor (PDGF) -chain and PDGF receptor suggests autocrine growth stimulation. Proc. Natl. Acad. Sci. USA 85:7745–7752.

Hopewell, J.W. (1974) The late vascular effects of radiation (letter). Br. J. Radiol. 47:157–158.

Hughes, E.N., and D.A. Boothman (1991) Effect of caffeine on the expression of an X-ray-induced protein in human tumor cells. Radiat. Res. 125:313–317.

Iliakis, G. (1988) Radiation-induced potentially lethal damage: DNA lesions susceptible to fixation. Int. J. Radiat. Biol. 53:541–584.

Jaberaboansari, A., M.R. Landis, C.A. Wallen, et al. (1989) Alterations of neuronal nuclear matrix and chromatin structure after irradiation under aerobic and anoxic conditions. Radiat. Res. 119:57–72.

Jaberaboansari, A., G.B. Nelson, J.L. Roti Roti, et al. (1988) Post-irradiation alterations of neuronal chromatin structure. Radiat. Res. 114:94–104.

Kartasova, T., M. Ponec, and P. van de Putte (1988) Induction of proteins and mRNAs after UV irradiation of human epidermal keratinocytes. Exp. Cell Res. 174:421–432.

Kaston, M.B., O. Onyine, D. Sidransky, B. Vogelstein, and R.W. Craig (1991). Participation of p53 protein in the cellular response to DNA damage. Cancer Res. 51:6304-6311.

Kumar, P.P., R.R. Good, F.M. Skutety, et al. (1987) Radiation-induced neoplasms of the brain. Cancer 59:1274–1282.

Lampert, P.W., and R.L. Davis (1964) Delayed effects of radiation on the human central nervous system. Neurology 14:912–917.

Leibel, S.A., and G.E. Sheline (1987) Radiation therapy for neoplasms of the brain. J. Neurosurg. 66:1–22.

Ljungman, M. (1991) The influence of chromatin structure on the frequency of radiation-induced DNA strand breaks: A study using neuclear and nucloid monolayers. Radiat. Res. 126:58–64.

Ljungman, M., S. Nyberg, J. Nygren, et al. (1991) DNA-bound proteins contribute much more than soluble intracellular cmpounds to the intrinsic protection against radiation-induced DNA strand breaks in human cells. Radiat. Res. 127:171–176.

Marks, J.E., R.J. Baglon, S.C. Prassad, et al. (1981) Cerebral radionecrosis: Incidence and risk in relation to dose, time, fractionation, and volume. Int. J. Radiat. Oncol. Biol. Phys. 7:243–252.

Nackerdien, Z., J. Michie, and L. Bohm (1989) Chromatin decondensed by acetylation shows an elevated radiation response. Radiat. Res. 117:234–244.

Narayan, R.K., W.E. Heydorn, W.E. Creed, et al. (1984) Proteins in normal, irradiated, and postmortem human brain qualitatively compared by using two-dimensional gel electrophoresis. Clin. Chem. 30(112):1989–1995.

Patronas, N.J., G. DiChiro, and R.A. Brooks (1982) [18F] fluorodeoxyglucose and positron emission tomography in the evaluation of radionecrosis of the brain. Radiology 144:885–889.

Reinhold, H.S., A. Keyeux, D. Durijic, et al. (1974) The influence of radiation on blood vessels and circulation. Curr. Top. Rad. Res. Q. 10:9–185.

Rider, W.D. (1963) Radiation damage to the brain—a new syndrome. J. Ca. Assoc. Radiol. 14:67–69.

Ron, E., B. Modal, and J.D. Boice (1988) Tumors of the brain and nervous system after radiotherapy in childhood. N. Engl. J. Med. 319:1033–1039.

Ronai, Z.A., M.E. Lambert, and I.B. Weinstein (1990) Inducible cellular responses to ultraviolet light irradiation and other mediators of DNA damage in mammalian cells. Cell Biol. Toxicol. 6:105–126.

Sarasin, A.R., and P.C. Hanawalt (1978) Carcinogens enhance survival of UV irradiated simian virus 40 in treated monkey kidney cells: Induction of a recovery pathway. Proc. Natl. Acad. Sci. USA 75:346–350.

Schorpp, M., U. Mallick, H.J. Rahmsdorf, et al. (1984) UV-induced extracellular factor from human fibroblasts communicates the UV response to non-irradiated cells. Cell 37:861–868.

Singh, S.P., and M.F. Lavin (1990) DNA-binding protein activated by gamma radiation in human cells. Mol. Cell Biol. 10:5279–5285.

van der Kogel, A.J. (1986) Radiation-induced damage in the central nervous system: An interpretation of target cell responses. Br. J. Cancer 53(Suppl. 7):207–217.

Walker, G.C. (1985) Inducible DNA repair systems. Annu. Rev. Biochem. 54:425–457.

Warters, R.L., and B.W. Lyons (1990) Detection of ionizing radiation-induced double strand breaks by filter elution is affected by nuclear chromatin structure. Radiat. Res. 124:309–316.

Weichselbaum, R.R., D.E. Hallahan, V. Sukhatme, et al. (1991) Biological consequences of gene regulation after ionizing radiation exposure. JNCI 83:480–484.

Withers, J.R., J.M.G. Taylor, and B. Maciejewski (1988) The hazard of accelerated tumor clonogen repopulation during radiotherapy. Acta Oncol. 27:131–146.

Yamaguchi, N., T. Yamashima, and J. Yamashita (1991) A histologic and flow cytometric study of dog brain endothelial cell injuries in delayed radiation necrosis. J. Neurosurg. 74:625–632.

Young, D.F., J.B. Posner, F. Chu, et al. (1974) Rapid course radiation therapy of cerebral metastasis: Results and complications. Cancer 34:1069–1076.

Chapter 14

Human Chromosome Maps: Relevance to Identifying Genetic Changes in Tumors of the Nervous System

Olufunmilayo I. Olopade and Janet D. Rowley

INTRODUCTION

Many of the recent advances in molecular genetics have resulted from the continuing expansion of our knowledge of the chromosomal location of human genes. Mapping genes to particular chromosomes and to well-defined regions of the chromosomes has been made possible using a combination of somatic cell genetics, *in situ* hybridization and chromosome sorting, and, more recently, fluorescence *in situ* hybridization. In fact, one of the goals of the Human Genome Project is to construct a high-resolution physical map of chromosomes in which contiguous regions of at least 2 mega base pairs (bp) are unambiguously ordered and identified by sequence tagged sites (STS) spaced about 100,000 base pairs apart. Mapping the human genome is a formidable task because the haploid genome contains 3×10^9 base pairs or 3,000 megabases of DNA. Progress in mapping human genes as well as anonymous cloned DNA segments has been remarkable. We already have some information about the location of many genes involved in more than 600 clinical disorders (Frezal et al., 1991). Detailed summaries of this progress are contained in the reports of the Human Genome Mapping Workshops, of which the most recent is Human Gene Mapping 11, 1991. The information contained in these reports is available, on line, though the Genome Data Base and On-line Mendellian Inheritance in Man (GDB and OMIM) at Johns Hopkins Medical Center. More recently, groups of investigators have formed a consortium to assume responsibilities for correlating and reconciling more detailed mapping data on individual chromosomes. Reports of these chromosome mapping groups are published in the appropriate journals; for example, a complete report of the chromosome 21 workshop can be found in *Cytogenetics and Cell Genetics* (vol. 59, 1992).

In addition to the information derived from localizing the genes associated with many genetic diseases, the study of recurring chromosome abnormalities in leukemia and lymphoma has been very fruitful (Rowley et al., 1990a,b; Sandberg, 1991; Solomon et al., 1991). Because many of these recurring abnormalities in leukemia and lymphoma involve previously uncharacterized regions of the human genome, their molecular elucidation has resulted in the discovery of several novel candidate cancer genes (Rowley et al., 1990; Solomon et al., 1990). For several of the rearrangements, some of the changes in gene structure and function have been defined and therefore some general principles that may be applicable to all chromosome rearrangements in human malignant disease, including central nervous system tumors, are beginning to emerge. The importance of chromosome maps is demonstrated in this review by using examples of important advances that resulted from knowledge of the chromosome location of the genes involved. A brief review of some of the currently available mapping technologies is presented.

Much of the detailed information regarding the relevant chromosome rearrangements is contained in a number of recent reviews, and only a general summary is presented here (Heim and Mitelman, 1987; Rowley, 1990a; Solomon et al., 1991). A complete compendium of chromosome abnormalities is available in the second edition of *The Chromosomes in Human Cancer and Leukemia* (Sandberg, 1990). Mitelman has published four editions of his *Catalog of Chromosome Aberrations in Cancer* (Mitelman, 1983, 1985, 1988, 1991). The most current one, published in 1991, contains 14,141 abnormal karyotypes, of which about 550 are tumors of the nervous system. A comparison of the number of abnormal karyotypes in the catalogs according to the type of neoplasia is summarized in Figure 1. Although

Molecular Genetics of Nervous System Tumors, pages 179-194 ©1993 Wiley-Liss, Inc.

carcinomas account for the greatest proportion of malignant disease, they represent only about 20% of the karyotypic data, whereas most information is available for leukemia and lymphoma.

A number of international meetings over the last 25 years have led to the establishment of a universally accepted system for chromosome nomenclature. The most authoritative document, *An International System for Human Cytogenetic Nomenclature* (1985) has recently been revised with regard to the nomenclature of the acquired abnormalities in cancer cells (ISCN, 1991). The simplest change is either a gain or a loss of a whole chromosome. Common structural alterations are translocations, which involve the exchange of material between two or more chromosomes, and deletions, which involve loss of DNA from a chromosome and thus from the affected cell (Fig. 2). Chromosome inversions have been observed; in this rearrangement, a single chromosome is broken in two places, and the central portion is inverted and rejoined to the ends of the chromosome. In describing the karyotype, the total chromosome number is followed by the sex chromosomes, and gains and losses of whole chromosomes are identified by a + or - before the chromosome number. A gain or loss of part of a chromosome is identified by a + or - after the chromosome arm involved; p and q represent the short and long arm, respectively. Translocations are indicated by t; the chromosomes involved are noted in the first set of brackets, and the breakpoints in the second set of brackets (see glossary of terminology, Table 1). Banding of chromosomes is essential to cytogenetic investigations because it allows identification of individual chromosomes and of parts of the chromosome as well. A band is defined as a chromosome segment that is distinguished from adjacent segments by appearing darker or lighter by one or more banding techniques. Each chromosome band is num-

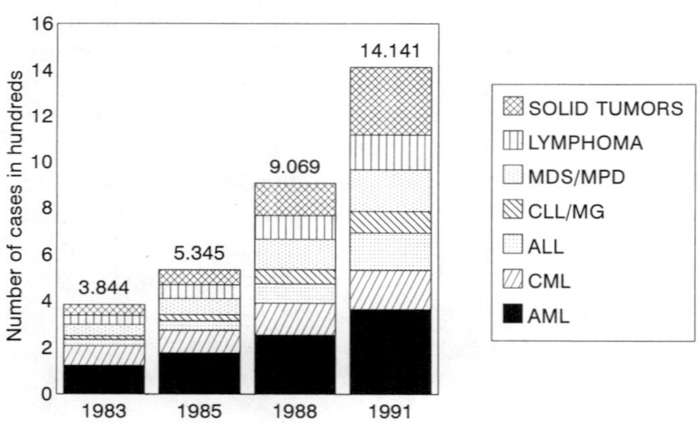

Fig. 1. Graph showing the proportion of abnormal karyotypes in the four *Catalogs* by disease type. The number in parentheses below the date is the total number of abnormal karyotypes in each edition. CML patients with only a t(9;22) are not included.

TABLE 1. Glossary of Cytogenetic Terminology

Centromere	The constriction along the length of the chromosome that is the site of the spindle fiber attachment. The position of the centromere determines whether chromosomes are metacentric (X-shaped, e.g., chromosomes 1, 3, 16, 19, 20) or acrocentric (inverted V-shaped, e.g., chromosomes 13–15, 21, 22, Y). During mitosis the two exact copies of the DNA in each chromosome are separated by shortening of the spindle fibers attached to opposite sides of the dividing cell.
Karyotype	Arrangement of chromosomes from a particular cell according to a well-established system such that the largest chromosomes are first and the smallest ones are last (see Fig. 3). Normal female karyotype is 46,XX; normal male karyotype is 46,XY.
Translocation	A break in at least two chromosomes with exchange of material; in a reciprocal translocation, there is no obvious loss of chromosomal material.
Deletion	A segment of a chromosome is missing as the result of two breaks and loss of the intervening piece (see Fig. 2).
Inversion	Two breaks occur in the same chromosome with rotation of the intervening segment. If both the breaks are on the same side of the centromere, it is called a paracentric inversion. If they are on opposite sides it is called a pericentric inversion.
Isochromo	A chromosome that consists of identical copies of one chromosome arm with loss of the other arm. Thus an isochromosome for the long arm of 17 [i(17q)] contains two copies of the long arm (separated by the centromere) with loss of the short arm of the chromosome.
Clone	In the cytogenetic sense, this is defined as two cells with the same additional or structurally rearranged chromosome or three cells with loss of the same chromosome.
Diploid	Normal chromosome number and composition of chromosomes.
Hyperdiploid	Additional chromosomes; therefore the modal number is 47 or greater.
Hypodiploid	Loss of chromosomes with modal number 45 or less.
Haploid	Only one-half the normal compliment, i.e., 23 chromosomes.
Karyotype symbols:	
p	Short arm
q	Long arm
+	If before the chromosome, indicates a gain of a whole chromosome (e.g., +8) and if after the chromosome, indicates gain of part of the chromosome (e.g., 14q+, added material at the end of the long arm of 14)
-	If before the chromosome, indicates a loss of a whole chromosome (e.g., -7) and if after the chromosome indicates loss of part of the chromosome (e.g., 5q-, loss of part of the long arm of 5)
?	Indicates uncertainty about the identity of the chromosome or band listed just after the ?.
t	Translocation
del	Deletion
inv	Inversion
i	Isochromosome

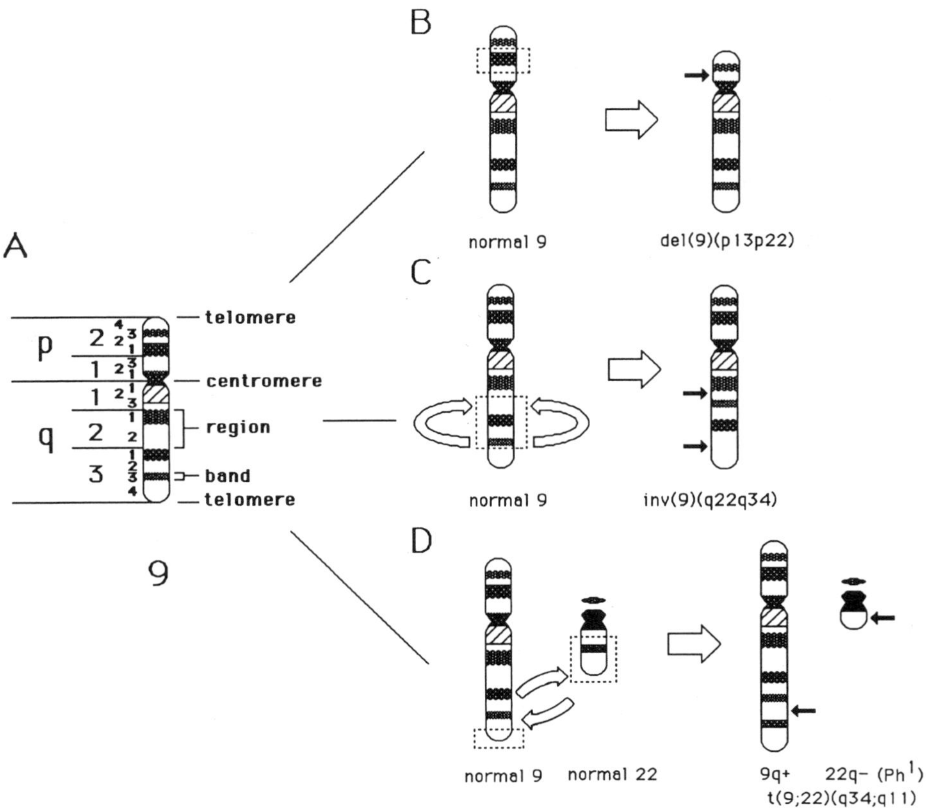

Fig. 2. Schematic diagram illustrating a normal chromosome and three chromosomal abnormalities observed in human neoplasms. **A**: Diagram of the banding pattern of a normal chromosome 9. The chromosome arms (p, short arm; q, long arm), regions, and band numbers are indicated on the left of the chromosome; specific chromosome structures are indicated on the right of the chromosome. **B**: Diagram of the mechanism of an interstitial deletion of the short arm of chromosome 9, a common abnormality in malignant gliomas. Chromosome breaks occur in bands 9p13 and 9p22, and the intervening chromosomal segment (band 9p21 and parts of bands 9p13 and 9p22) is lost [del(9)(p13p22)]. **C**: Diagram of the mechanism of a paracentric inversion. Chromosome breaks occur in two bands within a single chromosome arm, in this case, within 9p22 and 9q34; the intervening segment is inverted and the chromosome breaks are repaired [inv(9)(q22q34)]. **D**: Diagram of the mechanism of the reciprocal translocation involving chromosomes 9 and 22, t(9;22)(q34;q11), which gives rise to the Philadelphia (Ph) chromosome in the malignant cells of patients with chronic myelogenous leukemia. Breaks occur in bands q34 and q11 of chromosomes 9 and 22, respectively, followed by a reciprocal exchange of chromosomal material. This rearrangement results in the translocation of the *ABL* oncogene, normally located at 9q34, adjacent to the *BCR* gene on chromosome 22, giving rise to a chimeric *BCR-ABL* gene, whose protein product plays a role in the transformation of myeloid cells.

bered. A chromosome band contains approximately 5 to 15 megabases of DNA. Various banding methods are currently utilized, including quinacrine mustard (Q bands) and Giemsa stain (G bands). A more detailed description of cytogenetic methods is contained in the chapter by Poulsgård (Chapter 19, this volume). Molecular geneticists interested in nervous system tumors should be familiar with the cytogenetic literature because the application of several new molecular biologic techniques important for chromosome mapping requires a basic understanding of this literature.

STRATEGIES OR TECHNIQUES FOR CHROMOSOME MAPPING

Genetic Linkage or Meiotic Mapping

The construction of fine structure genetic linkage maps is important for the mapping of genes involved in genetic disorders. It is important for prenatal diagnosis and carrier detection in hereditary disorders and ultimately is useful for sequencing the human genome (Dunlop et al., 1991; Melki et al., 1992). The generation of comprehensive linkage maps of each human chromosome has been greatly facilitated by the study of the segregation of linked DNA markers in large pedigrees. The observation of infrequent recombination events between closely linked markers is more likely, if the segregation of the markers can be studied through many meioses. The Centre d'Etude du Polymorphisme Humain (CEPH) families in Paris have been used as a common linkage mapping panel by several investigators (Dausset, 1986; Donis-Keller et al., 1987). These families provide large reference pedigrees consisting of both parents and at least eight offspring so that a maximum of 16 meioses can be scored in the average family. CEPH distributes DNA from these 40 families to collaborating investigators, who, after completion of genotyping, submit the data to CEPH. In this manner, a common database of genetic markers is created that allows analysis and construction of maps from a single source, although multiple investigators contribute to the database. Linkage maps of several chromosome regions have been constructed using DNA polymorphic markers in these CEPH pedigrees (Warren et al., 1992). The Utah kindreds and Venezuelan kindreds are other families that are often used as mapping panels (Keats et al., 1991; Tanzi et al., 1988). Genetic linkage maps are good for ordering of loci and the relative distances between them. Recombination frequencies per megabase of DNA vary considerably by sex and by chromosomal region; telomeric regions appear to have a proportionately higher frequency of recombination in male meiosis, while centromeric regions have a higher frequency of recombination in female meiosis. The genetic linkage maps for females are longer because of the higher frequency of recombination in female meiosis. Interval distances of genetic linkage maps are based on the amount of breakage or exchange of DNA occuring during meiosis and is measured in terms of centimorgans (cM) with accompanying logarithm of the adds score (LOD) (Keats et al., 1991; Tanzi et al., 1988). The closer two markers are, the more informative meiosis that would have to be examined to determine their order. Thus, constructing a genetic linkage analysis with closely spaced markers requires a lot of work and is very time consuming. The use of restriction fragment lengths polymorphisms (RFLP) in conjunction with genetic linkage analysis has allowed the construction of fragmented meioitic linkage maps for each of the 23 human chromosomes, with an average resolution of 10–15 cM. It is not possible to give a comparison of genetic and physical maps at the present time except for a few chromosomal regions in which this has been worked out, e.g., Huntington disease locus on chromosome 4 and chromosome 21 (Beutow et al., 1991; Falk, 1992).

Somatic cell hybrids

Chromosome segregation in human–mouse somatic cell hybrids provides a valuable subdivision of the genome for many aspects of its analysis. Once a unique sequence is assigned to a specific chromosome, more precise physical localization on that chromosome is possible using a chromosome-specific somatic cell hybrid mapping panel (Gardiner and Patterson, 1992; Van Keuren et al., 1986). Such a panel consists of a collection of hybrids that retain different portions of that chromosome. This mapping method relies on the cytogenetic characterization of each hybrid with respect to the normal banding pattern of that chromosome to determine which segments are present and which are not. The accuracy of gene localization using this technique depends on the accuracy of the karyotype obtained. The development of panels of hybrids with only a single chromosome and also panels with parts of chromosomes derived using naturally occurring translocations has greatly aided our ability to localize unique sequence probes to defined regions of the particular chromosome (Kaiser-McCain et al., 1975). It has been shown by several investigators that for the most part, naturally occurring rearrangements of human chromosome isolated in Chinese hamster–human hybrid cell lines appear by Giemsa banding to be faithful copies of the parent chromosomes (Cox et al., 1989; Gardiner and Patterson, 1992). These panels of hybrids provide a useful and reliable resource for determining regional localization of unique DNA probes. A potential pitfall is related to the uncertainty that small submicroscopic deletions, inversions, or duplications might have occurred in the hybrids at the time of construction or during subsequent culture. Gardiner and Patterson (1992) have used a panel of somatic cell hybrids, each member of which contains a different naturally occurring rearrangement of chromosome 21 to regionally localize unique sequences on the long arm of chromosome 21. Hundreds of human loci have been assigned to specific human chromosomes with this technique. Thus, somatic cell genetics remains a powerful tool in chromosome mapping.

Radiation-reduced hybrids

Determination of the order and distance between DNA markers spanning 10 to 20 megabases of DNA at the 500 kb level of resolution is difficult with somatic cell hybridization and *in situ* hybridization. Radiation-reduced hybrid mapping (RH) is well suited for construction of high-resolution, long-range contiguous maps of mammalian chromosomes at a resolution of about 500 kb (Boehnke, 1992a,b; Boehnke et al., 1991; Cox, 1992; Cox et al., 1990; Lawrence et al., 1991). In this approach, based on earlier studies by Gross and Harris (1977), Chinese hamster–human somatic cell hybrid containing hamster chromosomes and a single copy of a human chromosome of particular interest is exposed to a lethal dose of X-rays, breaking each chromosome into several fragments 8–10 megabases in size. Following irradiation, the human and hamster chromosomal fragments rapidly join with one another, resulting in complex chromosomal rearrangements. The irradiated hybrid cell is then rescued by fusion with nonirradiated hamster cell under selective conditions such that only somatic cell hybrids formed by fusion of a nonirradiated cell to an irradiated cell grow to form viable colonies. Although this technique does not select for the retention of human chromosomal material in the resulting viable hybrid, each hybrid retains 20–60% of the human chromosome of interest as several independent human chromosome fragments that are often inserted into the middle of the hamster chromosome (Cox et al., 1990). The frequency of retention of the various human chromosomal fragments is surprisingly constant, irrespective of their chromosomal location, although some variation in retention frequency of individual markers is observed (Boehnke, 1992). Because the hybrids contain several noncontiguous fragments of human chromosome, it is not possible to use any single hybrid for mapping until extensive characterization of the retained human material is performed. Statistical methods have been developed to aid in the analysis of RH so that the order and distance between markers can be determined (Boehnke et al., 1991; Cox et al., 1990). The frequency of breakage between two markers is used to estimate the distance between them. This results in a centiRad (cR) score between markers with accompanying LOD scores. RH mapping involves a single copy of the human chromosome of interest and it can be used to map both polymorphic and nonpolymorphic markers in contrast to genetic linkage mapping, which can only map polymorphic markers. In addition, one can vary the resolution of the map by varying the dose of irradiation. At a dose of 8,000 rad, however, RH mapping fills a resolution gap between genetic linkage mapping on the one hand and metaphase *in situ* hybridization and pulse field gel electrophoresis (PFGE) on the other (Burmeister et al., 1991). It is most powerful when used in conjunction with other techniques.

Isolating Large DNA Probes

Using the technique of pulse field gel electrophoresis (PFGE) and restriction enzyme digestion with rare cutting enzymes, large fragments of DNA covering regions of the genome that are 1,000 kb or longer can now be readily resolved. PFGE can resolve fragments from less than 2 kb to greater than 6 megabase pairs in size, and thus can link probes separated by large physical distances (Carle et al., 1986; Smith and Cantor, 1986). This allows the generation of physical maps several megabase pairs in size (for example, Burmeister et al., 1988; Fulton et al., 1989). To study the particular chromosome region in more detail, overlapping clones can be obtained by screening recombinant cosmid library clones or yeast artificial chromosome (YAC) library clones. The largest cosmid clones contain inserts of approximately 40 kb; a 1,000 kb region can be covered by approximately 25 cosmid clones. Larger pieces of genomic human DNA up to 1,000 kb are now clonable in YACs. Several total genomic YAC libraries as well as phage and cosmid libraries from flow-sorted human chromosomes are now available for screening to obtain overlapping recombinant clones that cover the region of interest. At the Lawrence Livermore Laboratory, chromosomes 9, 12, 19, 21, 22, and Y cosmid libraries are available; from the Los Alamos Laboratory, chromosomes 4, 5, 6, 8, 11, 13, 16, and 17 libraries are available. YAC libraries are available from the Imperial Cancer Research Fund (ICRF), CEPH, and St. Louis University. The mapping resolution of YAC cloning in combination with PFGE is on the order of 100–500 kb.

Fluorescence *In Situ* Hybridization (FISH) and Interphase Mapping

FISH is a relatively new technique that has revolutionized chromosome mapping analysis by the use of nonradiolabeled DNA probes (for example, biotin-labeled probes) that are detected by fluorescence microscopy. A series of unique DNA probes can be ordered physically by FISH on metaphase chromosomes and/or interphase nuclei. This does not require that the probes be polymorphic. For this technique, phage DNA, cosmid DNA, or YAC DNA inserts are usually well resolved because of their larger sizes. However, inserts as small as 500–900 base pairs can now be detected by using more sophisticated set-ups such as coupled device camera (CCD), confocal or epifluorescent microscopes, and computer programs to merge images (Lemieux et al., 1992). FISH is usually performed using a DNA probe labeled with a single fluorochrome; two or more probes can be detected simultaneously when they are labeled with at least two fluorochromes of different colors, singly or in combination. DNA sequences that are spaced further apart than 1–2 megabases can usually be resolved using metaphase chromosomes, and their telomere–centromere orientation order can be determined. To obtain a higher resolution map, one can use these probes to perform FISH on interphase nuclei that has a resolution of 20–50 kb (Lawrence et al., 1988; Lichter et al., 1988; Traski et al., 1989). Interphase mapping relies on the linear relationship between the physical distance of the hybridization signals of two probes in interphase nuclei and their distance in base pairs at the DNA level. A near-linear relationship between these parameters exists in the range of 50 to 1,000 kb. Using different colors for different probes, one is able to rapidly determine the relative order of these probes and their approximate distance. Thus interphase mapping has become a powerful tool that will allow the efficient ordering of closely spaced probes derived from a band-specific library and will also yield an estimate of the physical distance between these probes.

A large number of chromosome-specific centromere probes are now available that unequivocally mark a pair of chromosomes. Using these probes, gain or losses of chromosomes can be detected not only in metaphase cells but also in interphase cells Lemieux et al. (1992). With the increasing availability of different fluorochromes and the remarkable developments in sensitive detection of combinations of fluorochromes, it is possible to screen cells for abnormalities of at least six different chromosomes simultaneously. Large-sized DNA probes that contain specific genes or anonymous DNA sequences, e.g., YACs or cosmids, can be used to screen the recurring translocations and to identify those probes that are split by the translocation breakpoints. This technique has been used by several groups in their search for genes involved in different translocations (Rowley et al., 1991).

Chromosome Microdissection

Another important new technique in chromosome mapping is the use of chromosome microdissection. This technique takes advantage of the natural division of the human genome into 23 distinct chromosome pairs and their subdivision into clearly distinct bands. Using an extremely fine glass rod, the DNA contained in a single specific band can be removed, combined with material from the same band from about a dozen other chromosomes, amplified with the polymerase chain reaction (PCR), and then used for cloning DNA from the band (Bohlander et al., 1992; Ludecke et al., 1989; Meltzer et al., 1992). With chromosome microdissection, it is now possible to obtain DNA from specific regions of the chromosome and construct band-specific libraries. Using the same strategy, one can microdissect unidentified marker chromosomes, amplify the DNA, and hybridize it to banded normal chromosomes.

We have provided only a brief overview of the different techniques for chromosome mapping. Each one of these techniques can be adapted to study nervous system tumors. Figure 3 shows the degree of resolution that can be achieved with each mapping method.

CHROMOSOME LOCALIZATION OF GENES AS KEY IN THE MOLECULAR ANALYSIS OF RECURRING CHROMOSOME ABNORMALITIES, PARTICULARLY TRANSLOCATIONS

At present, only a few of the genetic changes present in tumors of the nervous system are known in any molecular detail. However, the successful analysis of these changes in leukemia and lymphoma provide examples of the important biological insights that such research can reveal. The first consistent chromosome abnormality in any malignant disease was the Philadelphia or Ph chromosome identified in chronic myeloid leukemia (CML) (Nowell and Hungerford, 1960). The abnormality was thought to be a deletion of chromosome 22 (22q-). The correct defect was later shown to be a translocation involving chromosomes 9 and 22 [t(9;22)(q34;q11)] (Rowley, 1973). In 1972, Manolov and Manolova discovered that the malignant cells of patients with Burkitt's lymphoma had an additional band at the end of the long arm of one chromosome 14 (14q+). Zech and co-workers in 1976 first observed that the end of one chromosome 8 was consistently absent, and they suggested that the missing part of chromosome 8 was translocated to number 14 [t(8;14)(q24;q32)]. The t(8;14) has also been observed in nonendemic Burkitt's tumors from America, Europe, and Japan. Thus, it is a highly characteristic chromosome anomaly in Burkitt's lymphoma. This translocation has also been observed in other lymphomas, particularly those of the diffuse large-cell type. Two other related translocations were later identified in Burkitt's tumors. All three translocations involve chromosome 8, with a break in the same band, 8q24. One variant translocation involved chromosome 2, with a break in the short arm [t(2;8)(p12;q24)], and the other involved number 22, with a break in the long arm in band (22q11).

The reciprocal nature of the translocation in CML was established only when the Abelson proto-oncogene, *ABL*, was mapped to chromosome 9. The *ABL* gene is the homologue of a viral oncogene isolated from a mouse pre-B-cell leukemia. The initial localization of *ABL* to chromosome 9 prompted de Klein and colleagues to investigate the involvement of this gene in the t(9;22), and they found that *ABL* is translocated to the other chromosome (de Klein et al., 1982). This finding led Heisterkamp et al. (1984) to screen samples from patients with CML until one was identified that had an apparent DNA rearrangement in the vicinity of the *v-abl* homologous sequences. This led to cloning of the breakpoint involved in the t(9;22). The site on the Ph chromosome was called bcr, for breakpoint cluster region; in the majority of translocations, the breaks cluster in a small 5.8 kb region (Groffen et al., 1984). In contrast, the breaks on chromosome 9 occur over an incredible distance of more than 200 kb (Rubin et al., 1988). The genetic consequences of the standard t(9;22) or the complex translocation involving at least three chromosomes is to move the *ABL* proto-oncogene on number 9 next to the gene on number 22, called *BCR*. This translocation leads to the formation of a fusion BCR-ABL message and protein. The tyrosine kinase activity of ABL is unmasked by fusion to BCR sequences, thus giving the chimeric protein a more potent transforming ability (Gishizky et al., 1991; Kelliher et al., 1991; Manu and Witte, 1991; Shtivelman et al., 1985). In CML, the BCR-ABL fusion p210 protein is found, while in Ph-positive acute lymphoblastic lymphoma (ALL), a variant p185 protein formed due to a different breakage in BCR intron is seen in half the cases (Chan et al., 1987). Daley and co-workers as well as other investigators have now shown that this $P210^{bcr\text{-}abl}$ product is able to induce a CML-like disease when murine bone

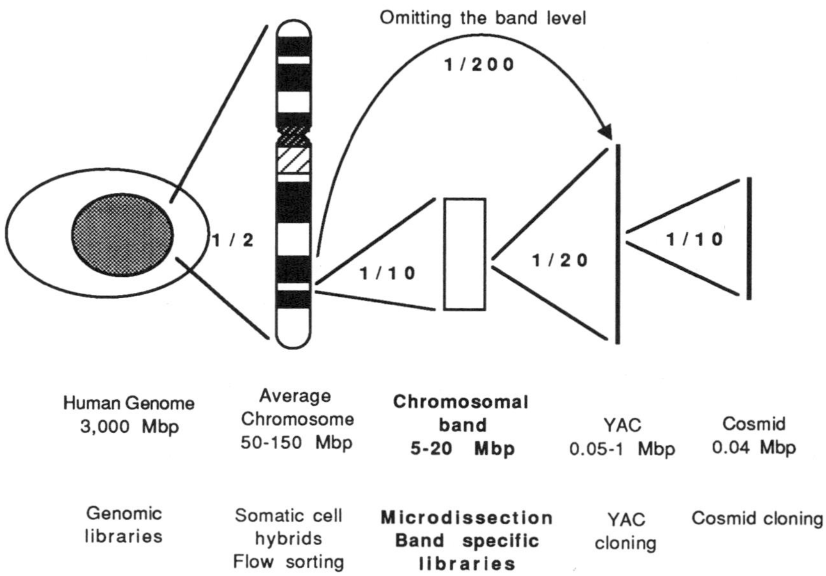

Fig. 3. A schematic diagram of the level of resolution of each mapping technique. Chromosome microdissection bypasses the level. (Prepared by Stefan Bohlander.)

marrow was infected with a retrovirus encoding P210$^{bcr\text{-}abl}$ and transplanted into irradiated syngeneic recipients (Daley et al., 1990; Groffen et al., 1992). Thus it is likely that expression of this fusion protein by bone marrow cells can induce CML.

For Burkitt's lymphoma, the specific translocations involved chromosome 8 with chromosomes 14, 2, and 22, on which were located the immunoglobulin heavy chain and kappa and lambda light chain genes, respectively. The gene for the proto-oncogene *MYC* (the cellular homologue of the avian myelocytomatosis virus) is on chromosome 8(q24). The immunoglobulin genes are located at the breakpoints (14q23, 2p12, and 22q11) on the three chromosomes other than 8 that are involved in the translocations in Burkitt's lymphoma. These translocations result in the aberrant juxtaposition of *MYC* and one of the immunoglobulin genes; this in turn leads to abnormal regulation of *MYC* expression (Caccia et al., 1985; Leder et al., 1983).

The observations that important genes lie at translocation breakpoints and that these rearranged genes contribute to the transformed phenotype have generated a great deal of interest in recurring translocations in both hematologic malignancies and solid tumors. It is clear that translocations pinpoint the location of important genes. In fact, patients with balanced translocations who are members of families with heritable diseases have helped in identifying the putative locus of cancer-associated genes. For example, the gene for *NF1* was recently identified by cloning the translocation breakpoints of two NF1 patients with constitutional translocations in the NF1 region on chromosome band 17q11.2 (Fountain et al., 1989).

While translocations can pinpoint important genes, in order to characterize them one must identify a breakpoint clone from among the 5 to 10 million base pairs suggested by the cytogenetic assignment of the breakpoint to a particular band (Solomon et al., 1991). The translocations involving the immunoglobulin genes were easy to clone because these DNA probes detected the rearranged fragments. The isolation of translocation breakpoints in human acute myeloid leukemia and in solid tumors has proceeded extremely slowly, partly because DNA probes for a particular chromosome region are not available in sufficient numbers to offer a high likelihood of identifying a DNA rearrangement on Southern blot analysis, either by conventional electrophoresis or by PFGE. The use of cosmids or YACs as probes to screen much larger fragments of DNA for rearrangements provides an additional strategy for the analysis of chromosomal abnormalities (Rowley et al., 1990c). We and others have used radiolabeled probes for *in situ* hybridization to identify genes that bracket the translocation breakpoints, although in some cases they showed no rearrangement on Southern blot analysis. With this technique, the detailed molecular analysis is focused on a particular DNA fragment. This technique is also useful in defining the limits of chromosome deletions. We have used this technique in identifying the 11q23 breakpoint in acute myeloid and lymphoid leukemia (Rowley et al., 1990).

One of the very few recurring translocations in solid tumors was recently cloned: namely, the t(11;22) in neuroepithelioma/Ewing's sarcoma (Delattre et al., 1992;

Fiedler et al., 1991). Whang-Peng et al. in 1984 described a t(11;22)(q24;q12) in two cases of peripheral neuroepithelioma (PNE); the same translocation has also been reported in more than 90% of Ewing's sarcoma (EWS) tumors (Aurias et al., 1983). Further, a comparison of EWS and PNE suggests that the two tumors are histogenetically related. It has recently been shown that the neuronal phenotype of EWS and PNE is the same. The similarity has been further substantiated by molecular analysis in which specific high expression of the MIC2 antigen occurs in both EWS and PNE (Ambros et al., 1991). The discovery of the apparently identical translocation in both tumor types has changed the treatment strategy in PNE. The genes involved in this translocation have just recently been identified by positional cloning because none of the candidate genes showed any rearrangement on standard Southern analysis or PFGE.

The breakpoint on chromosome 22 had been previously bracketed by two markers, D22S1 and D22S15, mapped to chromosome 22 (Zhang et al., 1990). An extended mapping panel of somatic cell hybrids, each containing only a small segment of chromosome 22, was developed and a number of other DNA probes were identified that mapped between these two loci. One small region of approximately 2 megabases in size termed 8.2 was immediately telomeric to the breakpoint. Differential screening of a chromosome 22-specific library with dispersed repetitive DNA sequences polymerase chain reaction (ALU-PCR) products generated from hybrids that defined region 8.2 led to the isolation of three independent clones. *In situ* hybridization on interphase nuclei, PFGE data, and generation of cosmid contigs ordered these three clones within region 8.2. Overlapping walking clones adjacent to the one closest to the breakpoint were identified until the breakpoint was reached. Screening of a cosmid library made from a PNE cell line identified cosmids containing the junction fragment of the der(22) and der(11) chromosomes as well as a 100 kb contig on chromosome 11 spanning the breakpoint. Determination of the position of the breakpoints in 20 EWS and PNE tumors showed that the chromosome 22 breakpoints are always within a small 7 kb region, termed EWSR1. In contrast, on chromosome 11, the breakpoints are scattered over a 40 kb region, termed EWSR2. cDNA from the involved genes have been cloned and sequenced. The gene on chromosome 22 is called *EWS1*. The 5' portion codes for a protein with homology to RNA polymerase 2, whereas the 3' portion codes for an RNA binding domain. The gene on chromosome 11 shows very high homology to the mouse *Fli1* gene; the 3' portion codes for a DNA binding domain. Both genes are oriented 5' to 3' centromere to telomere. As a consequence of the translocation, the 5' portion of *EWS1* and 3' portion of *FLI1*, including the DNA binding domain, form a chimeric gene on the der(22) chromosome. There is some variability in the fusion product due to the variable inclusion of one exon from either *EWS1* or *FLI1*. All the fusion transcripts can be detected with a single set of PCR primers (Dellatre et al., 1992).

The chromosome rearrangements for which the molecular alterations of the involved genes is known are summarized in Table 2. In Figure 4, we have also indicated the chromosome position of recurring, yet uncloned, translocation breakpoints.

CHROMOSOME LOCATION OF PROTO-ONCOGENES

One of the most surprising revelations in the recent past has involved the cellular oncogenes and their chromosome location. Much of the excitement derives from the observation that many proto-oncogenes are located in the bands that are involved in consistent translocations (Adams et al., 1985; Bishop, 1988; Rowley, 1990a; Solomon et al., 1991) (Fig. 3). Proto-oncogenes have their effects at many points in the cascade of information that flows from a signal outside a cell to the nuclear transcription factors that ultimately regulate the stage of cell cycle and degree of differentiation. For several of the translocation breakpoints that have been cloned, the genes identified act at different steps in this cascade (Table 3). Thus several of these genes are ligand receptors (*RARA*) or transcription regulating factors (*RBNT2*, *MYC*) and play a role in the cell cycle (Solomon et al., 1991; Table 4). An exception is *BCL2*, which is localized in the inner mitochondrial membrane and has been shown to prolong cell survival by blocking apoptosis. For a more detailed discussion of oncogenes, see chapter by Ruley (Chapter 7, this volume).

Several investigators have explored the possibility that different oncogenes may contribute to the malignant transformation due to their amplification in the tumors. Classic examples are the *EGFR* localized on 7p, which is amplified in gliomas, and *MYCN*, which is amplified in neuroblastoma (Bigner et al., 1990; Brodeur et al., 1984; Schwab et al., 1983; Wong et al., 1992). High levels of amplification often are marked cytogenetically by double minute chromosomes and homogeneously staining regions that are found in several CNS tumors. On balance, however, it seems that in the majority of solid tumors, including those of the CNS, chromosome deletions are going to be the predominant abnormality and tumor suppressor genes will be the dominant players (Tables 5, 6).

CHROMOSOME LOCATION OF TUMOR SUPPRESSOR GENES

Deletion of genomic material is usually suggestive of the presence of a tumor suppressor gene whose inactivation is important in the initiation or progression of malignancy.

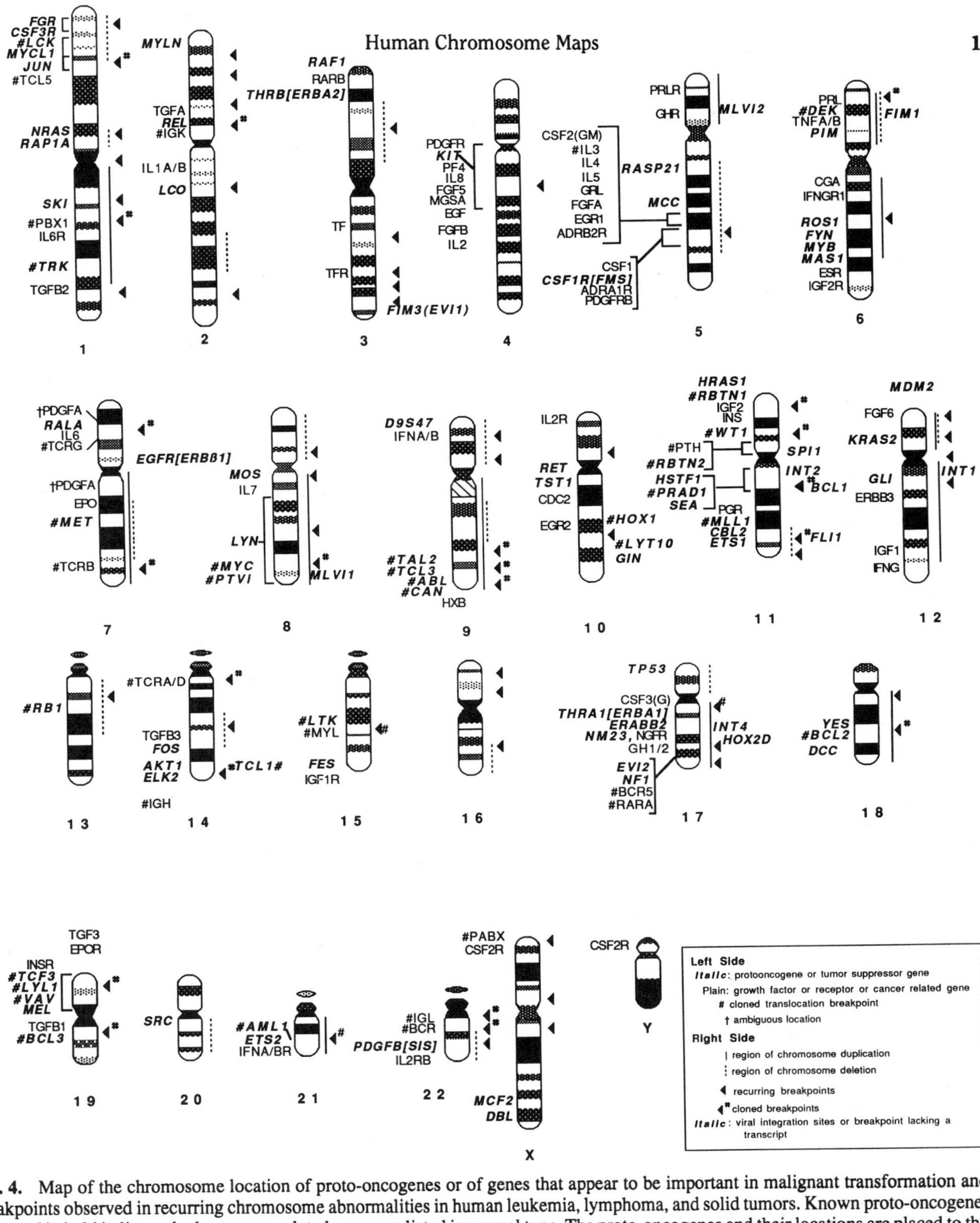

Fig. 4. Map of the chromosome location of proto-oncogenes or of genes that appear to be important in malignant transformation and the breakpoints observed in recurring chromosome abnormalities in human leukemia, lymphoma, and solid tumors. Known proto-oncogenes are indicated in bold italics and other cancer-related genes are listed in normal type. The proto-oncogenes and their locations are placed to the left of the appropriate chromosome band or region (indicated by a bracket). The breakpoints in recurring translocations, inversions, etc., are indicated with an arrowhead to the right of the affected chromosome band. The solid vertical lines on the right indicate regions frequently present in triplicate; the dashed lines indicate recurring deletions. Recurring viral integration sites and cloned translocation breakpoints with no identified transcripts are indicated to the right of the appropriate band. Genes or recurring breakpoints that have been cloned are identified by #. The locations of the cancer-specific breakpoints are based on the Report of the Committee on Structural Chromosome Changes in Neoplasia (Human Gene Mapping 11, 1991). Note that any map of this sort involves selection of the genes that should or should not be included; we have been relatively conservative. Also, for recurring breakpoints and deletions, we have included only those listed as status I or II in HGM11.

TABLE 2. Genes Involved in Neoplasia-Associated Chromosomal Rearrangements

Phenotype	Rearrangement	Involved genes	
Acute lymphoblastic leukemia			
Pre-B	t(1;19)(q23;p13)	*PBX*	*E2A*
B(SIg+)	t(8;14)(q24;q32)	*MYC*	*IGH*
(also NHL)	t(2;8)(p11–12;q24)	*IGK*	*MYC, PVT1*
	t(8;22)(q24;q11)	*MYC, PVT1*	*IGL*
	t(5;14)(q31;q32)	*IL3*	*IGH*
	dic(9;12)(p11;p12)		
B or B-myeloid	t(9;22)(q34;q11)	*ABL*	*BCR*
	t(4;11)(q21;q23)		
Other hyperdiploidy (50–60 chromosomes)			
	del(9p),t(9p)	*IFNA/B*	
	del(12p),t(12p)		
	t(8;17)(q24;q22)	MYC	BCL3*
T	t(11;14)(p15;q11)	*TAL1*/TTG#*	*TCRD*
	t(11;14)(p13;q11)	*TCL2#*	*TCRD*
	t(8;14)(q24;q11)	*MYC*	*TCRA*
	inv(14)(q11q32.3)	*IGH*	*TCRA*
	inv(14)(q11q32.1) /t(14;14)(q11;q32)	*TCL1*	*TCRA*
	t(10;14)(q24;q11)	*TCL3#*	*TCRD*
	t(1;14)(p32;q11)	*TCRD*	*TAL1*/SCL*
	t(7;9)(q35;q32)	*TCRB*	*TCL4*
	t(7;9)(q35;q34)	*TCRB*	*TCL3**
	t(7;7)(p15;q11)	*TCRG*	
	t(7;14)(p15;q11)	*TCRG*	
	t(7;14)(q35;q11)		
	t(7;19)(q35;p13)	*TCRB*	*LYL1*
	del(9p), t(9p)	*IFNA/B?*	
Non-Hodgkin's lymphoma			
B	t(8;14)(q24;q32)	*MYC*	*IGH*
	t(2;8)(p11–12;q24)	*MYC*	*IGK*
	t(8;22)(q24;q11)	*MYC*	*IGL*
	t(14;18)(q32;q21)	*IGH*	*BCL2*
	t(11;14)(q13;q32)	*IGH*	*BCL1#*
T	See T-cell ALL		
Chronic lymphocytic leukemia			
B	t(11;14)(q13;q32)	*IGH*	*BCL1#*
	t(14;19)(q32;q13)	*IGH*	*BCL3**
	t(2;14)(p13;q32)	*IGH*	
	t(18;22)(q21;q11)	*IGL*	*BCL2*
	t(14q32)		
	+12		
T	t(8;14)(q24;q11)	*MYC*	*TCRA*
	inv(14)(q11q32)	*TCRA*	*TCL1*
Multiple myeloma			
B	t(11;14)(q13;q32)	*BCL1*	*IGH*
	t(14q32)	IGH	
Adult T-cell leukemia			
	t(14;14)(q11;q32)		
	inv(14)(q11q32)		
	+3		
Retinoblastoma	del(13)(q14q14)	*RB1*	
Wilms' tumor	del(11)(p13p13)	*WT1*	
Ewing's sarcoma	t(11;22)(q24;q12)	*FLI1*	*EWS1*

*Same name used for different translocation breakpoints.
#No expressed gene identified at present.

Thus consistent and specific chromosomal deletions observed in tumors helped to delineate specific regions of the genome that harbor tumor suppressor genes; several different chromosome regions may be deleted in some malignant tumors. The use of highly polymorphic probes that have been mapped on different chromosomes have further defined regions showing extensive loss of heterozygosity, which indicates presence of a tumor suppressor gene (Table 7).

The tumor suppressor genes have been postulated to encode proteins that regulate normal growth and thus indirectly suppress neoplastic proliferation. In cells, these genes may act recessively, so that both maternal and paternal copies of the gene product must be inactivated in order to eliminate the tumor suppressor function. This model for tumor suppressor genes was originally proposed by Knudson (1971) and it has been confirmed by the study and molecular cloning of the retinoblastoma gene (*RB1*) on chromosome band 13q14. Patients with retinoblastoma have two patterns of inheritance, one as an autosomal dominant and one as a sporadic mutation. The age of the patient when the tumor was detected as well as other observations led Knudson to propose the "two-hit" hypothesis. According to this hypothesis, in patients who inherited one copy of a mutant allele, only one change in the other allele was needed for transformation of retinal cells to tumor cells, and in these individuals, multiple tumors were diagnosed at a very early age. In sporadic cases, two independent mutations had to affect both alleles in the same cell for transformation to occur; since this event was relatively uncommon, the tumors were unifocal and developed at an older age. Based on the consistent association of a deletion of 13q (later defined as 13q14) in rare patients with sporadic retinoblastoma, this band was thought to be the locus for the *RB1* gene (Dryja et al., 1984). Families without visible deletions were studied by using the esterase-D polymorphism, which had been assigned to chromosome 13 using somatic cell genetic techniques. Patients' tumors were found to have reduced or absent levels. Finally, Cavenee and colleagues were able to confirm that 13 was the critical chromosome by showing that tumor cells frequently had loss of heterozygosity (LOH) for polymorphic DNA markers mapped on chromosome 13, especially 13q14 (Cavenee et al., 1983; Murphree and Benedict, 1984). A gene from this region was molecularly cloned that had properties consistent with the *RB1* gene (Friend et al., 1986; Lee et al., 1987b). A 4.7 kb mRNA transcript of this gene was present in all normal tissues examined but was absent or altered in retinoblastoma cells, and deletions within the gene have been detected in many retinoblastoma tissues. The protein product of the *RB1* gene is a nuclear phosphoprotein of about 110 kd that has DNA binding activity (Lee et al., 1987a). This protein was shown to associate with large T antigen and E1A protein, the transforming proteins of DNA tumor viruses SV40 and adenovirus, respectively. These

TABLE 3. Functional Classification of Transforming Genes at Translocation Junctions

SRC family (TYR protein kinases)			
ABL	9q34	t(9;22)	CML/ALL
LCK	1p34	t(1;7)	T-ALL
Serum protein kinase			
BCR	22q11	t(9;22)	CML/ALL
Cell surface receptor			
TAN1	9q34	t(7;9)	T-ALL
Growth factor			
IL3	5q31	t(5;14)	PreB-ALL
Inner mitochondrial membrane protein			
BCL2	18q21	t(14;18)	NHL
Transcriptional regulating factors			
BCL3	19q13	t(14;19)	B-CLL
PBX	1q23	t(1;19)	PreB-ALL
E2A	19p13	t(1;19)	PreB-ALL
CAN	9q34	t(6;9)	AML
HOX11	10q24	t(10;14)/t(7;10)	T-ALL
LYL1	19p13	t(7;19)	T-ALL
MYC	8q24	t(8;14)	B-ALL/T-ALL
PML	15q22	t(15;17)	APL
RARA	17q12	t(15;17)	APL
TAL1(SCL)	1p34	t(1;14)	T-ALL
TAL2	9q32	t(7;9)	T-ALL
RBTN1(TTG1)	11p15	t(11;14)	T-ALL
RBTN2	11p13	t(11;14)	T-ALL

studies indirectly suggest that the RB protein has a role in regulating expression of other cellular genes, and may also mediate the oncogenic effects of some viral-transforming proteins. The protein product of the *RB1* gene has been implicated in the regulation of the cell cycle (De Caprio et al., 1989). Further discussion on the role of the *RB1* gene in the cell cycle and viral oncogenesis can be found in the chapter by Levine (Chapter 10, this volume).

An important tumor suppressor gene for CNS tumors (the *NF1* gene) was identified by cloning the translocation breakpoint (using markers that map close to the region) in the constitutional cells of patients from two families with neurofibromatosis (Fountain et al., 1989; Wallace and Collins, 1991). The hallmark of NF1 is the occurrence of multiple neurofibromas. In addition, there is an association with *café-au-lait* spots, skin-fold freckles, iris Lisch nodules, gliomas, including optic gliomas, etc. The locus for *NF1* had been previously mapped to 17q. This gene occupies over 300 kb on 17q and encodes a 13 kb mRNA that is translated into a 2,818 amino acid protein now referred to as neurofibromin (Xu et al., 1990). Sequence analysis shows that this protein has a region of homology with GTPase activating proteins (GAP), and functional activity has been demonstrated by the ability of the neurofibromin GAP domain to complement yeast IRA mutations (Martin et al., 1990; Xu et al., 1990). The GAP domain comprises only a small portion of the neurofibromin molecule; functions of other regions of the protein are unknown. A wide variety of mutations have been found in persons affected with neurofibromatosis, including chromosome translocations, deletions, an ALU insertion, and numerous point mutations (Cawthon et al., 1991; Wallace et al., 1990; Xu et al., 1991). The nature of lesions associated with NF1 are compatible with the hypothesis that this *NF1* gene is a tumor suppressor gene. *NF1* is discussed in greater detail in the chapter on tumor suppressor genes (Chapter 10, this volume).

Several tumor suppressor genes are implicated in the initiation and progression of gliomas. Reports on the cytogenetic abnormalities of these tumors cover all the histologic subtypes of gliomas, including astrocytomas, oligodendroglioma, and glioblastoma multiforme (Bigner et al., 1988). One of the largest series published in 1971 by Mark reported that 37 of 50 gliomas had near-diploid stem lines and that 26% contained double minutes (dmin). This study was done prior to the use of banding techniques. Jenkins et al. (1989) have also reported on 53 gliomas. No specific abnormalities have been detected but the most frequent findings have been dmin, structural abnormalities of chromosome 9 [del(9p) or translocation of 9p], trisomy 7, and loss of chromosomes 10, 18, and 22 (Jenkins et al., 1989). Cavenee has proposed a multistep process in gliomas in which several genetic mutations occur in the process of transformation of a normal astrocyte to a grade IV astrocyte. The earliest step involves inactivation of a gene on the short arm of chromosome 17 (*TP53*), an intermediate step involves loss of a gene on 9p (*?IFN*) and amplification of *EGFR* on 7p; loss of chromosome 10 material appears to be a late event (Mikkelsen et al., 1992).

In a report by Bigner et al. (1986), eight of 22 tumors contained marker chromosomes derived from chromosome 9; in three tumors, both chromosomes 9 participated in

TABLE 4. Alterations of Transcription Factors in Translocations

Deregulated expression		Fusion protein	
TCRA/D (14q11)	RBTN1 (11p15)*	E2A/PBX1*	t(1;19)
	RBTN2 (11p13)*	PML/RARA*	t(15;17)
	HOX11 (10q24)*	DEK/CAN*	t(6;9)
	TAL1 (1p34)*		
TCRB (7q34)	TAL2 (9q32)*		
	LYL1 (19p13)*		
	RBTN2 (11p13)*		
	HOX11 (10q24)*		
IGH/K/L (14q32) (2p12) (22q11)	MYC (8q24)		

*Gene identified by cloning breakpoint.

TABLE 5. Nonrandom Chromosomal Abnormalities in Solid Tumors

Disease	Chromosomal abnormalities
A. *Benign Tumors*	
Pleomorphic adenomas of the salivary glands	t(3;8)(p21;q12)
Meningioma and acoustic neuroma	Monosomy 22 or del(22q)
Lipoma chromosome	t(3;12)(q27–q28;q13–q14) ring
Ovarian tumors	Trisomy 12
Leiomyoma	t(12;14)(q14–15;q23–24)
B. *Adult Cancers*	
Breast	del(1p), del(3p)
Colon	del(17p), -18
Small-cell lung carcinoma	del(3)(p14–23)
Non-small-cell lung cancer	del(9p)
Renal carcinoma	del(3)(p11–p22)
Bladder cancer	i(5p), monosomy 9, del(19q)
Prostate	del(10)(q26), del(7)(q22)
Liposarcoma	t(12;16)(q13;p11)
Synovial sarcoma	t(X;18)(p11.2;q11.2)
Rhabdomyosarcoma (alveolar)	t(2;13)(q37;q14)
Malignant melanoma	del(1)(p11–22), del(6)(q11q27), i(6p)
Testicular tumors	i(12p)
Glioma	-10, del(9p)
C. *Involving Embryonic Cells*	
Neuroblastoma	del(1)(p32 to p36)
Ewing's sarcoma/peripheral neuroepithelioma	t(11;22)(q24;q12)
Wilms' tumor	del(11)(p13)*
Trisomy 1q	
Retinoblastoma	del(13)(q14)*
Trisomy 1q	

*Observed as a constitutional abnormality as well as in some tumors.

marker formation with different breakpoints, so that there were a total of 11 structural rearrangements of this chromosome. The most frequent breakpoints were at the centromere or in 9p. A region of chromosome 9 band p13–24 is also frequently deleted in ALL, melanomas, and lung cancer (Diaz et al., 1988; Olopade et al., 1992). The minimum region of overlap of these deletions involves band p22. We and others have observed homozygous loss of these genes in several glioma-derived cell lines and in primary patient samples (Miyakoshi et al., 1990; James et al., 1991; Olopade et al., 1990). We have used cell lines and tumors with submicroscopic deletions to identify the minimum region of overlap of the deletions. By comparing the presence of various genes in each deletion, we have determined that the order of these DNA probes on 9p is: telomere-*IFNB1*-*IFNA* gene cluster-*MTAP*-D9S3-centromere. Among the deletions that we have detected in the various cell lines, some have breakpoints within the *IFN* gene cluster and were used to obtain a deletion map of the *IFN* gene cluster (Olopade et al., 1992). We have used YAC clones, which contain large human DNA inserts that include the *IFNA* and *IFNB1* genes to complete a contiguous map of the cluster and the surrounding region. We have defined the minimum region of overlap of the deletions in the gliomas to the region between the centromeric end of the *IFN* gene cluster and the *MTAP* gene locus. We have prepared a phage library from 9p13–22 by microdissection. The unique probes obtained from the library in addition to the YAC contigs will be used to delimit the critical region to a size of a few hundred kilobases. Then, the search for normal transcripts in this critical region that are expressed in normal glial cells, but not in gliomas, may eventually lead to the identification of the tumor suppressor gene in this locus. This tumor suppressor gene may contribute to the malignant process in gliomas.

Although much of the discussion thus far implies that chromosome aberrations are equivalent to a malignant phenotype, there are a number of exceptions. Several benign tumors have clonal abnormalities of which the meningiomas described by Mark et al. (1972) and by Zankl and Zang have been most extensively studied. Recurent somatic alterations of chromosome 22 have been associated with many tumor types embryologically derived from the neural crest. Mark et al. (1972) first described a loss of one chromosome 22 in meningioma. Acoustic neuroma, which is another closely related benign tumor of the CNS, has also been shown to have complete or partial loss of the long arm of chromosome 22 (Seizinger et al., 1986). This observation was confirmed by molecular analysis of tumors for loss of constitutional heterzygozity of chromosome 22 alleles. In menigiomas, this approach has localized the critical region distal (telomeric) to the D22S1 locus. In addition, neurofibromatosis type 2 (*NF2*), a dominantly inherited disease that predisposes to the development of both meningiomas and acoustic neuromas, by genetic linkage analysis maps to the long arm of chromosome 22 (Rouleau et al., 1987; Seizinger et al., 1987). The

TABLE 6. Neoplastic Diseases Associated with Putative Tumor Suppressor Genes

Malignant diseases	Gene locus
Neuroblastoma	1p32–ter
Renal cell carcinoma	3p
Small-cell lung cancer	3p14–21
Familial adenomatous polyposis	
Colorectal carcinoma	5q21–22
ALL, NHL	9p22
Glioblastoma	9p13-ter
Wilms' tumor	11p13
Embryonal rhabdomyosarcoma	11p15-ter
Transitional cell bladder carcinoma	11p
Retinoblastoma	
Osteosarcoma	13q14.1
Ductal breast carcinoma	13q14.1
Meningioma	22q

recent discovery of a meningioma tumor specimen with a translocation on chromosome 22 has provided a potential shortcut to the isolation of the *NF2* gene (Lekanne Deprez et al., 1991). The translocation breakpoint maps to the region on chromosome 22 identified as the NF2 region, based on tumor deletion and linkage analysis. This translocation breakpoint has been cloned. Markers in the vicinity of the translocation breakpoint identify constitutional aberrations in patients with NF2 and multiple meningiomas. There is now a major effort to identify transcripts and the possible tumor suppressor gene in this locus. Refer to the chapter on meningiomas for more details (Chapter 22, this volume).

SEQUENCE AND SPECIFICITY OF GENETIC CHANGES

We do not know how consistent structural rearrangements occur, but there are at least two possibilities. The rearrangements may be random, but selection may act to eliminate the vast majority that do not provide the cell with a proliferative advantage. Alternatively, certain changes may occur preferentially and thus may be the ones we see. Some tantalizing data show an association of chromosome rearrangements in tumor cells from patients with fragile sites affecting one of the chromosome bands broken in the tumor cells (Sutherland and Hecht, 1985; Yunis, 1983). However, much more research is required to clarify the role of fragile sites as a predisposing factor to malignant transformation.

An equally important question is when in the multistage process of malignant transformation of a particular cell do translocations or other chromosome aberrations occur? Some chromosome changes occur as part of the further evolution of the malignant phenotype, e.g., blast crisis of CML, and they are, therefore, relatively late events. But what about the occurrence of the t(9;22) in CML, for example? Does the Ph occur in a single normal cell that becomes the progenitor of the leukemic clone, or is there expansion of a clone, possibly a leukemic one, in which a translocation occurs in one of these already abnormal cells? Fialkow and his colleagues (1987) have presented detailed evidence supporting the latter proposal. More recently, Adams et al. (1985) have constructed transgenic mice whose cells all have a vector containing the *myc*/*IgH* junction from a murine plasmacytoma. All cells contain this construct; however, the B-cell tumors that occurred in every animal were clonal, indicating that one or more additional changes occurred in one cell, resulting in clonality. Vogelstein and co-workers have elegantly shown that several genetic changes occur during the malignant process and that it is the progressive accumulation of these changes in the malignant cells that is most important (Goyette et al., 1992; Vogelstein et al., 1988). The evidence presented in this chapter clearly demonstrates the remarkable specificity of certain chromosome rearrangements for particular subtypes of tumors. The mechanism or mechanisms by which this specificity is achieved are unknown; however, a number of investigators have shown that certain proteins required for promotion of gene expression are synthesized in a very cell-type-specific manner (Nomiyama et al., 1987). These proteins are only present in the appropriate cell type and therefore the particular gene is activated only in that cell type. The chromosome rearrangements affecting *MYC* in B-cell and T-cell tumors strongly support the interpretation that the specificity resides in the gene that is uniquely active in a particular cell type. Thus the immunoglobulin genes are highly regulated in B cells and they can therefore serve as the switch or activator mechanism for *MYC* in B cells; on the other hand, *TCRA* is an active gene in T cells with a strong enhancer/promotor and it clearly is an activator for *MYC* in T cells. A reasonable paradigm is that translocations bring together in an inappropriate manner a growth factor or growth factor receptor gene (the proto-oncogene in the examples defined to date) adjacent to an active cell-specific gene (Adams et al., 1985).

TABLE 7. Chromosomal Deletions and Loss of Heterozygosity in CNS Tumors

Tumor	Chromosomal deletion in tumor	Loss of heterozygosity
Neuroblastoma	1p32-p34 14q	1p36.1-p36.2
Glioma	9p24-p13 10 17p	*IFNA*/*IFNB1* 10p; 10q 17p11-pter (*TP53*) 22q 8p21–23 6p15–q27
Retinoblastoma	13q14	13q (*RB1*)
Peripheral neuroepithelioma (PNE)		17p13 (not *TP53*)
Acoustic neuroma		22q11.2-q12 (*NF2*)
Neurofibrosarcoma	None	17p12-p13.1; 17q (*NF1*)
Meningioma	22q12–13 22	22q11.2–12 9 (*NF2*)

As has been described in this chapter, the analysis of various tumors for alterations in proto-oncogenes or tumor suppressor genes has revealed that a number are abnormal as a result of translocations, amplification, deletions, or mutations. In some situations the relationship of the change in the proto-oncogene to the multistage process of malignant transformation is unclear. Such ambiguity is not a problem with chromosome translocations; the evidence is overwhelming that the t(8;14) in Burkitt's lymphoma and the t(9;22) in CML are an integral component of the cascade of events leading to the transformation of a normal to a malignant cell. In the last decade, the definition of genetic lesions involved in the oncogenic process has advanced through the identification of multiple dominant oncogenes and tumor suppressor genes. Some of these genes, such as *MYC*, *RB1*, and

TP53, seem to participate in oncogenesis in a large variety of tumor types, while others such as *BCL2* or *GLI* seem to participate in oncogenesis in more restricted types of tumor. The challenge is to continue to advance the chromosome maps, to isolate more translocation breakpoint junctions and identify the genes involved, to identify the genes that are associated with other chromosomal aberrations, and then to determine the change in gene function that occurs as a consequence of the abnormalities. The ultimate measure of success, however, will be in the application of these new insights in the development of new, more effective treatments for cancer. In the future, each particular subtype of tumor will be treated in a uniquely defined way that is most appropriate for the specific genetic defect present in that tumor. This should lead to a new era of cancer therapy that is both more effective and less toxic.

REFERENCES

Adams, J.M., A.W. Harris, C.A. Pinkert, et al. (1985) The c-myc oncogene driven by immunoglobulin enhancers induces lymphoid malignancy in transgenic mice. Nature 318:533.

Ambros, I.M., P.F. Ambros, S. Strehl, et al. (1991) MIC2 is a specific marker for Ewing's sarcoma and peripheral primative neuroectodermal tumors. Evidence for a common histogenesis of Ewing's sarcoma and peripheral primitive neuroectodermal tumors from MIC2 expression and specific chromosome aberration. Cancer 67:1886–1893.

Aurias, A., C. Rimbaut, D. Buffe, et al. (1983) Chromosomal translocations in Ewing's sarcoma. N. Engl. J. Med. 309:496–497.

Bigner, S.H., P.A. Humphrey, A.J. Wong, et al. (1990) Characterization of the epidermal growth factor receptor human glioma cell lines and xerografts. Cancer Res. 50:8017–8022.

Bigner, S.H., J. Mark, D.E. Bullard, et al. (1986) Chromosomal evolution in malignant glioma starts with specific and usually numerical deviations. Cancer Genet. Cytogenet. 22:121–135.

Bigner, S.H., J. Mark, P.C. Burger, et al. (1988) Specific chromosomal abnormalities in malignant human gliomas. Cancer Res. 48:405–411.

Bishop, J.M. (1988) The molecular genetics of cancer. Leukemia 2:199–208.

Boehnke, M., K. Lange, and D.R. Cox (1991) Statistical methods for multipoint radiation hybrid mapping. Am. J. Hum. Genet. 49:1174–1188.

Boehnke, M. (1992a) Genetic analysis workshop 7: Radiation hybrid and somatic cell hybrid mapping of chromosome 21. Cytogenet. Cell Genet. 52:74–76.

Boehnke, M. (1992b) Radiation hybrid mapping by minimization of the number of obligate chromosome breaks. Cytogenet. Cell Genet. 59:96–98.

Bohlander, S.K., R. Espinosa, M.M. LeBeau, et al. (1992) A method for the rapid sequence-independent amplification of microdissected. Genomics 13:1322-1324.

Brodeur, G.M., A.A. Green, F.A. Hayes, et al. (1981) Cytogenetic features of human neuroblastomas and cell lines. Cancer Res. 41:4678.

Brodeur, G.M., R.L. Seeger, M. Schwab, et al. (1984) Amplification of N-myc in untreated neuroblastoma correlates with advanced disease stage. Science 224:1121.

Buetow, K.H., and A. Chakravati (1987) Multipoint gene mapping using seriation. I. General methods. Am. J. Hum. Genet. 41:180–188.

Buetow, K.H., R. Shiang, P. Yang, et al. (1991) A detailed multipoint map of human chromosome 4 provides evidence for linkage heterogeneity and position-specific recombination rates. Am. J. Hum. Genet. 48:911–925.

Burmeister, M., S. Kim, E.R. Price, et al. (1991) A map of the distal region of the long arm of human chromosome 21 constructed by radiation hybrid mapping and pulsed field gel electrophoresis. Genomics 9:19–30.

Burmeister, M., A.P. Monaco, E.F. Gillard, et al. (1988) A 10-mega base physical map of human Xp21, including the Ducherne muscular dystrophy gene. Genomics 2:198–202.

Caccia, N., G.A. Bruns, I.R. Kirsch, et al. (1985) T cell receptor-chain genes are located on chromosome 14 at 14q11-14q12 in humans. J. Exp. Med. 161:1255–1260.

Carle, G.F., M. Frank, and M.V. Olsen (1986) Electrophoretic separations of large DNA molecules by inversion of the electric field. Science 232:65.

Cavenee, W.K., T.P. Dryja, R.A. Phillips, et al. (1983) Expression of recessive alleles by chromosomal mechanisms in retinoblastoma. Nature 305:779–784.

Cawthon, R.M., L.B. Anderson, A.M. Buchberg, et al. (1991) cDNA sequence and genomic structure of EV12B, a gene lying within an intron of the neurofibromatosis type 1 gene. Genomics 9:446–460.

Chan, L.C., K.K. Karhi, S.I. Rayter, et al. (1987) A novel *abl* protein expressed in Philadelphia chromosome positive acute lymphoblastic leukaemia. Nature 325:635–637.

Cox, D.R. (1992) Radiation hybrid mapping. Cytogenet. Cell Genet. 59:80–81.

Cox, D.R., M. Burmeister, E.R. Price, et al. (1990) Radiation hybrid mapping: A somatic cell genetic method for constricting high resolution maps of mammalian chromosomes. Science 250:245–250.

Cox, D.R., C.A. Pritchard, E. Ughim, et al. (1989) Segregation of the Huntington disease region of chromosome 4 is a somatic cell hybrid. Genomics 4:397–407.

Daley, G.Q., R.A. Van Etter, and D. Baltimore (1990) Induction of chronic myelogenous leukemia in mice by p210 bcr/abl gene of the philadelphia chromosome. Science 247:824.

Dausset, J. (1986) Le Centre el'Etude du polymorphism Humain. La Press Medicale 15:1801–1802.

de Klein, A., A.G. van Kessel, G. Grosveld, et al. (1982) A cellular oncogene is translocated to the Philadelphia chromosome in chronic myelocytic leukemia. Nature 300:765–767.

De Caprio, J.A., J.W. Ludlow, D. Lynch, et al. (1989) The product of the retinolastoma gene has properties of a cell cycle regulatory element. Cell 58:1085.

Delattre, O., J. Zucmar, B. Plousgastel, et al. (1992) Cloning of the recurrent t(11;22) translocation of Ewing's sarcoma and peripheral neuroepithelioma. AACR Proc. 33:604.

Diaz, M.O., S. Zemin, M.M. LeBeau, et al. (1988) Homozygous deletions of the α- and β-interferon genes in human leukemia and derived cell lines. Proc. Natl. Acad. Sci. USA 85:5259–5263.

Donis-Keller, H., P. Green, C. Helms, et al. (1987) A genetic linkage map of the human genome. Cell. 51:319–337.

Dryja, T.P., W.K. Cavenee, R. White, et al. (1984) Homozygosity of chromosome 13 in retinoblastoma. N. Engl. J. Med. 310:550–553.

Dunlop, M.G., Wyllie, C.M. Steel, et al. (1991) Linked DNA markers for presymptomatic diagnosis of familial adenomatous polyposis. Lancet 337:314–316.

Falk, C.T. (1992) Multilocus mapping strategies on chromosome 21 data sets: Comparison of results from family data, radiation hybrids and somatic cell hybrids. Cytogenet. Cell Genet. 59:119–121.

Fialkow, P.J., J.W. Singer, W.H. Raskind, et al. (1987) Clonal development, differentiation and clinical remissions in acute nonlymphocytic leukemia. N. Engl. J. Med. 317:468–473.

Fong, C., P.S. White, K. Peterson, et al. (1992) Loss of heterozygosity for chromosomes 1 or 14 defines subsets of advanced neuroblastomas. Cancer Res. 52:1780–1785.

Fountain, J.W., M.R. Wallace, M.A. Bruce, et al. (1989) Physical mapping of a translocation breakpoint in neurofibromatosis. Science 244:1085–1087.

Frezal, J., and A. Schinzel (1991) Human Gene Mapping II. Report of the Committee on Clinical disorders. Chromosome aberrations and uniparent disomy. Cytogenet. Cell Genet. 58:986–1052.

Friend, S.H., R. Bernards, S. Rogelj, et al. (1986) A human DNA segment with properties of the gene that predisposes to retinoblastoma. Nature 323:643–646.

Fulton, T.R., A.M. Bowcollc, D.R. Smith, et al. (1989) 12 megabase restriction map at the cysticfibrosis loans. Nucl. Acids Res. 17:271–284.

Gardiner, K., and D. Patterson (1992) The use of somatic cell hybrids in physical mapping. Cytogenet. Cell Genet. 59:82–85.

Genetic Analysis Workshop 7 (1992) Issues in gene mapping and detection of major genes. MacAuer, J.W., A. Chakravarth, D. Cox, et al. (eds.). Cytogenet. Cell Genet. 59:65–240.

Gishizky, M.L., J. McLaughlin, A.M. Pendergast, et al. (1991) The 5' non-coding region of the BCR/ABL oncogene augments its ability to stimulate the growth of immature lymphoid cells. Oncogene 6:1299–1306.

Goyette, M.C., Chok, C.L. Fasching, et al. (1992) Progression of colorectal cancer is associated with multiple tumor suppressor gene defects but inhibition of tumorigenicity is accomplished by correction of any single defect via chromosome transfer. Mol. Cell. Biol. 12:1387–1395.

Groffen, J., J.R. Stephenson, H. Heisterkamp, et al. (1984) Philadelphia chromosomal breakpoints are clustered within a limited region, bcr, on chromosome 22. Cell 36:93–99.

Groffen, J., J.N. Vorcken, H. van Schiack, et al. (1992) Animal models for chronic myeloid leukemia and acute lymphoblastic leukemia. Leukemia (Suppl. 6) 1:44–46.

Gross, S.J., and H. Harris (1977) Gene transfer by means of cell fusion. II. The mapping of 8 loci on human chromosome 1 by statistical analysis of gene assortment in somatic cell hybrids. J. Cell Sci. 25:39–57.

Heim, S., and F. Mitelman (1987) Cancer Cytogenetics. New York: Alan R. Liss.

Heisterkamp, W., J.R. Stephenson, J. Groffen, et al. (1984) Localization of the C-abl oncogene adjacent to a translocation in chronic myelogenic leukemia. Nature 305:239–242.

Human Gene Mapping 11 (1991) Cytogenet. Cell Genet. 58:1–2200.

ISCN (1985) An international system for human cytogenetic nomenclature—high resolution banding. Cytogenet. Cell Genet. 31:1–23.

ISCN (1991) Guidelines for cancer cytogenetics supplement to *An International System for Human Cytogenetic Nomenclature*. F. Mitelman (ed.). Basel: S. Karger.

James, C.D., J. He, E. Carlbom, et al. (1991) Chromosome 9 deletion mapping reveals interferon A and interferon B1 gene deletions in human glial tumors. Cancer Res. 51:1684–1688.

Jenkins, R.B., D.W. Kimmel, C.A. Moertel, et al. (1989) A cytogenetic study of 53 human gliomas. Cancer Genet. Cytogenet. 39:253–279.

Kaiser-McCaw, B., F. Hecht, D.G. Harnden, et al. (1975) Somatic rearrangement of chromosome 14 in human lymphocytes. Proc. Natl. Acad. Sci. USA 72:2071.

Keats, B., S. Sherman, N. Morton, et al. (1991) Guidelines for human linkage maps (ISLM, 1990). Genomics 9:557–5.

Kelliher, M., A. Knott, J. McLaughlin, et al. (1991) Differences in oncogenic potency but not target cell specificity distinguish the two forms of the BCR/ABL oncogene. Mol. Cell. Biol. 11:4710–4716.

Knudson, A.G. (1971) Mutation and cancer: Statistical study of retinoblastoma. Proc. Natl. Acad. Sci. USA 68:820–823.

Lawrence, J.B., C.A. Villnave, and R.H. Singer (1988) Sensitive, high resolution chromatin and chromosome mapping in situ: Presence and orientation of closely integrated copies of EBV in a lymphoma line. Cell 42:51–61.

Lawrence, S., N.E. Morton, and D.R. Cox (1991) Radiation hybrid mapping. Proc. Natl. Acad. Sci. USA 88:7477–7480.

Leder, P., J. Battey, G. Lenoir, et al. (1983) Translocations among antibody genes in human cancer. Science 222:765–771.

Lee, W.H., J.Y. Slew, F.D. Hong, et al. (1987a) The retinoblastoma susceptibility gene encodes a nuclear phosphoprotein associated with DNA binding activity. Nature 329:642–645.

Lee, W.H., R. Bookstein, F. Hong, et al. (1987b) Human retinoblastoma susceptibility gene: Cloning, identification, and sequence. Science 235:1394–1399.

Lekanne Deprez, R.H., N.A. Groen, N.A. van Biezen, et al. (1991) A t(4;22) in a meningioma points to the localization of a putative tumor-suppressor gene. Am. J. Hum. Genet. 48:783–790.

Lemieux, N., D.B., and E. Vegas-Pequigout (1992) A simple method for simultaneous R- or G-banding and fluorescence in situ hybridization of single-copy genes. Cytogenet. Cell Genet. 59:311–312.

Lichter, P.T., T. Cremer, J. Borden, et al. (1988) Delineation of individual human chromosomes in metaphase and interphase cells by in situ suppression hybridization using recombinant DNA libraries. Hum. Genet. 80:224–234.

Ludecke, H.-J., G. Senger, U. Clausen, et al. (1989) Cloning defined regions of the human genome by microdissection of banded chromosomes and eurymatic amplifications. Nature 338:348–350.

Manolov, G., and Y. Manolova (1972) Marker band in one chromosome 14 from Burkitt lymphomas. Nature 237:33–34.

Mark, J. (1971) Chromosomal characteristic of neurogenic tumors in adults. Hereditas 68:61–99.

Mark, J., G. Levan, and F. Mitelman (1972) Identification by fluorescence of the G chromosome lost in human meningiomas. Hereditas 71:163–168.

Martin, G.A., Ulschit

Maru, Y., and O.N. Witte (1991) The BCR gene encodes a novel serine/threonine kinase activity within a single exon. Cell 67:459–468.

Melki, J., S. Abdelhak, P. Burlet, et al. (1992) Prenatal prediction of Werdnig-Hoffman disease using linked polymorphic DNA probes. J. Med. Genet. 29:171–174.

Meltzer, P.S., X.Y. Guan, A. Burgess, et al. (1992) Generation of region specific probes by chromosome microdissection: A novel approach to identify

Mikkelsen, T., J.G. Cairncross, and W.K. Cavanee (1991) Genetics of the malignant progression of astrocytoma. J. Cell Biochem. 46:3–8.

Mitelman, F. (1983) Catalog of chromosome aberrations in cancer. Cytogenet. Cell Genet. 36:1–515.

Mitelman, F. (1985) Catalog of chromosome aberrations in cancer. In: A.A. Sandberg (ed.). Progress and Topics in Cytogenetics, vol. 5. New York: Alan R. Liss.

Mitelman, F. (1988) Catalog of Chromosome Aberrations in Cancer. New York: Alan R. Liss.

Mitelman, F. (1991) Catalog of Chromosome Aberrations in Cancer, 4th ed. New York: Wiley-Liss.

Miyakoshi, J., K.D. Dobler, J. Allalunis-Turmer, et al. (1990) Absence of IFN1 and IFNB1 genes from human malignant glioma cell lines and lack of correlation with cellular sensitivity to interferons. Cancer Res. 50:278–283.

Murphree, A.L., and W.F. Benedict (1984) Retinoblastoma: Clues to human oncogenesis. Science 219:1028–1033.

Nomiyama, H., C. Fromental, and J.H. Xiao (1987) Cell-specific activity of the constituent elements of the simian virus 40 enhancer. Proc. Natl. Acad. Sci. USA 84:7881–7885.

Nowell, P.C., and D.A. Hungerford (1960) A minute chromosome in human granulocytic leukemia. Science 132:1497.

Olopade, O.I., R.B. Jenkins, D.T. Ransom, et al. (1992) Molecular analysis of deletions of the short arm of chromosome 9 in human gliomas. Cancer Res. 52:2523–2529.

Rowley, J.D. (1973) A new consistent chromosomal abnormality in chronic myelogenous leukemia. Nature 243:290–293.

Rowley, J.D. (1990a) Molecular cytogenetics: Rosetta stone for understanding cancer. Twenty-ninth GHA Clowes Memorial Award Lecture. Cancer Res. 50:3816.

Rowley, J.D. (1990b) Chromosome abnormalities in leukemia and lymphoma. Sem. Hematol. 27:122–136.

Rowley, J.D., M.O. Diaz, R. Espinosa III, et al. (1990) Mapping chromosome band 11q23 in human acute leukemia with biotingrated probes: Identification of 11q23 translocation breakpoints with a yeast artificial chromosome. Proc. Natl. Acad. Sci. USA 87:9358–9362.

Rubin, C.M., J.J. Carrino, M.N. Dickler, et al. (1988) Heterogeneity of genomic fusion of *BCR* and *ABL* in Philadelphia chromosome-positive acute lymphoblastic leukemia. Proc. Natl. Acad. Sci. USA 85:2795–2799.

Sandberg, A.A. (1990) The Chromosomes in Human Cancer and Leukemia, 2nd ed. New York: Elsevier.

Schwab, M., K. Alitalo, K.-H. Klempnauer, et al. (1983) Amplified DNA with limited homology to myc cellular oncogene is shared by human neuroblastoma cell lines and a neuroblastoma tumour. Nature 305:245–248.

Shtivelman, E., B. Lifshitz, P. Robert, et al. (1985) Fused transcript of abl and bcr genes in chronic myelogenous leukaemia. Nature 315:550–554.

Smith, C.L., and C.R. Cantor (1986) Pulsed field gel electrophoresis of large DNA molecules. Nature 319:701.

Solomon, E., J. Borrow, and A.D. Goddard (1991) Chromosome aberrations and cancer. Science 254:1153–1160.

Sutherland, G.R., and F. Hecht (1985) Fragile Sites on Human Chromosomes. New York: Oxford University Press.

Tazi, R.E., J.L. Haines, P.C. Watkins, et al. (1988) Genetic linkage map of human chromosome 21. Genomics 3:129–136.

Traski, B.J., D. Pinkel, and G. Van den Engh (1989) The proximity of DNA sequences in interphase cell nuclei is correlated to genomic distance and permits ordering of cosmids spanning 250 kilobase pairs. Genomics 5:710–717.

Van Keuren, M.L., P.C. Watkins, H.A. Drabkin, et al. (1986) Regional localization of DNA sequences on chromosome 21 using somatic cell hybrids. Am. J. Hum. Genet. 38:793–804.

Vogelstein, B., E.R. Fearon, S.R. Hamilton, et al. (1988) Genetic alterations during colorectal-tumor development. N. Engl. J. Med. 319:525.

Wallace, M.R., and F.S. Collins (1991) Molecular genetics of von Rocklinghauser neurofibromatosis. Adv. Hum. Genet. 20:267–307.

Wallace, M.K., D.A. Marchik, L.S. Andersen, et al. (1990) Type 1 neurofibromatosis gene: Identification of a large transcript disrupted in three NFI patients. Science 249:181–186.

Warren, A.C., Antonarakis, and Chakravati (1992) Chromosome 21 genetic linkage data set based on CEPH pedigrees. Cytogenet. Cell Genet. 59:86–87.

Whang-Peng, J., T.J. Triche, T. Knutsen, et al. (1984) Chromosome translocation in peripheral neuroepithelioma. N. Engl. J. Med. 311:584–585.

Wong, A.J., J.M. Ruppert, S.H. Bigner, et al. (1992) Structural alterations of the epidermal growth factor receptor gene in human gliomas. Proc. Natl. Acad. Sci. USA 89:2965–2969.

Xu, G., L. Nelson, P. O'Connell, et al. (1991) An alu polymorphism intragenic to the neurofibromatosis type 1 gene (NF1). Nucl. Acids Res.

Xu, G.F., P. O'Connell, D. Viskochil, et al. (1990) The neurofibromatosis type 1 gene encodes a protein related to GAP. Cell 62:599–608.

Yunis, J.J. (1983) The chromosomal basis of human neoplasia. Science 221:227.

Yunis, J.J., and A.L. Soreng (1984) Constitutive fragile sites and cancer. Science 226:1199.

Zech, L., U. Haglund, K. Nilsson, et al. (1976) Characteristic chromosomal abnormalities in biopsies and lymphoid-cell lines from patients with Burkitt and non-Burkitt lymphomas. Int. J. Cancer 17:47.

Zhang, F.R., A. Awias, O. Delattre, et al. (1990) Mapping of human chromosome 22 by in situ hybridization. Genomics 7:318–324.

Chapter 15

Strategies for Isolating Genes in Hereditary and Sporadic Tumors

Lois M. Mulligan and Sara E. Mole

INTRODUCTION

The development of a malignant neoplastic clone from a normal or predisposed precursor cell is attributable to the accumulation of multiple genetic events that contribute to the loss of normal cellular controls. These events may reflect the inappropriate activation (oncogenes) or inactivation (tumor suppressor genes) of genes involved in normal cell functions. As a general rule, oncogenes are dominantly acting genes that contribute to tumorigenesis by their expression, while tumor suppressor genes are recessive, with inactivation of both normal copies of the gene required for the development of neoplasia. The identification and characterization of genes that contribute to the tumorigenic phenotype has increased our understanding of neoplasia as an aberration of the normal cellular processes of development and differentiation, which, in turn, has significant implications for diagnosis and treatment of human cancer.

The genetic events contributing to tumorigenesis encompass two distinct although related processes: initiation, the conversion of a normal cell to a predisposed or neoplastic entity, and progression, the conferring of a selective advantage to an already neoplastic clone. While the former events reflect the direct relationship between alteration of a specific gene and the resultant tumor type, progressional events, which accumulate during neoplastic growth, are frequently common to tumors of diverse types.

Initiation events may occur somatically, resulting in sporadic tumors, but may also be inherited in the form of germline mutations of a "cancer" gene. The vast majority of human neoplasia occur sporadically, outside the context of a familial disease. The population at risk through hereditary or familial cancers is relatively small, representing only about 5% of cancer incidence. Individuals who have inherited a germline mutation are said to have a hereditary predisposition to cancer and are at high risk of developing that disease in their lifetime (Fig. 1). In general, the genes mutated in hereditary cancer predisposition have tumor-suppressing properties. Somatically acquired mutations that contribute to neoplasia may affect either tumor suppressor or oncogenes. Many human cancers occur in both sporadic and hereditary forms, reflecting either somatic or germline mutations of the same predisposing gene (e.g., retinoblastoma, RB1). Consequently the disparate types of information that may be gleaned from sporadic and heritable disease in these cases may be pooled to facilitate our identification and understanding of human cancer genes and the mutations to which they are subject.

How Do We Identify a Cancer Gene?

Genes that contribute to human cancer exhibit a broad range of expression patterns and functions (Table 1). Consequently, there are few clues provided by the previously identified genes as to the characteristics of a new cancer gene. Since the human genome comprises an estimated

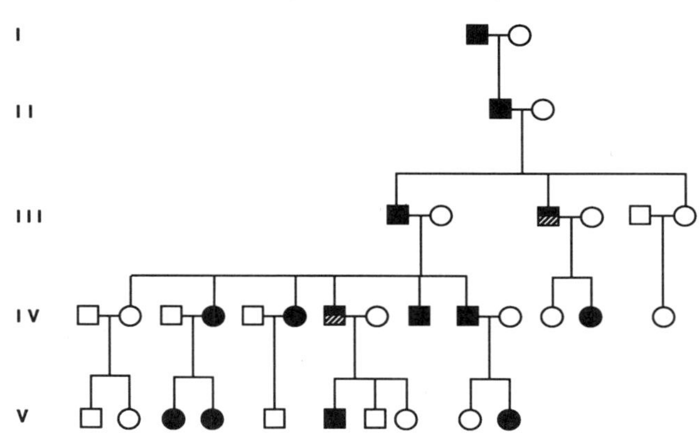

Fig. 1. Pedigree of a neurofibromatosis type 1 (NF1) family. ○- unaffected female; □-unaffected male; ●■-affected female and male; ▨ - male with NF1 and pheochromocytoma.

TABLE 15.1. Cancer Gene Characteristics and Their Contribution to Human Neoplasia

Gene product	Gene	Disease	Tumor-associated mutations	Localization
A. Oncogenes				
Tyrosine kinase	ABL	Chronic myelogenous leukemia	Rearrangement	9q34
	BCL1	Chronic B-cell lymphocytic leukemia	Rearrangement	11q13
	BCL2	Follicular lymphoma	Rearrangement/ overexpression	18q21
	RET	Papillary thyroid carcinoma	Rearrangement	10q11.2
Nuclear protein	MYC	Burkitt's lymphoma	Rearrangement/ overexpression	8q24
	MYCN	Neuroblastoma	Amplification/ overexpression	2p24.1
	MYCL	Small-cell lung cancer	Amplification/ overexpression	1p32
GTP binding protein	HRAS	Colon carcinoma	Point mutation	11p15.5
	KRAS	Pancreas carcinoma	Point mutation	12p12
Receptor	NEU	Breast carcinoma	Amplification/ overexpression	17q11.2-q12
	PML-RARA	Acute promyelocytic leukemia	Rearrangement	15q22
B. Tumor Suppressor Genes				
DNA binding protein	RB1	Retinoblastoma Osteosarcoma	Deletion, point mutation	13q14
	P53	Li-Fraumeni-associated tumors Colon carcinoma Breast carcinoma	Point mutation	17p13
		Lung carcinoma Soft tissue sarcoma	Deletion, point mutation	
Zinc finger protein	WT1	Wilms' tumor	Deletion, rearrange-ment, point mutation	11p13
GTPase activat-ing protein	NF1	Neurofibromatosis-assoc-iated tumors (neuro-fibrosarcoma glioma pheochromocytoma)	Deletion, point mutation	17q11.2
Cell adhesion molecule	DCC	Colon carcinoma	Deletion, insertion, point mutation	18q21.3
Other	MCC	Colon carcinoma	Rearrangement, point mutation	5q21
	APC	Colon carcinoma	Deletion, point mutation	5q21

30,000–100,000 genes, a prohibitive number of candidate genes exist that might play a role in tumorigenesis. Therefore, it is necessary to narrow the field of inquiry using a combination of techniques to identify first the general region in which a cancer gene may be located and then to map it more precisely within this general region. This approach is often referred to as *reverse genetics* or, more precisely, *positional cloning*.

STRATEGIES FOR MAPPING HUMAN CANCER GENES

The process of identifying genes with a role in human neoplasia encompasses four complementary phases.

1. Establishing a minimum regional assignment for a gene.
2. Physical analysis and mapping of this region of interest.
3. Identifying candidate genes.
4. Confirming the identity of the cancer gene by characterization of disease-specific mutations.

Some of the strategies used to address each of these phases (Fig. 2), together with examples of their contribution to cancer gene identification and characterization, are described below.

Regional Assignment of a Cancer Gene

The search for an unknown gene that plays a role in human cancer requires the defining and refining of a region that has the potential of containing such a gene. A combination of "low-resolution" gene mapping techniques such as cytogenetic and genetic linkage analyses, loss of heterozygosity, or amplification studies may be used to analyze material from sporadic tumors and patients with heritable cancers to identify the DNA boundaries of an interval encompassing a cancer gene. Once a limited region containing the gene is identified, more precise "high-resolution" physical mapping techniques may be employed to refine the interval further.

Cytogenetic analyses

A gross chromosomal alteration consistently associated with a specific neoplasm may indicate the location of the disease gene. In the case of familial cancers, these chromosomal aberrations may be inherited, appearing in all cells of the predisposed individual. Although the majority of inherited cancers are not associated with germline cytogenetic abnormalities, tumor-specific somatically acquired alterations occur consistently in many sporadic and heritable tumors. These changes may be *numerical*—representing a loss or duplication of a whole chromosome—or *structural*,

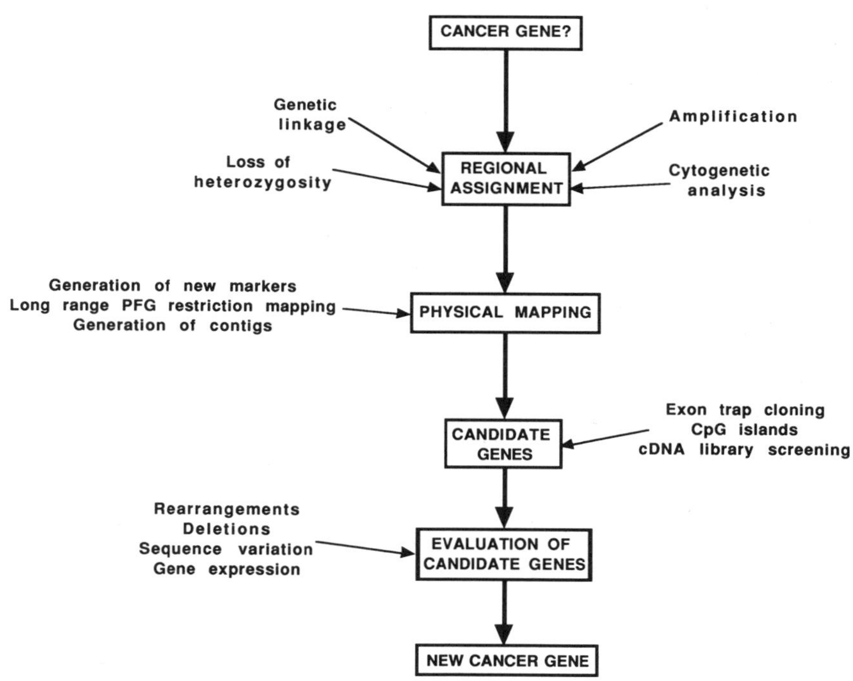

Fig. 2. Strategies for isolating genes in hereditary and sporadic tumors. Summary of the approaches taken and the techniques used in genetic and physical mapping to isolate cancer genes.

TABLE 2. Molecularly Characterized Rearrangements in Human Neoplasia

Disease	Rearrangement	Affected gene
Chronic myelogenous leukemia	t(9;22)(q34;q11)	ABL BCR
Chronic B-cell lymphocytic leukemia	t(11;14)(q13;q32)	BCL1
Follicular lymphoma	t(14;18)(q32;q21)	BCL2
Burkitt's lymphoma	t(8;14)(q24;q32) t(2;8)(p12;q24) t(8;22)(q24;q11)	MYC
Acute promyelocytic leukemia	t(15;17)(q22;q11.2-q12)	PML RAR
Acute myeloid leukemia	t(6;9)(p23;q34)	DEK CAN
T-cell acute lymphocytic leukemia	t(7;19)(q35;p13) t(1;14)(p32;q11) t(10;14)(q24;q11)	LYL1 TCL5 HOX-11
Non-Hodgkin's lymphoma	ins(2;2)(p13;p11.2–14)	REL NRG
Papillary thyroid carcinoma	inv(10)(q11.2q21)	RET

including translocations, deletions, and insertions of chromosomal material. Numerical alterations, such as monosomy for chromosome 22 in meningioma and chromosome 10 in advanced stage glioblastoma, suggest a possible chromosomal localization for a cancer gene or genes. In molecular terms, however, this leaves a prohibitively large region requiring analysis in order to identify that gene.

The relatively precise regional assignment provided by structural alterations has been more useful in the identification of genes implicated in tumorigenesis. Rearrangements that juxtapose normally unrelated sequences and the isolation of sequences flanking these breakpoints has led to identification of the cancer genes involved in chronic myelogenous leukemia (BCR-ABL), Burkitt's lymphoma (MYC), and numerous hematopoetic cancers (reviewed in Rabbitts, 1991; Sawyers et al., 1991). Such rearrangements can lead to the activation of an oncogene by fusing elements of two genes with very different functions or controls (Table 2). Examples of this are the deregulation of MYC (8q24) seen in Burkitt's lymphoma caused by enhancers from the immunoglobulin gene (2p12; 14q32.33; 22q11) to which it is fused and the translocation of chromosomes 9 and 22 in chronic myelogenous leukemia, which juxtaposes the GTPase domain of the BCR gene (22q11) and the tyrosine kinase domain of ABL (9q34).

TABLE 3. Cytogenetically Detected Deletions or Rearrangements of Tumor Suppressor Loci

Tumor	Chromosomal region	Tumor suppressor gene
Retinoblastoma	13q14	RB1
Neurofibromatosis 1	17q11.2	NF1
Wilms' tumor	11p13	WT1
Colon carcinoma	17p	TP53
	18q	DCC
	5q21–31	MCC
		APC

Cytogenetically detectable deletions and rearrangements have also indicated the location of putative tumor suppressor genes, most frequently in solid tumors (Table 3) (reviewed in Solomon et al., 1991). Often these events involve relatively large chromosomal regions in which only a small portion encodes the tumor suppressor gene, making further regional analyses obligatory for identification of that gene. The initial localization of the RB1 gene was based on cytogenetically detectable deletions of chromosome 13q14 found in the normal tissues of 3–5% of children with the heritable form of the disease. Deletions of chromosome 17p and 18q provided the initial clues that loci in these regions contributed to tumor progression in colon carcinoma (reviewed in Fearon and Vogelstein, 1990). Deletions of chromosome 1p32-pter in neuroblastoma and rearrangement of 3p26 in von Hippel-Lindau syndrome-associated renal cell carcinoma suggest mutations at loci in these regions may contribute to tumorigenesis. As yet, no appropriate candidate genes in these latter regions have been reported.

In summary, while in the case of oncogene activation cytogenetically detectable alterations have led directly to identification of genes implicated in tumorigenesis, the events recognized cytogenetically in tumor suppressor gene inactivation have generally provided a less specific gene assignment (reviewed in Solomon et al., 1991). Nevertheless, in each case a consistent chromosomal alteration that is inferred to alter or abrogate the normal function of a putative cancer gene provides the regional localization for that gene.

Gene amplification

The duplication of DNA sequences resulting in multiple copies of a gene that may contribute to the neoplastic phenotype is called *gene amplification*. The duplicated unit (*amplicon*) may be cytogenetically recognizable as a homogeneously staining region (HSR) within a chromosome or as small extrachromosomal units called double minute chromosomes (DM). Frequently the amplified unit is not found at the genomic position of the normal sequence. Therefore it is necessary to characterize the amplicon, identify the duplicated sequences, and extrapolate to the position, structure, and function of the corresponding sequences in the normal genome.

Examples of gene amplification are infrequent in primary tumors, affecting a limited number of loci and cancers. However, in advanced-stage neuroblastomas with poor prog-

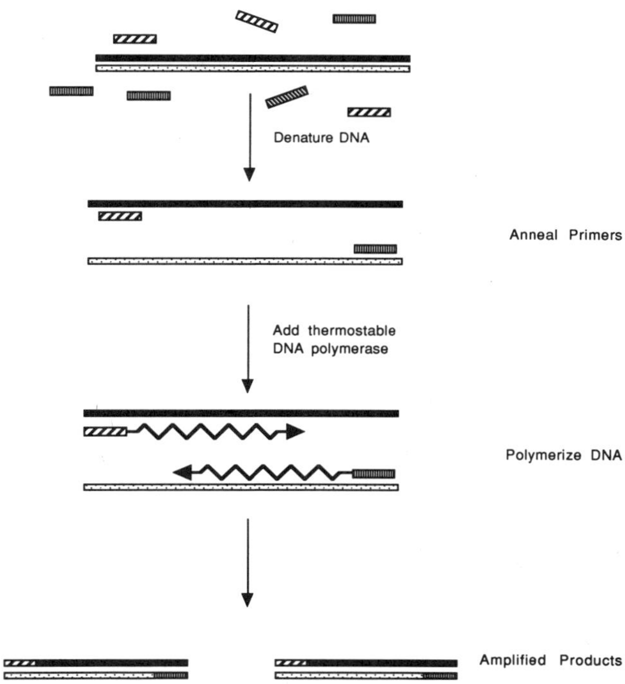

Fig. 3. The polymerase chain reaction (PCR). Double-stranded target DNA is mixed with primers (▭ ▭) homologous to DNA sequences within the target DNA and flanking the DNA sequence of interest. The DNA is denatured and primers annealed to it at an appropriate temperature. The small double-stranded segments thus created act as initiating points for polymerization of new complementary DNA strands by a thermostable DNA polymerase. Repeated rounds of these steps leads to an exponential amplification of the DNA sequence of interest.

nosis the MYCN oncogene may be amplified and highly expressed and the epidermal growth factor receptor can be amplified and expression increased in glioma.

Genetic linkage analyses

In familial cancers a mutant copy or allele of the cancer predisposition gene is inherited through the germline, providing an alternative method for gene localization termed *genetic linkage analysis*. This technique takes advantage of the frequent and normal DNA sequence variation (polymorphism) that affects nonessential DNA sequences throughout the genome. These variations can be detected using restriction enzymes which cut DNA at specific recognition sites (4–8 base pairs [bp]). DNA polymorphisms can disrupt these sequences, creating or eliminating sites, which results in changes in the resultant restriction fragment length. These polymorphic sites can be detected by hybridization of a radiolabeled DNA segment (probe) to digested genomic DNA that has been electrophoretically separated and transferred to a membrane support (Southern blot). Recently, methods have been developed for identifying highly polymorphic microsatellites, which are variable copy numbers of small (usually 1–4 bp) tandem repeats, using the polymerase chain reaction (PCR) (Fig. 3). In this technique, oligonucleotide primers (15–25 bases) complementary to DNA flanking the region of interest, in this case the microsatellite, are annealed to denatured genomic or other template DNA. Repeated cycles of denaturation, annealing, and DNA polymerization result in an exponential amplification of the sequence between the primers. Variation in microsatellite copy number is detected by separation of radiolabeled PCR amplification products by denaturing polyacrylamide gel electrophoresis.

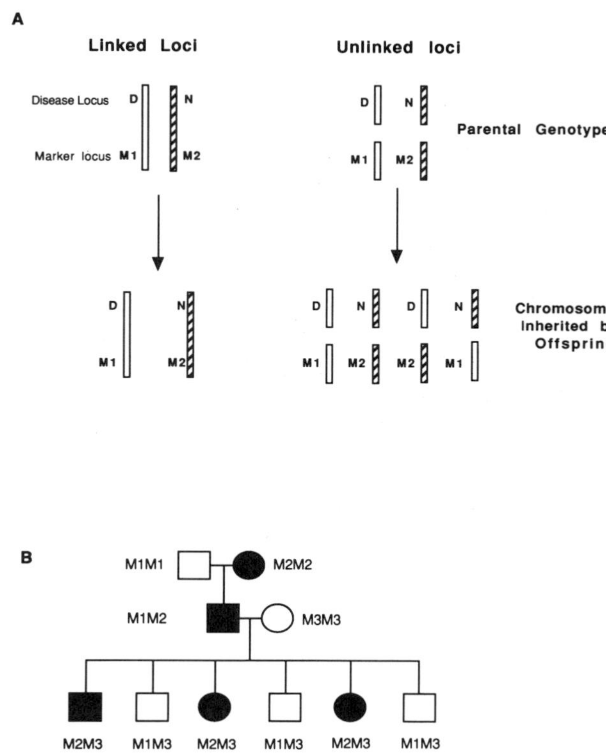

Fig. 4. Genetic linkage. **A:** Segregation of alleles at a disease locus with respect to alleles at a marker locus. The original parental genotype is shown and the potential chromosomes that might be inherited by the individual's offspring are indicated. If the loci are tightly linked (**left**) the parental disease and marker allele association will generally be maintained between generations. If the loci are unlinked (**right**) the disease and marker alleles may be inherited independently. D, Disease allele; N, normal allele at disease locus; M1, M2, marker alleles. **B:** Genotypes at marker locus (M) in a family with a hereditary cancer syndrome. In each individual the disease has been inherited with the M2 marker allele. ● ■ - affected female and male; ○ □ - unaffected female and male.

Restriction fragment length polymorphisms (RFLPs) and microsatellites allow us to distinguish the maternal and paternal copies (alleles) inherited by an individual at that locus. Linkage analyses rely on the fact that alleles for loci which are physically close are inherited together more frequently than alleles for loci which are widely separated in the genome (Fig. 4A). Therefore, a polymorphic locus close to the disease locus could act as a marker for the region containing a cancer gene. By identifying the alleles inherited by each member of a family at a polymorphic marker locus and comparing that pattern with the inheritance of the disease in the same family, we can look for coincidence of inheritance for marker and disease alleles (Fig. 4B). If a marker is found whose inheritance coincides with that of the cancer predisposition in the majority of a large sample of individuals, the marker must be close to the disease gene. As we test marker loci closer to the disease gene, the frequency of discordance, or cases where disease and marker allele are not co-inherited, should decrease. The identification of markers close (tightly linked) to the disease locus and, if possible, flanking the locus can provide a small genetic region in which the cancer predisposition gene may lie. Genetic linkage analyses have contributed to the localization of a number of human cancer genes (e.g., NF1, RB1, VHL, NF2).

Although genetic linkage may refine our localization of a gene, it rarely, by itself, allows us to identify the disease gene. It is very difficult to refine a linkage map for DNA markers that are separated by less than a genetic distance of 1 centiMorgan (cM) or approximately 1,000,000 bp (1 Mb).

Loss of heterozygosity studies

An alternative method of determining the location of a cancer gene is to look for tumor-specific alterations that may establish an interval containing the cancer gene. If the disease gene is a tumor suppressor gene, then a single normal copy of the gene is generally required for appropriate cell growth and function. If, however, a mutant copy of that gene is inherited through the germline or becomes mutated somatically, a subsequent somatic loss of the remaining copy unmasks that mutation conferring tumorigenic potential on the cell. Somatic loss may be generated by precise events such as gene conversion or point mutation or by mechanisms involving larger regions, including deletion, nondisjunction, and subsequent loss of the normal chromosome, or mitotic recombination and segregation, which results in one daughter cell with identical mutant copies of the disease gene (Fig. 5). These latter events are frequently recognized as a somatic loss of constitutional heterozygosity (LOH) in a tumor. In

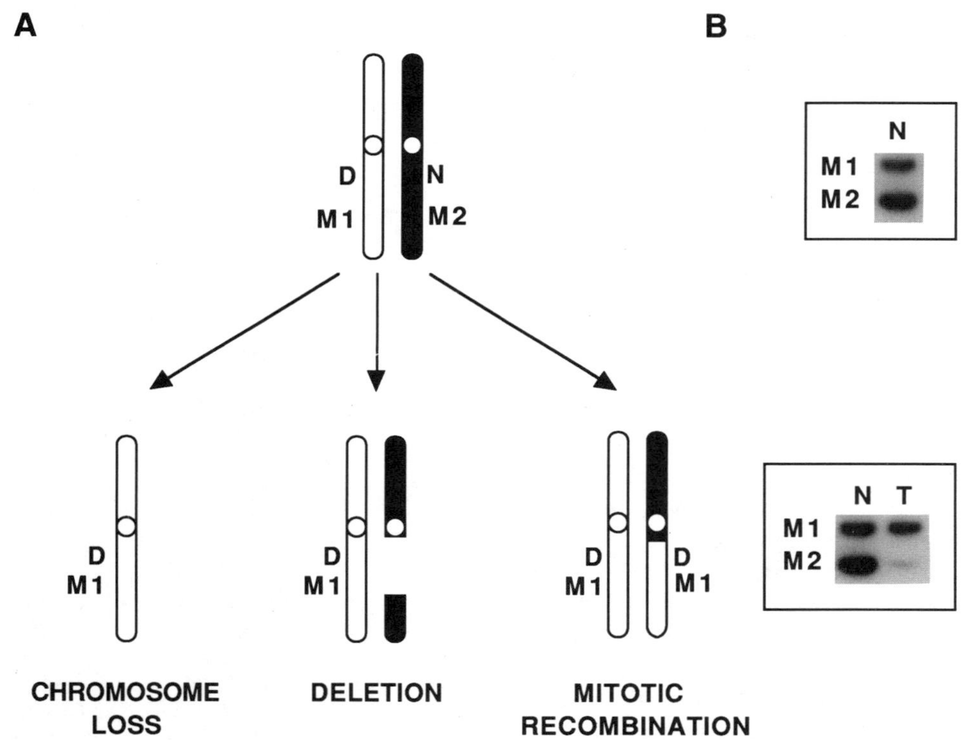

Fig. 5. Loss of heterozygosity. **A**: Schematic representation of events leading to loss of normal alleles (N) at a tumor suppressor gene locus in a cancer cell. The loss of the normal allele (N) and retention of the disease allele (D) may be inferred from the loss of alleles at the tightly linked marker locus (M). **B**: Allele loss at a locus on chromosome 1p in a pheochromocytoma from an individual with von Hippel-Lindau syndrome. The signal observed for allele M2 in the tumor DNA sample indicates the presence of contaminating normal tissue in the tumor sample. N, DNA from peripheral lymphocytes; T, DNA from a pheochromocytoma; M1, M2, marker alleles.

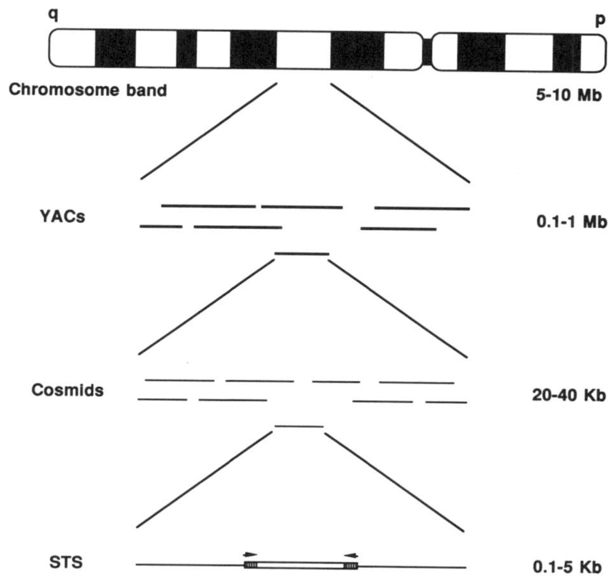

Fig. 6. Relative DNA content of genomic segments. The relationship between the DNA content of a chromosome band and three different cloning systems used in physical mapping. A chromosome band can encompass a number of overlapping YACs. Each YAC in turn may correspond to a number of overlapping cosmids, and each cosmid can be represented by one or more STSs. The approximate size of genomic DNA contained in each type of clone is indicated.

order to recognize these events, DNA samples from tumor and normal tissue from the same individual are compared using DNA markers closely linked to the disease gene. If one of these mechanisms of somatic loss has occurred, the alleles associated with the missing "normal" gene copy will be present in the normal tissue and absent in the tumor DNA samples (Fig. 5B). As in genetic linkage analyses, LOH studies require highly polymorphic marker loci. The characterization of allelic losses has proved useful in identifying the predisposing genes for retinoblastoma (RB1), Wilms' tumor (WT1), and, most recently, genes involved in colon carcinoma on chromosome 5q (APC, MCC) (reviewed in Seizinger et al., 1991; Weinberg, 1991). Unfortunately, LOH, like linkage analysis, does not usually provide us with a precise mapping tool for defining the gene of interest.

Physical Analysis and Mapping of the Chromosomal Region

Once a cancer gene has been mapped to a chromosomal region within the limits of resolution of techniques such as those described above, new approaches must be taken to refine the interval further. A genetically small region (1 cM) has been estimated to comprise on average 1 Mb of DNA, which may contain multiple genes, and often the regional assignment for a cancer gene is greater than this (Fig. 6). In order to characterize these large DNA segments and to provide a framework for the identification of candidate genes, it is necessary to develop a physical map of the region. A physical map describes the relationship between the known physical features that have been localized, or mapped, to the region of interest, such as chromosomal breakpoints, known genes, and the loci characterized by genetic linkage. As with any map, the greater the detail provided by the number of defined genetic loci on the physical map, the more information provided on the composition and size of the regions between these loci (Fig. 6). Physical mapping requires the generation and localization of *markers* as reference points in order to establish the arrangement of sequences within the region of interest, and to facilitate further subdivision of that region. A DNA marker is a fragment of genomic DNA that corresponds specifically to the chromosomal region from which it is derived. A sequence tagged site (STS) is a type of marker specific for only one chromosomal location (Fig. 6), the sequence of which is often used to derive primers for DNA amplification using the PCR (Fig. 3). This type of marker can therefore be defined by the DNA sequence of the primers. Many markers represent polymorphic loci, recognizing sequence variation (RFLPs or microsatellites) between chromosomes as described above.

Generation of markers

Once a region containing a putative cancer gene has been defined, a maximal density of markers within the region greatly increases the probability of identifying the cancer gene. It is estimated that there are approximately 30,000–100,000 genes in the human haploid genome, which consists of approximately 3,000 Mb of DNA. To be sure of finding any gene, a marker would be required at the same frequency as genes in that region of the genome, that is, on average, one marker for every 100 kb. Both genetic and physical mapping of a chromosomal region containing a putative cancer gene require a panel of well-defined DNA markers. This usually involves generating a *library*. A library is a collection of DNA segments, or clones, which is generated in a random or directed fashion from a variety of more or less specific DNA sources and can be used to isolate DNA markers (Table 4).

Genomic or chromosomal libraries. Construction of this

Fig. 7. Schematic representation of a yeast artificial chromosome (YAC). The essential elements for YAC maintenance and replication, TEL (telomeric sequences), ARS (autonomous replication sequences), CEN (centromere), and the position of a cloned genomic insert, are indicated. Hatched boxes represent yeast selectable markers.

TABLE 4. Libraries Used for the Generation of DNA Markers

Library	Starting material	Size cloned fragment (approx.)	Application
Genomic	Whole genome	0.5–40 kb	General
Chromosomal	One or more chromosomes, e.g., somatic cell hybrid, radiation hybrid	0.5–40 kb	General
YAC	Whole genome, one or more chromosomes	200 kb–1 Mb	Cloning large genomic fragments
Alu-PCR	Chromosome, e.g., somatic cell hybrid, radiation hybrid, flow-sorted chromosome	0.1–4 kb	Enrichment of gene sequences
Jumping/Linking	Whole genome, chromosome	0.5–5 kb	Enrichment of regions close to genes
Microdissection	Chromosome band	0.1–0.4 kb	Enrichment of small sequences

type of library involves restriction enzyme digestion of a source DNA followed by insertion into a suitable vector (plasmids, cosmids, or phage; Fig. 6). If the source DNA is limited to a defined region of the genome, then a large percentage of the clones generated will be derived from this region. However, highly specific material is frequently not available and a less defined source including one or more chromosomes is used. In this case, many of the clones generated will be derived from outside the region of interest and a screening procedure must be employed to map the clones with respect to that region. Commonly used DNA sources include radiation hybrids or single chromosomes isolated by fluorescence-activated chromosome sorting.

Yeast artificial chromosomes (YACs). YAC vectors allow the cloning of genomic DNA fragments 100 to 1,000 kb in length (Fig. 6). A YAC vector (Fig. 7) contains the elements essential for the maintenance of an artificial chromosome in a yeast cell, including a functional origin of replication (ARS) for the inserted foreign DNA segment, a region which acts as a centromere (CEN), allowing the attachment of microtubules during cell division, and two physical chromosome ends (TEL), which permit the replication and maintenance of a linear DNA molecule (Anand, 1992; Burke et al., 1987). Large DNA fragments from the source DNA are generated by cleavage using restriction enzymes that cut very infrequently in the genome, and ligated to the two arms that make up the YAC vector. YAC libraries that contain DNA from the whole genome or a defined region of the genome (e.g., a single chromosome) have been constructed. These libraries can be screened for the presence of DNA markers by direct hybridization or by PCR, providing a rapid increase in rate of physical mapping. While YACs provide large segments of cloned DNA and greatly expedite the development of the physical map, they are cumbersome to use as DNA markers themselves. Con-

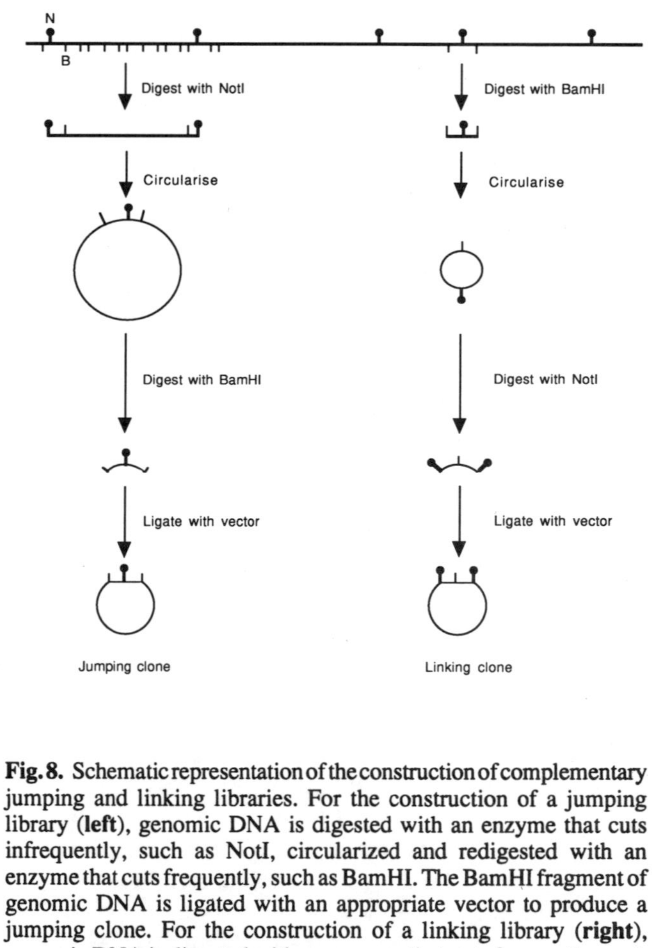

Fig. 8. Schematic representation of the construction of complementary jumping and linking libraries. For the construction of a jumping library (**left**), genomic DNA is digested with an enzyme that cuts infrequently, such as NotI, circularized and redigested with an enzyme that cuts frequently, such as BamHI. The BamHI fragment of genomic DNA is ligated with an appropriate vector to produce a jumping clone. For the construction of a linking library (**right**), genomic DNA is digested with an enzyme that cuts frequently, such as BamHI, circularized and redigested with an enzyme that cuts infrequently, such as NotI. The NotI fragment of genomic DNA is ligated with an appropriate vector to produce a linking clone.

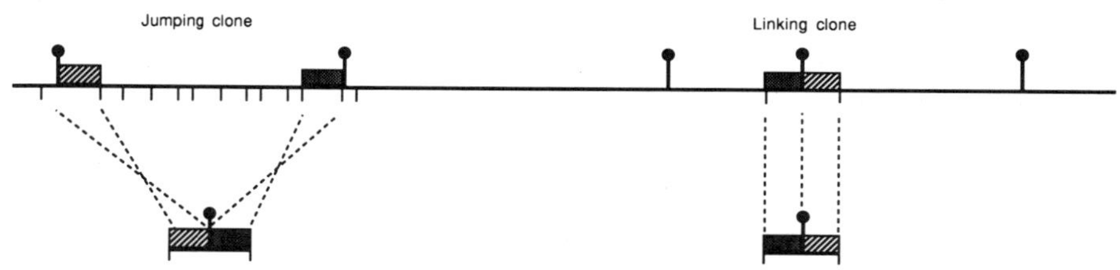

Fig. 9. Schematic representation of the relationship between the sequences contained within jumping and linking clones and the physical map.

sequently, YACs are often used as source material for generation of mini-libraries or for isolation of smaller DNA segments (e.g., YAC end clones), which are preferable as DNA markers.

***Alu*-PCR generated micro-libraries.** Primers to human-specific repetitive elements can be used for the direct amplification of human DNA by the PCR (Cole et al., 1991; Nelson et al., 1989; Fig. 3). The amplified material may be cloned in vectors, as described above, to construct a library of sequences found adjacent to these repetitive elements in the genome. These in turn are mapped by one or more of the techniques described below and those that map to the region of interest can be converted into STSs. The most efficient application of this approach uses a DNA template representing a very defined region such as a somatic cell or radiation hybrid, a single human chromosome, or a small segment of genomic DNA cloned in a vector such as a YAC. Although the genome contains a number of families of short interspersed repeat elements, the *Alu* repeat has been most frequently employed in this technique. *Alu* repeats are located at an average distance of 4 kb in the human genome which is a suitable distance for DNA amplification by PCR. Since *Alu* repeats are frequently clustered in or near genes, this type of library can enrich for coding sequences which could greatly facilitate the identification of candidate genes.

Microdissection libraries. Microdissection libraries are generated by physical dissection of a defined region of a metaphase chromosome, PCR amplification (Fig. 3) of the dissected DNA, and insertion into a suitable plasmid vector (Greulich, 1992; Lüdecke et al., 1990). This approach can produce numerous independent clones derived from a small region of a chromosome. Identification of those which are single copy followed by DNA sequencing to obtain suitable primer sequences for PCR generates a series of STSs spanning a defined region of the genome. In addition, these markers can be used to screen genomic DNA libraries to identify larger genomic fragments derived from the region of interest to expand the information available from each marker.

Jumping and linking libraries. A basic problem in identifying cancer genes is the need to explore large distances of genomic DNA for potential candidate genes. Two complementary strategies for library generation have been developed to address this difficulty (Poustka and Lehrach, 1986). An approach exploiting rare restriction enzyme digestion sites has been developed for chromosome "*jumping*," which allows large intervals in either direction from an initial marker to be spanned in a single "jump" (Fig. 8). The restriction enzymes used in this kind of analysis cut infrequently due to the length (8–12 bp) and composition (C- and G-rich) of their recognition sites, producing, on average, fragments of 100 kb or more in length. A jumping clone is constructed by first generating very large DNA fragments by restriction with one of these infrequently cutting restriction enzymes, such as NotI, and then recircularizing the large fragment to bring the unrelated ends together. A further digestion with a more frequently cutting enzyme and insertion into a suitable vector leads to deletion of the intervening DNA sequences. The resultant clone, therefore, has sequences adjacent to neighboring NotI restriction sites juxtaposed in a single clone (Fig. 9).

A *linking* library is composed of clones containing small regions of DNA immediately surrounding and including a restriction site for an infrequently cutting enzyme (Fig. 8). A linking clone is constructed by digestion of genomic DNA with a frequently cutting restriction enzyme, recircularization of the fragments, followed by a second digestion with an infrequently cutting restriction enzyme such as NotI. Fragments are then ligated with a vector that allows the cloning of NotI fragments but not those fragments undigested by NotI. Alternate and repeated screening of jumping and linking libraries permits directional and ordered movement through the genome (Fig. 9). This combined strategy generates markers on average every 100 kb from the initial marker locus and in particular at CpG islands, small CG-rich sequences that are often associated with genes (see below) and that are rich in restriction sites for infrequently cutting enzymes. This strategy has been used with great effectiveness in the isolation of the cystic fibrosis gene (Rommens et al., 1989).

Localization of DNA markers

Intrinsic to physical mapping is the derivation of new markers in the region of interest. All new markers, in particular those randomly isolated, need to be mapped in order to identify those that are within the region containing the gene of interest. This may be accomplished by a number of different methods.

Somatic cell hybrid mapping. A basic tool for mapping is the somatic cell hybrid, which carries a subset of human chromosomes within a rodent cell (Bentley et al., 1988). These cell lines can be generated by a variety of methods involving fusion of a human cell with a rodent cell, followed by screening for cells containing the desired human chromosome. Various selection methods are used to maintain the human chromosome complement of these cell lines. Human parent cells containing chromosomes with naturally occurring translocations, insertions, or deletions can be used to establish more precise mapping tools. Alternatively, subfragments of a human chromosome can be generated by artificially inducing chromosome breaks using radiation and selecting hybrids with the minimal human region retained (*radiation hybrids*). A panel of hybrids that unambiguously defines intervals along the chromosome can be used to map new markers as they are generated.

***In situ* hybridization.** The chromosomal location of new markers can be determined by *in situ* hybridization of [^{3}H]-labeled probes to banded metaphase preparations of human chromosomes. This method is slow and requires the pooling of data to give a statistically significant result. However, the recent development of *fluorescence* in situ *hybridization* (FISH), in which biotin-labeled markers are hybridized to suitable preparations of metaphase chromosomes, has proved to be an effective and rapid method of mapping markers (Korenberg et al., 1992). The biotin moiety is detected using streptavidin, which has a high affinity for biotin, coupled to a fluorescent dye. The simultaneous detection of more than one marker is possible using different fluorescent dyes, thus enabling the ordering of close markers with respect to each other to a resolution of 500–1,000 kb (Korenberg et al., 1992).

Fluorescence-activated chromosome sorting. Each chromosome in a cell can be distinguished by size and DNA content. By staining metaphase chromosome preparations with DNA binding fluorescent stains such as Chromomycin A3 and Hoechst 33258, a single human chromosome or small number of chromosomes can be isolated using a fluorescence-activated cell sorting apparatus (FACS) (Bentley et al., 1988). This material may be transferred to a membrane support and hybridized to radiolabeled probes or used directly as a template in PCR analysis to localize markers. Translocated, deleted, or rearranged chromosomes can also be separated by FACS and new markers localized with respect to the breakpoints of these derivative chromosomes.

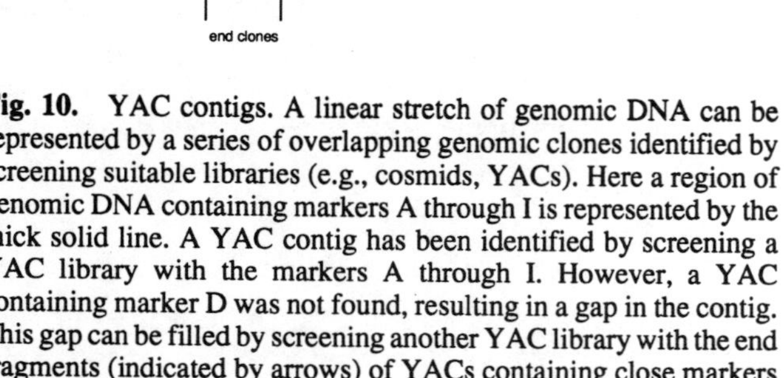

Fig. 10. YAC contigs. A linear stretch of genomic DNA can be represented by a series of overlapping genomic clones identified by screening suitable libraries (e.g., cosmids, YACs). Here a region of genomic DNA containing markers A through I is represented by the thick solid line. A YAC contig has been identified by screening a YAC library with the markers A through I. However, a YAC containing marker D was not found, resulting in a gap in the contig. This gap can be filled by screening another YAC library with the end fragments (indicated by arrows) of YACs containing close markers (C and E).

Construction of the physical map

Once a panel of markers that map to the region of interest has been generated, the physical relationship of these markers to each other and to other physical signposts such as translocation chromosome breakpoints must be established. Markers can be placed on this physical map using a number of techniques.

Pulsed field gel electrophoresis. Pulsed field gel electrophoresis (PFGE) is a method used for separating large DNA molecules, ranging in size from 20–2,000 kb. Genomic DNA immobilized in agarose blocks is digested with restriction enzymes which cut genomic DNA infrequently, fragments are separated using complex gel electrophoresis conditions in which the migration of these large fragments is related to their size, and Southern blots prepared. Hybridization to a radiolabeled probe permits the identification of the restriction fragment that corresponds to each DNA marker (Gemmill et al., 1987). By using a variety of suitable restriction enzymes, in isolation and in combinations of two or three enzymes, an overlapping set of restriction fragments to which a given marker will hybridize can be identified and ordered in relation to each other. In this way a restriction map of genomic DNA surrounding markers can be generated. If markers for the region of interest are sufficiently close, then these isolated maps will overlap and a long-range physical map can be constructed.

Chromosome walking and establishing contigs. Chromosome walking uses a DNA segment derived from one or both ends of a mapped DNA marker to identify overlapping genomic clones (Little, 1992). Each end of a cloned DNA segment may be used to screen a relevant genomic library to isolate a further set of cloned segments (Fig. 10). End segments from these overlapping clones can in turn be used in additional rounds of library screening and in this way it is possible to "*walk*" along the chromosome. The overlapping clones represent a continuous region of genomic DNA, often termed a *contig*. Generally a chromosome walk using DNA segments in phage or cosmid vectors, with cloning capacities of 20 kb and 40 kb, respectively, results in an average advance of approximately 20 kb per step. The advent of YAC technology has made steps of 100–500 kb possible. YAC contigs can be constructed by isolating clones containing the end fragments of the genomic DNA included in the YAC and using these to isolate new clones from the YAC library. Long-range restriction maps generated using PFGE of human DNA can be integrated with the contigs generated by chromosome walking simply by coordinating the restriction enzymes used in the analyses.

Identification of Candidate Genes

The isolation and characterization of genes that map to the interval of interest and may, therefore, be considered as candidates for a putative cancer gene is usually pursued in parallel with physical mapping of that region. New candidate genes may be identified directly through information from the physical map, providing simultaneous characterization and mapping data, or by less directed approaches, isolating all coding sequences from the interval of interest with subsequent mapping and characterization. Methods used to identify possible gene encoding sequences are described below.

Identification of CpG islands

There are small areas of the genome where the percentage of CpG dinucleotides (65%) is greater than in the remainder of the genome. CpG islands are associated with the 5' flanking sequences of some genes and hence the presence of CpG islands can be an indication of potential gene sequences (Lindsay and Bird, 1987). These sites are unmethylated and therefore accessible to cleavage by methylation-sensitive enzymes (e.g., HpaI). Digestion of genomic DNA or genomic clones with such restriction enzymes can lead to the identification of clusters of these cleavage sites, indicating the presence of CpG islands. In this way previously unknown genes can be recognized within the existing physical map of the chromosome. Markers associated with these islands can be used to screen cDNA libraries to isolate the gene(s) in the vicinity.

Screening cDNA libraries

Libraries composed of cDNAs (constructed using RNA sequences as templates for the synthesis of complementary DNA copies by the enzyme reverse transcriptase) cloned into appropriate vectors can be screened with any DNA sequence thought to contain coding sequences in order to identify the corresponding gene. The probe used may be DNA known to contain coding sequence based on other analyses or DNA fragments with unknown gene content. Methods have been developed for screening such libraries using large genomic fragments cloned in YACs (Brownstein et al., 1989). In this way each member of a contig of YACs derived from a region can be used to screen an appropriate cDNA library to identify genes in that region.

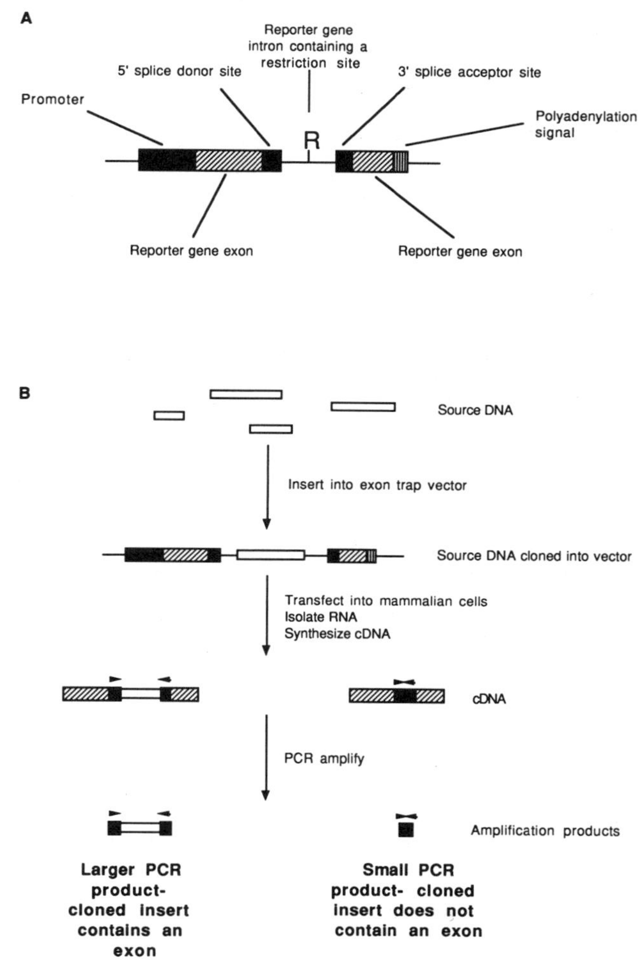

Fig. 11. Exon trapping. **A**: Schematic drawing of an exon trap vector. Essential elements are indicated. **B**: Exon trapping is performed by ligating digested fragments of source DNA (open box) into the exon trap vector and transfecting these constructs into mammalian cells. RNA is isolated and used as a template for cDNA synthesis. This is analyzed by PCR to identify cloned inserts that contain an exon and those that do not.

Exon trapping

A strategy, termed *exon trap cloning*, has been developed for the selective cloning of exon sequences from large genomic fragments (Auch and Reth, 1991; Fig. 11). Genomic DNA is cloned into the exon trap vector at a restriction site that is flanked by artificial 5' and 3' splice donor and acceptor sites and transfected into mammalian cells. RNA expressed from these cells is used as a template for cDNA synthesis, which is then analyzed by PCR amplification to identify the spliced product derived from the vector using primers close to the artificial splice sites. If genomic sequence containing an exon has been cloned, when a splice event occurs intronic sequences are deleted, leaving only the exon. In this way exons can be identified simply by the increase in size of the amplified product and previously unidentified genes cloned and investigated.

Characterization of Disease-Specific Mutations

The diverse characteristics of the known tumor suppressor and oncogenes (Table 1) make it difficult to distinguish a cancer gene from other genes that map to the region of interest. While the properties of cancer genes vary considerably (Table 1), the criterion by which they are defined is constant—the gene must be subject to disease-specific mutations. In the case of sporadic tumors these alterations are tumor-specific and should not be recognizable in the constitutional genotype of the patient. In heritable cancers, the mutation should segregate with the disease phenotype in the family being observed. In this case the disease mutation is present in all cells.

The mutations that result in the conversion of a normal gene to a cancer gene are diverse, ranging from gross, cytogenetically detectable changes to very specific events affecting one or very few base pairs of DNA. Consequently a number of approaches may be required to exclude or confirm the presence of mutations in a candidate gene that contribute to the neoplastic phenotype.

Gross chromosomal alterations (rearrangement/deletion)

Cytogenetic recognition of chromosomal translocations, rearrangements, or deletions, as described above, may be confirmed at the molecular level. Sequences flanking or deleted in such rearrangements may include genes that are altered structurally or functionally in the disease. Disruption of many oncogenes and tumor suppressor genes has been recognized by aberrant restriction fragment patterns using both PFGE and conventional Southern blot analyses. In these cases it is generally possible to detect such changes in fragments generated by multiple restriction enzymes. Comparison of the altered fragments with the physical or long-range restriction map of the region surrounding the candidate gene using multiple markers may also provide information as to the precise location and extent of sequence disruption. For example, altered sizes of PFGE fragments in two patients with balanced translocations involving 17q11.2 were instrumental in defining a 60 kb region that included part of the NF1 gene (Fountain et al., 1989).

PFGE can also be used to detect submicroscopic events undetectable cytogenetically. Interstitial nested deletions on chromosome 5q were detected by PFGE in two patients with familial adenomatous polyposis coli (APC) (Groden et al., 1991; Joslyn et al., 1991). This region was isolated in a YAC contig and the APC gene identified. Conventional Southern analyses are useful in identifying small deletions or insertions not detected by PFGE and to confirm and characterize the sequences juxtaposed by larger deletions.

Sequence variation

Cancer gene mutations that convert a normal cell to a neoplastic cell may be as simple as a single base substitution. In some cases these changes affect specific positions, as in the case of HRAS, where mutations of codons 12, 13, or 61 in a single copy of the gene can confer oncogenic potential on the cell. In other genes, such as p53 or RB1, point mutations may occur at a broader range of codons spanning the coding sequence.

Numerous techniques have been developed for efficiently screening a gene for DNA sequence variations (reviewed in Cotton, 1989).

Single strand conformational polymorphism (SSCP). SSCPs are recognized using radiolabeled PCR products amplified from the gene of interest using cDNA or genomic DNA as template. The two strands of the amplified DNA duplex are separated electrophoretically on a neutral polyacrylamide gel. A sequence variant, such as a base pair substitution or deletion, is recognized by a shift in mobility of the corresponding DNA fragment with the concomitant appearance of new bands on the gel.

Denaturing gradient gel electrophoresis (DGGE). In DGGE, digested DNA is electrophoresed through a polyacrylamide gel containing a gradient of increasing concentrations of denaturants (urea and formamide). As the DNA migrates the melting point of the DNA duplex is reached and the two DNA strands begin to dissociate, altering or stopping their migration through the gel. A single base pair difference in sequence is sufficient to alter this melting point and distinguish a fragment containing a mutation.

Chemical cleavage. In these analyses, DNA is denatured and hybridized to a radiolabeled probe. Chemical modifiers of DNA bases (e.g., osmium tetroxide, hydroxylamine) are added that interact with the DNA at positions that remain single stranded, such as sites of point mutation, and modify the DNA base. Subsequent addition of piperidine leads to chemical cleavage of the DNA fragment at these modified sites, recognized as an altered DNA fragment pattern by polyacrylamide gel electrophoresis.

DNA sequencing. These techniques are used to establish the exact sequence of a DNA segment. In general, any of the above methods for identifying sequence variation would be subsequently confirmed and further analyzed by sequencing that segment and comparing it to a normal control.

It should be emphasized that not all sequence variation represents a disease mutation. Naturally occurring polymorphism or conservative substitutions that do not affect the gene product will also be detected by these methods.

Gene expression

RNA. An alternative approach to evaluating a candidate gene is to examine expression of the gene products. RNA may be transferred to a membrane support (Northern blot) in a similar manner to DNA for investigation of product size and level of expression. Mutation of a candidate gene may result in the production of transcripts of altered size, reflecting deletions, sequence mutation, resulting in premature transcription termination, or juxtaposition of coding sequences not normally cotranscribed to make a chimeric mRNA. The expression of the candidate gene in tumor tissue may differ from the pattern or level of expression in the normal cognate cell or tissue. Abrogation or alteration of cancer gene transcripts is observed in many human tumors, including retinoblastoma (RB1), NF1 pheochromocytomas (NF1), and Wilms' tumor (WT1), while many oncogenes such as HRAS and MYCN are overexpressed in their active, oncogenic, forms.

Protein. The protein products of a mutated cancer gene can be detected using antibodies specific to the normal or mutated forms of that protein. However, the expense and time required for these analyses are prohibitive in evaluation of candidate genes. Characterization of these products is more efficiently left until the other techniques described identify the appropriate cancer gene.

CANCER GENES

A number of cancer genes have now been identified or are being actively pursued using combinations of the described techniques. Examples of the application of these strategies to the identification of genes contributing to three human cancers are detailed below.

Retinoblastoma (RB1)

The paradigm for identification of a human cancer predisposition gene is retinoblastoma. This is a neoplastic disease of the embryonic neural retina characterized by the early onset of malignant tumors of the immature retinoblast. The disease occurs in sporadic (unilateral) and heritable (usually bilateral or multifocal) forms, arising through mutation of the same gene either somatically or germinally. An initial gene localization to chromosome 13q14 was based on cytogenetically detectable deletions or translocations in peripheral blood lymphocytes from patients with familial retinoblastoma. Linkage analysis using markers such as Esterase D (ESD) confirmed that the retinoblastoma gene (RB1) must lie in this region. Tumor-specific loss of heterozygosity for the 13q13-qter interval (Cavenee et al., 1983) in approximately 70% of sporadic and hereditary tumors was also detected, suggesting a common genetic origin for these tumors. Using DNA markers isolated from a chromosome 13 library and mapped by somatic cell hybrid panel, a sequence, H3-8, was identified that detected submicroscopic deletions of 13q14 in retinoblastoma tumors. A genomic sequence from this region which was conserved between species was used to screen a cDNA library and a 4.7 kb cDNA was identified as an RB1 candidate (Friend et al., 1986). Southern blot analyses of retinoblastoma tumor DNAs using this cDNA as a probe have detected altered restriction fragments reflecting deletions or sequence alterations in approximately 20% of cases. The majority of these tumors expressed either no transcript or only truncated forms when analyzed by Northern blot. Direct sequencing of genomic and cDNA clones derived from RB1 tumors has identified minute mutations including base pair deletions or substitutions and disruption of splice donor and acceptor sites (Yandell et al., 1989). RB1 and its products have been characterized in numerous retinoblastomas. In each case, mutation of the gene consistent with its functional inactivation has been detected (Dunn et al., 1988).

Neurofibromatosis Type 1 (NF1)

NF1 is an autosomal dominant disease that seems predominantly to affect neural crest-derived cell lineages. It is characterized by *café-au-lait* spots, multiple neurofibromas that increase in size and number with age, and a small, but significantly increased, risk of several tumor types, including gliomas, pheochromocytomas, and neurofibrosarcomas. Genetic linkage in NF1 families suggested the disease gene lay close to the centromere of chromosome 17 at 17q11.2 (Goldgar et al., 1989). Localization of the disease was subsequently confirmed by cytogenetic identification of balanced translocations in two NF1 patients with breakpoints in 17q11.2. A long-range physical map of the region spanning the two breakpoints was established by PFGE using existing STSs and NotI-linking clones isolated from a sorted chromosome 17 library and localized by somatic cell hybrid panel (Fountain et al., 1989). Cosmids and YACs isolated for this region suggested that the translocation breakpoints were separated by approximately 60 kb. Two genes that mapped to this interval, EVI2 and NF1-C2, were initially considered as candidate genes; however, mutations were not detected at either locus in NF1 patients. A cDNA library was screened using a YAC spanning the 60 kb interval and a single copy sequence from close to one of the translocation breakpoints that was conserved between species. A candidate cDNA, NF1LT, was isolated (Wallace et al., 1990). The gene is interrupted by the two translocations and by an interstitial deletion of coding sequences detected in an NF1 patient by Southern blot analysis. SSCP has been used to identify several point mutations in the NF1 gene that in subsequent sequence analyses were shown to represent amino acid substitution or protein truncation (Cawthon et al., 1990), confirming the identification of NF1LT as the NF1 gene.

Neuroblastoma

Neuroblastoma is a pediatric tumor of the peripheral sympathetic nervous system. The disease occurs sporadically and, less frequently (approximately 1%), in heritable form. The most consistent cytogenetic abnormality is partial monosomy for chromosome 1p, most commonly the 1p32-pter region. This deletion has been confirmed by loss of heterozygosity analyses using DNA markers spanning chromosome 1, and the region of interest restricted to 1p36.1-36.2. Taken together, these data suggest the location of a tumor suppressor gene(s) inactivated in neuroblastoma at 1p36 (reviewed in Brodeur, 1990).

STSs have been isolated for loci mapped to chromosome 1p36 and used to screen a YAC library. A physical map of the region is currently being constructed using these STSs and new ones developed from the isolated YACs. Studies are underway to identify candidate genes within this region. However, to date no gene from chromosome 1p, which contributes to tumorigenesis in neuroblastoma, has been reported.

REFERENCES

Anand, R. (1992) Yeast artificial chromosomes (YACs) and the analysis of complex genomes. Trends Biotech. 10:35–40.

Auch, D., and M. Reth (1991) Exon trap cloning: using PCR to rapidly detect and clone exons from genomic DNA fragments. Nucl. Acids Res. 18:6743–6744.

Bentley, K.L., A.C. Ferguson-Smith, and F.H. Ruddle (1988) A review of genomic physical mapping. Cancer Surveys 7:267–294.

Brodeur, G.M. (1990) Neuroblastoma: clinical significance of genetic abnormalities. Cancer Surveys 9:673–688.

Brownstein, B.H., G.A. Silverman, R.D. Little, et al. (1989) Isolation of single-copy human genes from a library of yeast artificial chromosome clones. Science 244:1348–1351.

Burke, D.T., G.F. Carle, and M.V. Olson (1987) Cloning of large segments of exogenous DNA into yeast by means of artificial chromosome vectors. Science 236:806–812.

Cavenee, W.K., T.P. Dryja, R.A. Phillips, et al. (1983) Expression of recessive alleles by chromosomal mechanisms in retinoblastoma. Nature 305:779–784.

Cawthon, R.G., R. Weiss, G. Xu, et al. (1990) A major segment of the Neurofibromatosis type 1 gene: cDNA sequence, genomic structure and point mutations. Cell 62:193–201.

Cole, C.G., P.N. Goodfellow, M. Bobrow, et al. (1991) Generation of novel sequence tagged sites (STSs) from discrete chromosomal regions using *Alu*-PCR. Genomics 10:816–826.

Cotton, R.G. (1989) Detection of single base changes in nucleic acids. Biochem. J. 263:1–10.

Dunn, J.M., R.A. Phillips, A.J. Becker, et al. (1988) Identification of germline and somatic mutations affecting the retinoblastoma gene. Science 241:1797–1800.

Fearon, E.R., and B. Vogelstein (1990) A genetic model for colorectal tumourigenesis. Cell 61:759–767.

Fountain, J.W., M.R. Wallace, M.A. Bruce, et al. (1989) Physical mapping of a translocation breakpoint in neurofibromatosis. Science 244:1085–1087.

Friend, S.H., R. Bernards, S. Rogelj, et al. (1986) A human DNA segment with properties of the gene that predisposes to retinoblastoma and osteosarcoma. Nature 323:643–646.

Gemmill, R.M., J.F. Coyle-Morris, F.D. McPeek, et al. (1987) Construction of long-range restriction maps in human DNA using pulsed field gel electrophoresis. Gene Anal. Tech. 4:119–131.

Goldgar, D.E., P. Green, D.M. Parry, et al. (1989) Multipoint linkage analysis in neurofibromatosis type 1: An international collaboration. Am. J. Hum. Genet. 44:6–12.

Greulich, K.O. (1992) Chromosome microtechnology: Microdissection and microcloning. Trends Biotech. 10:48–51.

Groden, J., A. Thliveris, W. Samowitz, et al. (1991) Identification and characterization of the familial adenomatous polyposis coli gene. Cell 66:589–600.

Joslyn, G., M. Carlson, A. Thliveris, et al. (1991) Identification of deletion mutations and three new genes at the familial polyposis locus. Cell 66:601–613.

Korenberg, J.R., T. Yang-Feng, R. Schreck, et al. (1992) Using Fluorescence in situ hybridization (FISH) in genome mapping. Trends Biotech. 10:27–32.

Lindsay, S., and A.P. Bird (1987) Use of restriction enzymes to detect potential gene sequences in mammalian DNA. Nature 327:336–338.

Little, P.F.R. (1992) Generating 'cloned DNA maps'. Trends Biotech. 10:33–35.

Lüdecke, H.-J., G. Senger, U. Claussen, et al. (1990) Construction and characterization of band-specific DNA libraries. Hum. Genet. 84:512–516.

Nelson, D.L., S.A. Ledbetter, L. Corbo, et al. (1989) *Alu* polymerase chain reaction: A method for rapid isolation of human-specific sequences from complex DNA sources. Proc. Natl. Acad. Sci. USA 86:6686–6690.

Poustka, A., and H. Lehrach (1986) Jumping libraries and linking libraries: The next generation of molecular tools in mammalian genetics. Trends Genet. 2:174–179.

Rabbitts, T.H. (1991) Translocations, master genes, and differences between the origins of acute and chronic leukemias. Cell 67:641–644.

Rommens, J.M., M.C. Iannuzzi, B.-S. Kerem, et al. (1989) Identification of the cystic fibrosis gene: Chromosome walking and jumping. Science 245:1059–1065.

Sawyers, C.L., C.T. Denny, O.N. Witte (1991) Leukemia and the disruption of normal hematopoiesis. Cell 64:337–350.

Seizinger, B.R., H.P. Klinger, C. Junien, et al. (1991) Report of the committee on chromosome and gene loss in human neoplasia. Cytogenet. Cell Genet. 58:1080–1096.

Solomon, E., J. Borrow, and A.D. Goddard (1991) Chromosome aberrations and cancer. Science 254:1153–1160.

Wallace, M.R., D.A. Marchuk, L.B. Andersen, et al. (1990) Type I Neurofibromatosis gene: Identification of a large transcript disrupted in three NF1 patients. Science 249:181–186.

Weinberg, R.A. (1991) Tumour suppressor genes. Science 254:1138–1146.

Yandell, D.W., T.A. Campbell, S.H. Dayton, et al. (1989) Oncogenic point mutations in the human retinoblastoma gene: Their application to genetic counseling. N. Engl. J. Med. 321:1689–1695.

Chapter 16

Clonal Origin of Nervous System Tumors

Lee B. Jacoby

INTRODUCTION

A basic issue in the study of neoplasia is whether abnormal growth results from mutation in a single cell or occurs in many cells in response to an external stimulus. A clonal origin suggests that the initiating event in tumorigenesis is a genetic change or somatic mutation that confers a growth advantage on one parent cell. Daughter cells inherit this change and may, in addition, accumulate further genetic changes, each providing additional growth advantages. Most tumors do, in fact, appear clonal in nature, confirming the mutational theory of carcinogenesis (Fialkow, 1976; Knudson, 1973). Other tumors appear to be polyclonal, suggesting that growth is stimulated in multiple original cells by an exogenous or endogenous stimulus such as an infectious agent or humoral factor. Determination of clonal origin has both theoretical and practical implications. It is a first step toward the recognition of somatic mutations initiating carcinogenesis. Characterization of these mutations and identification of the gene products should lead to more rational treatment strategies.

Several methods for determining clonal origin have been developed in recent years that are applicable to a wide variety of solid tumors, including those of the nervous system (Wainscoat and Fey, 1990). These techniques, listed in Table 1, are based on one of two general approaches to distinguishing progeny of different cells, one involving X chromosome inactivation and the other involving somatic cell mutation.

X CHROMOSOME INACTIVATION

In females, one X chromosome is inactivated early in embryonic development and this inactivation is inherited in a stable fashion in all progeny cells, including tumor cells. In normal tissue, the paternal copy will be inactivated in approximately half the cells and the maternal copy in the other half. In contrast, in monoclonal cell populations, e.g., some tumors, the same X chromosome will be inactive in all cells. The active and inactive copies of an X-linked gene can be distinguished in females heterozygous for a particular X-linked polymorphism (that is, females having two different alleles at a genetic locus on the X chromosome) by differences in gene expression or by differences in DNA methylation, since changes in activity of many genes are accompanied by changes in DNA methylation.

Glucose-6-phosphate Dehydrogenase (G6PD)

In females heterozygous for a polymorphism in the X-linked enzyme glucose-6-phosphate dehydrogenase (G6PD), normal tissues will contain both type B and either type A or A^- isozymes, while clonally derived tissues will contain one isozyme or the other. Using this technique, Fialkow and his co-workers (Fialkow, 1976) showed that a variety of human tumors were clonal; other tumors, including neurofibromas, appeared to be polyclonal (Fialkow et al., 1971). The usefulness of G6PD analysis, however, is limited by the rarity of G6PD heterozygosity in most population groups.

Restriction Fragment Length Polymorphisms (RFLPs)

The use of X-linked DNA polymorphisms has made clonal analysis based on X chromosome inactivation more widely applicable. In the strategy described by Vogelstein and co-workers (Vogelstein et al., 1985, 1987), active and inactive copies of X chromosome genes are distinguished by differences in their methylation patterns. Normal and tumor DNAs from informative females are first digested with an appropriate restriction endonuclease that distinguishes the maternal and paternal copies of the gene through an RFLP. A second endonuclease then distinguishes the active from inactive copies of the gene through differences in methylation. If a tumor is monoclonal, the paternal copy of the gene will be cleaved by the second enzyme in a different manner

TABLE 1. Clonal Markers in Nervous System Tumors

Tumor type	X chromosome polymorphism	Karyotype	Loss of heterozygosity	Mutation	Viral integrations
Meningioma	HPRT	-22	22q		
Schwannoma	HPRT	-22	22q		
Pituitary adenoma	HPRT, PGK				
Neurofibroma	G6PD, PGK		None		
Neurofibrosarcoma			17p	p53	
Astrocytoma		-10, 17p-	10, 17p	p53	
Neuroblastoma		-1p	1p		
Medulloblastoma		i(17q), 17p-	17p		
Ependymoma		-22	22q		SV40-like

See text for descriptions of clonal markers. Only the most frequently observed clonal markers are included. Abbreviations: HPRT, hypoxanthine phosphoribosyltransferase; PGK, phosphoglycerate kinase; G6PD, glucose-6-phosphate dehydrogenase; p, short arm of chromosome; s, long arm of chromosome; i, isochromosome; SV40, simian virus 40.

from the maternal copy, since the paternal gene will be active in all cells or inactive in all cells. If the tumor is polyclonal, one-half of the cells will have an active paternal gene and one-half an active maternal gene, so that both gene copies will be equally affected by the second enzyme. This approach is illustrated in Figure 1. Using RFLPs occurring in the X-linked genes hypoxanthine guanine phosphoribosyltransferase (HPRT) and phosphoglycerate kinase (PGK), this method is applicable to approximately 50% of females and this frequency will increase as other polymorphisms are identified close to sites that are differentially methylated in active versus inactive X chromosomes.

The limitations of this technique include the relatively large amounts of DNA needed for Southern blotting (5–10 μg) and also the concern that contamination by normal tissue will obscure a clonal pattern. A recent modification of X chromosome analysis uses the polymerase chain reaction (PCR) to partially circumvent these problems and makes possible the clonal analysis of limited numbers of cells (Gilliland et al., 1991). Since PCR does not preserve methylation, DNA is digested with a methylation-sensitive endonuclease such as *Hpa*II prior to amplification by PCR. The active unmethylated allele will be cleaved and cannot be amplified, while the inactive methylated allele will not be disrupted and so will be amplified. The PCR products are then digested with a second enzyme (*Bst*XI) to distinguish the two polymorphic alleles. In a polyclonal tissue, both alleles will be amplified, whereas in a monoclonal tissue one allele will be absent or greatly reduced in intensity. This technique will be especially useful in analyzing heterogeneous tumors such as neurofibromas or gliomas, since DNA can be amplified from small regions of histopathologically pure tumor samples obtained from cryostat sections. However, whenever very small pieces of tissue are tested for clonality, patch size must be considered. Thus, a tumor could arise from more than one cell yet not appear polyclonal if all of those cells arose from a clonal embryonic "patch" of tissue. Therefore, when such studies are performed, multiple samples should be tested before reaching a conclusion.

SOMATIC MUTATION

Cytogenetics

The morphological analysis of chromosomes in metaphase cells is probably the most widely used method for recognizing markers unique to tumor cells. Chromosome abnormalities include losses or gains of entire chromosomes or losses of specific chromosome regions through deletions or unbalanced translocations. In addition, some malignant tumors have a high proportion of double minute chromosomes, associated with amplification of genetic regions. Karyotypic analysis has proven useful in identifying clonal populations of cells and in monitoring their evolution; however, cytogenetic studies of solid tumors requires that the cells be cultured, which is often difficult. Indeed, for some tumors, the normal cells may grow more rapidly than the tumor cells. Moreover, it is often difficult to distinguish which abnormalities are the primary pathogenic events in the disease and which augment tumor growth as secondary events.

Loss of Heterozygosity

Direct analysis of tumor DNA obviates the need for cultured cells and any artifacts associated with cell culture. In a patient heterozygous for a particular DNA RFLP, Southern blotting can reveal loss of one allele. Polymorphisms in hypervariable loci, i.e., loci containing variable

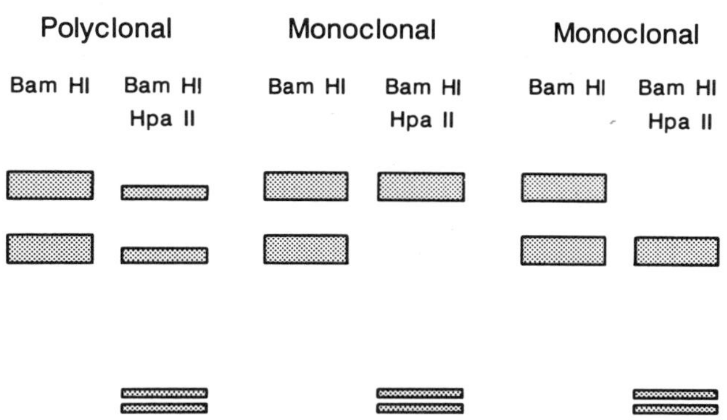

Fig. 1. X-linked chromosome analysis using the *Bam*HI polymorphism in the HPRT gene. DNA from a heterozygous female is digested with *Bam*HI to distinguish the maternal and paternal alleles and one aliquot is further digested with *Hpa*II to distinguish the active methylated allele from the inactive unmethylated allele. (The HPRT gene is an exception to the general rule that the active gene in unmethylated at critical sites.) The DNA is then Southern blotted and hybridized with a probe from the 5' end of the HPRT gene. In a polyclonal tissue, each allele is active in about half of the cells, so *Hpa*II digestion will reduce the density of both fragments about 50%. In a monoclonal tisssue, one allele is inactive in all cells and it will be selectively degraded by *Hpa*II in all cells.

numbers of tandem repeats of short oligonucleotide sequences, are useful for this analysis. In instances where specific chromosomal losses are associated with particular tumors, the presence of a tumor suppressor gene is strongly suggested, the absence of which promotes tumor progression.

DNA Fingerprinting

DNA fingerprinting represents another potential approach to distinguishing normal from tumor cell populations. DNA fingerprints represent multiple hypervariable fragments derived from a large number of autosomal loci dispersed throughout the human genome. In one study of gastrointestinal tumors, for example, about two-thirds of the tumors showed clonal somatic mutations not present in normal tissue DNA (Fey et al., 1989).

Point Mutation

Specific point mutations in oncogenes or tumor suppressor genes have been associated with certain human cancers and these can be used as tumor markers. Mutations in the *ras* oncogene, for example, occur frequently in some tumors such as adenocarcinomas of the pancreas, colon, and lung; however, they are rare in other tumor types (Bos, 1989). More generally useful for the study of nervous system tumors will be mutations in the p53 tumor suppressor gene. Such mutations occur with high frequency in many different types of human cancer and can occur at several different sites (Hollstein et al., 1991).

VIRAL INTEGRATION

The presence of a virus, whether integrated into the genome or not, provides an independent means of determining a clonal population. The homogeneous structure of the intracellular episomal Epstein-Barr virus genome in carcinomas of the nasopharynx and parotid gland, for example, indicated a clonal origin (Raab-Traub and Flynn, 1986). A recent report documenting simian virus 40-like DNA sequences in choroid plexus tumors and ependymomas suggests that this approach may be applicable to brain tumors (Bergsagel et al., 1992).

All of the above properties of tumors, including X chromosomal polymorphisms, chromosomal abnormalities, somatic mutations, and viral integration events, are useful clonal markers in solid tumors. Their uniform presence in a tumor cell population suggests a clonal origin but does not prove it. Rather it indicates that one cell (or a small number of cells with the same clonal marker) outgrew its neighbors during the process of neoplasia. Thus the possibility remains that one cell type was selected even though the tumor originated from many cells. Conversely, cells constituting a single clone are not necessarily identical since clonal evolution may occur within populations of cells following the primary neoplastic event (Nowell, 1976). Thus, a tumor cell population may be clonal in origin yet show heterogeneity in histologic or karyotypic preparations or even show different patterns of protein production by individual cells.

NERVOUS SYSTEM TUMORS

Meningioma

Meningiomas, which originate from the arachnoid layer of the brain, are the most common benign brain tumor (see Schmidek, Chapter 26, this volume). Extensive cytogenetic studies have been performed on cultured meningioma cells, revealing various chromosome abnormalities, most frequently the loss of one copy of chromosome 22 (Zang, 1982). DNA analyses have confirmed the loss of heterozygosity of alleles on chromosome 22 in about 40% of primary meningioma specimens, including tumors of the sporadic noninherited type and those associated with the inherited disorder neurofibromatosis type 2 (NF2) (Seizinger et al., 1987). These studies strongly implicate the presence of a tumor suppressor gene on the long arm of chromosome 22, the loss of which is a primary event in meningioma formation. They do not establish conclusively, however, whether a meningioma arises from a single cell.

The clonal origin of meningiomas was demonstrated using X chromosome inactivation analysis (Jacoby et al., 1990b). Among nine sporadic (not NF2) meningiomas from female patients heterozygous for the *Bam*HI polymorphism in the HPRT gene, all nine tumors appeared monoclonal; that is, one allele was selectively degraded by the methylation-sensitive enzyme *Hpa*II (see Fig. 2). These tumors included at least three showing loss of alleles on chromosome 22 and five that retained heterozygosity for chromosome 22 alleles. Thus meningiomas are monoclonal irrespective of whether one or two copies of chromosome 22 are present.

Schwannoma

Schwannomas are a second type of tumor that appears both as sporadic cases and also associated with NF2. Cytogenetic analyses have been less informative because of the relative difficulty in culturing schwannoma cells; however, analysis of schwannoma DNA shows loss of heterozygosity for genes on chromosome 22 in approximately half of the specimens (Seizinger et al., 1986).

Eight vestibular schwannomas (acoustic neuromas) from females were subjected to clonal analysis using the X chromosome inactivation approach (Jacoby et al., 1990b). As in the meningioma analysis, tumor DNAs were digested with *Bam*HI to distinguish the maternal from paternal HPRT alleles and then digested with *Hpa*II to distinguish the inactive from active alleles. Among eight schwannomas, at least seven of which retained heterozygosity for chromosome 22, seven were clonal. Thus two tumor types associated with NF2 show loss of heterozygosity for genes on chromosome 22 and are clonal in origin. This is consistent with the idea that loss of function of a tumor suppressor gene on the long arm of chromosome 22 is the primary event initiating tumor formation.

Pituitary Adenoma

Pituitary adenomas are among the most common intracranial tumors in adults and account for a diversity of clinical syndromes due to their hormone content and release. The pathogenesis of these tumors remains unclear. Multiple anterior pituitary cells could be stimulated by hypothalamic hormones, or alternatively pituitary adenomas could originate from the clonal expansion of a single altered cell. The clonal composition of several types of pituitary adenomas has been determined recently in several laboratories by

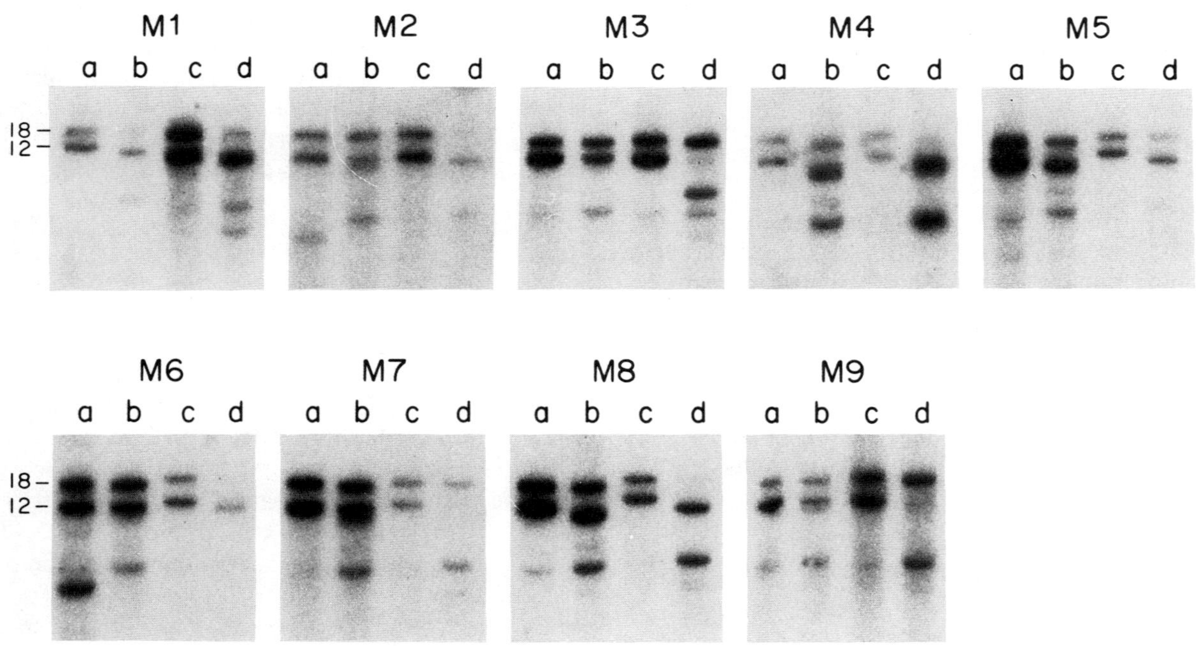

Fig. 2. Clonal analysis of meningiomas from nine patients using an HPRT gene probe. DNAs from blood leukocytes (lanes **a,b**) or tumors (lanes **c,d**) were digested with *Bam*HI and *Pvu*II. (*Pvu*II reduces the size of the larger *Bam*HI fragment from approximately 24 to 18 kb.) One aliquot was not digested further (lanes a,c), while the other was digested with *Hpa*II (lanes b,d). Leukocyte DNAs (lanes a,b) demonstrated the expected pattern for normal polyclonal tissue. Meningioma DNAs (lanes c,d) showed selective degradation of the 18 kb fragment in tumors M1, M2, M4, M5, M6, and M8 and degradation of the smaller 12 kb fragment in the remaining three meningiomas, indicating that all nine tumors were clonal in origin. (Reprinted with permission from Jacoby et al., 1990b.)

analyzing polymorphisms in the X-linked PGK and HPRT genes. Monoclonality has been demonstrated in at least 27 tumors, including somatotroph, lactotroph, corticotroph, gonadotroph, and nonsecretory adenomas (Alexander et al., 1990; Herman et al., 1990; Jacoby et al., 1990a; Schulte et al., 1991) (Table 2). A few tumors appeared polyclonal; in some cases, however, this result could be attributed to contamination of tumor tissue by normal pituitary tissue or blood. Thus somatic mutation precedes clonal expansion in the anterior pituitary and exogenous stimulatory factors alone are not sufficient for tumorigenesis, although they may facilitate the outgrowth of a mutated pituitary cell. It should be noted that even pituitary tumors containing cells heterogeneously expressing different hormones were still monoclonal in origin. The initial mutational event(s) are not known, although they may include mutations in the novel oncogene *gsp* described in a subset of growth hormone-secreting pituitary adenomas (Landis et al., 1989).

TABLE 2. Clonal Origin of Pituitary Adenomas

Tumor type	Mono-clonal	Poly-clonal	References
Somatotrophs	4	0	Herman et al., 1990; Jacoby et al., 1990a
Lactotrophs	4	2	Herman et al., 1990
Corticotrophs	9	4	Herman et al., 1990; Schulte et al., 1991
Gonadotrophs	1	0	Herman et al., 1990
Nonsecretory	9	0	Alexander et al., 1990; Jacoby et al., 1990a
Plurihormonal	0	1	Herman et al., 1990

Neurofibroma and Neurofibrosarcoma

Neurofibromas are one hallmark of von Recklinghausen neurofibromatosis or neurofibromatosis type 1 (NF1). These benign tumors consist of several cell types, including Schwann cells, fibroblasts, and mast cells, and their etiology is unknown despite the fact that the NF1 gene has been identified. No cytogenetic abnormalities have been observed in cultured neurofibromas nor has loss of heterozygosity for alleles in the NF1 region of chromosome 17 been observed in neurofibroma DNA (Menon et al., 1990; Skuse et al., 1991). Using G6PD isozyme analysis, Fialkow and coworkers (Fialkow et al., 1971) showed that each of 14 neurofibromas from two unrelated NF1 patients heterozygous for G6PD types A and B contained both isozymes, implying that hereditary neurofibromas are multicellular in origin. Neurofibromas are made up of several cell types, however, and it is entirely possible that the fibroblasts, for example, are of multicellular origin but that the neural crest-derived Schwann cells are, in fact, of single cell origin.

Recently clonality in neurofibromas was reexamined using the newer molecular genetic techniques (Skuse et al., 1991). Neurofibroma DNAs from eight NF1 patients heterozygous for the X-linked PGK RFLP were digested with one set of restriction enzymes to reveal the polymorphism and then with *Hpa*II to distinguish the active from inactive alleles. All eight neurofibromas showed monoclonal patterns, in contrast to the earlier G6PD isozyme studies. It was somewhat unexpected to obtain a uniformly monoclonal pattern in tumor tissue known to contain diverse cell types. Possibly these neurofibromas were enriched for one cell type, for example, Schwann cells, whereas those in the previous study may have contained a larger proportion of non-Schwann cells. It will be important to confirm this initial report of monoclonality in neurofibromas. The PCR-based modification of the X-linked chromosome analysis should facilitate this analysis by allowing the study of small amounts of relatively pure tissue enriched for either Schwann cells or another cell type. DNA can be extracted from regions of cryostat sections rich in tumor cells and then digested and amplified as described above (Gilliland et al., 1991) to show selective loss of either the maternal or paternal allele following digestion with *Bst*XI.

A benign neurofibroma infrequently progresses to a malignant neurofibrosarcoma. The mechanism leading to this malignant transformation is not fully understood at the present time. Mutation in the NF1 gene on chromosome 17 may be required as a first step and a later step in tumor progression may involve mutation in the p53 gene, conferring a selective growth advantage on the mutated cells. Chromosome 17p deletions and p53 mutations have indeed been associated with some neurofibrosarcomas (Glover et al., 1991; Menon et al., 1990; Nigro et al., 1989). Thus malignant neurofibrosarcoma cells are likely to be clonal in origin, but this has yet to be proven experimentally.

Astrocytoma and Glioma

Astrocytomas typically progress through stages of increasing malignancy, including benign tumors, low-grade astrocytomas, anaplastic astrocytoma, and glioblastoma multiforme (see James and Collins, Chapter 21, this volume). Glioblastomas can also arise *de novo*. In astrocytomas, the primary mechanism of tumor initiation is not known, but one early step is loss of function of p53 or another tumor suppressor gene on chromosome 17p, while later steps include loss of a second tumor suppressor gene on chromosome 10 and amplification of the epidermal growth factor receptor gene.

Clonal expansion of astrocytoma cells has been monitored using point mutations in the p53 gene as a tumor cell marker (Sidransky et al., 1992). In three pairs of recurrent gliomas, the same p53 mutation was found in the recurrent tumor as in the primary tumor. In three pairs of low-grade tumors that had progressed to high-grade tumors, cells in the high-grade tumor were enriched for the same p53 mutations present in a subpopulation of cells present in the low-grade tumor. Thus astrocytic tumor progression was associated with a clonal expansion of cells with a p53 mutation that provided them with a selective growth advantage. Tumor initiation in astrocytic malignancies is less well understood than tumor progression and it is not established whether the lower grade astrocytomas and gliomas are clonal in origin.

Neuroblastoma

Neuroblastomas are another tumor of neural crest origin (see Brodeur and Bernards, Chapter 20, this volume). Several genetic mechanisms have been associated with initiation and progression of neuroblastomas, including loss of heterozygosity for chromosome 1p, amplification of the N-myc oncogene, hyperploidy, and abnormalities in the nerve growth factor receptor. These genetic features are useful in predicting clinical behavior, but again the initial events underlying tumorigenesis are not yet clear.

Other Glial Tumors

Less is known about the genetic mechanisms underlying other types of glial tumors, including ependymomas and medulloblastomas. Cytogenetic and molecular studies of medulloblastomas reveal frequent rearrangements of chromosome 17 and loss of sequences from chromosome 17p, although these may involve a tumor suppressor gene other than p53 (Bigner et al., 1990; Saylors et al., 1991). Amplification of c-*myc* has also been described, as well as other karyotypic changes in tumors and tumor cell lines. Relatively few ependymomas have been examined, but both karyotypic and DNA studies suggest that loss of genetic information on chromosome 22 may occur frequently (Bigner et al., 1990). These are probably heterogeneous groups of tumors with more than one mechanism leading to tumor formation. In some instances, they are likely clonal but the data are too premature to discuss their clonal origin at present.

A recent report that most ependymomas contain simian virus 40 (SV40)-like DNA sequences, however, suggests one additional approach toward determining their clonal origin (Bergsagel et al., 1992). Whether the viral DNA is present as free episomal DNA or integrated into the cellular DNA, its precise structure can be determined by molecular genetic techniques. Assuming the original multiplicity of infection to be low, a monoclonal population of ependymoma cells should have only one form of the viral episome or one viral integration site in the host genome. In contrast, a polyclonal population of tumor cells should have viral episomes with different fused termini or viral sequences integrated at different chromosomal sites.

CONCLUSION

Knowledge of whether nervous system tumors are of clonal origin should increase our understanding of the mechanisms by which they arise and progress. Determining that a tumor originates from a single cell is a first step toward defining the somatic mutation that gives rise to the disease. The determination that a tumor arises from a single intracellular event to produce monoclonality versus an extracellular stimulus to produce polyclonality is also important for planning appropriate therapeutic strategies.

ACKNOWLEDGMENTS

I am indebted to Dr. R.L. Martuza for critically reviewing the manuscript and to USPHS grant CA51410 for support.

REFERENCES

Alexander, J.M., B.M.K. Biller, H. Bikkal, et al. (1990) Clinically nonfunctioning pituitary tumors are monoclonal in origin. J. Clin. Invest. 86:336–340.

Bergsagel, D.J., M.J. Finegold, J.S. Butel, et al. (1992) DNA sequences similar to those of simian virus 40 in ependymomas and choroid plexus tumors of childhood. N. Engl. J. Med. 326:988–993.

Bigner, S.H., J Mark, D.D. Bigner (1990) Cytogenetics of human brain tumors. Cancer Genet. Cytogenet. 47:141–154.

Bos, J.L. (1989) *ras* oncogenes in human cancer: A review. Cancer Res. 49:4682–4689.

Fey, M.F., C. Hesketh, J.S. Wainscoat, et al. (1989) Clonal allele loss in gastrointestinal cancers. Br. J. Cancer 59:750–754.

Fialkow, P.J. (1976) The clonal origin of human tumors. Biochim. Biophys. Acta 458:283–321.

Fialkow, P.J., R.W. Sagebiel, S.M. Gartler, et al. (1971) Multiple cell origin of hereditary neurofibromas. N. Engl. J. Med. 284:298–300.

Gilliland, D.G., K.L. Blanchard, J. Levy, et al. (1991) Clonality in myeloproliferative disorders: Analysis by means of the polymerase chain reaction. Proc. Natl. Acad. Sci. USA 88:6848–6852.

Glover, T.W., C.K. Stein, E. Legius, et al. (1991) Molecular and cytogenetic analysis of tumors in von Recklinghausen neurofibromatosis. Genes Chromosomes Cancer 3:62–70.

Herman, V., J. Fagin, R. Gonsky, et al. (1990) Clonal origin of pituitary adenomas. J. Clin. Endocrinol. Metab. 71:1427–1433.

Hollstein, M., D. Sidransky, B. Vogelstein, et al. (1991) p53 mutations in human cancers. Science 253:49–53.

Jacoby, L.B., E.T. Hedley-Whyte, K. Pulaski, et al. (1990a) Clonal origin of pituitary adenomas. J. Neurosurg. 73:731–735.

Jacoby, L.B., K. Pulaski, G.A. Rouleau, et al. (1990b) Clonal analysis of human meningiomas and schwannomas. Cancer Res. 50:6783–6786.

Knudson, A.G. (1973) Mutation and human cancer. Adv. Cancer Res. 17:317–352.

Landis, C.A., S.B. Masters, A. Spada, et al. (1989) GTPase inhibiting mutations activate the alpha chain of Gs and stimulate adenylyl cyclase in human pituitary tumours. Nature 340:692–696.

Menon, A.G., K.M. Anderson, V.M. Riccardi, et al. (1990) Chromosome 17p deletions and p53 gene mutations associated with the formation of malignant neurofibrosarcomas in von Recklinghausen neurofibromatosis. Proc. Natl. Acad. Sci. USA 87:5435–5439.

Nigro, J.M., S.J. Baker, A.C. Preisinger, et al. (1989) Mutations in the p53 gene occur in diverse human tumour types. Nature 342:705–708.

Nowell, P.C. (1976) The clonal evolution of tumor cell populations. Science 194:23–28.

Raab-Traub, N., K. Flynn (1986) The structure of the termini of the Epstein-Barr virus as a marker of clonal cellular proliferation. Cell 47:883–889.

Saylors, R.L. III, D. Sidransky, H.S. Friedman, et al. (1991) Infrequent p53 gene mutations in medulloblastomas. Cancer Res. 51:4721–4723.

Schulte, H.M., E.H. Oldfield, B. Allolio, et al. (1991) Clonal composition of pituitary adenomas in patients with Cushing's disease: Determination by X-chromosome inactivation analysis. J. Clin. Endocrinol. Metab. 73:1302–1308.

Seizinger, B.R., S. de la Monte, L. Atkins, et al. (1987) Molecular genetic approach to human meningioma: Loss of genes on chromosome 22. Proc. Natl. Acad. Sci. USA 84:5419–5423.

Seizinger, B.R., R.L. Martuza, J.F. Gusella (1986) Loss of genes on chromosome 22 in tumorigenesis of human acoustic neuroma. Nature 322:644–647.

Sidransky, D., T. Mikkelsen, K. Schwechheimer, et al. (1992) Clonal expansion of p53 mutant cells is associated with brain tumour progression. Nature 355:846–847.

Skuse, G.R., B. Kosciolek, P.T. Rowley (1991) The neurofibroma in von Recklinghausen neurofibromatosis has a unicellular origin. Am. J. Hum. Genet. 49:600–607.

Vogelstein, B., E.R. Fearon, S.R. Hamilton, et al. (1985) Use of restriction fragment length polymorphisms to determine the clonal origin of human tumors. Science 227:642–645.

Vogelstein, B., E.R. Fearon, S.R. Hamilton, et al. (1987) Clonal analysis using recombinant DNA probes from the X-chromosome. Cancer Res. 47:4806–4813.

Wainscoat, J.S., M.F. Fey (1990) Assessment of clonality in human tumors: A review. Cancer Res. 50:1355–1560.

Zang, K.D. (1982) Cytological and cytogenetical studies on human meningioma. Cancer Genet. Cytogenet. 6:249–274.

Chapter 17

Pediatric Brain Tumors

Pamela S. Karnes and Corey Raffel

INTRODUCTION

Primary central nervous system tumors are common childhood neoplasms, accounting for 9–20% of all malignancies in the pediatric age group. These tumors are often poorly responsive to treatment, with an overall five-year survival rate of about 50% (Finlay et al., 1987). For other tumors, cytogenetic studies have proven valuable in guiding searches for putative oncogenes and tumor suppressor genes, in pathologic identification, and as an aid in guiding treatment.

REPORTS ON RESEARCH

Cytogenetic and molecular biologic studies of pediatric brain tumors have been hindered by a number of factors. First, pediatric brain tumors are a clinically and pathologically heterogeneous group of neoplasms consisting of multiple tumor types and subtypes. In one large single-institution series of these tumors documented over seven years, astrocytomas represented the most common tumor in the series but accounted for less than one-third of the tumors (Rorke and Schut, 1989). In addition, astrocytomas can be classified into at least four different subtypes (Kernohan et al., 1949; Rorke et al., 1985). The incidence of each of these tumor types is low and few centers treat enough patients with these tumors to generate significant data. At Children's Hospital Los Angeles, about 60 new pediatric brain tumor cases are seen annually, distributed among at least 10 different tumor types.

Second, heterogeneity within an individual central nervous system (CNS) neoplasm is common, so that multiple cells must be examined in order to determine a pathologic diagnosis and karyotype. Third, tumors must be cultured in order to provide enough material for study. Pediatric CNS neoplasms are notoriously difficult to grow in culture. To date, only three primitive neuroectodermal tumor cell lines have been established and characterized (Friedman et al., 1985, 1988; Jacobsen et al., 1985). In addition, the karyotypes of CNS neoplasms may evolve and change during early passages in culture (Wikstrand et al., 1983). Certain chromosomal abnormalities may be overrepresented in cultured cell lines since they may offer a growth advantage (Wasson et al., 1990). Finally, many of these tumors occur in functionally important parts of the brain, resulting in small biopsies and limited availability of tissue for laboratory study. Despite these obstacles, progress has been made in both the cytogenetic and molecular study of pediatric CNS tumors.

ASTROCYTOMAS

Astrocytomas are the most common pediatric CNS neoplasm (Woo et al., 1988). These tumors are further subclassified into two types of low-grade benign astrocytomas, diffuse and pilocytic, and two types of malignant astrocytomas, anaplastic and glioblastoma multiforme (Rorke et al., 1985). Griffin and colleagues (1988) described cytogenetic findings in four astrocytomas (Table 1). Of two benign astrocytomas, one was diploid with a normal karyotype, and the other was near-diploid with more than one stem-line karyotype identified. Of two anaplastic astrocytomas, one was diploid with a normal karyotype and one was near-diploid with an extra copy of chromosomes 2 and 7, loss of a chromosome number 14, and a complex chromosome rearrangement [der(1q15q)].

Karnes and associates (1992) have reported cytogenetic analysis of 21 pediatric astrocytomas. Two of four pilocytic astrocytomas were abnormal; one had unidentified marker chromosomes, and the other was hyperdiploid (2N + X), with an extra copy of chromosomes 5, 7, 10, 11, 12, 20, and 21. Five of 14 anaplastic astrocytomas were cytogenetically abnormal. One of these was pseudodiploid, 46,XY,+12,-22. The other four of these cytogenetically abnormal tumors had structural abnormalities. The structural abnormalities involved both arms of chromosomes 1, 2, 3, 6, 10, and 11, as

Molecular Genetics of Nervous System Tumors, pages 217-221 ©1993 Wiley-Liss, Inc.

TABLE 1. Pediatric Astrocytomas and Their Karyotypes*

Author	Tumor	Modal karyotype
Karnes et al., 1992	PA	53,XY,+5,+7,+10,+11,+12, +20,+21
Karnes	PA	46,XY (2 of 10 cells with markers)
Karnes	PA	46,XX
Karnes	PA	46,XY
Griffin et al., 1988	PA	None
Karnes	AA	69,XXYY,+1,+2,+2,+4,+4,+7, +7,+8,+8,+9,+12,+13,+14, +15,+15,+17,+19, +22, +i(11p),+i(11q)
Karnes	AA	80,XX,+1,+1,+4,+4,+5,+5,+6, +7,+7,+9,+9,+11,+12,+13, +13,+14,+14,+15,+15,+16, +17,+17,+19,+20,+20,+21, +21,+22,+22,+del(3) (q13.2),+i(6p),+del(6)(p12), +der(8)t(3;8)(p21;q24.1), +t(3;8)(q21;q24)
Karnes	AA × 11	46,XX or 46,XY
Karnes	AA	91,XXXX,-3,+6,+6,+7,+7,+8, +8,+11,+11,+12,+12,+13, +13,+15,+16,+16,+17,+17, +17,+19,+20,+20,+21,+21, +2der(1)t(1;22)(p12;q11), +2der(1)t(1;3)(q21p36::q21 →qter),der(2)t(2;4)(p11.2; q12),+2der(2)t(2;?)(q11;?), +der(3)t(2;3)(pter→p11::q21 →pter),+2der(3)t(3;?)(pter→ q21::?),+der(4)t(4;5)(q13; q33),+2del(5)(q31.1),+2der (9)t(9;?)(p21;?),+2der(10) t(10;?;1)(qter→p15::?::p11 →qter)+2der(14)t(1;14)(q11; p11)+2mar
Karnes	AA	70,XX,+2,+3,+4,+5,+7,+7,+8, +9,+9,+10,+12,+12,+14, +14,+18,+18,+19,+20,+21, +22,+2der(16)t(7;16)(p15; p13.3)
Griffin	AA	48,XY,+2,+7,-14,der(1q15q)
Griffin	AA	46,XY
Griffin	A(other)	46,XX

*Dashes are substituted with arrows in production.
PA, pilocytic astrocytoma; A, astrocytoma; AA, anaplastic astrocytoma; AA × 11 = 11 anaplastic astrocytomas.

well as 4q, 5q, 7p, 8q, 9p, 14p, 16p, and 22q. Breakpoints occurred on chromosomes 3p and 3q in two tumors.

Five of six cytogenetically abnormal anaplastic astrocytomas described in the literature possessed structural abnormalities as well. Rearrangements involving chromosome regions 1p, 1q, 3p, 3q, and 7p have been reported in more than one astrocytoma. Examination of these regions may provide a clue to the genesis of some of these tumors.

These results illustrate that a number of numerical and structural abnormalities occur in pediatric astrocytomas, including chromosomal duplications, losses, deletions, insertions, and translocations, but no single characteristic anomaly has been identified. Low-grade tumors tend to exhibit diploid or near-diploid karyotypes, while high-grade tumors may exhibit pseudopolyploidy (Tran et al., 1990). Restriction fragment length polymorphism (RFLP) analysis shows that astrocytomas in adults exhibit loss of genetic material from chromosome 17p in 20–40% of tumors (Fults et al., 1989; James et al., 1990). We have looked for similar deletions in 11 pediatric astrocytomas using three RFLP markers and have found a 17p deletion in two cases (Karnes and Raffel, unpublished data). The tumor suppressor gene, p53, is located on 17p at 17p13.1. In adult tumors, p53 mutations have been reported in many tumor types (Lane and Benchimol, 1990). Sequencing the p53 gene in tumors that show allelic loss may aid the investigation of the role of p53 in pediatric astrocytomas.

PRIMITIVE NEUROECTODERMAL TUMORS

Primitive neuroectodermal tumors (PNET), also called medulloblastomas when they occur in the posterior fossa, account for about 20% of pediatric primary CNS neoplasms (Finlay et al., 1987). Histologic criteria have not provided prognostically useful data which help to subclassify PNET (Jenkin et al., 1989). Chromosome analyses have provided several consistent findings, but no apparent correlation exists between cytogenetic findings and tumor course. This suggests that the most fundamental molecular events have yet to be described. Cytogenetic results have, however, provided clues to the pathogenesis of at least some PNET. The tumors tend to be near-diploid or near-tetraploid, the near-tetraploid cells likely representing persistence of an endoreduplicated state. Griffin and colleagues (1988) reported near-diploidy in six of 10 PNET and pseudotetraploidy in one PNET (Table 2). Biegel and associates (1989) reported near-diploidy in 17 PNET and near-tetraploidy in four PNET. Karnes and co-workers (1992) reported near-diploidy in each of 10 PNET. Multiple numeric and structural abnormalities are also reported in these tumors.

Several chromosomes appear to be nonrandomly affected in PNET, including chromosomes 1, 7, 11, 17, and 22. Marker chromosomes have been noted in seven PNET. Gene amplification has been noted in a number of different types of tumors and has been prognostic in some, such as neuroblastoma (Seeger et al., 1985). Double minute chromatin bodies (dmins) have been reported in four PNET, and c-*myc* amplification has been identified in two tumors, although c-*myc* amplification is not a common event in PNET (Friedman et al., 1988; Raffel et al., 1990). Raffel et al. (1990) detected no amplification of N-*myc*, L-*myc*, or epidermal growth factor receptor. A common structural abnormality, iso(17q), has been reported in about one-third of these tumors (Biegel et al., 1989; Karnes et al., 1992). In

TABLE 2. Primitive Neuroectodermal Tumors (PNET) and Karyotypes

Author	Tumor type	Modal karyotype
Karnes et al., 1992	PNET	44,XX,-11,-17
Karnes	PNET × 4	46,XX, or 46,XY
Karnes	PNET	46,XX/46,XX,del(7q32)
Karnes	PNET	46,XY/47,XY,+22
Karnes	PNET	46,XY,-11,+satellited marker
Karnes	PNET	48,X–Y,-8,-9,-12,+21,-22, +5mar,dup(2)(q11→q31), del(6)(q12), +t(1;11;i(17q)),i(17q)
Karnes	PNET	46,XX/46,XX,del(7)(q33)
Griffin et al., 1988	PNET	46,XY/46,XY,+mar
Griffin	PNET	46,XX
Griffin	PNET	46,XX,6q-?
Griffin	PNET	None
Griffin	PNET	46,XY,der(1)t(1;?)(q44:?)
Griffin	PNET	45,XY,-6/45,XY,3q-,-6
Griffin	PNET	46,XY/51,XY,+3,der(5)t(1;5) (q21;q35),6q-,+13,+13, +2i(17q), der(18)t(18;?) (q21;?)
Griffin	PNET	42–46,XX,del(1)(p13→21), der(1)t(1;?)(p36;?)
Griffin	PNET	45–47,XY,der(1)t(1;?) (q44;?),Dq+
Griffin	PNET	78–80,XXYY,der(1)(1;?) (p3;?),der(4)t(1;4)(q21; q35),der(5)t(5;?)(p13;?), inv(7)(p13q11),del(9) (p13–21),del(11)(p13), +i(17q),+mar
Biegel et al.	PNET	46,XY/46,XY,9q+
Biegel	PNET	46,XX/47,XX,+mar
Biegel	PNET	46,XY,del(9)(q22),+dmin
Biegel	PNET	84,XX,-X,-X,-3,-8,-10,(-13), -15,-20,-21,i(17q),+der (21)t(21;?)(q22;?)
Biegel	PNET	45,X–Y,i(17q)
Biegel	PNET	45,XX,-6
Biegel	PNET	46,XY,del(16)(q11.2)
Biegel	PNET	45,X–X,-5,-22,+der(5)t(5;?) (q35;?),del(11)(q14), +der(22)t(22;?)(q13:?)
Biegel	PNET	91–101,XXYY,+13,+17
Biegel	PNET × 3	46,XY, or 46,XX
Biegel	PNET	45,XY,-5,-8,-11,+der(5) t(q35;q11),+der(8)t(8;?) ?)(p23;?),i(17q)
Biegel	PNET	83–89,XXYY,-3,-4,+7,-8, -10,-11,-11,-13,-20, +der(7)t(7;11)(p11;q11), del(16)(q11.2),2i(17q), +mar,+dmin
Biegel	PNET × 2	46,-X,+i(17q)
Biegel	PNET	89,XXX–X,-4,-4,-5,+7,+8, -11,-14,-14,-16,-17,-17, -18,-18,-19,-19,+del(1) (p13p32–35),+der(11) t(11;?)(p15;?),+2der(19) t(19;?)(q13;?),+3mar
Biegel	PNET	46,XY,-4,-5,-6,-6,-12,-12, -16,-17,+18,-19,del(1) (-21)+del(1)(q31),3p+q+, +der(4)t(4;?)(q35;?),+der (5)t(5;?)(q35;?),+der(16) t(16;?)(q12→21;?),+der (17)t(17;?)(p13:?),del(22) (q11),+4mar
Biegel	PNET	near-triploid,+7,-11,+14,-16, +der(11)t(11;?)(q23;?) +der(16)t(16;?)(q11→12;?)
Biegel	PNET	46,XY,t(16;20)(q13→22; q13.3),i(16q)/46,XY,del (2q2),t(16;20)(q13→22; q13.3),i(17q)
Biegel	PNET	46,XY,-10,+der(10)t(4;10) (q23→25;q23→24),+dmin
Biegel	PNET	53,XY,+4,+4,+5,+7,+7,+12, +13,i(17q),+dmin

Biegel's study, 18 tumors showed complex numerical and structural abnormalities (Biegel et al., 1989). The most common structural abnormality was i(17q), seen in seven of 21 tumors. Of note, there was no correlation between cytogenetic results and histology, size, metastasis, or extent of surgical resection (Biegel et al., 1989). Vagner-Capodano and colleagues (1989) reported loss of chromosome 22 in four PNET. Karnes and associates (1992) reported loss of chromosome 17 or i(17q) in three of 11 tumors.

MOLECULAR ANALYSIS

Molecular studies of PNET have confirmed the cytogenetic findings. Thomas and Raffel (1991) reported an RFLP analysis of 23 tumors and showed a loss of genetic material from 17p in six. Five of 11 PNET examined by James and coworkers (1990) using RFLP analysis demonstrated loss of genetic information on 17p as well. Similarly, Cogen and colleagues (1990) discovered loss of heterozygosity on chromosome 17p in four of nine tumors. Using 11 probes specific for 17p, the area of common loss was restricted to 17p11.2 to 17pter (Griffin et al., 1988). This region contains the p53 gene; a role for this gene in PNET has been hypothesized. Raffel and associates sequenced p53 cDNA from eight PNET and discovered no mutations in the conserved region (exons 5–8) in seven of the eight tumors (C. Raffel, J. Allem, A. Pellicer, and S. Lassoff, unpublished observation). This region is reported to contain more than 90% of p53 mutations in human tumors (Hollstein et al., 1991). Recent published work has also documented an absence of p53 mutations in PNET (Saylors et al., 1991). This strongly suggests that another, as yet unidentified gene on 17p may be involved in this tumor. Interestingly, one tumor with known loss of heterozygosity on 17p showed evidence of a rearrangement on the remaining 17p with a probe mapping to 17p13.3. No other 17p probe revealed this rearrangement. A novel tumor suppressor gene on 17p may be operative in PNET (C. Raffel, unpublished observation).

Cogen and colleagues (1990) suggest that loss of heterozygosity on chromosome 17p in PNET may be associated with a more aggressive phenotype; all four of their patients with 17p loss had experienced tumor recurrence or metastasis, while only one of five patients without this finding had recurrence or metastasis. This has not been true in our series of patients (C. Raffel, unpublished observations).

Cytogenetic work on PNET has revealed other areas of nonrandom changes. Biegel and colleagues (1989) mention chromosomes 5, 6, 11, and 22 as frequent targets of deletions and rearrangements. A der(11)t(8;11) was reported in two tumors. Loss of chromosome 22 was present in four of six PNET reported by Vagner-Capodano and associates (1989). Molecular studies have also revealed nonrandom loss of genetic material in other locations in PNET. Thomas and Raffel (1991) reported loss of heterozygosity in five of 23 tumors on chromosome 6q and three of 23 tumors on chromosome 16q. No loss was noted on chromosomes 1, 7, 10, 11, 13, or 22. Interestingly, no patient had loss on both 6q and 17p. The age distributions of these latter two groups did not overlap. Perhaps PNETs can be divided into two groups on this basis, each with different gene(s) involved in their formation. James and co-workers (1990) reported loss of heterozygosity on chromosome 10 in two of 11 PNET, loss of heterozygosity of chromosome 11 in three of 11 PNET, and loss of heterozygosity of 22 in two of 11 PNET. Loss on chromosome 11 was noted in association with loss of 10 in one tumor, and in association with loss of 17 in two tumors (one of which also had loss on 16). No significant difference in age distribution among tumors exhibiting different areas of loss was noted. Cogen and associates (1990) also showed loss on chromosome 6p using HSP70. These results suggest that several different molecular mechanisms are likely responsible for the genesis of pediatric PNET. All of the tumors reported by James and colleagues (1990) were classified as grade IV, suggesting as well that different mechanisms are capable of producing clinically aggressive tumors.

Cytogenetic data on other glial neoplasms occurring in children have been reported (Biegel et al., 1989; Wasson et al., 1990; Wikstrand et al., 1983). Too few of these tumors have been studied to allow identification of nonrandom changes.

CRANIOPHARYNGIOMAS

Craniopharyngiomas represent about 7% of pediatric CNS neoplasms and arise from remnants of Rathke's cleft. Only four of these suprasellar tumors have been studied. Karnes and colleagues (1992) described normal karyotypes in two tumors and complex structural abnormalities in a third. In this tumor, a translocation occurred between chromosomes 2 and 19 at 2q34 and 19q13.3 and between chromosomes 9 and 21 at 9q13 and 21q22, as well as deletion of chromosome 12 at 12q24.1, loss of chromosome 13, and an addition of an unidentified marker chromosome. One craniopharyngioma reported by Vagner-Capodano et al. (1989) exhibited loss of a Y chromosome.

RHABDOID TUMORS

Rhabdoid tumors are clinically aggressive, poorly understood tumors that occur in many different organs. Karnes and colleagues (1992) reported an undifferentiated rhabdoid tumor of the brain with multiple complex chromosome abnormalities, including a translocation involving chromosomes 1 and 19 at 1q25 and 19q13.4, a translocation of chromosome 2 at 2p11, deletion of chromosome 16 at 16q21.1, an insertion on chromosome 17 involving the long arm at 17q21, and a deletion of chromosome 20 at 20q13.1. One rhabdoid tumor of the brain and rhabdoid tumors at other sites exhibit abnormalities of chromosome 22 (Biegel et al., 1990; Karnes et al., 1992). This suggests that the rhabdoid tumor of the brain may be etiologically heterogeneous and may differ from rhabdoid tumors located elsewhere.

CONCLUSION

The cytogenetic study of pediatric primary CNS neoplasm is in its infancy. Results from study of astrocytomas, the most common group of tumors, have not yet been helpful in the understanding of these tumors. Study of PNET has allowed identification of regions of the genome that may contain genes important in the development of these tumors. Of special interest is the suggestion of a second recessive oncogene other than p53 on chromosome 17p. In addition, separation of PNET into two or more groups based on loss of heterozygosity in different locations has implications for the development of these tumors. Clearly, more data on all of these tumors needs to be accumulated, hopefully leading to a better understanding of the development and clinical behavior of pediatric central nervous system neoplasms.

REFERENCES

Biegel, J.A., L.B. Rorke, R.J. Packer, et al. (1989) Isochromosome 17q is the most common structural abnormality in CNS primitive neuroectodermal tumors. Pediatr. Neurosci. 14:153.

Biegel, J.A., L.B. Rorke, R.J. Packer, et al. (1990) Monosomy 22 in rhabdoid or atypical tumors of the brain. J. Neurosurg. 73:710–714.

Cogen, P.H., L. Danashvar, A.K. Metzger, et al. (1990) Deletion mapping of the medulloblastoma locus on chromosome 17p. Genomics 8:279–285.

Finlay, J.L., S.C. Goins, R. Uteg, et al. (1987) Progress in the management of childhood brain tumors. Hematol. Oncol. Clin. North Am. 1(4):753–776.

Friedman, H.S., P.C. Burger, S.H. Bigner, et al. (1985) Establishment and characterization of the human medulloblastoma cell line and transplantable xenograft D283 Med. J. Neuropathol. Exp. Neurol. 44(6):592–605.

Friedman, H.S., P.C. Burger, S.H. Bigner, et al. (1988) Phenotypic and genotypic analysis of a human medulloblastoma cell line and transplantable xenograft (D341 Med) demonstrating amplification of c-myc. Am. J. Pathol. 130(3):472–484.

Fults, D., R.H. Tippets, G.A. Thomas, et al. (1989) Loss of heterozygosity for loci on chromosome 17p in human malignant astrocytoma. Cancer Res. 49:6572–6577.

Griffin, C.A., A.L. Hawkins, R.J. Packer, et al. (1988) Chromosome abnormalities in pediatric brain tumors. Cancer Res. 48:175–180.

Hollstein, M., D. Sidransky, B. Vogelstein, et al. (1991) p53 mutations in human cancers. Science 253:49–53.

Jacobsen, P.F., D.J. Jenkyn, and J.M. Papadimitriou (1985) Establishment of a human medulloblastoma cell line and its heterotransplantation into nude mice. J. Neuropathol. Exp. Neurol. 44(5):472–485.

James, C.D., J. He, E. Carlbom, et al. (1990) Loss of genetic information in central nervous system tumors common to children and young adults. Genes Chromosomes Cancer 2:94–102.

Jenkin, D., K. Goddard, D. Armstrong, et al. (1989) Childhood medulloblastoma—prognostic factors. Pediatr. Neurosci. 14:151.

Karnes, P.S., T.N. Tran, M.Y. Cui, et al. (1992) Cytogenetic analysis of 39 pediatric central nervous system tumors. Cancer Genet. Cytogenet. 59:12-19.

Karnes, P.S., T.N. Tran, M.Y. Cui, et al. (1991b) Establishment of a rhabdoid tumor cell line with a specific chromosomal abnormality, 46,XY,t(11;22)(p15.5;q11.23). Cancer Genet. Cytogenet. 56:31–38.

Kernohan, J.W., R.F. Mabon, H.J. Svien, et al. (1949) A simplified classification of the gliomas. Proc. Staff Mtg. Mayo Clin. 24:71–75.

Lane, D.P., and S. Benchimol (1990) p53: Oncogene or anti-oncogene? Genes Dev. 4:1–8.

Raffel, C., F.H. Gilles, and K.I. Weinberg (1990) Reduction to homozygosity and gene amplification in central nervous system primitive neuroectodermal tumors of childhood. Cancer Res. 50:587–591.

Rorke, L.B., F.H. Gilles, R.L. Davis, et al. (1985) Revision of the World Health Organization classification of brain tumors for childhood brain tumors. Cancer 56:1869–1886.

Rorke, L.B., and L. Schut (1989) Introductory survey of pediatric brain tumors. In: R.L. McLaurin, L. Schut, J.L. Venes, F. Epstein (eds.). Pediatric Neurosurgery: Surgery of the Developing Nervous System. Philadelphia: W.B. Saunders, pp. 335–337.

Saylors, R.L. III, D. Sidransky, H.S. Friedman, et al. (1991) Infrequent p53 gene mutations in medulloblastomas. Cancer Res. 51:4721–4723.

Seeger, R., G. Brodeur, H. Sather, et al. (1985) Association of multiple copies of the n-myc oncogene with rapid progression of neuroblastomas. N. Engl. J. Med. 313:1111–1116.

Thomas, G.A., and C. Raffel (1991) Loss of heterozygosity on 6q, 16q, and 17p in human central nervous system primitive neuroectodermal tumors. Cancer Res. 51:639–643.

Tran, T.N., M.Y. Cui, P.S. Karnes, et al. (1990) Cytogenetic characterization of pediatric astrocytomas. Pediatr. Neurosci. 15A:152.

Vagner-Capodano, A.M., J.C. Gentet, M. Choux, et al. (1989) Chromosome abnormalities in sixteen pediatric brain tumors. Pediatr. Neurosci. 14:159.

Wasson, J.C., R.L. Saylors III, P. Zeltzer, et al. (1990) Oncogene amplification in pediatric brain tumors. Cancer Res. 50(10):2987–2990.

Wikstrand, C.J., S.H. Bigner, and D.D. Bigner (1983) Demonstration of complex antigenic heterogeneity in a human glioma cell line and eight derived clones by specific monoclonal antibodies. Cancer Res. 43:3327–3334.

Woo, S.Y., S.S. Donaldson, and R.S. Cox (1988) Astrocytomas in children: 14 years' experience at Stanford University Medical Center. J. Clin. Oncol. 6(6):1001–1007.

Chapter 18

Cytogenetic Analysis of Common Nervous System Tumors in Adults

Lars Poulsgård and Mogens Rønne

INTRODUCTION

Almost all human cancers have chromosome complement with nonrandom numerical and structural changes in the karyotype. Even small karyotypic deviation from normal diploid constitution may indicate an altered cellular behavior compatible with that of neoplasia. Hence karyotyping constitutes an important tool in the diagnosis of neoplasia. Solid tumors usually contain a number of different karyotypic changes and complex chromosomal rearrangements, making it very difficult to distinguish between early and late cytogenetic events. However, based on comparative studies, the observed aberrations can be subdivided into primary and secondary changes.

Primary aberrations are nonrandom changes consistently found in a tumor cell population and shared by a tumor entity from unrelated individuals. They are frequently found as the sole abnormalities in certain cancers, and often specifically associated with particular tumor types.

Secondary aberrations are often numerous and heterogeneous not only from tumor to tumor but also from cell to cell in a given tumor. They are rarely the only change occurring within the cell. These aberrations often occur in addition to the primary aberration and frequently dominate the karyotypes in later stages of the disease.

Although less specific than primary changes, secondary aberrations also demonstrate nonrandomness. The distribution patterns appear to be dependent both on the primary abnormality and on the type of the tumor. Secondary changes often become apparent only late in the diseases and therefore don't initiate neoplasia, but may play a role in tumor progression and metastasis. Primary aberrations occur most frequently very early in carcinogenesis. They are normally found at the time of diagnosis preceding the clinical manifestation of disease. It is not likely that a single chromosomal change by itself is sufficient to transform a healthy cell into a malignant one. It is more likely that carcinogenesis is multistaged, and that primary aberrations may be one out of several necessary events. Furthermore, although malignant tumors and leukemias display abnormal cells with primary or secondary aberrations or both, benign tumors also display nonrandom chromosomal changes. Clonal abnormalities such as trisomy 7, trisomy 10, and loss of the Y chromosome have been observed in short-term cultures established from normal kidney tissue (Elfving et al., 1990). Hence, the presence of clonal abnormality and secondary aberrations does not automatically imply malignancy, only that cell lines with altered karyotypes are present in the examined tissue.

ESTABLISHING TUMOR KARYOTYPE IN SOLID TUMORS

Chromosome banding is essential to cytogenetic investigations and detection of primary and secondary aberrations in tumor karyotypes. Banding allows identification of whole chromosomes and chromosomal segments. The nomenclature for the human karyotype is based on published reports from several international conferences (ISCN, 1985).

Chromosomes from solid tumors were first observed more than 100 years ago (Sandberg, 1990). Remarkable advances have been made in cytogenetic techniques since that time, but aberrant karyotypes in solid tumors are still difficult to examine in detail. There are two techniques used to establish tumor karyotype: (1) direct preparation of tumor biopsies, or (2) by cultures established from such biopsies.

Chromosome slides made from direct preparations of solid tumors often exhibit low mitotic indices and poor chromosome quality within metaphases. To counteract this, Rønne (1985) devised a direct preparation technique in which tumor biopsies were immediately transferred to complete culture medium supplemented with colcemid, which

Molecular Genetics of Nervous System Tumors, pages 223-228 ©1993 Wiley-Liss, Inc.

arrests cell division at the metaphase–anaphase border. The resultant single-cell suspension was incubated for 2–3 hours at 10–12°C before harvest. The tumor cells were able to enter mitosis at this temperature, but chromosome contractions were delayed or partially inhibited. Colcemid exposure at low temperature after hypotonic treatment and fixation produced slides that display metaphases with well-spread chromosomes of good quality. The advantage of this technique is that direct preparations ensure a true picture of cell division *in vivo*, provided that the samples are exposed to colcemid during the total incubation and preparation time. As mentioned earlier, direct preparations have severe limitations. In slow-growing tumors with a lot of connective tissue, the very low metaphase indices obtained make a representative karyotyping almost impossible. The use of cell cultures and synchronization have improved the quality of preparations and thereby increased the possibility of detecting minor chromosomal aberrations. Cell cultures established from tumor tissue present other difficulties in establishing tumor karyotypes. Cells that may not be representative of the *in vivo* condition may grow to dominance under these circumstances. *In vitro* selection and *in vitro* acquired chromosome aberrations may further confuse the overall picture. Long-term cultures tend to yield cells with high chromosome count, which seem to adapt better to *in vitro* conditions than the near-diploid cells of the same tumor. Thus, if tumor biopsies are cultured for cytogenetic analysis of primary and secondary alterations specific for that tumor, culture time should be kept at an absolute minimum to keep the *in vitro* condition as close as possible to the *in vivo* condition.

Yunis (1981) combined several previously published techniques to develop an effective short-term culture and synchronization procedure that provides sufficient material for cytogenetic analysis of tumor cells. Several types of primary tumors were examined with this technique. After Wright staining, the chromosomal material displayed metaphases in the 320–550 G-band stage. Yunis (1981) strongly advocated the use of cell synchronization with methotrexate (MTX) and a short colcemid exposure time. However, this technique may create new pitfalls. Short-term cultures established from tumor biopsies often contain several karyotypically different cell lines. Subjected to MTX synchronization and short colcemid exposure time, only cell lines with growth kinetics corresponding to the experimental set-up will be represented among the obtained metaphases. When synchronized and unsynchronized cultures are compared, one or more cell lines can be missing in the material prepared from the synchronized cultures (Rønne and Elberg, 1987).

Short-term cultures established from tumor biopsies run the risk of contamination from stroma cell proliferation. Several staining techniques should be employed and normal tissue examined for polymorphism and constitutive chromosomal abnormalities. One way to avoid stroma contamination is to use soft agar cultures (Yunis, 1981). This method yields preparations devoid of stroma cells since *in vitro* growth of normal fibroblast requires glass or plastic surfaces for attachment and proliferation. The drawbacks of this technique are the rather low cell yield obtained and the extra preparation procedures required. Another way to escape stroma contamination is to culture the tumor cells in athymic mice. Direct preparations or short-term cultures can then be established from these mice. When athymic mice are used as host for tumor transplants, invasion of host cells into the transplants may take place. Regular karyotyping in connection to subculturing is therefore highly recommended when transplants are used to establish *in vitro* cultures. There is a high risk that the original tumor preparation can be overgrown by contaminating host cells.

Leibovitz (1986) modified the standard growth media with several agents that sustain tumor growth *in vitro*. Leibovitz (1986) found the "mechanical spill out" disaggregation technique the least traumatic method of obtaining tumor cells with a minimum of stromal contamination. The enzymatic technique using collagenase disaggregation of tumor biopsies also provides high yield of viable cells (Leibovitz, 1986). However, toxic enzymes from disintegrated cells may be released during the process. Therefore, precautions must be taken when using the enzyme techniques. Rønne and Elberg (1987) also suggested mechanical disaggregation and supplemented growth medium to establish and sustain short-term culture from tumor biopsies. They coat the inner surfaces of the culture bottles with collagen; free tumor cells and biopsy fragments adhere and proliferate better on collagen-coated surfaces.

Trent and associates (1986) summarized several technical protocols about culture, harvest, and staining procedures from six laboratories, all of which successfully karyotyped a variety of human solid tumors. As mentioned earlier, technical difficulties in obtaining analyzable chromosomes from many of the most common human tumors are to some degree the consequences of the low mitotic indices, and the low quality of chromosome morphology and banding in the obtained metaphases. Eiseman and colleagues (1988) used phorbol-12,13-dibutyrate (POBU), a tumor promoter, as a mitogen in primary cultures of colon tumor cells and obtained at least 10 times more metaphases from stimulated cultures than from unstimulated controls. Chromosome morphology was also improved. With the addition of mitogenic calcium ionophore A23187, they obtained analyzable metaphases from the peripheral blood of normal individuals, chronic lymphatic leukemia patients, and a nodular small-cell lymphoma patient. According to Eiseman et al. (1988), the use of tumor promoters as mitogens may solve several of the technical problems of tumor cytogenetics.

Experiments performed by Rønne and co-workers (unpublished data) indicate, however, that these agents do not act as mitogens in all short-term cultures, so several modifications of these methods are needed before a general stimulation technique for tumor cells in short-term cultures is developed. The development and use of a method for *in vitro* stimulation of human tumor cells may be one of the ways to solve some of the current technical problems in cancer cytogenetics.

GLIOMA

Gliomas are a heterogeneous group of neoplasms that comprise nearly one-half of all intracranial tumors. They derive from cells of neuroectodermal origin; most of them arise from astrocytes, oligodendrocytes, and ependymal cells.

It is difficult to obtain an overview of the cytogenetic data concerning gliomas because of the great karyotypic diversity not only from tumor to tumor but sometimes within the same tumor. The largest published series of chromosome studies of gliomas are those of Rey et al. (1987b,c), Bigner et al. (1988), and Jenkins et al. (1989). In a cytogenetic study of 22 human low-grade gliomas, Rey and colleagues (1987c) found normal diploid stemline karyotype in all nine grade I–II astrocytomas and in seven of nine oligodendrogliomas examined. The presence of normal karyotypes in low-grade gliomas has also been found in the series of Jenkins and coworkers (1989), where nine of 16 low-grade gliomas had normal karyotypes.

The most frequent chromosomal aberration, observed in about 50% of all examined gliomas, is the presence of double minute chromatin bodies (dmins). Since dmins represent cytogenetic manifestation of gene amplification, it may be possible that amplification of an oncogene(s) could be involved in the initiation and progression of gliomas. Amplification and increased expression of the epidermal growth factor receptor gene (EGFR), which has marked structural similarities with the *erbB* oncogene located on chromosome 7, has been reported in a group of glioma patients (Libermann et al., 1985).

Besides the presence of dmins, the most consistent early karyotypic change in astrocytomas is numerical deviation of whole chromosomes, with gains of chromosome 7 and losses of chromosome 10. This aberration is found in both low-grade and high-grade astrocytomas, as well in the highly malignant glioblastoma multiforme, suggesting that these tumors arise in a uniform pattern. In an evaluation of 54 malignant gliomas, Bigner and colleagues (1988) found that 42 tumors yielded abnormal stemlines; 80% of these were near-diploid and 20% were either near-triploid or near-tetraploid. Statistically significant abnormalities of the near-diploid tumors were trisomy 7, monosomy 10, structural abnormalities of 9p and 19q, and the presence of dmins. Tumors containing only +7, -10, and dmin abnormalities suggest that these changes are the earliest chromosomal deviation in near-diploid highly malignant gliomas. Structural abnormalities, consisting mainly of deletions and unbalanced translocations, are superimposed on these primary changes. However, only structural changes at 9p and 19q were involved with a frequency that reaches statistical significance. The polyploid tumors had lower incidence of trisomy 7 and a higher incidence of monosomy 22 than the near-diploid, suggesting that these polyploid tumors may have followed an evolutionary route separate from the near-diploid ones.

Jenkins and associates (1989) obtained cytogenetic results in 50 of 54 (94%) gliomas. Abnormal clones were found in 66% of the analyzed specimens. Some karyotypes were complex, suggesting that clones with multiple chromosome abnormalities may exist in well-differentiated tumors. Fourteen tumors contained a near-triploid or near-tetraploid cell line. In addition to abnormalities involving 1p, 7q, 8p, and 9p, Jenkins et al. (1989) confirmed that trisomy 7 and monosomy 10 are the most frequently observed aberrations in diffuse astrocytomas. Other numerical gains and losses also occur frequently. Loss of chromosomes 14, 22, X, and gain of chromosomes 6, 15, 18, 19, 20, and X are reported (Sandberg, 1990). In contrast, oligodendrogliomas display karyotypes with frequent loss of the Y chromosome in addition to 1p, 6q, and 11q abnormalities (Jenkins et al., 1989).

In gliomas, associations between certain karyotypic abnormalities and histological subtypes have been suggested (Jenkins et al., 1989; Sandberg, 1990). No clearcut relationship between certain aberrant karyotypes and survival of patients with gliomas has been shown, although the longer survival of patients with anaplastic astrocytoma as compared to patients with glioblastoma multiforme may be related to the fact that most patients with anaplastic astrocytoma display normal karyotypes or lack only one sex chromosome in their tumors.

SCHWANNOMAS AND NEUROFIBROMATOSIS

Schwannomas or neurofibroma are benign encapsulated tumors that arise from Schwann cells on cranial, spinal, and peripheral nerves. These tumors lay close to the nerve without invading it. Neurofibromas may be solitary or multiple; in the latter case they constitute the feature of neurofibromatosis (NF). Neurofibromatosis consists of two distinct disorders, NF-1 and NF-2. NF-1 is an autosomal dominant condition with a prevalence of one in 3,000. Affected individuals have multiple benign neurofibromas and are at risk of developing glioma, meningioma, and neurofibrosarcoma. The gene responsible for the NF-1 condition is on chromosome 17 (Barker et al., 1987).

NF-2, also called bilateral acoustic neurofibromatosis, is a rare autosomal dominant disorder involving development of multiple tumors of the nervous system, including meningiomas, gliomas, neurofibromas, and particular bilateral acoustic neuromas. Using genetic linkage analysis with DNA markers, Rouleau and co-workers (1987) have shown that the defective gene causing NF-2 is on chromosome 22 and distinct from the gene for NF-1. Rey and associates (1987a) reported a case of a plexiform recurrent neurofibroma in a patient with a *de novo* constitutional reciprocal translocation, t(1;22)(p32;q11). The tumor showed presence of five cytogenetic clones, characterized by different chromosomal rearrangement. Chromosome aberrations included t(1;22), the marker, and chromosome pairs 1, 2, 3, 5, 8, 9, 10, 12, 16, and X. Although the patient did not display multiple lesions, the condition was considered to be a *de novo* neurofibromatosis, since plexiform neurofibromas are almost unknown in absence of NF-1. A possible association between NF-1 and a constitutional chromosomal translocation has previously been reported (Alvarez and Villapando, 1971) but is not a normal phenomenon (Kao et al., 1984).

Cytogenetic studies of neurofibromas have revealed either normal tumor karyotype or karyotype with loss of chromosome 22 (Couturier et al., 1990; Krone and Högemann, 1986; Rey et al., 1987). The monosomy 22 karyotype has been found in acoustic, spinal, and peripheral neurofibromas, as well as in sporadic neurofibromas, neurofibromas in NF-1, and in NF-2 (Couturier et al., 1990). Karyotypic changes other than monosomy 22 have been observed in neurofibromas but they share no similarities and seem to be random chromosomal changes.

Riccardi and Elder (1986) reported a cytogenetic study of neurofibrosarcomas from four unrelated patients with NF-1. They found multiple cytogenetic aberrations and highly abnormal karyotypes with loss of chromosomes and/or polysomy. Multiple structural abnormalities of diverse types were also observed. There were no consistently abnormal chromosome or chromosomal segments, indicating that the transformation of benign neurofibroma to a neurofibrosarcoma does not follow a well-defined nonrandom chromosomal pathway.

Evidence of chromosomal instability has also been reported (Hafez et al., 1985). Chromosome aberration in lymphocyte cultures, stimulated with the mitogen phytohemagglutinin from patients with neurofibromatosis and normal controls, were compared after *in vitro* irradiation of unstimulated whole blood. The test group displayed significantly higher frequencies of dicentrics and breaks and gaps than the controls, demonstrating chromosome instability (Hafez et al., 1985). The relationship between chromosome instability and cancer remains uncertain. Nevertheless, the incidence of neurosarcoma, glioma, meningioma, medullary thyroid carcinoma, rhabdomyosarcoma, and Wilms' tumor is much higher in patients with NF-1 than in the general population.

FRAGILE SITES AND NERVOUS SYSTEM TUMORS

According to Hecht and colleagues (1990), there are 107 fragile sites in the human genome. Twenty-four of these are rare, and are called heritable fragile sites. The majority of the rare fragile sites are folate-sensitive, but BUdR and ligands with preferential nonintercalate binding to dA-dT-rich DNA also unmask some of the rare sites. There is a remarkable concordance between the chromosomal location of oncogenes, chromosomal breakpoints in neoplasia, and location of fragile sites, which may be the targets for mutagens and carcinogens (Yunis et al., 1987). For example, the putative meningioma locus is within 22q12.3→qter, a region that also harbors a rare folate-sensitive site in 22q13, a common aphidicolin-sensitive site in 22q12.2, and the locus for the C-*sis* oncogene in 22q12.3→q13.1. In meningiomas, which display structural rearrangements of the tumor karyotype, this is more likely to occur in chromosome 22 with a constant involvement of the region around bands 22q11→12. To our knowledge no study has yet examined the correlation between chromosomal fragility at 22q11→12 and the risk of either developing meningiomas or of tumor recurrence after excision, so there is no direct evidence that chromosome 22 fragility is causally related to the development and recurrency of meningiomas.

In neuroblastoma, a common solid tumor in children, partial 1p monosomy is found in about 80% of the primary tumors and cell lines. The region most commonly deleted is between 1p32 and 1pter. This deletion is thought to represent the loss of the putative neuroblastoma suppressor gene (Brodeur and Fong, 1989). In addition to the 1p abnormalities, about two-thirds of all investigated neuroblastoma including cell lines display homogeneously staining regions (HSR) and/or dmins. The HSR and dmins formation in neuroblastoma appear to be linked to the amplification of the cellular oncogene N-*myc*, which is found not only in neuroblastoma cell lines but also in a substantial proportion of fresh tumors. The extent of N-*myc* amplification seems closely correlated with the stage of the disease and is especially pronounced in advanced cases and when disease progression is rapid. Since N-*myc* amplification is not a feature of all or even most primary neuroblastomas, it is unlikely to be a primary causative factor in tumor development, but may constitute a major mechanism in tumor progression. Two independent studies of chromosome fragility in peripheral blood lymphocytes from patients with neuroblastoma show interesting relationships between the N-*myc* amplification and fragile site expression. Vernole and colleagues (1988) found a higher expression of fragile sites after aphidicolin and, to a lesser extent, after methotrexate treatment in lymphocytes from neuroblastoma patients as compared with normal donors. The analysis showed that the patients had an increase in expression of fra(1)(p32), the same region where breakpoints, deletions, and rearrangements frequently occur in neuroblastoma cells. Rudolph et al. (1988) used methotrexate to unmask mostly folate-sensitive fragility sites. They examined lymphocytes from patients with neuroblastoma and their relatives. In about 45% of these patients they observed an unknown fragility in the band 1p13.1. This fragility was also observed in the mother's cell of one of the patients. The presence of fra(1)(p13.1) was also found in three members of a family with a history of neuroblastoma. Thus the fra(1)(p13.1) must be classified as a folate-sensitive hereditary fragile site. This fragile site might be of special interest in the development of neuroblastoma because 1p aberrations are specific for cells of primary neuroblastomas and cell lines. Most aberrations involve the loss of structural materials, distal to 1p13.1. Furthermore, two different breakpoints on the short arm of chromosome 2 were located to 2p23 and 2p24, the same region that harbors the N-*myc* gene at 2p24.1 (Rønne et al., 1990). The amplification and expression of that oncogene is an intrinsic biological property of neuroblastomas with a poor prognosis.

There are a number of reports on familial neuroblastoma that suggest that a gene inherited in an autosomal dominant manner confers genetic predisposition to the development of this tumor, as seen in retinoblastoma (RB) and in Wilms' tumor (Sandberg, 1990). In some patients with retinoblastoma and Wilms' tumor, a constitutional interstitial deletion involving band 13q14 (RB) and 11p13 (Wilms' tumor) have been identified. Children with these constitutional deletions develop multiple neoplasms early in life and have a variety of birth defects. On the other hand, patients whose normal cells did not contain cytogenetic abnormalities display minimal or no birth defects whether or not their tumor cells exhibited the above-mentioned deletions. Partial monosomy for chromosome 1p is the most consistent cytogenetic abnormality found in human neuroblastoma (Sandberg, 1990) and the loss of heterozygosity (LOH), which is the molecular equivalent of a deletion for the putative neuroblastoma locus, has been demonstrated for about 30% of all examined cases (Brodeur et al., 1988; Fong et al., 1989). According to Sandberg (1990) the 1p- deletion, which is restricted largely to neuroblastoma and other tumors of neuroectodermal origin, may in some cases be a genetic lesion analogous to the 13q14 deletion in retinoblastoma or the 11p13 deletion in Wilms' tumor (Knudson et al., 1976; Riccardi et al., 1978). However, no constitutional 1p- defect has yet been described in connection to neuroblastoma. It is likely that deletions large enough to be cytogenetically visible are incompatible with life, probably due to expression of embryonic lethal mutations (Sandberg, 1990).

As in cases of retinoblastoma and Wilms' tumor, patients with neuroblastoma whose normal cells do not contain cytogenetic abnormalities display minimal or no birth defects at all. To explain this, Pathak and co-workers (1989) suggested that a genetic or chromosomal defect characteristic of a particular neoplasm could be aquired postzygotically in a specific tissue, through specific somatic mutations or chromosomal rearrangement in all tissues. These aberrations may occur in a relatively low frequency resulting in somatic mosaicism, inviting the development of specific somatic cancers. Thus certain individuals may not inherit a genetic or chromosomal defect but an ability to acquire such a defect in either pre- or postnatal life. The acquired defect can be caused by the presence of an inherited fragile site, the presence of a mutated anti-oncogene, transmission of a clastogene, or exposure to environmental carcinogens. This hypothesis may be applicable to both hereditary and sporadic types of cancer. These suggestions seem at present compatible with the observed chromosome instability and the correlations between the location of inheritable fragile sites, LOH in the putative neuroblastoma locus, deletions, break, and rearrangements in 1p and amplification of N-*myc* as demonstrated in neuroblastoma.

REFERENCES

Alvarez, M.I., and J. Villapando (1971) Neurofibromatosis y translocación cromosómica informe de un caso 46,XX,t(2p-;22p+). Rev. Invest. Clin. 23:311–316.

Barker, D., E. Wright, K. Nguyen, et al. (1987) Gene for von Recklinghausen's neurofibromatosis is in the pericentric region of chromosome 17. Science 236:1100–1102.

Bigner, S.H., J. Mark, P.C. Burger, et al. (1988) Specific chromosomal abnormalities in malignant human gliomas. Cancer Res. 88:405–411.

Brodeur, G.M., and C.T. Fong (1989) Molecular biology and genetics of human neuroblastoma. Cancer Genet. Cytogenet. 41:153–174.

Brodeur, G.M., C.T. Fong, M. Morilla, et al. (1988) Molecular analysis and clinical significance of N-myc amplification and chromosome 1p monosomy in human neuroblastomas. Prog. Clin. Biol. Res. 271:3–15.

Couturier, J., O. Dalattre, M. Kujas, et al. (1990) Assessment of chromosome 22 anomalies in neurinomas by combined karyotype and RFLP analysis. Cancer Genet. Cytogenet. 45:55–62.

Eiseman, E., J.B. Luck, A.S. Mills, et al. (1988) Use of Phorbol-12,13-dibutyrate as a mitogen in the cytogenetic analysis of tumors with low mitotic indexes. Cancer Genet. Cytogenet. 34:165–175.

Elfving, P., J.C. Cigudosa, R. Lundgren, et al. (1990) Trisomy 7, trisomy 10 and loss of the Y chromosome in short term cultures of normal kidney tissue. Cytogenet. Cell Genet. 53:123–125.

Fong, C.T., N.C. Daracopoli, P.S. White, et al. (1989) Loss of heterozygosity for chromosome 1p in human neuroblastomas: Correlation with N-myc amplification. Proc. Natl. Acad. Sci. USA 86:3753–3757.

Hafez, M., L. Sharaf, S.M.A. El-Nabi, et al. (1985) Evidence of chromosomal instability in neurofibromatosis. Cancer 55:2434–2436.

Hecht, F., K.H. Ramesh, and D.H. Lockwood (1990) A guide to fragile sites on human chromosomes. Cancer Genet. Cytogenet. 44:37–45.

ISCN (1985) An International System for Human Cytogenetic Nomenclature. New York: Karger, pp. 1–118.

Jenkins, R.B., D.W. Kimmel, C.A. Moertel, et al. (1989) A cytogenetic study of 53 human gliomas. Cancer Genet. Cytogenet. 39:253–279.

Kao, Y.S., C.S. Kao-Shan, T. Knutsen, et al. (1984) Neurofibromatosis: No chromosomal defect by prophase banding technique. Cancer Genet. Cytogenet. 13:281–282.

Knudson, A.G., Jr., A.T. Meadows, W.W. Nichols, et al. (1976) Chromosome deletion and retinoblastoma. N. Engl. J. Med. 295:1120–1123.

Krone, W., and I. Högemann (1986) Cell culture studies on neurofibromatosis (von Recklinghausen's). V. Monosomy 22 and other chromosomal anomalities in cultures from peripheral neurofibromas. Hum. Genet. 74:453–455.

Leibovitz, A. (1986) Development of tumor cell lines. Cancer Genet. Cytogenet. 19:11–19.

Libermann, T.A., H.R. Nusbaum, N. Razon, et al. (1985) Amplification, enhanced expression and possible rearrangement of EGF receptor gene in primary human brain tumors of glial origin. Nature 313:144–147.

Pathak, S., M.K. Dhaliwal, and V.L. Hopwood (1989) Genetic susceptibility, somatic mosaicisms and predisposition to human cancer. Anticancer Res. 9:17–20.

Rey, J.A., M.J. Bello, J.M. de Campos, et al. (1987a) Cytogenetic clones in a recurrent neurofibroma. Cancer Genet. Cytogenet. 26:157–163.

Rey, J.A., M.J. Bello, J.M. de Campos, et al. (1987b) Chromosomal patterns in human malignant astrocytomas. Cancer Genet. Cytogenet. 29:201–221.

Rey, J.A., M.J. Bello, J.M. de Campos, et al. (1987c) Chromosomal composition of a series of 22 human low-grade gliomas. Cancer Genet. Cytogenet. 29:223–237.

Riccardi, V.M., and D.W. Elder (1986) Multible cytogenetic aberrations in neurofibrosarcomas complicating neurofibromatosis. Cancer Genet. Cytogenet. 23:199–209.

Riccardi, V.W., E. Sujansky, A.C. Smith, et al. (1978) Chromosomal imbalance in the aniridia Wilms' tumor association: 11p interstitial deletion. Pediatrics 61:604–610.

Rønne, M. (1985) An easy method for instant preparation of chromosome slides from solid tumors. Anticancer Res. 4:45–46.

Rønne, M., and J.J. Elberg (1987) Banding and preparational techniques in normal and neoplastic tissue. II. Short-term cultures from human solid tumors. Anticancer Res. 7:1209–1212.

Rønne, M., F. Kirpekar, Y. Shibasaki, et al. (1990) R-banding and in situ hybridization. Localization of single copy genes on high resolution banded chromosomes. Anticancer Res. 10:539–542.

Rouleau, G.A., W. Wertelecki, J.L. Haines, et al. (1987) Genetic linkage of bilateral acoustic neurofibromatosis to a DNA marker on chromosome 22. Nature 329:246–248.

Rudolph, B., J. Harbott, and F. Lampert (1988) Fragile sites and neuroblastoma: Fragile site at 1p13.1 and other points on lymphocyte chromosomes from patients and family members. Cancer Genet. Cytogenet. 31:83–94.

Sandberg, A.A. (1990) The Chromosome in Human Cancer and Leukemia, 2nd ed. New York: Elsevier.

Trent, J., K. Crickard, Z. Gibas, et al. (1986) Methodologic advances in the cytogenetic analysis of human solid tumors. Cancer Genet. Cytogenet. 19:57–66.

Vernole, P., B. Tedeschi, D. Caporossi, et al. (1988) Common fragile sites and human cancer. A study on lymphocytes from neuroblastoma patients. Cancer Genet. Cytogenet. 36:13–23.

Yunis, J.J. (1981) New chromosome technique in the study of human neoplasia. Hum. Pathol. 12:540–549.

Yunis, J.J., A.L. Soreng, and A.E. Bowe (1987) Fragile sites are target of diverse mutagens and carcinogens. Oncogene 1:59–69.

Chapter 19

Biology of Tumors of the Peripheral Nervous System

Garrett M. Brodeur and Rene Bernards

INTRODUCTION

Tumors of the peripheral nervous system include neuroblastomas, pheochromocytomas, and neuroepitheliomas. Neuroblastomas and pheochromocytomas are adrenergic in origin and share certain genetic features, whereas neuroepitheliomas are thought to be cholinergic and are characterized by distinct genetic features. Neuroblastomas are characterized by deletion of the short arm of chromosome 1 (1p), amplification of the N-*myc* proto-oncogene, and hyperdiploidy in subsets of tumors. All three of these genetic features have prognostic value in subsets of patients. Allelic loss of 14q also occurs with increased frequency, but the prognostic importance of this abnormality is not known yet. Pheochromocytomas have not been studied as extensively, but allelic loss for 1p appears to be a frequent change, and no clear examples of oncogene activation have been identified. Neuroepitheliomas are characterized by translocation between chromosomes 11 and 22. Although they have a characteristic pattern of proto-oncogene expression, it is not clear that any of these oncogenes are activated specifically, and no sites of allelic loss have been identified to date. Thus, cytogenetic and molecular analysis of neuroblastomas, pheochromocytomas, and neuroepitheliomas is useful in distinguishing them from each other and from other tumors in selected cases. Furthermore, certain genetic markers help predict a tumor's clinical behavior, especially for neuroblastoma.

NEUROBLASTOMA

Considerable progress has been made in recent years in advancing our knowledge of human neuroblastoma at the cellular and molecular level (Brodeur, 1990a; Brodeur and Fong, 1989; Evans et al., 1988, 1991; Pochedly, 1990). These advances have led to the identification of specific genetic abnormalities that are characteristic of subsets of tumors. These abnormalities include amplification of the *MYCN* (N-*myc*) proto-oncogene, deletion of the short arm of chromosome 1 (1p), and hyperdiploid or near-triploid tumor cell DNA content. In addition, other abnormalities have more recently been identified whose clinical significance is not yet known. These observations have contributed to our understanding of tumor heterogeneity, malignant transformation, and progression. In this chapter, the biologic and clinical significance of these genetic changes are reviewed.

Cytogenetic Analysis and Assessment of DNA Content

Neuroblastomas are characterized cytogenetically by 1p deletion, double minute chromatin bodies (dmins), and homogeneously staining regions (HSRs) (Brodeur et al., 1981; Gilbert et al., 1984). The former is thought to be the site of a suppressor gene that is lost or inactivated in neuroblastomas, whereas the latter two abnormalities are cytogenetic manifestations of gene amplification. No other cytogenetic abnormality has emerged as yet that is characteristic of neuroblastomas. The majority of neuroblastomas have a near-diploid chromosome number or DNA content, as detected by flow cytometric analysis, whereas some are clearly hyperdiploid, near-triploid or even tetraploid.

Flow cytometric analysis of the DNA content of human neuroblastoma cells was first reported by Look and colleagues (1984) in a series comprised of infants. In this analysis, abnormally high DNA content was associated with lower stages of tumor and a response to initial chemotherapy, whereas those with a "normal" DNA content were likely to have advanced stages of tumor (especially stage 4) and a poor response to this chemotherapy. This latter group of tumors likely had subtle genetic abnormalities that were not detectable by flow cytometry, such as dmins or 1p

Molecular Genetics of Nervous System Tumors, pages 229-239 ©1993 Wiley-Liss, Inc.

deletions. Subsequent studies by Look et al. (1991) and others have confirmed the prognostic importance of hyperdiploidy by flow cytometric measurement of DNA content (Bourhis et al., 1991; Cohn et al., 1990; Taylor and Locker, 1990) or by karyotype analysis (Christiansen and Lampert, 1988; Hayashi et al., 1988; Kaneko et al., 1987), although the utility of this marker may be limited to patients less than one or two years of age (Look et al., 1991).

Loss of heterozygosity and mapping of putative suppressor genes

Deletion of 1p has emerged as the most characteristic cytogenetic abnormality in primary human neuroblastomas and tumor-derived cell lines (Brodeur et al., 1981; Brodeur and Fong, 1989; Gilbert et al., 1984). These deletions usually lead to partial 1p monosomy. The region most commonly deleted in these tumors is between 1p32 and 1pter (Brodeur et al., 1981; Brodeur and Fong, 1989). A panel of polymorphic DNA probes for 1p was used to assess chromosome deletion or somatic loss of heterozygosity (LOH). Based on our analysis of a panel of normal and tumor DNA pairs from individual patients (Fong et al., 1989), the common region of LOH is between 1p36.1 and proximal 1p36.3 (Fig. 1).

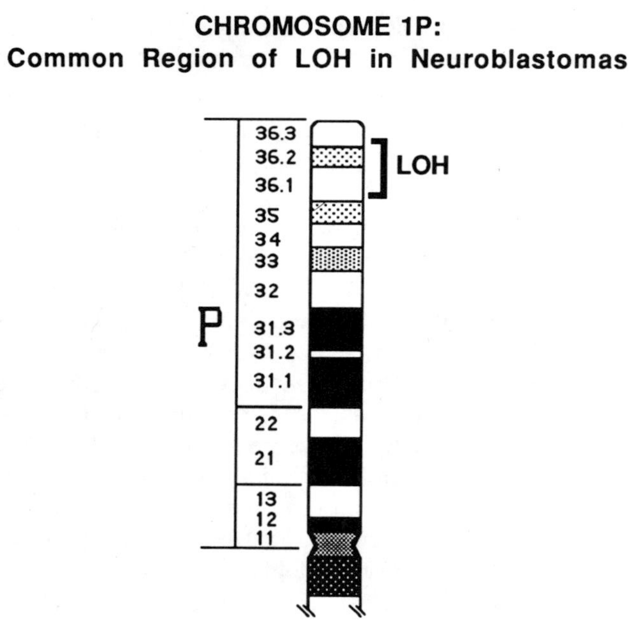

Fig. 1. Consistent region of LOH on chromosome 1p in neuroblastomas. Shown is a diagram of the short arm of chromosome 1, with the bands and subbands indicated on the left. Also shown is a bracket indicating the region that is consistently deleted in human neuroblastomas, based on molecular analysis with polymorphic probes (Fong et al., 1989; Weith et al., 1989). A neuroblastoma suppressor gene is probably encoded in this region. (Reproduced with permission from Brodeur, 1991b.)

Similar results have been obtained independently by other investigators (Weith et al., 1989). Loss or inactivation of a putative neuroblastoma suppressor gene at this site may be critical for the development or progression of neuroblastoma. Furthermore, 1p deletion or LOH may be an adverse prognostic marker (see below).

Mutation in the critical 1p36 locus on one chromosome, followed by deletion of the same region on the homologous chromosome (as manifested by LOH), could be an important mechanism in the malignant transformation or progression of some or all human neuroblastomas. Indeed, there is recent evidence that the introduction of 1p into a neuroblastoma cell line caused neural differentiation followed by cessation of cell growth and ultimately cell death (Bader et al., 1991). This provides functional evidence to support the structural evidence for a neuroblastoma suppressor gene on the short arm of chromosome 1.

There is recent evidence that LOH for the long arm of chromosome 14 also occurs with increased frequency in neuroblastomas (Suzuki et al., 1989). In this report by Suzuki, LOH for chromosome 14 was found in six of 12 informative cases, whereas LOH for chromosome 1p was found in two of nine cases. One patient had LOH for both regions. In addition, Suzuki and co-workers found LOH for chromosome 13q in two of 11 cases. Although the exact frequency and significance of these abnormalities is not yet clear, the finding of LOH on other chromosome arms besides 1p provides further evidence for genetic heterogeneity in this tumor.

Oncogene Activation

Oncogene amplification

A substantial number of neuroblastomas have either extrachromosomal dmins or chromosomally integrated HSRs (Brodeur et al., 1981; Brodeur and Fong, 1989). These two abnormalities are cytogenetic manifestations of gene amplification, but the nature of the amplified sequences was not known until recently. Schwab and colleagues (1983) identified an oncogene called N-*myc* that was amplified in all the neuroblastoma cell lines with dmins or HSRs, but not in other cell lines with these abnormalities. The amplified N-*myc* sequence was mapped to the HSRs on different chromosomes in neuroblastoma cell lines, and the normal single-copy locus was mapped to the distal short arm of chromosome 2 (Schwab et al., 1984).

We determined that N-*myc* was amplified from three- to 300-fold in a substantial number of primary neuroblastomas from untreated patients (Brodeur et al., 1984). N-*myc* amplification was highly correlated with advanced stages of disease, rapid progression, and a poor outcome (Brodeur et al., 1984; Seeger et al., 1985). These studies have been extended to more than 1,200 patients with neuroblastoma enrolled in protocols of the Children's Cancer Study Group (CCSG) and the Pediatric Oncology Group (POG) (Brodeur, 1990a; Brodeur and Fong, 1989). Examples of N-*myc* amplification seen in some of the primary tumors are shown in Figure 2. The strong association between N-*myc* amplification and advanced stages of disease seen in our earlier studies were borne out by the larger study (Table 1). However, about 5–10% of patients with more localized disease have tumors with N-*myc* amplification. Tumors with amplification are associated with more rapid progression and a worse outcome than their nonamplified counterparts, and this was especially true for patients with lower stages of disease (Brodeur, 1990a; Brodeur and Fong, 1989). This association between N-*myc* amplification and a poor prognosis is supported by studies from Japan and Europe (Bartram and Berthold, 1987; Bourhis et al., 1991; Nakagawara et al., 1987; Tsuda et al., 1987).

TABLE 1. Correlation of N-*myc* Copy Number and Stage in 1,200 Patients with Neuroblastoma*

Stage at diagnosis	Frequency of N-*myc* amplification	
Benign ganglioneuromas	0/26	(0%)
Low stages (A, B; I, II)	10/309	(3%)
Stage D-S or IV-S	6/85	(7%)
Advanced stages (C, D: III, IV)	230/780	(31%)

*Modified from previous publications (Brodeur and Fong, 1989; Brodeur, 1990a,b).

We have sought evidence for amplification of a number of other oncogenes in a panel of neuroblastoma cell lines and primary tumors, but none was found (Brodeur and Fong, 1989). Thus, N-*myc* amplification of N-*myc* appears to be characteristic of a subset of neuroblastomas.

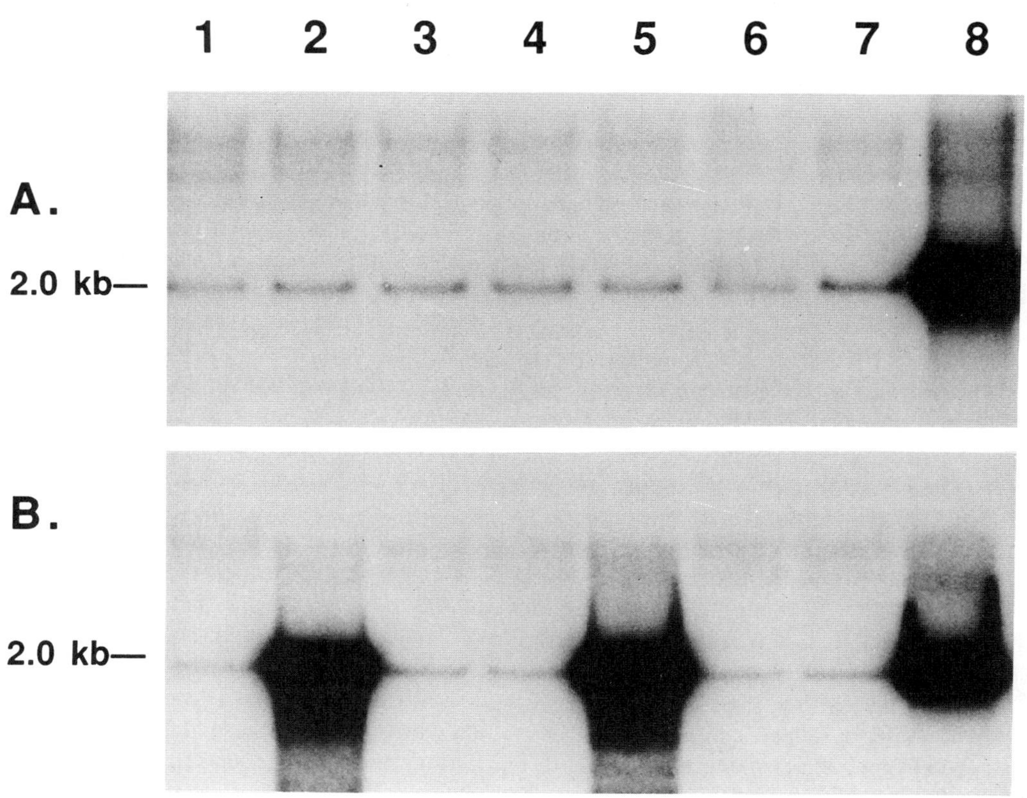

Fig. 2. Southern blots showing N-*myc* amplification. Equal amounts of DNA were digested to completion with the restriction enzyme EcoRI, electrophoresed, blotted, and hybridized to a radioactive probe for the N-*myc* oncogene. In both rows, **lane 1** represents DNA from a normal lymphoblastoid cell line as a single-copy control, and **lane 8** represents DNA from the NGP cell line, with 150 copies of N-*myc* per haploid genome, as determined by serial dilution and laser densitometry. **A: Lanes 2–7** represent six neuroblastomas with a single copy of N-*myc* per haploid genome. **B: Lanes 2 and 5** show examples of tumors with N-*myc* amplification, whereas the other tumors have the normal single-copy signal. (Reproduced with permission from Brodeur, 1990a.)

We analyzed the N-*myc* copy number in multiple simultaneous or consecutive samples of neuroblastoma tissue from 60 patients (Brodeur et al., 1987) to determine whether or not the presence or absence of N-*myc* was consistent in different tumor samples from a given patient, or if single-copy tumors ever developed amplification at the time of recurrence. Indeed, we found a consistent pattern of N-*myc* copy number (either amplified or unamplified) in different tumor samples taken from an individual patient, either simultaneously or consecutively (Brodeur et al., 1987).

These results suggest that N-*myc* amplification is an intrinsic biological property of a subset of aggressive neuroblastomas. Tumors that develop N-*myc* amplification generally do so by the time of diagnosis, and so far no cases of neuroblastoma with a single copy (per haploid genome) of N-*myc* at the time of diagnosis have developed amplification subsequently.

Our recent studies showed a very strong correlation between N-*myc* amplification and chromosome 1p LOH, indicating that 1p LOH was common in patients with amplification (Fong et al., 1989). Both N-*myc* amplification (Brodeur, 1990a; Brodeur et al., 1984; Bourhis et al., 1991; Seeger et al., 1985) and deletion of chromosome 1p (as detected by cytogenetic analysis) (Christiansen and Lampert, 1988; Hayashi et al., 1989; Kaneko et al., 1987) appear strongly correlated with a poor clinical outcome and with each other, but it is not yet clear if they are independent prognostic variables. Nevertheless, they appear to characterize a genetically distinct subset of very aggressive neuroblastomas. Our data showing the consistency of N-*myc* copy number over time (Brodeur et al., 1987) would suggest that N-*myc* amplification is an intrinsic biologic property of a subset of tumors, so it must occur relatively early in these cases. Because cases with N-*myc* amplification represent a subset of patients with chromosome 1p deletion, we suspect that the 1p deletion may precede the development of amplification.

Recent studies have compared tumor cell ploidy and N-*myc* amplification in neuroblastomas (Bourhis et al., 1991; Cohn et al., 1990; Look et al., 1991; Taylor and Locker, 1990). There was a general correlation between N-*myc* amplification and near-diploidy, although the correlation was not absolute. It appears that each is the predominant prognostic factor for a given subset of patients. The majority of infants with localized disease had hyperdiploid tumors, and hyperdiploidy identified a subset of infants with more advanced disease who had a good prognosis.

On the other hand, while N-*myc* amplification does not identify all patients who are destined to fail, it does identify a substantial number of infants, as well as older patients with nonmetastatic disease, who have a very poor prognosis. Thus, analysis of N-*myc* copy number, as well as tumor cell ploidy, provides complementary genetic information that is extremely useful in predicting response to therapy and outcome.

Oncogene expression

About 25–30% of the children with neuroblastoma have N-*myc* amplification in their tumors, and virtually all of these children have rapidly progressive and fatal disease. However, in patients with single-copy tumors, there is not yet a biological marker or explanation why half of the remaining do not survive. A general correlation has been demonstrated between N-*myc* copy number and expression (Bartram and Berthold, 1987; Grady-Leopardi et al., 1986; Nisen et al., 1988; Seeger et al., 1988; Slavc et al., 1990). Also, it has been shown that a substantial number of tumors without N-*myc* amplification overexpress this gene, but single-copy tumors with high levels of N-*myc* expression do not appear to have a particularly poor outcome (Bartram and Berthold, 1987; Grady-Leopardi et al., 1986; Nisen et al., 1988; Seeger et al., 1988; Slavc et al., 1990). It is still possible that activation of N-*myc* by mechanisms other than amplification or overexpression may play an important role. In addition, it is likely that either activation of other oncogenes, deletion of suppressor genes, or other genetic lesions may contribute to the poor outcome in these patients.

The N-*myc* oncogene was identified first in 1983 by virtue of its homology to v-*myc* and c-*myc*. The structural homology between c-*myc* and N-*myc* proteins suggests that these proteins also are similar functionally. Indeed, to date no substantial differences have been found between these two myc gene family members, except that N-*myc* is more tissue-restricted in its expression than c-*myc*. Probably as a result of this tissue-specific expression, amplification of N-*myc* is found primarily in tumors of neuroectodermal origin (neuroblastoma, retinoblastoma, and small-cell lung carcinoma), whereas c-*myc* is amplified sporadically in a variety of human tumors.

An important clue to understanding myc protein function was the observation that myc proteins shared common structural motifs with several transcription factors: a stretch of basic amino acids followed by a domain with a basic helix–loop–helix structure (bHLH motif) and a number of evenly spaced leucines (leucine zipper motif). These motifs have been shown to be involved in DNA binding and in protein dimerization (Luscher and Eisenman, 1990). Recently, Eisenman and co-workers (Blackwood and Eisenman, 1991) isolated a gene coding for a protein named MAX, which can form a specific complex with both c-*myc* and N-*myc* through these bHLH and leucine zipper motifs. Furthermore, it was found that the myc/MAX heterodimer could bind to DNA in a sequence-specific fashion. These data strongly suggest that myc proteins act by altering the expression of certain key cellular genes, the altered expression of which is responsible for the myc-induced changes in cellular phenotype.

We have demonstrated that the N-*myc* protein can form a specific complex with the product of the retinoblastoma gene, $pp105^{Rb}$ (Rustgi et al., 1991). Because the retinoblastoma protein generally is thought to be a negative regulator of cell growth, it is possible that $pp105^{Rb}$ suppresses the activity of the N-*myc* protein through direct binding. Amplification of N-*myc* would result in a large excess of N-*myc* protein relative to $pp105^{Rb}$, leaving much of the N-*myc* protein free to disrupt cellular gene expression. We have studied the effects of N-*myc* expression on neuroblastoma phenotype by transfecting a rat neuroblastoma cell line with the N-*myc* oncogene. We found that N-*myc* elicits a number of responses in these cells that are reminiscent of advanced stage human neuroblastoma, such as increased growth rate and increased metastatic ability. We compared the pattern of cellular gene expression in the parental rat neuroblastoma cells with that of the N-*myc*-transfected derivatives and found that N-*myc* suppresses the expression of major histocompatibility comples (MHC) class I antigens (Bernards et al., 1986), neural cell adhesion molecule (NCAM) (Akeson and Bernards, 1990), and the δ isoform of protein kinase C (Bernards, 1991) in these cells. These data suggest that N-*myc* is a pleiotropic factor that can perturb a variety of cellular processes in neuroblastoma, including immune recognition, cell adhesion, and signal transduction.

We have sought evidence for amplification or overexpression of a number of other oncogenes in a panel of neuroblastoma cell lines and primary tumors, but none was found (Brodeur and Fong, 1989). Although N-*ras* was first identified as the transforming gene of a human neuroblastoma cell line, subsequent studies of primary neuroblastomas by ourselves and others (Ballas et al., 1988; Ireland, 1989; Moley et al., 1991) indicate that ras activation by mutation of codons 12, 13, 59, or 61 is rare. Recent studies of H-*ras* and c-*src* expression in neuroblastomas suggest that higher levels of expression were associated with differentiated histology and with a good prognosis (Matsunaga et al., 1991; Tanaka et al., 1988). Thus, in the patients whose tumors lack N-*myc* amplification, patterns of oncogene expression may identify a subset with a more favorable outcome.

Summary

Neuroblastomas are heterogeneous in terms of their genotype and their clinical behavior. Recent studies suggest that these two features are related, and that the genotype frequently is predictive of response to treatment or the outcome of the patient. The genetic abnormalities that are characteristic of certain neuroblastomas include: (1) LOH for the short arm of chromosome 1, including band 1p36; (2) amplification of the N-*myc* proto-oncogene; and (3) hyperdiploidy, or near-triploidy determined by flow cytometry or karyotype. Hyperdiploidy is associated with lower stages of disease and with a favorable outcome in infants. However, LOH for 1p36 and N-*myc* amplification are more common in patients over one year of age with advanced stages of disease.

TABLE 2. Clinical/Genetic Types of Neuroblastoma*

Feature	Type 1	Type 2	Type 3
Ploidy	Hyperdiploid Near-triploid	Near-diploid Near-tetraploid	Near-diploid Near-tetraploid
1p LOH	Absent	±Present	±Present
N-*myc* Copy	Normal	Normal	Amplified
Age	<12 months	Any age	Any age
Stage	1, 2, 4S	3, 4	Any stage
Outcome	Good	Intermediate	Bad

*Modified from previous publications (Brodeur, 1990a,b).

Indeed, patterns are emerging, based on cytogenetic, molecular, and flow cytometric analysis, that suggest that neuroblastomas may be assigned to three genetically distinct groups (Table 2). The first comprises those with hyperdiploid or triploid modal karyotypes (or compatible DNA content by flow cytometry). LOH of 1p and N-*myc* amplification are rarely seen. These patients are more likely to be infants with low stages of disease (stages 1, 2, or 4-S by the International Neuroblastoma Staging System) (Brodeur et al., 1988), and they generally have a very favorable prognosis. The second group consists of tumors that generally have a near-diploid or tetraploid modal chromosome number or DNA content. These tumors lack 1p LOH and N-*myc* amplification, but the patients are more likely to be over one year of age and have advanced stages of disease (stages 3 or 4). While their overall outcome is intermediate, their tumors generally respond to initial treatment, and they may survive for years before succumbing to their disease. The third group consists of tumors that are also generally near-diploid or tetraploid, with 1p LOH and/or N-*myc* amplification. The patients also are more likely to be over one year of age and have advanced stages of disease (stages 3 or 4). In contrast to the second group, they respond to treatment only transiently, if at all, before they have rapid tumor progression and die within six to 12 months. It remains to be determined if tumors in one group ever evolve or "progress" into a less unfavorable group, but current evidence would suggest that they are distinct genetically. Thus, genetic analysis of neuroblastoma cells provides prognostic information beyond that provided by age and stage alone. Consideration of the genotype of neuroblastomas can assist in making the most appropriate choice of treatment for a given patient.

PHEOCHROMOCYTOMAS

Pheochromocytoma is a rare tumor derived from chromaffin cells (Carney et al., 1976). This tumor generally occurs in the adrenal medulla, but may also arise in the organ of Zuckerkandl or in the retroperitoneal area. About 20% of pheochromocytomas occur in the pediatric age group, and about one-third have multiple primary tumors. Since this tumor excretes urinary catecholamines, some patients have hypertension due to excessive catecholamine secretion. Since urinary catecholamine metabolites are almost always increased, the tumor must be distinguished from neuroblastoma by clinical, radiographic, and histological features (Bravo and Gifford, 1984). Although the majority of pheochromocytomas are benign, some are locally invasive and metastatic. The distinction between malignant and benign tumors is based on clinical behavior.

Pheochromocytomas can occur sporadically or as part of several genetic syndromes, such as multiple endocrine neoplasia (MEN) type 2A, MEN-2B, neurofibromatosis type 1 (NF-1), or von Hippel-Lindau (vHL). Interestingly, the predisposition locus for MEN-2A and -2B has been mapped to the pericentromeric region of chromosome 10 (Jackson et al., 1988; Mathew et al., 1987a; Simpson et al., 1987), whereas NF-1 has been mapped to chromosome 17q (Seizinger et al., 1987) and vHL has been mapped to 3p (Seizinger et al., 1988). This suggests that predisposition to develop this tumor is heterogeneous genetically. Thus, either there are several different genetic pathways to develop pheochromocytoma, or else what appears to be the same tumor histologically may actually be several different tumors genetically. Frequently it is difficult to predict the clinical behavior of these tumors based on the histologic appearance alone. A recent study showed a correlation between expression of neuropeptide Y benign clinical behavior of pheochromocytomas (Helman et al., 1989). However, this association was not confirmed in another report (Grouzmann et al., 1990). A better understanding of the genetic features of these different forms of neuroendocrine neoplasia may provide useful markers to predict clinical behavior.

Cytogenetic Analysis and Assessment of DNA Content

There is very limited information about genetic features of sporadic or hereditary pheochromocytomas. This is due in part to its infrequency, less malignant behavior, and its low mitotic rate. In one study of three pheochromocytomas in patients with vHL, one had a normal karyotype, one had trisomy 7, and a third had several clones, one of which had abnormalities of chromosome 3p (Kiechle-Schwartz et al., 1989). In another report there was partial monosomy for chromosome 22 (Szucs et al., 1989). However, no consistent karyotypic abnormalities have been identified to date, and there have not been any flow cytometric studies of the DNA content of pheochromocytomas.

Loss of heterozygosity and mapping of putative suppressor genes

In patients with hereditary retinoblastoma, all of the normal (or constitutional) cells of the body are heterozygous for the retinoblastoma locus on the long arm of chromosome 13. That is, they have one copy of a mutated gene, and one copy of the normal gene. The normal gene is expressed in a dominant fashion, so these cells do not exhibit malignant features. The retinoblastomas that develop in these patients, on the other hand, are homozygous for a mutated retinoblastoma gene. That is, they have lost the normal, wild-type gene, which allows expression of the retinoblastoma phenotype and formation of a tumor. However, LOH at the predisposing gene locus has not been found in tumors of the MEN-2A syndrome. Three recent reports noted allelic loss on chromosome 10 in only 5% of cases (4 of 84 informative tumors) (Landsvater et al., 1989; Nelkin et al., 1989; Okazaki et al., 1989). Therefore, malignant transformation in MEN-2A is probably caused by some mechanism other than homozygous loss or mutation of the predisposition locus, although homozygous submicroscopic mutations cannot be ruled out as yet.

Several studies using polymorphic probes for the short arm of chromosome 1 have noted LOH involving 1p in about half the cases studied (Khosla et al., 1991; Mathew et al., 1987b; Moley et al., 1992; Takai et al., 1987; Tsutsumi et al., 1989; Yang et al., 1990). In cases in which it was examined, these deletions were not noted in the constitutional (normal) DNA of the patients, indicating that the deletion does not reflect the site of the inherited mutation. LOH has been noted on other chromosomes also, including 3p, 11p, 17p, and chromosome 22q, but LOH for 1p has been the most consistent finding to date (Khosla et al., 1991; Mathew et al., 1987b; Moley et al., 1992; Takai et al., 1987; Tsutsumi et al., 1989; Yang et al., 1990). This suggests that an important suppressor gene resides on this chromosomal arm, and its loss or inactivation plays a role in the development or progression of pheochromocytomas, at least in a substantial subset of patients. It is possible that allelic loss occurs with increased frequency on some other chromosomal arm or arms, since most studies have involved small numbers of cases and/or polymorphic probes for a limited number of chromosomal arms. Nevertheless, 1p LOH has been seen frequently in every study in which it was examined.

Oncogene Activation

One group used *in situ* hybridization of paraffin-imbedded tissue sections and noted high levels of expression of the proto-oncogene N-*myc* in six of 21 medullary thyroid cancers (MTCs) analyzed (Boultwood et al., 1988). In addition, a recent study demonstrated expression (without gene amplification or rearrangement) of c-*myc* and c-*fos* in six pheochromocytomas, suggesting that these cells are in a state of growth stimulation (Goto et al., 1990). These studies suggest that increased expression of N-*myc*, c-*myc*, c-*fos*, and possibly other oncogenes may be characteristic of at least some pheochromocytomas. Whether or not this increased expression represents activation of these oncogenes, or merely reflects their active state of growth, remains to be determined.

The ras family of oncogenes (N-*ras*, H-*ras*, K-*ras*) has been implicated in the development of up to 30% of human tumors. Given the prevalence of ras mutations in human cancers, and the association of ras mutations with early changes and lower-grade malignancies, it was reasonable to suspect that ras mutations might occur in pheochromocytomas and MTCs. Two recent studies have examined the ras genes of pheochromocytomas and medullary thyroid cancers for mutations of codons 12, 13, 59, 60, and 61 (Moley et al., 1991; Okazaki et al., 1989). Both studies concluded that mutational activation of ras genes rarely if ever occurs in these tumors.

Insulinlike growth factor I (IGF-I) and IGF-II are polypeptide hormones that are structurally and functionally similar to insulin. There is some evidence that IGF-II is expressed at high levels in fetal tissues, and it stimulates growth of certain cells *in vitro*. The growth-stimulating activity of IGF-I and IGF-II is mediated by the IGF receptor (IGFR), which is a heterotetrameric protein that is similar to the insulin receptor (Rechler and Nissley, 1985). Increased levels of IGF-II (but not IGF-I) have been demonstrated in fetal adrenal medulla, compared to the adult adrenal (Brice et al., 1989). Interestingly, increased levels of IGF-II were detected in 11/11 pheochromocytomas from two studies (El-Badry et al., 1989; Haselbacher et al., 1987). A neuroblastoma cell line has been described that shows autonomous growth in serum-free medium mediated by IGF-II (El-Badry et al., 1989). Indeed, some other neuroblastoma cell lines show increased growth rates in response to IGF-II in the medium (El-Badry et al., 1991). These results suggest that the IGF-II/IGFR pathway may represent an autocrine or paracrine pathway of growth stimulation in neuroblastomas and possibly pheochromocytomas.

Summary

Pheochromocytomas may occur sporadically or in a variety of genetic backgrounds. The most characteristic genetic feature of these tumors is allelic loss for 1p, but other sites of allelic loss may occur as well. There have been several preliminary studies to detect activation of selected oncogenes and growth factors by amplification, gene rearrangement, overexpression, or base-pair mutation, but there are no unequivocal examples of genetic alterations leading to activation of a particular oncogene in these tumors. It is reasonable to examine more tumors for patterns of allelic loss or oncogene activation, but no particular oncogene or mechanism of activation has been identified so far that is characteristic of these tumors.

TABLE 3. Distinguishing Features of Neuroblastoma and Neuroepithelioma*

Characteristic	Neuroblastoma	Neuroepithelioma
Location	Adrenal, paraspinal	Extremity, Chest wall
Patient age	Usually <5 yrs	Usually >5 yrs
Neurotransmitter	Norepinephrine	Acetylcholine
HLA expression	Absent	Present
Cytogenetics	Del 1p; dmins, HSR	Trans (11;22)
Oncogene amplification	Common (N-*myc*)	Uncommon (c-*myc*)
Oncogene expression	N-*myc*	c-*myc*

*Modified from previous publication (Brodeur, 1991a).

NEUROEPITHELIOMAS

Peripheral neuroepithelioma (also known as primitive neuroectodermal tumor, or PNET) is an uncommon malignant tumor of the peripheral nervous system (Askin et al., 1979; Lagervist et al., 1969; Miser et al., 1987; Voss et al., 1984). While this tumor can resemble neuroblastoma histologically, there are distinguishing clinical and biological features (Table 3) (Brodeur, 1991a; Thiele et al., 1987a). These tumors tend to occur in older patients and arise more commonly in the chest wall or extremities from peripheral nerves. Biologically these tumors are characterized by a specific translocation between chromosomes 11 and 22 (see below), and have a cholinergic neurotransmitter enzyme phenotype. Recent evidence from chromosome analysis and oncogene expression pattern suggests that they may be related closely to Ewing's sarcomas.

Cytogenetic Analysis and Assessment of DNA Content

The majority of neuroepitheliomas that have been karyotyped have a reciprocal translocation involving chromosomes 11 and 22: t(11;22)(q24;q12) (Whang-Peng et al., 1984, 1986). However, they do not have 1p deletions, and they seldom have the cytogenetic manifestations of gene amplification (dmins, HSRs), which are characteristic of many neuroblastomas. Thus, the cytogenetic features serve to distinguish neuroepitheliomas from neuroblastomas (Table 3). There have not been any reports of analysis of tumor cell DNA content, but the majority of neuroepitheliomas have karyotypes that are pseudodiploid or hyperdiploid (47–53 chromosomes) (Whang-Peng et al., 1984, 1986).

Loss of heterozygosity and mapping of putative suppressor genes

There have been no studies documenting consistent deletions or LOH of any chromosomal regions. It is possible that there is a suppressor gene whose expression is altered by the translocation between chromosomes 11 and 22. However, it seems more likely that this rearrangement results in activation of an oncogene, as is seen in most translocation breakpoints that have been cloned in leukemias. There have been very few molecular studies of neuroepitheliomas other than those focused on the 11;22 translocation. Thus, it is likely that more information will be obtained concerning LOH and putative suppressor gene loci once additional tumors are studied.

Oncogene Activation

There have been several studies of oncogene expression in neuroepitheliomas. In one study, patterns of oncogene expression served to distinguish this tumor from neuroblastomas, which it resembles histologically (Thiele et al., 1987a). In another study, the pattern of oncogene expression of neuroepitheliomas and Ewing's sarcoma, a tumor of bone, were indistinguishable (McKeon et al., 1988). Both had moderate expression of c-*myc* (and no expression of N-*myc*). Occasionally neuroepithelioma cell lines have c-*myc* amplification (Kohl et al., 1983; Reynolds et al., 1988). In addition, both had moderate expression of HLA class I antigens, whereas neuroblastomas generally have very low HLA antigen expression, low or absent c-*myc* expression, and moderate to high N-*myc* expression. Even though the c-*ets1* gene maps near the breakpoint on chromosome 11, and c-*sis* maps near the breakpoint on chromosome 22, neither appears to be activated by the 11;22 translocation in neuroepitheliomas or Ewing's sarcomas (Bechet et al., 1984; de Taisne et al., 1984; Emanuel et al., 1985; Thiele et al., 1987b). Thus, despite a characteristic pattern of oncogene expression in these tumors, no oncogene has been shown to be activated specifically or consistently. The most promising candidate at this time (other than the presumptive gene activated by the 11;22 translocation) is c-*myc*, which is expressed frequently and amplified occasionally.

Summary

Neuroepitheliomas resemble neuroblastomas histologically, although they are readily distinguishable on the basis of clinical and biological features. However, neuroepitheliomas do share a number of features with Ewing's sarcomas, such as a characteristic 11;22 translocation and similar patterns of expression of oncogenes, HLA antigens, and other histochemical markers. Although this translocation was identified more than seven years ago, the breakpoint has not yet been cloned, and no one has identified as yet a specific suppressor gene that is inactivated or an oncogene that is activated.

CONCLUSION

Neuroblastoma, pheochromocytoma, and neuroepithelioma are neuroectodermal tumors of the peripheral nervous system. Although the former two share a similar genetic lesion-deletion of 1p, it is not clear if the same gene or locus is involved in the pathogenesis of both diseases. Indeed, given the rather large deletions of 1p in most pheochromocytomas, it is possible that a more proximal locus is involved, or possibly two distinct loci on 1p. Furthermore, these two tumors have very different clinical behaviors. Neuroblastomas are more common in infants than in adults and frequently are metastatic at diagnosis, whereas pheochromocytomas are more common in adults, and have a much more benign clinical behavior.

Neuroepitheliomas resemble neuroblastomas histologically, but they can be distinguished readily based on histochemical, genetic, and clinical features. Although a characteristic translocation and pattern of oncogene expression have been identified, very little is known about the molecular genetic basis of this disease. It is not clear if the group of tumors that shares the 11;22 translocation (neuroepithelioma, Ewing's sarcoma, Askin's tumor, esthesioneuroblastoma) is identical, or if they have variations on a common rearrangement, as is seen in chronic myelogenous leukemia and acute lymphoblastic leukemia with a 9;22 translocation and different rearrangements of BCR and c-abl. Furthermore, there may be other genetic lesions (loss of suppressor genes, activation of oncogenes) that serve to distinguish these diseases. Finally, the clinical significance of genetic changes in these tumors remains to be determined.

Neuroblastomas have been characterized more thoroughly than the other two tumors by both cytogenetic and molecular analysis. Based on genetic features, these tumors can be categorized into at least three groups that have distinctive clinical behaviors. Hyperdiploidy, with whole chromosome gains, characterizes the first group, which has the least aggressive clinical behavior. Tumors that are near-diploid without N-*myc* amplification have an intermediate behavior, whereas tumors with N-*myc* amplification are very aggressive, regardless of other clinical or genetic features. The contribution of 1p LOH or 14q LOH to clinical behavior remains to be determined, although the former has been associated with a poor outcome. It is not clear if this is due to an adverse impact of 1p LOH itself or to its association with N-*myc* amplification. However, genetic analysis indicates clearly that neuroblastomas are more complex genetically than was thought originally.

ACKNOWLEDGMENTS

Some of the material in this review was published previously (Brodeur, 1990a,b; Brodeur et al., 1991a,b; Brodeur and Fong,

1989; Brodeur and Moley, 1991; Brodeur, 1992). This work was supported in part by National Institutes of Health grants RO1-CA-39771, PO1-CA49712, and U10-CA-05887.

REFERENCES

Akeson, R., and R. Bernards (1990) N-*myc* downregulates NCAM expression in rat neuroblastoma. Mol. Cell. Biol. 10:2102–2016.

Askin, F.B., J. Rosai, R.K. Sbley, et al. (1979) Malignant, small cell tumor of the thoracopulmonary region in childhood. Cancer 43:2438–2451.

Bader, S., C. Fasching, G.M. Brodeur, et al. (1991) Dissociation of suppression of tumorigenicity and differentiation in vitro effected by transfer of single human chromosomes into human neuroblastoma cells. Cell Growth Diff. 2:245–255.

Ballas, K., J. Lyons, J.W.G. Jannsen, et al. (1988) Incidence of ras gene mutations in neuroblastoma. Eur. J. Pediatr. 147:313–314.

Bartram, C.R., and F. Berthold (1987) Amplification and expression of the N-*myc* gene in neuroblastoma. Eur. J. Pediatr. 146:162–165.

Bechet, J.M., G. Bornkamm, and G.M. Lenoir (1984) The c-sis oncogene oncogene is not activated in Ewing's sarcoma. N. Engl. J. Med. 310:393.

Bernards, R. (1991) N-*myc* disrupts protein kinase C-mediated signal transduction in neuroblastoma. EMBO J. 10:1119–1125.

Bernards, R., S.K. Dessain, and R.A. Weinberg (1986) N-*myc* amplification causes down-modulation of MHC class I antigen expression in neuroblastoma. Cell 47:667–674.

Blackwood, E.M., and R.N. Eisenman (1991) Max, a helix-loop-helix protein that forms a sequence-specific DNA binding complex with myc. Science 251:1211–1217.

Boultwood, J., F.S. Wyllie, E.D. Williams, et al. (1988) MYCN expression in neoplasia of human thyroid C-cells. Cancer Res. 48:4073–4077.

Bourhis, J., F. DeVathaire, G.D. Wilson, et al. (1991) Combined analysis of DNA ploidy index and N-*myc* genomic content in neuroblastoma. Cancer Res. 51:33–36.

Bravo, E.L., and R.W. Gifford (1984) Pheochromocytoma: Diagnosis, localization and management. N. Engl. J. Med. 311:1298–1303.

Brice, A.L., J.E. Cheetham, V.N. Bolton, et al. (1989) Temporal changes in the expression of the insulin-like growth factor II gene associated with tissue maturation in the human fetus. Development 106:543–554.

Brodeur, G.M. (1990a) Neuroblastoma: Clinical significance of genetic abnormalities. Cancer Surveys 9:673–688.

Brodeur, G.M. (1990b) Neuroblastoma—Clinical applications of molecular parameters. Brain Path. 1:47–54.

Brodeur, G.M. (1991a) Neuroblastoma and other peripheral neuroectodermal tumors. In: D.J. Fernbach and T.J. Vietti (eds.). Clinical Pediatric Oncology, 4th ed. St. Louis: Mosby Year Book, pp. 437–464.

Brodeur, G.M. (1991b) Patterns and significance of genetic changes in neuroblastomas. In: T.P. Pretlow and T.G. Pretlow (eds.). Biochemical and Molecular Aspects of Selected Tumors. Orlando: Academic Press (in press).

Brodeur, G., C. Azar, M. Brother, et al. (1992) Neuroblastoma: Effect of genetic factors on prognosis and treatment. Cancer 70:1685-1694.

Brodeur, G.M., and C.T. Fong (1989) Molecular biology and genetics of human neuroblastoma. Cancer Genet. Cytogenet. 41:153–174.

Brodeur, G.M., A.A. Green, F.A. Hayes, et al. (1981) Cytogenetic features of human neuroblastomas and cell lines. Cancer Res. 41:4678–4686.

Brodeur, G.M., F.A. Hayes, A.A. Green, et al. (1987) Consistent N-*myc* copy number in simultaneous or consecutive neuroblastoma samples from sixty individual patients. Cancer Res. 47:4248–4253.

Brodeur, G.M., and J.F. Moley (1991) Biology of tumors of the peripheral nervous system. Cancer Metas. Rev. 10:321-333.

Brodeur, G.M., R.C. Seeger, A. Barrett, et al. (1988) International criteria for diagnosis, staging, and response to treatment in patients with neuroblastoma. J. Clin. Oncol. 6:1874–1881.

Brodeur, G.M., R.C. Seeger, M. Schwab, et al. (1984) Amplification of N-*myc* in untreated human neuroblastomas correlates with advanced disease stage. Science 224:1121–1124.

Carney, J.A., G.W. Sizemore, and G.M. Tyre (1976) Adrenal medullary disease in multiple endocrine neoplasia type 2: Pheochromocytoma and its precursor. Am. J. Clin. Pathol. 66:279–290.

Christiansen, H., and F. Lampert (1988) Tumour karyotype discriminates between good and bad prognostic outcome in neuroblastoma. Br. J. Cancer 57:121–126.

Cohn, S.L., A.W. Rademaker, H.R. Salwen, et al. (1990) Analysis of DNA ploidy and proliferative activity in relation to histology and N-*myc* amplification in neuroblastoma. Am. J. Pathol. 136:1043–1052.

de Taisne, C., A. Gregonne, D. Stehelin, et al. (1984) Chromosomal localization of the human proto-oncogene c-ets. Nature 310:581–583.

El-Badry, O.M., L.J. Helman, J. Chatten, et al. (1991) Insulin-like growth factor II-mediated proliferation of human neuroblastoma. J. Clin. Invest. 87:648–657.

El-Badry, O.M., J.A. Romanus, L.J. Helman, et al. (1989) Autonomous growth of a human neuroblastoma cell line is mediated by insulin-like growth factor II. J. Clin. Invest. 84:829–839.

Emanuel, B.S., P.C. Nowell, C. McKeon, et al. (1985) Translocation breakpoint mapping: Molecular and cytogenetic studies of chromosome 22. Cancer Genet. Cytogenet. 19:81–92.

Evans, E., G.J. D'Angio, A.G.J. Knudson, et al. (eds.) (1991) Advances in Neuroblastoma Research III. New York: Wiley-Liss.

Evans, E., G.J. D'Angio, A.G.J. Knudson, et al. (eds.) (1988) Advances in Neuroblastoma Research II. New York: Alan R. Liss.

Fong, C.T., N.C. Dracopoli, P.S. White, et al. (1989) Loss of heterozygosity for chromosome 1p in human neuroblastomas: Correlation with N-myc amplification. Proc. Natl. Acad. Sci. USA 86:3753–3757.

Gilbert, F., M. Feder, G. Balaban, et al. (1984) Human neuroblastomas and abnormalities of chromosome 1 and 17. Cancer Res. 44:5444–5449.

Goto, K., A. Ogo, T. Yanase, et al. (1990) Expression of c-*fos* and c-*myc* proto-oncogenes in human adrenal pheochromocytomas. J. Clin. Endocrinol. Metab. 70:353–357.

Grady-Leopardi, E.F., M. Schwab, A.R. Ablin, et al. (1986) Detection of N-*myc* expression in human neuroblastoma by in situ hybridization and blot analysis: Relationship to clinical outcome. Cancer Res. 46:3196–3199.

Grouzmann, E., C. Gicquel, P.F. Plouin, et al. (1990) Neuropeptide Y and neuron-specific enolase levels in benign and malignant pheochromocytomas. Cancer 66:1833–1835.

Haselbacher, G.K., J.C. Irminger, J. Zapf, et al. (1987) Insulin-like growth factor II in human adrenal pheochromocytomas and Wilms tumors: Expression at the mRNA and protein level. Proc. Natl. Acad. Sci. USA 84:1104–1106.

Hayashi, Y., T. Inabada, R. Hanada, et al. (1988) Chromosome findings and prognosis in 15 patients with neuroblastoma found by VMA mass screening. J. Pediatr. 112:67–71.

Hayashi, Y., N. Kanda, T. Inaba, et al. (1989) Cytogenetic findings and prognosis in neuroblastoma with emphasis on marker chromosome 1. Cancer 63:126–132.

Helman, L.J., P.S. Colten, S.D. Averbuch, et al. (1989) Neuropeptide Y expression distinguishes malignant from benign pheochromocytoma. J Clin. Oncol. 7:1720–1725.

Ireland, C.M. (1989) Activated N-ras oncogenes in human neuroblastoma. Cancer Res. 49:5530–5533.

Jackson, C., R. Nourm, L. O'Neal, et al. (1988) Linkage between MEN-IIb and chromosome 10 markers linked to MEN-IIa. Am. J. Hum. Genet. 43:A147.

Kaneko, Y., N. Kanda, N. Maseki, et al. (1987) Different karyotypic patterns in early and advanced stage neuroblastomas. Cancer Res. 47:311–318.

Khosla, S., V.M. Patel, I.D. Hay, et al. (1991) Loss of heterozygosity suggests multiple genetic alterations in pheochromocytomas and medullary thyroid carcinomas. J. Clin. Invest. 87:1691–1699.

Kiechle-Schwartz, M., H.P.H. Neumann, H.J.J. Decker, et al. (1989) Cytogenetic studies on three pheochromocytomas derived from patients with von Hippel-Lindau syndrome. Hum. Genet. 82:127–130.

Kohl, N.E., N. Kanda, R.R. Schreck, et al. (1983) Transposition and amplification of oncogene-related sequences in human neuroblastomas. Cell 35:359–367.

Lagervist, B., B. Ivemark, and B. Sylven (1969) Malignant neuroepithelioma in childhood. Acta Chir. Scand. 135:641–645.

Landsvater, R.M., C.G.P. Mathew, B.A. Smith, et al. (1989) Development of multiple endocrine neoplasia type IIa does not involve substantial deletions of chromosome 10. Genomics 4:246–250.

Look, A.T., F.A. Hayes, R. Nitschke, et al. (1984) Cellular DNA content as predictor of response to chemotherapy in infants with unresectable neuroblastoma. N. Engl. J. Med. 311:231–235.

Look, A.T., F.A. Hayes, J.J. Shuster, et al. (1991) Clinical relevance of tumor cell ploidy and N-myc gene amplification in childhood neuroblastoma: A pediatric oncology group study. J. Clin. Oncol. 9:581–591.

Luscher, B., and R.N. Eisenman (1990) New light on myc and myb. 1. Myc. Genes Dev. 4:2025–2035.

Mathew, C.G.P., B.A. Smith, K. Thorpe, et al. (1987a) Deletion of genes on chromosome 1 in endocrine neoplasia. Nature 328:524–526.

Mathew, C.G.P., K.S. Chin, D.F. Easton, et al. (1987b) A linked genetic marker for multiple endocrine neoplasia type 2A on chromosome 10. Nature 328:527–528.

Matsunaga, T., H. Takahashi, N. Ohnuma, et al. (1991) Expression of N-*myc* and c-*src* protooncogenes correlating to the undifferentiated phenotype and prognosis of primary neuroblastomas. Cancer Res. 51:3148–3152.

McKeon, C., C.J. Thiele, R.A. Ross, et al. (1988) Indistinguishable patterns of protooncogene expression in two distinct but closely related tumors: Ewing's sarcoma and neuroepithelioma. Cancer Res. 48:4307–4311.

Miser, J.S., T.J. Kinsella, T.J. Triche, et al. (1987) Treatment of peripheral neuroepithelioma in children and young adults. J. Clin. Oncol. 5:1752–1758.

Moley, J.F., M.B. Brother, S.A. Wells, et al. (1990) Prevalence of chromosome 1p deletions and ras gene activation in neuroblastomas, pheochromocytomas and medullary thyroid cancers. Surg. Forum 41:466–468.

Moley, J.F., M.B. Brother, C.-T. Fong, et al. (1992) Consistent association of 1p loss of heterozygosity with pheocromocytomas from patients with multiple endocrine neoplasia type 2 syndromes 52:770-774.

Nakagawara, A., K. Ikeda, T. Tsuda, et al. (1987) Amplification of N-*myc* oncogene in stage II and IVS neuroblastomas may be a prognostic indicator. J. Pediatr. Surg. 22:415–418.

Nelkin, B.D., Y. Nakamura, R.W. White, et al. (1989) Low incidence of loss of chromosome 10 in sporadic and hereditary human medullary thyroid carcinoma. Cancer Res. 49:4114–4119.

Nisen, P.D., P.G. Waber, M.A. Rich, et al. (1988) N-*myc* oncogene RNA expression in neuroblastoma. J. Natl. Cancer Inst. 80:1633–1637.

Okazaki, M., A. Miya, N. Tanaka, et al. (1989) Allele loss on chromosome 10 and point mutation of ras oncogenes are infrequent in tumors of MEN 2A. Henry Ford Hosp. Med. J. 37:112–115.

Pochedly, C. (ed.) (1990) Neuroblastoma: Tumor Biology and Therapy. Boca Raton: CRC Press.

Rechler, M.M., and S.P. Nissley (1985) The nature and regulation of the receptor for insulin-like growth factors. Annu. Rev. Physiol. 47:425–442.

Reynolds, C.P., M.M. Tomayko, L. Donner, et al. (1988) Biological classification of cell lines derived from human extra-cranial neural tumors. In: A.E. Evans, G.J. D'Angio, A.G. Knudson, and R.C. Seeger (eds.). Advances in Neuroblastoma Research II. New York: Alan R. Liss, pp. 291–306.

Rustgi, A.K., N. Dyson, and R. Bernards (1991) Amino-terminal domains of c-*myc* and N-*myc* proteins mediate binding to the retinoblastoma gene product. Nature 352:541–544.

Schwab, M., K. Alitalo, K.H. Klempnauer, et al. (1983) Amplified DNA with limited homology to myc cellular oncogene is shared by human neuroblastoma cell lines and a neuroblastoma tumour. Nature 305:245–248.

Schwab, M., H.E. Varmus, J.M. Bishop, et al. (1984) Chromosome localization in normal human cells and neuroblastomas of a gene related to c-*myc*. Nature 308:288–291.

Seeger, R.C., G.M. Brodeur, H. Sather, et al. (1985) Association of multiple copies of the N-*myc* oncogene with rapid progression of neuroblastomas. N. Engl. J. Med. 313:1111–1116.

Seeger, R.C., R. Wada, G.M. Brodeur, et al. (1988) Expression of N-*myc* by neuroblastomas with one or multiple copies of the oncogene. Progr. Clin. Biol. Res. 271:41–49.

Seizinger, R.B., G.A. Rouleau, L.J. Ozelius, et al. (1987) Genetic linkage of von Recklinghausen neurofibromatosis to the nerve growth factor receptor gene. Cell 49:589–594.

Seizinger, R.B., G.A. Rouleau, L.J. Ozelius, et al. (1988) Hippel-Lindau disease maps to the region of chromosome 3 associated with renal cell carcinoma. Nature 332:268–269.

Simpson, N.E., K.K. Kidd, P.J. Goodfellow, et al. (1987) Assignment of multiple endocrine neoplasia type 2A to chromosome 10 by linkage. Nature 328:528–530.

Slavc, I., R. Ellenbogen, W.H. Jung, et al. (1990) Myc gene amplification and expression in primary human neuroblastoma. Cancer Res. 50:1459–1463.

Suzuki, T., J. Yokota, H. Mugishima, et al. (1989) Frequent loss of heterozygosity on chromosome 14q in neuroblastoma. Cancer Res. 49:1095–1098.

Szucs, S., R. Herzog, H. Dralle, et al. (1989) Chromosome 22 deletion in pheochromocytoma. Cancer Genet. Cytogenet. 38:189.

Takai, S.I., H. Tateishi, I. Nishisho, et al. (1987) Loss of genes on chromosome 22 in medullary thyroid carcinoma and pheochromocytoma. Jpn. J. Cancer Res. (Gann) 78:894–898.

Tanaka, T., D.J. Slamon, H. Shimoda, et al. (1988) Expression of Ha-ras oncogene products in human neuroblastomas and the significant correlation with a patient's prognosis. Cancer Res. 48:1030–1034.

Taylor, S.R., and J.A. Locker (1990) A comparative analysis of nuclear DNA content and N-*myc* gene amplification in neuroblastoma. Cancer 65:1360–1366.

Thiele, C.J., C. McKeon, T.J. Triche, et al. (1987a) Differential protooncogene expression characterizes histopathologically indistinguishable tumors of the peripheral nervous system. J. Clin. Invest. 80:804–811.

Thiele, C.J., J. Whang-Peng, C.S. Kao-Shan, et al. (1987b) Translocation of c-sis proto-oncogene in peripheral neuroepithelioma. Cancer Genet. Cytogenet. 24:119–128.

Tsuda, T., M. Obara, H. Hirano, et al. (1987) Analysis of N-*myc* amplification in relation to disease stage and histologic types in human neuroblastomas. Cancer 60:820–826.

Tsutsumi, M., J. Yokota, T. Kakizoe, et al. (1989) Loss of heterozygosity on chromosomes 1p and 11p in sporadicpheochromocytoma. J. Natl. Cancer Inst. 81:367–370.

Voss, B.L., T.J. Pysher, and G.B. Humphrey (1984) Peripheral neuroepithelioma in childhood. Cancer 54:3059–3064.

Weith, A., T. Martinsson, C. Cziepluch, et al. (1989) Neuroblastoma consensus deletion maps to 1p36.1-2. Genes Chromosomes Cancer 1:159–166.

Whang-Peng, J., T.J. Triche, T. Knutsen, et al. (1984) Chromosome translocation in peripheral neuroepithelioma. N. Engl. J. Med. 311:584–585.

Whang-Peng, J., T.J. Triche, T. Knutsen, et al. (1986) Cytogenetic characterization of selected small round cell tumors of childhood. Cancer Genet. Cytogenet. 21:185–208.

Yang, K.P., C.V. Nguyen, S.G. Castillo, et al. (1990) Deletion mapping on the distal third region of chromosome 1p in multiple endocrine neoplasia type IIA. Anti Oncogene Res. 10:527–534.

Chapter 20

Glial Tumors

C. David James and V. Peter Collins

INTRODUCTION

Tumorigenesis is a dynamic process that occurs through a series of events that confer increasing *in vivo* growth advantage to affected cells. "Growth advantage" refers to all events associated with increasing tumor malignancy (e.g., deregulation of cell proliferation controls, invasion, immunosuppression, etc.). Nowell (1976) proposed a model frequently used to explain this process based upon genetic alterations. Viewed from the perspective of this model (Fig. 1), the *in vivo* evolution of tumor cell populations occurs through the stepwise accumulation of growth-advantageous genetic changes (T1–6) in affected cells that were initially of normal (N) genotype and phenotype. Several recent molecular genetic studies of solid tumors have shown that the most malignant neoplasms of a specific tissue origin consistently have more detectable mutations than the less aggressive neoplasms from the same tissue. Consequently, the tumor evolution model has some prognostic significance: the more growth advantageous mutations accumulated within the cells of a tumor, the more aggressive or malignant will be its behavior.

CATEGORIZATION OF GENETIC ALTERATIONS IN TUMOR CELLS

The types of genetic changes that occur during the evolution of tumor cells can be grouped in either of two categories: those involving the functional loss or inactivation of gene products having a negative growth-regulatory function, and those involving the activation, through overexpression and/or mutation, of gene products conferring growth advantage.

Cellular-Recessive Genetic Alterations

In solid tumors, the localization and identification of genes whose products have a negative growth-regulatory function (tumor suppressor genes: TSGs) (Weinberg, 1991) has been accomplished using two molecular genetic approaches, both of which rely upon restriction fragment length polymorphisms (RFLPs). RFLPs are a consequence of normal DNA sequence variation that exists among the members of a species that provide a means with which to distinguish maternally and paternally inherited genetic information based upon minimal sequence differences in the two copies of chromosomal DNA inherited from one's parents. Sequence differences between homologous regions of homologous chromosomes (corresponding maternally and paternally inherited chromosomes) can be detected by the treatment of an individual's cellular DNA with a sequence-specific restriction endonuclease and hybridization of a labeled, human chromosome-specific DNA probe to the filter-bound cellular DNA following its gel electrophoresis and Southern transfer. Using this procedure, sequence variations between chromosome homologues appear as different-sized restriction fragments, referred to as restriction fragment length alleles. Since one person's allele pattern for a given chromosomal region can frequently be distinguished from another's, RFLP analysis can show the different segregation patterns of parental genetic information. This application is extremely important to the development of human gene maps. Furthermore, the nonrandom, familial cosegregation of an allele defined by a chromosomal marker with the presentation of a specific disease has been of fundamental importance to the identification of genes whose alteration are associated with disease predisposition.

This approach of linkage analysis has been successfully applied to the identification of predisposing genetic lesions for certain cancers, and through the study of syndromes associated with central nervous system neoplasia such as neurofibromatosis type 2 and von Hippel-Lindau, linkage analysis may contribute important information regarding specific gene involvement and the genetic pathways involved in the development of glial tumors. Comparisons of inherited and sporadic development of retinoblastoma, neurofibromatosis type 1, and other inherited diseases for which specific gene targets have been identified suggest that the same genes are targeted for functional inactivation in

Molecular Genetics of Nervous System Tumors, pages 241-248 ©1993 Wiley-Liss, Inc.

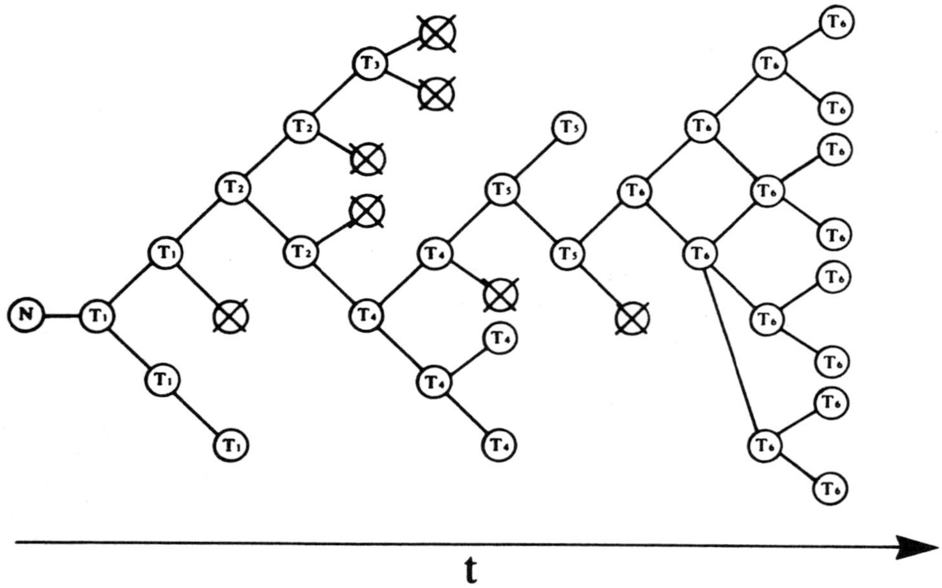

Fig. 1. Tumor cell evolution model. Tumor initiation is a consequence of a genetic alteration in a normal (N) cell that confers selective *in vivo* growth advantage to an affected daughter cell, T1. Tumor evolution or progression is a consequence of subsequent genetic alterations in the daughter cells that confer increasing growth advantage (T2–T6). These alterations, associated with tumor progression, cause tumor genetic heterogeneity. The relative proportions of tumor cell subpopulations (T1–T6) are dependent upon the cumulative growth advantage of each combination of alterations and the time (t) that has elapsed since their occurrence. The crossed out circles represent genetic alterations that are lethal to the tumor cells.

both forms of the disease. The discovery of genes predisposing to glial tumor development could therefore be of great value toward understanding the etiology of all gliomas and may identify the "T1" events depicted in Figure 1.

There is a second application of RFLP analysis to the localization of TSGs that involves the comparison of a cancer patient's corresponding normal and tumor DNA patterns as revealed on Southern filters by chromosome-specific probes. This type of analysis was initially applied to the study of retinoblastoma. The comparison of patient normal and tumor DNA patterns frequently showed a two-allele-to-one-allele reduction (loss of heterozygosity or LOH, Fig. 2) for markers on chromosome 13. This suggests the existence of a gene on chromosome 13 whose loss may be important to the development of retinoblastoma (Cavenee et al., 1983). LOH analysis has subsequently been applied to many other tumors and has been useful in establishing several specific chromosomal LOH–tumor associations (Weinberg, 1991). Although it is not clear whether such losses mark events associated with tumor initiation (T1) or progression (T2–T6), the apparent tumor cell clonality (detection of a common event in a DNA specimen derived from $>10^7$ tumor cells) indicated by LOH implies a biological consequence associated with these alterations that are of fundamental importance to the molecular mechanisms of tumorigenesis.

Once a chromosome is identified as harboring a TSG whose loss is involved in the etiology of a specific type of tumor, several markers (human DNA probes) that collectively span the chromosome of interest can be used to examine a panel of corresponding normal–tumor DNA pairs to determine a minimum region of chromosomal deletion among the tumors and therefore sublocalize the position of the TSG. Although several candidate tumor suppressor genes have been implied and identified through the utilization of this type of analysis, conclusive evidence supporting a tumor suppression function exists for only two genes: RB1 (Huang et al., 1988) and TP53 (Finlay et al., 1989). Interestingly, each of these has been shown to be involved with inherited cancers: predisposition to retinoblastoma in the case of inherited RB1 gene defects (Cavanee et al., 1983), and predisposition to any of several tumors (including brain tumors) associated with Li-Fraumeni syndrome in the case of inherited TP53 gene defects (Malkin et al., 1990). The assay used to demonstrate the property of tumor suppression involves the introduction of an expressible, wild-type TSG into tumor cells having corresponding inactivated, endogenous genes. The exogenous, wild-type gene product diminishes the growth potential of the host cells. Such tests therefore establish a biological significance of the observed genetic alterations and act as a bridge between descriptive and mechanistic studies.

Cellular-Dominant Genetic Alterations

Dominant genetic alterations elicit a proliferative response in affected cells as a consequence of increased production of associated protein and protein functional modification through gene mutation. These alterations involve the class of cellular genes called oncogenes (Bishop, 1987). Such alterations vary from subtle, single base substitutions (point mutations) of oncogene coding sequences that activate the growth-promoting function of the corresponding protein to gross increases in gene copy number (more than 100-fold in some instances), resulting in the overproduction of associated protein. In the former case, such alterations can only be detected using molecular methods. Gene amplification can be detected under a light microscope as double minute chromosomes or homogenously staining regions.

Two methods of analysis are used to describe specific oncogene–tumor associations. One involves determination of oncogene gene dosage (gene amplification) in the DNA extracted from tumor tissue. Increasing gene dosage is a cellular mechanism for increasing the amount of corresponding gene product by increasing the number of transcribable genes. The second approach involves measuring oncogene transcript and/or protein levels in tumor tissue and comparing these levels with those of corresponding normal tissue. Observations of increased oncogene transcripts and/or proteins in tumor tissue have served as catalysts promoting the analysis of corresponding genes for types of alterations that would increase gene transcription.

Elevated, tumor-specific gene transcription does not necessarily involve corresponding gene alteration. Presumably, the transcriptional activity of many cellular genes is altered during tumorigenesis as a consequence of quantitative and/or functional changes in proteins that either enhance or suppress their transcription. Although many such transcriptional changes are of fundamental importance to tumor development, they are nonetheless secondary to primary, mutational events; in this chapter the objects of interest are the genes whose direct alteration promote tumorigenesis.

NEUROPATHOLOGIC CLASSIFICATION

For all solid tumors, the histopathologic indicators of biological aggressiveness have been empirically determined through the comparison of tumor morphology with corresponding patient clinical outcome. For gliomas, classification is based upon the type of normal glial cell precursor the tumor cells most closely resemble, as well as the tumor's malignancy grade (Chapter 14, this volume). Because of the obligate nature of tumor histopathologic classification within a clinical setting, its well-documented positive correlation with tumor biological/clinical behavior, and due to its prognostic significance, it is reasonable to use glioma histopathology as a standard with which to compare results obtained from their molecular genetic analysis.

The diversity of glial tumor classification makes it difficult to compare the results of different studies directed at the same types of tumors. The tumors described in Table 1 are classified according the World Health Organization criteria. Low malignancy grade gliomas are indicated by the use of nomenclature reflecting cellular differentiation only. Gliomas of an intermediate grade of malignancy are "anaplastic" with their apparent cellular differentiation. Gliomas of highest grade of malignancy are classified as glioblastomas.

COMPARISON OF MOLECULAR GENETIC AND HISTOPATHOLOGIC ANALYSIS

We have examined glial tumorigenesis with respect to both the cellular dominant and recessive changes by applying LOH and oncogene amplification analysis to several series of gliomas. Here we present the results of these investigations in combination with data developed from similar studies performed by other investigatory groups.

The Identification of TSGs Involved in Glial Tumor Development

Multiple markers are used in LOH analysis to examine loci on each autosome and sex chromosome for loss of genetic information (Vogelstein et al., 1989). Our results with this type of analysis are summarized in Table 1. Several associations between detectable alterations and tumor histopathology are apparent. The most striking of these is the loss of genetic information from chromosome 10 and its highly specific association with glioblastoma. The total database available in the literature indicates detection of such loss in 59 of 84 glioblastomas, five of 33 gliomas of intermediate (anaplastic) malignancy, and none of 32 low malignancy grade gliomas (El-Azouzi et al., 1989; Fults et al., 1992; James et al., 1988; Venter and Thomas, 1991; Watanabe et al., 1990).

Further analysis has revealed that loss of the 9p-localized interferon genes is restricted to the anaplastic gliomas and glioblastomas and that loss of genetic information from chromosomes 13, 17, and 22 occurs at similar frequencies in each malignancy grade of glioma (Fig. 2). Of these changes, the loss of genetic information from the short arm of chromosome 17 has been investigated and confirmed by several additional groups, with the cumulative data indicating loss in five of 19 low-grade astrocytomas, seven of 24 anaplastic astrocytomas, and 19 of 49 glioblastomas (El-Azouzi et al., 1989; Fults et al., 1992; James et al., 1989; Venter and Thomas, 1991). Three additional chromosomes for which we have determined frequent loss of genetic information

TABLE 1. Genetic Alterations in Gliomas

	Chromosomes retaining tumor heterozygosity	Chromosomes losing heterozygosity	Gene amplification
Glioblastomas			
GB1	1,7,11,13,15,17,18,19,20,X,Y	9*,10,22	None
GB2	3,4,5,9,11,15,16,18,20,21,X,Y	10,13,17,22	None
GB3	1,3,5,8,11,12,14,16,21,22,X,Y	4,7,10,18	EGFR
GB4	1,2,13,15,17,18,20,22	10	EGFR
GB5	1,2,7,11,13,14,15,16,17,20,22,X,Y	9*,10	EGFR
GB6	1,2,3,4,7,8,13,15,17,18,19,21,X,Y	9*,10,20,22	EGFR
GB7	2,3,5,8,9,19,22,X	7,9*,10,11,13,15,17	EGFR
GB8	1,2,9,11,14,15,20,21,X,Y	10,13,17,22	None
GB9	1,4,8,11,13,16,17,19	10	None
GB10	1,2,3,11,12,13,17,18,19,20,21,X,Y	9,10,22	EGFR
GB11	1,4,5,6,9,11,12,15,16,17,18,20,22,X,Y	10	EGFR
GB12	2,3,4,7,8,9,11,13,14,15,17,18,19,22,X,Y	9*,10	EGFR
GB13	1,2,8,9,11,14,17,19,20,22,X,Y	10,13	EGFR
GB14	5,7,8,9,11,13,15,16,20,21,22,X	10	EGFR
GB15	1,3,7,8,13,14,16,18,19,22,X,Y	10,17	None
GB16	2,3,4,7,9,11,12,13,15,20,22,X,Y	10,17	None
Anaplastic Gliomas			
AA1	1,4,5,9,10,11,13,17,19,21,X,Y	None	None
AA2	1,3,4,5,7,8,9,10,11,13,14,15,16,19,20,21,22,X 9,17	None	
AA3	7,9,10,11,13,14,17,19,22,X,Y	None	None
AA4	1,2,3,5,7,9,10,13,16,18,19,20,21,22,X,Y	9,17	None
AA5	3,4,7,9,13,14,15,19,21,22,X,Y	None	None
AA6	1,2,3,10,11,12,13,14,15,17,22,X	9	EGFR
AO1	3,8,9,10,12,13,15,16,17,22,X	1,4,9,21	None
AO2	1,2,5,8,9,10,14,17,18,19,21,22,X,Y	13	None
AOA1	1,2,4,7,8,9,10,11,12,15,17,19,21,22,X,Y	None	None
AE1	3,5,6,9,10,11,12,15,17,19,21,X	14,22	None
AE2	1,2,7,9,10,12,15,16,20,21,X,Y	1,13,17,22	None
Gliomas			
A1	1,5,7,9,10,11,12,13,18,19,20,21,22,X,Y	17	None
A2	1,2,5,9,10,11,13,14,15,18,19,22,X,Y	3,17	None
A3	1,7,9,10,12,15,16,17,20,22	13	None
A4	4,5,9,10,13,14,17,18,20,21,22	None	None
A5	2,3,9,10,11,13,15,17,22,X,Y	None	None
O1	1,4,5,7,9,10,11,13,15,17,20,X,Y	None	None
O2	2,3,8,10,11,13,14,16,17,20,X,Y	22	None
O3	1,2,3,5,8,9,10,11,13,16,17,18,20,22,X,Y	None	None
O4	1,2,8,9,10,11,12,13,15,16,17,19,21,22,X	None	None
OA1	1,2,3,4,5,7,9,10,11,13,14,15,17,20,22,X	None	None
OA2	2,7,10,12,13,15,22	11	None
OA3	2,7,9,10,11,13,17,18,19,20,21,22,X	None	None
E1	3,8,9,10,11,15,16,17,20	13,21,22,X	None

Tumor types: GB, glioblastoma; AA, anaplastic astrocytoma; AO, anaplastic oligodendroglioma; AOA, anaplastic oligoastrocytoma; AE, anaplastic ependymoma; A, astrocytoma; O, oligodendroglioma; OA, oligoastrocytoma; E, ependymoma.

*Cases in which chromosome 9 interferon nullizygosity was indicated.

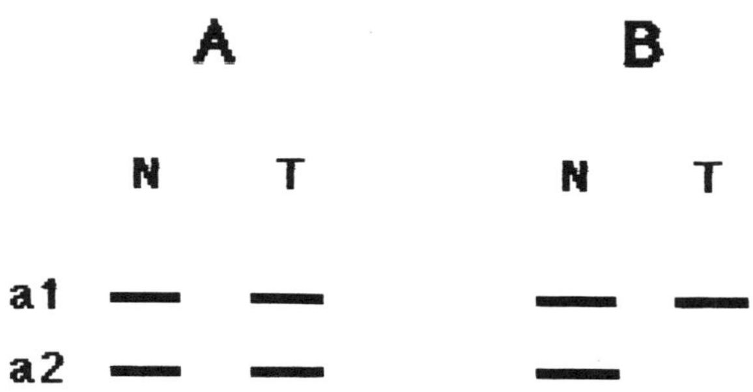

Fig. 2. Patterns revealed by maintenance and loss of constitutional heterozygosity. Alleles (a1 and a2) detected by markers revealing tumor (T) DNA maintenance (**A**) and loss (**B**) of heterozygosity identified in corresponding normal (N) DNA. Such results represent the basis for the data presented in Table 1 under the headings "Chromosomes Retaining Tumor Heterozygosity" (A) and "Chromosomes Losing Heterozygosity" (B). In instances where LOH is determined, the marker revealing loss identifies a chromosome harboring a TSG whose deletion is involved in the etiology of the tumor. Several markers from the same chromosome for which a TSG has been identified can be used to sublocalize the position of the TSG through deletion mapping, in which marker proximity to the TSG is estimated by the frequency of loss it reveals among the members of a particular type of tumor. Since these genetic alterations are determined relative to corresponding normal (N) DNA, they occur as somatic events and do not represent inherited gene defects.

have yet to be confimed and/or addressed by others. Of these, the chromosome 9 interferon gene deletions have been determined in none of 10 low-grade gliomas, four of eight anaplastic gliomas, and six of 12 glioblastomas (James et al., 1991). Data from a study involving analysis of the 9p-localized interferon genes in cell lines established from malignant gliomas (primarily glioblastomas; Miyakoshi et al., 1990) supports their deletion as a common event of glial tumorigenesis. For chromosome 13, frequent loss of genetic information was determined in two of 13 low-grade tumors, two of nine anaplastic gliomas, and four of 14 glioblastomas. On chromosome 22, loss was determined in two of 12 low-grade gliomas, two of 10 anaplastic gliomas, and five of 15 glioblastomas.

Alterations of Oncogenes in Glial Tumors

With regard to the oncogene amplification analysis of this series of tumors (Table 1), the results presented are consistent with the previously established association between epidermal growth factor receptor (EGFR) gene amplification and advanced glial tumor malignancy. A partial survey of the literature reveals this event in 63 of 136 glioblastomas, 10 of 31 anaplastic gliomas, and one of 36 low malignancy grade gliomas (Diedrich et al., 1991; Ekstrand et al., 1991; Libermann et al., 1985; Strommer et al., 1990; Venter and Thomas, 1991; Wong et al., 1987). Although other genes, including MYCN, CMYC, and GLI, have been reported as being amplified in primary malignant gliomas and/or cell cultures of such tumors, their amplifications appear to represent rare events in glioma development.

The data presented in Table 1 indicate that glioblastomas have nearly twice the average number of detectable genetic alterations as the anaplastic gliomas, which have nearly twice the average number of alterations as the low malignancy grade tumors. These data support the model proposed by Nowell (1976) and suggest that increased tumor malignancy is associated with increased numbers of genetic alterations. With regard to possible associations between genetic alterations and cellular differentiation, there are suggestions of some specific relationships. For example, there appears to be an association between chromosome 17p deletions and tumors of astrocytic differentiation, while deletion of sequences from chromosome 22 may be primarily associated with tumors of ependymal origin. However, there is insufficient data at present to draw any conclusions regarding these possible associations.

IDENTIFYING SPECIFIC GENE ALTERATIONS IN GLIAL TUMOR DEVELOPMENT

The LOH investigations represent preliminary attempts to identify, characterize, and understand the function of the genes whose alterations are causally associated with glial cell neoplasia. LOH analysis is a technique that has been utilized to help accomplish the first of these goals and is useful only for the identification of genes whose products provide negative regulation of cell proliferation (Weinberg, 1991).

The LOH studies conducted thus far have suggested a minimum of five tumor suppressor genes at five corresponding genomic locations whose loss is associated with the development of glial tumors. Of these, chromosome 10 is perhaps the most significant due to the frequency of its loss and its association with highest histopathologic grade of malignancy. Molecular genetic studies of chromosome 10 suggest that loss of an entire chromosome is the preferred mechanism by which genetic information is eliminated from this chromosome (Fults et al., 1992; James et al., 1988; Venter and Thomas, 1991; Watanabe et al., 1990), data in good accord with the frequent cytogenetic determination of monosomy 10 in the cells derived from glioblastomas (Bigner et al., 1988). However, there is some data that reveal the infrequent partial loss of genetic information from this chromosome and that the loss can occur on either its short (p) or long arm (q) (Fults et al., 1992; Watanabe et al., 1990). These results suggest that there may be at least one TSG on each arm of this chromosome whose loss promotes glial tumor progression, and that loss of the entire chromosome is a convenient mechanism to allow for their simultaneous elimination.

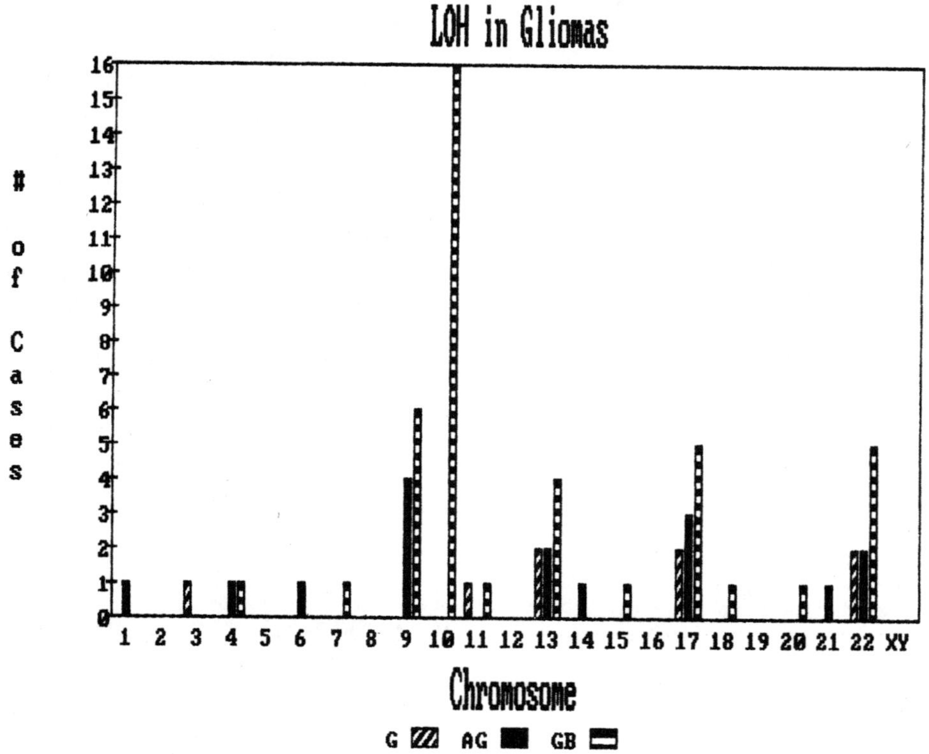

Fig. 3. Analysis of loss of genetic information in gliomas. LOH analysis was applied to the 40 tumors listed in Table 1 by examining patient normal–tumor DNA patterns at one or more loci from each chromosome. The results of such analysis are displayed in a histogram format to allow comparison of genetic alterations determined in low malignancy gliomas (G), intermediate malignancy or anaplastic gliomas (AG), and high malignancy glioblastomas (GB).

Less often involved, but at seemingly nonrandom frequencies in glial tumor LOH, is a locus or loci on chromosome 22 (James et al., 1988). Instances of loss from this chromosome have been determined in each malignancy grade of glioma. The locus involved may be the as yet unidentified and putative TSG implicated in predisposition to neurofibromatosis type 2 (Seizinger et al., 1986) or the TSG involved in the development of meningioma (Dumanski et al., 1987; Seizinger et al., 1987). However, there is no deletion mapping data available at this time to suggest a specific subregion of chromosome 22 as harboring a TSG whose loss is involved in glial tumorigenesis.

The losses of heterozygosity on chromosome 13 show some specificity for eliminating an RB1 tumor suppressor gene (Venter et al., 1991). Chromosome 13 losses of heterozygosity, including the region of the RB1 gene, have been observed in other cancers in association with a missing or mutant form of the remaining RB1 allele. Although a thorough study directed at the analysis of the remaining RB1 allele in gliomas showing an LOH-associated RB1 gene deletion has yet to be conducted, the retinoblastoma gene stands as an attractive candidate for the target of the chromosome 13 deletions observed in glial tumors.

The analysis of chromosome 9 has revealed the frequent loss of one (hemizygous deletion) or both (homozygous deletion) type I interferon gene complements (IFN-b and/or IFN-a) on the short arm of the chromosome (James et al., 1991). Although the region of 9p loss has not been well-characterized with respect to LOH deletion mapping, it is reasonable to speculate that glioma cells might eliminate the genes whose products have been shown to inhibit their own growth in *in vitro* systems. Data that further support this idea have recently been revealed through the transfer into and expression of an IFN-b gene in the cells of cultured gliomas with resulting and marked growth retardation of the recipient cells (Mizuno et al., 1990).

Frequent loss of genetic information from chromosome 17p was determined at a relatively early stage during the LOH analysis of adult astrocytomas. This loss has been observed at similar frequencies in each malignancy stage of this tumor, suggesting that this may be an early event during the formation of at least a subgroup of astrocytomas. Subsequently it was shown that the 17p-localized TP53 gene is frequently mutated in primary malignant gliomas and xenografts of such tumors (Nigro et al., 1989). Furthermore, these mutations tend to be coupled with 17p deletions, supporting the notion that normal TP53 genes are the targets of these losses of genetic information revealed by the LOH analysis (Chung et al., 1991). Additional evidence in favor of the involvement of TP53 mutations in glial tumorigenesis stems from the determination of the suppression of malignant glioma cell growth by introduction and expression of exogenous, wild-type TP53 gene (Mercer et al., 1990). There is, however, no direct evidence that supports TP53 gene mutations as early events of glial tumor development as implied by the chromosome 17 deletion mapping data (El-Azouzi et al., 1989; Fults et al., 1992; James et al., 1989; Venter and Thomas, 1991).

The identification and characterization of the dominant components of glial tumorigenesis are being directed at the growth factor systems. By virtue of its frequent amplification in malignant gliomas, the EGFR gene has received the majority of attention. Although there is some disparity between the results of different studies, this molecular genetic alteration is frequently associated with both intermediate and advanced grades of glial malignancy. In addition to its increasing gene dosage in these tumors, recent studies have indicated a second aspect of the EGFR gene in relation to the malignant evolution of glial tumors. Alteration of amplified EGFR genes in glioblastomas was initially reported in 1985 (Libermann et al., 1985). Subsequently, several groups characterized EGFR transcripts and receptor proteins associated with gene rearrangements. It now appears that there are multiple "hot spots" within this large gene for the occurrence of intragene rearrangements (Fig. 4). For three types of rearrangements the associated mRNA transcripts have been characterized (Ekstrand et al., 1992; Humphrey et al., 1991; Sugawa et al., 1990). For two of these there is evidence of physically and functionally altered receptors (Humphrey et al., 1991; Yamazaki et al., 1990). One of these alterations is of particular interest since it appears to occur in 20–25% of all glioblastomas. Its corresponding receptor presents a unique tumor antigen at the cell surface. Such a cell surface antigen provides an attractive target for monoclonal antibodies that may be useful in tumor imaging or the precise delivery of conjugated toxic agents to tumor cells.

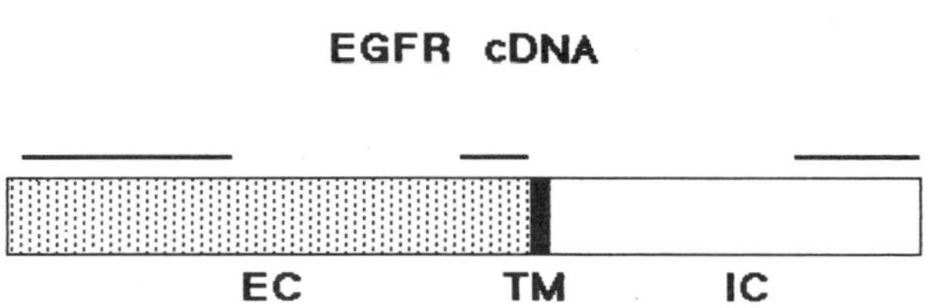

Fig. 4. Alterations of the EGFR oncogene. The cDNA of the epidermal growth factor receptor gene is shown and has been divided into portions that encode the extracellular (EC), transmembrane (TM), and intracellular (IC) domain of the receptor. Coding sequences that are eliminated as a result of intragene rearrangements are indicated by lines above the cDNA.

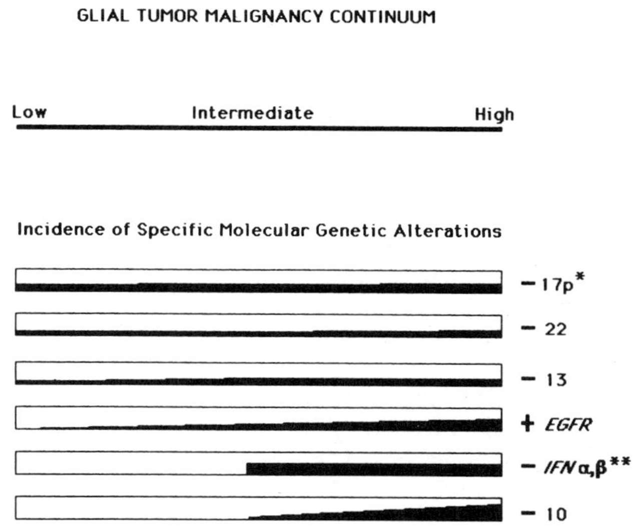

Fig. 5. Glial tumor progression through the accumulation of specific genetic alterations: Specific genetic alterations that have been identified in glial tumors are indicated to the far **right** of the diagram, representing a malignancy continuum for glial tumors. In this format, the association of genetic alterations with specific histopathologic categories is deemphasized while emphasizing the continuous nature of tumor evolution. The shaded portion of each bar represents the frequency of occurrence for the corresponding alteration. Their heights have been determined based upon the combined data cited in the text for low malignancy grade gliomas, anaplastic (intermediate) gliomas, and glioblastomas (high).

*Data is based on astrocytomas only.

**Alterations reported in intermediate malignancy grade gliomas have been exclusively IFN-a,b hemizygosity (loss of one chromosomal complement), while IFN-a,b nullizygosity (loss of both chromosomal complements) has been reported exclusively in glioblastomas.

AN INTEGRATED MODEL FOR GLIAL TUMORIGENESIS

Comparison of molecular genetic and histopathologic analysis of glial tumors suggests specific associations between tumor cellular differentiation and genetic alterations. For example, there appears to be an association between chromosome 17p deletions and tumors of astrocytic differentiation, while deletion of sequences from chromosome 22 may be primarily associated with tumors of ependymal origin. Given the relatively small database with which to evaluate such relationships, however, at this time it seems more appropriate to evaluate the cumulative data with respect to histopathologic malignancy (Fig. 3). These data are consistent in indicating accumulating genetic alterations with increasing tumor malignancy and show some consistency between the occurrence of specific alterations and tumor malignancy staging (Fig. 5). There is an underlying simplicity to this data that suggests that among glial tumors studies there is an endpoint (glioblastoma) and that progression toward this endpoint is a consequence of accumulated, growth-advantageous genetic alterations. It is not, however, intended to indicate an obligate, invariant genetic pathway toward high-grade malignancy, nor that all of the genomic locations of genetic alterations in glial tumors have been identified. Given the complexity of the human genome, the limitations of existing methods of analysis, and the complexity of molecular events within a cell, we have only taken the first step toward understanding the molecular genetic basis of glial neoplasia.

REFERENCES

Bigner, S.H., J. Mark, P.C. Burger, et al. (1988) Specific chromosomal abnormalities in malignant human gliomas. Cancer Res. 48:405–411.

Bishop, J.M. (1987) The molecular genetics of cancer. Science 235:305–311.

Cavenee, W.K., T.P. Dryja, R.A. Phillips, et al. (1983) Expression of recessive alleles by chromosomal mechanisms in retinoblastoma. Nature 305:779–784.

Chung, R., J. Whaley, N. Kley, et al. (1991) TP53 gene mutations and 17p deletions in human astrocytomas. Genes Chromosomes Cancer 3:323–331.

Diedrich, U., S. Soja, J. Behnke, et al. (1991) Amplification of the c-erb B oncogene is associated with malignancy in primary tumours of neuroepithelial tissue. J. Neurol. 238:221–224.

Dumanski, J.P., E. Carlbom, V.P. Collins, et al. (1987) Deletion mapping of a locus on human chromosome 22 involved in the oncogenesis of meningioma. Proc. Natl. Acad. Sci. USA 84:9275–9279.

Ekstrand, A.J., C.D. James, W.K. Cavenee, et al. (1991) Genes for epidermal growth factor receptor, transforming growth factor-alpha, and epidermal growth factor and their expression in human gliomas in vivo. Cancer Res. 51:2164–2172.

Ekstrand, A.J., N. Sugawa, C.D. James, et al. (1992) Amplified EGFR genes in human glioblastomas reveal mutations of sequences coding for extracellular and cytoplasmic regions which occur singularly or in combination. Proc. Natl. Acad. Sci. USA (in press).

El-Azouzi, M., R.Y. Chung, G.E. Farmer, et al. (1989) Loss of distinct regions on the short arm of chromosome 17 associated with tumorigenesis of human astrocytomas. Proc. Natl. Acad. Sci. USA 86:7186–7190.

Finlay, C.A., P.W. Hinds, and A.J. Levine (1989) The p53 protooncogene can act as a suppressor of transformation. Cell 57:1083–1093.

Fults, D., D. Brockmeyer, M.W. Tullos, et al. (1992) p53 mutation and loss of heterozygosity on chromosomes 17 and 10 during human astrocytoma progression. Cancer Res. 52:674–679.

Huang, H.S., J.K. Yee, J.Y. Shew, et al. (1988) Suppression of the neoplastic phenotype by replacement of the Rb gene in human cancer cells. Science 242:1563–1566.

Humphrey, P.A., L.M. Gangarosa, A.J. Wong, et al. (1991) Deletion-mutant epidermal growth factor receptor in human gliomas: Effect of type II mutation on receptor function. Biochem. Biophys. Res. Comm. 178:1413–1420.

James, C.D., E. Carlbom, J.P. Dumanski, et al. (1988) Clonal genomic alterations in glioma malignancy stages. Cancer Res. 48:5546–5551.

James, C.D., E. Carlbom, M. Nordenskjold, et al. (1989) Mitotic recombination of chromosome 17 in astrocytomas. Proc. Natl. Acad. Sci. USA 86:2858–2862.

James, C.D., J. He, E. Carlbom, et al. (1991) Chromosome 9 deletion mapping reveals interferon a and interferon b-1 gene deletions in human glial tumors. Cancer Res. 51:1684–1688.

Libermann, T.A., H.R. Nusbaum, N. Razon, et al. (1985) Amplification, enhanced expression and possible rearrangement of the EGF receptor gene in primary human brain tumors of glial origin. Nature 313:144–147.

Malkin, D., F.P. Li, L.C. Strong, et al. (1990) Germ line p53 mutations in a familial syndrome of breast cancer, sarcomas, and other neoplasms. Science 250:1233–1238.

Mercer, W.E., M.T. Shields, M. Amin, et al. (1990) Negative growth regulation in a glioblastoma tumor cell line that conditionally expresses human wild-type p53. Proc. Natl. Acad. Sci. USA 87:6166–6170.

Miyakoshi, J., K.D. Dobler, J. Allalunis-Turner, et al. (1990) Absence of IFNA and IFNB genes from human malignant glioma cell lines and lack of correlation with cellular sensitivity to interferons. Cancer Res. 50:278–283.

Mizuno, M., J. Yoshida, K. Sugita, et al. (1990) Growth inhibition of glioma cells transfected with the human by-interferon gene by liposomes coupled with a monoclonal antibody. Cancer Res. 50:7826–7829.

Nigro, J.M., S.J. Baker, A.C. Preisinger, et al. (1989) Mutations in the p53 gene occur in diverse human tumor types. Nature 342:705–708.

Nowell, P.C. (1976) The clonal evolution of tumor cell populations. Science 194:23–28.

Seizinger, B.R., De La Monte, S. Atkins, et al. (1987) Molecular genetic approach to human meningioma: Loss of genes on chromosome 22. Proc. Natl. Acad. Sci. USA 84:5419–5423.

Seizinger, B.R., R.L. Martuza, and J.F. Gusella (1986) Loss of genes on chromosome 22 in tumorigenesis of of human acoustic neuroma. Nature 322:644–647.

Strommer, K., M.F. Hamou, H. Diggelmann, et al. (1990) Cellular and tumoral heterogeneity of EGFR gene amplification in human malignant gliomas. Acta Neurochir. 107:82–87.

Sugawa, N., J. Ekstrand, C.D. James, et al. (1990) Identical splicing of aberrant epidermal growth factor receptor transcripts from amplified rearranged genes in human glioblastomas. Proc. Natl. Acad. Sci. USA 87:8602–8606.

Venter, D.J., K.L. Bevan, R.L. Ludwig, et al. (1991) Retinoblastoma gene deletions in human glioblastomas. Oncogene 6:445–448.

Venter, D.J., and D.G.T. Thomas (1991) Multiple sequential abnormalities in the evolution of human gliomas. Br. J. Cancer Res. 63:753–757.

Vogelstein, B., E.R. Fearon, S.E. Kern, et al. (1989) Allelotype of colorectal carcinomas. Science 244:207–211.

Watanabe, K., M. Nagai, S. Wakai, et al. (1990) Loss of constitutional heterozygosity in chromosome 10 in human glioblastoma. Acta Neuropath. 80:251–254.

Weinberg, R. (1991) Tumor suppressor genes. Science 254:1138–1146.

Wong, A.J., S.H. Bigner, D.D. Bigner, et al. (1987) Increased expression of the epidermal growth factor receptor gene in malignant gliomas is invariably associated with gene amplification. Proc. Natl. Acad. Sci. USA 84:6899–6903.

Yamazaki, H., Y. Ohba, N. Tamaoki, et al. (1990) A deletion mutation within the ligand binding domain is responsible for activation of the epidermal growth factor receptor gene in human brain tumors. Jpn. J. Cancer Res. 81:773–779.

Chapter 21

The Cytogenetic and Molecular Genetic Analysis of Meningiomas

Lars Poulsgård, Mogens Rønne, and Henry H. Schmidek

INTRODUCTION

Meningiomas are common, usually benign, primary tumors of the nervous system arising from the membranes covering the brain and spinal cord. These tumors are of major clinical significance due to their frequency and the morbidity that they produce (Schmidek, 1991). The analysis of the cytogenetic and molecular genetic properties of these tumors with current techniques has recently opened important vistas into trying to understand the mechanisms involved in the initiation and progression of these neoplasms (Martuza, 1983).

CYTOGENETIC ANALYSIS OF MENINGIOMAS

Meningiomas are the most extensively studied human solid tumor using cytogenetic techniques. To date, about 500 meningiomas have been characterized, with about 200 cases studied using banding techniques. These studies indicate that the incidence of consistent clonal chromosomal abnormality varies widely between laboratories (from 60–100%). Geographic and racial differences among patients may explain some of the statistical variation, but technical differences between laboratories also contribute.

The most consistent chromosome aberration in meningiomas is simple monosomy 22 (Atkinson and Schmidek, 1991; Zankl and Zang, 1971). Less frequently a deletion of 22q is found instead of complete monosomy. Monosomy 22 has been observed in about 70–80% of meningiomas studied so far. Since chromosomal abnormalities without monosomy 22 are very seldom found, a role for this chromosome in the pathogenesis of meningiomas is strongly indicated. This assumption is supported by the findings of a patient with a constitutional ring chromosome 22, in whom multifocal meningiomas were found at autopsy (Arinami et al., 1986). This patient had a constitutional Robertsonian translocation t(14;22)(14qter→cen→22qter) and a history of familial meningioma but no evidence of von Recklinghausen's disease (Bolger et al., 1985). The t(14;22) translocation was associated with a variant of the human homologue of the simian sarcoma virus transforming gene (C-sis), which has been mapped to the long arm of chromosome 22. The role of this oncogene in the pathogenesis of meningioma, acting either as a dominant oncogene in the tumorigenesis of meningioma or as a putative tumor suppressor gene, is unknown.

In addition to monosomy 22, loss of other chromosomes is often seen. Aberrant karyotypes in meningiomas may evolve through loss of chromosomes subsequent to the acquired monosomy 22. As evolution proceeds, the karyotype becomes progressively hypodiploid and eventually structural chromosomal rearrangements such as translocations, rings, dicentrics, and marker chromosomes may occur.

The chromosome most often involved in nonrandom loss or structural rearrangement subsequent to monosomy 22 is chromosome 14 (Al Saadi et al., 1987; Poulsgård et al., 1989, 1990; Rey et al., 1988a,b). In addition to monosomy 22, Al Saadi et al. (1987) found nonrandom involvement of chromosomes 1, 7, 14, and Y, and, to a lesser degree, chromosomes 18, 19, and 20. Maltby and co-workers (1988) observed that chromosome 17 and the Y chromosome were those most often lost subsequent to monosomy 22, and that chromosomes 1, 14, 10, and 19 were most often involved in structural rearrangement. Chromosomes 1 and 14 were most often lost subsequent to monosomy 22; structural rearrangement occurred most often between chromosomes 1, 14, 22, and the X chromosome (Poulsgård et al., 1989, 1990). Rey et al. (1988a,b) found loss of chromosomes 14, 8, and 1 as the most frequent numerical deviation in meningiomas, next to monosomy 22.

Molecular Genetics of Nervous System Tumors, pages 249-254 ©1993 Wiley-Liss, Inc.

In spite of lack of complete accordance concerning the cytogenetic data, there is no doubt that chromosome 14 is the second most involved chromosome in meningiomas, and chromosomes 1, 7, and Y are frequently involved in nonrandom losses in the hypodiploid evolution of meningiomas. Furthermore, chromosomes 1, 14, and 22 are also the chromosomes most often involved in structural abnormalities.

Hyperdiploidy is uncommon in meningiomas. In a survey of 267 meningiomas, only 15 were reported to be hyperdiploid (Zang, 1982). Poulsgård et al. (1989, 1990) analyzed 30 meningiomas and only one was hyperdiploid. Rønne and Poulsgård (1990) found a case of meningioma with putative multiclonal origin where one of the sublines was hyperdiploid and displayed a translocation between 8p and 22p, t(8;22)(:8p12→8pter::22pter→22qter).

The actual significance of hypodiploid evolution and structural changes in relation to tumor development are unclear. The majority of human malignant neoplasms are usually characterized either by specific chromosomal rearrangement or by hyperdiploidy. For some unknown reason, hypodiploidy is more often encountered in benign neoplasms. Some tumors are able to tolerate considerable chromosome losses without developing malignancy. We reported a case of meningioma with frequent relapses and a hyperhaploid karyotype of 32 chromosomes (Rønne et al., 1988), indicating that tumor behavior may change in connection to a karyotypic evolution. In the early stage the meningioma is likely to have a normal karyotype or a simple monosomy 22 and meningotheliomatous morphology. As the tumor evolves, the karyotype becomes progressively abnormal and the histopathology more fibromatous. Along with these karyotypic and histopathogical changes, the tumor becomes more aggressive (Al Saadi et al., 1987; Poulsgård et al., 1989, 1990; Zang, 1982). Thus karyotyping of meningiomas may be of use to identify a subgroup of patients at higher risk for recurrence who may need special follow-up and treatment.

Volpe (1988) suggested that tumors may be initiated by two methods: either by an error affecting one or several oncogenes or by an error affecting one or several genes controlling the stability of the genome. This group of genes is comprised of repair genes, recombination genes, replication genes, and genes involved in the control of nuclear divisions. Misexpression of oncogenes may lead to the formation of benign, chromosomally stable tumors. Misexpression of stability genes may lead to the formation of a chromosomally unstable cell line or benign tumor. Misexpression of both oncogenes and stability genes within the same cell may lead to the formation of a genetically unstable tumor that creates karyotypically diverse variants that allow extensive tumor cell evolution and development of malignant and metastatic properties. Inherent genetic instability is one characteristic separating metastatic and malignant tumors from benign tumors. However, abnormal karyotypes with nonrandom chromosomal changes are observed in meningiomas and several other benign tumors. Thus genomic instability is not incompatible with benign proliferation and behavior. Only the degree of instability, which according to Volpe (1988) is much more pronounced in malignant tumors than in benign ones, separates the benign from the malignant tumors.

Meningiomas exhibit chromosome abnormalities that are consistent with malignant behavior, but these tumors are considered benign on histological and behavioral grounds. This is a serious dilemma with regard to the role of chromosome abnormalities in malignancy. Strachan and colleagues (1989) reevaluated the claimed benign nature of meningiomas. They pointed out that some of these tumors invade other tissues, including brain, bone, orbital cavity, and paranasal sinuses. Several authors also note that the recurrence rate of completely excised tumors remains remarkably stable, around 13%. The recurrence rate seems therefore "operator independent" and may reflect a propensity for recurrence. These findings lead Strachan and colleagues (1989) to challenge the concept that meningiomas are benign lesions. They suggest that chromosome analysis may help explain and predict tumor behavior and rate of recurrence. Currently, no clear-cut correlation has been found between certain aberrant karyotypes and a meningioma's premalignant or malignant behavior.

The significance of chromosome abnormalities in benign tumors seems difficult to evaluate. In some cases such abnormalities may indicate that tumor proliferation has become premalignant or malignant. In other cases the presence of cytogenetic abnormalities is compatible with benign tumor proliferation and behavior (Sandberg, 1990). Thus the rigid definition of benign/malignant tumor behavior based on clinical and histological grounds seems to some extent contradicted by the cytogenetic data. In light of this, tumor classifications should be reconsidered in accordance with all available data to provide a more complete description of the tumor.

MOLECULAR GENETIC ANALYSIS OF MENINGIOMAS

Current molecular genetic techniques allow improved definition of the chromosomal abnormalities in meningiomas that were first demonstrated by classic cytogenetic techniques. In addition, molecular genetic analysis has some specific advantages over cytogenetic analysis (Martuza et al., 1988). Cytogenetic studies are based on the detailed analysis of a few tumor cells, while molecular genetic analysis examines the DNA from millions of tumor cells. In addition, cytogenetic techniques are most often applied to tumor cells in culture, whereas fresh tumor tissue obtained at surgery is used for molecular genetic analysis. The most important difference between these techniques, however, is that the level of resolution of molecular genetic analysis far exceeds that of cytogenetic techniques where sufficient probes are available.

Analysis of polymorphic DNA loci in DNA isolated from meningioma tumor specimens by Dumanski et al. (1987), Katsuyama et al. (1986), and Meese et al., (1987) has allowed the sublocalization of the region of chromosome 22 involved in the tumorigenesis of meningioma. Polymorphic DNA loci are sites at which small mutation in the DNA alters the recognition site of a restriction endonuclease (RE), an enzyme that cleaves DNA at sequence-specific sites (Martuza et al., 1988). These mutations are stably inherited, normal variations in a DNA sequence that cause different allele fragment lengths to be produced by redigestion of homologous chromosomes. These fragment length variants are referred to as restriction fragment length polymorphisms (RFLPs) and are used to detect chromosomal loss or mutation in patients who are heterozygous for the DNA marker being studied.

Hybridization of radiolabeled chromosome-specific DNA probes to redigested DNA is used to compare the genetic composition of DNA isolated from a tumor specimen against normal DNA from the same patient's leukocytes. If the patient is constitutionally heterozygous for a given chromosomal marker, and if examination of the tumor DNA reveals a loss of heterozygosity at that locus, then loss of genetic information in the tumor is established. In patients who are homozygous for a given chromosomal marker, the paternal and maternal allele fragments cannot be distinguished, so that comparison of constitutional and tumor genotypes is not informative.

Seizinger and co-workers (1987) studied 51 patients with meningiomas not associated with neurocutaneous syndromes. Forty patients were constitutionally heterozygous for at least one chromosome 22 DNA marker. Of these 40 informative patients, 43% (17 patients) demonstrated loss of heterozygosity for at least one chromosome 22 DNA marker. The four polymorphic DNA markers examined were (1) *sis* (the platelet-derived growth factor beta-chain locus that is homologous to the c-sis oncogeny 22q 12.3→13.1; (2) D22S1 (anonymous DNA segment at 22q12→13); (3) D22S9 (anonymous DNA segment at 22q11); and (4) IGLC (gamma-chain IgG-constant regions of 22q11) (Fig. 1). All of these markers identify regions on the long arm of chromosome 22. All meningiomas that showed loss of genetic information at one chromosome 22 marker locus showed loss of genetic information at all chromosome 22 marker loci tested. This finding corresponds to karyotype data in six of these tumors, which confirmed loss of one copy of chromosome 22. Application of these techniques involving use of probes for other chromosomes (4, 10, 11, 13, 14, and 19) has revealed loss of heterozygosity in some tumors, but at much lower frequency than is seen for chromosome 22.

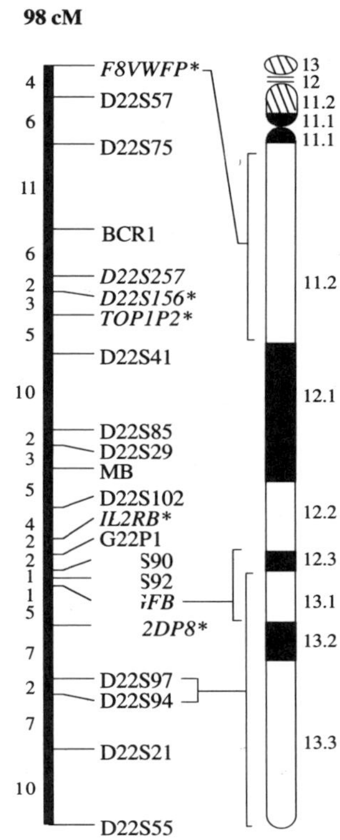

Fig. 1. Human chromosome 22 genetic map. The sex-averaged genetic map (male recombination—female recombination) is shown beside maps constructed on the basis of only male or female meioses, respectively. Optimal estimates of recombination frequencies were obtained by multipoint analysis of five to six loci using ILINK. The maps are in centiMorgans (using Haldane's mapping function to convert recombination frequency). The bracketed regions indicate the physical regions to which particular loci have been assigned. (Rouleau, G.A., J.L. Haines, A. Bazanowski, et al. (1989) A genetic linkage map of the long arm of human chromosome 22. Genomics 4:1–6. Used with permission of Academic Press.)

A similar study of meningiomas that occur in association with bilateral acoustic neurofibromatosis (BANF), carried out by Seizinger and colleagues, has demonstrated loss of genetic information in the same region of chromosome 22 as seen in sporadic cases of meningioma (i.e., those not associated with neurofibromatosis) (Seizinger et al., 1987). This finding supports the hypothesis that loss of a tumor suppressor gene is involved in the etiologic mechanisms of meningiomas. In patients with BANF, who already have one abnormal copy of chromosome 22, loss of genetic information on the other copy of chromosome 22 may confer a monozygous deletion of a putative tumor suppressor gene. This theory would explain the increased incidence and earlier peak age of meningiomas in patients with BANF as compared with those with sporadic meningiomas.

Dumanski and associates (1987, 1990) examined 22 polymorphic DNA loci on chromosome 22 in 81 cases of sporadic meningioma. Fifty-two percent of tumors examined showed loss of heterozygosity at all informative loci, which is consistent with monosomy 22 in the tumor DNA. Eleven percent of tumors showed loss of one constitutional allele at one or more chromosome 22 locus, but maintenance of heterozygosity at other chromosome 22 loci, which is consistent with variable terminal deletions of the long arm of one chromosome 22 in the tumor DNA. This finding suggests that the meningioma locus is located distal to the myoglobin locus, within the region 22q 12.3→qter. Recently a new marker, D22516, allowed sublocalization within the band region 22q 12→13.1—in the vicinity of the gene for the platelet derived growth factor (PDGFB); however, pulsed-field gel electrophoresis analysis of the two loci (PDGFB and D22516) failed to reveal major rearrangements in meningioma (Herzog et al., 1991). Further attempts to identify the meningioma locus indicated that in a t(4:22) translocation, the gene was mapped between D2251 and D22515, a distance of 1 cM on the linkage map of this chromosome. The area in which the translocation is located is within the region where the gene predisposing to neurofibromatosis II has been mapped (Lekanne Deprez et al., 1991).

Densitometric scanning confirms the ratio of copy number of chromosome 22 between tumor and normal DNA as approximately 1:2, indicating that loss of alleles from one chromosome 22 is due to a simple loss of genetic material from one copy of chromosome 22 and not to mitotic recombination, or to loss and reduplication of the remaining alleles.

Multiple loci on other chromosomes have also been examined (Dumanski et al., 1987). A loss of heterozygosity is seen in tumor DNA on chromosomes 1, 2, 3, 6, 7, 10, 11, 14, 17, 18, and X, but these changes are seen much less frequently than is the loss of chromosome 22 alleles. All tumor DNAs that demonstrated a loss of genetic information on another chromosome also showed loss of genetic information from chromosome 22, which suggests that this loss may be the primary etiologic mechanism for meningioma.

The establishment of loss of genetic information on chromosome 22 in meningiomas supports the hypothesis that the loss of a tumor suppressor gene from the long arm of chromosome 22 may be the initiating genetic event in the genesis of this type of tumor (Dumanski et al., 1987). However, this hypothesis requires an explanation of the absence of demonstrable loss of genetic information in at least half of the meningiomas studied. One explanation is that the resolution of RFLP analysis is not precise enough to reveal small regions of loss of genetic information, or that a mutation in the critical locus may make the tumor suppressor gene nonfunctional although no genetic information is lost. Alternative hypotheses are based on the speculation that other types of genetic alterations may be present in those tumors that do not have loss of chromosome 22. These alterations could include loss of genetic information on these chromosomes or possibly amplified expression of dominant oncogenes.

ONCOGENE EXPRESSION IN HUMAN MENINGIOMAS

Carstens and colleagues (1988) investigated oncogene expression in human meningiomas. They studied human meningioma cells in culture, and they used normal human fibroblasts and meningeal cells *in vitro* as negative controls. Tumors with known oncogenic amplification were used as positive controls. The meningiomas studied had either normal karyotypes or monosomy 22 detected by banding analysis. RNA extracted from the tumors was analyzed for oncogene-related transcripts from a panel of 16 different oncogenes. The only oncogene whose expression was found to be enhanced was k-*ras*; this enhancement was sixfold to eightfold greater in the meningiomas than in the patient's meningeal cells or fibroblasts. Although it revealed enhanced expression of this oncogene, DNA analysis did not reveal any major structure changes in the k-*ras* gene, or any evidence of increased copy numbers of the k-*ras* gene (i.e., gene amplification). No difference in k-*ras* expression was noted between tumor cells with monosomy 22 and those with a normal karyotype. The authors hypothesized that increased expression of k-*ras* may be caused by loss of function of a tumor suppressor gene, as a result of either losses of or deletions within particular chromosomes that house tumor suppressor genes (for example, chromosome 22). Another possibility raised was that a mutation of the k-*ras* gene too small to be detected by RFLP analysis could be the cause of altered k-*ras* expression.

Although c-*sis* has not been shown to be amplified in meningiomas, there is evidence that it may be implicated in their tumorigenesis. The c-*sis* locus lies within the region of the putative meningioma locus on chromosome 22, 22q12.3→qter (Bolger et al., 1985). Bolger reported a pedigree with familial meningioma without evidence of neurofibromatosis. Three siblings with meningioma carried a constitutional translocation t(14;22)(14qter→cen→22qter) in their peripheral blood leukocytes, where the two centromeric ends of the long arms of one chromosome 22 and one chromosome 14 are joined together to form a morphologically abnormal hybrid chromosome. Two of these patients had a variant of the c-*sis* gene in their leukocyte DNA revealed by RFLP analysis using a simian sarcoma virus probe. The variant of the oncogene was felt to be located on the morphologically normal chromosome 22. To further elucidate the molecular basis of meningioma development in these individuals, investigators have recently cloned, mapped, and sequenced the abnormal c-*sis* allele from one member of this family (Bolger et al., 1985). Analysis of the structural abnormality using several different restriction enzyme digestions demonstrated DNA fragments D.1kb smaller than those in DNA from unaffected individuals. The abnormality in the c-*sis* gene therefore represents a deletion, and this deletion is located within an intron and an Alu sequence. The finding of an identical deletion in the familial and in a sporadic case of meningioma suggests that the c-*sis* deletion is associated with meningioma development, possibly by activating the expression of the c-*sis* gene (Smidt et al., 1990). The activation of c-*sis*/PDGF-2 finds expression through binding to cell-surface PDGF receptors. *In situ* hybridization and immunocytochemistry have localized the expression of c-*sis*/PDGF-2 and PDGF-R on mRNAs and their protein products in tumor cells of meningiomas. The coexpression of a potent mitogenic growth factor and its receptor in meningiomas suggests an autocrine mechanism contributing to the growth of meningiomas (Maxwell et al., 1990).

Enhanced expression of 5- to 20-fold c-*myc* has been found in 63% of 19 human meningiomas studied. Southern blot hybridization of meningioma DNAs with the c-*myc* probe showed neither amplification nor rearrangement of this gene in any of the specimens. However, in the cases with loss of heterozygosity on chromosome 22, the levels of c-*myc* expression were higher than those without loss, again suggesting the loss of a tumor suppressor gene in the genesis of this type of tumor (Kazumoto et al., 1990).

In the study by Westphal and co-workers (1989) on the spectrum of karyotypic aberrations in human meningiomas, two cases were reported with normal chromosome 22s with specific aberrations involving chromosome 7. One of these patients had trisomy 7 and the other had a t(4;7) translocation. The gene for the epidermal growth factor receptor (EGFR) is located on the short arm of chromosome 7 and the gene encoding the receptor for insulinlike growth factor 2 is on the long arm of chromosome 7. EGFR expression has been examined in meningiomas by several investigators (Jones et al., 1990; Liberman et al., 1985; Weisman et al., 1987). Elevated levels of the EGFR have been reported in some meningiomas, although amplification of the EGFR gene appears to be rare in these tumors. It was not detected in any of 11 meningiomas in one study (Helseth et al., 1988) and in three other meningiomas only single copies of the EGFR gene was found in all of them. However, increased expression of the receptor protein has been detected and the increased expression of this receptor in the absence of gene amplification has been reported in other tumors. Both high- and low-affinity EGF binding sites have been detected in meningiomas. When EGFR is present, it is localized to tumor cells (Jones et al., 1990).

Recently meningiomas have also been found to express both the insulin growth factor (IGF) I and insulin growth factor II in RNAs. Only a low level of IGFI and IGFII mRNA expression was detectable in normal human brain and in non-neoplast meninges. The coexpression of these on RNAs and of the c-*sis* in RNA suggests a significant contribution of PDGF and the IGFs to the unregulated growth of these tumors (Antoniades et al., 1992).

Fibroblast growth factors (FGFs) are potent mitogens, differentiation factors, and angiogenic agents and may be involved in tumor progession as autocrine growth agents. Significant amounts of mRNAs for basic FGFs have been detected in meningiomas. *In situ* hybridization shows that the main source of basic FGF in RNA is tumor cells and immunocytochemistry shows that basic FGF peptide is produced in meningioma cells (Takahashi et al., 1990). FGFs may be involved in neoplastic growth from the following evidence: oncogenes such as hst/k-fgf, int-2, FGF-5, and hst2/FGF-6 encode FGF-related proteins; certain cells can acquire transformed phenotypes after transfection of the basic FGF gene; and many kinds of tumor cells produce FGFs as autocrine growth factors. Recently the cDNAs for two types of human FGF receptors, flg and bck, were isolated. These receptors are able to act as high-affinity binding sites for FGFs. Takahashi and colleagues (1990) demonstrated that meningiomas produce FGFs *in vivo* while these are not detected in normal brain or meninges and that meningiomas express either flg or bck genes *in vivo*, suggesting that tumor-derived FGFs may play a role in tumor genesis through an autocrine mechanism.

CONCLUSION

Research in the molecular genetics of meningiomas has moved from the observations of cytogenetically visible chromosomal abnormalities to the meningioma locus in the region 22q12.3-qter. It remains for the gene to be isolated, cloned, and characterized and the apparatus that controls the gene deciphered, which would then lead to another major insight into the etiology of meningiomas. Subsequently, based on this information, we may be able to develop diagnostic and therapeutic strategies likely to revolutionize the management of these neoplasms.

REFERENCES

Al Saadi, A., F. Latimer, M. Madercic, et al. (1987) Cytogenetic studies of human tumors and their clinical significance II: Meningioma. Cancer Genet. Cytogenet. 26:127–141.

Antoniades, H.N., T. Galanpoulos, J. Neville-Golden, et al. (1992) Expression of IGFI and IGFII genes in primary human astrocytomas and meningiomas. Int. J. Cancer 50(2):215–222.

Arinami, T., I. Kondo, H. Hamaguchi, et al. (1986) Multifocal meningiomas in a patient with a constitutional ring chromosome 22. J. Med. Gen. 23:178–180.

Atkinson, L., and H. Schmidek (1991) Genetic aspects of meningiomas. In: H. Schmidek (ed.). Meningiomas and Their Surgical Management. Philadelphia: W.B. Saunders, pp. 42–47.

Bolger, G.B., J. Stamberg, I.R. Kirsch, et al. (1985) Chromosome translocation t(14;22) and oncogene (c-sis variant in a pedigree with familial meningioma. N. Engl. J. Med. 312:564–567.

Carstens, C., E. Messe, K.D. Zang, et al. (1988) Human KRAS oncogene expression in meningioma. Cancer Lett. (1–2)37–41.

Dumanski, J.P., E. Carlbom, V.P. Collins, et al. (1987) Deletion mapping of a locus on human chromosome 22 involved in the oncogenesis of meningioma. Proc. Natl. Acad. Sci. USA 84:9275–9279.

Dumanski, J.P., G.A. Rouleau, M. Nordenskjold, et al. (1990) Molecular genetic analysis of chromosome 22 in 81 cases of meningioma. Cancer Res. 50(18):5863–5867.

Helseth, E., G. Unsgaard, A. Dalen, et al. (1988) Amplification of EGFR gene in biopsy specimens from human intracranial tumors. Br. J. Neurosurg. 2:217–225.

Herzog, R., E. Gottert, W. Henn, et al. (1991) Large-scale physical mapping within the region 22 q 12.3-13.1 in meningioma. Genomics 19(4):1041–1046.

Jones, N.R., M.L. Rossi, M. Gregoriou, et al. (1990) Epidermal growth factor receptor expression in 72 meningiomas. Cancer 66:152–155.

Katsuyama, J., P.R. Papenhausen, F. Herz, et al. (1986) Chromosome abnormalities in meningiomas. Cancer Genet. Cytogenet. 22:63–68.

Kazumoto, K., M. Tamura, H. Hoshino, et al. (1990) Enhanced expression of the sis and c-myc oncogenes in human meningiomas. J. Neurosurg. 72:786–791.

Lekanne Deprez, R.H., N.A. Groen, N.A. van Biezen, et al. (1991) A t(4;22) in a meningioma points to the localization of a putative tumor-suppressor gene. Am. J. Hum. Genet. 48(4):785–790.

Liberman, T.A., H.R. Nussbaum, N. Razon, et al. (1985) Amplification, enhanced expression and possible rearrangement of EGF receptor gene in primary human brain tumors of glial origin. Nature 313:144–147.

Maltby, E.L., J.W. Ironside, and R.D. Battersby (1988) Cytogenetic studies in 50 meningiomas. Cancer Genet. Cytogenet. 31:199–210.

Martin, G.A., D. Viskochil, G. Bollag, et al. (1990) The GAP-related domain of the neurofibromatosis type 1 gene product interacts with ras p21. Cell 63:843-849.

Martuza, R.L. (1983) Genetics in neuro-oncology. Clin. Neurosurg. 31:417-440.

Martuza, R.L., B.R. Seizinger, L.V. Jacoby, et al. (1988) The molecular biology of human glial tumors. Trends Neurosci. 11(1):22–27.

Maxwell, M., T. Galanopoulos, E.T. Hedley-White, et al. (1990) Human meningiomas co-express platelet-derived growth factor (PDGF) and PDGF-receptor genes and their protein products. Int. J. Cancer 46:16–21.

Meese, E., N. Blin, and K.D. Zang (1987) Loss of heterozygosity and the origin of meningioma. Hum. Genet. 77:349–351.

Poulsgård, L., M. Ronne, and H.D. Schroder (1989) Cytogenetic studies of 19 meningiomas and their clinical significance. Anticancer Res. 9:109–112.

Poulsgård L., H.D. Schrøder, and M. Rønne (1990) Cytogenetic studies of 11 meningiomas and their clinical significance. II. Anticancer Res. 10:535–538.

Rey, J.A., M.J. Bello, J.M. de Campos, et al. (1988a) Incidence and origin of dicentric chromosomes in cultured meningiomas. Cancer Genet. Cytogenet. 35:55–60.

Rey, J.A., M.J. Bello, J.M. de Campos, et al. (1988b) Chromosomal involvement secondary to -22 in human meningiomas. Cancer Genet. Cytogenet. 33:275-290.

Rønne, M., L. Poulsgård, and J.J. Elberg (1988) A case of meningioma with frequent relapses and a hyperhaploid stemline. Anticancer Res. 8:545–548.

Rønne, M. and L. Poulsgård (1990) A case of meningioma with multiclocal origin. Anticncer Res. 10:539-542.

Rouleau, G.A., W. Wertelecki, J.L. Haines et al. (1987) Genetic linkage of bilateral acoustic neurofibromatosis to a DNA marker on chromosome 22. Nature 329:246-248.

Schmidek, H.H. (ed.) (1991) Meningiomas and Their Surgical Management. Philadelphia: W.B. Saunders.

Seizinger, B.R., R.L. Martuza, J.F. Gussella (1986) Loss of genes on chromosome 22 in tumorgenesis of human acoustic neuroma. Nature 322:644-647.

Seizinger, B.R., G. Rouleau, L.J. Ozelius, et al. (1987) Common pathogenetic mechanism for three tumor types in bilateral acoustic neurofibromatosis. Science 236:317–319.

Smidt, M., I. Kirsch, and L. Ratner (1990) Deletion of Alu sequences in the fifth c-sis intron in individuals with meningiomas. J. Clin. Invest. 86:1151–1157.

Strachan, R., C. Clarke, M. Nurbhai, et al. (1989) Will chromosome karyotyping of meningioma aid prediction of tumour recurrence? Br. J. Neurosurg. 3:583–590.

Takahashi, J.A., H. Mori, M. Fukumoto, et al. (1990) Gene expression of fibroblast growth factors in human glicomas and meningiomas: Demonstration of cellular source of basic fibroblast growth factor mRNA and peptide in tumor tissues. Proc. Natl. Acad. Sci. USA 87:5710–5714.

Volpe, J.P.G. (1988) Genetic instability of cancer. Why a metastatic tumor is ustable and a benign tumor is stable. Cancer Cytogenet. 34:125-134.

Weisman, A.S., S.S. Rawuet, P.A. Kelly, et al. (1987) Characterization of the EGFR in human meningioma. Cancer Res. 47:2172–2176.

Westphal, M., M. Hansel, R. Kunzmann, et al. (1989) Spectrum of Karyotypic aberrations in cultured human meningiomas. Cytogenet. Cell Genet. 52:45–49.

Zang, K.D. (1982) Cytological and cytogenetical studies on human meningioma. Cancer Genet. Cytogenet. 6:249-274.

Zankl, H., and K.D. Zang (1971) Cytological and cytogenetical studies on brain tumors. IV. Identification of the missing G chromosome in human meningiomas as no. 22 by fluorescence technique. Hum. Genet. 14:167–169.

Chapter 22

Pathogenesis of Pituitary Tumors

Shlomo Melmed

INTRODUCTION

Pituitary tumors are benign adenomas with well-encapsulated margins. Although they may be locally invasive, they very rarely metastasize. The transformation and subsequent proliferation of the pituicyte is accompanied by autonomous hypersecretion of the peptide hormone gene product. Several genetic, endocrine, and paracrine factors have been implicated in the pathogenesis of both the disordered cell proliferation and the hormone hypersecretion that is the hallmark of these tumors (Reichlin, 1991). This putative linkage between dysregulated hormone transcription and secretion, cell differentiation, and tumor formation has recently been elucidated and several molecular and hypothalamic mechanisms for pituitary tumor pathogenesis proposed. This chapter describes the etiology and pathogenesis of these tumors but does not deal with their diagnosis, presentation, or management.

Anterior Pituitary Cell Types

The anterior pituitary gland is composed of several distinct cell types, each of which expresses a unique gene product (Fig. 1). The pituitary trophic hormones are synthesized and secreted by highly differentiated cells, each of whose gene expression is under specific endocrine and paracrine hormonal control. The somatotroph cells express growth hormone (GH) and lactotrophs secrete prolactin (PRL). Both of these cells appear to arise from a common progenitor cell, the acidophil stem cell (Mulcahey and Jaffe, 1988). Furthermore, a mature mammosomatotroph cell has been identified that expresses both GH and PRL. The glycoprotein pituitary hormones thyrotropin (TSH), follicle stimulating hormone (FSH), and luteiniaing hormone (LH) are secreted by the thyrotroph and gonadotroph cells, respectively. These cells express both the common alpha and unique beta subunits that comprise the glycoprotein hormones. The corticotroph cell expresses the products of the pro-opiomelanocortin gene, including adrenocorticotrophic hormone (ACTH) and beta-endorphin. In addition to these functional hormone-secreting cells, the pituitary gland also contains cells that express well-characterized growth factors, including the folliculo-stellate cell, as well as other cell types whose gene products are not fully elucidated.

Each of the hormone-secreting cells may give rise to functional anterior pituitary tumors with resultant hypersecretion of PRL, GH, ACTH, FSH/LH, or TSH. Pituitary tumors previously recognized as nonfunctional all probably express either gonadotropin hormone or their hormonally inactive alpha subunits.

Hypothalamic Regulation of Pituitary Function

The regulation of pituitary function is under both hypothalamic and peripheral hormonal control (Table 1). The hypothalamus secretes releasing and inhibiting hormones, each of which bind to specific receptors on the anterior pituitary cells and elicit their respective hormonal signal. GHRH stimulates GH synthesis and secretion, while SRIF inhibits GH and TSH release (Frohman and Jansson, 1986). GnRH stimulates FSH and LH, TRH stimulates both TSH and PRL secretion, and POMC expression is induced by CRH. Hypothalamic dopamine acts to suppress PRL transcription and secretion. Peripheral hormones also regulate the expression and secretion of their respective pituitary trophic hormones, usually by exerting a potent negative feedback regulation. Thyroid hormone, corticosteroids, and IGF-I inhibit the transcription of TSH, ACTH, and GH, respectively.

The role of hypothalamic and peripheral hormones in regulating anterior pituitary hormone secretion has been relatively well defined. However, the role of disordered hypothalamic or peripheral endocrine function in the pathogenesis of pituitary cell transformation has only recently

Molecular Genetics of Nervous System Tumors, pages 255-263 ©1993 Wiley-Liss, Inc.

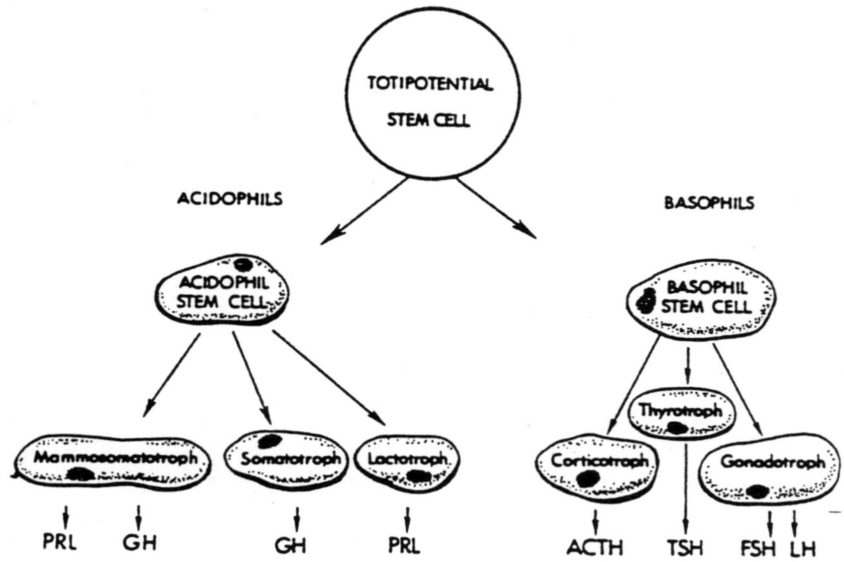

Fig. 1. Descriptive figure of anterior pituitary cell types. Immuno-peroxidase staining, immunoelectron microscopy, and *in situ* hybridization techniques have confirmed at least five functional cell types. True nonfunctioning adenomas may arise from undifferentiated stem cells or degranulated mature cells. No basophil stem cell has been identified. Mammosomatotrophs, somatotrophs, and lactotrophs are assumed to arise directly from the acidophil stem cell. The corticotroph cell contains proopiomelanocortin, which is the source of ACTH, and related opioid peptides. (Reproduced from Melmed et al., 1983.)

become apparent, and is reviewed below.

Pituitary-Specific Cell Differentiation

A totipotential stem cell appears to be the progenitor for the differentiated pituitary cell types. Differentiation of these cells occurs early in embryogenesis, giving rise to clearly specialized cell populations. Tissue-specific transcription factors confer differentiated gene expression. Pit-1, a pituitary-specific nuclear protein, a member of the homeobox Pou-domain family, binds to several sites on the 5' flanking region of the GH and PRL genes. Pit-1 appears to be an essential factor for phenotypic pituitary cell commitment and GH and PRL gene expression (Ingraham et al., 1988). TSH-secreting cells also contain Pit-1, which may also regulate its expression. Transgenic mice bearing mutated regions of Pit-1, rendering it dysfunctional, exhibit hypoplasia of pituitary somatotrophs, lactotrophs, and thyrotrophs (Castrillo et al., 1991). These animals also have diminished GH, PRL, and TSH secretion with resultant growth retardation and attenuated thyroid function. In another model, inhibition of Pit-1 synthesis by antisense oligonucleotide techniques results in decreased proliferation of GH-secreting cells. Other cell-specific factors, including repressors of gene expression, are clearly present in the pituitary, but their structure has not been elucidated.

PITUITARY ADENOMAS

Prolactinomas

These common pituitary tumors usually arise as pure PRL-secreting cell neoplasms. Occasionally, they may be associated with excess GH secretion, if the tumors arise from a mixed or dedifferentiated cell population. Other causes of hyperprolactinemia associated with pituitary tumors include nonfunctioning adenomas or craniopharyngiomas and similar parasellar neoplasms. These lesions, by impinging on the pituitary stalk, cause impaired dopamine delivery to the lactotroph cell. This results in release of the lactotroph from tonic dopaminergic inhibition and resultant hypersecretion of PRL. Prolactinomas cause amenorrhea, galactorrhea, and infertility in females, and impotence, infertility, and galactorrhea in males.

Gonadotroph Cell Adenomas

Until recently, these tumors expressing FSH, LH, or their

TABLE 1. Hypothalamic–Pituitary Axis

Hypothalamic hormone	Pituitary target	Hormone expression	
		Trophic	Peripheral
Stimulatory			
Growth hormone-releasing hormone	Somatotroph	Growth hormone	Insulinlike growth factor I
Corticotropin-releasing hormone	Corticotroph	ACTH and related peptides	Cortisol
Thyrotropin-releasing hormone	Thyrotroph	Thyrotropin	Thyroid hormones
Gonadotropin-releasing hormone	Gonadotroph	Luteinizing hormone Follicle-stimulating hormone	Gonadal steroids
Inhibitory			
Somatostatin	Somatotroph	Growth hormone	
Dopamine	Lactotroph	Prolactin	

Reproduced from Melmed et al., 1986.

alpha or beta subunits were considered rare. It is now apparent that the majority of pituitary tumors previously classified as "nonfunctional" with no demonstrable pituitary hormone hypersecretion do in fact express one of the glycoprotein hormone subunits. The adenomas secreting FSH or LH actually cause hypogonadism and the high levels of gonadotropin hormones appear to down-regulate gonadal function.

Thyrotroph Cell Adenomas

These rare tumors arising from TSH-secreting cells are usually locally invasive. The TSH hypersecretion is associated with thyroid stimulation and resultant elevated thyroid hormone levels.

Acromegaly

Acromegaly is caused by either excess GH or GHRH secretion (Melmed, 1990). Excess GH secretion may be due to a pituitary adenoma or an extrapituitary tumor secreting ectopic GH. GH-secreting pituitary adenomas include densely or sparsely granulated GH-cell adenomas, mixed GH- and PRL-cell adenomas, mammosomatotroph adenomas, acidophil stem cell adenomas, and plurihormonal adenomas. GH-cell carcinoma is extremely rare. Excess GHRH secretion may be due to hypothalamic tumors hypersecreting GHRH, or ectopic secretion of GHRH by an extrapituitary tumor (Guillemin et al., 1982; Thorner et al., 1982). These latter tumors are usually abdominal or thoracic carcinoids, pancreatic tumors, or other endocrine tumors. These high levels of GHRH cause somatotroph hyperplasia and resultant acromegaly.

Regardless of the etiology, all of these conditions cause hypersecretion of GH and its target hormone, IGF-I. If the tumor hypersecretion starts prior to epiphyseal closure, patients present with gigantism. If, as usually occurs, the tumor manifests later in life, patients present with signs of hypersomatotrophism, including acral bony changes, soft tissue swelling, and glucose intolerance.

Cushing's Disease

Cushing's disease is characterized by hypercortisolism, and is usually due to excessive ACTH secretion by a corticotroph-cell adenoma that is usually small and often undetectable by routine pituitary imaging. Other causes of Cushing's syndrome include ectopic ACTH production by lung carcinoma, thymoma, or other malignancies. Rarely, ectopic CRH production by an extrapituitary neoplasm may cause secondary hypersecretion of pituitary ACTH. These conditions are characterized by high levels of cortisol, skin changes, hirsutism, bruising, striae, osteopenia, glucose intolerance, and hypertension.

Plurihormonal Tumors

Functional pituitary tumors predominantly secrete a single trophic hormone (Table 2). Although the normal adult human pituitary consists mainly of well-differentiated cells, multidifferentiated cells expressing more than one hormone within the same cell have also been identified (Kovacs et al.,

TABLE 2. Classification of Pituitary Adenomas

Adenoma cell origin	Hormone product	Clinical syndrome
Lactotroph	PRL	Hypogonadism, Galactorrhea*
Somatotroph	GH	Acromegaly/ gigantism
Corticotroph	ACTH	Cushing's disease
Mixed growth hormone cell and prolactin cell	GH, PRL	Acromegaly hypogonadism
Other plurihormonal cell	Any	
Gonadotroph	FSH, LH, subunits	Hypogonadism
Acidophil stem cell	PRL, GH	Hypogonadism, acromegaly
Mammosomatotroph	PRL, GH	Hypogonadism, acromegaly
Thyrotroph	TSH	Hyperthyroidism
Null cell	None	Pituitary failure
Oncocytoma	None	Pituitary failure

Hormone-secreting tumors are listed in decreasing order of frequency.
*Females present with amenorrhea and infertility and males present with impotence and/or infertility. All tumors may cause local pressure effects, including visual disturbances, cranial nerve palsy, and headache.

1989). Occasionally, GH, PRL, and even TSH may be cosecreted from the same tumor, probably reflecting a common transcriptional control mechanism for these hormones. As GH- and PRL-secreting cells both arise from a common progenitor cell, tumors arising from these more primitive cells hypersecrete the two hormones. Cosecretion of apparently unrelated hormones, including GH, PRL, TSH, glycoprotein subunits, and ACTH, may imply that these tumors originate from an earlier poorly differentiated stem cell. Alternatively, a mature fully differentiated cell could dedifferentiate to a more primitive form after its transformation, resulting in cosecretion of more than one hormone.

PATHOGENESIS

Several endocrine and other factors have been implicated in the pathogenesis of pituitary tumors (Table 3) (Molitch, 1987). Experimental and clinical evidence for these causes are discussed below.

Pituitary Angiogenesis

Hypothalamic dopamine normally exerts a tonic inhibitory effect on PRL secretion and lactotroph cell division. The concentration of hypothalamic–pituitary portal vessel dopamine is higher than in the systemic circulation. Spontaneously occurring or estrogen-responsive pituitary arteriolar shunt formation results in foci of lactotroph cells being released from the normal tonic dopaminergic inhibition and exposed to relatively lower concentrations of dopamine (Elias and Weiner, 1984). This may be followed by subsequent formation of prolactin-secreting adenomas. Estrogen-induced angiogenesis appears to be mediated by reorganization of pituitary interstitial cells and basement membrane, including folliculo-stellate cells, and local release of basic FGF. The increased levels of local FGF may act to stimulate new blood vessel formation, lactotroph cell proliferation, and possibly induce PRL secretion. These adenomas retain their sensitivity to dopamine and are amenable to suppression of both hormone secretion and tumor shrinkage by dopamine agonists. Prolactinomas may therefore develop as a consequence of estrogen stimulation, either directly on the lactotroph or indirectly via induction of new vessel formation, leading to a diminished dopaminergic inhibition.

Loss of Feedback Regulation

Thyroid hormone, corticosteroids, and gonadal steroids exert powerful negative feedback regulatory effects on the pituitary. Chronic failure of the thyroid, adrenal, or gonadal target gland leads to trophic hormone hypersecretion, hyperplasia, and eventually adenoma formation of the respective trophic cell (Furth et al., 1955). Patients with longstanding thyroid failure exhibit high levels of TSH and pituitary enlargement due to thyrotroph cell hyperplasia. Very rarely, TSH-secreting adenomas may occur in these patients. Most TSH-adenomas are however primary pituitary lesions and are not associated with thyroid failure.

Chronic gonadal insufficiency in men may lead to development of gonadotroph-cell pituitary adenomas. Most of these tumors are clinically silent, as they express heterogenous forms of glycoprotein hormones or their respective

TABLE 3. Etiology of Pituitary Adenomas

	Adenoma cell type
Hypothalamic	
Excess GHRH secretion	GH
Excess CRH secretion	ACTH
Defective dopamine secretion	PRL
Pituitary	
Monoclonal expansion	All
gsp oncogene mutation	GH
MEN-1 deletion	PRL
New angiogenesis	PRL
Peripheral	
Hypothyroidism	TSH
Hypogonadism	FSH/LH/subunit

subunits. Most of these tumors are not associated with target organ hypogonadism, and arise as primary pituitary lesions. In fact, primary ovarian failure occurring in menopause is not associated with an increased incidence of gonadotroph cell adenomas.

Intrinsic Pituitary Etiology of Pituitary Adenomas

Pituitary adenomas characteristically exhibit autonomous hormone secretion and are usually independent of physiologic hypothalamic control. Compelling clinical evidence in favor of an intrinsic lesion of the pituicyte leading to adenoma formation has been provided by commonly encountered successful surgical "cures" occurring after adenoma resection. Assuming that all abnormal tissue, including microscopic tumor rests, are removed during a well-demarcated surgical dissection, the subsequent benign natural history of patients with acromegaly, Cushing's disease, and sometimes prolactinomas supports the presence of an intrinsic pituitary defect. However, biochemical or clinical disease persistence or recurrence after apparently successful surgery, as is often seen with larger tumors, may indicate incomplete resection and tumor tissue regrowth, or new tumor growth. The occurrence of true recurrent *de novo* tumor growth suggests persistent hypothalamic dysfunction, and would be difficult to distinguish from incomplete resection at surgery. The tissue surrounding an adenoma is not hyperplastic, strongly mitigating against a generalized hypothalamic overstimulation, and would favor an intrinsic pituitary cell defect leading to local adenoma formation.

Clonal Origin of Pituitary Adenomas

The often unrestrained pituitary hormone secretion in patients harboring functional adenomas is usually associated with relatively intact hormone responsiveness to hypothalamic stimulation, and paradoxical responses to other stimuli. These latter abnormal dynamic responses often return to normal after surgical resection, suggesting that removal of the tumor results in functional and anatomic cure (Ross et al., 1987). These observations may imply that these tumors are clonal expansions of a single, disordered cell. The determination of the clonal origin for pituitary adenomas was therefore important for understanding the pathogenesis of these tumors. This would determine whether an intrinsic cellular defect is associated with development of pituitary adenomas. Excessive pituitary stimulation by hypothalamic hormones would be expected to result in a generalized polyclonal pituitary cell hyperplasia. Alternatively, a genomically mutated cell would be expected to undergo clonal proliferation and form a monoclonal tumor. Clonality of pituitary tumors was determined by X-chromosomal inactivation analysis in female patients heterozygous for hypoxanthine phosphoribosyl transferase (HPRT) and phosphoglycerate kinase (PGK). When clonal analysis was performed on tumor DNA derived from heterozygous female patients with acromegaly, Cushing's disease, prolactinomas, and nonfunctioning tumors, a preferential loss of the inactive X-chromosome allele was observed after digestion with methylation-sensitive enzymes (Herman et al., 1990). Monoclonality of these tumors was confirmed by the high allelic cleavage ratio of the visualized demethylated vs. nondemethylated DNA bands (Table 4).

TABLE 4. Clonality of Functional Pituitary Adenomas

DNA source	Allelic cleavage ratio	Clonality
Normal pituitary	1	Polyclonal
GH-cell	11.2 ± 5.00	Monoclonal
PRL-cell adenomas	6.60 ± 4.40	Monoclonal
ACTH-cell adenomas	>30	Monoclonal

X-chromosomal inactivation analysis of normal pituitary tissue and adenomas. Allelic cleavage ratio was determined on DNA samples derived from 10 hormone-secreting adenomas by comparing the densitometric intensities of the X-linked HPRT or PGK DNA restriction fragments after methylation-sensitive digestion. A ratio approximating 1 is expected for polyclonal tissue, indicating the relatively nonpreferential cleavage of either of the two alleles.
Adapted from Herman et al., 1990.

These results indicate that functional and nonfunctional pituitary tumors probably arise from a single precursor cell. The monoclonality of pituitary tumors suggests that they arise as the result of an intrinsic defect that results in either activation of a cell stimulator or inactivation of an inhibitor of cell proliferation (Alexander et al., 1984; Klibanski, 1990). Although this conclusion would appear to imply a secondary role for hypothalamic hormones in triggering adenoma formation, CRH and GHRH may play a facilitatory role by stimulating clonal expansion of a genomically altered corticotroph or somatotroph cell.

Pituitary Tumor-Promoting Oncogenes

Dominant structural genetic mutations in genes controlling cell growth may lead to activation of growth-promoting events, or deletion of a growth inhibitory or tumor suppressor gene may occur. Both these events have been suggested as possible etiologic factors in the pathogenesis of pituitary tumors.

Constitutive activation of hypothalamic–pituitary receptor units will putatively be shown to induce hormone hypersecretion and pituitary proliferation. Potential tumors whose pathogenesis may thus be determined include prolactinomas (TRH or VIP receptors), Cushing's (CRH receptor), and gonadotroph cell adenomas (GnRH receptor).

After ligand-receptor binding, the GTP-binding (G) proteins transduce the cell membrane signal. GHRH regulates GH gene transcription after binding to the surface and inducing G_s activation of adenyl cyclase. Thus, the membrane-associated receptors are coupled to the cyclic AMP effector system and facilitate transduction of growth-promoting and hormone-secretory signals from the hypothalamic ligand. Thus far, constitutive activation of a hypothalamic–pituitary signaling unit has only been demonstrated in a subset of GH-cell adenomas. Two groups of GH-cell adenomas have been identified (Landis et al., 1989). The first group consists of patients harboring GH-cell tumors with intact adenylate cyclase response to GHRH stimulation. The second group of tumors contain a constitutively activated high basal level of adenyl cyclase, rendering the cell relatively unresponsive to further GHRH stimulation. These latter tumors contain somatic point mutations of the G_s alpha chain. The stimulatory effects of GHRH on GH secretion and somatotroph proliferation are mimicked by this dominant mutation of the G_s alpha chain gene, which causes a constantly GTP-bound state, leading to constitutive adenyl cyclase activation. The mutation, termed gsp, consists of Arg 201 being replaced with either Cys or His, or Gln 227 being replaced with Arg (Table 5).

In vitro, tumors bearing the gsp mutation have markedly increased intracellular cAMP as well as basal GH secretion. Clinically, these tumors are smaller, have higher basal GH levels, and are less responsive to GHRH stimulation, as compared to nonmutant tumors (Landis et al., 1990).

Somatotroph proliferation and GH secretion are both clearly stimulated by GHRH and cyclic AMP, but additional non-G_s-mediated mechanisms may also be involved in the pathogenesis of GH cell adenomas. Mice expressing a nonphosphorylatable CREB mutant transgene had resultant inactivated cAMP activity. These animals exhibited hypoplasia of somatotroph cells with resultant dwarfism (Struthers et al., 1991). Conversely, mice bearing an irreversible activator of the G_s transgene driven by a pituitary-specific promoter developed somatotroph hyperplasia, elevated serum GH levels, and gigantism (Burton et al., 1991). Functional somatotroph hyperplasia may therefore be induced by *in vivo* manipulation of intracellular pituitary cyclic AMP levels. No clear-cut somatotroph adenomas developed during the time period of these experiments. In contrast, discrete pituitary adenomas developed in mice bearing a GHRH transgene for a longer period (>1 yr) (Asa et al., 1990). Cyclic AMP alone therefore appears to induce GH secretion and somatotroph cell proliferation, but may require a prolonged exposure time as well as additional hormonal or other cofactors to initiate pituitary tumor formation.

TABLE 5. gsp Mutations in Pituitary GH-Cell Adenomas

	Adenylyl cyclase (pmol cAMP/ mg.min)		Adenoma-encoded amino acids in G_s alpha	
	Tumor	Basal	Codon 201	Codon 227
Group 1 patients	1	13	Arg	Gln
	2	6	Arg	Gln
	3	16	Arg	Gln
	4	43	Arg	Gln
Group 2 patients	5	170	Cys	Gln
	6	480	His	Gln
	7	190	Cys	Gln
	8	180	Arg	Arg

Mutations of G_s alpha were determined in DNA derived from two groups of acromegalic patients. Group 1 patients all encode wild-type Arg 201 and Gln 227, while group 2 tumors express a mutation in either codon 201 or 227. Adapted from Landis et al., 1989.

Recently, prolactinoma DNA was shown to contain transforming activity in an *in vitro* transformation assay (Gonsky et al., 1991). The transforming DNA contained coding regions for the hst gene, a member of the FGF family. hst also appears to be expressed in prolactinoma tissue, supporting the role of FGF-related genes in development of PRL-cell tumors.

Pituitary Tumor Suppressor Genes

Deletions or damage to putative growth suppressor genes may result in abnormal cell proliferation. Such a mechanism has been proposed for endocrine tumor development in patients with the multiple endocrine neoplasia 1 syndrome. This is an autosomal dominant disorder with any combination of parathyroid, pancreatic, and/or pituitary adenomas. The susceptibility locus for MEN-1 has been mapped to chromosome 11q13 and may undergo recessive mutations in affected individuals. The MEN-1 gene probably functions as a tumor suppressor gene and a few prolactinomas have shown loss of heterozygozity at the MEN-1 site, suggesting deletion of a tumor suppressor gene (Bystrom et al., 1990). Similar allelic losses have not been demonstrated in sporadic GH-cell pituitary adenomas. In the future, precise sequencing of the MEN-1 gene locus may allow characterization of discrete mutations in other sporadic pituitary adenomas.

HYPOTHALAMIC ETIOLOGY OF PITUITARY TUMORIGENESIS

Despite the molecular, biochemical, and clinical evidence in favor of a pituitary etiology for these tumors, a hypothalamic contribution to development of pituitary tumors is supported by several observations (Faglia et al., 1991).

Excess Hypothalamic Releasing Hormone

Pituitary hormone secretion is controlled by the hypothalamic hormones, which also exert positive trophic effects on

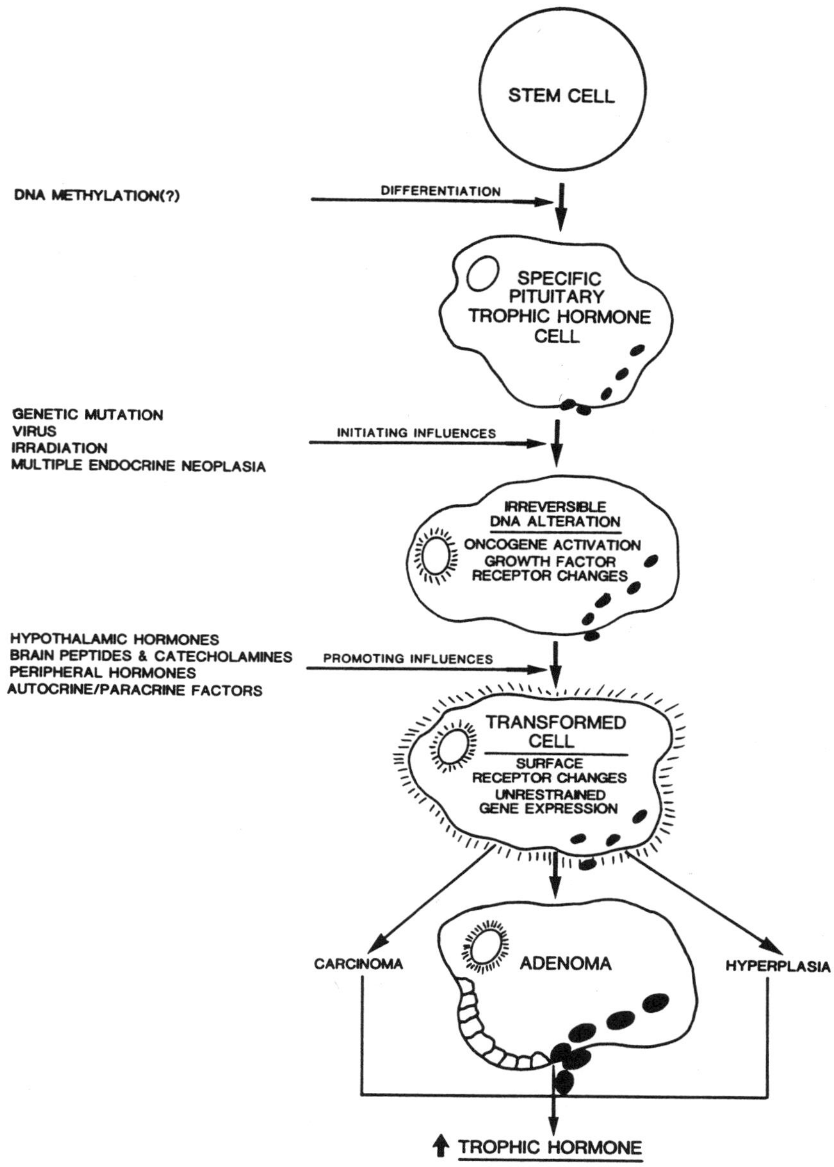

Fig. 2. Pathogenesis of a pituitary adenoma is schematically depicted. After an irreversible cellular initiation, several endocrine or paracrine factors may promote subsequent tumor formation and unrestrained hormone secretion. (Reproduced from Melmed et al., 1986.)

anterior pituitary cells. Excessive secretion of GHRH or CRH may cause proliferation and adenoma formation of the somatotroph. GHRH stimulates DNA synthesis in rat somatotroph cells, and also induces expression of c-fos, an early response signal-transducing oncogene (Billestrup et al., 1986, 1987). SRIF blocks both the hormonal as well as the trophic stimulatory effects of GHRH. Rats receiving chronic CRH administration also exhibit an increase in corticotroph cell number. These trophic actions of hypothalamic hormones have been further studied in a variety of animal and human models.

GH-cell hyperplasia, elevated GH secretion, and gigantism are found in mice bearing a GHRH transgene (Mayo et al., 1988; Stefaneanu et al., 1989). Similarly, patients harboring tumors secreting high levels of ectopic GHRH develop pituitary somatotroph cell hyperplasia and acromegaly, and a few cases of well-defined GH-cell adenomas have been documented. A patient harboring a hypothalamic gangliocytoma hyperexpressing GHRH also developed a pituitary GH-cell adenoma (Asa et al., 1984; Sano et al., 1988). Similarly, ectopic CRH secretion by an intrasellar gangliocytoma and a prostatic carcinoma were associated with pituitary corticotroph hyperplasia and Cushing's syndrome. The aggressive ACTH-secreting tumors developing after bilateral adenomectomy for Cushing's disease is probably due to chronic hypersecretion of CRH. However, in patients with well-defined primary GH-cell or ACTH-cell adenomas, the nontumorous somatotroph or corticotroph cells situated outside the demarcated tumor capsule are not hyperplastic, implying that no generalized hypersecretion of GHRH or CRH is present. Paradoxical GH responses to glucose, TRH, and dopamine, which characterize both pituitary adenomas as well as ectopic GHRH secretion, are normalized after excision of the ectopic GHRH source (Melmed and Rushakoff, 1987). These abnormal GH responses therefore appear to be associated with a higher GHRH milieu. The presence of GHRH or CRH receptors on adenoma tissue, and the lack of *in vitro* down-regulation of pituitary hormone secretion by these releasing hormones, also supports their pathogenetic role in tumor formation or promotion. Finally, the promising results of biochemical and anatomic therapy of acromegaly with somatostatin analogs (Lamberts, 1988) is also compelling evidence in favor of disordered hypothalamic control leading to pituitary tumor formation.

The isolation of GHRH, SRIF, and TRH within functional pituitary tissue and their presumed synthesis within the adenoma also favors a paracrine mechanism for their action (Pagesy et al., 1990). Tumors containing the highest concentrations of SRIF actually appear to express lower amounts of GH.

The sequence of events leading to pituitary adenoma formation is as yet unclear, and may involve multiple intracellular and extracellular disruptions. For example, the gsp mutation should render the pituitary relatively nonresponsive to GHRH, probably minimizing a potential role for GHRH in the initiation of these tumors. Both CRH and GHRH, however, may confer growth advantages on their respective target cells and they may in fact promote the proliferation of a more aggressive clone of transformed cells.

CONCLUSIONS

Experimental and clinical evidence for either an intrinsic pituitary or hypothalamic pathogenesis of pituitary tumors has shown that pituitary tumorigenesis probably involves multiple steps (Fig. 2). Structural gene mutations, including gsp, possibly MEN-1, and other putative growth control genes may be required for pituicyte transformation. This primary initiating event may only be followed by tumor formation if appropriate endocrine or paracrine promotion of cell proliferation occurs. GHRH and CRH are also involved in growth promotion of the transformed somatotroph or corticotroph cell and they may function as progression factors for tumor cell replication. It is unlikely that GHRH, CRH, or TRH are primary initiators of pituitary tumorigenesis but rather act as growth promoters of the already disordered cell. This multistep etiology for pituitary tumor formation, involving genetic endocrine and paracrine factors, will be further clarified as fundamental knowledge of subcellular pituitary mechanisms improves.

REFERENCES

Alexander, J.M., B.M.K. Biller, H. Bikkal, et al. (1984) Clinically nonfunctioning pituitary tumors are monoclonal in origin. J. Clin. Endocrinol. (Oxf.) 21:709.

Asa, S.L., K. Kovacs, L. Stefaneanu, et al. (1990) Pituitary mammosomatotroph adenomas develop in old mice transgenic for growth hormone-releasing hormone. Proc. Soc. Exp. Biol. Med. 193:232.

Asa, S.L., B.W. Scheithauer, J.M. Bilbao, et al. (1984) A case for hypothalamic acromegaly: A clinicopathological study of six patients with hypothalamic gangliocytomas producing growth hormone-releasing factor. J. Clin. Endocrinol. Metab. 58:796.

Billestrup, N., R.L. Mitchell, W. Vale, et al. (1987) Growth hormone-releasing factor induces c-fos expression in cultured primary cells. Mol. Endocrinol. 1:300.

Billestrup, N., L.W. Swanson, and W. Vale (1986) Growth hormone-releasing factor stimulates proliferation of somatotrophs in vitro. Proc. Natl. Acad. Sci. USA 83:6854.

Burton, F.H., K.W. Hasel, F.E. Bloom, et al. (1991) Pituitary hyperplasia and gigantism in mice caused by a cholera toxin transgene. Nature 350:74.

Bystrom, C., C. Larsson, C. Blomberg, et al. (1990) Localization of the MENI gene to a small region within chromosome 11q13 deletion mapping in tumors. Proc. Natl. Acad. Sci. USA 87:1968.

Castrillo, J.-L., L.E. Theill, and M. Karin (1991) Function of the homeodomain protein GHF1 in pituitary cell proliferation. Science 253:197.

Elias, K.A., and R.I. Weiner (1984) Direct arterial vascularization of estrogen-induced prolactin-secreting anterior pituitary tumors. Proc. Natl. Acad. Sci. USA 81:4549.

Faglia, G., A. Spada, B. Ambrosi, et al. (1991) What is the role of the hypothalamus in pituitary adenoma formation? In: S. Melmed and R. Robbins (eds.). Molecular and Clinical Advances in Pituitary Disorders. Boston: Blackwell, p. 18.

Frohman, L.A., and J.O. Jansson (1986) Growth hormone-releasing hormone. Endocr. Rev. 7:223.

Furth, J., N.N. Dent, and W.T. Burnett (1955) The mechanism of induction and the characteristics of pituitary tumors induced by thyroidectomy. J. Clin. Endocrinol. Metab. 15:81.

Gonsky, R., V. Herman, S. Melmed, et al. (1991) Transforming DNA sequences present in human prolactin-secreting pituitary tumors. Mol. Endocrinol. 5:1687.

Guillemin, R.R., P. Brazeau, P. Bohlen, et al. (1982) Growth hormone-releasing factor from a human pancreatic tumor that caused acromegaly. Science 218:585.

Herman, V., J. Fagin, R. Gonsky, et al. (1990) Clonal origin of pituitary adenomas. J. Clin. Endocrinol. Metab. 71:1427.

Ingraham, H.A., R. Chen, H.J. Mangalam, et al. (1988) A tissue-specific transcription factor containing a homeodomain specifies a pituitary phenotype. Cell 55:519.

Klibanski, A. (1990) Further evidence for a somatic mutation theory in the pathogenesis of human pituitary tumors. J. Clin. Endocrinol. Metab. 71:1415A.

Kovacs, K., E. Horvath, S.L. Asa, et al. (1989) Pituitary cells producing more than one hormone. Human pituitary adenomas. Trends Endocrinol. Metab. 1:104.

Lamberts, S.W.J. (1988) The role of somatostatin in the regulation of anterior pituitary hormone secretion and the use of its analog in the treatment of human pituitary tumors. Endocr. Rev. 9:417.

Landis, C.A., G. Harsh, J. Lyons, et al. (1990) Clinical characteristics of acromegalic patients whose pituitary tumors contain mutant G3 protein. J. Clin. Endocrinol. Metab. 71:1416.

Landis, C.A., S.B. Masters, A. Spada, et al. (1989) GTPase inhibiting mutations activate the alpha chain of Gs and stimulate adenylyl cyclase in human pituitary tumors. Nature 340:692.

Mayo, K.E., R.E. Hammer, L.W. Swanson, et al. (1988) Dramatic pituitary hyperplasia in transgenic mice expressing a human growth hormone-releasing factor gene. Mol. Endocrinol. 2:606.

Melmed, S. (1990) Acromegaly. N. Engl. J. Med. 322:966.

Melmed, S., G.D. Braunstein, E. Horvath, et al. (1983) Pathophysiology of acromegaly. Endocr. Rev. 4:271.

Melmed, S., G. Braunstein, J. Chang, et al. (1986) Pituitary tumors secreting growth hormone and prolactin. Ann. Intern. Med. 105:238.

Melmed, S., and R.J. Rushakoff (1987) Ectopic pituitary and hypothalamic hormone syndromes. Endocr. Metab. Clinics 16:805.

Molitch, M.E. (1987) Pathogenesis of pituitary tumors. Endocrinol. Metab. Clin. North Am. 166:503.

Mulcahey, J.J., and R.B. Jaffe (1988) Detection of a potential progenitor cell in the human fetal pituitary that secretes both growth hormone and prolactin. J. Clin. Endocrinol. Metab. 66:24.

Pagesy, P., J.Y. Li, F. Rentier-Delrue, et al. (1990) Growth hormone and somatostatin gene expression in pituitary adenomas with active acromegaly and minimal plasma growth hormone elevation. Acta Endocrinol. (Copenh.) 122:745.

Reichlin, S. (1991) Pathogenesis of pituitary tumors. In: G. Faglia, P. Beck-Oeccizm, B. Ambrosi, et al. (eds.). Pituitary Adenomas: New Trends in Basic and Clinical Research. New York: Elsevier, p. 113.

Ross, D.A., and C.B. Wilson (1987) Results of transsphenoidal resection of growth hormone-secreting pituitary adenomas. In: R.J. Robbins and S. Melmed (eds.). Acromegaly. New York: Plenum, p. 229.

Sano, T., S.L. Asa, and K. Kovacs (1988) Growth hormone-releasing hormone-producing tumors: Clinical, biochemical, and morphological manifestations. Endocr. Rev. 9:357.

Stefaneanu, L., K. Kovacs, E. Horvath, et al. (1989) Adenohypophysial changes in mice transgenic for human growth hormone-releasing factor: A histological, immunocytochemical and electron microscopic investigation. Endocrinology 125:2710.

Struthers, R.S., W.W. Vale, C. Arias, et al. (1991) Somatotroph hypoplasia and dwarfism in transgenic mice expressing a non-phosphorylation CREB mutant. Nature 350:622.

Thorner, M.O., R.L. Perryman, M.J. Cronin, et al. (1982) Somatotroph hyperplasia: Successful treatment of acromegaly by removal of a pancreatic islet tumor secreting a growth hormone-releasing factor. J. Clin. Invest. 70:965.

Chapter 23

Cytogenetic Analysis of Pituitary Adenomas

Jack P. Rock, V.R. Babu, and Jacob Chason

INTRODUCTION

Pituitary adenomas are usually slow-growing neoplasms that may be hormonally active. The natural history of an individual tumor is unpredictable and, as a result, optimal treatment remains unknown. Treatment, therefore, can vary considerably from patient to patient. Identification of indicators of a tumor's biologic potential could greatly improve the management of these tumors in patients.

Many investigators have sought to gain insight into the nature of pituitary adenomas by the cytogenetic analysis of these tumors. In 1969, Mark reported that aneuploid chromosome numbers can be seen in pituitary tumors. Although the study by Mark was from the prebanding era, it became clear that benign tumors can harbor chromosomal abnormalities. Subsequently, others utilizing flow cytometry to study the DNA content of pituitary adenomas noted correlations between aggressive tumor growth and abnormal DNA content (Anniko et al., 1981, 1983; Anniko and Wersall, 1988). Although there were no definitive correlations between nuclear DNA content and clinical behavior, there were suggestions that deviations from the normal chromosome content could denote aggressive growth (Anniko and Wersall, 1988). After the authors analyzed proliferation rates, they found higher growth rates in GH tumors. Landolt et al. (1987) reported similar results using Ki-67, a monoclonal antibody that recognizes a proliferation-associated nuclear antigen not present in cells in the resting (G_0) phase of the cell cycle. In this study, invasive tumor growth and growth hormone secretion were associated with a higher proportion of cells in the growth fraction. The growth fraction was considered to be the percentage of Ki-67-positive nuclei in the cell population. This antibody test can be performed immediately at the time of surgery and the authors concluded that the test may prove a useful estimate of prognosis. More recently, Knosp and colleagues (1989) reported findings using Ki-67, but did not note a correlation between endocrinologic subtype and proliferation rate. They did, however, find a higher proliferation rate with invasive tumors.

These studies are based on DNA characteristics but were not on the detailed morphology of the individual chromosomes. The study of chromosomal morphology has revealed many clinically relevant facts (Figs. 1, 2). In 1960, Nowell and Hungerford discovered the marker chromosome in chronic myelogenous leukemia, the Philadelphia chromosome (Ph). Levan (1966) and van Steenis (1966) then demonstrated that certain chromosomal abnormalities are commonly found in association with specific tumors. This suggests that certain chromosome aberrations may be causally related to neoplasia. Shortly thereafter, the nonrandom (clonal) nature of some chromosomal abnormalities was demonstrated in some solid tumors (Sandberg, 1980). By 1970, Caspersson et al. had developed the technique of chromosome banding. This revolutionized cytogenetics by demonstrating that while the histologic appearance of certain tissues could be homogenous, the genetic structure could show considerable diversity. Emanating from these results, subsets based on chromosomal morphology were noted and correlations between these cytogenetic subsets and clinical behavior could be recognized. These karyotypic changes have been well established in association with various leukemias. Although the hematopoietic malignancies were relatively easy to raise in culture, improvements in tissue culture techniques were required before a similar ability to discern changes in solid tumors evolved. As early as 1967, Lamb reported a relationship between chromosome number, histologic appearance, and tumor invasiveness in patients with bladder cancer. This study demonstrated that all noninvasive bladder tumors analyzed had a near-diploid chromosome count, whereas their invasive counterparts had near-tetraploid counts. Subsequently, Sandberg (1981) demonstrated a correlation between patient survival and the presence or absence of marker chromosomes utilizing banding techniques for the cytogenetic analysis of bladder cancer. These investigators found that for noninvasive bladder tumors, with and without markers, the five-year survival rate was 55% and 95%, respectively.

The study of malignant tumors is facilitated by their complex karyotypes, their higher mitotic indices, and the

Molecular Genetics of Nervous System Tumors, pages 265-269 ©1993 Wiley-Liss, Inc.

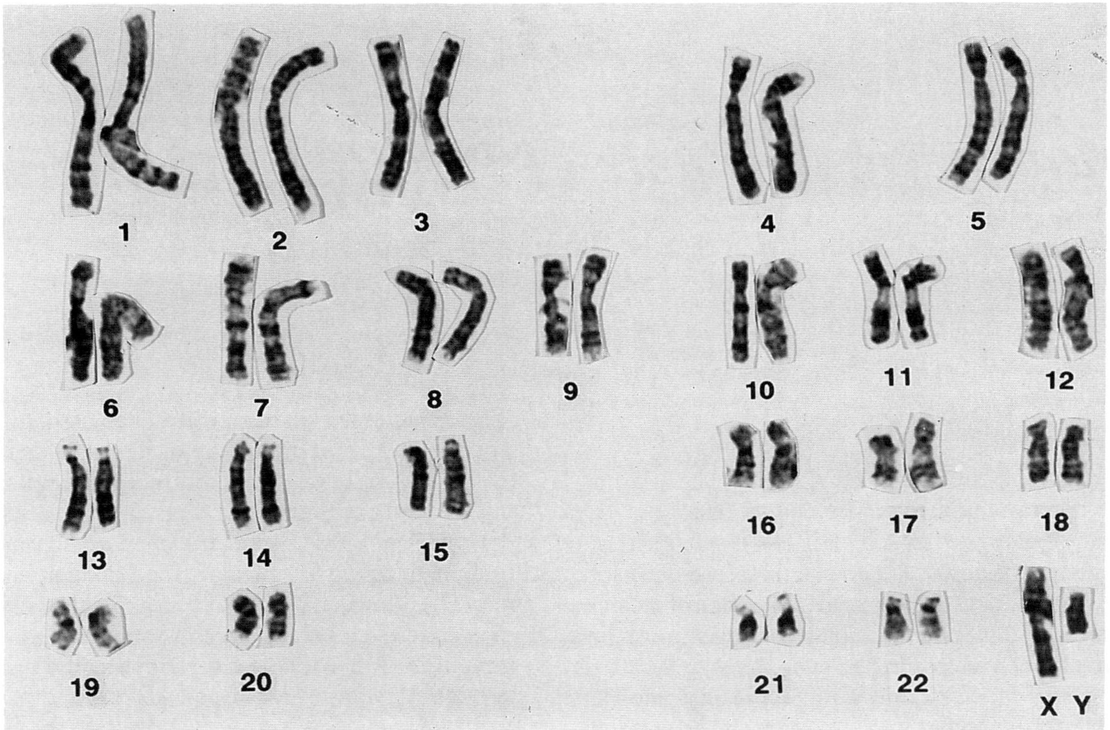

Fig. 1. Normal male karyotype. Note banding pattern, which defines subregions on the chromosome, thereby allowing specific identification.

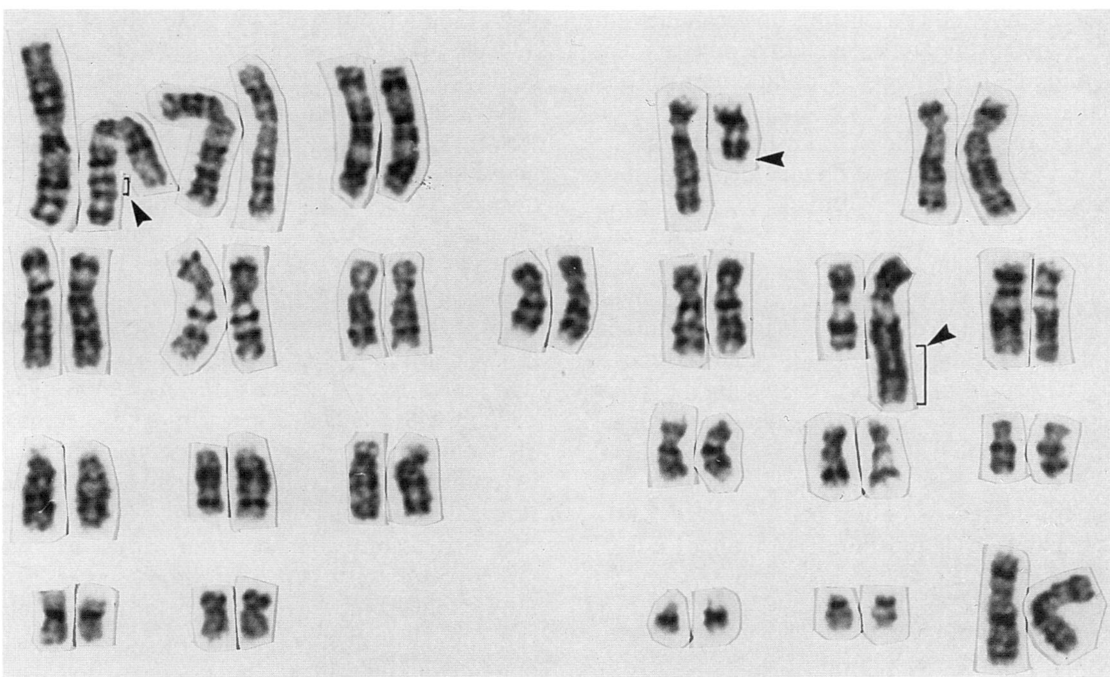

Fig. 2. Abnormal pseudodiploid karyotype noted in patient 11 with a duplication (arrow) on the long arm of chromosome 1 and a rearrangement (double arrow) of genetic material between chromosome 4 and 11. Chromosome pairs are numbered as in Figure 1.

relative ease with which these tumors can be raised in culture. Benign tumors are harder to raise in culture and are slower growing. Several benign neoplasms, however, have been analyzed cytogenetically using refinements in tissue culture techniques (Sandberg and Turc-carel, 1987). The most extensively studied tumor to date is the meningioma, of which more than 50% have a specific karyotypic change: loss of one of the alleles of chromosome 22 (Zang, 1982). While this abnormality correlates poorly with biologic behavior, other chromosome changes appear to correlate with the tumor's invasiveness. This does not signify the potential for malignant transformation. Other benign tumors studied include the mixed tumors of the salivary glands, lipomas, and colonic adenomas. From these studies, it is clear that even benign tumors may harbor abnormal karyotypes, the significance of which remains unknown. In addition, benign neoplasms may appear to be histologically homogenous while subsets with relevance to treatment and prognosis are being noted (Sandberg, 1986).

MATERIAL AND METHODS

Tumor samples from 29 patients were submitted and 18 survived in short-term culture. All specimens were collected in the operating room, in centrifuge tubes with sterile F10 medium supplemented with 15% fetal bovine serum, penicillin–streptomycin, and L-glutamine. The specimens were then prepared according to the technique of Gibas with slight modification (Gibas et al., 1984). The specimens were mechanically disaggregated and then dissociated by immersion in 0.8% collagenase for 2 to 3 hours at 37°C. For short-term cultures, the dissociated material was transferred to culture flasks and bathed in fresh F10 medium containing 15% fetal bovine serum and antibiotics. These cultures were observed on a daily basis for clonal growth. When the culture clones were adequate for harvesting, colcemid was added at a final concentration of 0.01 g/ml for 2 to 3 hours, and the cells were released with 2 cc trypsin EDTA 1×. The sample was then centrifuged and the supernatant discarded. The cells were suspended in 5–8 cc of 0.75 M KCl and incubated for 15 minutes in a 37°C water bath. Next, 4 cc of fresh 3:1 absolute methanol:glacial acetic acid fixative was added to the cell suspension and mixed vigorously by pipetting. The cells were then washed in fresh fixative. Slides were prepared by dropping several drops of cell suspension onto a clean, cold, and wet slide. The slide was dried overnight at 65°C. For identification of chromosome rearrangements, the slide was prebanded by 0.25% trypsin using Giemsa.

On each suspension, at least 30 cells were analyzed. The analysis included number of chromosomes, presence or absence of marker chromosomes, and specific rearrangements. For defining a clonal abnormality, guidelines established by the International Workshops on Chromosomes in Leukemia were followed. In order to be considered clonal, structural abnormalities were present in at least two cells. In order to define a numerical abnormality, at least three cells harbored the changes. All chromosome breakpoints in the rearrangements were recorded using the ISCN 1985 (ISCN, 1985).

RESULTS

Clinical Data

The patients ranged from 22–68 years in age and significant differences in sex were not noted (Table 1). Most patients with null-cell-type tumors had visual loss secondary to suprasellar extension of tumor, and those with hormone-secreting tumors had manifestations of the underlying endocrinopathy. In patient 12, who complained of headache, the tumor was discovered upon examination. In patient 14, the work-up of abdominal carcinoid disclosed an elevated serum prolactin that was followed by magnetic resonance imaging (MRI) of the brain. Patient 9 had Cushing's disease unresponsive to medical, surgical, and radiation treatment (5,000 R).

Pathological Studies

All surgical specimens were submitted for routine histologic and immunocytochemical analysis. Note was made of loss of normal glandular structure, presence of necrosis, hemorrhage, fibrosis, hypercellularity, granularity, number

TABLE 1. Clinical Data at Time of Surgical Exploration

Patient	Age	Sex	Symptoms
1	49	M	Visual loss
2	56	F	Visual loss
3	57	F	Visual loss
4	44	M	Visual loss
5	51	F	Visual loss
6	57	M	Visual loss
7	68	M	Visual loss
8	42	M	Acromegaly
9	46	M	Cushing's disease
10	46	M	Visual loss
11	44	F	Acromegaly
12	52	F	Headache
13	22	F	Acronomegaly
14	58	F	Carcinoid*
15	40	F	Acromegaly
16	45	F	Visual loss
17	41	F	Acromegaly
18	37	F	Cushing's disease

*Endocrine analysis disclosed an elevated prolactin value and subsequent MRI of the brain demonstrated the pituitary adenoma.

of mitoses per high-power field, and nuclear pleomorphism. No correlation could be made between karyotypic detail and cell morphology.

Criterion of Invasiveness

All patients had computed tomographic (CT) and MRI examination of the brain. Invasive behavior was thought to be present if CT demonstrated destruction of surrounding bone and/or growth of tumor into the adjacent sphenoid sinus. Invasion of the cavernous sinus was thought to exist if irregular enhancing tissue extended beyond the medial wall up to and around the carotid artery. Intraoperative invasion of the optic chiasm was noted in patient 2.

Cytogenetic Analysis

The predominant chromosomal abnormalities are noted in Table 2. All karyotypes are metaphase preparations and, accordingly, two copies of each chromosome are present. Five of seven null-cell tumors had normal karyotypes, while two were abnormal. Of the hormonally active tumors, three had normal karyotypes, while the remaining eight were abnormal. Seven tumors were of the mixed growth hormone–prolactin secreting type and six of these had abnormal karyotypes. Rearrangements involving chromosomes 17 and 19 were noted in four tumors, and trisomy 7 was noted in two tumors. Numerous changes were noted in one tumor that had been previously irradiated (patient 9). These abnormalities were noted in other tumors but were present in only one cell under examination. They therefore cannot be considered clonal in origin.

TABLE 2. Summary of Invasive, Histologic, and Karyotypic Data

Point	Invasion*		Immunochemistry (normal/abnormal)	Karyotype description
1	-	Null	+	46,XY
2	+	Null	+	46,XX,19q+
3	-	Null	+	46,XX,TA(17q;19q)
4	+	Null	+	46,XY
5	-	Null	+	46,XX
6	-	Null	+	46,XY
7	+	Null	+	46,XY
8	+	GH-prl	+	46,XY,TA(17;19q)
9	+	ACTH**	+	30,XY multiple marker chromosomes including 19q+
10	-	GH-prl	+	44,XY,-2,-10,-13,-14,+der (2)t(2;10;14) (q35;q23; q11)
11	+	GH-prl	+	46,XX,dup(1)(q32q42),dir ins(11;4)(q22;q13q31.3)
12	-	GH-prl	+	Near diploid, trisomy 7
13	-	GH-prl	+	44,XX,-1,-4,-4,-11,+der (4)t(1;4)(q1;q35),+mar***
14	+	Prl	+	46,XX
15	-	GH-prl	+	60–65,XXX,+X,-1,+4,-6, +7,+12,-15,-16,+17,-18, -21,-22
16	+	Prl	+	51,X,-X,-1,+4,-6,+7,+9, -18,-21,-22
17	+	GH-prl		46,XX
18	+	ACTH	+	46,XX

*CT/MRI or intraoperative evidence of invasiveness.
**Patient previously irradiated.
***Trisomy 7 was seen in one cell; TAs of 19q with different chromosomes were seen in several cells.
TA, Telomeric association; dwr, derivative chromosome; mar, marker chromosome of unknown origin; dup, duplication; dir ins, direct insertion.

DISCUSSION

To our knowledge, there have been no karyotype studies of pituitary adenomas in the postbanding era. In the current study, five of seven null-cell tumors had normal karyotypes, while only three of 11 hormonally active tumors had normal chromosome composition. Among the hormonally active tumors, six of seven GH-prl adenomas had abnormal karyotypes. Similar abnormalities have been noted previously with this hormonal subtype, but the significance of this finding is unclear (Anniko and Wersall, 1988; Landolt et al., 1988).

With respect to specific chromosomal abnormalities, four tumors had trisomy 7 (nonclonal in patients 13 and 16). Three of these tumors were GH-prl and one was prl type. Other nonrandom changes were noted on chromosome 19 in three tumors, two of which secreted GH-prl and the other ACTH. One tumor had a hypodiploid count (less than 46 chromosomes), and one tumor had changes on chromosomes other than 1, 4, 7, and 19. Numerous changes including a chromosome 19 abnormality were noted in one ACTH tumor that had been previously irradiated. Both of the null-cell tumors with abnormal karyotypes had changes involving chromosome 19 on the long arm. When the karyotypes of the GH-prl tumors are examined, four patterns are demonstrated. The tumors had the following: (1) a normal karyotype; (2) abnormalities of chromosomes 1, 4, and 7; (3) abnormality of chromosome 19, or (4) other chromosome changes. This subgrouping of chromosomal abnormalities is reminiscent of the subgrouping described in leukemia and some solid tumors. Thus, the concept of genetic diversity despite histologic homogeneity is noted (Sandberg and Turc-carel, 1987). Review of several histologic parameters, including vascularity, necrosis, cellularity, number of mitoses, nuclear pleomorphism, and cytoplasmic granularity, failed to reveal significant correlation. This is in agreement with the literature and further demonstrates the need for more definitive methods with which to study anaplasia.

We have attempted to correlate invasive tumor growth with chromosomal changes. We have chosen to define invasiveness as the capacity of the tumor to infiltrate surrounding tissues, such as the sphenoid sinus and adjacent bony structures, cavernous sinus, and the optic chiasm. Of eight invasive tumors (a number that based on recent reviews is probably underestimated) five had abnormal karyotpyes while four of 10 non-invasive tumors had abnormal karyotypes. (Sandberg, 1980). Although the numbers are small, the findings suggest that chromosomal abnormalities might correlate with the potential to aggressive growth (Anniko et al., 1983; Knosp et al., 1989; Landolt et al., 1987).

CONCLUSION

This study demonstrates that some pituitary tumors have nonrandom chromosome changes involving chromosomes 1, 4, 7, and 19. In addition, the karyotypes of 71% of the nonsecreting tumors were normal, while only 28% of hormone-secreting tumors were normal. This finding bears further investigation. Tumors of the GH-prl type appear to have a high incidence of chromosomal abnormalities, and subgroups among these abnormalities might be noted as a greater number of tumors are studied. It is also likely that with documentation of the genetic characteristics of other central nervous system tumors, our ability to make important differential diagnoses will be improved. By utilizing cytogenetic techniques to point the way for more detailed molecular biological analysis, we might increase our understanding of the nature of pituitary adenomas and neoplasia in general.

ACKNOWLEDGMENTS

The authors gratefully acknowledge the assistance of Henry H. Schmidek with the preparation of this manuscript.

REFERENCES

Anniko, M., L.E. Holm, B. Silversward, et al. (1981) DNA measurements of human pituitary tumors. Acta Otolaryngol. (Stockh.) (Suppl.) 379:1–28.

Anniko, M., L. Holm, and J. Wersall (1983) Aggressive pituitary tumor growth. Arch Otorhinolaryngol. 238:53–62.

Anniko, M., and J. Wersall (1988) DNA studies for prediction of prognosis of pituitary adenomas. In: A. Landolt, P.U. Hertz, J. Zapf, et al. (eds.). Advances in Pituitary Adenoma Research. New York: Pergamon Press, pp. 45–52.

Caspersson, T., L. Zech, and C. Johansson (1970) Differential binding of alkylating flourocarbons in human chromosomes. Exp. Cell Res. 60:315–319.

Gibas, L.M., Z. Gibas, and A.A. Sandberg (1984) Technical aspects of cytogenetic analysis of human solid tumors. Karyogram 10:25–27.

ISCN (1985) D.G. Harnden and H.P. Klinger (eds.). An International System for Human Cytogenetic Nomenclature. Basel: Karger.

Knosp, E., K. Kitz, and A. Perneczky (1989) Proliferation activity in pituitary adenomas; measurement by monoclonal antibody Ki-67. Neurosurgery 25(6):927–929.

Lamb, D. (1967) Correlation of chromosome counts with histological appearances and prognosis of transitional cell carcinoma of the bladder. Br. Med. J. 1:273–277.

Landolt, A.M., T. Shibata, and P. Kleihues (1987) Growth rate of human pituitary adenomas. J. Neurosurg. 67:803–806.

Landolt, A.M., T. Shibata, P. Kleihues, et al. (1988) Growth of human pituitary adenomas: Facts and speculations. In: A.M. Landolt, F.U. Hertz, J. Zapf, et al. (eds.). Advances in Pituitary Adenomas Research. New York: Pergamon Press, pp. 53–67.

Levan, A. (1966) Nonrandom representation of chromosome types in human tumor stemlines. Hereditas 55:28–38.

Mark, J. (1969) Two benign intracranial human tumors with an abnormal chromosome picture. Acta Neuropath. 14:174–189.

Nowell, P.C., and D.A. Hungerford (1960) A minute chromosome in human chronic granulocytic leukemia. Science 132:1497-.

Sandberg, A.A. (1980) The Chromosomes in Human Cancer and Leukemia. New York: Elsevier/North Holland.

Sandberg, A.A. (1981) Chromosome studies in bladder cancer. In: J.G. Connolly (ed.). Carcinoma of the Bladder. New York: Raven Press.

Sandberg, A.A. (1986) Chromosome changes in bladder cancer: Clinical correlations. Cancer Genet. Cytogenet. 19:163–175.

Sandberg, A.A., and C. Turc-carel (1987) The cytogenetics of solid tumors. Cancer 57:387–395.

van Steenis, H. (1966) Chromosomes and cancer. Nature 209:819–821.

Zang, D.K. (1982) Cytological and cytogenetic studies on human meningioma. Cancer Genet. Cytogenet. 6:249–274.

Chapter 24

Molecular Mechanisms Involved in the Multiple Endocrine Neoplasia Syndromes

Robert F. Gagel, Sara Peleg, Marion F. Robinson, and Aubrey E. Boyd III

INTRODUCTION: OVERVIEW OF THE MULTIPLE ENDOCRINE NEOPLASIA SYNDROMES

The identification of genetic disorders that cause neoplasia of certain cell types within the endocrine system has fascinated several generations of endocrinologists since their formal description less than 40 years ago. These genetic syndromes have received attention out of proportion to their frequency in the general population, because of the distinctive clinical syndromes caused by hormone-producing neoplasms. Our understanding of the multiple endocrine neoplasia (MEN) syndromes has evolved through a predictable series of stages. The first was the description of the syndrome and its general clinical features. This was followed by a several-decade period during which the diagnosis, management, and natural history of each syndrome was characterized. It has been only during the past five years that progress toward the elucidation of the molecular defects in these genetic syndromes has occurred. This chapter focuses only on selected aspects of the MEN syndromes that are relevant to the elucidation of the genetic defects.

The Classification of the Hereditary Endocrine Neoplasia Syndromes

Multiple endocrine neoplasia type 1 (MEN 1) is a multiglandular syndrome recognized since the turn of this century (Erdheim, 1903). The patients develop neoplasia of the parathyroid glands, endocrine pancreas, and the anterior pituitary. Wermer was the first to point out that these tumors are inherited in an autosomal dominant manner with a high degree of penetrance (Wermer, 1954) (Table 1). Multiple endocrine neoplasia type 2 (MEN 2), the association of medullary thyroid carcinoma with parathyroid and adrenal medullary neoplasia, was first formulated as a clinical syndrome by Sipple (1961), MEN 1 and 2 were differentiated first as distinct clinical syndromes by Steiner et al. in 1968, leading to the current classification system (Table 1). The identification of a variant of MEN 2 in which the mucosal neuromas are associated with medullary thyroid carcinoma and adrenal medullary neoplasia has led to subcategorization of MEN 2 into types A and B. MEN 2A has been further subcategorized by the identification of a variant of MEN 2A with cutaneous lichen amyloidosis (Gagel et al., 1989; Nunziata et al., 1989a).

The Importance of the Multiple Endocrine Syndromes

The MEN syndromes are important because the predisposition genes on chromosomes 10 and 11 are clearly different from any other known cancer predisposition genes. Identification and cloning of these genes will provide additional information about molecular events leading to carcinogenesis in neural crest cell types. In addition, neoplastic transformation of multiple cell or tissue types occurs in each of the MEN syndromes. Because these cell types are thought to derive from the neural crest, it is possible the predisposing genes are involved in early differentiation of the neural crest. Finally, early identification of gene carriers by DNA-based diagnostic techniques is likely to result in improved management of clinical features of these syndromes.

MULTIPLE ENDOCRINE NEOPLASIA TYPE 1

Clinical Features of MEN 1

Hyperparathyroidism is the most common clinical manifestation of MEN 1, with parathyroid hyperplasia occurring in 90–95% of the families; pancreatic tumors are found in about 75–80%; and pituitary tumors occur in about 65% of

TABLE 1. Classification of Multiple Endocrine Neoplasia Syndromes

Multiple endocrine neoplasia type 1	
Pituitary neoplasia	
Nonfunctioning	
Functioning with secretion of:	
Prolactin	
Adrenocorticotropic hormone	
Growth hormone	
Parathyroid neoplasia	
Pancreatic islet cell neoplasia	
Nonfunctioning	
Functioning with secretion of:	
Gastrin	
Vasoactive intestinal peptide	
Insulin	
Glucagon	
Pancreatic polypeptide	
Multiple endocrine neoplasia type 2A	
Medullary thyroid carcinoma	
Parathyroid neoplasia	
Adrenal medullary neoplasia	
± cutaneous lichen amyloidosis	
Multiple endocrine neoplasia type 2B	
Medullary thyroid carcinoma	
Adrenal medullary neoplasia	
Mucosal neuromas	

the patients. Less frequently, other tumors, including carcinoids of the foregut, bronchus and thymus, and adrenalcortical or thyroid adenomas, or nonfunctional tumors, including lipomas or liposarcomas, are diagnosed. The incidence of malignant transformation in insulinoma is 20% and as high as 50–60% in gastrinoma. The potential to develop malignant pancreatic tumors is seen in some families and not in others. Prolactinomas are the most common pituitary tumors (Carlson et al., 1978). Less frequent are pituitary tumors producing growth hormone (15%), ACTH, combinations of hormones (growth hormone and prolactin), or no secretory peptides. Clinical management of a single clinical syndrome in these patients is complicated; the frequent simultaneous occurrence of more than one clinical syndrome in these patients makes their clinical managment even more challenging.

The Genetic Abnormality in MEN 1

There are two major hypotheses to explain the development of the transformation in MEN 1. The first argues that the loss of a "tumor suppressor" gene on chromosome 11 initiates transformation. The evidence for this hypothesis is considerable. The second suggests that production of a growth factor is an important factor, especially for the parathyroid glands.

Identification of a loss of genes on chromosome 11 derived from the normal parent in several pancreatic tumors obtained from MEN 1 patients led Nordenskӧld and his collaborators to suspect the presence of a predisposing gene for MEN 1 on chromosome 11 (Larsson et al., 1988). Studies by this same group led to the establishment of genetic linkage of MEN 1 to several polymorphic DNA loci on chromosome 11 (Fig. 1). These results were confirmed by several investigators (Friedman et al., 1989; Nakamura et al., 1989a; Radford et al., 1990; Thakker et al., 1989). More recent efforts have led to the identification of flanking markers for the MEN 1 predisposition gene and further narrowed the region containing the tumor suppressor gene to 11q11–13 by combining results obtained from meiotic recombination and mapping of allelic losses in tumors (Fig. 1). It is generally believed that the region spanned by the flanking sequences is 8–12 megabases and is flanked by the DNA sequences for PYGM and INT2.

The molecular observations in MEN 1 have led to the belief that the MEN 1 gene is likely to belong to the category of genes known as tumor suppressor genes. This hypothesis predicts that a mutated MEN 1 gene is inherited from an affected parent and a normal copy from an unaffected parent. Loss of the normal copy in a single cell sets in motion a series of molecular events that lead to transformation of the cell.

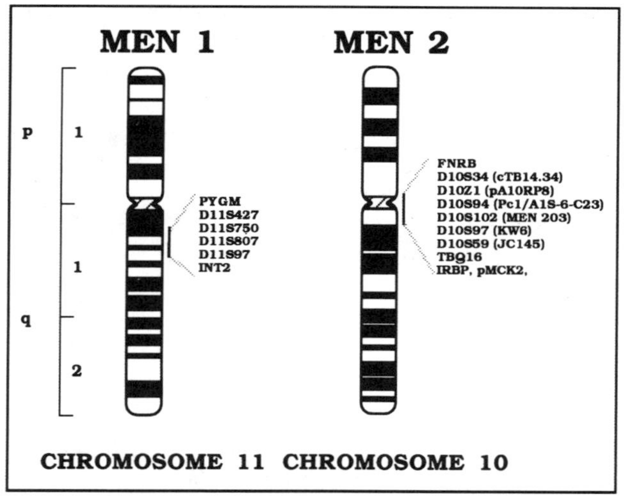

Fig. 1. Chromosomal location of MEN 1 and MEN 2 genes. The MEN 1 gene has been mapped by a combination of linkage and tumor deletion approaches to the proximal long arm of chromosome 10. It is generally agreed that PYGM and INT2 flank the disease gene. Other loci have been tentatively mapped to the region between these flanking markers (Larsson et al., 1992; Pang et al., 1992). The MEN 2 gene has been mapped by linkage approaches to centromeric chromosome 10; however, assignation to the long or short arm is unclear at this time. The DNA sequences for the subunit of the fibronectin receptor (FNRB) and the interstitial retinal binding protein (IRBP) flank the MEN 2 locus (Simpson, 1991). It has been difficult to map the region between these two flanking loci because of the relatively low recombination rate in this region of chromosome 10.

The clinical findings in MEN 1 support the tumor suppressor hypothesis. Although some of the MEN 1 patients may have what appears grossly to be an adenoma, the surrounding pituitary, parathyroid, or pancreatic tissue shows hyperplasia of the same cell type as that in the apparent adenoma. Furthermore, removal of the primary tumor is frequently associated with development of a new tumor derived from a different clonal population of the same cell type. Virtually none of the patients with MEN 1 have a long-term cure following surgical resection. In contrast, patients with sporadic disease usually have discrete adenomas without evidence of hyperplasia. The incidence of recurrence of small, sporadic growth hormone-secreting pituitary adenomas is low, about 3% in five years, and 14% at 5–10 years (Molitch, 1987). Similar high rates of permanent cure occur following surgical removal of sporadic parathyroid and islet cell tumors.

Genetic Abnormalities in Specific Tumors Associated with MEN 1 and in Sporadic Tumors of the Same Tissue Type

Insulinomas

The MEN 1 locus was first mapped to chromosome 11 in pancreatic tumors derived from a MEN 1 kindred (Larsson et al., 1988). In four MEN 1 families they showed showed tight linkage to the phosphorylase gene located on the long arm of chromosome 11. These results have been confirmed by all other groups working in this field. In addition to pinpointing the primary molecular defect in MEN 1, the allelic loss has been used as indirect evidence for monoclonality of the tumors. The allelic loss or "loss of heterozygosity" has been found in pancreatic islet cell hyperplasia as well (Bale et al., 1991; Larsson et al., 1988), suggesting that the tumors are monoclonal in origin. The sporadic islet cell neoplasms that have been studied to date do not usually show inactivation of the region of chromosome 11, which appears to contain the MEN 1 gene (Bale et al., 1991).

Parathyroid glands

There is apparent contradictory evidence regarding the development of parathyroid hyperplasia and adenomas in MEN 1. Reports from several groups have shown allelic loss in parathyroid tumors (Friedman et al., 1989; Friedman, 1987; Radford et al., 1990; Thakker et al., 1989) identical to those found in pancreatic islet cell tumors in MEN 1. These studies would suggest a monoclonal pattern of cell growth with a particular tumor derived from a single cell mutation. Earlier reports demonstrated the presence of a parathyroid growth factor in the serum of patients with MEN 1 (Brandi et al., 1989), suggesting a mechanism for polyclonal expansion of parathyroid tissue, although widespread confirmation of these findings has not been forthcoming. More recent studies have demonstrated this factor stimulates parathyroid endothelial cell growth (Zimering et al., 1990). Although it seems difficult to reconcile these two differing hypotheses, there are some clinical observations that suggest elements of both could be important. Primary hyperparathyroidism in MEN 1 is the first clinical manifestation to occur. Almost all affected individuals will develop hyperparathyroidism caused by multigland involvement, whereas a smaller percentage of affected individuals will develop the pituitary and pancreatic abnormalities, usually later than the parathyroid manifestation. It is reasonable, therefore, to hypothesize that evolution of the parathyroid neoplastic process differs from development of pituitary and pancreatic islet neoplasia, perhaps by combining elements of these two hypotheses. It is possible that a mutational event (loss of the normal MEN 1 allele on chromosome 11) is required for initiation of the monoclonal neoplastic transformation, but that expansion of the clone is dependent upon growth factor stimulation.

Sporadic parathyroid tumors. About 85% of patients with sporadic hyperparathyroidism have overactivity in a single parathyroid gland, resulting in a solitary parathyroid adenoma. Initial studies to determine clonality of parathyroid adenomas were performed using isozymes of the enzyme glucose-6-phosphate dehydrogenase (G-6-PD) and suggested that the tumors were polyclonal (Jackson et al., 1982). More recent studies that used two different independent molecular techniques (clonal gene rearrangement and clonality of an X-linked marker in adenomas obtained from women) established that most parathyroid adenomas were monoclonal (Arnold et al., 1988). The patterns were equivocal in some tumors, but it appears that some and probably most parathyroid adenomas are monoclonal in origin. In Arnold's study none of five hyperplastic glands that were subjected to a similar analysis had a monoclonal pattern. Thus, these studies suggest that mutations are also involved in the pathogenesis of the more common form of parathyroid tumors, the adenomas.

Pituitary tumors

MEN 1 pituitary tumors. There are few molecular studies of pituitary tumors from patients with MEN 1. The high rate of recurrence after surgical removal and the availability of alternative approaches for treating these tumors have made surgery an infrequent treatment for pituitary tumors in this syndrome, thereby reducing the amount of tissue available for study.

Sporadic pituitary tumors. The clonal origin of nonfunctional and functional sporadic pituitary adenomas has been established in several studies (Alexander et al., 1990; Herman et al., 1990), suggesting that these tumors contain one or more genetic alterations that lead to a clonal population. Since the prolactin-secreting lactotroph and the growth hormone-secreting somatotroph arise from a common ancestral stem cell, a mutation(s) in this stem cell could lead to secretory tumors producing both growth hormone and prolactin.

A series of mutations have been described in a small subset of patients with acromegaly (Vallar et al., 1987) in the GTP-binding proteins, which couple ligand-receptor signaling to cellular effector systems. The G proteins are heterotrimers, composed of α, β, and γ subunits. The signaling system cycles between an inactive GDP-bound state and the active GTP-bound configuration. On binding to GTP, the alpha subunit is activated, dissociates from the β/γ subunits and associates with the effector system, thereby activating the effector system. The system is returned to the inactive state by the hydrolysis of the bound GTP.

The hypothalamic peptide, growth hormone-releasing factor (GH-RF), is coupled to growth hormone synthesis, secretion, and somatotroph growth through G_s by activating adenylyl cyclase and increasing cAMP levels in the cell. To determine if this signal transduction pathway was activated in acromegaly, adenylyl cyclase activity was analyzed in a series of growth hormone-secreting pituitary adenomas. Two different patterns were found. In some adenomas, enzymatic activity was low and responded to stimulation, whereas others had high, fixed cyclase activity that could not be further activated. Other studies (Landis et al., 1990; Lyons et al., 1990) showed point mutations that resulted in single amino acid substitutions (missense mutations) for tumors in which adenylate cyclase was continually active. The mutation alters the ability of the G protein to inhibit its intrinsic GTPase activity. Thus, the activated alpha subunit would remain chronically bound to GTP, the alpha subunit of the protein would be associated with the effector system, and adenylyl cyclase and the enzyme would be constitutively activated. The G_s mutation would continuously turn on the growth-promoting adenylyl cyclase signaling pathway. This candidate oncogene has been termed gsp for G_s-protein. In contrast, none of the patients with regulated adenylyl cyclase activity were found to have such mutations. The studies on the G_s alpha mutations have been confirmed (Landis et al., 1990; Spada et al., 1990). In Spada's study, 29 of of 80 patients had activating mutations and in Landis's study 10 of 25 tumors removed by surgery (40%) contained activating mutations. It was observed in both studies that patients with these mutations have smaller tumors and represent 30–40% of tumors causing acromegaly. The importance of this observation is that it builds upon a known ligand/receptor coupling system involved in the dual functions of cell growth and growth hormone production.

Although activating mutations of the G protein p21 ras occur frequently in many human neoplasms, there is no evidence that this oncogene plays a role in human pituitary tumorigenesis. Recently a mutation that activates the ras oncogene was reported in one of 19 spontaneous pituitary adenomas (Karga et al., 1992). This tumor was highly invasive and ultimately led to the patient's death. Thus, although the *ras* mutations were uncommon in pituitary adenomas, such mutations may prove predictive of highly invasive tumors.

THE GENETIC ABNORMALITIES IN MULTIPLE ENDOCRINE NEOPLASIA TYPE 2

Clinical Findings in MEN 2

Three types of tumors are found in the MEN 2A syndrome. Medullary thyroid carcinoma is the most common tumor type and is found in greater than 90% of gene carriers. Clinically recognized pheochromocytomas develop in approximately 50% of gene carriers, although more careful examination indicates that a much higher percentage have histologic evidence of adrenal medullary hyperplasia. Clinically evident parathyroid disease caused by hyperplasia or adenomas are found in 10–20% of patients. In MEN 2B, medullary thyroid carcinoma has been found in all affected patients and pheochromocytoma is found in approximately 75% of affected individuals. The most characteristic feature of MEN 2B is the presence of mucosal neuromas on the tongue, under the eyelids, and throughout the gastrointestinal tract, which provides a distinctive clinical phenotype. A variant of MEN 2A in which there is localized cutaneous amyloidosis located over the upper back has been recently identified and appears to be a distinct genetic syndrome. This is discussed in further detail below.

Morbidity and mortality in MEN 2 is related to metastasis of medullary thyroid carcinoma or cardiovascular death caused by pheochromocytoma. Prospective screening of affected kindreds has resulted in the identification of medullary thyroid carcinoma prior to the development of metastasis in most cases, and ongoing studies indicate death from metastatic tumor is likely to be prevented by early thyroidectomy. Routine annual screening for pheochromocytoma has similarly improved the clinical outcome in these patients and death from unsuspected pheochromocytoma is a rare occurrence in kindreds screened routinely.

In contrast, death from metastatic disease in MEN 2B is more common, a result of early metastasis of the medullary thyroid carcinoma and the more aggressive nature of the tumor. It is not yet known whether thyroidectomy shortly after birth will modify the clinical course in the these patients.

Genetic Linkage of the MEN 2 Syndrome to a Centromeric Locus on Chromosome 10

The effort to develop genetic linkage for MEN 2 dates to the early part of the 1980s, when the use of genetic linkage techniques for genes was revived by the identification of frequent polymorphism of DNA sequence within the human genome. Early efforts were hampered by the lack of polymorphic DNA sequences spread equidistant across the entire genome, but persistence led to the linkage of MEN 2A to the interstitial retinal binding protein locus on chromosome 10 by two groups in 1987 (Mathew et al., 1987a; Simpson et al., 1987). Efforts by several additional groups have confirmed these earlier findings and a large number of additional closely linked DNA sequences have been identified (Bowden et al., 1989; Goodfellow et al., 1990; Lairmore et al., 1991; Myers et al., 1988; Nakamura et al., 1989b; Narod et al., 1991; Schuffenecker et al., 1991; Simpson, 1991; Wu et al., 1990). Studies from several groups now indicate that MEN 2A, MEN 2B, the syndrome of hereditary medullary thyroid carcinoma without pheochromocytoma, and the MEN 2A/cutaneous lichen amyloidosis syndrome are all linked to the same centromeric chromosome 10 locus. Although it is possible a second MEN 2 locus exists, there is currently no compelling evidence.

Flanking DNA sequences have been identified for the MEN 2 gene encompassing a region estimated to be up to 10 megabases in size (Fig. 1). Further use of linkage approaches to narrow the search region for the MEN 2 gene is dependent upon finding recombination between the MEN 2 locus and linked DNA sequences. Unfortunately the rate of meiotic recombination in this region is low and little or no recombination has been observed between recently isolated DNA sequences and the MEN 2 locus (Simpson, 1991). This has made it difficult to use this strategy to further map the MEN 2 locus. Current efforts are focused on the development of collaboration to increase the number of potential meiotic events studied for a particular DNA sequence and physical mapping of the MEN 2 locus. Chromosomal walks have been initiated from several loci within this region, with the exclusion of at least one candidate gene. Progress toward identification of the MEN 2 gene has, however, proceeded slowly.

The Nature of the Genetic Defect in MEN 2

Loss or mutation of a tumor suppressor gene at the MEN 2 locus on chromosome 10 was considered the most likely mechanism for initiation of transformation in MEN 2. Studies by several groups, however, have failed to demonstrate a loss of heterozygosity on chromosome 10 in either medullary thyroid carcinoma or pheochromocytoma (Landsvater et al., 1989; Nelkin et al., 1989b). Because the region of search is still large, spanning between 3 and 10 megabases, a smaller deletion has not been completely excluded. Cytogenetic analysis of primary tumor tissue or human cell lines derived from medullary thyroid carcinoma, however, has demonstrated a number of consistent genetic abnormalities on chromosomes 1p, 3p, 9, 11p, and 22 (Taylor et al., 1989). Studies by several groups have now demonstrated convincing evidence for loss of heterozygosity on chromosomes 1p (Mathew et al., 1987b) and 22 (Takai et al., 1987).

The failure to find allelic loss at the centromeric chromosome 10 locus suggests that loss or mutation of a tumor suppressor gene is not the primary mechanism for initiation of transformation, making the molecular events in MEN 2 likely different from that observed for MEN 1. The precise nature of the initiating event in MEN 2 is not clear, although some investigators have hypothesized an initial event that causes polyclonal expansion of a cell type with superimposition of other molecular abnormalities (perhaps those observed on chromosomes 1p and 22) to further drive the transformation process (Nelkin et al., 1989a).

The MEN 2A/Cutaneous Lichen Amyloidosis Syndrome

The cutaneous lichen amyloidosis/MEN 2A syndrome is a unique variant of MEN 2A (Table 1). This syndrome was first identified in two families in 1989 (Gagel et al., 1989; Nunziata et al., 1989a). An additional six kindreds with this variant have been subsequently described (Chabre et al., 1991; Robinson et al., 1992). The key finding in this variant is the presence of a cutaneous form of lichen amyloidosis over the upper back of affected patients that is associated with intense intermittent pruritus. The finding that the skin lesion preceded the development of clinical evidence of medullary thyroid carcinoma in at least one child (Nunziata et al., 1989b) suggests that the skin lesion does not result from dermal deposition of a tumor product. Biopsy of advanced skin lesions has shown the presence of cutaneous amyloid, which stains positive for keratin but not calcitonin gene products, making this skin lesion analogous to other forms of hereditary and sporadic cutaneous lichen amyloidosis (Masu, 1981). Biopsies of less extensive skin lesions have not shown evidence of amyloid, suggesting that amyloid deposition may be the result of the chronic pruritus associated with this skin lesion (Chabre et al., 1991).

The evidence that this is a distinct genetic variant of MEN 2 came with the identification of an additional six kindreds with this syndrome. We have studied five kindreds with this syndrome (44 affected individuals) and found that 31 of 44 individuals with MEN 2A also had the cutaneous lesion. Each individual with cutaneous lichen amyloidosis also had MEN 2A, with one exception, a young child genetically at risk for MEN 2. We believe the skin lesion is not related to deposition of a tumor product in the skin because the skin lesion has preceded the development of any evidence of tumor in several patients.

The nature of the genetic defect in the MEN 2A/cutaneous lichen amyloidosis syndrome is unclear. In an attempt to utilize this variant to localize the MEN 2 locus, we performed cytogenetic analyses of peripheral white blood cells from two patients in the Texas kindred; these studies showed no evidence of deletion or translocation (Fig. 2A,B), even when analyzed by techniques that elongate interphase chromosomes (data not shown). We have also studied two tumors from kindred members to determine whether there is allelic loss at the MEN 2 locus. Neither tumor showed evidence of allelic loss by Southern analysis using DNA sequences (interstitial retinal binding protein and pMCK2) closely linked to the MEN 2 locus. Finally, in a collaborative effort we have used flanking DNA sequences for the MEN 2 locus (FNRB, D10S94, pMCK 2, and H4.IRBP, Fig. 1) to link this syndrome to the MEN locus, with a maximum lod score obtained by multipoint analysis of 3.9, indicative of linkage to the MEN 2 locus (Robinson et al., 1991).

We have interpreted these results to indicate that the MEN 2A/cutaneous lichen amyloidosis syndrome is a distinct clinical syndrome and, like the other three variants of MEN 2 (MEN 2A, MEN 2B, and MTC-only syndrome), the predisposition gene is located at the centromeric chromosome 10 locus. It is not possible to predict the nature of the genetic defect from our current analysis; however, the clinical finding of pruritus, limited to a particular dermatome distribution, suggests the possibility of a defect of the neural differentiation.

The Characterization of a Cell-Specific Transcription Factor for the Calcitonin Gene

In addition to linkage approaches and studies of tumor tissue to identify allelic loss, other approaches have been employed to determine the molecular defects present in endocrine tumors. One example, the role of G_s mutations in pituitary tumors, has already been described. Overexpression of the calcitonin gene is a frequent observation in medullary thyroid carcinoma and may represent another example. In an effort to determine the regulatory processes involved in control of calcitonin gene expression, we and others have identified the presence of a cell-specific transcription factor that may also be important for differentiation of the normal calcitonin-producing cells of the thyroid gland.

The calcitonin gene encodes for two protein products, calcitonin and calcitonin gene-related peptide. The gene is expressed uniquely in a subset of neuroendocrine cells of neural crest type, including the thyroid C-cell, the Kulchitsky cell of the lung, certain neuronal cell types, and other scattered neuroendocrine types of cells (Gagel, 1983).

Studies from this laboratory and others (de Bustros et al., 1990; Fredman et al., 1990) have demonstrated the presence of a cell-specific enhancer of transcription in the medullary thyroid carcinoma cells that interacts with a type of enhancer sequence (E-box) shown to be important for tissue-specific regulation of gene expression and differentiation of certain tissue and cell types (Garrell and Modelell, 1990; Lo et al., 1991; Sassoon et al., 1989).

TABLE 2. Sequence Comparison of Cell-Specific Regulatory Elements for the Calcitonin Gene and Their Location Relative to the Transcription Start Site

Element	Sequence	Type
(-1,049 to -1,041)	CACACCTGG	E-box
(-941 to -933)	GGCAGCTGT	E-box
(-974 to -963)	GCCAAGATGGCCG	E-box
(-1,008 to -997)	ACCCCCACCC	AP-2

CT, calcitonin; AP-2, activator protein 2.

These studies were performed by cloning 1,450 bp of 5' flanking DNA from the calcitonin gene into a contruct containing the human growth hormone gene. This construct was transfected into a human medullary thyroid carcinoma cell line. Transcriptional activity of the construct was determined by measurement of growth hormone in cells and medium from cells transfected with this construct. Initial studies demonstrated there was enhanced growth hormone production by the construct containing calcitonin 5' flanking DNA compared to a construct containing only promoter sequences (Peleg et al., 1989). This enhancement was specific for calcitonin-producing cells (medullary thyroid carcinoma and small-cell carcinoma of the lung) and was not observed in a variety of other cell types (Peleg et al., 1990). Deletional mapping of this construct identified a region between -1,060 and -905 responsible for cell-specific expression of this construct. Sequence analysis identified four potential regulatory sequences within this region, one with a consensus sequence for activator protein II and three others with a consensus response element (CANNTG) for the helix–loop–helix group of transcriptional factors. Additional deletional and mutation studies confirmed the role of two E-box sequences (-1,049 to -1,041 and -941 to -933) to mediate the cell-specific transcription (Fig. 3) (Peleg et al., 1990). These studies also demonstrated clearly that the -941 to -933 E-box sequence was mandatory for cell-specific expression (Table 2). Use of a restriction site within this sequence (Pvu I site in Fig. 3) or point mutation of this sequence (data not shown) completely destroyed cell-specific expression. Additional studies that utilized gel mobility shift techniques demonstrated specific binding of a nuclear factor from human medullary thyroid carcinoma cell lines to the cell-specific enhancer (Peleg et al., 1990) and efforts are currently underway to purify and clone this factor.

There is potential significance for these findings in the pathogenesis of medullary thyroid carcinoma. It is known that the helix–loop–helix class of transcription factors function not only as cell- or tissue-specific enhancers of tran-

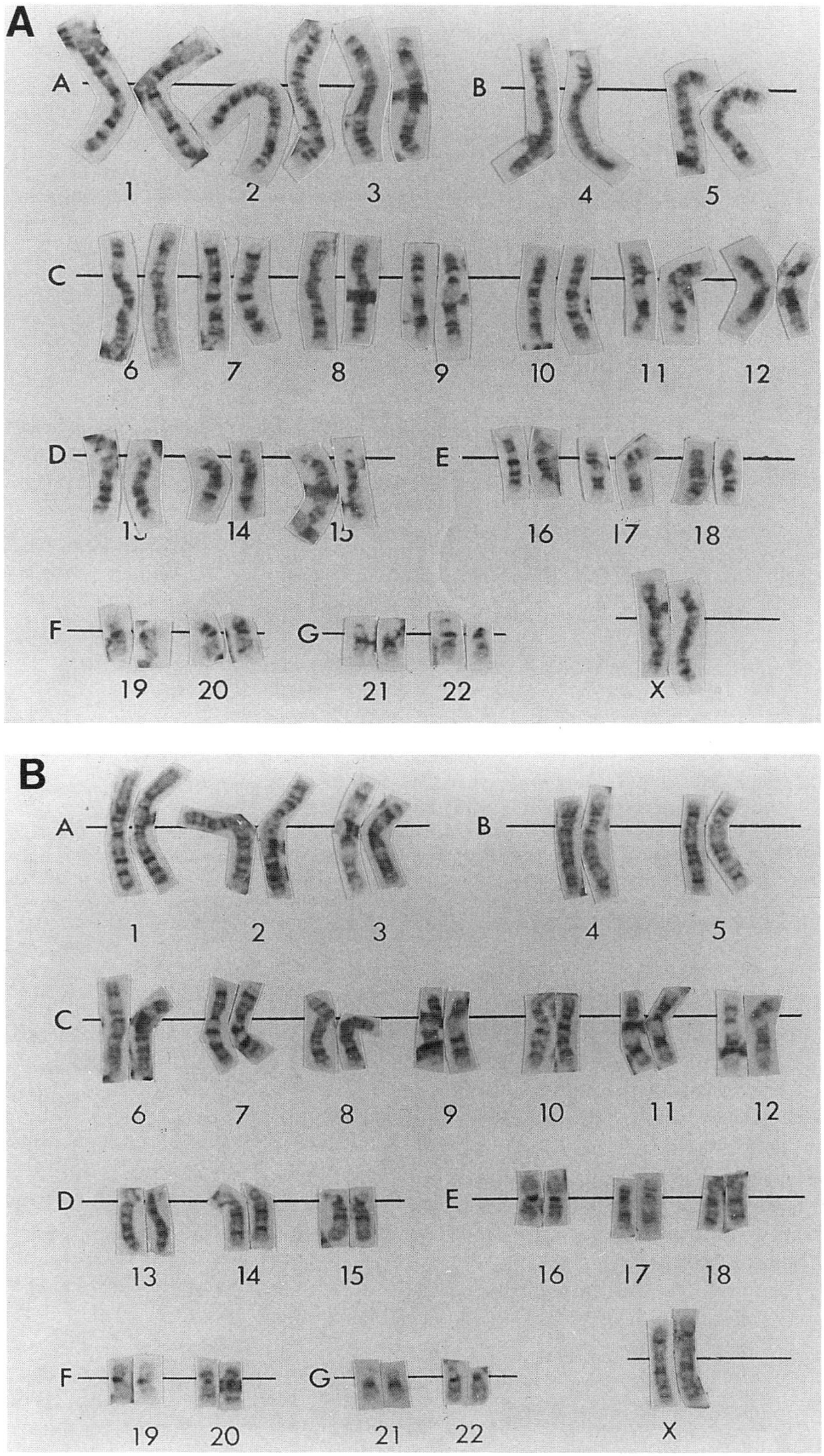

Fig. 2. Cytogenetic analysis of peripheral blood chromosomes in two patients with the MEN 2A/cutaneous lichen amyloidosis syndrome. Cytogenetic analysis of peripheral blood chromosomes in two patients with the MEN 2A/CLA syndrome showed no evidence of deletion or translocation. Specific studies performed to focus on the centromeric region of chromosome 10 showed no abnormalities (data not shown).

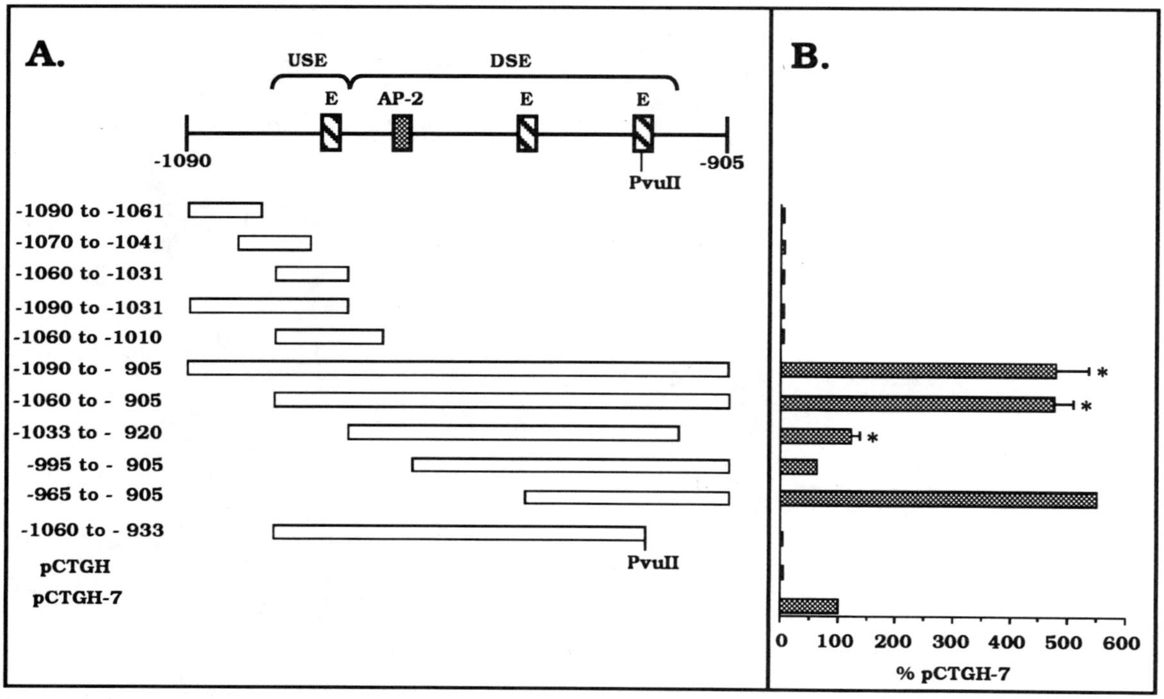

Fig. 3. Characterization of the cell-specific enhancer sequences for the calcitonin gene. A series of fragments and double-stranded oligonucleotides spanning the enhancer region were prepared and inserted individually into the growth hormone reporter construct in an upright orientation. Panel **A** illustrates the full-length enhancer. Within the enhancer is shown the location of three potential E-boxes and an AP-2 (activator protein 2) binding site. Aligned underneath the enhancer are the additional oligonucleotides and fragments relative to their position in the full-length enhancer. Panel **B** shows growth hormone production by TT cells caused by transfection of each construct expressed as a percentatge of pCTGH-7, a construct that contained the sequence -1,333 to -731 cloned into the reporter construct pCTGH and was included in each experiment. Each bar shows the mean ± SEM from three to seven transfections. * indicates a p value < 0.001. (Modified from Peleg et al., 1990.)

scription, but are involved in differentiation of specific tissues. One well-characterized factor is myoD, a helix–loop–helix transcription factor that activates muscle creatine kinase and is required for normal muscle differentiation. A second example is the achaete–scute locus in *Drosophila*, important for nervous system development. Mammalian homologues (MASH 1 and 2) for this locus have been identified, although the biologic role of these homologues has not been completely elucidated (Johnson et al., 1990). We believe it reasonable to hypothesize that the cell-specific transcription factor for the calcitonin gene may be involved in differentiation of the C-cell. In this context there is potential for a role in the transformation process. Further studies to characterize this unique transcription factor are likely to provide additional information of importance for understanding normal and abnormal development of the C-cell.

Clinical Implications of Identification of the Genetic Abnormalities for MEN 1 and 2

The identification of flanking markers for the MEN 1 and 2 loci has made it possible to utilize closely linked markers for the genetic diagnosis of these cancer syndromes. The availability of these markers makes it possible to predict gene carrier status with a certainty of approximately 90–95% in MEN 1 (Skogseid et al., 1991) and greater than 95% in MEN 2 (Gagel, 1991; Sobol et al., 1989). Of immediate impact will be the ability to counsel young couples in affected families of gene carrier status, thereby permitting each couple to make an informed decision about childbearing. Identification and cloning of the disease genes for MEN 1 or 2 will make it possible to designate gene carrier status with certainty in affected kindreds.

Of equal or greater importance is the impact the cloning of disease genes will have for management of patients with sporadic tumors. For example, a frequently asked question from a patient with medullary thyroid carcinoma is whether family members should be screened. Although such testing is regularly suggested, it is rare for a family to be screened adequately to exclude hereditary disease. The identification of DNA sequence abnormalities for MEN 1 and 2 will make it possible to directly address the question of whether a germline mutation exists in individuals with these tumors. It is likely that the ability to reassure patients there is *not* a hereditary component will be one of the most important by-products of this work.

In MEN 1, the greatest impact of genetic screening will be to reduce the frequency of screening in individuals who are not gene carriers. It will also focus attention on predicted gene carriers and enable clinicians to utilize expensive hormonal and radiographic techniques in the most efficient manner. It is not likely, however, that genetic diagnosis will alter the management approach in those identified as gene carriers. Identification of specific tumors in MEN 1 will still depend upon clinical and laboratory evaluation. In families where there is a high incidence of pancreatic islet cell carcinoma, positive identification of gene carrier status may lead to an early total pancreatectomy. Since such families are in the minority, this type of aggressive therapy is not likely to be applied in a widespread manner. Long-term morbidity from diabetes mellitus produced by total pancreatectomy is likely to be more life-threatening than the potential consequences of malignancy. There are some clinical situations where early identification of gene carrier status may alter management. For example, the early treatment of prolactinomas with dopamine agonists may prevent the tumors from interfering with pubertal development or growing to a size that threatens vision or other pituitary function.

The impact of genetic screening in MEN 2 will clearly be greater. Total removal of the thyroid gland prior to the development or spread of medullary thryoid carcinoma is mandatory to prevent death from metastasis at a later age (Gagel et al., 1988). We have used genetic screening to identify children at risk for development of medullary thyroid carcinoma. For example, identification of a two-year-old member of a MEN 2A kindred by genetic screening led us to perform provocative studies of calcitonin release at age 2½ and 3 years that were abnormal and to perform a total thyroidectomy at age 3½ years. The finding of bilateral C-cell hyperplasia in the thyroid gland provided confirmation of diagnosis diagnosis. Prophylactic total thyroidectomy is likely to be recommended for children who are gene carriers once the predisposing MEN 2 gene has been cloned and genetic diagnosis can be established with 100% certainty, because early metastasis of the thyroid tumor has been described. Management of pheochromocytoma in MEN 2 is not likely to be affected by genetic diagnosis. The pheochromocytomas in this syndrome are rarely malignant, they usually do not manifest until long after the diagnosis of medullary thyroid carcinoma, and removal of adrenal glands would not be contemplated in most patients until necessary because of the potential morbidity associated with corticosteroid and mineralocorticoid replacement.

REFERENCES

Alexander, J.M., B.M.K. Biller, and H. Bikkal (1990) Clinically nonfunctioning pituitary tumors are monoclonal in origin. J. Clin. Invest. 86:336–340.

Arnold, A., C.E. Staunton, H.G. Kim, et al. (1988) Monoclonality and abnormal parathyroid hormone genes in parathyroid adenomas. N. Engl. J. Med. 318:658–662.

Bale, A.E., J.A. Norton, E.L. Wong, et al. (1991) Allelic loss on chromosome 11 in hereditary and sporadic tumors related to familial multiple endocrine neoplasia type 1. Cancer Res. 51:1154–1157.

Bowden, D.W., T.C. Gravius, P. Green, et al. (1989). A genetic linkage map of 32 loci on human chromosome 10. Genomics 5:718–726.

Brandi, M.L., M.B. Zimering, S.J. Marx, et al. (1989). Multiple endocrine neoplasia type 1: Role of a circulating growth factor in parathyroid hyperplasia. In: M. Kleerekoper and S. Krane (eds.). Clinical Disorders of Bone and Mineral Metabolism. New York: Mary Ann Liebert.

Carlson, H.E., G.A. Levine, N.J. Goldberg, et al. (1978) Hyperprolactinemia in multiple endocrine adenomatosis, type 1. Arch. Intern. Med. 138:1807–1808.

Chabre, O., F. Labat, F. Berthod, et al. (1991) Cutaneous manifestation asociated with multiple endocrine neoplasia type 2A. La Presse Medicale (in press).

de Bustros, A., R.Y. Lee, D. Compton, et al. (1990) Differential utilization of calcitonin gene regulatory DNA sequences in cultured lines of medullary thyroid carcinoma and small-cell lung carcinoma. Mol. Cell Biol. 10:1773–1778.

Erdheim, J. (1903) Zur normalen und pathologischen Histologie der Glandula Thyreoidea, Parathyreoidea und Hypophysis. Beitr. Pathol. Anat. 33:158–236.

Fredman, L.S., S.E. Leff, E.S. Klein, et al. (1990) A tissue specific enhancer in the rat calcitonin/CGRP gene is active in both neural and endocrine cell types. Mol. Endocrinol. 4:497–504.

Friedman, E., K. Sakaguchi, A.E. Bale, et al. (1989) Clonality of parathyroid tumors in familial multiple endocrine neoplasia type 1. N. Engl. J. Med. 321:213–218.

Friedman, G. (1987) Omeprazole. Am. J. Gastroenterol. 82:188–191.

Gagel, R. (1983) Tumor markers of medullary thyroid carcinoma. In: W. Fishman (ed.). Oncodevelopmental Markers: Biologic, Diagnostic and Monitoring Aspects. New York: Academic Press, pp. 222–239.

Gagel, R.F. (1991) The impact of gene mapping techniques on the diagnosis of multiple endocrine neoplasia type 2. Trends Endocrinol. Metab. 2:19–25.

Gagel, R.F., M.L. Levy, D.T. Donovan, et al. (1989) Multiple endocrine neoplasia type 2a associated with cutaneous lichen amyloidosis. Ann. Intern. Med. 111:802–806.

Gagel, R.F., A.H. Tashjian, Jr., T. Cummings, et al. (1988) The clinical outcome of prospective screening for multiple endocrine neoplasia type 2a. An 18-year experience. N. Engl. J. Med. 318:478–484.

Garrell, J., and J. Modelell (1990) The Drosophila extramacroachaetae locus, an antagonist of proneural genes that like these genes encodes a helix-loop-helix protein. Cell 61:39–48.

Goodfellow, P.J., S. Myers, L.L. Anderson, et al. (1990) A new DNA marker (D10S94) very tightly linked to the multiple endocrine neoplasia type 2A (MEN2A) locus. Am. J. Hum. Genet. 47:952–956.

Herman, V.S., J.A. Fagin, R. Gonsky, et al. (1990) Clonal origin of secretory and non-secretory pituitary adenoma. J. Clin. Endocrinol. Metab. 71:1427–1433.

Jackson, C.E., J.C. Cerny, M.A. Block, et al. (1982) Probable clonal origin of aldosteronomas versus multicellular origin of parathyroid adenomas. Surgery 92:875–879.

Johnson, J.E., S.J. Birren, and D.J. Anderson (1990) Two rat homologues of Drosophila achaete-scute specifically expressed in neuronal precursors. Nature 346:858–861.

Karga, H.J., J.M. Alexander, E.T. Hedley-White, et al. (1992) Ras mutations in human pituitary tumors. J. Clin. Endocrinol. Metab. 74:914–919.

Lairmore, T.C., J.R. Howe, J.A. Korte, et al. (1991) Familial medullary thyroid carcinoma and multiple endocrine neoplasia type 2B map to the same region of chromosome 10 as multiple endocrine neoplasia type 2A. Genomics 9:181–192.

Landis, C.A., G. Harsh, J. Lyons, et al. (1990) Clinical characteristics of acromegalic patients whose pituitary tumors contain mutant Gs protein. J. Clin. Endocrinol. Metab. 71:1416–1420.

Landsvater, R.M., C.G.P. Mathew, B.A. Smith, et al. (1989) Development of multiple endocrine neoplasia type 2a does not involve substantial deletions of chromosome 10. Genomics 4:246–250.

Larsson, C., B. Skogseid, K. Oberg, et al. (1988) Multiple endocrine neoplasia type 1 gene maps to chromosome 11 and is lost in insulinoma. Nature 332:85–87.

Larsson, C., G. Weber, and M. Janson (1992) Sublocalization of the multiple endocrine neoplasia type 1 gene. Henry Ford Hosp. J. (in press).

Lo, L.C., J.E. Johnson, C.W. Wuenschell, et al. (1991) Mammalian achaete-scute homolog 1 is transiently expressed by spatially restricted subsets of early neuroepithelial and neural crest cells. Genes Dev. 5:1524–1537.

Lyons, J., C.A. Landis, G. Harsh, et al. (1990) Two G protein oncogenes in human endocrine tumors. Science 249:655–659.

Masu, K. (1981) Amyloid in localized cutaneous amyloidosis: Immunofluorescence studies with antikeratin antiserum especially concerning the difference between systemic and localized cutaneous amyloidosis. Acta Derm. Venereol. (Stockh.) 61:381.

Mathew, C.G., K.S. Chin, D.F. Easton, et al. (1987a) A linked genetic marker for multiple endocrine neoplasia type 2A on chromosome 10. Nature 328:527–528.

Mathew, C.G., B.A. Smith, K. Thorpe, et al. (1987b) Deletion of genes on chromosome 1 in endocrine neoplasia. Nature 328:524–526.

Molitch, M. (1987) Pathogenesis of pituitary tumors. Endocrinol. Metab. Clin. North Am. 16:503–525.

Myers, S., J. Wu, P.J. Goodfellow, et al. (1988) A linkage map of the pericentric region of chromosome 10 including the locus for multiple endocrine neoplasia (MEN 2A). Genome 30:239.

Nakamura, Y., C. Larsson, C. Julier, et al. (1989a) Localization of the genetic defect in multiple endocrine neoplasia type 1 within a small region of chromosome 11. Am. J. Hum. Genet. 44:751–755.

Nakamura, Y., C.G. Mathew, H. Sobol, et al. (1989b) Linked markers flanking the gene for multiple endocrine neoplasia type 2A. Genomics 5:199–203.

Narod, S.A., H. Sobol, I. Schuffenecker, et al. (1991) The gene for MEN 2A is tightly linked to the centromere of chromosome 10. Hum. Genet. 86:529–530.

Nelkin, B.D., .A.C. de Bostros, M. Mabry, et al. (1989a) The molecular biology of medullary thyroid carcinoma. A model for cancer development and progression. JAMA 261:3130–3135.

Nelkin, B.D., Y. Nakamura, R.W. White, et al. (1989b) Low incidence of loss of chromosome 10 in sporadic and hereditary human medullary thyroid carcinoma. Cancer Res. 49:4114–4119.

Nunziata, V., G. di Giovanni, A.M. Lettera, et al. (1989a) Cutaneous lichen amyloidosis associated with multiple endocrine neoplasia type 2A. Henry Ford Hosp. J. 37:144–146.

Nunziata, V., R. Giannattasio, G. di Giovanni, et al. (1989b) Hereditary localized pruritus in affected members of a kindred with multiple endocrine neoplasia type 2A (Sipple's syndrome). Clin. Endocrinol. 30:57–63.

Pang, J.T., M.A. Pook, J.H. Eubanks, et al. (1992) Molecular genetic mapping of the multiple endocrine neoplasia type 1 (MEN 1) locus. Henry Ford Hosp. J. (in press).

Peleg, S., R.V. Abruzzese, G.J. Cote, et al. (1990) Transcription of the human calcitonin gene is mediated by a C cell-specific enhancer containing E-box-like elements. Mol. Endocrinol. 4:1750-1757.

Peleg, S., G.J. Cote, R.V. Abruzzese, et al. (1989) Transcription regulation of the human calcitonin gene: A progress report. Henry Ford Hosp. Med. J. 37:194–197.

Radford, D.M., S.W. Ashley, S.A.J. Wells, et al. (1990) Loss of heterozygosity of markers on chromosome 11 in tumors from patients with multiple endocrine neoplasia syndrome type 1. Cancer Res. 50:6529–6533.

Robinson, M.F., E.J. Furst, V. Nunziata, et al. (1991) The multiple endocrine neoplasia type 2/cutaneous lichen amyloidosis syndrome is linked to the MEN 2 locus. Abstract presented at the Fourth International Workshop on Multiple Endocrine Neoplasia, October 3–5, 1991, Houston, Texas.

Robinson, M.F., E.J. Furst, V. Nunziata, et al. (1992) Characterization of the clinical features of five families with hereditary primary cutaneous lichen amyloidosis and multiple endocrine neoplasia type 2. Henry Ford Hosp. J. (in press).

Sassoon, D., G. Lyons, W.E. Wright, et al. (1989) Expression of two myogenic regulatory factors myogenin and MyoD1 during mouse embryogenesis. Nature 341:303–307.

Schuffenecker, I., S.A. Narod, R.A. Ezekowitz, et al. (1991) The gene for mannose-binding protein maps to chromosome 10 and is a marker for multiple endocrine neoplasia type 2. Cytogenet. Cell Genet. 56:99–102.

Simpson, N.E. (1991) The exploration of the locus or loci for the syndromes associated with medullary thyroid cancer (MTC) on chromosome 10. In: M.L. Brandi and R. White (eds.). Hereditary Tumors. New York: Raven Press, pp. 55–67.

Simpson, N.E., K.K. Kidd, P.J. Goodfellow, et al. (1987) Assignment of multiple endocrine neoplasia type 2A to chromosome 10 by linkage. Nature 328:528–530.

Sipple, J.H. (1961) The association of pheochromocytoma with carcinoma of the thyroid gland. Am. J. Med. 31:163–166.

Skogseid, B., C. Larsson, K. Oberg (1991) Genetic and clinical characteristics of multiple endocrine neoplasia type 1. Acta Oncol. 30:485–488.

Sobol, H., S.A. Narod, Y. Nakamura, et al. (1989) Screening for multiple endocrine neoplasia type 2a with DNA-polymorphism analysis. N. Engl. J. Med. 321:996–1001.

Spada, A., M. Arosio, D. Bochicchio, et al. (1990) Clinical, biochemical, and morphological correlates in patients bearing growth hormone-secreting pituitary tumors with or without constitutively active adenylyl cyclase. J. Clin. Endocrinol. Metab. 71:1421–1426.

Steiner, A.L., A.D. Goodman, and S.R. Powers (1968) Study of a kindred with pheochromocytoma, medullary carcinoma, hyperparathyroidism and Cushing's disease: Multiple endocrine

neoplasia, type 2. Medicine 47:371–409.
Takai, S., H. Tateishi, I. Nishisho, et al. (1987) Loss of genes on chromosome 22 in medullary thyroid carcinoma and pheochromocytoma. Jpn. J. Cancer Res. 78:894–898.
Taylor, L.D., F.B. Elder, A. Knuth, et al. (1989) Cytogenetic characterization of two human and three rat medullary thyroid carcinoma cell lines. Henry Ford Hosp. Med. J. (abstr., in press).
Thakker, R.V., P. Bouloux, C. Wooding, et al. (1989) Association of parathyroid tumors in multiple endocrine neoplasia type 1 with loss of alleles on chromosome 11. N. Engl. J. Med. 321:218–224.
Vallar, L., A. Spada, and G. Giannattasio (1987) Altered Gs and adenylate cyclase activity in human GH-secreting pituitary adenomas. Nature 330:566–568.
Wermer, P. (1954) Genetic aspects of adenomatosis of endocrine glands. Am. J. Med. 16:363–371.
Wu, J.S., S. Myers, N. Carson, et al. (1990) A refined linkage map for DNA markers around the pericentromeric region of chromosome 10. Genomics 8:461–468.
Zimering, M.B., M.L. Brandi, D.A. deGrange, et al. (1990) Circulating fibroblast growth factor-like substance in familial multiple endocrine neoplasia type 1. J. Clin. Endocrinol. Metab. 70:149–154.

Chapter 25

Some Current Concepts Regarding Medulloblastomas

Henry H. Schmidek

INTRODUCTION

Neoplasia may be viewed as a disease in which the normal mechanisms leading to the maturation and terminal differentiation of stem cells are perturbed at a molecular level. These cells continue to proliferate and are prevented from achieving their fully differentiated, growth-arrested state, maintaining the capacity to divide that most cells lose as they mature.

Stem cells are cells whose daughter cells may pass through a process of differentiation that leads either to a more specialized cell or to continued growth and division and the production of additional stem cells. Stem cells are either unipotent (engendering only one type of endstage differentiated cell) or pluripotent (giving rise to two or more types of differentiated cells). It is from among the stem cells that cells capable of malignant transformation arise: Neoplastic change often occurs in a committed but not fully differentiated stem cell. Thus growth and differentiation are coupled by a linkage between cellular differentiation and cessation of growth. Anaplasia contrasts with cellular differentiation in being characterized by the progressive accumulation of mutations in some tumor cells. These cells then emerge to form the tumor's dominant proliferative cell population. The process results in progressive changes manifested as increasing aneuploidy, loss of tissue-specific enzymes, a variability of phenotype, and increasing chromosomal changes (Schmidek, 1987).

In the investigation of the etiologic mechanisms underlying a particular type of tumor, much is to be learned by identifying the stem cell which gives rise to that neoplasm as well as the mechanisms that are responsible for its malignant transformation. In the last decade, many of the most advanced techniques of molecular genetics have been focused on gaining an insight into identifying the stem cells of the developing nervous system and in studying changes in gene expression in neural neoplasms. As a result, more has been learned about the biology of the medulloblastoma in this decade than the previous 57 years since its identification as a specific neuropathiologic and clinical entity (Bailey and Cushing, 1925). Some of these issues are reviewed in this chapter.

THE MEDULLOBLASTOMA

The medulloblastoma is a common and important group of nervous system neoplasms that shares sufficient clinical and neuropathologic features to warrant discussion as an entity. Most medulloblastomas originate in the cerebellar vermis during the first two decades of life and appear as a rapidly growing brain tumor located in the posterior cranial fossa. This tumor has many properties that suggest membership in a family of tumors that includes neuroblastoma, retinoblastoma, pineoblastoma, and esthesioneuroblastoma. For example, 50% of 16 cerebellar medulloblastomas examined exhibit retinal S antigen (Bonnin and Perentes, 1988) and 30% of 66 medulloblastomas possessed S antigen and rod-opsin cells (Czerwionka et al., 1989), proteins that suggest a shared lineage with retinoblastomas and pineal cell tumors. The medulloblastoma is rarely found in association with autosomal dominant disorders such as familial adenomatous polyposis or the nevoid basal cell carcinoma syndrome, suggesting the expression of an underlying genetic defect in neural and extraneural tissues (Evans et al., 1991; Kropilak et al., 1989; Lacombe et al., 1990).

Among the discrete histologic variants of the medulloblastoma are: (1) the undifferentiated medulloblastoma, (2) the medulloblastoma with neuronal differentiation, which may demonstrate Homer-Wright rosettes, unipolar or bipolar cells that contain neurosecretory granules, and occasionally ganglion cells at various stages of differentiation, and (3) medulloblastomas with glial differentiation that includes astrocytes, oligodendrocytes, and rarely ependymal cells. The medulloblastomas also present with a number of transitional and mixed lesions, and "variants" that contain striated muscle (medullomyoblastoma) and melanin-producing cells (melanotic medulloblastoma). In 30 cases of medulloblastoma classified on immunochemical characteristics, 15 of 30 cases showed evidence of glial differentiation, 17 of 30 showed neurofilament-positive cells indicating neuronal differentiation, and eight of 30 tumors exhibited both proteins. In five cases the cells were nonreactive to both antibodies (Tohyama et al., 1988).

IN SEARCH OF THE "MEDULLOBLAST" AND IDENTIFYING THE STEM CELLS OF THE NERVOUS SYSTEM

Until recently, attempts at identification of the stem cells of the nervous system has been hampered by the lack of tools allowing the homogeneous-appearing cells of the early nervous system to be segregated and their differentiation examined in an orderly manner. These techniques are now available. "Immortalization" of stem cells by retroviruses incorporating an oncogene arrests the differentiation program and is one such technique. The establishment of a cell line from primary cells using retroviruses leads to the growth of such cells continuously in culture and serves an establishment or "immortalizing" function shown to arrest the differentiation program of several cell types, including myoblasts, chondrocytes, retinal melanoblasts, and neuroretinal cells. This approach has been used by Frederiksen and colleagues (1988) to establish cell lines from the developing rat cerebellum. These investigators also used a panel of monoclonal antibodies to recognize the stages of neuronal and glial differentiation (Hayes and McKay, Chapter 2, this volume). Included in this panel are reagents to nestin, an intermediate filament gene specifically expressed in central nervous system stem cells but not in differentiated cells of the adult brain, and to troponin T, an antigen characteristic of muscle differentiation, as well as antibodies to vimentin, neurofilament, and GFAP.

The SV15 cell line developed by Frederiksen and coworkers produced by the immortalization of the postnatal rat cerebellum with a temperature-sensitive variant of the SV40 large T antigen in a retrovirus vector is of particular interest. This clonal population of cell line proliferates and is nestin-positive and vimentin-positive when grown at 33°C. However, when the temperature is shifted to 39°C, a temperature at which the large T protein is rapidly degraded, the cells lose the nestin antigenicity and become GFAP-positive (i.e., glial) cells. The particular pathway of differentiation assumed by these stem cells depends on the microenvironmental influences to which they are subjected. For example, a serum-free media promotes neuronal differentiation. Media containing fetal calf serum promotes glial differentiation of these cells. Depending on these influences the cells differentiate into either neurons, astrocytes, oligodendrocytes, or muscle cells. These are the properties predicted for the "medulloblast," a stem cell whose neoplastic transformation can produce a tumor with the potential to differentiate along neuronal, glial, oligodendroglial, myoblastic lineages, or combinations thereof.

STUDIES IN HUMAN MEDULLOBLASTOMA CELL LINES

The experimental study of the medulloblastoma has been hindered by the difficulties encountered in maintaining these tumors' culture. Prior to 1989, much of the experimental work on the properties of these tumors was performed on the readily available human cell line TE 671. These tumors are now grown in athymic mice as serially transplantable intracranial and subcutaneous xenografts.

The nature of the TE 671 cell line fell into suspicion when Stratton and colleagues (1989a) transfected the TE 671 DNA into NIH 3T3 cells and demonstrated the presence of an activated N-ras gene. Using oligonucleotide probes, the investigators then showed that the N-ras gene is activated by a point mutation at the third base of codon 61 resulting in the substitution of histidine for glutamine in the p21 ras gene product. This uncommon mutation is also present in the human rhabdomyosarcoma cell line RD. Cytogenetic analysis revealed the presence of marker chromosomes common to both TE 671 and RD. DNA fingerprinting using both locus-specific and multilocus core probes subsequently showed that the band patterns of the two cell lines TE 671 and RD are indistinguishable, with the TE 671 cell line exhibiting muscle-type nicotinic acetylcholine receptors and intermediate filament protein desmin suggestive of a myoid origin. These studies suggest that the TE 671 and RD are probably the same rhabdomyosarcoma line.

The TE 671 cell line was of particular interest in our laboratory because treatment in culture with dibutyryl cyclic AMP changed the cellular morphology from small spindle-shaped cells to cells having the classic appearance of neurons with extremely long axonal-appearing processes. Coincident with these morphologic changes, the cells become permanently growth-arrested. Taken together, the morphologic and kinetic data suggested the terminal differentiation of this cell line. The data was not published in view of the uncertainties concerning the nature of the TE 671 cell line. The same type of experiment was carried out by Maria and associates (1990) on non-TE 671 medulloblastoma cells. In the tumor section from which the cells were derived, more than 80% of the cells contained vimentin, while less than 1% of cells contained GFAP. When these cells were cultured with dibutyryl cyclic AMP there was a progressive increase in GFAP expression in the cells, which reached maximal values over 72 hours, suggesting the induced differentiation of a predominantly astrocytic medulloblastoma from its precursor cell and the potential for inducing the terminal differentiation of such a neoplasm and therefore its growth-arrest.

We examined two human medulloblastoma cell lines, 282 Med and DAOY, using the same panel of antibodies as used on the ST 15 cell line. These two lines are composed of anti-vimentin and anti-neurofilament-positive and anti-GFAP-negative cells. Tamura and colleagues (1989) reported on the immunochemical studies of two cell lines of human medulloblastoma, ONS-76 and ONS-81, which they established. Both of these lines express neurofilament protein, neuron-specific enolase without GFAP, or S-100 protein, indicating neuronal but not glial differentiation.

Valtz and co-workers (1991) recently reported the examination of surgical samples of medulloblastoma for evidence of muscle development. These primary cerebellar cells stain with troponin T antibodies. The troponin-positive cells have the morphologic characteristics of muscle. In addition, examination of five surgical specimens of medulloblastoma revealed nestin-positive cells in each of the tumors, although their numbers differed in additional support for the concept of their derivation from a neuroectodermal stem cell.

ONCOGENES AND TUMOR SUPPRESSOR GENES IN MEDULLOBLASTOMA

The medulloblastoma and its variants can be hypothesized to arise following the neoplastic transformation of stem cells at various stages of their maturation. Little is known regarding the cellular events leading to the transformation of the stem cell, although there has been intense interest in scrutinizing neural tumors triggered by the observations of alterations of N-myc oncogene expression in neuroblastomas and by the loss of the Rb tumor suppressor gene and its product in retinoblastoma (Friend et al., 1986, 1988; Lee et al., 1987).

Although medulloblastoma and neuroblastoma share many common features, the frequency of N-myc amplification differs markedly between the two tumors. In a study by Garson et al. (1989), the N-myc gene was not amplified in any of the nine medulloblastoma samples analyzed. This lack of amplification in medulloblastomas was confirmed by the studies of Badiali and associates (1991), who investigated both N-myc and c-myc amplification in 27 medulloblastomas. Their study showed no evidence of N-myc amplification and only one case in 27 of c-myc amplification. Wasson et al. (1990) examined 20 primary medulloblastomas with probes for 11 known oncogenes (erb B1, gli, neu, myc, L-myc, N-myc, H-ras, K-ras, N-ras, sis, src) and found only one tumor to have erb B1 amplification. In contrast, four medulloblastoma cell lines showed c-myc amplification in one, erb B1 amplification in one, and N-myc in one case. Taken together, these studies suggest that oncogene amplification is relatively uncommon in medulloblastomas but that the frequency and diversity of oncogene amplification is greater in tumors that can be established as cell lines and in these cases may provide the cells with a growth advantage.

Although studies of medulloblastoma for evidence of loss of genetic information has been negative for a loss of the RB gene, several investigators have been able to detect a loss of genetic information in these tumors (Raffel et al., 1990; Thomas and Raffel, 1991). This loss occurs most frequently from the short arm of chromosome 17 (Biegel et al., 1989; James et al., 1990). Cogen and colleagues (1990), in an examination of the surgical specimens of nine patients with medulloblastomas, showed that there was a loss of alleles from DNA sequences mapped to the distal part of chromosome 17 in four of the nine patients. A further examination revealed a loss of p53 sequences in two of the specimens. Saylors et al. (1991) searched for p53 mutations in 12 medulloblastoma tumors, eight xenografts, and three permanent cell lines and was only able to find a mutation of the p53 gene in one of three cell lines tested and in none of the xenografts or in any of the primary tumors. They suggest that p53 is mutated in an unusual way or that a second tumor suppressor gene is located on the short arm of chromosome 17 that is involved in the pathogenesis of medulloblastoma.

CONCLUSION

In the last decade and particularly the last five years, the techniques of molecular genetics (such as monoclonal antibodies, immortalized cell lines whose differentiation program can be regulated and examined, and the restriction fragment length polymorphism (RFLP) analysis of tumors for loss or deletion of genetic information) has allowed the identification of a stem cell population in the developing cerebellum whose properties are consistent with those expected of putative "medulloblast." The loss or mutation of a tumor suppressor gene may provide the triggering event leading to the formation of a nervous system neoplasm having the neuropathologic hallmarks of the medulloblastoma.

REFERENCES

Badiali, M., A. Pession, G. Basso, et al. (1991) N-myc and c-myc oncogenes amplification in medulloblastoma. Tumori 77(2):118–121.

Bailey, P., and H. Cushing (1925) Medulloblastoma cerebelli: A common type of midcerebellar tumor of childhood. Arch. Neurol. Psychiatry 14:192–224.

Biegel, J.A., L.B. Rorke, R.J. Packer, et al. (1989) Isochromosome 17q in primitive neuroectodermal tumors of the central nervous system.

Genes Chromosomes Cancer 1(2):139–147.

Bonnin, J.M., and E. Perentes (1988) Retinal S-antigen immunoreactivity in medulloblastoma. Acta. Neuropathol. 76(2):204–207.

Cogen, P.H., L. Daneshvar, L., A.K. Metzger, et al. (1990) Deletion mapping of the medulloblastoma locus on chromosome 17p. Genomics 8(2):279–285.

Czerwionka, M., H.W. Korf, O. Hoffman, et al. (1989) Differentiation in medulloblastomas: Correlation between the immunocytochemical demonstration of photoreceptor markers (S antigen, rod-opsin) and the survival rate in 66 patients. Acta Neuropathol. 78(6):629–636.

Evans, D.G., P.A. Farndon, L.D. Burnell, et al. (1991) The incidence of Gorlin syndrome in 173 consecutive cases of medulloblastoma. Br. J. Cancer 64(5):959–961.

Frederiksen, K., J.S. Jat, N. Valtz, et al. (1988) Immortalization of precursor cells from mammalian CNS. Neuron 1:439–448.

Friend, S.H., R. Bernards, S. Rogelj, et al. (1986) A human DNA segment with properties of the gene that predisposes to retinoblastoma and osteosarcoma. Nature 323:643–647.

Friend, S.H., T.P. Dryja, and R.A. Weinberg (1988) Oncogenes and tumor-suppressor genes. N. Engl. J. Med. 318:618–622.

Garson, J.A., L.F. Pemberton, P.W. Sheppard, et al. (1989) N-myc gene expression and oncoprotein characterization in medulloblastoma. Br. J. Cancer 59(6):889–894.

James, C.D., J. He, E. Carlbom, et al. (1990) Loss of genetic information in central nervous system tumors common to children and young adults. Genes Chromosomes Cancer 2(2):94–102.

Kropilak, M., D.G. Jagelman, and V.W. Fazio (1989) Brain tumors in familial polyposis. Dis. Colon Rectum 32(9):778–782.

Lacombe, D., J.F. Chateil, D. Fontan, et al. (1990) Medulloblastoma in the nevoid basal cell carcinoma syndrome: Case reports and review of the literature. Genet. Couns. 1(3–4):273–277.

Lee, W.H., J.Y. Shew, F.D. Hong, et al. (1987) The retinoblastoma susceptibility gene encodes a nuclear phosphoprotein associated with DNA binding activity. Nature 329:642–645.

Maria, B.L., D. Wong, and V.I. Kalnins (1990) Dibutyryl cyclic AMP induces vimentin and GFAP expression in cultured medulloblastoma cells. Can. J. Neurol. Sci. 17(1):15–20.

Raffel, C., F.E. Gilles, and K.I. Weinberg (1990) Reduction to homozygosity and gene amplification in central nervous system primitive neuroectodermal tumors of childhood. Cancer Res. 50(3):587–591.

Saylors, R.L., D. Sidransky, H.S. Friedman, et al. (1991) Infrequent p53 gene mutations in medulloblastoma. Cancer Res. 51(17):4721–4723.

Schmidek, H.H. (1987) The molecular genetics of nervous system tumors. J. Neurosurg. 67:1–16.

Stratton, M.R., J. Darling, G.J. Pilkington, et al. (1989a) Characterization of the human cell line TE 671. Carcinogenesis 10(5):89–905.

Stratton, M.R., B.R. Reeves, and C.S. Cooper (1989b) Misidentified cell (letter). Nature 337(6205):311–312.

Tamura, K., K. Shimizu, M. Yamada, et al. (1989) Expressions of major histocompatibility complex on human medulloblastoma cells with neuronal differentiation. Cancer Res. 49(19):5380–5384.

Thomas, G.A., and C. Raffel (1991) Loss of heterozygosity on 6q, 16q, and 17p in human central nervous system primitive neuroectodermal tumors. Cancer Res. 51(2):639–643.

Tohyama, T., O. Kubo, M. Katahira, et al. (1988) Glial fibrillary acidic protein and neurofilament protein in medulloblastoma. Neurol. Surg. 16(11):1243–1250.

Valtz, N.L., T.E. Hayes, T. Norregard, et al. (1991) New Biologist 3(4):364–371.

Wasson, J.C., R.L. Saylors, P. Zeltzer, et al. (1990) Oncogene amplification in pediatric brain tumors. Cancer Res. 50(10):2987–2990.

Chapter 26

Tumors of the Choroid Plexus

Terry A. Van Dyke

THE NORMAL CHOROID PLEXUS

Structure and Function

The choroid plexus is located in the ventricles of the mammalian brain; together with the arachnoid membrane, it forms a barrier between the blood and cerebrospinal fluid (CSF) (Netsky and Shuangshoti, 1975; Spector and Johanson, 1989). The choroid plexus produces about 80–90% of the CSF and maintains its composition through active transport mechanisms. The total mass of the choroid plexus represents only about 0.25% of the adult mammalian brain; the lateral and fourth ventricles contain approximately equal amounts and the third ventricle contains only 10% of the total. Macroscopically, the choroid plexus appears as frondlike structures protruding from the roofs of the third and fourth ventricles and from the walls of the lateral ventricles. Each frond consists of a vascularized core surrounded by a single layer of highly specialized epithelium (hereafter referred to as CPE). The cuboidal cells of the CPE exhibit a polarity, with a ciliated apical surface facing the ventricular side and a basolateral surface resting on basement membrane in contact with the vasculature (Fig. 1).

Fenestrated capillaries, which supply the choroid plexus, permit the passage of solutes, but free exchange between the blood and CSF is prevented by the tight junctions of the epithelium. Nutrients such as vitamins and nucleosides are selectively extracted, concentrated, and passed into the CSF by the CPE. The ependymal cells, which line the ventricles and are loosely connected by gap junctions, permit diffusion between the CSF and the interstitial fluid that bathes the brain. Metabolic by-products are removed from the CSF via the choroid plexus. For example, antibiotics, small inorganic ions such as iodide, heavy metals, and neurotransmitter derivatives are all concentrated or removed by the choroid plexus (Spector and Johanson, 1989; Zheng et al., 1991). The presence of many specific protease activities in the CPE suggests that it also plays a key role in the breakdown and removal of certain proteins (Bourne et al., 1989).

During embryonic development, the choroidal epithelium and the ependymal cells derive from a common precursor in the primary layer of neural epithelium lining the primitive neural tube. In the human, the first sign of choroid plexus appears in the form of small epithelial protrusions extending into the ventricle. Choroid villi are apparent by stage 19 (7 weeks) in the fourth ventricle and by stage 20 in the lateral ventricles. By stage 23, the epithelium is composed of a single layer and displays cuboidal characteristics. The third ventricle choroid plexus is not apparent until the fetal period (O'Rahilly and Muller, 1990).

In the mouse, morphologic studies detect the presumptive lateral choroid plexus at embryonic day 11 (E11) as a bilateral ridge in the roof of the interventricular foramen (fourth ventricle choroid plexus appears slightly earlier). By E12 this ridge is well defined and subsequently rapidly increases in size. At E13 the previously largely pseudostratified epithelium becomes more columnar distally and by E14 it becomes more extensively folded, with the epithelial cells appearing morphologically well differentiated (Sturrock, 1979). By this time, few cells of the forebrain are similarly well differentiated. Based on studies in rodents, the CPE cells proliferate rapidly until birth, at which time their division rate slows. Autoradiographic analyses of ^{3}H-thymidine incorporation suggest that much of the cell division ceases by two weeks after birth (Mares and Lodin, 1974).

Differentiation early in development, occupation of a large portion of the embryonic ventricles relative to the adult, and the presence of large amounts of glycogen not present postnatally may indicate that the choroid plexus plays a critical role in nourishing the brain during development. To test this interesting hypothesis, morphological

Molecular Genetics of Nervous System Tumors, pages 287-301 ©1993 Wiley-Liss, Inc.

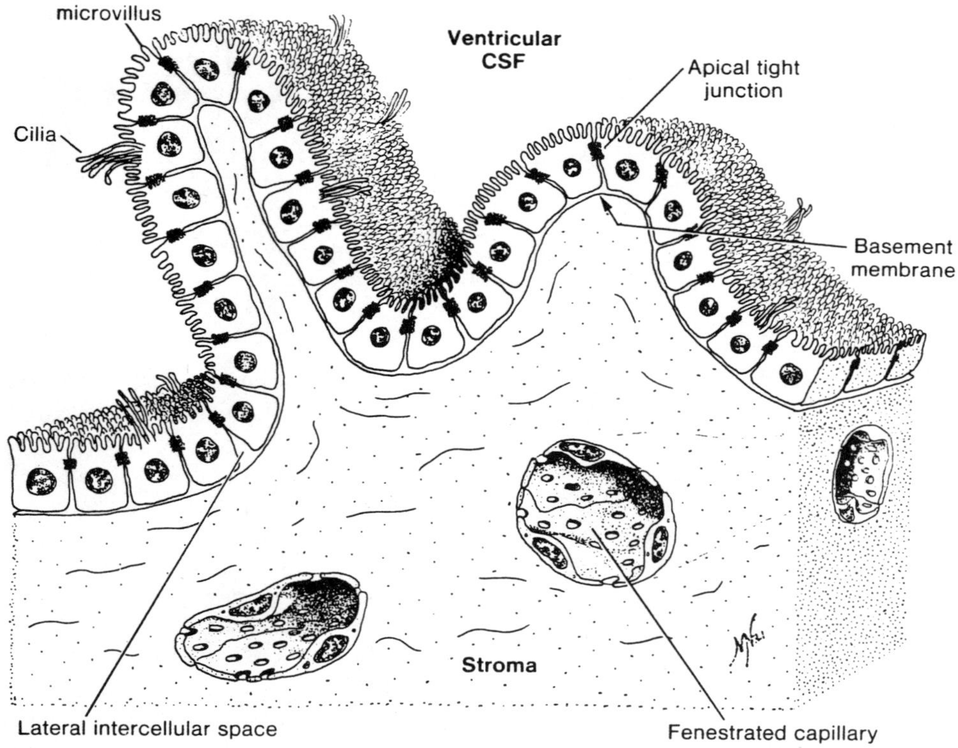

Fig. 1. Diagram of a choroid plexus villus covered by a single layer of cuboidal epithelium with apical microvilli protruding into the ventricular CSF. The base of these cells rests upon a basement membrane. Tight junctions connecting apical regions of the epithelial cells constitute the blood–CSF barrier. The underlying connective tissue stroma contains capillaries with fenestrations (Reproduced from Carpenter and Sutin, 1983, with permission of the publisher).

studies must be accompanied by functional studies in order to determine when the choroid plexus becomes *functionally* differentiated and to assess whether it has a unique role in the embryonic period. As molecular studies progress, and the genes encoding specific functions are characterized, correlation of gene expression and function with morphological changes during development will be possible.

Recent Molecular Studies

Much of what we currently understand about choroid plexus function has come from morphological studies and physiological analyses of *in vivo* and organ culture systems. A large body of work on the physiology of the choroid plexus has recently been reviewed and is not described here. I have focused instead on highlighting some recent molecular genetic studies; the reader is encouraged to explore the additional readings list at the end of the chapter for excellent reviews on other aspects of choroid plexus function. Molecular genetic studies have only recently begun to provide insight into regulatory processes within the choroid plexus. These studies have previously been hampered because of the difficulty in obtaining significant quantities of purified tissue, and because a cell line has not been available. However, several genes have recently been cloned that have been shown to be expressed specifically in the choroid plexus, and the advent of technology for introducing genes into the animal has provided an approach for studying their regulation and function.

Many genes for which expression in the choroid plexus has been characterized are listed in Table 1. Expression of several of these genes within the central nervous system (CNS) is largely limited to the choroid plexus, while some of them are expressed in multiple CNS sites (Table 1). This list represents a start to examining molecular regulation in the choroid plexus and is sure to expand rapidly as more genes are cloned and as more is learned about choroid plexus

TABLE 1. Genes for Which Choroid Plexus Epithelium Is a Prominent CNS Site for Expression

Gene	Expression in other CNS sites*	References
AE2 (anion exchanger)	-	Lindsey et al., 1990
Ceruloplasm		Aldred et al., 1987
IGF binding protein (BP-3A)	-	Tseng et al., 1989
IGF-II	Leptomeninges	Han et al., 1987; Stylianopoulou et al., 1988
Transferrin	Glial cells	Dickson et al., 1985
Transthyretin	Retina pigment epithelium	Soprano et al., 1985

*Based on current *in situ* hybridization data.
Additional CNS expression in embryonic life.
Note: Only genes for which CPE is one of the few, or the only, major site for expression in the CNS, and for which transcripts have been detected, are listed. Only one reference for each is listed for further reading, although several independent reports often exist.

function. With respect to intracellular transport mechanisms, expression of a specific anion exchanger and of the Na,K–ATPase have recently been examined. AE2, a homologue of the erythrocyte anion exchanger BAND3, was recently cloned and shown to encode a Cl/HCO_3 exchanger (Lindsey et al., 1990). Its expression in brain is limited to the basolateral surface of the choroid plexus, a finding that provides some insight into the mechanism of HCO_3 secretion across this tissue. Moreover, the availability of a functional clone provides a molecular tool for further examination of the role of anion exchange in CSF production.

Unlike AE2, the Na,K–ATPase is found associated with both the apical and basolateral surfaces of the CPE. This protein is comprised of two polypeptide chains, an α and β chain, each of which can be encoded by more than one gene. The existence of two α and three β genes creates the potential for several different isoforms of the enzyme, depending on which subset of genes is expressed in a given tissue. Using probes for each gene, *in situ* hybridization experiments show that the $\alpha1$, $\beta1$, and $\beta2$ genes are transcribed in the choroid plexus (Watts et al., 1991). The potential to produce two isoforms in the same cell raises the possibility that different isoforms may associate with the two different surfaces of the CPE. Whether this is in fact the case, and whether the different isoforms are needed for differential localization or function, will be targets for future molecular studies.

Although only recently appreciated, it is now clear that the choroid plexus synthesizes many of the proteins that are found in the CSF. Since many of the CSF proteins are also found in the serum, it was previously thought that their presence in the CSF indicated transport or leakage from the blood supply. This assumption was first shown to be incorrect when transthyretin (TTR; also called prealbumin), a protein that transports thyroxine and retinol binding protein, was found to be highly expressed in the CPE (Soprano et al., 1985). (TTR gene expression is discussed in more detail below.) Transferrin (Dickson et al., 1985) and ceruloplasm (Aldred et al., 1987), transporters of iron and copper, respectively, are also synthesized *de novo* in the choroid plexus, as are the protease inhibitors, cystatin C and $\alpha2$ macroglobulin (Thomas et al., 1989). Thus, CPE not only regulates the composition of CSF through active transport, it synthesizes and secretes a number of important CSF proteins.

Another area of current choroid plexus research employing molecular genetic approaches is the analysis of receptors that are associated with this tissue and the elucidation of their various roles. The first choroid plexus receptor gene to be cloned encodes the serotonin 5-HT_{1C} receptor, a receptor found at very high densities in the choroid plexus. The initial cloning utilized the choroid plexus papillomas from a transgenic mouse model (described further below) as an abundant source of transcripts (Lubbert et al., 1987). Although this gene was originally isolated from choroid plexus, it has subsequently been shown to be expressed in several CNS sites (Molineaux et al., 1989). Binding studies have identified several other high-affinity receptors in the choroid plexus (Table 2). Included among this group are receptors for vasoactive intestinal peptide (Nilsson et al., 1991), atrial and brain natriuretic factor (Zorad et al., 1992), angiotensin

TABLE 2. Receptors With High-Affinity Sites in Choroid Plexus

Ligand	References
Angiotensin II	Gehlert et al., 1986
C-atrial natriuretic peptide	Zorad et al., 1992
Growth hormone	Lai et al., 1991
IGF-II/mannose-6-phosphate	Valentino et al., 1988
IL1	Takao et al., 1990
Prolactin	Di Carlo et al., 1992
Serotonin 5-HT_{IC}	Yagaloff and Hartig, 1985; Lubbert et al., 1987
Vasoactive intestinal peptide	Nilsson et al., 1991

Note: Receptors that have been demonstrated to be present in the choroid plexus are listed. Not all examples have been tested for expression at the level of RNA. Representative references are listed; in several cases other reports exist.

II (Gehlert et al., 1986), insulinlike growth factor II (IGF-II) (Valentino et al., 1988), prolactin (Di Carlo et al., 1992), growth hormone (Lai et al., 1991), and IL1 (Takao et al., 1990). For some of these ligands, including the latter three, the choroid plexus appears to represent the site of highest specific binding within the brain. While few molecular studies on these choroid plexus receptors have yet been carried out, future characterization should provide insight into the regulation of CSF production and perhaps also uncover other aspects of choroid plexus function.

Essentially nothing is known about the factors that regulate gene expression specifically in the choroid plexus. However, the availability of several CPE-specific genes has provided an avenue to explore this area. As the regulation of these differentiated gene products becomes understood, it should be possible to explore the molecular basis for differentiation of this tissue. This type of approach has been productive in several systems, such as muscle and erythrocyte differentiation, where examining the factors involved in the regulation of several cell-specific gene products has begun to provide clues about the mechanism of differentiation. Beyond the basic interest in development, insight into the mechanism of choroid plexus differentiation should provide tools for characterizing the development of neoplasia within the CPE. Our studies on the regulation of TTR expression in the choroid plexus represent the first example of this type of analysis for the choroid plexus and are summarized here. Since CPE cell lines have not been available, our first line of approach was to utilize transgenic mice.

In the early 1980s, the technology was developed for the stable introduction of foreign DNA into the mouse germline, and this technique has since been widely used for a variety of aims (reviewed in Jaenisch, 1988; Palmiter and Brinster, 1986). The most common method for generating "transgenic mice" is the microinjection of naked DNA into the pronuclei of fertilized mouse eggs. The transgenic mouse is a powerful experimental system because the introduced gene becomes exposed to every cell type in all developmental stages of the animal. This approach has been used widely for the elucidation of cis-acting genetic elements required for appropriate tissue-specific regulation of several genes (Jaenisch, 1988; Palmiter and Brinster, 1986). Additionally, experiments using hybrid constructs in which reporter genes are linked to heterologous regulatory signals reveal that expression of a particular gene product can be directed to a specific subset of tissues in transgenic mice with the appropriate regulatory signals (Jaenisch, 1988). This versatility provides a powerful experimental system for studying tumorigenesis, as well as other aspects of gene function. We and others have utilized this method to study choroid plexus tumorigenesis and these studies are also described later in this chapter. For a more detailed description of the methodology and applications of transgenic mice, refer to Chapter 12 (this volume).

Analysis of Transthyretin Gene Expression: A Model for CPE-Specific Gene Regulation

As mentioned above, TTR, a thyroid hormone binding protein, is expressed abundantly in the choroid plexus and is secreted into the CSF. Estimates suggest that TTR transcripts in the CPE account for as much as 10% of the total mRNA (Soprano et al., 1985). While the role of TTR in brain development and function has not yet been directly examined, it is clear that thyroid hormone, acting through a receptor to regulate transcription, is a critical factor in normal CNS development and regulation. Choroid plexus represents the only site in the brain for TTR synthesis.

Beyond exploring the regulation of genes in the choroid plexus, examination of TTR gene regulation in transgenic mice addresses the broader problem of temporal and spatial regulation of a gene in multiple cell types. In the adult mammal, in addition to the CPE, TTR is expressed abundantly in liver hepatocytes and the retina pigment epithelium (RPE) of the eye. In the developing rat, TTR is expressed in the visceral yolk sac, foregut endoderm (where expression is subsequently silenced), fetal liver, the RPE, and the epithelial layer of the choroid plexus primordia (Makover et al., 1989). By 11 days of gestation, TTR RNA is detected in the fourth ventricle tela choroidea (the structural forerunner of the choroid plexus), and thus represents a very early marker for CPE cell differentiation.

To locate the cis-regulatory elements, we analyzed in transgenic mice TTR minigenes that carry various amounts of upstream DNA: 7 kbp, 3 kbp, and 0.3 kbp, which contains only the minimal sequences required for expression in hepatoma cells in culture (an enhancer from -1.86 to -1.96 kbp and a promoter–proximal region from +1 to -200 bp; Costa et al., 1988). The minigenes were constructed so that their expression could be distinguished from that of the endogenous TTR gene at the RNA level. Transcripts from both the endogenous gene and from the transgenes in adult and embryonic tissues were measured quantitatively by RNase protection studies and were further localized by *in situ* RNA hybridization.

Using this approach, we demonstrated that the minimal hepatoma cell-specific elements are sufficient for quantitatively normal expression in hepatocytes of transgenic mice, but not for restricted expression in the CPE cells of the brain. Instead, they direct aberrant widespread expression in regions of the brain outside the CPE (Fig. 2; Yan et al., 1990). While the minimal liver-specific elements do not direct expression to the CPE, they do function in the RPE. However, ectopic expression in a neuronal cell population of the retina is evident. The aberrant ectopic expression in both the brain and the eye indicate that negative elements not present in this short transgene may normally function to silence expression of TTR in these cells. Indeed with 3 kbp of upstream sequence the TTR minigene is expressed specifi-

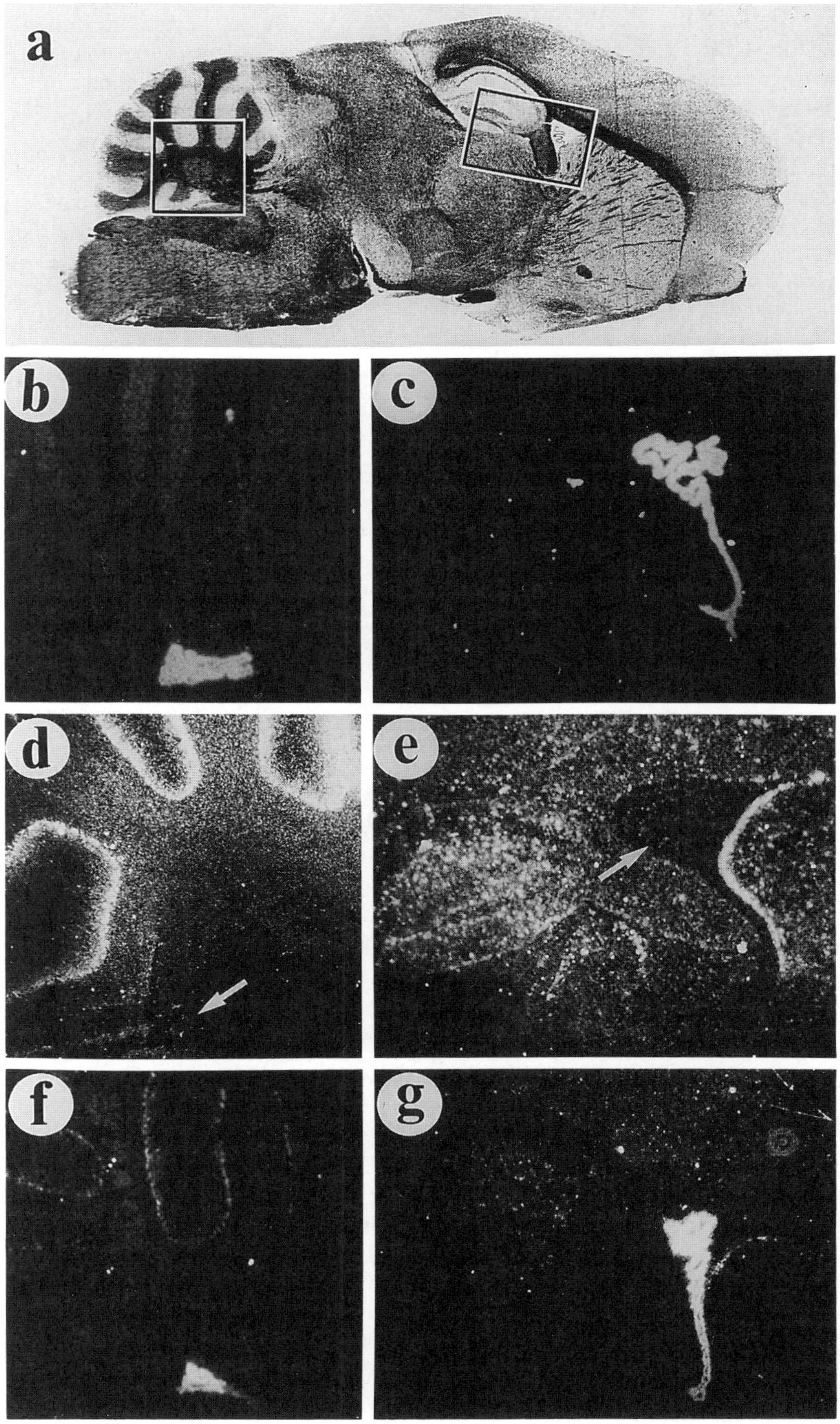

Fig. 2. In situ localization of brain-specific TTR transgene expression. The bright field image of a mouse brain stained with hematoxylin and eosin is shown in **(a)**. The location of regions present in the dark field images of b-g, which include the fourth ventricle choroid plexus **(b,d,f)** and lateral ventricle choroid plexus **(c,e,g)**, are indicated approximately with boxes in (a). Sections b and c, which demonstrate the normal restriction of TTR expression to the CPE, were hybridized with a probe that only detects the endogenous TTR RNA. A transgene which is regulated by the minimal liver specific elements is expressed widely in the brain, but not in the CPE (d and e). In contrast, with -3Kbp of upstream sequence, the transgene is expressed specifically in the CPE (f and g). Arrows in d and e indicate the location of the choroid plexus (Reproduced from Yan et al., 1990, with permission of the publisher).

cally in the choroid plexus (Fig. 2; Yan et al., 1990), the liver (Yan et al., 1990), and the RPE (Yan and Van Dyke, unpublished), demonstrating the normal cell-type specificity for TTR. These results suggest that the 3 kbp upstream region contains positive element(s) required for adult CPE expression that are distinct from those utilized in the hepatocyte, and also contains negative element(s) that function to suppress transcription in other brain and eye cell types.

Analysis of embryonic expression indicates that additional elements in the -3 kbp region are required for fetal liver expression. Furthermore, the -3 to -7 kbp region harbors silencers for the shut-off of TTR expression in foregut endoderm, as well as additional silencer(s) required in the developing CNS (Yan, McGaffin, and Van Dyke, unpublished). This type of analysis has led to the discovery of both positive and negative regulatory mechanisms for the accurate temporal and spatial expression of TTR. Based on these initial results, further molecular definition of both the positive and negative elements involved in CPE expression will provide our first look at choroid plexus gene regulation. As the genes for other products of differentiated CPE are examined in similar ways, a more complete analysis of the molecular mechanisms controlling choroid plexus differentiation should emerge.

Regulation of IGF-II: Another Level of Gene Control

The mitogenic factor, insulinlike growth factor II (IGF-II), is expressed in the choroid plexus and the leptomeninges in adult animals (Han et al., 1987; Stylianopoulou et al., 1988). During embryonic development, however, it is expressed in many other tissues as well. Recent experiments in which the mouse gene was disrupted by embryonic stem cell-mediated gene targeting (see Chapter 12, this volume) have demonstrated a role for IGF-II in embryonic growth. Moreover, these experiments demonstrated that the embryonic-specific expression of this gene is subject to a level of gene control referred to as parental imprinting. Interestingly, the IGF-II gene in choroid plexus and leptomeninges somehow escapes this control.

Parental imprinting is the differential modification of paternal and maternal genomes that results in differential expression of particular genes in the offspring, depending on parental origin. Estimates suggest that 10–25% of mammalian genes are subject to this effect (Surani et al., 1990). The mechanism and regulation of this epigenetic phenomenon and their importance in development are currently subjects of intense interest. Imprinting of the IGF-II gene was uncovered in heterozygous mice that carry one disrupted copy and one good copy of the gene (DeChiara et al., 1991). Only the offspring that receive the disrupted copy from the father show the growth-defective phenotype. When received from the mother alone, the mutant locus exerts no effect. Furthermore, the mutant phenotype in heterozygous mice is the same as that of homozygous mice carrying the disruption at both alleles. This result indicates that the maternal gene is normally inactive. Indeed, when expression of IGF-II is examined in the embryo, it is only detected when the intact gene comes from the father. Choroid plexus expression, however, is unaffected by the parental origin of the gene.

The IGF-II disruption indicates that imprinting can result in tissue-specific differences, and provides the basis for further examination of the molecular mechanisms involved. The observation that the choroid plexus escapes the epigenetic imprint is intriguing and raises the question as to whether this is a general property of the choroid plexus (or CNS) or is specific to this gene. It is interesting that the IGF-II receptor, which is also expressed in the choroid plexus, has recently been shown to be maternally imprinted (Barlow et al., 1991). Examination of the choroid plexus for the imprinting effect has not yet been reported.

TUMORIGENESIS OF THE CHOROID PLEXUS

Human Tumors

Tumors arising from the ventricular choroid plexus epithelium, first reported in 1832 by Guerard, represent about 0.5% of intracranial tumors but 1–4% of brain tumors in children (Lena et al., 1990; Zulch, 1986). Within the pediatric group, various studies report that the vast majority of cases (60–82%) occur before age two (reviewed in Lena et al., 1990). The risk factors for developing choroid plexus tumors are presently unknown. Because the tumor is rare, it has been difficult to obtain highly significant data concerning its etiology. Most children present with an increase in intracranial pressure or head enlargement. Hydrocephalus is a common feature and is believed to be secondary to obstruction and/or hypersecretion of CSF (which can reach 4–5 times higher than normal). Preoperative diagnosis of choroid plexus tumors is usually based on computerized tomography (CT) and magnetic resonance imaging (MRI) scans. With the knowledge of patient age and location of the tumor in the ventricles, the differential diagnosis can be narrowed to a few possibilities (Jelinek et al., 1990). By cerebral angiography, choroid plexus tumors, especially if malignant, appear as highly vascularized masses fed by the choroidal arteries.

Choroid plexus tumors can arise in any of the ventricles. In children, the predominant location (60–78% of cases) is in the lateral ventricles (Lena et al., 1990). Bilateral tumors have been reported, but are very rare. In adults the most common site of these tumors is the fourth ventricle (about 45% of cases; Palazzi et al., 1989). Third ventricle tumors arise much less frequently in both groups. Choroid plexus tumors can be further classified into two groups, choroid

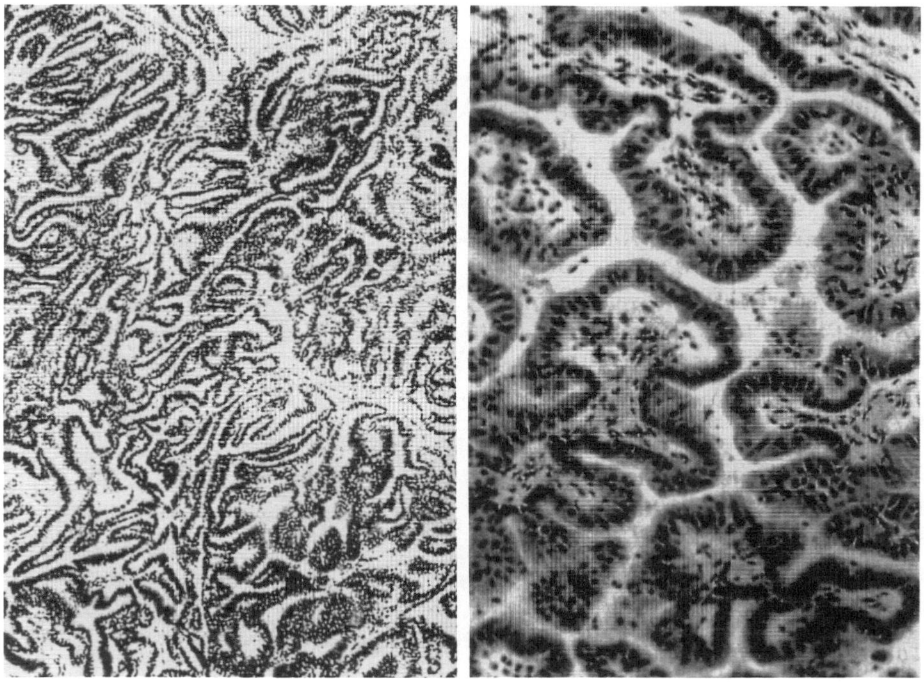

a b

Fig. 3. **a**: Distinct formation of papillae with a central stroma. The papilla is surrounded by the parietal cells in a regular formation. Cresyl violet, ×25. **b**: Higher magnification allows a more distinct view of the papillae. Most contain a capillary vessel. The epithelial cells are pavement-shaped or cylindrical. Cresyl violet, ×112. (Reproduced from Zulch, 1986, with permission of the publisher).

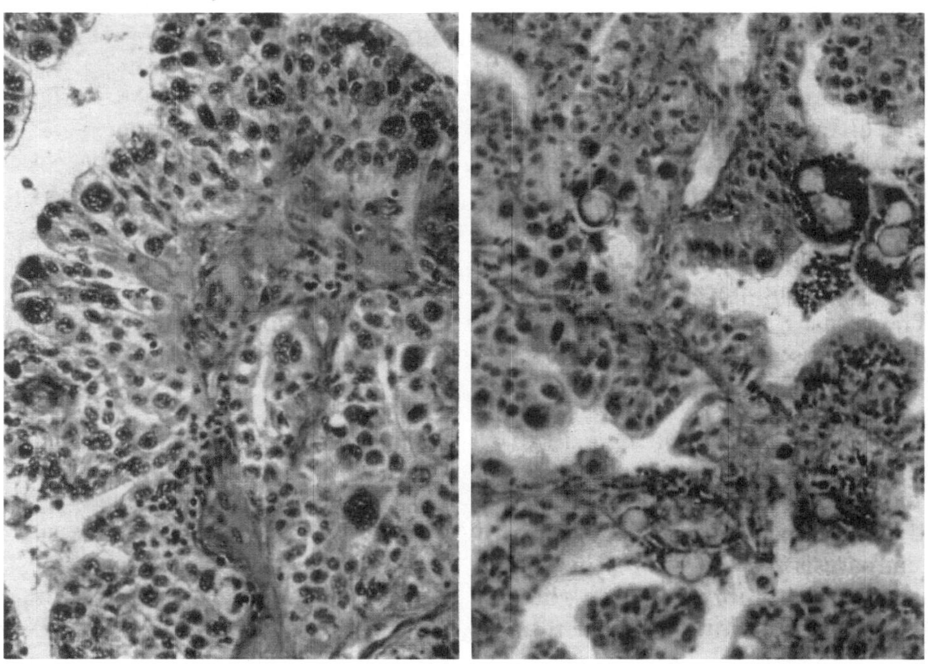

a b

Fig. 4. Anaplastic choroid plexus papilloma. The papillary type of basic architecture is present in other locations of the same tumor. Psammoma bodies are visible. In some locations, the epithelium is even more irregular and in places the papillary structure is not retained, the epithelial cells and polymorphic giant cells being interspersed. H & E, ×125 (**a**) and ×200 (**b**) (Reproduced from Zulch, 1986, with permission of the publisher).

plexus papilloma (CPP) and choroid plexus carcinoma (CPC), based on their neuropathologic features (Figs. 3, 4). According to Russell and Rubenstein (1977), the CPP is characterized by histologic features that recapitulate that of the normal choroid plexus: papillae composed of a single layer of epithelium surrounding a vascular core (grade I, WHO). The features that characterize CPC, the malignant variant of CPP, are local invasiveness, disappearance of regular papillary architecture, presence of a transitional zone between normal and neoplastic tissue, the existence of prominent mitoses, and variable nuclear size (grades III and IV, WHO; Zulch, 1986). The distinction between CPP and CPC is not always obvious, and histologic characteristics are not always predictive of the tumor's biologic behavior. This problem is important because the CPP is curable with surgery, whereas CPC is associated with a poor prognosis. One report estimated that the mean time from onset of symptoms to death in CPC with treatment is about one year, with few cases surviving five years (Bennedbaek and Therkildsen, 1990). Hence, better predictors of the biologic behavior of these neoplasms are needed. Such predictors should emerge as we learn more about the molecular basis of choroid plexus tumor progression (see Studies in an Animal Model, below). Recent studies have examined several markers by immunohistochemical staining in an attempt to correlate them with prognosis. Although progress has been achieved in distinguishing choroid plexus tumors from other papillary neoplasms, no reliable markers exist that predict the prognosis with these tumors. Among the markers that are useful in discriminating choroid plexus tumors from other tumors with similar histological features are glial fibrillary acidic protein (GFAP), vimentin (VM), and cytokeratin (three intermediate filament proteins), S100 P (a calcium-binding protein first identified in neural tissue), and carcinoembryonic antigen (CEA). In choroid plexus tumors, GFAP and VM (less studied) are generally either absent or expressed focally. Cytokeratin is generally expressed. In contrast, ependymomas frequently show widespread expression of GFAP and VM, but do not express cytokeratin. While these markers usually allow distinction of these tumor types, exceptions to these patterns have been reported, as have tumors with profiles characteristic of both choroid plexus and ependymal tumors (Mannoji and Becker, 1988). Recently these markers were examined in several types of papillary tumors, including metastatic papillary tumors, papillary ependymomas, myxopapillary ependymomas, papillary pituitary adenomas, astroblastomas, and choroid plexus tumors (Ang et al., 1990). S100 P, which is usually expressed in choroid plexus tumors, was absent from the metastatic papillary tumors examined (most of which originated from the gut), and this finding is supported by other studies. Astrocytoma and astroblastoma showed staining characteristics similar to ependymoma. While the sample size was small, it is clear that a panel of immunohistochemical stains is useful in establishing the nature of some intraventricular tumors.

Recently, TTR has been examined in several immunohistochemical studies of choroid plexus tumors (Herbert et al., 1990). The majority of choroid plexus tumors demonstrate TTR staining; furthermore, some tumors of nonchoroid plexus origin have occasionally shown a positive reaction. CPP usually shows TTR reactivity, whereas TTR has not been demonstrated in several CPC examined; however, exceptions exist. Some tumors diagnosed as CPP have not expressed TTR, and TTR reactivity has been demonstrated in some CPC tumor cells (reviewed in Paulus and Janisch, 1990). Whether TTR reactivity will prove valuable in discriminating choroid plexus tumors from other tumor types, or in classifying the choroid plexus tumors, requires further analysis of larger series of cases. A recent examination of both the histological characteristics and marker studies of 49 choroid plexus tumors, including 14 cases of CPC (Paulus and Janisch, 1990), showed a poor prognosis correlated significantly with several features of these tumors. In order of decreasing significance, these features are: fewer than 50% tumor cells are strongly S100 P positive, existence of mitoses, absence of TTR-positive tumor cells, invasion of brain by neoplastic cell nests, absence of marked stromal edema, and presence of necrotic areas. In the absence of known specific markers for tumor progression, additional studies such as this one should help to provide more accurate criteria for the diagnosis of malignant CPC.

The ideal treatment for CPP is its complete surgical removal, which can result in a cure. No single recommended treatment program exists for CPC, although the current management involves surgery, chemotherapy, and radiation therapy. Recent studies (Gianella-Borradori et al., 1992; Palazzi et al., 1989) have reviewed the literature with regard to the effectiveness of radiation and chemotherapy, and definitive conclusions on the effectiveness of these treatments are difficult to draw.

Mechanism of CPE Tumorigenesis: Studies in an Animal Model

Considerable evidence suggests that the natural induction of neoplasia involves multiple genetic events (Hunter, 1991). Alterations both in cellular proto-oncogenes (dominant events) and in tumor suppressor genes (recessive events) appear to be key steps in tumor induction (Bishop, 1991; Marshall, 1991). For many tumor types, correlation with mutations in these genes can lead to hypotheses as to which molecular events lie on the pathway to tumorigenesis. Because CPE tumors are rare in humans, even preliminary correlative results on which to base hypotheses are difficult to obtain. Choroid plexus tumors have been observed, albeit infrequently, in two familial syndromes. Of 24 kindreds with Li-Fraumeni syndrome recently examined, five contained a single individual that developed a choroid plexus tumor (Bergsagel et al., 1992; Garber et al., 1990). Individuals with this dominantly inherited disorder develop a variety of neoplasms. Six individuals with Aicardi's syndrome, a congenital malformation syndrome of females, have been reported with choroid plexus tumors (Bergsagel et al., 1992; Robinow et al., 1984). While there is clearly not enough data to suggest an association of choroid plexus tumor etiology with these syndromes, the possibility that the affected loci could contribute to the disease is worth investigating. Interestingly, mutations in the p53 gene (see below) have been found in patients with Li-Fraumeni syndrome (Malkin et al., 1990). One recent genetic analysis of several CPE tumors suggests the presence of viral sequences, leading to speculation that these tumors may arise, at least in part, by an infectious mechanism. This study is described further below.

Animal models in which tumors can be reproducibly induced are key in the elucidation of molecular events that may participate in the genesis of any particular human tumor. In the absence of genetic clues, such a model provides an approach for understanding events that can lead to a tumor; subsequent analyses of human tumors are then required to determine if these events are associated. Ultimately even hypotheses based on correlative data must be tested by demonstrating cause and effect relationships with the introduction of a particular genetic lesion or activated oncogene into an experimental animal. Many years ago, experiments in which live virus preparations were injected intracerebrally into hamsters showed that the polyomaviruses SV40 (a monkey virus that infects kidney), JC virus (JCV, a human virus associated with progressive multifocal leukoencephalopathy), and BK virus (BKV, also a human virus) could induce choroid plexus papillomas and ependymomas when introduced into neonates (reviewed in Bergsagel et al., 1992). JCV also induced a number of other tumor types. An unrelated virus, adenovirus type 3, has also induced CPP by these methods (Motoi and Ogawa, 1985). These results are intriguing but did not produce a viable animal model for studying the genesis of choroid plexus tumors or elucidating the mechanism by which these viruses induced the tumors.

The first example of direct induction of tumors by a purified oncogene utilized the tumor (T) antigen genes of SV40 (Brinster et al., 1984). When the large T antigen gene was introduced into transgenic mice, it induced tumors of the choroid plexus. This gene encodes a multifunctional phosphoprotein of 708 amino acids, which has a role in cellular transformation, the activation of cellular DNA and RNA synthesis, the initiation of viral DNA replication, the negative regulation of its own expression, and the activation of capsid gene expression. A diagram of T antigen and its biochemical and biological activities is shown in Figure 5. Although T antigen has long been known to efficiently transform cells in culture and to induce tumors upon virus infection of test animals (Monier, 1986), the mechanism(s) by which it subverts normal growth control is largely unknown.

One hypothesis is that T antigen induces transformation by interacting with key cellular proteins that normally function to regulate cell proliferation. Two candidate proteins are the retinoblastoma susceptibility gene product (pRB) and p53, each of which is found in a physical complex with T antigen in SV40-transformed cells (reviewed in Levine, 1990). Because p53 and pRB have been shown to block the growth or transformation of cells in culture and are frequently mutated or lost in tumors, they are considered to be "tumor suppressors." The transforming proteins of several DNA tumor viruses, including the large T antigens of several polyomaviruses, are thought to neutralize or inactivate the growth-suppressive functions of one or both of these proteins (Green, 1989; Levine, 1990).

The T antigen-induced choroid plexus tumors have been extensively studied in recent years. In the early studies, the apparent specificity of T antigen for the choroid plexus was shown to be due to restricted expression of the transgene. Mutation of the transcriptional regulatory signals resulted in loss of tumor induction (Palmiter et al., 1985), and replacement of these signals with other cell-specific promoters resulted in the induction of different tumor types (reviewed in Hanahan, 1988, 1989). While many cell types appear to

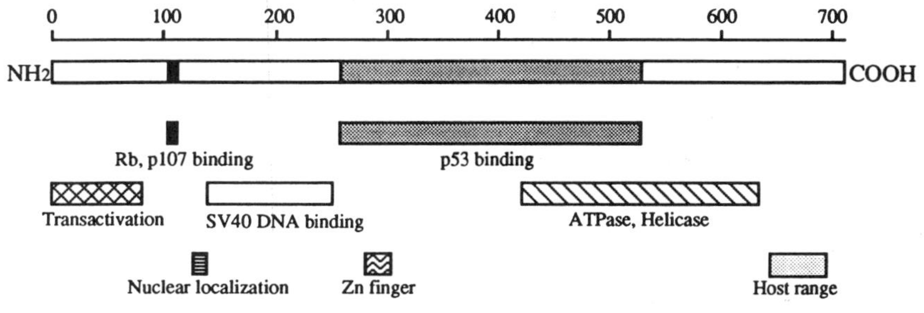

Fig. 5. Diagram of SV40 T antigen. The regions to which biochemical and biological activities have been mapped are indicated. The scale at the top is in amino acids.

respond to the effects of T antigen, the choroid plexus appears to be among the most sensitive (see below). The reason for this apparent sensitivity remains to be determined.

When SV40 T antigen is driven by its natural regulatory elements, tumors of the CPE develop in founder animals at around 2–6 months of age (Brinster et al., 1984). (Founder mice arise from the initial introduction of DNA; hence each founder mouse carries the transgene at different genomic locations.) Another SV40 tumor antigen, the small t antigen, is dispensable for this phenotype (Palmiter et al., 1985; Van Dyke et al., 1985). Within a given line of mice established from a founder animal, the timing of tumorigenesis is remarkably consistent, and penetrance of the phenotype is 100% (Fig. 6; Brinster et al., 1984; Van Dyke et al., 1987). In the SV11 line, one of the better characterized lines (Fig. 6), tumors consistently become life-threatening by 104 + 11 days (Van Dyke et al., 1985, 1987). The SV11 mice and similar lines provide a continual resource for the study of choroid plexus tumorigenesis.

Analysis of the histological progression of tumors in the SV40 mice show that the tumors develop focally out of a background of normal CPE (Fig. 6, bottom; Messing et al., 1988; Van Dyke et al., 1987). In the SV11 line, focal abnormalities are first observed at around 36–41 days of age (Fig. 6a). Histologically, these early foci are characteristic of grade I (WHO). In much older animals (90–110 days) and in some younger mice (56 days), tumors with charateristics of grades III and IV are apparent, and local invasiveness is often present (Fig. 6d). Thus, histological characteristics observed during the progression of these tumors appear to recapitulate properties found in human choroid plexus tumors.

As with human tumors, the mouse tumors are not generally metastatic, although rarely these tumors do metastasize within the CNS (Messing et al., 1988; Van Dyke et al., 1987). Furthermore, chromosome analysis indicates that normal ploidy is maintained in primary tumors (Messing et al., 1988). The presence of specific markers has not been systematically correlated with the grade of tumor in these mice, although expression of the 5-HT_{1C} receptor and TTR have been demonstrated in these neoplasms (Marks et al., 1988; Yagaloff et al., 1986). Correlative marker analyses similar to the ones performed on human tumors may prove informative. The advantage of the animal model is that large numbers of samples can be examined at several stages during tumor progression.

The observation that CPE tumors develop focally out of the background of normal CPE, although low levels of T antigen can be detected in apparently normal tissue, led to the early suggestion that T antigen may not be sufficient to produce the tumors, and that other genetic changes may be required (Messing et al., 1988; Van Dyke et al., 1987). However, our recent studies suggest that the activities contained within this protein appear to be sufficient to disturb the growth of the CPE. In the lines of mice generated in early studies, such as SV11, T antigen is regulated by its natural transcriptional control elements, and expression is limited to scattered choroid plexus cells prior to tumor development (Messing et al., 1988; Van Dyke et al., 1987). However, when high-level expression is directed to the majority of CPE cells using a control region from the related lymphotropic papovavirus (LPV), tumors result from nonfocal expansion of the CPE and are fatal by a much younger age (usually by 25–42 days; Chen et al., 1989; Chen and Van Dyke, 1991). Under these conditions, about 30% of the cells in fully developed tumors, as well as in the CPE at earlier times, express the S-phase marker, proliferating cell nuclear antigen (PCNA; Chen and Van Dyke, 1991). Nuclear PCNA expression is not detectable in the CPE of nontransgenic age-matched animals, consistent with reports in which ^{3}H-thymidine incorporation shows a dramatic decrease in CPE cell proliferation by birth and then cessation within the first two weeks of life (Mares and Lodin, 1974).

Extending the analysis of S-phase cells to mice that were mosaic for the SV40 transgene provided further insight into the mechanism that T antigen employs to transform the CPE (Chen and Van Dyke, 1991). Since normal and transgenic cells are present within the same choroid plexus in mosaic mice, it is possible to determine whether the effect of T antigen is limited to the cells that express it by immunohistochemical analyses of both T antigen and PCNA. In this experiment, about 30% of the T antigen-positive cells also express PCNA, as is the case in tumors. The adjacent morphologically normal CPE does not express either protein. Hence, the targets of T antigen involved in loss of growth control exert their effects within the cell; that is, T antigen is not activating a paracrine mechanism by, for example, inducing the elaboration of growth factors.

Mechanism of T Antigen-Induced CPE Tumorigenesis

The possibility that tumor virus-transforming proteins such as T antigen act to disrupt the normal function of the tumor suppressor proteins p53 and pRB provides an opportunity to explore the action of these cellular proteins in growth regulation of cells within the animal. The transgenic mouse system described above can be used to determine which key activities may be involved in CPE tumorigenesis. Since these studies are carried out under normal physiological conditions, they potentially address which pathways could be targets for the induction of similar tumors in humans.

Two approaches have begun to explore the properties of T antigen with a role in choroid plexus tumorigenesis. First, some insight is offered by comparing the family of related polyomavirus T antigens. The polyomavirus T antigens encoded by LPV (Chen et al., 1989; Chen and Van Dyke, 1991) and JCV (Feigenbaum et al., 1992) have recently been

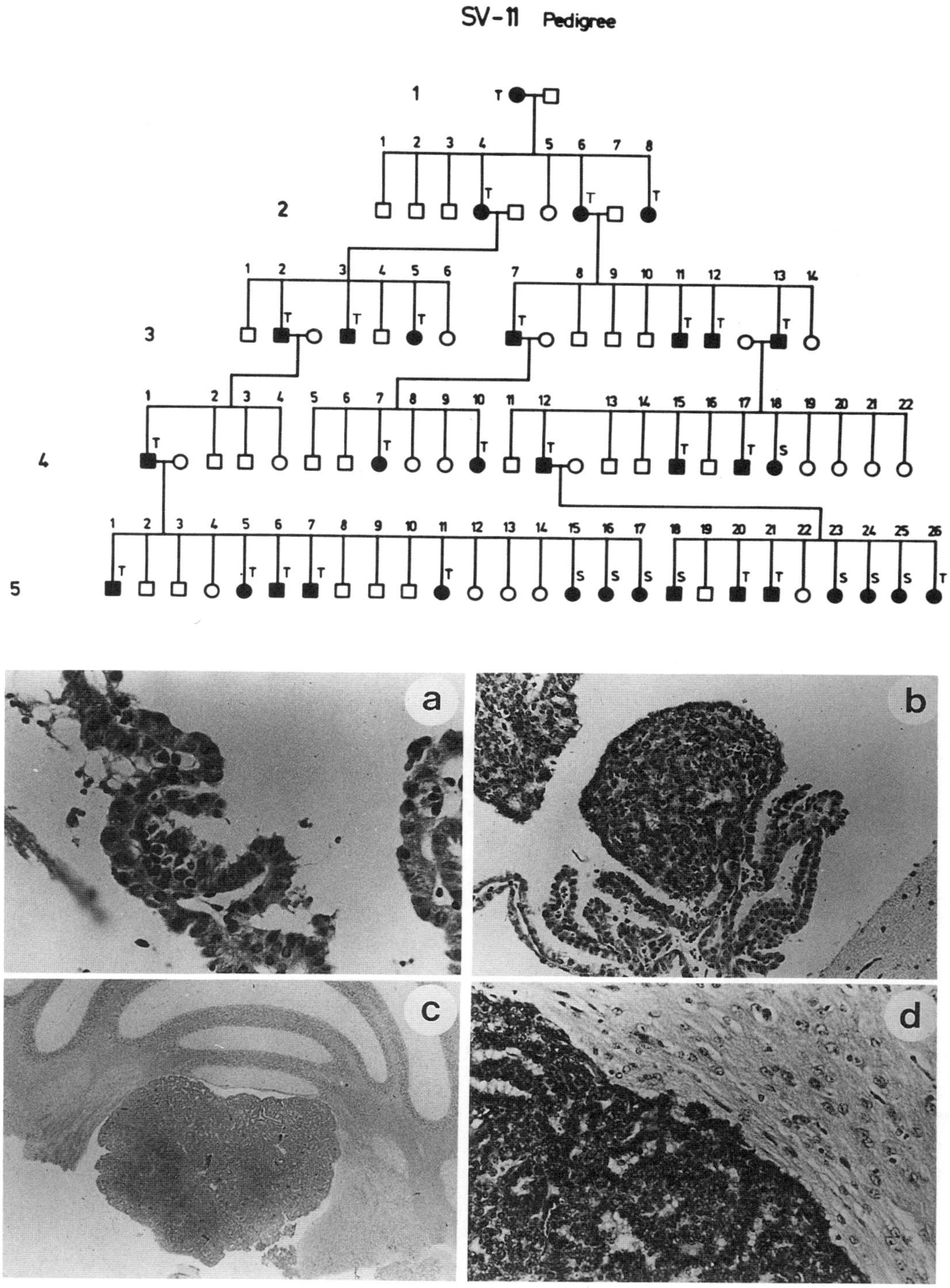

Fig. 6. Choroid plexus tumorigenesis in the SV11 transgenic mouse line. **Top:** The early pedigree of the SV11 mice shows the consistent development of tumors in all transgenic mice. Closed symbols represent transgenic mice, open symbols, non-transgenic mice. Females are indicated by squares and males by circles. T indicates that the animal developed a choroid plexus tumor; S indicates that the animal was sacrificed for analysis before it developed a tumor. **Bottom:** Progression of choroid plexus tumors in the SV11 mice. Brains from SV11 mice were analyzed by hematoxylin and eosin staining at different times after birth (Van Dyke et al., 1987). **(a)** Normal choroid plexus containing a Grade I focus (magnification, X396), **(b)** Grade II nodular tumor (magnification, X100); **(c)** Grade III tumor which is detectable by gross analysis and fills the ventricle, but retains some papillary architecture and does not invade the surrounding parenchyma (magnification, X20), **(d)** Grade IV tumor which fills the ventricle and invades the subependymal parenchyma (magnification, X198).

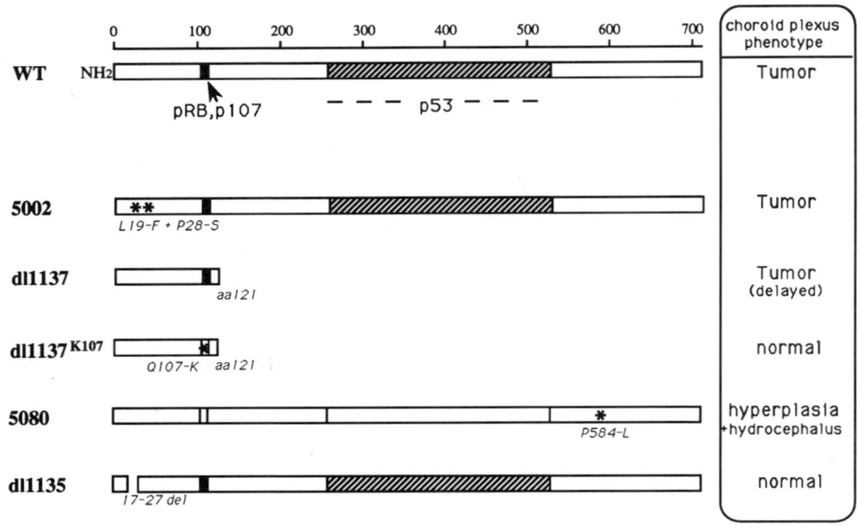

Fig. 7. SV40 T antigen mutant activities in the choroid plexus of transgenic mice. Mutant proteins which have been analyzed in transgenic mice are diagrammed. The regions of T antigen involved in pRB/p107 and p53 binding are indicated in the maps and are highlighted (black or stippled) if still active in the protein. Single amino acid changes are indicated by asterisks. The abnormality induced by each protein in the choroid plexus is indicated in the box to the right.

shown to induce choroid plexus tumors in transgenic mice. This observation suggests that the biochemical activities involved in the induction of these tumors are common among these proteins. Both JCV and LPV T antigens form a complex with pRB *in vitro* (Dyson et al., 1990) and LPV T antigen complexes with and stabilizes p53 (Symonds et al., 1991). Based on this comparison, then, both pRB and p53 binding activities remain candidates for important steps in the production of choroid plexus tumors.

In a second, more direct approach, we have analyzed several mutant forms of T antigen for their ability to induce choroid plexus tumors in transgenic mice (Fig. 7). These mutant proteins, which differ in their ability to form complexes with pRB and p53, were analyzed under conditions where wild-type T antigen induces choroid plexus papillomas as a result of uniform proliferation of the entire choroid plexus epithelium (described above). The results show that the p53 binding function of T antigen is not absolutely required for tumor induction in the choroid plexus, although it may improve the efficiency of this process. This conclusion is based on the result that a form of T antigen truncated at amino acid 121, and thus lacking the p53 binding domain, causes CPE tumors (Chen et al., 1992). Tumorigenesis in CPE does appear to require the pRB-binding site, however, since further mutation of amino acid 107, which abrogates pRB binding, results in loss of tumor induction (Van Dyke, unpublished). These studies also demonstrate that an additional activity, probably resident in the amino-terminal one-fifth of the protein, is required for CPE tumorigenesis. A mutant form of T antigen that does not bind either pRB or p53 fails to induce tumors but does cause hyperplasia. Animals harboring this mutant die of hydrocephalus (Chen et al., 1992). Furthermore, a T antigen missing amino acids 17–27, which retains the ability to bind both tumor supressors, fails to cause CPE tumors (Van Dyke and Pipas, unpublished). These data are summarized in Figure 7. The examination in animals of T antigen mutants with different defects in the interaction with tumor suppressor gene products has provided a way to determine which targets are involved in the control of cell proliferation and tumorigenesis in the CPE.

How does the transgenic mouse model improve our understanding of choroid plexus tumorigenesis in humans? Learning the targets for T antigen activity in the CPE will, at the very least, suggest which cellular activities or genes should be examined in greater detail in human choroid plexus tumors. One recent study in humans suggests that this animal model may be even more directly relevant to the human case. Based on the observation that papovaviral T antigens induce CPE tumors in transgenic mice, and that virus preparations induce CPP and ependymomas, Bergsagel et al. (1992) examined several human CPPs and ependymomas for the presence of papovaviral nucleic acid sequences. Utilizing the polymerase chain reaction (PCR) assay, they determined that 10 of 20 CPPs and 10 of 11 ependymomas tested positive for SV40-like sequences. Furthermore, reactivity to T antigen-specific antibodies was demonstrated in seven of 11 tumor samples tested. Importantly, none of the 10 neuroblastomas examined showed evidence of the SV40-like sequences, suggesting that the results were tumor specific. Less sensitive methods have also indicated that SV40-like sequences may be associated with CPPs and ependymomas (reviewed in Bergsagel et al., 1992). While SV40 or a related virus may cause CPE tumorigenesis in humans, the association of viral sequences with a tumor, like the association of alterations in oncogenes and tumor supressor genes, does not prove an etiological role. However, this evidence, together with the results of animal studies, suggests that the possibility warrants investigation.

Possibly choroid plexus tumors arise from several alternative pathways, making correlative data for rare tumors even more difficult to obtain and interpret. If viral infection were one route to tumor production, accumulating mutations in the cellular genes that are the relevant targets for the virus could be a second route. Transgenic mice will be valuable tools by which to test this possibility. In summary, interesting clues are emerging that may prove instrumental in determining the etiology for CPE tumors in humans.

FUTURE PERSPECTIVES

Molecular genetic approaches to understanding the regulatory processes that govern normal differentiated function of the CPE have begun to yield interesting information. As these studies expand and progress, we can expect to learn much more about developmental processes in the origin of this important tissue. Avenues for understanding the aberrations that lead to unregulated proliferation and tumor formation should also emerge. However, the systems needed for complete examination of molecular events in the CPE have not yet been fully developed. Molecular studies, for example, have been slowed due to a lack of appropriate cell lines. Preliminary progress has been made in establishing a candidate cell line from the SV40-induced transgenic tumors which may prove promising for future molecular studies (Costa and Van Dyke, unpublished). When a cell line is available for genetic and biochemical manipulation, whole animal approaches can be complimented by more detailed studies *in vivo*.

A transgenic mouse model for CPE tumorigenesis has been established and these mice are being utilized to understand the molecular basis for tumor formation. Results from these studies may provide insight into the etiology of the human tumors. The compelling observation describing SV40-like sequences associated with human tumors invites more studies on human samples. For example, more detailed analysis will determine the nature and state of the sequences that are detected by the PCR approach. If these results indicate consistent correlations, the animal model already developed may prove directly applicable to the study of the human disease.

In the diagnosis and treatment of human tumors, two areas require progress. First, reliable indicators of malignancy and prognosis must be developed further. Perhaps help in this effort can come from existing animal model systems. Additionally, as more is known about normal CPE development and function, new markers specific for CPE (both differentiated and undifferentiated) will emerge. Second, progress is also required in developing an optimal treatment for malignant CPC. Although these tumors are rare, the effects are devastating. Unfortunately, the small number of these cases precludes large clinical trials. Perhaps the progress with transgenic mice, and the potential to produce transgenic models in other mammalian species, will provide a much-needed avenue of approach. Finally, as molecular genetisists and clinicians become mutually familiar with molecular mechanisms and research models and with clinical observations and needs, the ultimate goal of the prevention and cure for CPC will become more attainable.

REFERENCES

Aldred, A.R., A. Grimes, G. Schreiber, et al. (1987) Rat ceruloplasmin. Molecular cloning and gene expression in liver, choroid plexus, yolk sac, placenta, and testis. J. Biol. Chem. 262:2875–2878.

Ang, L.C., A.R. Taylor, D. Bergin, et al. (1990) An immunohistochemical study of papillary tumors in the central nervous system. Cancer 65:2712–2719.

Barlow, D.P., R. Stoger, B.G. Herrmann, et al. (1991) The mouse insulin-like growth fFactor type-2 receptor is imprinted and closely linked to the Tme locus. Nature 349:84–87.

Bennedbaek, O., and M.H. Therkildsen (1990) Choroid plexus carcinoma—report of a case with metastases within the central nervous system. Acta Oncol. 29:241–243.

Bergsagel, D.J., M.J. Finegold, J.S. Butel, et al. (1992) DNA sequence similar to those of simian virus 40 in ependymomas and choroid plexus tumors of childhood. N. Engl. J. Med. 326:988–993.

Bishop, M.J. (1991) Molecular themes in oncogenesis. Cell 64:235–248.

Bourne, A., K. Barnes, B.A. Taylor, et al. (1989) Membrane peptidases in the pig choroid plexus and on other cell surfaces in contact with the cerebrospinal fluid. Biochem. J. 259:69–80.

Brinster, R.L., H.Y. Chen, A. Messing, et al. (1984) Transgenic mice harboring SV40 T-antigen genes develop characteristic brain tumors. Cell 37:367–379.

Carpenter, M., and Sutin (1983) Human Neuroanatomy, 8th Ed. Baltimore: Williams and Wilkins Co., p. 22, Fig. 1-18.

Chen, J., K. Neilson, and T. Van Dyke (1989) Lymphotropic papovavirus early region is specifically regulated in transgenic mice and efficiently induces neoplasia. J. Virol. 63:2204–2214.

Chen, J., G. Tobin, J.M. Pipas, et al. (1992) T antigen mutant activities in transgenic mice: Roles of p53 and pRB-binding in tumorigenesis of the choroid plexus. Oncogene 7:1167-1175.

Chen, J., and T. Van Dyke (1991) Uniform cell-autonomous tumorigenesis of the choroid plexus by papovavirus large T antigens. Mol. Cell. Biol. 11:5968-5976.

Costa, R.H., E. Lai, D.R. Grayson, et al. (1988) The cell-specific enhancer of the mouse transthyretin (prealbumin) gene binds a common factor at one site and a liver-specific factor(s) at two other sites. Mol. Cell Biol. 8:81–90.

DeChiara, T.M., E.J. Robertson, and A. Efstratiadis (1991) Parental imprinting of the mouse insulin-like growth factor II gene. Cell 64:849–859.

Di Carlo, R., G. Muccioli, M. Papotti, et al. (1992) Characterization of prolactin receptor in human brain and choroid plexus. Brain Res. 570:341–346.

Dickson, P.W., A.R. Aldred, P.D. Marley, et al. (1985) High prealbumin and transferrin mRNA levels in the choroid plexus of rat brain. Biochem. Biophys. Res. Commun. 127:890–895.

Dyson, N., R. Bernards, S.H. Friend, et al. (1990) Large T antigens of many polyomaviruses are able to form complexes with the retinoblastoma protein. J. Virol. 64:1353–1356.

Feigenbaum, L., S.H. Hinrichs, and G. Jay (1992) JC virus and simian virus 40 enhancers and transforming proteins: Role in determining tissue specificity and pathogenicity in transgenic mice. J. Virol. 66:1176–1182.

Garber, J.E., E.M. Burke, B.L. Lavally, et al. (1990) Choroid plexus tumors in the breast cancer-sarcoma syndrome. Cancer 66:2658–2660.

Gehlert, D.R., R.C. Speth, and J.K. Wamsley (1986) Distribution of [^{125}I] angiotensin II binding sites in rat brain: A quantitative autoradiographic study. Neuroscience 18:837–856.

Gianella-Borradori, A., P.M. Zeltzer, B. Bodey, et al. (1992) Choroid plexus tumors in childhood. Cancer 69(3):809–816.

Green, M.R. (1989) When the products of oncogenes and anti-oncogenes meet. Cell 56:1–3.

Guerard, M. (1832) Tumeor fonguesedans le bentricle droit du cerveau chez une petite fille de trois ans. Bull. Soc. Anat. Paris 8:211-214.

Han, V., K.M. D'Ercole, A. Joseph, et al. (1987) Cellular localization of somatomedin (insulin-like growth factor) messenger RNA in the human fetus. Science 236:193–197.

Hanahan, D. (1988) Dissecting multistep tumorigenesis in transgenic mice. Ann. Rev. Genet. 22:479–519.

Hanahan, D. (1989) Transgenic mice as probes into complex systems. Science 246:1265–1275.

Herbert, J., T. Cavallaro, and A.J. Dwork (1990) A marker for primary choroid plexus neoplasms. Am. J. Pathol. 136:1317–1325.

Hunter, T. (1991) Cooperation between oncogenes. Cell 64:249–270.

Jaenisch, R. (1988) Transgenic animals. Science 240:1468–1474.

Jelinek, J., J.G. Smirniotopoulos, J.E. Parisi, et al. (1990) Lateral ventricular neoplasms of the brain: Differential diagnosis based on clinical, CT, and MR findings. Am. J. Roentgenol. 155:365–372.

Lai, Z., M. Emtner, P. Roos, et al. (1991) Characterization of putative growth hormone receptors in human choroid plexus. Brain Res. 546:222–226.

Lena, G., L. Genitori, J. Molina, et al. (1990) Choroid plexus tumours in children. Review of 24 cases. Acta Neurochir. 106:68–72.

Levine, A.J. (1990) The p53 protein and its interactions with the oncogene products of the small DNA tumor viruses. Virology 177:419–426.

Lindsey, A.E., K. Schneider, D.M. Simmons, et al. (1990) Functional expression and subcellular localization of an anion exchanger cloned from choroid plexus. Proc. Natl. Acad. Sci. USA 87:5278–5282.

Lubbert, H., B.J. Hoffman, T.P. Snutch, et al. (1987) cDNA cloning of a serotonin 5-HT 1C receptor by electrophysiological assays of mRNA-injected Xenopus oocytes. Proc. Natl. Acad. Sci. USA 84:4332–4336.

Makover, A., D.R. Soprano, M.L. Wyatt, et al. (1989) An in situ-hybridization study of the localization of retinol-binding protein and transthyretin messenger RNAs during fetal development in the rat. Differentiation 40:17–25.

Malkin, D., F.P. Li, L.C. Strong, et al. (1990) Germ line p53 mutations in a familial syndrome of breast cancer, sarcomas, and other neoplasms. Science 250:1233–1238.

Mannoji, H., and L.E. Becker (1988) Ependymal and choroid plexus tumors. Cytokeratin and GFAP expression. Cancer 61:1377–1385.

Mares, V., and Z. Lodin (1974) An autoradiographic study of DNA synthesis in adolescent and adult mouse forebrain. Brain Res. 76:557–561.

Marks, J., J. Lin, D. Miller, et al. (1988) The expression of viral and cellular genes in papillomas of the choroid plexus induced in transgenic mice. Prog. Clin. Biol. Res. 284:163–186.

Marshall, C.J. (1991) Tumor suppressor genes. Cell 64:313–326.

Messing, A., C.A. Pinkert, R.D. Palmiter, et al. (1988) Developmental study of SV40 large T antigen expression in transgenic mice with choroid plexus neoplasia. Oncogene. Res. 3:87–97.

Molineaux, S.M., T.M. Jessell, R. Axel, et al. (1989) 5-HT1c receptor is a prominent serotonin receptor subtype in the central nervous system. Proc. Natl. Acad. Sci. USA 86:6793–6797.

Monier, R. (1986) The Papovaviridae. New York: Plenum Press.

Motoi, M., and K. Ogawa (1985) Choroid plexus papilloma and giant cell glioblastoma induced in hamsters with bovine adenovirus type 3. Acta Neuropathol. 66:218–222.

Netsky, M.G., and S. Shuangshoti (1975) The Choroid Plexus in Health and Disease. Bristol: John Wright.

Nilsson, C., J. Fahrenkrug, M. Lindvall-Axelsson, et al. (1991) Epithelial cells purified from choroid plexus have receptors for vasoactive intestinal polypeptide. Brain Res. 542:241–247.

O'Rahilly, R., and F. Muller (1990) Ventricular system and choroid plexuses of the human brain during the embryonic period proper. Am. J. Anat. 189:285–302.

Palazzi, M., A. Di Marco, T. Campostrini, et al. (1989) The role of radiotherapy in the management of choroid plexus neoplasms. Tumori 75:463–469.

Palmiter, R.D., and R.L. Brinster (1986) Germ line transformation of mice. Ann. Rev. Genet. 20:465–499.

Palmiter, R.D., H.Y. Chen, A. Messing, et al. (1985) SV40 enhancer and large-T antigen are instrumental in development of choroid plexus tumours in transgenic mice. Nature 316:457–460.

Paulus, W., and W. Janisch (1990) Clinicopathologic correlations in epithelial choroid plexus neoplasms: A study of 52 cases. Acta Neuropathol. 80:635–641.

Robinow, M., G.F. Johnson, and P.A. Minella (1984) Aicardi syndrome, papilloma of the choroid plexus, cleft lip, and cleft of the posterior palate. J. Pediatr. 104:404–405.

Russell, D.S., and L.J. Rubenstein (1977) Pathology of Tumors of the Nervous System. Baltimore: Williams and Wilkins.

Soprano, D.R., J. Herbert, K.J. Soprano, et al. (1985) Demonstration of transthyretin mRNA in the brain and other extrahepatic tissues in the rat. J. Biol. Chem. 260:11793–11798.

Spector, R., and C.E. Johanson (1989) The mammalian choroid plexus. Sci. Am. 261:68–74.

Sturrock, R.R. (1979) A morphological study of the development of the mouse choroid plexus. J. Anat. 129:777–793.

Stylianopoulou, F., J. Herbert, M.B. Soares, et al. (1988) Expression of the insulin-like growth factor II gene in the choroid plexus and the leptomeninges of the adult rat central nervous system. Proc. Natl. Acad. Sci. USA 85:141–145.

Surani, M.A., R. Kothary, N.D. Allen, et al. (1990) Genome imprinting and development in the mouse. Development (Suppl.) 110:89–98.

Symonds, H., J. Chen, and T. Van Dyke (1991) Complex formation between the lymphotropic papovavirus large tumor antigen and the tumor suppressor protein, p53. J. Virol. (in press).

Takao, T., D.E. Tracey, W.M. Mitchell, et al. (1990) Interleukin-1 receptors in mouse brain: Characterization and neuronal localization. Endocrinology 127:3070–3078.

Thomas, T., G. Schreiber, and A. Jaworowski (1989) Developmental patterns of gene expression of secreted proteins in brain and choroid plexus. Dev. Biol. 134:38–47.

Tseng, L.Y., A.L. Brown, Y.W. Yang, et al. (1989) The fetal rat binding protein for insulin-like growth factors is expressed in the choroid plexus and cerebrospinal fluid of adult rats. Mol. Endocrinol. 3:1559–1568.

Valentino, K.L., H. Pham, I. Ocrant, et al. (1988) Distribution of insulin-like growth factor II receptor immunoreactivity in rat tissues. Endocrinology 122:2753–2763.

Van Dyke, T., C. Finlay, and A.J. Levine (1985) A comparison of several lines of transgenic mice containing the SV40 early genes. Cold Spring Harbor Symp. Quant. Biol. 50:671–678.

Van Dyke, T.A., C. Finlay, D. Miller, et al. (1987) Relationship between simian virus 40 large tumor antigen expression and tumor formation in transgenic mice. J. Virol. 61:2029–2032.

Watts, A.G., G. Sanchez Watts, J.R. Emanuel, et al. (1991) Cell-specific expression of mRNAs encoding Na+,K(+)-ATPase alpha- and beta-subunit isoforms within the rat central nervous system. Proc. Natl. Acad. Sci. USA 88:7425–7429.

Yagaloff, K.A., and R.R. Hartig (1985) ^{125}I-Lysergic acid diethylamide binds to a novel serotonergic site on rat choroid plexus epithelial cells. J. Neurosci. 5:3178–3183.

Yagaloff, K.A., G. Lozano, T. Van Dyke, et al. (1986) Serotonin 5-HT (1C) receptors are expressed at high density on choroid plexus tumors from transgenic mice. Brain Res. 385:389–394.

Yan, C., R.H. Costa, J.E. Darnell, et al. (1990) Distinct positive and negative elements control the limited hepatocyte and choroid plexus expression of transthyretin in transgenic mice. Oxford: Oxford University Press. EMBO J. 9:869–878.

Zheng, W., D.F. Perry, D.L. Nelson, et al. (1991) Choroid plexus protects cerebrospinal fluid against toxic metals. FASEB J. 5:2188–2193.

Zorad, S., K. Tsutsumi, and J.M. Saavedra (1992) Selective localization of C atrial natriuretic peptide receptors in the rat brain. Brain Res. 570:149–153.

Zulch, K.J. (1986) Brain Tumors: Their Biology and Pathology. New York: Springer-Verlag.

FURTHER READINGS

Johanson, C.E. (1988) The choroid plexus-arachnoid-cerebrospinal fluid system. In Boulton, A., Baker, G., and Walz, W. (eds.) Neuromethods, Vol. VII: The Neuronal Microenvironment-Electrolytes and Water Speces. Clifton, NJ: Human Press, pp 33-104.

Netsky, M.G., and S. Shuangshoti (1975) The Choroid Plexus in Health and Disease. Bristol: John Wright.

Nilsson, C., M. Lindvall-Axelsson, and C. Owman (1992) Neuro-endocrine regulatory mechanisms in the choroid plexus-cerebral spinal fluid system. Brain Res. Rev. 17:109-138.

Spector, R., and C.E. Johanson (1989) The mammalian choroid plexus. Sci. Am. November.

Chapter 27

Tuberous Sclerosis: Hamartomas, Subependymal Giant Cell Astrocytomas, and Other Central Nervous System Tumors

Jonathan L. Haines and M. Priscilla Short

INTRODUCTION

The original pathologic observations of tuberous sclerosis (TSC) were provided by von Recklinhausen in 1862, and in 1880, Bournville reported the central nervous system (CNS) pathology and the associated neurologic symptoms. The multisystemic involvement of tuberous sclerosis has been appreciated since Vogt's observation in 1908 of the association between the CNS pathology, neurologic symptoms, and the cutaneous manifestations; these became known as Vogt's triad.

Diagnostic criteria that encompass all patients with TSC do not exist, given the variable expressivity and less than full penetrance of disorder. However, characteristic skin lesions are seen in 96% of affected individuals (Gomez, 1988), and provide the easiest and earliest ascertained diagnostic criteria. The findings of facial angiofibromas, ungual fibromas, and forehead/scalp fibrous plaques are diagnostic. The finding of hypomelanotic patches known as "ash leaf" spots, a Shagreen patch, gingival fibromas, and dental enamel pits are also diagnostic in a person with a known affected relative or who has one of the other presumptive diagnostic criteria (Gomez, 1988). Diagnostic neuro-imaging studies, such as computed tomography (CT) and magnetic resonance imaging (MRI), can reveal CNS manifestations including cortical tubers and subependymal glial nodules, which are hamartomas of the CNS. In some cases calcifications of these lesions can be seen on plain skull radiographs.

The presence of periventricular calcifications is the second most common feature in individuals with definite TSC, occurring in 90% of patients (Gomez, 1988). Other neurologic symptoms associated with the hamartomas and hamartias include seizures, mental retardation, learning disabilities, and autism.

Previously it was thought that TSC was invariably associated with severe mental retardation and seizures. However, recent information suggests that the finding of Vogt's triad (seizures, mental retardation, and adenoma sebaceum) is true for only one in three affected individuals (Gomez, 1988). Seizures occur in 84% of patients with TSC, making it the third most common clinical feature. TSC is one of the leading differential diagnoses in a child with infantile spasms, a generalized neonatal seizure disorder with a characteristic electroencephalogram (EEG) tracing.

Between one-quarter to one-half of children with infantile spasms will have diagnostic criteria of TSC. The probability of mental retardation occurring in a child with infantile spasms is 87% (Pampiglione and Pugh, 1975). Affected individuals who are severely mentally retarded invariably have recurrent seizures, but there is no clear relationship between the seizure disorder severity and mental retardation. Recent information suggests a correlation between MRI findings and the severity of TSC in children (Roach et al., 1987).

The fourth most common clinical feature of patients with TSC is retinal hamartoma or phakoma visible on fundoscopic examination. Multiple kidney hamartomas, angiomyolipomas, and cysts are seen in 54% of patients in a clinical series and 100% in an autopsy series (Gomez, 1988). Renal involvement often underlies significant morbidity, causing the secondary development of hypertension and renal failure in severe cases. The incidence of cardiac rhabdomyoma is 43% in a series of children with TSC (Watson, 1991). These may lead to symptoms of outflow obstruction or cause arrhythmias, and represent the leading cause of death in infants with TSC.

Hamartomas or benign tumors are a hallmark of tuberous sclerosis and are seen in almost every tissue. The most

Molecular Genetics of Nervous System Tumors, pages 303-310 ©1993 Wiley-Liss, Inc.

common sites of hamartomas are the CNS, the heart, and the kidneys. In the CNS, the hamartomas or tubers are most commonly located in the subependymal region, as well as along the cortical surface. In some cases, tubers located in the region of the foramen of Monro can produce symptoms by causing obstruction of the outflow of the lateral ventricle. When these tubers become enlarged, they are reclassified as subependymal giant cell astrocytomas. In a retrospective series from the Mayo Clinic, 20 out 23 patients with tuberous sclerosis and brain tumors had subependymal giant cell astrocytoma. There did not appear to be a correlation between the clinical course and the histological profile (Shepherd et al., 1991). There have been rare single reports of co-occurrence of giant cell astrocytomas and other CNS tumors (neuroblastoma, oligodendroglioma) and a single report of the occurrence of an extracranial meningioma in a child with typical features of TSC (Irving et al., 1991). The typical renal lesions, angiomyolipomas, are histologically benign, but there are reports of the co-occurrence of angiomyolipoma and renal cell carcinoma by pathologic examination. On the basis of routine screening (ultrasonography, CT, and angiography) they may be difficult to distinguish (Yamanaka et al., 1990). In some cases renal angiomyolipomas have been described as invading retroperitoneal lymph node involvement, suggesting a continuum between angiomyolipoma and renal cell carcinoma. Single renal angiomyolipomas do occur in the general population over the age of 30; the majority of cases occur in women (Bernstein and Robbins, 1991), and in an autopsy series, occur as frequently as 12% of the population.

Two-dimensional echocardiography is useful to detect cardiac rhabdomyomas in TSC. Watson (1991) recently reported a significant difference of prevalence of these tumors in children with TSC versus adults with TSC. Two of four cardiac rhabdomyomas seen prenatally were demonstrated to regress in subsequent studies. Regression has been reported by others as well, arguing for conservative management (Smith et al., 1989). Pericardial mesothelioma, lipoma, and angiomyoma have all been reported in association with TSC (Naramoto et al., 1989).

The retinal phakoma has to be distinguished from retinoblastoma, with which it can be confused when first observed. Its lack of growth over time clearly points to the benignity of the TSC retinal phakoma (Robertson, 1991). Benign tumors of the gastrointestinal tract, the liver, bone, uterus, and lung have all been reported in association with TSC. In some cases, adenoma sebaceum has occurred in clusterlike growths (Park and Hann, 1989).

GENETIC ETIOLOGY

Although TSC is clearly inherited as an autosomal dominant disease (first noted by Berg, 1913), there is a continuum of varied expressivity from severely affected through quite mild expression all the way to apparent total nonpenetrance (Gomez, 1988). Often the clinical diagnosis of other relatives occurs only when the family of a severely affected proband is investigated further. The new mutation rate is presumed to be rather high, as the majority of new patients have parents who are clinically normal. However, the accuracy of these assessments may be questionable given the variable expressivity of this disease. Germline mosaicism might account for cases in which apparently unaffected parents have more than one affected child. These complications make it difficult to determine an accurate mutation rate for this disease, which has been variously estimated to be from 5 to 40×10^{-5} per gamete per generation (Gomez, 1988).

Tissue-Specific Expression

The clinical variability of TSC within families, assuming that those affected have inherited the same TSC mutation, requires the effect of tissue-specific modifying factors. Whether or not the molecular analysis of lesions associated with TSC will yield additional mutations in other loci, as has been seen in astrocytomas associated with neurofibromatosis type 1, is not known (Mashiyama et al., 1991). It is not even known whether or not the lesions of TSC are clonal in origin, as has been noted in pituitary adenomas (Alexander et al., 1990), meningiomas, and schwannomas, the latter two being associated with neurofibromatosis type 2 (Jacoby et al., 1990). Whether or not similar genetic mechanisms will be true for the clinical variability seen in TSC (as are being demonstrated in neurofibromatosis types 1 and 2 and von Hippel-Lindau disease, neurocutaneous disorders with which TSC is most commonly compared) awaits further investigation.

There have been a number of hypotheses proposed to account for the pathology of TSC. Many of these lesions are described as disorganized, with large embryonic-type cells suggesting a halt in the maturation process. The lack of involvement of the spinal cord has been used to suggest that the normal temporal effect of the TSC gene is between the third and fourth week of gestation (Hallervorden and Krucke, 1956). Inglis hypothesized in 1950 that the lesions all arose from single cells or rests of immature neural cells or in response to some unknown neurotrophic substance. Though he later abandoned this theory, the subsequent culturing of neuronal-like "N" cells from a number of different TSC lesions (Johnson et al., 1991) has revived his theory.

Ishibashi and colleagues (1991) have described in cultures of adenoma sebaceum biopsy material, as well as *in situ*, cells that are more glial-like than fibroblastic, on the basis of staining for vimentin and glial fibrillary acid protein. In addition, they noted an abnormal histogram pattern of DNA in culture of this diagnostic lesion of TSC and in some cases abnormal chromosomal arrangement at metaphase, suggesting an abnormality of cellular cytoskeleton as a possibly underlying TSC. Whether or not this observation is characteristic of all lesions of TSC awaits further investigation. Other investigators have noted abnormal amount of extracellular matrix components (Khalkhali-Ellis et al., 1991). A unifying theory that takes into account the various observations noted in TSC may need to await the isolation of at least one of the genes causing TSC.

GENETIC STUDIES

The research outlined above has not allowed us to discern the underlying causes of TSC or to develop any effective cures or preventions. At best only the symptoms can sometimes be alleviated using standard treatments. Thus another research strategy needs to be tried, one based on genetics. If we can identify the underlying genetic etiology of TSC, then we have the substrate necessary to learn directly about the physiological pathways through which the defective TSC gene creates its havoc. Two approaches to understand TSC etiology are currently being used. The first is to specifically test known genes that are hypothesized to play a role in TSC. This falls under the wider purview of the second approach: positional cloning.

Paradigm of Positional Cloning

The term positional cloning (previously referred to as reverse genetics) refers to the identification of a gene from its location in the genome rather than from its function. Positional cloning of genes causing disease requires several steps. The first step is to identify a set of families that clearly segregate the disease in question. Blood samples from all available members of the family must be collected so that DNA samples can be used for study. The second step is to test various DNA markers for genetic linkage to the disease trait (as discussed below). Markers dispersed throughout the human genome are tried until a linkage is found. The third step is to refine this linkage once it has been established by testing other markers in that region of the genome. This also requires testing for genetic heterogeneity, since more than one gene may be involved. The final step is to actually clone the gene using molecular genetic techniques. Once the candidate genomic region has been narrowed to 1–2 million base pairs, techniques such as exon trapping, pulsed-field gel electrophoresis, and tests for expressed sequences can be used to identify the defective gene. The positional cloning methodology has been partially successful in TSC. Currently, researchers are working on steps two and three, since linkage for one form of TSC has been found, but explains only about one-third of families. In order to understand the current state of research, it is necessary to have an understanding of the basic principles of linkage.

Genetic Linkage

Linkage analysis is a statistical technique that compares within a family the pattern of inheritance of a disease with any number of other genetic loci (generally referred to as marker loci). If the disease and the marker locus segregate together throughout the family, then they are likely to reside close to each other in a chromosomal segment. Conversely, if they segregate differently, then they probably reside far apart, most likely on different chromosomes. The likelihood, or probability, that the disease and marker locus reside close to each other can be measured using a test designated the lod score (Ott, 1991), which provides an estimate of the odds in favor of linkage. A lod score of +3.0 is generally accepted as significant evidence for linkage, and a lod score of -2 as significant evidence against linkage. Not only can we tell if two loci reside on the same chromosomal segment, we can also measure the distance between them in units named centiMorgans (cM). In theory, cM are calculated by observing the ratio of the number of meiotic events in which the two loci do not segregate together over the total number of meiotic events. In practice, it is often not possible to determine with certainty the exact number of meiotic events being observed, and thus maximum likelihood estimation is used to generate both the lod score and distance estimates. Lod scores are meaningful only when given along with a specific recombination fraction, since they are totally dependent on the distance being assumed between the two loci.

Figure 1 presents an example of a simple pedigree with both disease and marker data. From the grandparents (individuals 1 and 2), we can tell that the disease gene must have been passed down to individual 3 with marker allele A. If the disease and the marker are linked, then allele A and the disease should show up together in the third generation. This is the case, since affected individuals 6, 7, and 8 have all inherited the A allele from their father, and the two unaffected individuals 5 and 9 have inherited the B allele. The number of observable meiotic events is 5, and there are no

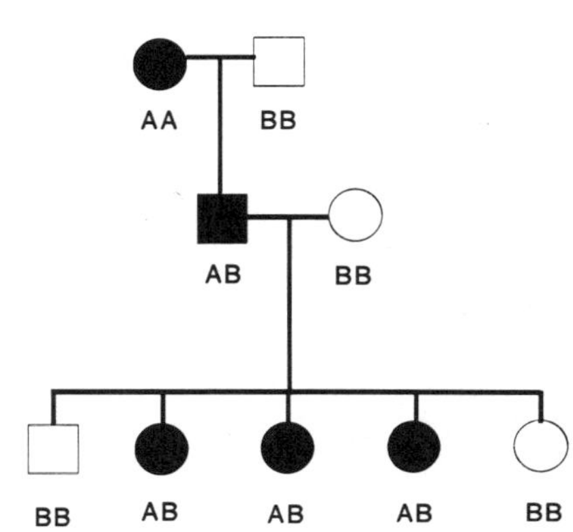

Fig. 1. Example of an autosomal dominant trait.

recombinants, so the estimated recombination fraction is 0 (the lod score is +1.5).

Although linkage can be detected when considering only two loci (two-point analysis), analyses can be extended to include any number of loci. In practice it is rarely possible to consider at once more than five loci in addition to the disease. This type of multipoint analysis has two major advantages in disease studies. The first is that it maximizes the amount of segregation information available from the pedigree. No matter how informative a genetic marker may be, there are likely to be at least a few individuals who are homozygous (having the same allele on each chromosome), and thus provide no segregation information relative to that marker. However, another closely linked marker may be informative and provide the needed segregation information. If these markers are examined simultaneously, then each individual chromosome is more likely unique, and more segregation information is recovered. The second advantage is that one can recover from the data information on the order of the disease relative to the tested markers. Thus it may be possible to identify markers that reside on either side of the disease gene (flanking markers), a process that is difficult when using only two-point lod score results.

One obvious question that arises is which genetic markers to use for testing against the disease gene. Only 12 years ago, human geneticists relied primarily on 20–30 expressed polymorphic markers, such as the ABO blood group system or the HLA transplantation antigens. These markers covered less than 15% of the genome (Conneally and Rivas, 1980). Fortunately, recombinant DNA technology provided the means for identifying a plethora of DNA polymorphisms in the form of restriction fragment length polymorphisms (RFLPs) (Botstein et al., 1980). The great advantage of RFLPs was (and still is) that they need not be biologically expressed, and are located throughout the genome. Thousands of RFLPs have now been described.

Several recent advances in genetic marker technology have further enhanced our ability to test for and identify genetic linkage. The first was the description of variable number of tandem repeat (VNTR) markers (Nakamura et al., 1987). These markers are highly polymorphic, often having as many as 10 alleles (most RFLPs have only two). The high level of polymorphism allows the researcher to capture more of the probable segregation information with less effort. However, VNTRs are more technically difficult to genotype, and also tend to cluster toward telomeres; thus not all of the genome can be examined.

Another type of marker holds the greatest potential for future linkage studies. Microsatellite repeat markers (Weber and May, 1989) are highly polymorphic, are spread evenly throughout the genome, and can be tested using polymerase chain reaction (PCR) technology, thus vastly reducing the amount of DNA that needs to be tested. The majority of microsatellite markers so far described are of the type (GT)n, where n indicates the number of repeated units of (GT), and is usually greater than 10 in highly polymorphic systems.

Considerations in applying linkage analysis

When applying linkage analysis in a positional cloning strategy, it is important to consider the effect of family collection. Both size and availability of families are important, and are used to determine which families to collect. In general, it is better to examine fewer larger families rather than a multitude of smaller ones, since genetic heterogeneity (as discussed below) may be an issue. For detecting heterogeneity, it is more powerful to have a high lod score (greater than +3) in a single family, even if other clinically similar families have negative lod scores, than it is to have many small families, several of which are positive. In TSC, family collection has been a somewhat difficult process, since families tend to be small and difficult to ascertain. In the entire National Tuberous Sclerosis Association (NTSA) collaborative database of 117 families from around the world, only five are sufficiently large to generate a lod score of +3 (Haines et al., 1991a).

The key to understanding linkage results in TSC is accurate definition of the underlying heterogeneity. There are three major types of heterogeneity. Allelic heterogeneity involves different mutations at a single locus, and can only be confirmed by examining the exact polypeptide or DNA sequence. Nonallelic heterogeneity (mutations at two different loci, both giving rise to the disease phenotype) is testable through linkage analysis. Finally, nongenetic heterogeneity (that is, some cases arising from nongenetic etiologies) is also testable through linkage analysis. The procedure is straightforward. Lod scores for linked marker(s) are obtained on each family. Using maximum likelihood estimation, it is possible to generate the likelihood of the data fitting either the homogeneity or the heterogeneity hypothesis. If the likelihood under the heterogeneity hypothesis is significantly greater than under the homogeneity hypothesis (tested using the likelihood ratio test), then the homogeneity hypothesis is rejected.

Two heterogeneity models are commonly tested: (1) one linked and one unlinked set of families, and (2) two linked sets of families, at either different recombination fractions from the same locus or to two separate unlinked loci. Two results arise from applying homogeneity tests. First, the proportion of linked (to locus 1) families (a) is estimated. If more than two loci are being hypothesized, then the proportion of each type is estimated. Additionally, posterior probabilities of linkage to each locus are generated for each family, aiding in the classification of these families by linkage type.

Three different statistical tests of linkage heterogeneity have been proposed. Morton's test is the oldest, but is generally not used, as it has been shown to be sensitive to various assumptions, and can give false-positive results. The admixture test is generally superior, and under most conditions produces somewhat conservative results. The newly proposed b-test functions similarly, and is perhaps less sensitive to certain assumptions (Ott, 1991). Both can be used to generate heterogeneity statistics.

Chromosome 9 linkage

The application of linkage analysis to TSC has provided some interesting findings. Two independent research groups have reported that a tuberous sclerosis gene resides on chromosome 9 (Connor et al., 1987; Fryer et al., 1987), closely linked to the ABO locus. This work engendered a massive effort by other researchers to confirm the linkage. However, several other groups have found substantially negative lod scores with ABO and other markers on this region of chromosome 9 (Kandt et al., 1989; Northrup et al., 1987; Smith et al., 1990), suggesting either heterogeneity or that the earlier linkage results were false-positive. Testing the heterogeneity hypothesis has been difficult for individual research groups, as the positive and negative families seem to segregate by study group, an interesting but inexplicable result. Another factor is the small size (less than 10 families, in most cases) of the sample sizes being studied. However, several recent studies have concluded that there is a chromosome 9-linked locus, and that genetic heterogeneity exists. In 1989, Sampson and associates found in a small group of families significant evidence for linkage. They then tested the hypothesis of heterogeneity, and obtained a marginally significant result. In 1990, three different groups, working with an international collaborative data set, also found significant evidence of heterogeneity using a variety of statistical methods (Haines et al., 1991a; Janssen et al., 1991; Povey et al., 1991). Finally, Haines and co-workers (1991b) found significant evidence of heterogeneity while examining a set of 22 TSC families. In all cases, the estimated proportion of chromosome 9-linked families was 30–40%.

The initial linkage was to the ABO blood group, indicating a position in 9q34 for the TSC gene. In order to localize the gene even further, two complementary and similar approaches are taken. The first is to test other polymorphic markers in the region of ABO, both centromeric and telomeric, and use multipoint analysis to determine the most likely location of the TSC gene relative to the tested markers. Although this makes maximal use of the segregation information, it requires genotyping every person in every pedigree, and often requires unmanagable amounts of computer time. The second approach is to examine the segregation information visually by aligning the markers in chromosomal order and assigning specific alleles to specific chromosomes (generating haplotypes). Haplotype analysis identifies the individual recombination events in each family, but depending on the specific pedigree structure this may be possible for all the chromosomes, for only a few chromosomes, or for none of the chromosomes. Thus it may be possible to identify a small subset of individuals that provide much of the ordering information, and these are the only people that need to be genotyped at additional loci. These two approaches are complementary, and are being used together to localize the TSC gene on chromosome 9 as quickly as possible.

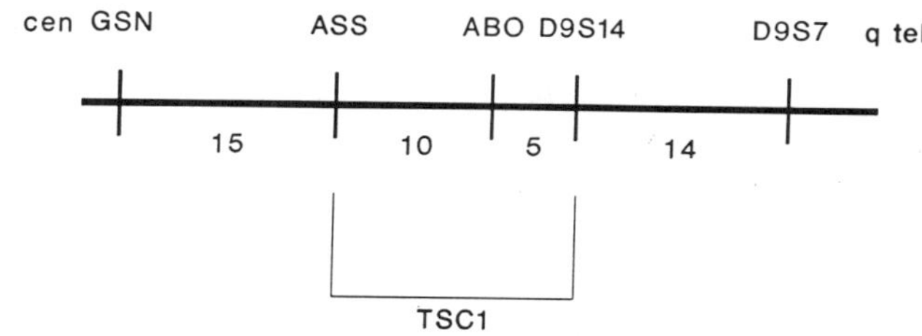

Fig. 2. Location of TSC1 on chromosome 9q.

Care must be taken to include only those families likely to be chromosome 9-linked. Families with TSC caused by another gene elsewhere in the genome will provide misleading ordering information since the chromosome 9 segregation is independent of the disease segregation. For TSC, at least two recombinants in well-defined 9-linked families have been seen with GSN, and at least one recombinant has been seen with ASS, all indicating a position distal to these genes (Fig. 2). Many recombinants with D9S7 have been seen, and at least one with D9S14, indicating a position proximal to these loci. Thus the TSC gene lies within an approximate 15 cM region of chromosome 9. Current studies are aimed at narrowing this region even further. However, given the relatively few families and recombination events left to examine, it may not be easy to reduce this region below about 5 cM.

Linkage to other chromosomes

Given the fact that only 30–40% of families are chromosome 9-linked, at least one other TSC gene must be segregating elsewhere in the genome. Several regions have been looked at extensively, but no confirmed linkage has yet been identified.

The first region to be examined was chromosome 11q. This region was looked at because of an 11:22 chromosomal translocation patient detected upon amniocentesis. The child was stillborn and had several features consistent with TSC (Clark, 1991). However, the child was an unbalanced carrier, and several relatives also carried the translocated chromosome and were clinically normal. Markers on chromosome 11q were examined and substantial positive lod scores for several markers on chromosome 11 were found (Smith et al., 1990). These results were isolated to a few pedigrees, none of which individually generated a lod score greater than 3. Recent analysis using the tuberous sclerosis international collaborative data set (Haines et al., 1991a; Janssen et al., 1991; Povey et al., 1991), and including these families studied by Smith and co-workers, found good evidence for linkage to chromosome 11q23, in the region near D11S35. However, the evidence is not as convincing as that for chromosome 9, and little support was found for this locus aside from Smith's pedigrees. Thus only a small number (5–15%) of TSC families could possibly be explained by this locus. Additional markers up and down chromosome 11 have been examined by several research groups (Haines et al., 1991b; Kandt et al., 1990) but no further evidence of linkage to 11q has been found. At this point, it seems unlikely that a chromosome 11q locus, if it exists at all, can explain more than a very few families with TSC.

A second region currently being examined is on chromosome 12, near the PAH gene. Again, this region was targeted because of a sporadic patient who had a 3:12 translocation (Fahsold et al., 1991). Markers on both chromosomes near the breakpoint were tested, and positive lod scores for two markers on chromosome 12 were found in a set of seven informative families (Fahsold et al., 1991). This work has not yet been confirmed. Not every family tested was positive for these markers (primarily those already shown to have chromosome 9 linkage), so that the proportion of chromosome 12-linked families, if any, is not yet known. One recent study, examining a collaborative data set, could not find significant evidence for chromosome 12 linkage, suggesting that only a small portion of families are linked to this locus, or that the result is spurious.

A locus residing on chromosome 14 has also been suggested (Kandt et al., 1990). They obtained a positive lod score of about 2 (at $\theta = 0.15$) in a set of 11 families. Again, this result has not been confirmed by any other group, and examination of any collaborative data set has not found significant evidence of linkage to this locus. It is important to point out that there are several large pedigrees (Haines et al., 1991b; Kandt et al., 1989) that have not shown linkage to any of these confirmed or proposed locations for TSC loci. Thus it is highly likely that a TSC locus exists elsewhere in the genome, in regions not yet examined.

FUTURE DIRECTIONS

Genetic Studies

It is likely, based on the genetic linkage work already performed, that the majority of TSC families segregate a gene that has not yet been localized. Of course, markers on chromosome 11q and 12q need to be examined further, but given the lack of confirmation up to this time, it is unlikely that loci in these regions will account for more than 10–20% of TSC families, if any at all.

Therefore, one immediate goal is to find the location of the other TSC gene(s). To do this requires screening markers throughout the genome. Given the estimated size of the genome (more than 3,000 cM) and the resolution of detection for most markers (no more than 10 cM), more than 300 markers will likely need to be tested. In the near future, it will be possible to do this in a coordinated and efficient manner, due to the emergence of "index marker maps" as part of the Human Genome Initiative. Index marker maps are being made for each chromosome, and will consist of markers with very high heterozygosities (greater than 70%), and spaced no further than 10 cM apart. Additionally, every effort is being made to have all markers use the same laboratory technology (PCR of microsatellite repeats). It is expected that such index marker maps will be ready before the end of 1994, although preliminary maps are likely to be published in the interim. Once this other locus (or loci) are found, then approaches similar to that outlined for chromosome 9 can be used to clone the aberrant gene(s).

In the case of the chromosome 9q-linked TSC gene, it will be important to minimize the region as much as possible. There are two crucial matters that must be taken into account. The first is to maximize the number of informative families to be studied. Since only 30–40% of any family collection is chromosome 9q-linked, each research group has only a limited amount of information. Thus it is necessary to pool this information, as is already being done through the NTSA-sponsored genetic linkage collaborative database. The second is to consider only those families that are chromosome 9q-linked. As already discussed, a recombination event gives ordering information only if it is truly segregating the disease gene. In general, this problem is solved by using data from only the larger families showing clear evidence of linkage to chromosome 9. One rule of thumb is to use either a lod score greater than 1.5 at some recombination fraction, or a posterior probability of linkage greater than 80%.

Chromosomal Aberration Studies

Given the high proposed mutation rate in TSC, and the fact that it causes generally benign tumor growth, deletions and/or rearrangements are prime candidates for being the TSC lesion. Thus it may seem surprising that few studies of tumor tissue have been done, since this has been so successful in identifying and delineating candidate region for other genetic tumor diseases such as neurofibromatosis type 1 (Mashiyama et al., 1991) and type 2 (Seizinger et al., 1986). The primary difficulty has been one of availability. Since most of the tumors are benign, there is usually little medical reason to remove them. Several groups are now mounting a large effort to obtain and study the rare tumor tissues as they become available. Standard cytogenetic studies of banded chromosomes have failed to reveal any abnormalities in more than 30 familial and 30 sporadic cases in our data. However, if the rearrangements and/or deletions are small, they are not likely to be seen at a resolution of 600–800 bands. Another approach holds more promise. Fluorescence *in situ* hybridization (FISH) provides a much higher resolution than standard banding, and thus might identify small deletions or rearrangements. However, it is necessary to have on hand clones (usually cosmids) of the region to be looked at, and it is generally only possible to examine one chromosome at a time. Thus FISH may be applied to the 9q TSC locus. If cosmids known to reside from the TSC region do not hybridize to one of the chromosomes in an affected patient, then a deletion is suspected. This region is immediately delineated since the cosmids have previously been mapped and ordered.

Clinical Correlations to Genetic Findings

Another interesting facet to TSC will be to determine if there are any specific clinical differences between the genetic subtypes of TSC. Is it possible that specific organ involvement is at least somewhat dependent on which TSC gene is segregating? There is currently no evidence to suggest that clinical correlations exist, but they might be subtle and remain undetected until all the TSC genes have been localized. This is unlike the situation for neurofibromatosis, where there has been a long appreciation that some families had involvement limited to the central nervous system, but most had a pattern of peripheral involvement. These two forms are now recognized as distinct genetic entities, NF2 and NF1, respectively. The genes for these have been localized (to chromosomes 22 and 17, respectively) and the NF1 gene has been cloned (Wallace et al., 1990). Winship and colleagues (1990) recently described a TSC family that is not chromosome 9-linked and has some unusual features. In our first 28 families analyzed for linkage we have not seen any distinction between our 9-linked families and the non-9-linked families with regard to clinical features (Haines et al., 1991b).

Another possibility to consider is that the wide clinical variation seen in TSC, even within families, also has a genetic etiology. There may be several genes that influence the expression of the TSC mutation, but are not the mutation themselves. These other genes could be considered susceptibility genes. Identifying such genes could provide insight not only into the functioning of the normal TSC gene but also to overall growth and regulation of cell and tissues.

Is it possible that two fairly common genes are required to manifest TSC? Such two-locus interactions have been described in animal models, and it is tempting to speculate on its viability in TSC. Suppose that two genes can cause TSC, but each by itself has a penetrance of only 30–40%, and expression is usually mild. However, if they happen to cosegregate, then the joint penetrance is near 100%, and expression is generally severe. This would predict that several collected families would have many affected individuals, more than the expected 50/50 ratio of an autosomal dominant trait. It would also predict that positive lod scores would be found on multiple chromosomes in the same family. In support of this hypothesis, several of the larger families that have been collected have a greater than expected number of affected individuals (Haines et al., 1991b; Pericak-Vance, personal communication). In addition, at least one family has provided a positive lod score on three different chromosomes. Family 1036 of the Coriell Mutant Cell Bank Repository gives a lod score of 1.8 on chromosome 9, 0.3 on chromosome 11, and 1.5 on chromosome 12 (all at $\theta = 0.00$). This family also has only one unaffected individual in seven meiotic events. Although this remains highly speculative, it will need to be considered once all the TSC gene(s) have been identified.

Prenatal Diagnosis

It is natural to wonder when a test for carrying the TSC gene(s) will be available. For a very few families under very strict conditions (e.g., large families linked to chromosome 9), it might be possible to offer prenatal diagnosis based on the segregation of the chromosome 9 markers. However, even here, there would be a probabilistic error rate of at least 5%. For the remaining vast majority of families, a test will not be available soon. The primary difficulty lies in the genetic heterogeneity, since the majority of families are not chromosome 9-linked. The problem for chromosome 9-linked families is showing that they belong to this class. Unless the family is large, it will not be possible to be confident about the chromosomal localization. To provide a test for TSC in familial cases, localization of all the TSC genes will be required. Even then, with small families, it often will not be possible to determine to which class they belong. Sporadic cases are even more difficult. Since there is no segregation data available, the only test will arise when the gene defects are identified, and can be directly tested. This is an unfortunate circumstance, but the challenges raised in studying TSC will generate new analytical methods, and provide greater insight into human physiology.

REFERENCES

Alexander, J.M., B.M.K. Biller, et al. (1990) Clinically nonfunctioning pituitary tumors are monoclonal in origin. J. Clin. Inv. 86:336–340.

Berg, H. (1913) Vererbung der tuberosen sklerose durch zwei bzw. drei generationen. Ges. Neurol. Psychiatr. 19:528–539.

Bernstein, J., and T.O. Robbins (1991) Renal involvement in tuberous sclerosis. Ann. N.Y. Acad. Sci. 615:36–49.

Botstein, D., R.L. White, M. Skolnick, et al. (1980) Construction of a genetic linkage map in man using restriction fragment length polymorphisms. Am. J. Hum. Genet. 32:314–331.

Bournville, D.M. (1880) Sclerose tubereuse des circonvolutions cerebrales: Idiotie et epilepsie hemiplegique. Arch. Neurol. (Paris) 1:81–91.

Clark, R. (1991) A cytogenetic abnormality in tuberous sclerosis. Ann. N.Y. Acad. Sci. 615:243–251.

Conneally, P.M., and M.L. Rivas (1980) Linkage analysis in man. In: H. Harris and K. Hirschhorn (eds.). Advances in Human Genetics, vol. 10. New York: Plenum Press, pp. 209–266.

Connor, J.M., L.S. Pirrit, J.R.W. Yates, et al. (1987) Linkage of the tuberous sclerosis locus to a DNA polymorphism detected by v-abl. J. Med. Genet. 24:544–546.

Fahsold, R., H.D. Rott, P. Lorenz, et al. (1991) A new gene for tuberous sclerosis maps to chromosome 12. Cytogenet. Cell. Genet. (in press).

Fryer, A.E., A. Chalmers, J.M. Connor, et al. (1987) Evidence that the gene for tuberous sclerosis is on chromosome 9. Lancet 1:659–661.

Gomez, M.R. (1988) Tuberous Sclerosis. New York: Raven Press, pp. 1–230.

Haines, J.L., J. Amos, J. Attwood, et al. (1991a) Genetic heterogeneity in tuberous sclerosis: Study of a large collaborative dataset. Ann. N.Y. Acad. Sci. 615:256–264.

Haines, J.L., M.P. Short, D.J. Kwiatkowski, et al. (1991b) Localization of one gene for tuberous sclerosis within 9q32-9q34 and further evidence for heterogeneity. Am. J. Hum. Genet. 49:764–772.

Hallervorden, J., and W. Krucke (1956) Die tuberose Hirnsklerose. In: Handbuch der speziellen pathologischen Anatomie and Histologie, vol. 13/4. Berlin: Springer-Verlag, pp. 602–663.

Inglis, K. (1950) Neurolemmoblastosis: The influence of intrinsic factors in disease when development of the body is abnormal. Am. J. Pathol. 26:521–536.

Irving, R.M., G.R. Ford, and N.S. Jones (1991) Tuberous sclerosis with primary meningioma of the maxillary antrum. J. Laryngol. Otl. 105:481–483.

Ishibashi, Y., R. Watanabe, T. Nogita, et al. (1991) Abnormal gene expressions of stroma cells in patients with tuberous sclerosis. Ann. N.Y. Acad. Sci. 615:228–242.

Jacoby, L.B., K. Pulaski, G.A. Rouleau, et al. (1990) Clonal analysis of haman meningiomas and schwannomas. Cancer Res. 50:6783–6786.

Janssen, L.A.J., S. Povey, J. Attwood, et al. (1990) A comparative study on genetic heterogeneity in tuberous sclerosis: Evidence for one gene on 9q34 and a second gene on 11q22-23. Ann. N.Y. Acad. Sci. 615:306–314.

Johnson, W.G., H. Yoshidome, E.S. Stenroos, et al. (1991) Origin of the neuron-like cells in tuberous sclerosis tissues. Ann. N.Y. Acad. Sci. 615:211–219.

Kandt, R.S., M.A. Pericak-Vance, W.-Y. Hung, et al. (1989) Absence of linkage of ABO blood group locus to familial tuberous sclerosis. Exp. Neurol. 104:223–228.

Kandt, R.S., M.A. Pericak-Vance, W.-Y. Hung, et al. (1990) Heterogeneity in tuberous sclerosis: Chromosome 9?, 11?, or maybe 14! Am. J. Hum. Genet. 47:A223.

Khalkhali-Ellis, Z., K. Henderson, and F.W. Hemming (1991) Glycoprotein and proteoglycan alterations in tuberous sclerosis. Ann. N.Y. Acad. Sci. 615:149–157.

Mashiyama, S., Y. Murakami, T. Yoshimoto, et al. (1991) Detection of p53 gene mutations in human brain tumors by single-strand conformation polymorphism analysis of polymerase chain reaction products. Oncogene 6:1313–1318.

Nakamura, Y., M. Leppert, P. O'Connell, et al. (1987) Variable number of tandem repeat (VNTR) markers for human gene mapping. Science 235:1616–1622.

Naramoto, A., N. Itoh, M. Nakano, et al. (1989) An autopsy case of tuberous sclerosis associated with primary mesothelioma. Acta Pathol. (Japan) 39:400–406.

Northrup, H., A.L. Beaudet, W.E. O'Brien, et al. (1987) Linkage of tuberous sclerosis to ABO blood group. Lancet 1:804–805.

Ott, J. (1991) Analysis of Human Genetic Linkage. Baltimore: Johns Hopkins University Press.

Pampiglione, G., and E. Pugh (1975) Infantile spasms and subsequent appearance of tuberous sclerosis syndrome. Lancet 2:1046.

Park, Y.K., and S.K. Hann (1989) Cluster growths in adenoma sebaceum associated with tuberous sclerosis. J. Am. Acad. Dermatol. 20:918–920.

Povey, S., B. Janssen, J. Attwood, et al. (1991) An attempt to map two genes for tuberous sclerosis using novel two-point methods. Ann. N.Y. Acad. Sci. 615:298–305.

Roach, E.S., D.P. Williams, and D.W. Laster (1987) Magnetic resonance imaging in tuberous sclerosis. Arch. Neurol. 44:301–303.

Robertson, D.M. (1991) Ophthalmic manifestations of tuberous sclerosis. Ann. N.Y. Acad. Sci. 615:17–25.

Sampson, J.R., J.R. Yates, L.A. Pirrit, P. Fleury, I. Winship, P. Beighton, and J.M. Connors (1989) Evidence for genetic heterogeneity in tuberous sclerosis. J. Med. Genet. 26:511-516.

Seizinger, B.R., R.L. Martuza, and J.F. Gusella (1986) Loss of genes on chromosome 22 in tumorigenesis of human acoustic neuroma. Nature 322:644–647.

Shepherd, C.W., B.W. Scheithauer, M.R. Gomez, et al. (1991) Subependyma giant cell astrocytoma: A clinical, pathological, and flow cytometric study. Neurosurgery 28:864–868.

Smith, H.C., G.H. Watson, R.G. Patel, et al. (1989) Cardiac rhabdomyomata in tuberous sclerosis: Their course and diagnostic value. Arch. Dis. Child 64:196–200.

Smith, M., S. Smalley, R. Cantor, et al. (1990) Mapping a gene determining tuberous sclerosis to human chromosome 11q14-11q23. Genomics 6:105–114.

Vogt, H. (1908) Zur Diagnostik der tuberosen Sklerose Z. Erforsch Behandle Jugeudl Schwachsinns 2:1–12.

von Recklinghausen, F. (1862) Ein Herz von einem neugeborenen welches mehrere tueils nach aussen, thiels nach den hohlen prominveude tumoren. (Myomen) trug. Verh Ges Geburtsh Monatschr Geburtsh 20:1–2.

Wallace, M.R., et al. (1990) Type I neurofibromatosis gene: Identification of a large transcript disrupted in three NF1 patients. Science 249:181–186.

Watson, G.H. (1991) Cardiac rhabdomyoma. Ann. N.Y. Acad. Sci. 615:50–57.

Weber, J.L., and P.E. May (1989) Abundant class of human DNA polymorphisms which can be typed using the polymerase chain reaction. Am. J. Hum. Genet. 44:388–396.

Winship, I.M., J.M. Connor, and P.H. Beighton (1990) Genetic heterogeneity in tuberous sclerosis: Phenotypic correlations. J. Med. Genet. 27:418–421.

Yamanaka, N., T. Imai, M. Fujisawa, et al. (1990) Renal cell carcinoma and angiomyolipoma in tuberous sclerosis. Case report. Nippon Hinyokika Gakkai Zasshi 81:304–307.

Chapter 28

Tumor Suppressor Genes and Hereditary Tumor Syndromes of the Human Nervous System: Isolation of the Primary Genetic Defect in von Hippel-Lindau Disease

Bernd R. Seizinger

INTRODUCTION

Von Hippel-Lindau disease (VHL) is a devastating hereditary tumor syndrome associated with a variety of neoplasms in multiple organ systems, including hemangioblastomas in the cerebellum, retina, and spinal cord, pheochromocytomas, and—a particularly frequent cause of death in VHL—renal cell carcinomas. For those hereditary disorders where no defective protein has been identified (the vast majority of the 4,000 or so hereditary diseases known in humans), the so-called "reversed genetics" approach has emerged as a particularly powerful strategy toward the cloning and identification of the defective gene. VHL has become a model system for the application of this approach. Using DNA linkage analysis in a number of families displaying VHL, we have recently shown that the primary defect in VHL maps to the short arm of chromosome 3. Based on the approximate knowledge of its chromosomal location, the putative "tumor suppressor" gene has been bracketed to a small region in chromosome 3p25-p26. The identification of closely linked flanking markers, together with the apparent genetic homogeneity of VHL, allows for the development of a reliable diagnostic genetic test. Furthermore, this has provided the starting point for directed chromosomal "walking" and "jumping," which is being pursued in conjunction with several possible shortcuts toward the isolation of the defective VHL gene.

The characterization of the VHL gene should have important implications not only for patients with VHL but for a much larger number of cancer patients in the general population who are afflicted with the sporadic counterparts of VHL-associated tumor types. Furthermore, the normal counterparts of those genes whose mutation or loss of function lead to tumor formation are likely to play a key role in regulating growth and development. Thus, the isolation and characterization of "tumor suppressor" genes, including the VHL gene, should ultimately be of great significance not only from a clinical point of view but also from a basic science perspective for our understanding of fundamental genetic mechanisms of growth control in the normal cell.

The "Reversed Genetics" Approach

Many human tumors are known to occur in two different forms: as sporadic tumors in the general population, and as hereditary tumors, where they are transmitted from generation to generation within families. There is accumulating evidence that the hereditary and sporadic forms of one and the same tumor type are caused by similar pathogenetic mechanisms affecting the same genes. Thus, hereditary tumor syndromes offer unique model systems for isolating genes whose mutations lead to cancer. The cloning and characterization of these genes may therefore have important implications for diagnosis and treatment not only of the relatively rare hereditary tumors but also for their much more common sporadic counterparts.

In almost all cases of hereditary tumor syndromes, the underlying biochemical defect is unknown, but presumably is related to fundamental mechanisms controlling growth and differentiation during development of various organ systems. The task of finding a specific protein defect among the 100,000 or so possible proteins encoded in the human genome is enormous unless additional information is available to aid the search. Thus, this approach has been unsuc-

Molecular Genetics of Nervous System Tumors, pages 311-318 ©1993 Wiley-Liss, Inc.

cessful in the vast majority of cases.

During the last decade, however, recombinant DNA technology has provided a powerful alternative strategy for investigating inherited diseases in which the defective protein is unknown. This strategy has been termed reversed genetics because the first goal is not to find the defective protein, but rather to isolate the defective gene based on the determination of its chromosomal location in the human genome (Fig. 1). The first crucial step in this approach, the chromosomal location of the gene defect, can be achieved by DNA linkage analysis in families displaying the disease phenotype. The high frequency and ease of detection of restriction fragment length polymorphisms (RFLPs), as well as other types of highly polymorphic elements in human DNA, such as the (CA)n repeats, has resulted in an abundant new supply of polymorphic DNA markers for linkage studies. These polymorphic markers allow testing of whether a particular DNA marker is linked to the disease gene, i.e., whether the marker and the disease gene are in close proximity on the same chromosome and therefore are co-inherited in disease families.

The chromosomal location of the defective gene provides the basis for its subsequent isolation and characterization. Several new techniques have been developed in recent years to bridge the gap between a genetically linked marker and the actual disease gene. However, the cloning of a disease gene based on the knowledge of its chromosomal location still represents a major enterprise, and has so far been successful in only relatively few cases. Thus, most genes that have been cloned thus far using the reversed genetics approach have actually been isolated with the aid of additional shortcuts, for example, with the help of cytogenetically visible constitutional chromosomal aberrations involving the disease locus, such as small deletions or translocations.

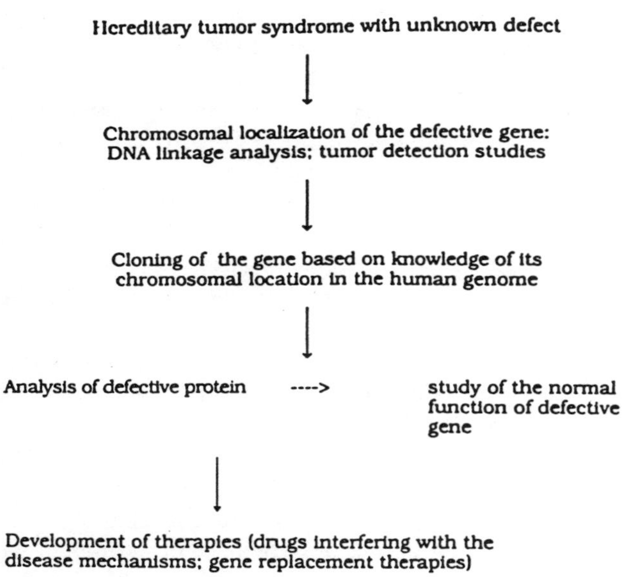

Fig. 1. "Reversed genetics" approach, used to isolate and characterize the defective genes for hereditary disorders if the protein defect is unknown.

The following paragraphs briefly outline strategies toward the isolation and functional characterization of the defective gene causing von Hippel-Lindau disease (VHL), a devastating hereditary tumor syndrome involving multiple organ systems. Similar strategies may be applied toward the isolation of other hereditary tumor syndromes.

VON HIPPEL-LINDAU DISEASE: CLINICAL FEATURES

VHL is a devastating disorder associated with various neoplasms in multiple organ systems (Go et al., 1984; Huson et al., 1986; Memon and Rosen, 1964). The most common manifestations of VHL include tumors in the central nervous system (hemangioblastomas in the cerebellum, retina, and spinal cord), pheochromocytomas, bilateral and multifocal renal cell carcinomas, pancreatic tumors, and cysts of the kidney, pancreas, epididymis, and other organs. These tumors frequently lead to serious clinical complications and death. VHL has been found in people of white, black, and Asian descent and follows an autosomal dominant trait of inheritance with high penetrance but variable expressivity. An individual with the VHL phenotype appears to be susceptible to any of its lesions in any combination and in any sequence, with a highly variable age of onset, usually between ages 15 and 50 years. Renal cell carcinomas are a particularly frequent cause of death in VHL. In comparison to their sporadic counterparts, familial renal cell carcinomas in VHL are usually bilateral and multifocal, and show an earlier age of onset (Go et al., 1984; Huson et al., 1986; Memon and Rosen, 1964).

Genetic Linkage Studies in VHL: Implications for Diagnostic Testing and Cloning of the Disease Gene

Since no clue about the underlying biochemical defect had been reported in VHL, we decided to approach this hereditary tumor syndrome with a reversed genetics approach, as outlined above. We used DNA linkage analysis to achieve the first crucial step in this approach, the chromosomal localization of the gene defect. Although we were prepared to search through the whole human genome to localize the VHL defect, our genetic linkage studies were particularly focused on chromosome 3 as a possible "candidate chromosome" for containing the VHL defect for two reasons. First, two pedigrees with hereditary renal cell carcinomas had been reported to exhibit transloctions involving chromosome 3p (a constitutional 3p;8 translocation in one family) (Cohen et al., 1979), and a 3p;11 translocation restricted to the tumor tissue in the other family (Pathak et al., 1982). This suggested that chromosome 3p may contain a gene possibly important for tumorigenesis of renal cell carcinomas. Despite the fact that the two translocation families did not have VHL, since they did not show any other symptoms of this tumor syndrome, they were associated with renal cell carcinoma, one of the hallmarks in VHL.

Second, karyotype studies (and, subsequently, RFLP studies) had shown chromosome 3p deletions in sporadic renal cell carcinoma, suggesting that this chromosomal region contains a tumor suppressor gene whose loss or inactivation leads to renal cell carcinoma (King et al., 1987; Yoshida et al., 1986; Zbar et al., 1987).

DNA linkage analysis provided conclusive evidence that the defective gene causing VHL maps to chromosome 3p, since we found VHL to be genetically linked to c-RAF1 in the region 3p25-p26 (Seizinger et al., 1988). Although c-RAF1 had been a potential candidate gene for causing VHL, this oncogene is not the VHL gene, due to the relatively high recombination frequency between VHL and c-RAF1 (Seizinger et al., 1988). c-RAF1 and the VHL gene are probably several million bases apart from each other.

Based on the finding that the gene causing VHL is genetically linked to c-RAF1 on chromosome 3p, we intended (1) to further narrow the region on chromosome 3p containing the defective VHL gene; (2) to determine whether VHL is genetically heterogenous, i.e., whether there is more than one gene for the same disease phenotype; and (3) to identify flanking markers for the gene. These three objectives are important for identifying genetic markers suitable for prenatal and presymptomatic diagnostic testing in VHL, and to clone the defective gene itself. The availability of a genetic test for the early and reliable detection of individuals carrying the defective gene would be beneficial for VHL patients and their relatives, since many of the manifestations of VHL can be successfully treated if detected in their early stages (e.g., hemangioblastomas in the nervous system), while the complications of undetected disease can be devastating and may lead to death.

To provide genetic markers for the development of a reliable diagnostic test, and to further narrow and eventually clone the VHL defect, we have generated new DNA markers for chromosome 3p. With these markers, we performed a multipoint genetic linkage analysis in 28 VHL pedigrees, comprising 470 individuals, 164 of whom were affected with VHL. We identified tightly linked markers, including flanking markers that bracket the VHL gene to a small region on chromosome 3p25-p26 (Seizinger et al., 1991a) (Fig. 2). In agreement with our findings, studies in Bert Zbar's laboratory have also shown that the VHL gene maps telomeric to c-RAF1 in 3p25-p26 (Hosoe et al., 1990).

We have meanwhile identified a number of additional polymorphic markers for the VHL region in 3p25-p26 (Klauck and Seizinger, in preparation). This includes genetic probes that detect (CA)n repeats, highly polymorphic repeat elements that are analyzed using polymerase chain reaction (PCR) technology and can be detected in great abundance in all parts of the human genome. These markers are useful for two purposes: (1) to improve presymptomatic and prenatal diagnostic testing in VHL, i.e., to generate probes that are progressively closer to the VHL gene and are more informative in VHL pedigrees, and (2) as a tool for the isolation of the disease gene itself. For example, these markers can be used to screen for deletions in VHL patients using Southern and pulsed-field blots.

In addition to providing the basis for the development of a genetic test for VHL, and for the isolation of the disease gene using chromosome-directed cloning techniques, these linkage studies have had direct implications for both the clinical manangement of VHL patients and for our understanding of the genetic mechanisms of tumorigenesis in this disorder.

While visceral cysts of the kidney, pancreas, and epididymis are commonly found in VHL and are considered a diagnostic criterion for this disorder (Memon and Rosen, 1964), they also have a significant occurrence in the general population. The presence of cysts, unaccompanied by other more typical lesions such as retinal and cerebellar hemangioblastoma, may therefore represent a major diagnostic problem, leading to errors in the assessment of the disease status. The application of flanking markers for the VHL gene for presymptomatic diagnostic testing confirms that epididymal cysts (and possibly other visceral cysts) are indeed not suitable as a diagnostic criterion in this disorder (Seizinger et al., 1991a). Since visceral cysts, as well as renal cell carcinoma and pheochromocytoma, can occur sporadically even in young people, and may also be associated with other tumor syndromes, the availability of flanking markers for the VHL gene will be useful in identifying VHL gene carriers, particularly among those individuals at risk in whom these are the only manifestations of disease.

Pheochromocytomas occur nonuniformly in VHL families, and may also be associated with other hereditary tumor syndromes. Our genetic studies imply that the phenotype in VHL families with and without pheochromocytomas is caused by defects within the same gene. This finding has been confirmed by other investigators (Hosoe et al., 1990; Vance et al., 1990).

Thus, there appears to be only one single gene for VHL in the human genome (in genetic terms, there is no evidence for "nonallelic genetic heterogeneity" in VHL). However, it is possible that there is genetic heterogeneity within the VHL gene itself (so-called "allelic genetic heterogeneity"). The observed clinical heterogeneity in VHL may therefore be due to different types of deletions within the VHL gene (e.g., larger deletions vs. point mutations; different functional

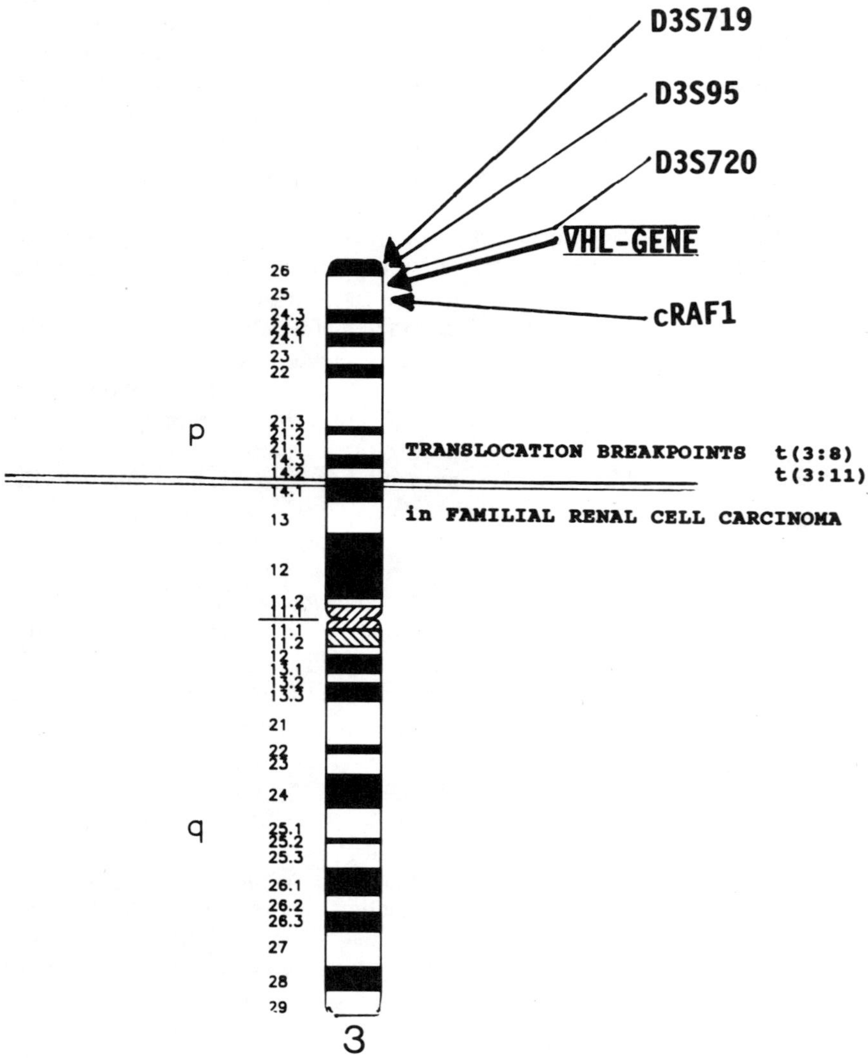

Fig. 2. Chromosomal location of the VHL gene in 3p25-p26, relative to the location of several chromosomal translocations breakpoints in families with "pure familial renal cell carcinoma."

domains of the gene affected, etc.).

Alternatively, other "modulatory genes" distinct from the VHL gene may also play a role in the expressivity of VHL. For example, the influence of such modulatory genes has been suggested to cause the variable expressivity in neurofibromatosis type 1 (NF1) (Ponder et al., personal communication). NF1 is—like VHL—apparently genetically homogeneous, i.e., there appears to be only a single NF1 gene that maps to chromosome 17q11.2.

The VHL Gene as a Tumor Suppressor Gene

In addition to genes that stimulate growth and differentiation (e.g., certain proto-oncogenes), most (if not all) cells seem to contain genes against excessive growth. The function of these genes may be to regulate (suppress) the expression of growth-promoting factors, thereby exercising growth control. The loss or inactivation of these growth-inhibiting genes has been associated with the development of cancer; hence, they have been termed tumor suppressor genes or anti-oncogenes (Ponder, 1988; Seizinger and Breakefield, 1990; Weinberg, 1991).

Several different avenues of research have provided evidence for the existence of tumor suppressor genes. For example, fusion of normal and malignant cells leads to the suppression of the tumorigenic phenotype in the majority of the resulting hybrids which retain a relatively complete chromosomal complement. Reappearance of tumorigenicity is often accompanied by chromosomal loss. Similarly, the discovery of deletions of genomic DNA in a variety of human cancers has strongly supported the concept that the human genome contains a limited number of tumor suppressor genes whose loss, deletion, or inactivation leads to tumor formation (Ponder, 1988; Seizinger and Breakefield, 1990; Weinberg, 1991). Indeed, the loss of tumor suppressor genes appears to be one of the most fundamental mechanisms of sporadic and hereditary cancer development, and has been implicated in numerous neoplasias from multiple organ systems, including the nervous system (Ponder, 1988; Seizinger and Breakefield, 1990; Weinberg, 1991).

As part of a new committee on Chromosome and Gene Loss in Human Neoplasia of the International Gene Mapping Conferences (HGM), we have recently attempted to generate a comprehensive list of consistently observed allele losses and deletions in human tumors (Seizinger et al., 1991b). Figure 3 shows the number of different solid tumor types known to be associated with chromosomal deletions for each individual chromosome. This suggests that the distribution of tumor suppressor genes may not be at random. Chromosome 3 is among those autosomes that are most frequently lost or deleted in human tumors. This is partly due to chromosome 3 deletions in various tumor types associated with VHL.

A recent collaborative study between Bert Zbar's laboratory and our laboratory revealed loss of regions on chromosome 3p in renal cell carcinomas, hemangioblastomas, and pheochromocytomas, suggesting a common mechanism of tumorigenesis for different tumor types in VHL (Tory et al., 1989). In all informative cases, the balancing wild-type chromosome 3p containing the wild-type copy of the VHL gene from the nonaffected parent was lost in the tumor, whereas the copy containing the defective VHL gene inherited from the affected parent was retained. Thus, in the tumor tissues both copies of the VHL gene were either defective or lost. This is consistent with a recessive mechanism of tumorigenesis in which both copies of a tumor suppressor gene are lost, deleted, or inactivated, similar to the retinoblastoma model originally postulated by Alfred Knudson (1971). According to this model, sporadic and hereditary retinoblastomas (and possibly other tumor types) are caused by the loss or inactivation of both alleles of the Rb gene. The sporadic tumors are the result of two somatic cell mutations, whereas the hereditary tumors are caused by a germline mutation in one of the two copies of the tumor suppressor gene, followed by the loss or inactivation of the balancing wild-type allele, due to a somatic cell mutation.

In fact, most if not all hereditary tumor syndromes appear to be caused by aberrations in tumor suppressor genes. Table 1 describes a number of hereditary disorders associated with neural tumors that are thought to be caused by mutations in tumor suppressor genes. In each case the chromosomal location of the gene defect has been determined using genetic linkage studies.

Interestingly, several of the above-mentioned hereditary tumor syndromes are not exclusively associated with neural tumors, but include a variety of neoplasms from other organ systems (e.g., renal cell carcinoma in VHL, or osteosarcoma in familial retinoblastoma). Thus, it is likely that a variety of organ systems and cell types share the same tumor suppressor systems, and use similar genes for the control of growth and differentiation.

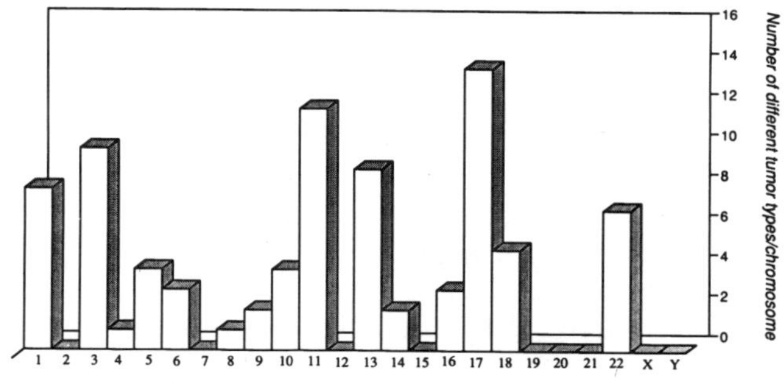

Fig. 3. The number of different solid tumor types associated with chromosomal loss or deletions for each individual human chromosome.

Not all tumor suppressor genes seem to follow a recessive mechanism of tumorigenesis. For example, the p53 tumor suppressor gene on chromosome 17p (and possibly the APC gene on 5q, which is associated with familial adenomatous polyposis coli) displays a "negative dominant" effect (Levine et al., 1991). In this mechanism, the mutation of only one copy may already contribute to tumorigenesis (Levine et al., 1991).

TABLE 1. Hereditary Disorders Associated with Neural Tumors Caused by Mutations in Tumor Suppressor Genes

Tumor syndrome	Typical tumor types	Chromo -somal location
Von Hippel-Lindau disease	Hemangioblastomas (spinal, cerebellar, retinal) pheochromocytoma, renal cell carcinoma	3p
Multiple endocrine neoplasia type 2 (MEN2)	Pheochromocytoma, medullary thyroid carcinoma	10
Multiple endocrine neoplasia type 1 (MEN1)	Pituitary tumors, parathyroid adenomas, tumors of the endocrine pancreas	11q
Familial retinoblastoma	Retinoblastoma, osteosarcoma	13q (cloned!)
Von Recklinghausen neurofibromatosis (NF1)	Neurofibroma	17q (cloned!)
Bilateral acoustic neurofibromatosis (NF2)	Acoustic neuroma, meningioma, spinal nerve root neurofibroma	22q

Alternative Mechanisms of Tumorigenesis in VHL-Associated Tumor Types

Renal cell carcinoma

The relatively telomeric position of the VHL gene in the region 3p25-p26 came as a surprise to us, since we expected the VHL gene to map more centromeric, close to the region in chromosome 3p14.2 known to contain the 3;8 and 3;11 translocation breakpoints in familial renal cell carcinoma. In fact, the VHL gene may be as far as 50 cM (approximately 50 million base pairs) apart from the translocation breakpoints in familial renal cell carcinoma. Thus, there may be more than one gene on chromosome 3p for renal cell carcinoma: one in chromosome 3p25-p26 associated with VHL and another in chromosome 3p14.2 associated with the translocation breakpoints in familial renal cell carcinoma. Alternatively, the translocations may be associated with a complex, cytogenetically invisible rearrangement on chromosome 3p that affects the VHL gene in chromosome 3p25-26, or these translocation breakpoints may affect the VHL locus through a position effect. However, although this possibility cannot be excluded, it appears to be less likely in light of the fact that the VHL gene is 50 cM apart from the translocation breakpoints in 3p14.2.

Moreover, Yusuke Nakamara's group has recently reported a third region in 3p21.3 whose loss possibly leads to the formation of some of the sporadic renal cell carcinomas. Thus, there may be as many as three distinct tumor suppressor genes on 3p associated with renal cell carcinoma.

Pheochromocytoma

As with renal cell carcinoma, there appear to be alternative genetic mechanisms of tumorigenesis for pheochromocytoma. This tumor type is a hallmark of VHL; as predicted, it shows deletions on 3p, including the VHL region in patients with VHL. However, pheochromocytomas can also be associated with other hereditary tumor syndromes, including multiple endocrine neoplasia type 2 on chromosome 10 (where it occurs together with medullary thyroid carcinoma) (Mathew et al., 1987; Simpson et al., 1987), a "pure" form of hereditary pheochromocytomas without other tumor types, and NF1, a disease associated with multiple aberrations in the nervous system including Schwann cell-derived tumors (Barker et al., 1987; Seizinger et al., 1987; Viscochil et al., 1990; Wallace et al., 1990). Pheochromocytomas from patients with NF1 show loss or deletions on chromosome 17q, including the NF1 locus in 17q11.2 (Ponder et al., 1992), suggesting that the NF1 gene may act as a tumor suppressor gene in this tumor type.

Thus, one and the same histologic tumor type such as pheochromocytoma or renal cell carcinoma may be caused by the loss of various distinct tumor suppressor loci. Conversely, mutations in one and the same tumor suppressor gene such as the VHL gene may lead to the formation of various seemingly unrelated tumor types (e.g., tumors histologically as diverse as renal cell carcinoma, pheochromocytoma, and hemangioblastoma).

Hemangioblastoma

In contrast with renal cell carcinoma and pheochromocytoma, there is no evidence for an alternative genetic mechanism of tumorigenesis in hemangioblastoma. This rare tumor usually occurs in patients with VHL and is associated with 3p deletions including the VHL locus (Tory et al., 1989).

Possible Shortcuts to the Isolation of the VHL Gene

As already mentioned earlier, the isolation of a disease gene using reversed genetics is still a very arduous approach. We are therefore exploring a variety of additional approaches that may provide possible shortcuts toward the cloning of the VHL gene. Many of these approaches are also applicable to other hereditary tumor syndromes.

The search for small deletions

Since the VHL gene appears to be a tumor suppressor gene, i.e., a gene that normally confers growth suppression in the cell and whose loss or deletion is associated with tumor formation, the search for deletions in both constitutional DNA and DNA from tumor tissues may provide a possible shortcut toward the identification of the disease gene. We are currently using both Southern analysis and pulsed-field gel electrophoresis to systematically search for such deletions. This approach has gained increased importance since we have identified close genetic markers for the VHL gene (e.g., the marker D3S720, which shows no recombinations with the VHL gene in genetic linkage studies; see Fig. 2). This and other markers may well be close enough to the VHL gene to be located on the same pulsed-field fragment (within approximately 1 megabase or less). If so, this technique should be able to recognize larger deletions in and around the VHL locus.

Somatic cell hybrids

Several of our strategies involve libraries that are enriched for markers mapping to the critical chromosomal region 3p25-p26 containing the VHL gene. Our collaborator Harry Drabkin has recently generated radiation reduction hybrids from cell lines containing human chromosome 3. One of these hybrid lines carries a small piece of the VHL region 3p25-p26 as its only human component; this hybrid contains the locus D3S720, our most closely linked marker with no recombination with the VHL gene (Seizinger et al., 1991a). Thus, it provides a highly enriched source of probes in close vicinity to the VHL locus, and may even contain the VHL gene. Markers generated from libraries constructed from this hybrid cell line will be used for genetic linkage, as well as for the search for small deletions in constitutional and tumor-derived DNA from VHL patients (see above).

A novel microdissection library/cDNA library approach

Based on several studies by our laboratory and other groups (Hosoe et al., 1990; Seizinger et al., 1988, 1991a), the VHL gene acts like a recessive tumor suppressor genes, which means both copies have to be lost, deleted, or inactivated in order to induce tumor formation. Thus, it is reasonable to assume that the VHL gene is either not expressed or aberrantly expressed in tumors from VHL patients. In contrast, the VHL gene is likely to be expressed in those normal tissues from which the VHL tumors arise, such as normal kidney and adrenal medulla. In order to obtain candidate genes that are expressed in those normal tissues and also specifically map to the VHL region on 3p25-p26, we have developed a novel subtractive hybridization approach. In this approach, a microdissection library exclusively containing DNA from the VHL region 3p25-p26 (constructed in collaboration with Bernhard Horsthemke and Uwe Klausen) is used in conjunction with cDNA libraries from kidney and adrenal medulla. Candidate cDNAs that are expressed in the normal tissues and map to the correct chromosomal region can then be tested in VHL tumors to search for aberrant or missing transcripts (Kley and Seizinger, in preparation). Furthermore, these candidate genes will be sequenced, and constitutional DNA from a variety of VHL pedigrees will be tested for point mutations and small deletions using PCR-SSCP and PCR sequencing (see below). This may lead to the identification of the VHL gene.

ANALYSIS OF CANDIDATE GENES AND IDENTIFICATION OF THE VHL GENE

Using a combination of the above strategies, we expect to obtain candidate genes for the VHL locus within the next few months. Candidate cDNAs will undergo a variety of tests.

Northern blots with RNA from VHL tumors, including renal cell carcinomas, pheochromocytomas, and hemangioblastomas, will be hybridized with candidate cDNA probes. The VHL gene is expected to be either not expressed, or aberrantly expressed in those tumor tissues. The candidate cDNAs will be analyzed on Southern blots and pulsed-field filters containing genomic DNA from both constitutional and VHL tumor tissues to search for deletions and rearrangements. Candidate genes will be sequenced, and the sequences compared with known sequences in various databases. This may provide some clue about their possible function.

For candidate genes that appear to be of particular interest based on the first three tests described above, we will perform a systematic screening for point mutations and small deletions in a variety of constitutional DNAs from various VHL pedigrees, using PCR-SSCP analysis (polymerase chain reaction-associated single-stranded conformation polymorphism analysis, a rapid screening technique for point mutations) and PCR sequencing. The VHL gene is expected to be mutant only in the affected individuals of each VHL pedigree, and should segregate with the disease phenotype from generation to generation.

If a particular candidate gene is identified, we will proceed to an in-depth analysis of the sequence and expression of this gene in normal and tumor tissues using any techniques that are necessary, including S1 analysis for determining the intron–exon structure or for delineating multiple transcripts, generation of fusion proteins for preparation of antibodies for structural studies and immunocytochemistry, determination of subcellular localization of the protein product, etc.

Reintroduction of the missing sequence into cultured tumor cells may be a method of finally proving the identity of the gene. For example, suppression of the neoplastic phenotype has recently been observed after replacement of the RB1 gene in human retinoblastoma and osteosarcoma cells.

The sequence of candidate genes will be compared with those sequences currently stored in the various databases to determine whether the protein product can be classified and ascribed a potential function based on a relationship to known protein families or structures. For example, it may have a zinc-finger motif like the Wilms' tumor gene, suggesting that it may be involved in transcriptional regulation, or it may be a membrane-bound cell surface protein like the DCC ("deletion in colon carcinoma") tumor suppressor gene on chromosome 18q, which has substantial homology with the family of cell adhesion molecules (Weinberg, 1991).

Once the VHL gene is cloned, we plan an in-depth characterization of the role of this gene in tumorigenesis of VHL-associated tumor types, e.g., its relation to other genes such as oncogenes. Moreover, we will investigate whether this gene is also involved in the development of other seemingly unrelated tumor types, analogous to the RB1 and p53 genes.

CONCLUSION

The elucidation of the biochemical disease mechanisms based on the knowledge of the protein sequence may ultimately lead to the design of specific drugs interfering with the disease mechanisms. Gene "replacement" therapy, directed by gene "targeting," may be an alternative therapeutic option, although this technology is still in its infancy.

The study of the normal function of this and other cancer-related genes may ultimately provide new insights into a conundrum of modern biology, the understanding of fundamental genetic mechanisms regulating the normal development and differentiation of cells in the nervous system and other organ systems.

ACKNOWLEDGMENTS

The author is the recipient of a Faculty Research Award from the American Cancer Society. Funds for this work were provided by grant ROI-CA 49455-03 from the National Cancer Institute.

REFERENCES

Barker, D., E. Wright, K. Nguyen, et al. (1987) Gene for von Recklingausen neurofibromatosis is in the pericentromeric region of chromosome 17. Science 236:1100–1102.

Cohen, A.J., F.P. Li, S. Berg, et al. (1979) Hereditary renal-cell carcinoma associated with a chromosomal translocation. N. Engl. J. Med. 301:592–595.

Go, R.C., J.M. Lamiell, Y.E. Hsia, et al. (1984) Segregation and linkage analyses of von Hippel Lindau disease among 220 descendants from one kindred. Am. J. Hum. Genet. 36:131–142.

Hosoe, S., H. Brauch, F. Latif, et al. (1990) Localization of the von Hippel Lindau disease gene to a small region of chromosome 3. Genomics 8:634–640.

Huson, S.M., P.S. Harper, M.D. Hourihan, et al. (1986) Cerebellar hemangioblastom and von Hippel Lindau disease. Brain 109:1297–1310.

King, C.R., R.N. Schmike, T. Arthur, et al. (1987) Proximal 3p deletion in renal cell carcinoma cells from a patient with von Hippel Lindau disease. Cancer Genet. Cytogenet. 27:345–348.

Knudson, A.G. (1971) Mutation and cancer: Statistical study of retinoblastoma. Proc. Natl. Acad. Sci. USA 68:820–824.

Levine, A.J., J. Momand, and C.A. Finlay (1991) The p53 tumour suppressor gene. Nature 351:453–456.

Mathew, C.G.P., B.A. Smith, K. Thorpe, et al. (1987) Deletion of genes on chromosome 1 in endocrine neoplasia. Nature 328:524–527.

Memon, K.L., and S.W. Rosen (1964) Lindau's disease. Am. J. Med. 36:595–617.

Pathak, S., L.C. Strong, R.E. Ferrell, et al. (1982) Familial renal cell carcinoma with a 3:11 chromosome translocation limited to tumor cells. Science 217:939–941.

Ponder, B.G. (1988) Gene losses in human tumours. Nature (London) 335:400–402.

Ponder, B., et al. (1992) Genes Chromosomes Cancer (in press).

Seizinger, B.R., and X.O. Breakefield (1990) The role of "tumor suppressor" genes in neural tumors. Trends Neurosci. 13:3–6.

Seizinger, B.R., H.P. Klinger, C. Junien, et al. (1991b) International Human Genome Conference 1991: Report of the Committee on Chromosome and Gene Loss in Human Cancer. Cytogenet. Cell Genet. (in press).

Seizinger, B.R., G.A. Rouleau, L.J. Ozelius, et al. (1987) Genetic linkage of von Recklinhausen neurofibromatosis to the nerve growth factor receptor gene. Cell 49:589–594.

Seizinger, B.R., G.A. Rouleau, L.J. Ozelius, et al. (1988) von Hippel-Lindau disease maps to the region of chromosome 3 associated with renal cell carcinoma. Nature (London) 332:269–270.

Seizinger, B.R., D.I. Smith, M.R. Filling-Katz, et al. (1991a) Genetic flanking markers refine diagnostic criteria and provide insights into the genetics of von Hippel Lindau disease. Proc. Natl. Acad. Sci. USA 88:2864–2868.

Simpson, N.E., K.K. Kidd, P.J. Goodfellow, et al. (1987) Assignment of multiple endocrine neoplasia type 2A to chromosome 10 by linkage. Nature 328:528–531.

Tory, K., H. Brauch, M. Linehan, et al. (1989) Specific genetic change in tumors associated with von Hippel Lindau disease. J. Natl. Cancer Inst. 81:1097–1107.

Vance, J.M., K.W. Samll, M.A. Jones, et al. (1990) Confirmation of linkage in von Hippel Lindau disease. Genomics 6:565–567.

Viscochil, D., A.M. Buchberg, G. Xu, et al. (1990) Deletions and a translocation interrupt a cloned gene at the neurofibromatosis type 1 locus. Cell 62:187–192.

Wallace, M.R., D.A. Marchuk, L.B. Anderson, et al. (1990) Type 1 neurofibromatosis gene: Identification of a large transcript disrupted in three NF1 patients. Science 249:181–186.

Weinberg, R.A. (1991) Tumor suppressor genes. Science 254:1138–1146.

Yoshida, M.A., K. Ohyashiki, H. Ochi, et al. (1986) Rearrangement of chromosome 3 in renal cell carcinoma. Cancer Genet. Cytogenet. 19:351–354.

Zbar, B., H. Brauch, C. Talmadge, et al. (1987) Loss of alleles of loci on the short arm of chromosome 3 in renal cell carcinoma. Nature 327:721–724.

Chapter 29

Neurofibromatosis 1

Richard Cawthon and Ray White

INTRODUCTION

Neurofibromatosis 1 (NF1) is the most common hereditary disease predisposing to neoplasia in people (Crowe et al., 1956; Riccardi and Eichner, 1986). The typical clinical features of NF1 are neurofibromas, benign tumors appearing just beneath the skin and along peripheral nerve tracts; *café-au-lait* macules (CLS), hyperpigmented flat patches of skin; and bilateral Lisch nodules, tiny glassy brown growths appearing on the surface of the iris. Microscopically, the neurofibromas are a mixture of cell types: primarily Schwann cells, along with fibroblasts, mast cells, and at least one nerve ending (Riccardi and Eichner, 1986). The CLS and Lisch nodules contain melanocytes that have proliferated excessively and/or overproduced melanin (Ishida and Jimbow, 1987; Perry and Font, 1982; Slater et al., 1986; Williamson et al., 1991).

The mode of inheritance of NF1 is autosomal dominant, and the incidence is approximately equal in men and women—about one in 3,500 (Crowe et al., 1956; Riccardi and Eichner, 1986). Genetic linkage studies indicate that a single locus, within band q11.2 of chromosome 17, accounts for essentially all cases of the disease (Goldgar et al., 1989). Since only half of the cases are inherited, the rate of new mutations is about one in 7,000, one of the highest mutation rates known for any human gene.

The disease is fully penetrant but highly variable in its expressivity, sometimes producing both mildly affected and severely affected individuals within a single family. In addition to the neurofibromas that are typical of the disease, NF1 patients are predisposed to other neoplasms, benign and malignant. These include malignant schwannomas, pheochromocytomas, rhabdomyosarcomas, optic gliomas, astrocytomas, and glioblastomas (Bataini et al., 1991; Blatt et al., 1986; Cohen and Rothner, 1989; Kibirige et al., 1989; Riccardi and Eichner, 1986; Skuse et al., 1989; Sørensen et al., 1986). The lifetime risk for malignant cancers among NF1 individuals compared to the risk in the general population is unknown; however, one study found that the mean age of NF1 patients at the time of diagnosis of malignancy was 38 years, compared to 65 years mean age at diagnosis of malignancy in the general population (Lefkowitz et al., 1990).

Recently, the gene responsible for neurofibromatosis 1 was identified and cloned (Cawthon et al., 1990; Viskochil et al., 1990; Wallace et al., 1990b). Since then, significant progress has been made toward understanding how the normal *NF1* gene product, neurofibromin, may be involved in the regulation of cell growth, how germline *NF1* mutations cause neurofibromatosis 1, and how somatic *NF1* mutations may contribute to the development of sporadic tumors in the general population.

IDENTIFICATION OF THE *NF1* GENE

The studies that culminated in the identification of the *NF1* gene have been reviewed in detail elsewhere (White, 1991). A brief overview is presented here. The *NF1* gene was mapped by linkage analysis to the pericentromeric region of chromosome 17 in 1987; further linkage studies narrowed the candidate region to 17q11.2. Two remarkable NF1 patients with independent balanced translocations involving 17q11.2 (one with a 1;17 translocation, the other with a 17;22 translocation) provided physical landmarks that greatly simplified the search for the gene. The simplest hypothesis was that interruption of *NF1* by a translocation break was the cause of neurofibromatosis 1 in each of these patients. If correct, this would mean that both translocation breakpoints lay within *NF1*. Separate mouse–human somatic cell hybrids were constructed containing different segments of the broken chromosomes 17 produced by the translocations. Following restriction-enzyme digestion of the DNA from these cell lines and agarose gel electrophoresis, Southern blots were made and probed with clones of human genomic DNA from the region in order to score each clone as lying proximal to, distal to, or actually crossing a translocation breakpoint. Along the normal chromosome 17, both the site of the 1;17 break and the site of the 17;22

Molecular Genetics of Nervous System Tumors, pages 319-327

break were found to lie within the same 600-kb NruI fragment.

Evi2 was a newly identified mouse gene that mapped to mouse chromosome 11, in a region known to be syntenic with the proximal part of human 17q. Therefore, the location of *EVI2*, the human homologue of *Evi2*, was sought. *EVI2* in fact lies between the two NF1 translocation breakpoints; furthermore, these experiments demonstrated that along the normal chromosome 17 the 1;17 breakpoint lies centromeric of the 17;22 breakpoint. However, no part of this gene appeared to cross the site of either translocation breakpoint, and no mutations in *EVI2* specific to NF1 patients were found. Cloned segments of *EVI2* used as probes onto cosmid libraries of human genomic DNA identified two overlapping cosmids that spanned the region between the two translocation breakpoints, approximately 60 kb. When DNA fragments from these cosmids were used to probe cDNA libraries, two additional genes were identified; like *EVI2* (later designated *EVI2A*), these genes, *EVI2B* and *OMGP*, lay entirely between the two NF1 translocation breakpoints but contained no mutations specific to NF1 patients.

A screen of Southern blots of DNA from approximately 80 unrelated NF1 patients, using probes from the translocation breakpoint region, identified three patients with deletions (Viskochil et al., 1990). The smallest deletion (11 kb) disrupted none of the previously identified genes, but removed a portion of a 3.8-kb EcoRI fragment previously shown to span the 17;22 translocation breakpoint and to contain a region with strong homology to murine genomic DNA, suggesting the presence of a gene. Using this human genomic fragment as probe onto a murine macrophage cDNA library, a single clone was identified. This mouse cDNA was then used to probe a randomly primed human fetal brain cDNA library, and a large set of overlapping positive clones was obtained.

Segments of the new cDNA contig mapped to both sides of the 17;22 translocation breakpoint, and in one NF1 patient sequences lying within the cDNA contig were removed from genomic DNA by an 11-kb deletion (see above); thus, the mutations of these two NF1 patients fell inside the newly identified gene, presumably disrupting its function. Furthermore, in screening a small portion of the open reading frame of the new gene, two additional NF1 patients were identified with single-base-pair changes expected to disrupt the function of the encoded protein (Cawthon et al., 1990). (One point mutation predicted a Leu to Pro change, the other an Arg to STOP truncation of the protein.) Whereas the translocations and even the 11-kb deletion could be considered large-scale changes that might alter the expression or function of more than one gene in the region, these NF1 patient-specific point mutations were likely to affect the functioning of only one gene. On the basis of these combined findings the new gene was identified as the *NF1* gene.

Further evidence bolstering this conclusion was provided by the concomitant description of two additional overlapping cDNA clones that hybridized to the 3.8-kb EcoRI fragment containing the 17;22 translocation breakpoint (Wallace et al., 1990b). In the same report, a sporadic (new mutation) NF1 patient was identified with a 500-bp insertion in this EcoRI fragment that was not present in the DNA of either parent. Subsequently, it became clear that the cDNA clones described by these workers neither spanned the 17;22 translocation breakpoint nor the site of the 500-bp insertion (Wallace et al., 1990a); however, the cDNA sequence was contained within that reported by Cawthon et al. (1990), proving that the same gene had been identified by both groups.

By Northern blot analysis the *NF1* mRNA is 11–13 kb long (Viskochil et al., 1990; Wallace et al., 1990b). The *NF1* gene also spans the 1;17 translocation breakpoint (Xu et al., 1990b). The protein-coding regions are spread over 51 exons and approximately 400 kb of genomic DNA (Y. Li, unpublished). The 5' end (beginning) of the gene is toward the centromere (Cawthon et al., 1990). The three genes identified along the way to finding *NF1* lie together in the order *EVI2A*, *EVI2B*, and *OMGP* within a single intron of *NF1*, and in the opposite orientation to *NF1* along the chromosome.

NEUROFIBROMIN: HOMOLOGY AND SHARED FUNCTION WITH GAP AND IRA PROTEINS

Clues regarding the function of neurofibromin came from running computer-assisted queries of the DNA and protein databases to look for amino acid sequences similar to that predicted by the open reading frame (ORF) of *NF1*. Even before the complete *NF1* ORF had been sequenced, these searches identified a domain of approximately 360 amino acids with significant homology to the catalytic domains of the mammalian GAP (ras-*G*TPase *A*ctivating *P*rotein) and yeast IRA1 and IRA2 proteins (*I*nhibitors of the *R*as–cyclic *A*MP activation pathway, *1* and *2*) (Xu et al., 1990b). As shown in Figure 1, significant amino acid sequence homology between the NF1 protein and GAP is restricted to this catalytic domain; however, homologies between the NF1 protein and the IRA1 and IRA2 proteins occur not only in the catalytic domain but also in additional peptide domains both upstream and downstream of this central region.

To test the putative catalytic domain of neurofibromin for functional activity, proteins consisting of the GAP-related domain of neurofibromin (NF1-GRD proteins) were synthesized and assayed for their ability to bind to and stimulate the GTPase of mammalian p21 ras, and for their ability to complement the heat-shock sensitivity and glycogen storage deficiency of yeast ira$^-$ mutants (Ballester et al., 1990; Martin

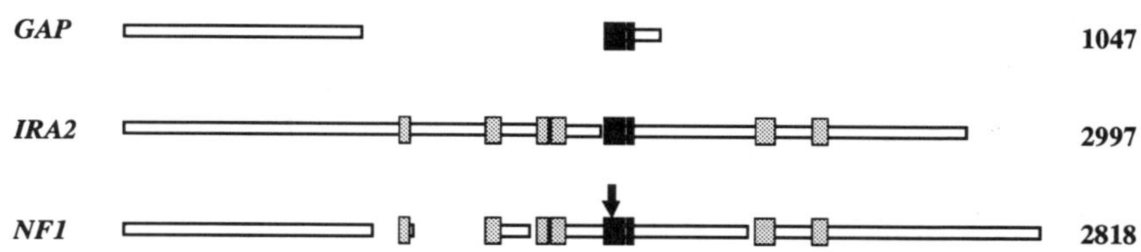

Fig. 1. Regions of similarity in amino acid sequence among neuro-fibromin-related proteins. The full length of each protein, represented by a series of open and filled blocks, is totaled at the right. Blocks filled in black are similar in all three proteins. Stippled blocks are similar only in neurofibromin and the IRA proteins. The arrow shows the site in the catalytic domain of neurofibromin where an additional 21 amino acids may appear due to alternative splicing of the *NF1* transcript (see text). The MACAW program (Schuler et al., 1991) was used to identify and align regions of similar amino acid sequence.

et al., 1990; Xu et al., 1990a). The normal NF1-GRD was active by all of these tests. These results were of profound importance, because they suggested that neurofibromin may be a component of ras-GTP signal transduction pathways, pathways already known to be of major importance in the regulation of growth and differentiation (Hall, 1990). Neurofibromin could be viewed as a down-regulator of ras proteins, accelerating the conversion of active ras^{GTP} to inactive ras^{GDP}. Alternatively, since mutant ras proteins with defective effector domains are impaired in their interactions with the NF1-GRD protein (Bollag and McCormick, 1991; Martin et al., 1990; Xu et al., 1990a), it is possible that neurofibromin normally transduces growth-regulating signals coming from ras.

Within this framework two mechanisms to explain how partially or completely inactivated neurofibromin might cause tumors were considered (Xu et al., 1990b). In cell types where ras^{GTP} is growth-promoting, the diminished capacity of mutant neurofibromins to stimulate GTP hydrolysis on ras may allow a rise in ras^{GTP} levels, causing growth. In cell types where ras^{GTP} appears to be growth-inhibiting, e.g., Schwann cells (Ridley et al., 1988) and pheochromocytomas (Bar-Sagi and Feramisco, 1985; Noda et al., 1985), mutant neurofibromin may disrupt a ras^{GTP}–neurofibromin complex that is required for transduction of the growth-inhibiting signal. The balance of growth-regulating signals would then shift toward growth. The latter mechanism is the one more likely to be involved in neurofibromatosis 1, since Schwann cells appear to proliferate excessively in the development of neurofibromas (Riccardi and Eichner, 1986) and pheochromocytomas have an increased incidence among NF1 patients (Cohen and Rothner, 1989; Riccardi and Eichner, 1986). Furthermore, a careful screen for activating mutations in *ras* genes in human pheochromocytomas revealed none (Moley et al., 1991), the expected result if ras^{GTP} inhibits growth in this cell type.

The full length of the *NF1* open reading frame (8,454 bp, encoding 2,818 amino acids) has been determined by cDNA walking and sequencing and the 5' end of the mRNA identified by primer extension (Marchuk et al., 1991; G. Xu, unpublished). Approximately 2 kb of 3' untranslated mRNA sequences also have been cloned and sequenced (G. Xu, unpublished); however, the 3' end of the *NF1* message has not been obtained at this writing. The predicted protein contains several potential cAMP-dependent protein kinase recognition sites and one potential tyrosine kinase phosphorylation site (Marchuk et al., 1991; Xu et al., 1990b). It is a hydrophilic protein with no apparent membrane-spanning domains.

REGULATION OF NEUROFIBROMIN–RAS INTERACTION

How does neurofibromin affect the growth and differentiation of different cell types at different stages of development? How are the activities of neurofibromin itself regulated? The answers to these questions are likely to be complex.

The NF1-GRD has a fivefold higher affinity for H-ras than for N-ras (Bollag and McCormick, 1991). Perhaps only one of the p21 ras proteins (H-, N-, or K-) normally interacts with neurofibromin; however, it is possible that different ras proteins interact with neurofibromin in different cell types, or even in the same cell, depending on the stimuli the cell is receiving from its environment. Another possibility is that none of the three ras proteins is a physiologic partner of neurofibromin: Instead, another member of the large family of ras-like proteins with intrinsic GTPases (Bourne et al., 1991) may be neurofibromin's true partner. Recently, a full-length (2,818 amino acids long) *NF1* cDNA was constructed (G. Xu, unpublished), permitting a full-length neurofibromin protein to be made and tested for its ability to interact with p21 ras proteins. In both its binding affinity for p21 ras and its ras-GTPase stimulating activity, the behavior of full-length neurofibromin was equivalent to that of the smaller NF1-GRD protein (G. Bollag, personal communication).

Neurofibromin's interaction with ras (or some other ras-like protein) may also be controlled in part by certain lipids, since it has been shown that phosphatidic acid and arachidonic acid can inhibit NF1-GRD's stimulation of the ras-GTPase (Bollag and McCormick, 1991; Golubic et al., 1991; Han et al., 1991).

Another avenue for control of neurofibromin activity is the inclusion or exclusion (by alternative RNA splicing) of a segment of 21 amino acids in the middle of the NF1-GRD (Nishi et al., 1991). No corresponding alternatively spliced segment has been found in other members of the GAP family of proteins (Fig. 1). Inclusion of this peptide interrupts a relatively hydrophobic segment of the NF1-GRD with a very hydrophilic segment. Compared to type I NF1-GRD (without the insert), the type II NF1-GRD (with the insert) has a twofold higher binding affinity for H-ras and a diminished ability to stimulate the H-ras GTPase; however, type I and type II NF1-GRD do not significantly differ in their interactions with N-ras (D. Viskochil, unpublished). If one of the functions of neurofibromin is to serve as an effector for p21 H-ras, then perhaps signaling through the H-ras–type II neurofibromin complex would be more frequent or of longer duration than signaling through the H-ras–type I neurofibromin complex, due to a diminished rate of signal termination by GTP hydrolysis on ras.

The potential phosphorylation sites on neurofibromin (see above) may provide additional points for control of neurofibromin–ras interactions.

To date, the focus of functional studies of neurofibromin, understandably, has been the interaction with p21 ras proteins. Neurofibromin is a very large protein, and the potential for additional proteins and other cell components to bind and regulate the interaction of neurofibromin with ras should not be overlooked. Furthermore, it is possible that other domains of neurofibromin mediate additional important functions independent of interactions with ras (or ras-like) proteins.

EXPRESSION IN TISSUES: mRNA, PROTEIN

The 11–13-kb *NF1* mRNA has been detected by Northern blot analysis of human (Viskochil et al., 1990; Wallace et al., 1990b) and mouse (Buchberg et al., 1990) tissues. *NF1* mRNA has also been detected by S1 nuclease analysis (Suzuki et al., 1991) and by RNA-PCR (Nishi et al., 1991; Suzuki et al., 1991; Wallace et al., 1990b), starting with total RNA from human tissues. (In the RNA-polymerase chain reaction [PCR] studies, RNA was primed with random hexamers or with a specific *NF1*-cDNA-based oligonucleotide; first-strand cDNA synthesis was then carried out using reverse transcriptase; finally, PCR was performed using two *NF1*-specific primers.) By RNA-blot analysis, the level of *NF1* expression varies widely in different tissues (Buchberg et al., 1990; Wallace et al., 1990b); by the sensitive RNA-PCR method, *NF1* message appears to be ubiquitously expressed (Wallace et al., 1990b; G. Xu, unpublished).

Type I and type II *NF1* mRNA (see above) were discovered by RNA-PCR of the NF1-GRD (Nishi et al., 1991) from various tissue sources. These authors provided three lines of evidence to support the idea that in the human nervous system the inclusion of this 63-bp exon by alternative splicing is associated with the progression of cells toward a more differentiated state. First, prior to 20 weeks' gestation, type I *NF1* mRNA (exon removed) predominates in human brain, whereas after 22 weeks' gestation type II *NF1* mRNA (exon included) predominates. Second, in relatively undifferentiated primitive neuroectodermal tumors (neuroblastoma, retinoblastoma, and medulloblastoma), type I *NF1* mRNA predominates, whereas in more differentiated tumors (fibrosarcoma, astrocytoma) type II *NF1* mRNA predominates. Third, treating a neuroblastoma cell line with retinoic acid to induce neuronal differentiation shifts the *NF1* mRNA pools from predominantly type I to predominantly type II with a time course that precedes, then parallels, the time course of cellular morphological change.

However, other authors (Suzuki et al., 1991), also studying the ratio of type I:type II *NF1* mRNA in human tissues, in contrast to the report of Nishi et al. (1991), have found that adult brain expresses predominantly the type I form of *NF1* mRNA. Suzuki et al. (1991) also found predominant expression of type II *NF1* mRNA in glioblastomas, astrocytomas, and meningiomas, with no clear correlation between the ratio of type I:type II and the grade of malignancy of the tumors. In normal tissues outside the nervous system (placenta, kidney, and lung) approximately equal expression of types I and II *NF1* mRNAs was observed. The role played by the alternate-splice forms in the cellular biology of *NF1* remains uncertain; however, the rapid change in the type I:type II ratio (within 30 min) in cultured cells in response to retinoic acid (Nishi et al., 1991) suggests that this ratio is tightly controlled in cells.

Antibodies to peptides and larger protein domains predicted by the open reading frame of the human *NF1* gene have been used to detect neurofibromin in mouse, rat, and human tissues. By Western blot analysis (with or without a preceding immunoprecipitation step), specific immuno-staining of a protein in the 220–280 kd size range has been achieved by three research groups (Daston et al., 1992; DeClue et al., 1991; Gutmann et al., 1991). The predicted size of neurofibromin based on the 2,818 amino acids predicted by the full-length cDNA sequence (Marchuk et al., 1991) is 327 kd. The discrepancy between this predicted size and that observed on the Western analysis may be due to posttranslational addition of negatively charged residues or the assumption of a nonglobular conformation during protein-folding.

In human cell lines, neurofibromin exists in part as monomer and in part associated with a 400–500 kd protein (DeClue et al., 1991). To locate neurofibromin intracellularly, NIH3T3 cells were grown in ^{35}S-labeled amino acids, then gently lysed and subjected to differential centrifugation (DeClue et al., 1991). The resulting subcellular fractions were solubilized, and neurofibromin was immunoprecipitated and analyzed by polyacrylamide gel electrophoresis. By this immunoassay, neurofibromin was detected in the 100,000*g* nonnuclear particulate fraction, indicating that it is associated with a cytoplasmic structure. Immunocytochemical staining has also located neurofibromin predominantly in the cytoplasm of cells, but occasionally some specific staining of the nucleus also has been seen (Daston et al., 1992).

To assay for neurofibromin expression in adult mouse, Gutmann et al. (1991) subjected tissue homogenates to standard SDS-PAGE, followed by electrophoretic transfer to polyvinylidene difluoride membranes and detection of neurofibromin with neurofibromin-fusion protein antiserum or peptide antiserum raised against a different domain of neurofibromin. By this assay neurofibromin was detected in roughly equivalent amounts in all tissues tested, including colon, muscle, lung, kidney, spleen, Schwann cells, fibroblasts, brain, and liver (Gutmann et al., 1991). A very different, more restricted pattern of neurofibromin expression was observed in adult rat, however (Daston et al., 1992). These workers first immunoprecipitated neurofibromin from solubilized tissue homogenates. They then subjected the immunoprecipitates to SDS-PAGE, followed by transfer to nitrocellulose and detection of neurofibromin with a second antiserum, raised against an immunologically distinct domain of the protein. By this assay the tissues with the highest concentrations of neurofibromin were brain, spinal cord, adrenal gland, and peripheral nerve; lower levels were found in liver, spleen, and pancreas; and barely detectable quantities were present in cardiac tissue. Neurofibromin was not detectable in skeletal muscle, lung, kidney, and skin.

Daston et al. (1992) have also studied the distribution of neurofibromin in rat and human nervous system by immunocytochemical staining. In the peripheral nervous system, the cells with the highest levels of neurofibromin by immunostaining were nonmyelinating Schwann cells and unmyelinated neurons. Low but detectable neurofibromin was present in myelinated neurons. In the rat and human spinal cord, neurofibromin was detected in both motor and sensory neurons and in oligodendrocytes, but not in astrocytes. In rat brain, expression was limited to neurons and oligodendrocytes.

NF1 MUTATIONS IN TUMORIGENESIS

Germline Mutations

NF1 mutations detected in the constitutional DNA of NF1 patients include translocations, deletions, and stop codons, all occuring within protein-coding regions of the gene (Cawthon et al., 1990; Estivill et al., 1991; Stark et al., 1991; Viskochil et al., 1990). In one sporadic (new mutation) NF1 case, a *de novo* Alu insertion in an *NF1* intron causes an exon to be skipped during RNA splicing, resulting in a frameshift and a premature stop (Wallace et al., 1991). In another family a Leu to Pro amino acid substitution in *NF1* co-inherits with NF1 disease (Cawthon et al., 1990). All of these mutations would be expected to inactivate neurofibromin, because they should truncate the protein and/or drastically change its overall shape. However, since these mutations lie 3' of the NF1-GRD, it is formally possible that stable, mutant neurofibromins with intact, functioning GRDs are made from these *NF1* alleles. At least two NF1 patients have been described, however, with deletions that remove the entire chromosomal region containing the NF1-GRD exons (K. Stephens, personal communication; D. Viskochil, unpublished). Furthermore, an NF1 patient with a CGA to TGA stop mutation early in the GRD and another NF1 patient with a single-base deletion, frameshift, and stop in the middle of the GRD have been identified recently (R. Cawthon, unpublished); the mutant neurofibromins predicted in these cases, even if they are synthesized and are stable, would not be expected to bind ras^{GTP} and stimulate the ras GTPase.

Taken together, these results constitute convincing evidence that the germline *NF1* mutations that lead to neurofibromatosis 1 are inactivating mutations. This conclusion, in turn, suggests that *NF1* is a tumor suppressor gene, but it is also consistent with the possibility that during tumor development in NF1 patients, the remaining normal *NF1* allele actually undergoes a gain-of-function mutation, creating a mutant neurofibromin that is actively growth-promoting. If *NF1* is truly a tumor suppressor gene, then during the development of at least some of the tumors of NF1 patients the remaining normal *NF1* allele would be expected to undergo inactivation. One mechanism of inactivation relatively easy to detect is allele loss.

Second Hits: Somatic Mutation of *NF1* in the Tumors of NF1 Patients

Therefore, investigators have examined both benign and malignant tumors from NF1 patients for evidence of allele loss on chromosome 17 in the region of the *NF1* gene. Extensive analyses of neurofibromas, the common benign tumors of NF1 patients, have never detected a loss of heterozygosity (LOH) (Skuse et al., 1989, 1991) anywhere along chromosome 17. Analyses of some malignant tumors from NF1 patients have shown LOH for DNA markers on 17p or along the entire chromosome 17 (Glover et al., 1991; Menon et al., 1990; Skuse et al., 1989), perhaps because the loss of a normal P53 allele on 17p contributes to tumor progression. However, other malignant tumors from NF1 patients have shown LOH specific to 17q. Skuse et al. (1989; Skuse, 1990) observed LOH specific to 17q in one NF1 astrocytoma and several NF1 neurofibrosarcomas; furthermore, the alleles retained in the tumor cells were those inherited from the NF1-affected parent. Also, Ponder et al.

(1990) found LOH specific to 17q in some pheochromocytomas from NF1 patients. These results strongly suggest that inactivation of both *NF1* alleles contributes to the development of several NF1-associated malignancies. Thus, *NF1* appears to behave as a true tumor suppressor gene in these tissues. The failure to find LOH in neurofibromas suggests that events other than inactivation of the second *NF1* allele are involved in the development of these benign tumors. However, inactivations of the second *NF1* allele by mechanisms more subtle than allele loss (e.g., point mutations) have not yet been ruled out.

Loss of alleles in the vicinity of *NF1* or encompassing the *NF1* locus leaves open the possibility that some other locus in the region, and not *NF1* itself, is the relevant contributor to tumor progression in these NF1 malignancies. The strongest evidence directly implicating *NF1* in this progression comes from studies of NF1 neurofibrosarcomas (Basu et al., 1992; DeClue et al., 1992). Cell lines from these malignant tumors have drastically diminished or undetectable levels of neurofibromin by immunoassay, and ras^{GTP} is elevated and growth-promoting in the cells. These results strongly suggest that a nearly complete loss of neurofibromin function contributed to the development of the tumors. Thus, *NF1* apparently acts as a tumor suppressor gene in this cell type, and most likely in all tissues affected by neurofibromatosis 1.

NF1 Mutations in Sporadic Cases of NF1-Associated Tumors

In the general population, sporadic cases of the tumor types that have an increased incidence among NF1 patients would be expected to occasionally harbor somatic mutations in the *NF1* gene. Astrocytoma, one of the most common malignant brain tumors, appears to have an increased incidence among NF1 patients (Blatt et al., 1986; Cohen and Rothner, 1989; Kibirige et al., 1989; Sørensen et al., 1986). Recently, paired samples of tumor DNA and accompanying constitutional DNA from the same individuals were screened for sequence variants in an *NF1* exon in the middle of the GRD (Li et al., 1992). Among 10 anaplastic astrocytomas from non-NF1 individuals, one somatic mutation was detected, a point mutation changing $codon_{1423}$ from AAG to CAG (Lys-to-Gln). This Lys residue is normally invariant among GAP-related proteins (Wang et al., 1991; Xu et al., 1990b) and likely, therefore, to be crucial for normal functioning. Indeed, the mutant NF1-GRD protein binds with normal affinity to ras^{GTP} but is severely impaired in stimulating the ras GTPase (Li et al., 1992). In the tumor sample the amount of mutant allele was approximately equal to the amount of normal allele, but the extent of admixture of normal tissue with tumor tissue in the tumor sample was uncertain. Therefore, it is possible that the tumor cells carrying the Gln_{1423} mutant allele have lost the normal *NF1* allele. It is also possible that the tumor cells have retained a second *NF1* allele that is normal in the exon screened but mutant elsewhere in the gene. This somatic mutation was detected in the first *NF1* exon screened, and 50 additional protein-encoding *NF1* exons remain to be scanned. Therefore, it is too early to accurately assess the frequency of *NF1* somatic mutations in sporadic astrocytomas, but the true incidence may be quite high.

NF1 Mutations in Tumor Types Associated with *ras* Activation

NF1 appears to be expressed in many tissues (Buchberg et al., 1990; Daston et al., 1992; Gutmann et al., 1991; Wallace et al., 1990b). Neurofibromin may be an important regulator of ras^{GTP} levels in many cell types. Somatic mutations that activate *ras* genes may contribute to tumorigenesis in up to 30% of human tumors (Bos, 1989). If somatic *NF1* inactivation can raise ras^{GTP} levels, then *NF1* mutations should promote growth in cell types where activated *ras* genes are known to promote growth. Screening was undertaken for somatic mutations in *NF1* in some of the tumor types where activated *ras* genes are frequently found. Colon cancers frequently contain activated K-*ras*, and myelodysplasias frequently contain activated N- or K-*ras* genes (Bos, 1989). Screening the same exon that was mutated in the astrocytoma sample described above led to detection of a somatic mutation in one of 22 colon adenocarcinomas; the change was also in $codon_{1423}$; in this case the normal AAG had become GAG (Lys-to-Glu). The same mutation was found in one of 28 myelodysplasias. The mutant Lys-to-Glu NF1-GRD, like the mutant Lys-to-Gln NF1-GRD, binds ras^{GTP} with normal affinity but shows greatly reduced activity toward the ras GTPase (Li et al., 1992).

Since the prevalence of colon cancers and myelodysplasias does not seem to be increased among NF1 patients as compared with the general population, it seemed possible that the mutant *NF1* allele encoding the Glu_{1423} neurofibromin, while capable of contributing to the development of certain sporadic tumors, might not be a cause of neurofibromatosis 1. However, the identical Lys-to-Glu mutation was found in one NF1 patient during a screen of 80 unrelated NF1 individuals. The mutation was found to co-inherit with NF1 disease in the patient's family. No history of myelodysplasia, leukemia, or colon cancer was present in any of the five NF1-affected family members for whom clinical data were available.

There is a precedent for this apparent paradox among the hereditary neoplastic syndromes. Somatic mutations in the retinoblastoma gene are found frequently in sporadic small-cell lung cancers, yet this tumor type does not show an increased incidence among individuals with hereditary retinoblastoma (Harbour et al., 1988; Yokota et al., 1988).

Thus, somatic mutations in the *Rb* gene in lung cells may contribute to—and may even be necessary for—the development of the tumors; however, the mutations are not rate-limiting for tumorigenesis in this cell type.

Mechanisms by Which *NF1* Mutations May Contribute to Tumor Development

These considerations suggest that *NF1* mutations might promote growth in NF1-associated tumor types by one mechanism and promote growth in other tumor types by a somewhat different mechanism. The great importance of neurofibromin in suppressing growth in the cell types affected by neurofibromatosis 1 may be based on the full utilization of both ras-regulator and ras-effector functions in these cell types. Alternatively, signal transduction through neurofibromin may be the only pathway by which these cell types can achieve their normal differentiated fate. Figure 2a shows a model for neurofibromin action in tissues affected by neurofibromatosis 1. The two downward-pointing arrows coming from the ras^{GTP}-neurofibromin complex represent the ras-regulator and ras-effector activities of neurofibromin. On the right, a growth-promoting pathway is shown, coexisting in the same cell. Normally in these tissues the left-hand ras-signaling pathway is dominant; ras^{GTP}'s net effect is growth-inhibiting. Figure 2b shows the effects of mutations inactivating *NF1*. Both ras-regulator and ras-effector activities are eliminated. ras^{GTP} is then free to promote growth via another signaling pathway, the existence of which is suggested by the results of Basu et al. (1992) and DeClue et al. (1992).

The less crucial, though still important, role of neurofibromin in suppressing growth in cell types not affected by neurofibromatosis 1 may be based on the primary utilization of only its ras-regulator function in these cell types. Thus, the model of Figure 2 would still apply, but the arrow going toward differentiation would not be involved (or would be much reduced) in these cell types. Normally in these tissues ras^{GTP} is growth-promoting. Here, inactivation of neurofibromin diminishes ras^{GTP} hydrolysis, leading to increased ras^{GTP} signaling via the growth-promoting pathway. Furthermore, ras regulation by neurofibromin may be only one of several means by which these cells control ras^{GTP} levels.

The Glu_{1423} *NF1* mutation would be expected to disrupt both of the proposed functions of neurofibromin (ras-regulator and ras-effector) in order to contribute to the development of colon tumors and myelodysplasias and to cause neurofibromatosis 1; i.e., the mutant neurofibromin should be functionally null. In addition to its diminished ability to stimulate the ras GTPase, it should be blocked in signal transduction. Alternatively, the levels of the mutant neurofibromin in the cell could be very low, if the mutation makes the protein unstable. As a functional null, the Glu_{1423}

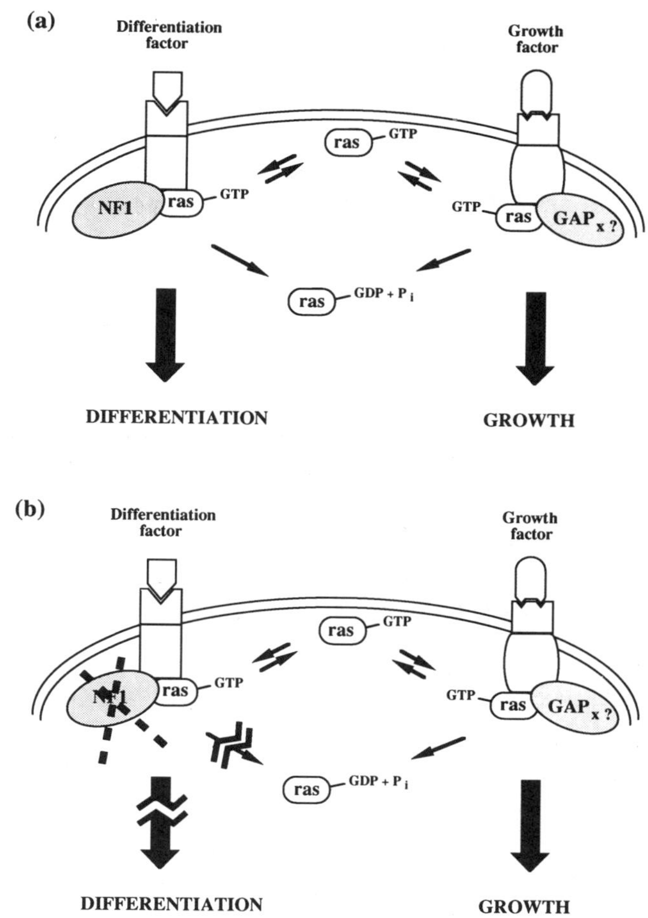

Fig. 2. A model for neurofibromin action in the tissues affected by neurofibromatosis 1. **a:** Normal cell. **b:** Cell with mutation inactivating neurofibromin. See text.

mutant allele would simply be one additional member of the growing class of null alleles identified in NF1 patients.

However, overlapping but distinct classes of mutant *NF1* alleles may yet prove to be involved in the development of different tumor types. For example, mutant neurofibromins, blocked in signal transduction but able to down-regulate ras^{GTP}, may cause neurofibromatosis 1 but be unable to contribute to tumor development in tissues where ras^{GTP} is growth-promoting. Other mutants, competent in signal transduction but blocked in the down-regulation of ras^{GTP}, may not cause neurofibromatosis 1 but might contribute to tumor development in tissues where ras^{GTP} promotes growth. In Figure 2b, such mutants could be viewed as blocking just one or the other, but not both, of the two downward-pointing arrows coming from the ras^{GTP}-neurofibromin complex.

CONCLUSION

How broad a role do somatic mutations at the *NF1* locus play in tumorigenesis? Three of 60 tumors have revealed functionally important missense mutations affecting a single normally invariant residue in a highly conserved NF1-GRD exon. However, less than 2% of the *NF1* open reading frame has been screened with methods sensitive to point mutations. If intron-based mutations affecting RNA splicing are also a significant source of tumorigenic mutations, then the sequences screened so far constitute only about 0.05% of the relevant target region (i.e., the entire *NF1* gene). Therefore, the true incidence of somatic mutations in the *NF1* gene in colon adenocarcinomas, myelodysplasias, and anaplastic astrocytomas may be quite high.

In light of these mutations, a possible role for *NF1* mutations should be considered in any tumor type in which activated *ras* genes are frequently found. Perhaps in these tumor types *NF1* mutations are the functional complement of *ras* mutations and will be found in those tumor samples lacking *ras* mutations. Since in neurofibromatosis 1 tumor incidence is elevated in at least some tissues where ras^{GTP} is growth-inhibiting, a role for *NF1* mutations must also be considered in any tumor type in which activated *ras* has been shown to inhibit growth (e.g., medullary thyroid carcinoma; Nakagawa et al., 1987). Perhaps in these tumor types *NF1* mutations will be found in those tumor samples composed of cells that continue to proliferate when activated *ras* is introduced. While many kinds of tumors have been screened for mutations activating *ras* genes (Bos, 1989), the effect on growth of introducing activated *ras* into tumor cell lines has been tested only rarely. Therefore, any tumor types in which activating *ras* mutations are strikingly absent would be good candidates for such an assay. Nervous system neoplasms in this category include neuroblastomas (Moley et al., 1991) and glioblastomas (Bos, 1989). The hopes raised by identification of the gene are being realized as characterization of *NF1* mutations continues to offer new insights into fundamental biological processes.

REFERENCES

Ballester, R., D. Marchuk, M. Boguski, et al. (1990) The NF1 locus encodes a protein functionally related to mammalian GAP and yeast IRA proteins. Cell 63:851–859.

Bar-Sagi, D., and J.R. Feramisco (1985) Microinjection of the *ras* oncogene protein into PC12 cells induces morphological differentiation. Cell 42:841–848.

Basu, T.N., D.H. Gutmann, J.A. Fletcher, et al. (1992) Aberrant regulation of ras proteins in malignant tumour cells from type 1 neurofibromatosis patients. Nature 356:713-715.

Bataini, J.P., S. Delanian, and D. Ponvert (1991) Chiasmal gliomas: Results of irradiation management in 57 patients and review of literature. Int. J. Radiat. Oncol. Biol. Phys. 21:615–623.

Blatt, J., R. Jaffe, M. Deutsch, et al. (1986) Neurofibromatosis and childhood tumors. Cancer 57:1225–1229.

Bollag, G., and F. McCormick (1991) Differential regulation of rasGAP and neurofibromatosis gene product activities. Nature 351:576–579.

Bos, J.L. (1989) *ras* oncogenes in human cancer: A review. Cancer Res. 49:4682–4689. [Published erratum appears in Cancer Res. (1990) 50(4):1352.]

Bourne, H.R., D.A. Sanders, and F. McCormick (1991) The GTPase superfamily: Conserved structure and molecular mechanism. Nature 349:117–127.

Buchberg, A.M., L.S. Cleveland, N.A. Jenkins, et al. (1990) Sequence homology shared by neurofibromatosis type-1 gene and IRA-1 and IRA-2 negative regulators of the RAS cyclic AMP pathway. Nature 347:291–294.

Cawthon, R.M., R. Weiss, G. Xu, et al. (1990) A major segment of the neurofibromatosis type 1 gene: cDNA sequence, genomic structure, and point mutations. Cell 62:193–201.

Cohen, B.H., and A.D. Rothner (1989) Incidence, types, and management of cancer in patients with neurofibromatosis. Oncology (Williston Park) 3:23–30.

Crowe, F., W. Schull, and J. Neel (1956) A Clinical, Pathological, and Genetic Study of Multiple Neurofibromatosis. Springfield, Ill.: Charles C. Thomas.

Daston, M.M., H. Scrable, M. Nordlund, et al. (1992) The protein product of the neurofibromatosis type 1 gene is expressed at highest abundance in neurons, Schwann cells, and oligodendrocytes. Neuron (in press).

DeClue, J.E., B.D. Cohen, and D.R. Lowy (1991) Identification and characterization of the neurofibromatosis type 1 protein product. Proc. Natl. Acad. Sci. USA 88:9914–9918.

DeClue, J.E., A.G. Papageorge, J.A. Fletcher, et al. (1992) Abnormal regulation of mammalian $p21^{ras}$ contributes to malignant tumor growth in von Recklinghausen (type 1) neurofibromatosis. Cell 69:265-273.

Estivill, X., C. Lazaro, T. Casals, et al. (1991) Recurrence of a nonsense mutation in the NF1 gene causing classical neurofibromatosis type 1. Hum. Genet. 88:185–188.

Glover, T.W., C.K. Stein, E. Legius, et al. (1991) Molecular and cytogenetic analysis of tumors in von Recklinghausen neurofibromatosis. Genes Chromosomes Cancer 3:62–70.

Goldgar, D.E., P. Green, D.M. Parry, et al. (1989) Multipoint linkage analysis in neurofibromatosis type I: An international collaboration. Am. J. Hum. Genet. 44:6–12.

Golubic, M., K. Tanaka, M.H. Tsai, et al. (1991) The GTPase stimulatory activities of the neurofibromatosis type 1 and the yeast IRA2 proteins are inhibited by arachidonic acid. EMBO J. 10:2897–2903.

Gutmann, D.H., D.L. Wood, and F.S. Collins (1991) Identification of the neurofibromatosis type 1 gene product. Proc. Natl. Acad. Sci. USA 88:9658–9662.

Hall, A. (1990) The cellular functions of small GTP-binding proteins. Science 249:635–640.

Han, J.-W., F. McCormick, and I.G. Macara (1991) Regulation of ras-GAP and the neurofibromatosis-1 gene product by eicosanoids. Science 252:576–579.

Harbour, J.W., S.L. Lai, J. Whang-Peng, et al. (1988) Abnormalities in structure and expression of the human retinoblastoma gene in SCLC. Science 241:353–357.

Ishida, O., and K. Jimbow (1987) A computed image analyzing system for quantitation of melanocyte morphology in café-au-lait macules of neurofibromatosis. J. Invest. Dermatol. 88:287–291.

Kibirige, M.S., J.M. Birch, R.H. Campbell, et al. (1989) A review of astrocytoma in childhood. Pediatr. Hematol. Oncol. 6:319–329.

Lefkowitz, I., A. Obringer, and A. Meadows (1990) Neurofibromatosis and cancer: Incidence and management. In: A. Rubenstein and B. Korf (eds.). Neurofibromatosis: A Handbook for Patients, Families and Health-Care Professionals. New York: Theime Medical Publishers, pp. 99–110.

Li, Y., G. Bollag, R. Clark, et al. (1992) Somatic mutations in the neurofibromatosis 1 gene in human tumors. Cell 69:275-281.

Marchuk, D.A., A.M. Saulino, R. Tavakkol, et al. (1991) cDNA cloning of the Type 1 Neurofibromatosis gene: Complete sequence of the *NF1* gene product. Genomics 11:931–940.

Martin, G.A., D. Viskochil, G. Bollag, et al. (1990) The GAP-related domain of the neurofibromatosis type 1 gene product interacts with ras p21. Cell 63:843–849.

Menon, A.G., K.M. Anderson, V.M. Riccardi, et al. (1990) Chromosome 17p deletions and p53 gene mutations associated with the formation of malignant neurofibrosarcomas in von Recklinghausen neurofibromatosis. Proc. Natl. Acad. Sci. USA 87:5435–5439.

Moley, J.F., M.B. Brother, S.A. Wells, et al. (1991) Low frequency of *ras* gene mutations in neuroblastomas, pheochromocytomas, and medullary thyroid cancers. Cancer Res. 51:1596–1599.

Nakagawa, T., M. Mabry, A. de Bustros, et al. (1987) Introduction of v-Ha-ras oncogene induces differentiation of cultured human medullary thyroid carcinoma cells. Proc. Natl. Acad. Sci. USA 84:5923–5927.

Nishi, T., P.S.Y. Lee, K. Oka, et al. (1991) Differential expression of two types of the neurofibromatosis type 1 (*NF1*) gene transcripts related to neuronal differentiation. Oncogene 6:1555–1559.

Noda, M., M. Ko, A. Ogura, et al. (1985) Sarcoma viruses carrying ras oncogenes induce differentiation-associated properties in a neuronal cell line. Nature 318:73–75.

Perry, H.D., and R.L. Font (1982) Iris nodules in von Recklinghausen's Neurofibromatosis. Electron microscopic confirmation of their melanocytic origin. Arch. Ophthalmol. 100:1635–1640.

Ponder, B., W. Xu, M. Ponder, et al. (1990) Allele losses in tumours from patients with neurofibromatosis type 1 (NF1). Am. J. Hum. Genet. 47:A14.

Riccardi, V.M., and J.E. Eichner (1986) Neurofibromatosis: Phenotype, Natural History, and Pathogenesis. Baltimore: Johns Hopkins University Press.

Ridley, A.J., H.F. Paterson, M. Noble, et al. (1988) *ras*-mediated cell cycle arrest is altered by nuclear oncogenes to induce Schwann cell transformation. EMBO J. 7:1635–1645.

Schuler, G., S. Altschul, and D. Lipman (1991) A workbench for multiple alignment construction and analysis. Prot. Struct. Funct. Genet. 9:180–190.

Skuse, G.R. (1990) Technical comments (letter). Science 250:1749.

Skuse, G., B.A. Kosciolek, and P.T. Rowley (1989) Molecular genetic analysis of tumors in von Recklinghausen neurofibromatosis: Loss of heterozygosity for chromosome 17. Genes Chromosomes Cancer 1:36–41.

Skuse, G.R., B.A. Kosciolek, and P.T. Rowley (1991) The neurofibroma in von Recklinghausen neurofibromatosis has a unicellular origin. Am. J. Hum. Genet. 49:600–607.

Slater, C., M. Hayes, N. Saxe, et al. (1986) Macromelanosomes in the early diagnosis of neurofibromatosis. Am. J. Dermatopathol. 8:284–289.

Sorensen, S.A., J.J. Mulvihill, and A. Nielsen (1986) Long-term follow-up of von Recklinghausen neurofibromatosis. Survival and malignant neoplasms. N. Engl. J. Med. 314:1010–1015.

Stark, M., G. Assum, and W. Krone (1991) A small deletion and an adjacent base exchange in a potential stem-loop region of the neurofibromatosis 1 gene. Hum. Genet. 87:685–687.

Suzuki, Y., H. Suzuki, T. Kayama, et al. (1991) Brain tumors predominantly express the neurofibromatosis type 1 gene transcripts containing the 63 base insert in the region coding for GTPase activating protein-related domain. Biochem. Biophys. Res. Commun. 181:955–961.

Viskochil, D., A.M. Buchberg, G. Xu, et al. (1990) Deletions and a translocation interrupt a cloned gene at the neurofibromatosis type 1 locus. Cell 62:187–192.

Wallace, M.R., L.B. Andersen, A.M. Saulino, et al. (1991) A de novo Alu insertion results in neurofibromatosis type 1. Nature 353:864–866.

Wallace, M.R., D.A. Marchuk, L.B. Andersen, et al. (1990a) Type 1 neurofibromatosis gene: Correction. Science 250:1749.

Wallace, M.R., D.A. Marchuk, L.B. Andersen, et al. (1990b) Type 1 neurofibromatosis gene: Identification of a large transcript disrupted in three NF1 patients. Science 249:181–186.

Wang, Y., M. Boguski, M. Riggs, et al. (1991) Sar1, a gene from Schizosaccharomyces pombe encoding a GAP-like protein that regulates ras1. Cell Reg. 2:453–465.

White, R. (1991) Identification of the neurofibromatosis gene. In: J. Brugge, T. Curran, E. Harlow, et al. (eds.). Origins of Human Cancer: A Comprehensive Review. Cold Spring Harbor, N.Y.: Cold Spring Harbor Laboratory Press, pp. 623–632.

Williamson, T.H., A. Garner, and A.T. Moore (1991) Structure of Lisch nodules in neurofibromatosis type 1. Ophthalmic Paediatr. Genet. 12:11–17.

Xu, G., B. Lin, K. Tanaka, et al. (1990a) The catalytic domain of the neurofibromatosis type 1 gene product stimulates ras GTPase and complements ira mutants of S. cerevisiae. Cell 63:835–841.

Xu, G., P. O'Connell, D. Viskochil, et al. (1990b) The neurofibromatosis type 1 gene encodes a protein related to GAP. Cell 62:599–608.

Yokota, J., T. Akiyama, Y.K. Fung, et al. (1988) Altered expression of the retinoblastoma (RB) gene in small-cell carcinoma of the lung. Oncogene 3:471–475.

Chapter 30

Tumors of the Lymphoreticular System: CNS Lymphoma and AIDS

Lee Ratner

INTRODUCTION

Within the first few years of the recognition of the acquired immunodeficiency syndrome (AIDS) epidemic, the frequent development of lymphomas was noted as an important feature of the disease (Ziegler et al., 1985). Approximately 3% of those diagnosed with AIDS had non-Hodgkin's lymphoma as their initial manifestation of the disease. However, lymphomas continued to develop throughout the course of AIDS, at an incidence rate 60-fold higher than that of the uninfected patient population (Beral et al., 1991). This has had profound impact on the epidemiology of lymphomas, such that AIDS-related non-Hodgkin's lymphomas will constitute between 8 and 27% of the 36,000 cases expected to be diagnosed in the United States in 1992 (Gail et al., 1991).

With new therapeutic strategies for both the primary human immunodeficiency virus type 1 (HIV-1) infection and opportunistic infections, increased survival is notable with AIDS. Thus, the proportion of individuals that are ultimately at risk for development of lymphoma has increased. A study by Pluda and colleagues (1990) evaluated 55 patients treated with antiretroviral therapies at the National Cancer Institute. All patients entered on to the study had AIDS, and thus this was a highly immunosuppressed patient population. In addition, the mean CD4 count was 75/mm^3 (normal CD4 lymphocyte count 1,000–2,000/mm^3. Over the 36-month follow-up period, 46% of the surviving patients developed lymphomas. The 95% confidence intervals for this finding were 20–75%. This included eight patients with lymphomas, of which five were primary central nervous system (P-CNS) lymphomas. In this population, the risk of lymphoma was 9% per year.

Other HIV-1-infected patient cohorts have also been examined for the development of lymphomas. In a study of 1,065 HIV-infected and 636 seronegative hemophiliacs, non-Hodgkin's lymphomas developed in 12 HIV-infected and no seronegative individuals (Rabkin et al., 1992). The incidence of lymphoma increased with the duration of HIV infection up to 0.75% per year after eight years. This finding agrees with the data of Pluda and colleagues (1990) that the incidence of lymphoma increases during HIV-1 infection. The lower overall incidence of lymphomagenesis in the hemophiliac cohort compared to the cohort followed at the National Cancer Institute is likely due to selection of a less immunosuppressed patient population.

At San Francisco General Hospital, lymphomas occurred in 2.5% of AIDS cases in 1985 and 5% of AIDS cases in 1987 (Kaplan et al., 1989), reaffirming the finding of a significant increase in the incidence of lymphoma among HIV-1-infected individuals. There remain some uncertainities as to the magnitude of this risk. More intensive monitoring techniques increase the detection of lymphomas, and are particularly important in the diagnosis of primary central nervous system lymphomas. The role of antiretroviral therapies in lymphomagenesis is also uncertain. However, preliminary data suggest that the development of lymphomas correlates more closely with CD4 count than the presence or absence of antiretroviral therapy, and is independent of the type of antiretroviral therapy (Pluda et al., 1991).

Further evidence to support the association of HIV-1 with lymphoma is derived from studies of the SIV_{mac}-rhesus macaque model system (Feichtinger et al., 1992). SIV_{mac} is related to HIV-1 with regard to genome structure and expression, cell-specific tropism, and pattern of replication. Infected animals develop an immunodeficiency syndrome over several months to years, characterized by preferential depletion of CD4+ lymphocytes. Opportunistic infections are noted in these animals as well. Aggressive non-Hodgkin's lymphomas develop in about 40% of these animals.

The increased incidence of non-Hodgkin's lymphoma in AIDS is reminiscent of its occurrence in other immunodeficiency disorders, including primary immunodeficiencies such as the Wiskott-Aldrich syndrome and ataxia

Molecular Genetics of Nervous System Tumors, pages 329-340 ©1993 Wiley-Liss, Inc.

telangiectasia, and secondary immunodeficiencies due to use of immunosuppressive medications for organ transplantation (Biemer, 1990). An increased incidence of non-Hodgkin's lymphoma is also noted in autoimmune diseases, including Hashimoto's thyroiditis, Sjogren's syndrome, celiac disease, rheumatoid arthritis, and systemic lupus erythematosis. In each case, the predominant forms of lymphoma are of the B lymphocytic lineage and 90% are of moderate- to high-grade types (Fig. 2). These include primarily diffuse large-cell lymphomas, immunoblastic lymphomas, and small non-cleaved cell lymphomas or Burkitt-like lymphomas. In contrast, these types of lymphomas account for only 10% of non-Hodgkin's lymphomas in the nonimmunosuppressed population (Lukes et al., 1978).

A notable feature of lymphomas developing in HIV-1-infected individuals is the involvement in 85% of patients of extranodal sites, including liver, gastrointestinal tract, and bone marrow (Knowles et al., 1988; Lowenthal et al., 1988). Extranodal involvment by non-Hodgkin's lymphoma in HIV-seronegative individuals occurs in only 40% of cases (Jones et al., 1973). About 65% of HIV-1-infected patients present with stage III disease (nodal involvement on both sides of the diaphragm) or stage IV disease (parenchymal lymphoma). In addition, central nervous system involvement with AIDS lymphomas is much more common than with lymphomas in the nonimmunosuppressed patient population. With AIDS-associated systemic lymphomas, central nervous system disease is present at the time of diagnosis in 25% of cases (Levine et al., 1991b). Central nervous system relapses occurring during the clinical course are found in 18% of cases. In addition, at autopsy 60% of patients are found to have central nervous system involvement with their systemic lymphomas.

P-CNS lymphomas are defined as those in which no involvement outside the central nervous system system is found at the time of diagnosis. Though P-CNS lymphomas account for only about 1% of lymphomas in a nonimmunosuppressed HIV-negative population, among HIV-1-infected patients, primary central nervous system lymphomas account for 10% of all lymphomas (Remick et al., 1990). A similar increased proportion of P-CNS lymphomas with immunosuppression is also seen with other primary and second immunodeficiency disorders. Of all patients presenting with P-CNS, approximately 1% were found to have an immunosuppressive syndrome (Heile et al., 1984). This number will undoubtedly increase with the continuation of the AIDS epidemic.

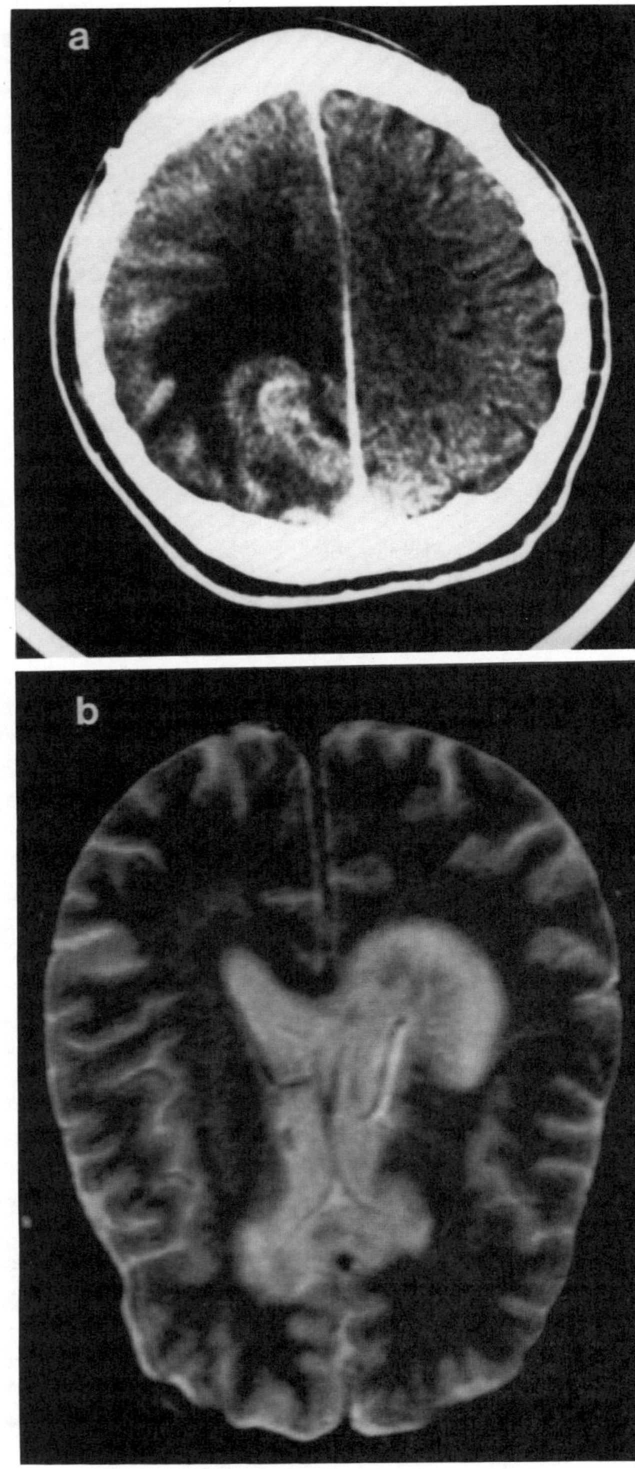

Fig. 1. Radiological findings in P-CNS lymphoma. **a**: Enhanced computer tomographic scan of a 36-year-old patient with P-CNS lymphoma demonstrating a low attenuation lesion in the right parieto-occipital area, compressing the ventricle and with surrounding edema. Partial enhancement is seen. **b**: T2 weighted MRI image demonstrating a large periventricular lesion.

CLINICAL FEATURES

AIDS-associated lymphomas occur in all risk group populations. This is to be contrasted with Kaposi's sarcoma, which occurs almost exclusively among homosexual or bisexual males with HIV-1 infection (Knowles et al., 1988).

The clinical features of AIDS-associated systemic lymphomas contrast sharply with those of P-CNS lymphomas in this population (Table 1). Individuals may develop systemic lymphoma at any stage of HIV-1 infection, and for 63% of these individuals, non-Hodgkin's lymphoma will represent the initial manifestation of AIDS. In contrast, P-CNS lymphoma occurs as a late manifestation of HIV-1 disease after the development of multiple other complications, including opportunistic infections, Kaposi's sarcoma, or HIV-1-associated encephalopathy in 73% of cases (Levine et al., 1991b). Thus, the mean CD4 lymphocyte count for newly diagnosed systemic lymphoma patients is 189/mm^3 compared to 30/mm^3 for newly diagnosed P-CNS lymphoma patients. In addition, the median survival is 2.5 months for P-CNS lymphoma patients compared to 6 months for systemic lymphoma patients. The poor prognosis of P-CNS lymphoma was also recognized in a study at Memorial Sloan Kettering Cancer Institute (Lowenthal et al., 1988). Of nine cases of P-CNS lymphoma diagnosed at that institution, seven were only diagnosed at autopsy.

Several clinical features have been shown to have prognostic significance for survival in patients with systemic AIDS-associated lymphoma (Levine et al., 1991b). These include the patient's performance status, previous opportunistic infections, and bone marrow involvement. Interestingly, factors that had no predictive value with regard to survival included leptomeningeal disease, pathologic type, size of mass lesions, level of lactic dehydrogenase, presence of B symptoms (fevers, weight loss, night sweats), and gastrointestinal involvement.

The primary clinical signs and symptoms of P-CNS lymphomas are nonlocalizing and include stupor, seizures, and headaches (Gill et al., 1985). Hemiparesis and cranial nerve palses are also frequent presenting signs. Thus, the presenting clinical signs are similar to those of non-AIDS P-CNS lymphomas. However, AIDS P-CNS lymphoma patients have a higher prevalence of B symptoms than non-AIDS P-CNS lymphoma patients.

The characteristic radiological picture is that of a contrast-enhancing mass lesion, more than 3 cm in diameter, which is usually a single lesion with associated edema (Fig. 1).

The primary differential diagnostic consideration is encephalitis caused by *Toxoplasma gondii*. This disorder is due to reactivation of latent organisms, and thus, almost all patients have detectable levels of IgG antibody against this organism. The radiological picture may be similar to that of P-CNS lymphoma, though ring enhancement of the lesion and multiple lesions are more common with *Toxoplasma* encephalitis than P-CNS lymphoma. Furthermore, *Toxoplasma* lesions are frequently less than 3 cm in diameter, and are found most commonly in the deep gray matter structures, whereas P-CNS lymphoma may be found in any area of the brain. Most cases of *Toxoplasma* encephalitis will respond within 2–4 weeks to therapy with sulfadiazine and pyrimethamine, and such a therapeutic trial may allow one to bypass a biopsy of deep-seated brain lesions.

TABLE 1. Comparison of HIV-1-Associated Primary CNS Lymphomas and Systemic Lymphomas

Characteristic	Primary CNS lymphoma	Systemic lymphoma
Prior AIDS (%)	73	37
B symptoms (%)	91	82
Median CD4 count (per dl)	30	189
Median Karnofsky (%) performance status	70	70
Median survival (months)	2.5	6

Data from Levine et al. (1991b).

Other considerations in the differential diagnosis of a mass lesion in an AIDS patient include abscesses caused by *Cryptococcus neoformans* or *Mycobacteria species*. Rarely, Kaposi's sarcoma or Hodgkin's disease may present as mass lesions in the brain. It is important to reconize that AIDS patients may occasionally have multiple coexistent pathological disorders.

B LYMPHOID PROLIFERATION

Lymphomas associated with AIDS tend to have a higher proliferative fraction than those examined from the HIV-seronegative population (McDunn et al., 1991). This is manifested by a higher proportion of tumor cells in S-phase from AIDS lymphomas (24%) compared to non-AIDS lymphomas (10%). There was also a trend toward less aneuploidy among the HIV-associated lymphomas. However, it is unclear whether the clinically aggressive behavior of AIDS-associated lymphomas is due to such differences in the biological properties or, more likely, to variations in therapeutic response, advanced stage, large tumor burden, intolerance of cytotoxic agents, underlying degree of immunosuppression, or coexistent opportunistic infections.

Several cytokines have been postulated to play a role in autocrine growth stimulation in B lymphoid cell tumors. These include interleukin (IL) 5, IL 6, and a soluble form of CD23 (Baumann and Paul, 1992; Tosato et al., 1990). In addition, IL6 may be an autocrine growth factor in multiple myeloma and chronic lymphocytic leukemia (Biondi et al., 1989; Kawano et al., 1989). Furthermore, IL6 is released by HIV-1-infected macrophages (Nakajima et al., 1989) and is capable of stimulating the growth of HIV-1-associated lymphomas *in vitro*. Recent data from the cohort of patients followed at the National Cancer Institute demonstrate a correlation between serum IL6 levels and the development of lymphoma (Pluda et al., 1991).

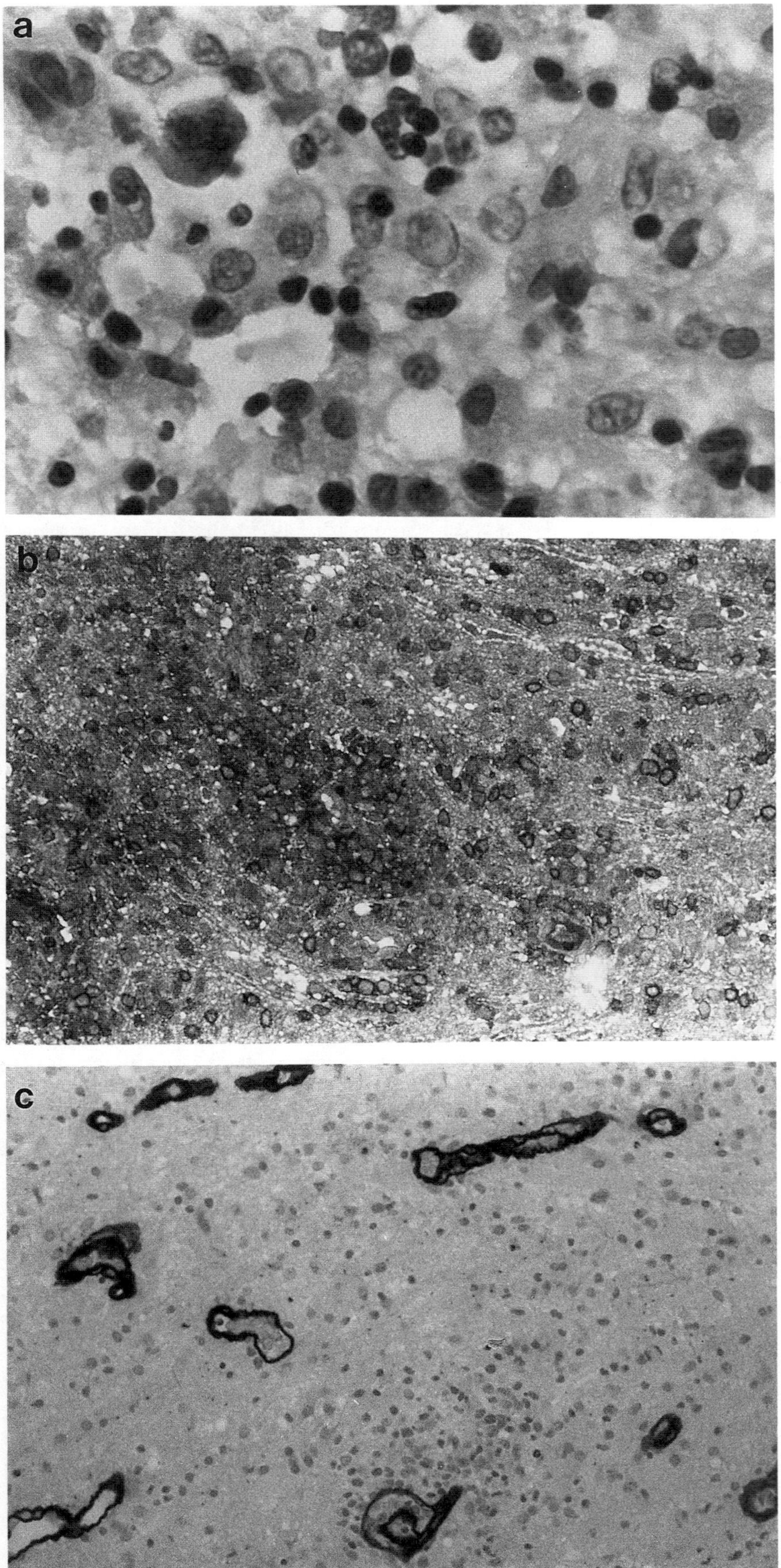

Fig. 2. Pathology of P-CNS lymphoma. Hematoxylin and eosin-stained biopsy material from the patient described in Figure 1b demonstrates karyorrhectic cells and large atypical lymphocytes (**a**, ×2,900), which also react with an antibody directed against leukocyte common antigen (**b**, ×730) primarily in a perivascular distribution, as further illustrated with immunoperoxidase labeling with an antibody directed against type IV collagen present in the basement membrane (**c**, ×730).

CLONALITY

Several different techniques have been used to assess the clonality of B lymphoid proliferative disorders. Expression of exclusively immunoglobulin (Ig) lambda or Ig kappa light chains may indicate clonality. Alternatively, specific immunoglobulin gene rearrangements are consistent with clonality. In the presence of Epstein-Barr virus (EBV) DNA, analysis of the structure of variable sequences such as those present at the termini of the genome may provide information about clonality (Fig. 3).

Three different patterns of AIDS-associated lymphomas have been defined (Table 2). P-CNS lymphomas tend to be monoclonal, immunoblastic, and associated with EBV and c-*myc* rearrangements. Systemic lymphomas have been described as either polyclonal or monoclonal tumors, analogous to those occurring in the solid organ transplant population (Hanto et al., 1983; Starzl et al., 1984). Polyclonal lymphomas tend to be immunoblastic, are often not associated with EBV, and also lack c-*myc* rearrangements. Monoclonal systemic lymphomas are of varied pathology, may be EBV-associated, and tend to exhibit c-*myc* rearrangements.

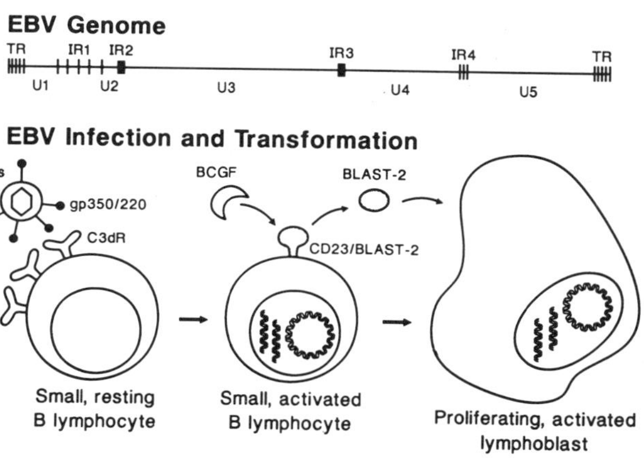

Fig. 3. Epstein-Barr virus infection and B lymphocyte proliferation. The **top** panel is a schematic illustration of the 170 kilobase double-stranded DNA genome of EBV. Positions of terminal repeat (TR), inverted repeat (IR), and unique (U) sequences are indicated. The **bottom** panel illustrates critical steps in EBV infection and transformation. EBV binds to small, resting B lymphocytes through an interaction between the gp350/220 protein complex on the virus and the C3d receptor on the cell. Within 12 hr, one circular and multiple linear copies of the EBV genome are present in the nucleus. The cells are activated, partially mediated by B-cell growth factor (BCGF) and express EBNA proteins and the CD23/BLAST-2 protein complex. By 48 hours after infection, the infected cells are proliferating, activated lymphoblasts, which contain one circular genome in the nucleus that is amplified after an additional 7–10 days. EBV LMP proteins are expressed, as well as adhesion molecules, LFA-1, LFA-3, and ICAM-1.

TABLE 2. Molecular Classification of Lymphomas

Type of lymphoma	Clonality	EBV (%)
Primary CNS lymphoma		
AIDS	Monoclonal	100
Posttransplant	Monoclonal	100
Nonimmunodeficiency	Monoclonal	<10
Systemic lymphoma		
AIDS	Polyclonal	22
AIDS	Monoclonal	50

Data from Meeker et al. (1991) and MacMahon et al. (1991).

In a study of allograft recipients with lymphoproliferative disorders, seven patients had polyclonal lesions, four had monoclonal lesions, and five had polyclonal and monoclonal lesions (Starzl et al., 1984). Several reports document remissions of some of the polyclonal lymphoproliferative lesions with either decreased immunosuppressive therapy or acyclovir therapy (Hanto et al., 1983). In contrast, monoclonal disorders rarely respond to such maneuvers.

One study reported clonal ambiguity in AIDS lymphomas in two patients with lambda light chain predominance in some clusters of lymphoblasts and kappa light chain predominance on other cell clusters (Lippman et al., 1988). In a series of 25 HIV-associated lymphomas examined for Ig heavy chain gene rearrangements, 14 were found to be monoclonal, and 11 were polyclonal (Table 2; Meeker et al., 1991). EBV DNA was found in seven of the 14 monoclonal tumors and in two of the nine polyclonal tumors available for study. It is unclear whether these polyclonal tumors are biologically or clinically distinct from monoclonal tumors in AIDS patients. Furthermore, it remains to be determined whether polyclonal tumors respond differently than monoclonal tumors to therapeutic maneuvers.

In addition, clonal HIV-1-associated lymphomas differ from those in seronegative individuals in that multiple B cell clones may be present simultaneously in AIDS lymphomas (Pelicci et al., 1986b). However, only one clone contains a *myc* gene rearrangement.

The previous evidence suggests that lack of clonality should not be equated with lack of malignancy. In addition, the presence of clonality does not equate with malignancy. In a study of immunoglobulin genes in individuals with progressive generalized lymphadenopathy, rearrangements were occasionally found in the absence of malignant tumors (Pelicci et al., 1986b).

ROLE OF EPSTEIN-BARR VIRUS (EBV)

EBV has been implicated in a wide range of neoplastic conditions, including nasopharyngeal carcinomas, Hodgkin's disease, thymomas, and parotid cancers. A common feature of lymphomas developing in the setting of immunodeficiency is the presence of EBV (Biemer, 1990).

EBV infects virtually all humans by adulthood, and remains latent throughout adult life (Miller, 1985). Children, particularly those of lower socioeconomic status, are frequently infected before the age of five years, and the infection is often silent or mild. Adolescents infected with EBV may develop acute infectious mononucleosis (IM), characterized by fever, pharyngitis, fatigue, cervical lymphadenopathy, hepatosplenomegaly, atypical lymphocytosis, or polyclonal hypergammaglobulinemia.

A chronic form of IM may follow the acute manifestations. This is characterized by continuation of the acute symptoms as well as headaches, neuropsychologic difficulties, and occasionally arthritis and rashes. However, in the immunocompromised individual, a much wider spectrum of disease manifestations may occur, including polyclonal B-cell lymphoproliferation, malignant B-cell lymphomas, virus-associated hemophagocytic syndrome, aplastic anemia, or acquired hypogammaglobulinemia.

One particularly striking form of EBV infection occurs in individuals with the X-linked lymphoproliferative disorder (XLP or Duncan's syndrome) (Purtillo, 1986). These individuals have a selective immune defect in EBV regulation and succumb in more than 85% of cases to complications of EBV infection.

EBV infects B lymphocytes utilizing the same receptor that mediates binding of the complement protein, C3d (Fig. 3; Thorley-Lawson, 1988). Although EBV can also infect epithelial cells, the receptor on this cell type has not been characterized. EBV activates resting small lymphocytes to become proliferating lymphoblasts. These cells produce a number of factors that may have autocrine growth stimulatory properties. They are related to the CD23 or BLAST-2 cell surface antigen. In addition, EBV produces at least six early replication cycle proteins that are found in the nucleus (EBNA antigens). These proteins have transcriptional activation functions toward cellular genes.

Active EBV infection is particularly common among homosexual men. In one study, 11 of 25 homosexual men had antibodies to EBV early antigen (EA) and 13 of 17 were excreting the virus in the saliva (Crawford et al., 1984).

Patients infected with HIV-1 manifest defective T-cell regulation of EBV replication (Birx et al., 1986). Similar defects in regulation of EBV-infected cells *in vitro* or *in vivo* are induced by a T-cell monoclonal antibody, OKT3, or by cyclosporin treatment (Ren and Chan, 1988; Thorley-Lawson, 1988). AIDS patients have elevated anti-EBV antibody titers, and impairments in the ability to suppress immunoglobulin secretion *in vitro* by EBV-infected B lymphocytes. This is consistent with the frequent presence of paraproteins in the serum of AIDS patients. Furthermore, higher numbers of circulating EBV-containing cells are found capable of giving rise to spontaneous B lymphoblastoid cell lines. This suggests that there is an expanded B-cell precursor compartment in HIV-infected individuals, thus increasing the likelihood for development of a subsequent genetic event(s) responsible for lymphomagenesis.

The EBV genome is found in 30–50% of cases of systemic lymphomas associated with AIDS (Table 2; Hamilton-Dutoit et al., 1991). The failure to identify EBV in the majority of cases is not due to the lack of sensitivity of techniques utilized. Similar results have been obtained by Southern blot hybridization studies and more sensitive *in situ* RNA hybridization analyses (Hamilton-Dutoit et al., 1991). In those cases in which EBV DNA is found in the tumor cells, it is present in a clonally expanded population of cells. This has been demonstrated by the presence of distinct types of EBV genomic terminal repeat sequences in each tumor (Fig. 3; Neri et al., 1991). These findings suggest that EBV infection precedes clonal expansion of tumor cells.

In contrast to systemic lymphoma in AIDS, in P-CNS lymphomas, EBV is found in the tumor cells in 100% of cases (Table 2; MacMahon et al., 1991). In both types of tumors, EBV is found in a latent state, with abundant expression of short, noncoding and nonpolyadenylated EBV transcripts (EBERs). Latent membrane protein (LMP) is also detected in these tumors. These results are analogous to the uniform presence of EBV DNA in P-CNS lymphomas arising in immunosuppressed allograft recipients (Nakleh et al., 1991). In contrast, P-CNS lymphomas in nonimmuno-suppressed individuals are not routinely associated with EBV.

One animal model for EBV-induced lymphomas utilizes cottontop tamarin monkeys (Cleary et al., 1985). In these animals, EBV produced oligoclonal or monoclonal B-cell lymphomas containing EBV DNA. These tumors expressed EBNA-1, EBNA-2, and LMP proteins. In contrast, in squirrel monkeys, EBV induces a lymphoproliferative disease only (Miller, 1985). However, lymphomas may be induced in these animals by co-infection with malarial parasites. EBV-related virus, Marek's disease virus, and *Herpes ateles* and *saimarii* also induced lymphomas in chickens and New World monkeys, respectively.

ROLE OF HIV AND *TAT*

HIV-1 is likely to be involved in the formation of B lymphoid cell tumors in at least three different ways. First, HIV-1 leads to a profound state of immunosuppression. This results in a defect in immunological mechanisms required for recognition and eradication of malignant cells and pathogens involved in induction and maintenance of malignancy (Birx et al., 1986). The role of defective immunity in allowing expansion of EBV-infected B-cell precursors has already been discussed.

Second, the HIV-1 surface envelope protein, gp120, is capable of stimulating B lymphoid proliferation and immunoglobulin secretion (Schnittman et al., 1986).

Third, the HIV-1 transcriptional *trans*-activator protein, *tat*, is capable of stimulating the replication of several different types of cells. It has been implicated as a released growth factor for Kaposi's sarcoma cells, and a possible stimulating role for B lymphocytes has also been suggested (Ensoli et al., 1990). Recent data has suggested that *tat* may activate interleukin 6 (IL6), a B lymphocyte proliferation factor (Miles et al., 1990). There is also evidence to implicate IL6 in the development of murine plasmacytomas, a tumor in which immunoglobulin–*myc* rearrangements have been demonstrated (Potter, 1990).

Interaction between EBV and HIV-1 in lymphomagenesis is suggested by recent studies of HIV-1 infection of EBV-positive lymphoblastoid cell lines in cultures (Henderson et al., 1991). This resulted in a decrease in expression of B-cell activation antigens, CD23 and CD39 expression, and a concomitant increase in the expression of surface CD10 and Bac-1 (De Rossi et al., 1990). Furthermore, the HIV-1-infected B lymphoblastoid cell lines no longer grew in tight clusters, but rather as disperse suspensions. Thus, the phenotype of HIV-1-infected B lymphoblastoid cell lines closely resembles that of B lymphomas *in vivo*.

However, a direct role for HIV-1 in the development of B lymphoid tumors is unlikely since HIV-1 DNA sequences are not found in the malignant B cells *in vivo* (Subar et al., 1988). An indirect role for HIV-1 is possible, as indicated by the finding that HIV-1 exposed lymphocytes from EBV-seropositive individuals are tumorigenic in nude mice (Laurence and Astrin, 1991). In tissue culture, several different types of HIV-1 and EBV interactions can occur. HIV-1 can stimulate the growth of EBV replication and EBV increases HIV-1 transcription *in vitro* (Dahl et al., 1987). Furthermore, HIV can enhance EBV-induced transformation, presumably by infection of regulatory T cells (Henderson et al., 1991).

HIV-1 can also stimulate the release of lymphokines and cytokines by infected lymphocytes, monocytes, and macrophages (Rosenberg and Fauci, 1990). Several such products, including IL6, have been implicated in B-cell activation and proliferation.

An alternative indirect role for HIV-1 in B-cell neoplasia is based on the finding of a single case in which a B lymphoma cell line was established and found to express on its surface an IgM antibody reactive with the HIV-1 envelope (Shiramizu and McGrath, 1991). This is similar to an experimental form of B-cell lymphoma that develops in mice after hyperimmunization with sheep red blood cells. In this case, immunoglobulins on the surface of the malignant cells are reactive with the antigen, sheep red blood cells. Furthermore, B-cell lymphomas arising in BALB/c mice in association with an endogenous retrovirus express surface immunoglobulins directed against retroviral products. These findings are consistent with an aberrant antigenic proliferative response that is critical for neoplastic development.

ROLE OF OTHER VIRUSES

Human T lymphotropic virus type 1 (HTLV-1) is the etiological agent of a distinct form of non-Hodgkin's lymphoma, designated adult T-cell leukemia-lymphoma (ATLL; Ratner and Poiesz, 1988). The virus causes a pleomorphic lymphoma with convoluted nuclei. Distinct clinical features include the frequent presence of hypercalcemia, peripheral blood involvement, and skin involvement. The primary target cell for infection *in vivo* is the CD4+ lymphocyte, though the virus can infect a wide range of different human cell types *in vitro*. Studies of B-cell lymphomas in AIDS patients for HTLV-1 antibodies or sequences have been uniformly negative (Subar et al., 1988). However, several cases of CD8 lymphoproliferative disorders in HIV-infected patients have been attributed to infection by HTLV-1 or a closely related virus, HTLV-2 (Harper et al., 1986; Kaplan et al., 1991).

A recent report demonstrated the presence of a type D retrovirus in a B-cell lymphoma from an AIDS patient (Bohannon et al., 1991). This virus resembled the Mason-Pfizer monkey virus (MPMV) and the simian AIDS (SAIDS)-related virus (SRV). The patient's serum contained antibodies reactive with the MPMV p27 major capsid antigen. This individual had no contact with monkeys, and thus the mode of acquisition of the virus is unclear. It will be important to determine whether such an MPMV-like virus is present in other B-cell lymphomas.

Other viruses with tropism for B lymphocytes include the newly described human herpesviruses type 6 (Jarrett et al., 1988). It selectively infects but does not immortalize B lymphocytes. Its role in lymphoproliferative disorders remains to be evaluated. Cytomegalovirus has also been reported to induce B-cell lymphoproliferation, but its role in lymphogenesis is not well understood (Feichtinger et al., 1992).

CHROMOSOMAL REARRANGEMENTS INVOLVING *MYC*

Genes that encode immunoglobulin molecules include the gene for the heavy chain molecule (*IgH* on chromosome 14), and genes for the light chain molecules (*Igkappa* on chromosome 2 and *Iglambda* on chromosome 22) (Potter, 1990). The *IgH* gene is composed of multiple variable (V), diversity (D), joining (J), and constant (C) regions. During B lymphocyte differentiation, one of each of these regions combines through the action of an enzyme designated recombinase. The large number of different combinations of elements are at least partially responsible for the tremendous diversity of immunoglobulins encoded by the lymphoid system.

In Burkitt's lymphoma, rearrangements occur in all cases involving an immunoglobulin gene being juxtaposed with the *myc* proto-oncogene (Fig. 4; Dalla-Favera et al., 1983). In 85% of cases, rearrangements involve the *IgH* gene, whereas the remainder involve the light chain genes. The *myc* gene is situated on chromosome 8, and includes three exons. The open reading frame for the protein product is confined entirely within exons 2 and 3, whereas exon 1 is noncoding. The *myc* gene product is a 62 kd nuclear protein that associates with specific DNA sequences through a second protein known as Max (Blackwood and Eisenman, 1991).

In endemic Burkitt's lymphoma occurring in Africa, the predominant forms of rearrangements are between the D element of *IgH* locus recombining within 1,000 kilobases 5' of the c-*myc* gene or between the J element of the light chain genes and a sequence over a span of at least 300 kilobases 3' of the c-*myc* gene (Pelicci et al., 1986a). In contrast, in the sporadic Burkitt's lymphoma cases occurring in developed countries and the aggressive lymphomas accompanying AIDS, the rearrangement involves the switch region of *IgH* and exon 1 sequences of c-*myc*. Such c-*myc* rearrangements are found in 75% of AIDS lymphoma cases (Subar et al., 1988). These translocations overcome a normal negative regulatory mechanism that results in premature transcriptional elongation within the first exon of the c-*myc* gene.

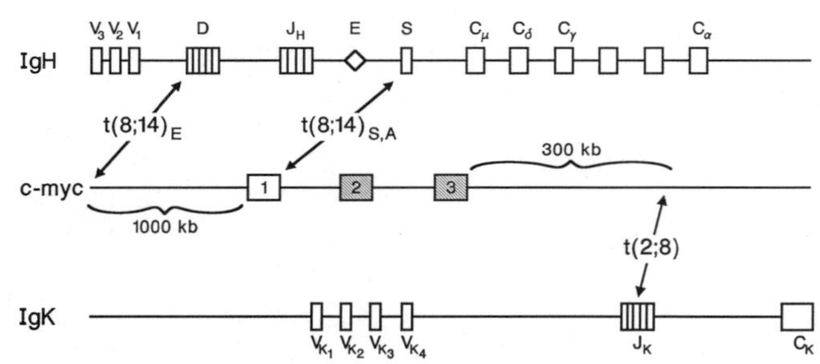

Fig. 4. Chromosomal rearrangements involving *IgH* and *myc* genes. The **top** figure shows the arrangement of elements in the *IgH* gene, including variable (V), diversity (D), joining (J_H), switch (S), and constant (C) regions, and the position of the transcriptional enhancer (E). The **middle** figure shows the arrangement of the three exons in the c-*myc* gene, with shading of exons 2 and 3 to indicate that the open reading frame is located in those positions. The **bottom** figure shows the arrangement of elements in the *IgK* gene, including variable (V_K), joining (J_K), and constrant (C_K) regions. Positions of elements are not drawn to scale. Translocations between chromosomes 8 and 14 in endemic, African Burkitt lymphomas (E) generally occur about 1,000 kilobases (kb) upstream of exon 1. Translocations between chromosomes 8 and 14 in sporadic (S) and AIDS-associated lymphomas (A) generally occur within the first exon or first intron of the c-*myc* gene. Translocations between chromosomes 2 and 8 generally occur about 300 kb downstream of the c-*myc* gene.

A critical role in the development of lymphomas of c-*myc* activation has been demonstrated in two different experimental systems. In tissue culture, EBV-infected B lymphocytes form clusters and are morphologically normal in appearance, and fail to form colonies in soft agar or in nude mice (Lombardi et al., 1987). After transfection of EBV-infected lymphocytes with a *myc* expression clone, the cells fail to form clusters and grow dispersed similar to those of Burkitt lymphoma cells in culture. The altered adhesive properties may be due to down-regulation of expression of cell surface adhesion proteins, such as LFA-1. Furthermore, the *myc*-transfected and EBV-infected lymphocytes form colonies in soft agar and tumors in nude mice.

The second model system includes transgenic mice with Ig-*myc* constructs expressing high levels of c-*myc* products (Adams et al., 1985). B lymphoid tumors arise in these mice in a stochastic fashion.

Another interesting model system for studying such chromosomal rearrangements has been reported by Mosier and colleagues (1989). For their experiments they utilized severe combined immunodeficiency (scid) mice, a strain of mice with a defective recombinase enzyme. Implantation of human B lymphocytes from EBV-positive humans results in stable engraftment. Immunoblastic lymphomas of donor cell origin arose in 63–120 days in 14 of 17 mice transplanted with 3×10^6 lymphocytes from eight different EBV-seropositive donors (Rowe et al., 1991). Tumors manifested clonal immunoglobulin gene rearrangements, and expressed L26 and CD23 pan-B-cell markers, as well as EBNA proteins and LMP. Furthermore, a large percentage of these lymphomas are noted to have rearrangements of the c-*myc* locus (Mosier et al., 1989). These rearrangements were shown to be present at low levels in the donor human lymphocytes and were positively selected in the scid mice.

The c-*myc* gene has also been shown to be activated in B lymphoid tumors in animals by either an analogous chromosomal rearrangement with immunoglobulin genes, or by retroviral insertion adjacent to the c-*myc* gene (Potter, 1990). In the latter case, c-*myc* expression is enhanced by the insertion of a strong retroviral promoter or enhancer that bypasses normal transcriptional regulatory mechanisms of the c-*myc* gene. Though similar retroviral insertions are theoretically possible in humans, none have yet been demonstrated.

INVOLVEMENT OF OTHER PROTO-ONCOGENES

The c-H-*ras* and c-K-*ras* proto-oncogenes are homologous to the v-*ras* oncogenes transduced by Harvey and Kirsten murine sarcoma viruses (Barbacid, 1987). An additional member of the gene family has been designated N-*ras*. Mutations in *ras* gene products are found in more than 50% of lung carcinomas, principally involving amino acid residues 12 and 61. Mutant forms of N-*ras* are found in more

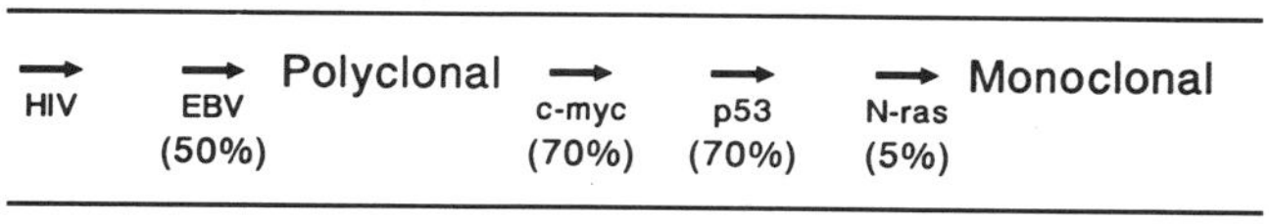

Fig. 5. Multistep process hypothesized for development of AIDS lymphomas. Both HIV and EBV can induce a polyclonal expansion of B lymphocytes. Subsequent molecular defects are suggested to be important for selection and expansion of a particular B-cell clone. The genes possibly involved are listed in descending order of frequency of occurrence of mutations in these loci. The order listed here does not necessarily imply a particular temporal order of occurrence.

than 50% of acute myeloid leukemias, with mutations principally involving amino acid residue 13. The mutant forms of *ras* have in common a defect in their ability to release guanosine triphosphate (GTP) and become deactivated. Mutations in N-*ras* have been reported in about 5% of AIDS-associated lymphomas (Fig. 5; R. Dalla-Favera, personal communication).

Rearrangements of other dominantly acting oncogenes have also been investigated in AIDS lymphoma cases. bcl-2 is a proto-oncogene situated on chromosome 18 that is activated by a rearrangement that juxtaposes the immunoglobulin heavy chain enhancer region from chromosome 14 (McDonnell and Korsmeyer, 1991). This rearrangement occurs in 85% of indolent lymphomas and in about 30% of higher grade lymphomas. Recent studies with transgenic mice expressing bcl-2 have demonstrated the etiological role of this genetic abnormality in lymphomagenesis, by the reproduction of a remarkably similar indolent lymphoid proliferative disease in mice to that present in humans. The lymphoid expansion in these mice is due to the ability of the *bcl-2* product to abrogate the normal program of B lymphoid cell death by a process known as apoptosis. In about 10% of the bcl-2 transgenic mice, more aggressive lymphomas develop. These are associated with c-*myc* rearrangements. Studies of bcl-2 rearrangements in AIDS lymphomas, however, have been negative (Subar et al., 1988).

One report has also noted an increased frequency of trisomy 12 in AIDS-associated lymphomas (Kornblau et al., 1991). The significance of this finding, however, remains unclear.

DEFECTS IN SUPRESSOR ONCOGENE, P53

A number of genes have been characterized as anti-oncogenes or tumor suppressor genes, with mutations occurring frequently in a variety of malignancies (Marshall, 1991). The most commonly mutated tumor suppressor gene is the p53 gene. It encodes a nuclear protein that is phosphorylated during the S-phase of the cell cycle and is a regulator of cell cycle progression. The normal p53 protein can complex with transforming simian virus 40 large T, human papillomavirus E6, and adenovirus E1B proteins. Mutations lead to a loss of function, or dominant negative mutations that inhibit normal p53 function by complex formation. In a study of Burkitt's lymphomas, mutations in p53 were noted in nine of 27 biopsies, 15 of 27 cell lines, and five of nine samples from the analogous L3 form of acute lymphoblastic leukemia (Burkitt's leukemia) (Gaidano et al., 1991). Mutations are present in the p53 gene in 70% of AIDS lymphomas (Fig. 5; R. Dalla-Favera, personal communication).

MULTISTEP MODEL FOR LYMPHOGENESIS IN AIDS

The presence of HIV-1 infection in all individuals with AIDS, the finding of EBV infection in some B-cell tumors, and the presence of various molecular defects in such B-cell tumors is consistent with a multistep model for B-cell transformation in AIDS patients (Fig. 5). Such a theory is supported by convincing evidence in other malignancies for multiple different oncogenic events. For example, in colon carcinomas, three to five genetic abnormalities are found, including mutations in the K-*ras* gene, the DCC (deleted in colon carcinoma) gene, familial adenoma polyposis (FAP) gene, and p53 (Fearon and Vogelstein, 1990).

IMMUNOLOGICAL TOLERANCE OF THE CENTRAL NERVOUS SYSTEM

The high frequency of P-CNS lymphomas in immunocompromised individuals may at least in part be due to the fact that the CNS is an immunologically tolerant site. The CNS has no lymphatic system and has a paucity of infiltrating mononuclear inflammatory cells. Furthermore, tissue allografts implanted in the brain are generally tolerated (Hasek et al., 1977). Tumors implanted into animal brains generally demonstrate more rapid growth rate than those implanted elsewhere in the body. These findings may further explain the high rate of P-CNS lymphomas in AIDS.

IMPLICATIONS FOR DETECTION AND DIAGNOSIS

The close association of EBV with P-CNS lymphomas suggests that assays for detection of EBV DNA in cerebrospinal fluid may prove useful in diagnosis of this malignancy. Identification of EBV sequences in enlarged lymph nodes has also been suggested to be a marker for subsequent lymphoma development (Shibata et al., 1991). Intriguing new data on the role of cytokines, such as IL6 in the development of AIDS lymphomas, may ultimately lead to specific diagnostic assays.

IMPLICATIONS FOR THERAPY AND PREVENTION

The current therapeutic strategy for treatment of systemic lymphoma is to utilize combination chemotherapeutic agents. A recent multicenter trial in 49 HIV-1-positive patients utilizing attenuated doses of mBACOD therapy (methotrexate, bleomycin, adriamycin, cyclophosphamide, oncovin, and dexamethasone) reported a 46% complete remission rate (Levine et al., 1991a). If active central nervous system lymphoma is found, intrathecal therapy combined with radiation therapy is utilized to eradicate disease in this sanctuary site. If no lymphoma is found by either CT scan or lumbar puncture, prophylactic intrathecal therapy is provided in light of the high rate of central nervous system relapse in this population.

The current treatment strategy of P-CNS lymphoma in AIDS patients relies on the use of radiation therapy. In a retrospective study of 55 patients with AIDS and P-CNS lymphoma, 29 patients received whole brain radiation therapy (Baumgartner et al., 1990). In this group, a partial radiological remission was seen in 52% of cases, and a complete remission in an additional 17% of cases. The mean duration of response was four months. Most deaths were due to opportunistic infections, and no deaths were due to uncontrolled CNS lymphoma. In another study of 10 cases of P-CNS lymphoma, a complete remission was obtained with radiation therapy in six individuals (Formenti et al., 1989). At the time of the report, two of these six patients were alive eight and 14 months after treatment, two individuals had died of opportunistic infections, and two individuals died of recurrent lymphoma. The median survival of the entire population was 5.5 months. Another study also reported occasional long-term remissions and survival with AIDS-associated P-CNS lymphoma (Gill et al., 1985). However, other investigators have reported poorer responses to radiation therapy (DeWeese et al., 1991).

Current trials in lymphomas in AIDS are attempting to improve results in two ways. First, therapeutic strategies are being evaluated that combine therapies for lymphoma with those for both HIV-1 and prophylaxis for opportunistic infections. Though zodivudine is poorly tolerated with chemotherapy, it is likely that the less hematopoietic suppressive agents, dideoxycytidine and dideoxyinosine, will be easier to combine with combination chemotherapeutic agents.

Second, treatment strategies are investigating whether higher doses of chemotherapeutic agents improve responses in AIDS-associated systemic lymphoma and whether combination of chemotherapy with radiation therapy improves results in AIDS-associated P-CNS lymphomas.

Other approaches in the future may involve suppression of EBV replication. In the case of polyclonal tumors in allograft recipients, withdrawal of immunosuppression may reverse B lymphoproliferative disorders. Do polyclonal tumors in HIV-infected individuals behave differently toward therapeutic maneuvers than monoclonal tumors? Continued treatment of HIV-1 infection possibly combined with immune replacement therapy may have a beneficial response in at least certain lymphomas.

Other approaches may involve the use of specific toxin-antibody conjugates directed against B-cell antigens (Grossbard et al., 1992), or agents capable of suppressing expression of certain cytokines, e.g., IL6. There is little question that further studies of AIDS-associated lymphomas will lead to the development of novel therapeutic approaches.

ACKNOWLEDGMENTS

This work was supported by NIH grants. The author is an American Cancer Society Research Professor.

REFERENCES

Adams, J.M., A.W. Harris, C.A. Pinkert, et al. (1985) The *c-myc* oncogene driven by immunoglobulin enhancer induces lymphoid malignancy in trangenic mice. Nature 318:533–538.

Barbacid, M. (1987) Ras genes. Ann. Rev. Biochem. 56:779–827.

Baumann, M.A., and C.C. Paul (1992) Interleukin-5 is an autocrine growth factor for Epstein-Barr virus-transformed B lymphocytes. Blood 79:1763–1767.

Baumgartner, J.E., J.R. Rachlin, J.H. Beckstead, et al. (1990) Primary central nervous system lymphomas: Natural history and response to radiation therapy in 55 patients with acquired immunodefiency syndrome. J. Neurosurg. 73:206–211.

Beral V., T. Peterman, R. Berkelman, et al. (1991) AIDS-associated non-Hodgkin's lymphoma. Lancet 337:805–809.

Biemer, J.J. (1990) Malignant lymphomas associated with immunodeficiency states. Ann. Clin. Lab. Sci. 20:175–191.

Biondi, A., V. Rossi, R. Bassan, et al. (1989) Constitutive expression of interleukin-6 gene in chronic lymphocytic leukemia. Blood 73:1279–1284.

Birx, D.L., R.R. Redfield, and G. Tosato (1986) Defective regulation of Epstein-Barr virus infection in patients with acquired immunodeficiency syndrome (AIDS) or AIDS-related disorders. N. Engl. J. Med. 314:874–879.

Blackwood, E.M., and R.M. Eisenman (1991) Max: A helix-loop-helix zipper protein that forms a sequence-specific DNA complex with *myc*. Science 251:1211–1217.

Bohannon, R.C., L.A. Donehower, and R.J. Ford (1991) Isolation of a type D retrovirus from B-cell lymphomas of a patient with AIDS. J. Virol. 65:5663–5672.

Cleary, M.L., M.A. Epstein, S. Finerty, et al. (1985) Individual tumors of multifocal EB virus-induced malignant lymphomas in tamarins arise from different B-cell clones. Science 228:722–724.

Crawford, D.H., I. Weller, V. Iliescu, et al. (1984) Epstein-Barr (EB) virus infection in homosexual men in London. Br. J. Vener. Dis. 60:258–264.

Dahl, K., K. Martin, and G. Miller (1987) Differences among human immunodeficiency virus strains in their capacities to induce cytolysis or persistent infection of a lymphoblastoid cell line immortalized by Epstein-Barr virus. J. Virol. 6:1602–1608.

Dalla-Favera, R., S. Martinotti, R.C. Gallo, et al. (1983) Translocation and rearrangement of the *c-myc* oncogene locus in human undifferentiated B-cell lymphomas. Science 219:963–967.

De Rossi, A., S. Roncella, M.L. Calabro, et al. (1990) Infection of Epstein-Barr virus transformed lymphoblastoid B cells by the human immunodeficiency virus: Evidence for a persistent and productive infection leading to B cell phenotypic changes. Eur. J. Immunol. 20:2041–2049.

DeWeese, T.L., M.B. Hazuka, D.J. Hommel, et al. (1991) AIDS-related non-Hodgkin's lymoma: The outcome and efficacy of radiation therapy. Int. J. Rad. Oncol. Biol. Phys. 20:803–808.

Ensoli, B., G. Barillari, S.Z. Salahuddin, et al. (1990) Tat protein of HIV-1 stimulates growth of cells derived from Kaposi's sarcoma lesions of AIDS patients. Nature 345:84–86.

Fearon, E.R., and B. Vogelstein (1990) A genetic model for colorectal tumorigenesis. Cell 61:759–767.

Feichtinger H., E. Kaaya, P. Putkonen, et al. (1992) Malignant lymphoma associated with human AIDS and with SIV-induced immunodeficiency in macaques. AIDS Research and Human Retroviruses 8:339–348.

Formenti, S.C., P.S. Gill, and E. Lean (1989) Primary central nervous system lymphoma in AIDS. Results of radiation therapy. Cancer 63:1101–1107.

Gaidano, G., P. Ballerini, J.Z. Gang, et al. (1991) p53 mutations in human lymphoid malignancy: Association with Burkitt lymphoma and chronic lymphocytic leukemia. Proc. Natl. Acad. Sci. USA 88:5413–5417.

Gail, M.H., J.M. Pluda, C.S. Rabkin, et al. (1991) Projections of the incidence of non-Hodgkin's lymphoma related to acquired immunodeficiency syndrome. Natl. Cancer Inst. 83:695–701.

Gill, P.S., A.M. Levine, P.R. Meyer, et al. (1985) Primary central nervous system lymphoma in homosexual men: Clinical, immunologic, and pathologic features. Am. J. Med. 78:742–748.

Grossbard, M.L., A.S. Freedman, J. Ritz, et al. (1992) Serotherapy of B-cell neoplasms with anti-B4-blocked ricin: A phase I trial of daily bolus infusion. Blood 79:576–585.

Hamilton-Dutoit, S.J., G. Pallesen, M.B. Franzmann, et al. (1991) AIDS-related lymphoma: Histopathology, immunophenotype, and association with Epstein-Barr virus as demonstrated by *in situ* nucleic acid hybridization. Am. J. Pathol. 138:149–163.

Hanto, D.W., K.J. Gajl-Peczalska, G. Frizzera, et al. (1983) Epstein-Barr virus (EBV) induced polyclonal and monoclonal B-cell lymphproliferative diseases occurring after renal transplantation. Clinical, pathologic, and virologic findings and implications for therapy. Ann. Surg. 198:356–369.

Harper, M.E., M.H. Kaplan, and L.M. Marselle (1986) Concomitant infection with HTLV-I and HTLV-III in a patient with a T8 lymphoproliferative disease. N. Engl. J. Med. 315:1073–1081.

Hasek, M., J. Chutna, M. Sladecek, et al. (1977) Immunologic tolerance and tumor allografts in the brains. Nature 268:68–69.

Heile, T.L., R.H. Britt, and Th.V. Colby (1984) Primary lymphoma of the central nervous system. Clincopathological experience at Stanford. J. Neurosurg. 60:94–103.

Henderson, E.E., Y.-Y. Yang, R.-D. Zhang, et al. (1991) Altered HIV expression and EBV-induced transformation in coinfected PBLs and PBL subpopulations. Virology 182:186–198.

Jarrett, R.F., S. Sledhill, F. Quereshi, et al. (1988) Identification of human herpes virus 6-specific DNA sequences in two patients with non-Hodgkin's lymphoma. Lancet 2:496.

Jones, S.E., Z. Fuks, M. Bull, et al. (1973) Non-Hodgkin's lymphoma: IV. Clinicopathological correlation in 405 cases. Cancer 31:806–823.

Kaplan, L.D., D.I. Abrams, E. Feigal, et al. (1989) AIDS-associated non-Hodgkin's lymphomas in San Francisco. J. Am. Med. Assoc. 261:719–724.

Kaplan, M.H., W.W. Hall, M. Susin, et al. (1991) Syndrome of severe skin disease, eosinophilia, and dermatopathic lymphadenopathy in patients with HTLV-II complicating human immunodeficiency virus infection. Am. J. Med. 91:300–309.

Kawano, M., T. Hinano, T. Matsuda, et al. (1988) Autocrine generation and requirement of BSF-2/IL-6 for human multiple myelomas. Nature 332:83–85.

Knowles, D.M., G.A. Chamulak, M. Subar, et al. (1988) Lymphoid neoplasia associated with the acquired immunodeficiency syndrome (AIDS). Ann. Intern. Med. 108:744–753.

Kornblau, S.M., A. Goodacre, and F. Cabanillas (1991) Chromosomal abnormalities in adult non-endemic Burkitt lymphoma and leukemia: 22 new reports and a review of 148 cases from the literature. Hematol. Oncol. 9:63–78.

Laurence, J., and S.M. Astrin (1991) Human immunodeficiency virus induction of malignant transformation in B lymphocytes. Proc. Natl. Acad. Sci. USA 88:7635–7639.

Levine, A.M., J.C. Wernz, L. Kaplan, et al. (1991a) Low dose chemotherapy with central nervous system prophylaxis and zodovudine maintenance in AIDS-related lymphoma: A prospective multi-institutional trial. J. Am. Med. Assoc. 266:84–88.

Levine, A.M., J. Sullivan-Halley, M.C. Pike, et al. (1991b) Human immunodeficiency virus-related lymphoma. Prognostic factors predictive of survival. Cancer 68:2466–2472.

Lippman, S.M., J.R. Volk, C.M. Spier, et al. (1988) Clonal ambiguity of human immunodeficiency virus-associated lymphomas. Arch. Pathol. Lab. Med. 112:128–132.

Lombardi, L., E.W. Newcomb, and R. Dalla-Favera (1987) Pathogenesis of Burkitt lymphoma: Expression of an activated *c-myc* oncogene causes the tumorigenic conversion of EBV-infected human B lymphoblasts. Cell 49:161–170.

Lowenthal, D., D. Straus, S. Wise Campbell, et al. (1988) AIDS related lymphoid neoplasia. The Memorial Hospital experience. Cancer 61:2325–2337.

Lukes, R.J., J.W. Parker, C.R. Taylor, et al. (1978) Immunologic approach to non-Hodgkin's lymphoma and related leukemias. Analysis of multiparameter studies of 425 cases. Semin. Hematol. 15:322–251.

MacMahon, E.M.E., J.D. Glass, S.D. Hayward, et al. (1991) Epstein-Barr virus in AIDS-related primary central nervous system lymphoma. Lancet 338:969–972.

Marshall, C.J. (1991) Tumor suppressor genes. Cell 64:313–326.

McDonnell, T.J., and S. Korsmeyer (1991) Progression from lymphoid hyperplasia to high-grade malignant lymphoma in mice transgenic for the t(14;18). Nature 349:254–256.

McDunn, S.H., J.N. Winter, D. Variakojis, et al. (1991) Human immunodeficiency virus-related lymphomas: A possible association between tumor proliferation, lack of ploidy anomalies, and immune deficiency. J. Clin. Oncol. 9:1334–1340.

Meeker, T.C., B. Shiramizu, L. Kaplan, et al. (1991) Evidence for molecular subtypes of HIV-associated lymphoma: Division into peripheral monoclonal, polyclonal, and central nervous system lymphoma. AIDS 5:669–674.

Miles, S.A., A.R. Rezai, J.F. Salazar-Gonzalez, et al. (1990) AIDS Kaposi sarcoma derived cells produce and respond to interleukin 6. Proc. Natl. Acad. Sci. USA 87:4068–4072.

Miller, G. (1985) Epstein-Barr Virus, In: B.N. Fields, D.M. Knipe, R.N. Channock, et al. (eds.). Virology. New York: Raven Press, pp. 563–590.

Mosier, D.E., R.J. Gulizia, S.M. Baird, et al. (1989) Studies of HIV infection and the development of Epstein-Barr human lymphocytes to mice with severe combined immunodeficiency. 5th International AIDS Conference, Montreal, June 1989.

Nakajima, K., O. Martinez-Mara, T. Hirano, et al. (1989) Induction of IL-6 (B cell stimulating factor-2/IFN-beta$_2$) producton by HIV. J. Immunol. 142:531–536.

Nakleh, K.E., J.C. Manivel, C.M. Copenhaver, et al. (1991) In situ hybridization for detection of Epstein-Barr virus in central nervous system lymphomas. Cancer 67:444–448.

Neri, A., F. Barriga, G. Inghirami, et al. (1991) Epstein-Barr virus infection precedes clonal expansion in Burkitt's and acquired immunodeficiency syndrome-associated lymphoma. Blood 77:1092–1095.

Pelicci, P.-G., D.M. Knowles, I. Magrath, et al. (1986a) Chromosomal breakpoints and structural alterations of the *c-myc* locus differ in endemic and sporadic forms of Burkitt lymphoma. Proc. Natl. Acad. Sci. USA 83:2984–2988.

Pelicci, P.-G., D.M. Knowles, Z.A. Zarlin, et al. (1986b) Multiple monoclonal B cell expansions and *c-myc* oncogene rearrangements in acquired immune deficiency syndrome-related lymphoproliferative disorders. Implications for lymphomagenesis. J. Exp. Med. 164:2049–2076.

Pluda, J.M., D. Benzon, G. Tosato, et al. (1991) Factors which predict for the development of non-Hodgkin's lymphoma (NHL) in patients with HIV infection receiving antiretroviral therapy. Blood 78:285a.

Pluda, J.M., R. Yarchoan, E.S. Jaffe et al. (1990) Development of non-Hodgkin lymphoma in a cohort of patients with severe human immunodeficiency virus (HIV) infection on long-term antiretroviral therapy. Ann. Intern. Med. 113:276–282.

Potter, M. (1990) Neoplastic development in B lymphocytes. Carcinogenesis 1–13.

Purtillo, D.T. (1986) Prevention and treatment of Epstein-Barr virus (EBV)-associated lymphoproliferative disease in immune deficient patients. AIDS Res. 2 (Suppl.): 177–181.

Rabkin, C.S., M.W. Hilgartner, K.W. Hedberg, et al. (1992) Incidence of lymphomas and other cancers in HIV-infected and HIV-uninfected patients with hemophilia. J. Am. Med. Assoc. 267:1090–1094.

Ratner, L., and B.J. Poiesz (1988) Leukemias associated with human T-cell lymphotropic virus type I in a non-endemic region. Medicine 67:401–422.

Remick, S.C., C. Diamond, J.A. Migliozzi, et al. (1990) Primary central nervous system lymphoma in patients with and without the acquired immunodeficiency syndrome. A retrospective analysis and review of the literature. Medicine 69:345–360.

Ren, E.C., and S.H. Chan (1988) Possible enhancement of Epstein-Barr virus infection by use of OKT3 in transplant recipients. Transplantation 45:988–989.

Rosenberg, Z., and A.S. Fauci (1990) Immunopathologic mechanisms of HIV infection: Cytokine induction of HIV expression. Immunol. Today 11:176–180.

Rowe, M., L.S. Young, J. Crocker, et al. (1991) Epstein-Barr virus (EBV)-associated lymphoproliferative disease in the *SCID* mouse model: Implications for the pathogenesis of EBV-positive lymphomas in man. J. Exp. Med. 173:147–158.

Schnittman, S.M., H.C. Lane, S.H. Higgins, et al. (1986) Direct polyclonal activation of human B lymphocytes by the acquired immunodeficiency syndrome virus. Science 233:1084–1086.

Shibata, D., L.M. Weiss, B.N. Nathwani, et al. (1991) Epstein-Barr virus in benign lymph node biopsies from individuals infected with the human immunodeficiency virus is associated with concurrent or subsequent development of non-Hodgkin's lymphoma. Blood 77:1527–1533.

Shiramizu, B., and M.S. McGrath (1991) Molecular pathogenesis of AIDS-associated non-Hodgkin's lymphoma. Hematol. Oncol. Clin. North Am. 5:323–330.

Starzl, T.E., M.A. Nalenik, K.A. Porter, et al. (1984) Reversibility of lymphomas and lymphoproliferative lesions developing under cyclosporin-steroid therapy. Lancet 1:583–587.

Subar, M., A. Neri, G. Inghirami, et al. (1988) Frequency of *c-myc* oncogene activation and infrequent presence of Epstein-Barr virus genome in AIDS-associated lymphoma. Blood 72:667–671.

Thorley-Lawson, D.A. (1988) Basic virological aspects of Epstein-Barr virus infection. Semin. Hematol. 25:247–260.

Tosato, G., J. Tanner, K.D. Jones, et al. (1990) Identification of interleukin-6 as an autocrine growth factor for Epstein-Barr virus-immortalized B cells. J. Virol. 64:3033.

Ziegler, J., J. Beckstead, P. Volberding, et al. (1985) Non-Hodgkin's lymphoma in 90 homosexual men. Relation to generalized lymphadenopathy and the acquired immunodeficiency syndrome. N. Engl. J. Med. 311:565–570.

Chapter 31

Extracellular Matrix in the Developing Nervous System

Arthur D. Lander and Anne L. Calof

INTRODUCTION

Extracellular matrix (ECM), in its broadest definition, refers to all the molecules that are immobilized in the spaces between cells. The most extensively studied ECMs are those found in connective tissues, such as bone, cartilage, tendons, ligaments, fascia, and the stromal components of organs. Connective tissues are typically cell-poor, i.e., the ECM comprises a major fraction of the tissue mass. Accordingly, connective tissue ECMs are comparatively easy to isolate in large quantities, permitting direct biochemical analysis.

In contrast, the ECMs of nonconnective tissues are difficult or in some cases impossible (so far) to isolate, and usually must be studied by more indirect means. For this reason most progress toward an understanding of nonconnective tissue ECMs has been made only in the last 15–20 years. Under the circumstances, it is not surprising that most generalizations that have been made about how ECMs are organized and how they function have been drawn from work on connective tissues, and are not necessarily applicable to the ECMs of nonconnective tissues. Nowhere is this fact more evident than when considering the ECM of the central nervous system (CNS). For example, it is traditional to think of ECMs as consisting largely of structural molecules, such as collagens and elastin, that confer bulk tissue properties such as resistance to stress, rigidity, elasticity, or compressibility. The CNS, however, is virtually devoid of these molecules (except in association with meninges and blood vessels), a fact that explains its jellylike consistency. Moreover, while it is traditional to expect ECMs to take up certain light-microscopic stains and to exhibit certain ultrastructural features (fibrils, electron-dense material), the ECM of the CNS shows virtually none of these characteristics.

Indeed, were it not for the fact that the presence of many known ECM molecules can be readily demonstrated in the brain, there would be serious debate as to whether an ECM existed there at all (as, in fact, there was for many years). Yet the existence of an ECM in the brain underscores the fact that ECMs must have functions beyond the structural and mechanical roles envisaged for them in connective tissues. Indeed, the explosion of research in the ECM field over the past 15 years has focused largely on the nonstructural properties of ECM molecules. The picture that has emerged (and is still changing) is of the ECM as a key regulator of virtually all aspects of cell behavior, including cell proliferation, shape, motility, cell–cell interactions, growth factor responses, even gene expression. During embryonic development, the ECM is uniquely suited, by virture of its immobility, to being a repository of morphogenetic information, such as the information that guides growing axons to their targets. During carcinogenesis, the ECM is ideally suited to regulate growth, differentiation, invasion, and metastasis.

Most of this new information about ECMs is being pieced together from studies of many types of cells, normal and transformed, *in vivo* and *in vitro*. Although cells of the nervous system have played an important role in this work, some of what is said in this chapter has had to be extrapolated from work on nonneural tissues. Clearly, our current knowledge of the ECM of the nervous system and its functions is at a very early stage, and much new information will undoubtedly be obtained in the next several years. Nevertheless, there is enough information now to conclude that the ECM plays a critical role in the formation of the nervous system, and its response to injury.

FORMS OF ECM IN THE NERVOUS SYSTEM

As stated above, cells of the CNS find themselves surrounded by an ECM that is very different from that of connective tissues. Those connective tissue elements that are present in the CNS are associated with meninges or surround blood vessels. Astrocytic processes that end on these structures apparently isolate them from direct contact with other CNS cells. This arrangement also isolates CNS cells from another type of ECM, the basement membrane (or, in the parlance of electron microscopists, the basal lamina). Basement membranes are specialized sheets of ECM almost universally found adjacent to the basal surfaces of epithelia (including the endothelial cells of blood vessels), as well as in a few other locations. During early development, cells and axons may be in contact with the basement membrane that surrounds the neural tube (originally an epithelium), but such contacts are not present in the mature brain.

In contrast to the cells of the CNS, cells of the peripheral nervous system (PNS) do contact conventional matrices, especially basement membranes, which surround Schwann cells and muscles. During peripheral nerve regeneration, axons come into direct contact with these basement membranes. During development, neural cells and axons also come into wide contact with interstitial ECMs, i.e., ECMs that will eventually become components of connective tissues. Thus, the PNS and CNS possess distinctly different ECMs. The change from one type to another occurs abruptly at the glia limitans.

MOLECULES OF THE NERVOUS SYSTEM ECM

Laminin

The term *laminin* was coined to refer to a molecule isolated from a murine tumor (the Engelbreth-Holm-Swarm, or EHS, sarcoma) that produces large amounts of basement membranes (for review, see Beck et al., 1990). This molecule is a large glycoprotein (~850 kd), consisting of three polypeptide chains, one long ("A," ~400 kd) and two shorter, related chains ("B1" and "B2," ~220 and ~210 kd, respectively). All three chains are encoded by separate genes. The organization of the chains into an intact molecule is shown in Figure 1. Like many of the large glycoproteins found in ECMs, laminin consists of many distinct structural motifs strung together. Some of these structural elements have been associated with the ability to bind certain cellular receptors, or have certain effects on cell behavior, especially the behavior of neurons (see below).

Polyclonal antibodies raised against laminin have been used to localize it in many tissues. The results imply that laminin is a ubiquitous and abundant component of basement membranes, and occasionally a much less abundant component of other ECMs. Usually, when laminin is found outside basement membranes its presence is only transient, i.e., it appears only during development, or as part of an injury response. Transiently expresssed laminin is known to be encountered by nervous system cells in several circumstances: In the CNS, laminin appears transiently along a small number of axon tracts, at or about the time that axons are navigating these pathways. Examples include the avian and rodent optic tract and the early-forming ventral longitudinal tract of the mouse hindbrain (Cohen et al., 1987; Letourneau et al., 1988). Laminin is also reexpressed in the CNS at sites of injury (Liesi et al., 1984). In the PNS, non-basement membrane laminin is also found at locations where developing axons grow (e.g., Rogers et al., 1986).

Recently, it has become clear that several isoforms of laminin exist. In these molecules, one or more of the chains of laminin is substituted with a novel, but related, polypeptide. Thus, the "A" chain may be replaced by a homologous "M" chain to generate a molecule referred to as merosin. The "B1" chain may be replaced by a homologous "S" chain to generate what has been called s-laminin. Since most early immunohistochemical studies on the localization of laminin *in vivo* did not use antibodies specific for individual chains, one cannot conclude from such studies which laminin isoforms are present. More recent studies, however, indicate that laminin isoforms have distinctly different distributions *in vivo*. Merosin and classical laminin, for example, are found in different types of basement membranes (Sanes et al., 1990). In the neuromuscular system, laminins containing the "S" chain are specifically localized to the ECM of the neuromuscular junction (the term s-laminin is short for synapse-specific laminin). The form of laminin present in the developing and injured CNS has yet to be determined.

Fibronectin

Fibronectin has probably been more intensively studied than any other ECM protein (for reviews, see Hynes, 1990; Yamada, 1990). Fibronectin is a ~440 kd glycoprotein that is composed of two disulfide-linked polypeptide chains (Fig. 1). Both chains are encoded by a single gene, but are not necessarily identical in amino acid sequence, as considerable alternative splicing of fibronectin transcripts occurs. Alternative splicing patterns are known to be both developmentally regulated and tissue-specific *in vivo*. At least some of the heterogeneity created by alternative splicing affects the ability of fibronectin to interact with certain cellular receptors and to form higher-order fibronectin aggregates and fibrils.

Fibronectin is an abundant component of many connective tissue ECMs, and sometimes may also be found in basement membranes. Fibronectin is also present in soluble form in plasma. In the PNS, fibronectin is widely distributed among matrices through which neural crest cells migrate, as well as in territories encountered by peripheral nerves. In the CNS, fibronectin can be found associated with meninges and blood vessels, but, like laminin, is otherwise expressed only transiently, and in a limited number of sites. Its localization in the brain has been best worked out in the forming mammalian cerbral cortex (Sheppard et al., 1991; Stewart

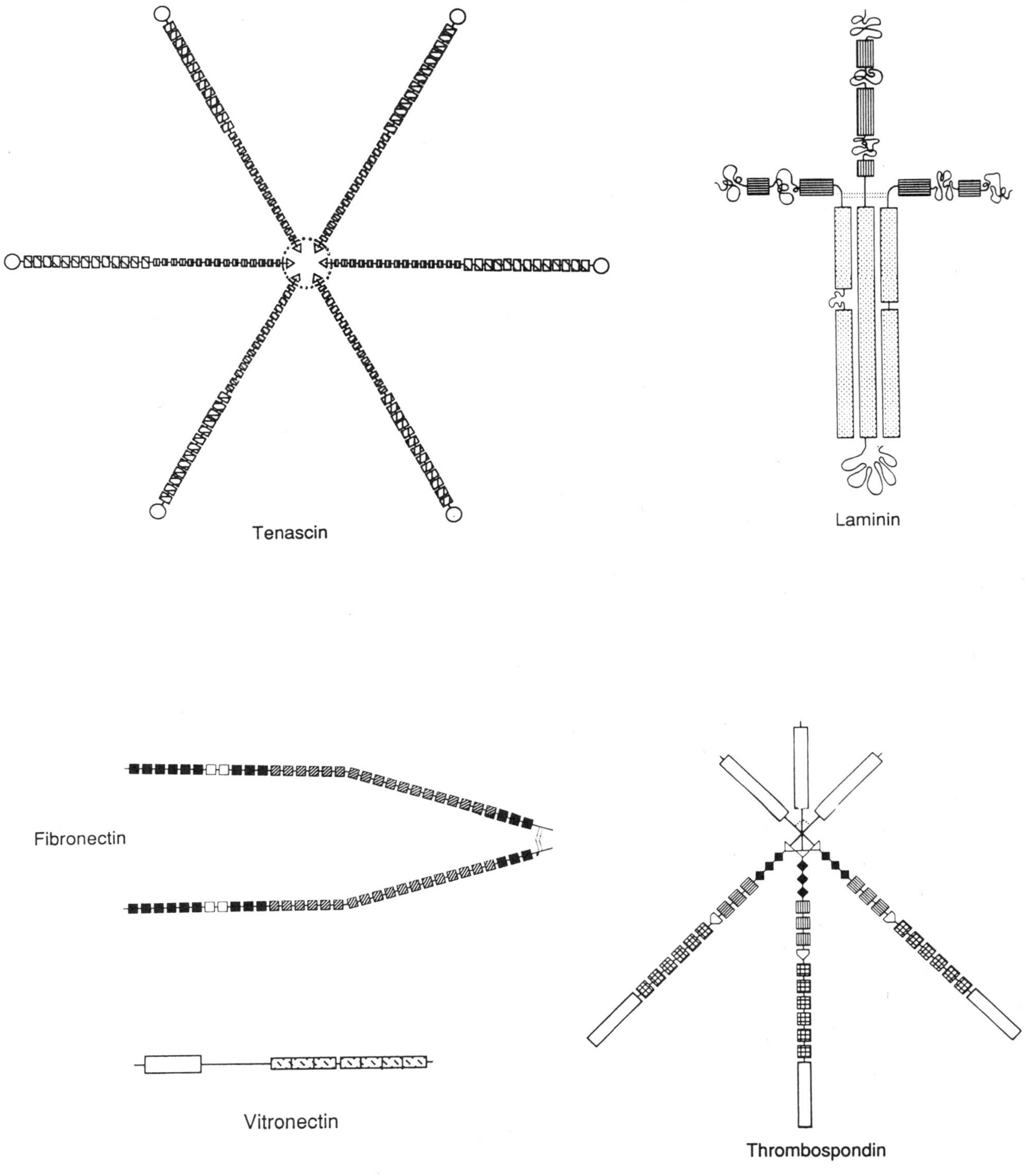

Fig. 1. Structures of ECM proteins found in the central nervous system. These schematic drawings illustrate the polypeptide components and repeating domain structures of five of the best understood ECM components in the nervous system. The rectangular symbols distributed along each polypeptide chain represent repeated amino acid motifs such as EGF-like repeats and fibronectin-type III repeats. Disulfide linkages between polypeptide chains are represented as dotted lines (for further details, see review by Lander, 1989).

and Pearlman, 1987). There it appears earliest in the ventricular zone, where neural precursors proliferate. Later, expression shifts to the subplate region, the layer of developing cortex into which cortical efferent and afferent fibers grow. As the cortex matures, more superficial neuron layers are added, and thalamocortical axons grow out from the subplate and innervate their final targets in the cortex. At about the same time, fibronectin immunoreactivity in the subplate disappears (as does the subplate itself shortly thereafter).

Tenascin

Tenascin, like fibronectin, is a large molecule comprised of identical (or closely related) polypeptide chains held together near their ends by disulfide bonds (for reviews, see Chiquet, 1989; Erickson and Bourdon, 1989). Typically, six polypeptides are linked together to form a six-armed structure known as a hexabrachion (Fig. 1), although forms with fewer arms have been observed in the electron microscope. Tenascin subunits are approximately 190–220 kd in size, and, like fibronectin subunits, vary in structure due to alternative splicing.

Tenascin was discovered more recently than fibronectin and laminin, and was found independently by a number of laboratories. Some of these groups first identified tenascin in nervous system tissue, and some of the early names given to this molecule, such as "cytotactin" and "J1," still persist in some of the neurobiological literature. Although the highest levels of tenascin in the body are found in connective tissue and other mesenchymal ECMs, the developing CNS also contains a considerable amount, especially during late developmental stages, in a wide variety of locations. In laminating CNS structures, such as the developing retina and cerebellum, tenascin can be found associated with particular cell layers and with the radial glial cells that are thought to guide cell migration into appropriate layers. In some CNS structures, such as the developing rodent somatosensory cortex, tenascin expression is transiently associated with the boundaries that separate functional units (such as "barrels"). In the developing PNS, tenascin is strongly associated with the pathways followed by neural crest cells, and later with peripheral nerves and ganglia.

Although the expression of tenascin in the developing CNS is much more widespread than that of laminin and fibronectin, it too largely disappears from the mature brain. The strong expression of tenascin by human glioma cells, however, suggests that reexpression of tenascin may be an important concomitant of neoplastic transformation. Indeed, tenascin is a prominent component of many neural and nonneural tumors. Among human gliomas, tenascin expression tends to correlate with a poorly differentiated histological appearance (Erickson and Bourdon, 1989).

Thrombospondin

Like fibronectin and tenascin, thrombospondin is a large (~420 kd) glycoprotein composed of multiple similar subunits (Lawler and Hynes, 1986). Three polypeptide chains are disulfide-bonded together to form an intact thrombospondin molecule (Fig. 1). Currently, the three chains in each molecule are thought to be identical, but the possibility of alternative splicing to produce molecular heterogeneity has not been ruled out. Recently, the existence of a second gene that encodes a thrombospondin isoform was reported (Bornstein et al., 1991). It is not yet known whether the product of this gene accounts for any of the thrombospondin that has been biochemically or immunohistochemically identified so far, nor whether thrombospondin trimers exist that contain a mixture of chains from the two thrombospondin genes.

Thrombospondin was first identified as a molecule released by activated platelets, and was later found to be a major ECM protein secreted by endothelial and smooth muscle cells. Although thrombospondin secretion by cultured human glial cells was noted relatively early (Asch et al., 1986), full appreciation of the widespread expression of thrombospondin in the nervous system awaited the development of antibodies that could be used on rodent and avian tissue. It is now known that thrombospondin appears early and abundantly througout the developing CNS. In the developing cerebellum, thrombospondin appears transiently in particular layers (O'Shea et al., 1990). Unlike laminin, fibronectin, and tenascin, thrombospondin remains expressed at high levels in many parts of the CNS, suggesting that thrombospondin may play a role in mature brain functions.

Vitronectin

Vitronectin is an ECM glycoprotein that, like fibronectin, is present at relatively high levels in plasma. It is an unusually small ECM component, consisting of a single polypeptide chain of ~78 kd (Fig. 1). Because of vitronectin's presence in plasma, immunolocalization studies must be carefully controlled to rule out artefactual staining arising from vascular leakage. So far, only one region of the CNS, the developing retina (along with the optic nerve), has been shown to contain vitronectin (Neugebauer et al., 1991). Whether vitronectin persists in this location in the mature brain, and whether it is present elsewhere in the CNS, remains to be determined.

J1-160/180

Among the many groups to independently identify tenascin (see above), one isolated a monoclonal antibody, termed "J1," that additionally immunoprecipitates something else from mouse brain. This additional material is composed of polypeptide chains of 180 and 160 kd and is clearly immunologically distinguishable from tenascin (for discussion, see Morganti et al., 1990). Like tenascin, however, electron microcgraphs of J1-160/180 preparations reveal multi-armed structures (usually one, two, or three arms, as opposed to tenascin's six) attached at a central point. Based on these facts, and some of the biological activities of J1-160/180 (see below), it is reasonable to speculate that J1-160/180 will turn out to be structurally related to tenascin. Clarification of this issue currently awaits molecular cloning efforts.

Despite the apparent relatedness of J1-160/180 to tenascin, its tissue distribution is quite different. In the rodent, J1-160/180 appears only after birth, and is found only in white matter of the CNS, where it is associated with oligodendrocytes and the myelin they produce. J1-160/180 appears not to be expressed outside the nervous system, an almost unique characteristic for an ECM molecule.

Agrin

The discovery and characterization of agrin were the result of a concerted effort to identify a molecule involved in neuromuscular synaptogenesis. Although muscle cells are surrounded entirely by an ECM (a basement membrane or basal lamina), the matrix lying between nerve and muscle at synapses has long been known to contain unique components and possess distinctive biological activities (reviewed by Sanes, 1989). Intriguingly, the synaptic basal lamina alone is capable of causing muscle to assemble a postsynaptic apparatus. One of the most striking features of this apparatus is a very dense clustering of cell-surface acetylcholine receptors. Through a series of investigations, a molecule, agrin, was identified that could trigger acetylcholine receptor clustering *in vitro*. Recent molecular cloning experiments indicate that that molecule is a fragment of a large (~215 kd) ECM protein that contains regions of amino acid similarity to laminin and to certain protease inhibitors (Rupp et al., 1991). *In situ* hybridization experiments indicate that agrin is expressed in the developing nervous system—especially by spinal motoneurons—as well as the developing gut and aorta. Antibody staining experiments confirm that agrin is present at, and localized to, the synaptic basal lamina of neuromuscular junctions.

Proteoglycans

Proteoglycans are proteins that possess covalently attached carbohydrate chains known as glycosaminoglycans (GAGs). Proteoglycans are ubiquitous components of ECMs, and are also found on virtually all cell surfaces and within intracellular vesicles (including synaptic vesicles). From the point of view of their protein components (their "cores"), proteoglycans are certainly not a single family of molecules. Proteoglycans cores fall into many different families, and vary from very large (e.g., 400 kd) to very small (e.g., 10 kd); from membrane-spanning, to lipid-tailed, to secreted (for reviews see Gallagher, 1989; Lander, 1992).

In contrast, from the point of view of GAG structure, proteoglycans may be classified into three basic categories: those bearing GAGs of the heparinlike family, those bearing GAGs of the chondroitinlike family, and those bearing the GAG keratan sulfate. Each of these GAG types is synthesized onto the protein core as an unbranched linear polymer consisting of a simple disaccharide repeat; which of three basic disaccharide repeats is used determines which of the three families the GAG will belong to. Following chain polymerization, GAGs are chemically modified in a variety of ways. Most of these modifications introduce sulfate groups at various positions into some—but rarely all—of the disaccharide subunits. These sulfate groups give GAGs a strong negative charge. Another modification isomerizes glucuronic acid to iduronic acid.

One result of the chemical modifications that occur during GAG biosynthesis is to generate a great deal of complexity within each GAG family. It is likely that this is of considerable biological significance, since the biological functions of proteoglycans are likely to depend strongly on the binding properties of their GAG chains. Unfortunately, the full range of GAG complexity *in vivo* is not well understood, nor does any simple nomenclature exist that adequately conveys such complexity. Instead, GAGs are usually referred to by broad terms such as heparan sulfate (all members of the heparinlike family with the exception of heparin itself—a highly modified species produced only by mast cells); chondroitin sulfate (members of the chondroitin family containing little to no iduronic acid); dermatan sulfate (members of the chondroitin family containing large amounts of iduronic acid); and keratan sulfate. Although these labels are convenient for some purposes, those reading the literature should keep in mind that the "same" GAG chain—e.g., heparan sulfate—on any two proteoglycans may be structurally very different.

Perhaps in view of the inadequacies of GAG nomenclature, it has become standard practice to name proteoglycans according to their core proteins only. Thus, perlecan is a heparan sulfate proteoglycan found in basement membranes, while syndecan is a cell-surface proteoglycan that contains both heparan sulfate and chondroitin sulfate. One potential difficulty with this practice is the fact that, in a few cases, a core protein may bear one class of GAG in one cell type, a different type of GAG in another cell type, and no GAG at all in yet another.

Although heparan sulfate, chondroitin sulfate, dermatan sulfate, and keratan sulfate are all present in the nervous system (the first two being the most abundant), the considerable number of core proteins that bear these chains are only just now being identified. Many of these exhibit interesting developmental regulation of their expression (Herndon and

Lander, 1990). Some are integral membrane proteins, and are therefore not true components of the ECM. The remainder include some that are not readily extractable from neural tissue (as might be expected for an ECM component) and others that are surprisingly easy to solubilize. Immunohistochemical studies suggest that even these latter proteoglycans are frequently found in the ECM, suggesting that some of interactions holding proteoglycans in the nervous system ECM may be very weak.

Examples of proteoglycans that appear to be components of the ECM of the CNS are T1-PG, a widely distributed chondroitin sulfate proteoglycan (Iwata and Carlson, 1991a); 1D1, 3F8, and 3H1, two chondroitin sulfate proteoglycans and one mixed chondroitin- and keratan sulfate proteoglycan, respectively (Rauch et al., 1991); cytotactin-binding proteoglycan, a chondroitin sulfate proteoglycan that copurifies with tenascin (Hoffman et al., 1988); claustrin, a keratan sulfate proteoglycan (Cole and McCabe, 1991); and a chondroitin sulfate proteoglycan recognized by monoclonal antibody Cat-301 (Zaremba et al., 1989). The molecule recognized by Cat-301 is intriguing in that its expression surrounding neurons is controlled by neural (i.e., electrical) activity during development. Interestingly, no heparan sulfate proteoglycans have been identified as components of the ECM of the CNS. Whether it is only a matter of time before some are, or whether the CNS generally lacks heparan sulfate in the ECM, remains to be determined.

Proteoglycans are also found in the ECMs encountered by the PNS. In addition to many that are common components of interstitial matrices, some are associated with basement membranes of the PNS (e.g., Iwata and Carlson, 1991b), and one appears to be localized to the neuromuscular junction (Anderson and Fambrough, 1983).

Hyaluronic Acid- and Hyaluronate-Binding Proteins

Whereas most types of glycosaminoglycans are found as side chains of proteoglycans, one glycosaminoglycan, hyaluronic acid or hyaluronan, is not found covalently attached to protein. Hyaluronic acid is the simplest of GAGs, consisting of a linear polymer of the disaccharide [D-glucuronic acid $\beta(1\rightarrow3)$ D-N-acetylglucosamine $\beta(1\rightarrow4)$] that is not further chemically modified (e.g., by sulfation) as are other GAGs. Although structurally simple, hyaluronic acid chains are very long (M_rs in the range of 1,000 kd are common), giving hyaluronic acid solutions some unique biophysical properties. Among these properties is the ability to bind large amounts of water, and hyaluronic acid has been suggested to play widespread roles in controlling tissue hydration and the sizes of cell-free spaces *in vivo*.

The developing brain is particularly rich in hyaluronic acid. In addition, several brain ECM proteins have been identified to be hyaluronic acid-binding proteins. These include brain-specific hyaluronectin, a 68 kd protein, and glial hyaluronate-binding protein, a 60 kd protein (Perides et al., 1989). Glial hyaluronate-binding protein appears to occur only in the white matter of brain and spinal cord. Its amino acid sequence suggests that it may be derived (by alternative splicing or proteolytic cleavage) from the hyaluronic acid-binding domain of the core protein of the proteoglycan versican.

Collagens

Collagens are a large family of ECM glycoproteins. Collagen molecules consist of homo- or heterotrimers of polypeptides that contain regions of "collagenous" structure (amino acid sequences in which every third residue is glycine, and high levels of proline, lysine, hydroxyproline, and hydroxylysine are found). Because the virtual absence of collagens is a hallmark of the ECM of the CNS, collagens have not been extensively studied by neurobiologists. They are, however, abundant components of the ECM encountered by peripheral neurons and axons. Type IV collagen, in particular, is an abundant component of basement membranes, including those contacted by peripheral nerves during development and regeneration.

Other ECM Molecules

In recent years, several ECM molecules of unknown function have been identified in the nervous system. One of these, P30, is a small heparin-binding protein found at high levels in the developing CNS and PNS (Daston and Ratner, 1991). Another, SC1, was isolated by expression cloning from a rat brain cDNA library using an antibody directed against synaptic junction proteins. The cDNA cloned encodes a protein with a predicted size of 71 kd (the apparent size of the molecule synthesized *in vivo* is 120 kd) that exhibits substantial similarity to the ECM glycoprotein known as osteonectin or SPARC (Johnston et al., 1990). SC1 appears to be expressed by many cells of the adult CNS, including neurons of the cerebellum, cerebral cortex, and hippocampus. Most recently, a molecule showing some similarity to thrombospondin has been cloned and localized to the floor plate of the developing spinal cord (Klar et al., 1992). This protein is predicted to be 91 kd in size, and is also detected in adult kidney and brain.

Matrix-Associated Molecules

Although the above discussion has focused on what might be termed the "structural" components of ECM—those molecules that are abundant enough to play a role in determining overall ECM structure—there are many other molecules that, while normally thought of as soluble proteins, probably spend much of their time *in vivo* physically bound to the ECM. These molecules include polypeptide growth factors, secreted proteases and protease inhibitors, and other extracellular enzymes. Examples of each type of molecule are found in the nervous system: These include the acidic and basic fibroblast growth factors, the protease thrombin, the protease inhibitor known as protease nexin-1 (also known as glial-derived nexin), and one form of the synaptic enzyme acetylcholinesterase (Brandan et al., 1985; Dihanich et al., 1991; Eckenstein et al., 1991; Gloor et al., 1986). In all cases, it is interaction with glycosaminoglycans, particularly heparan sulfate, that appears to bind these molecules to the ECM. The biological significance of matrix association for these molecules is unclear. Potentially, the ECM can act to precisely localize these molecules (as in a synapse) or act as a storage depot, permitting the molecules to be mobilized by local ECM degradation. Alternatively, the ECM may alter the intrinsic properties of these molecules. Heparin and heparan sulfate are known to have such an effect on many proteases and on protease nexin-1, and appear to play a critical role in the functioning of acidic and basic fibroblast growth factor as well. Finally, association with the ECM is thought to dramatically influence the sensitivity of these molecules to proteolytic degradation. Clearly, much remains to be learned about matrix-associated molecules in the nervous system.

AMYLOID: A PATHOLOGICAL ECM

Under the broadest definition of ECM—insoluble material found between cells—must be included those extracellular deposits that occur in certain disease states. The amyloids are a large group of such deposits, and they are charactertized by a relatively uniform histological appearance, even though their molecular composition may be quite different depending on the disease process involved. Amyloidosis involving the nervous system is a key feature of Alzheimer's disease, as well as the so-called prion encephalopathies (Creutzfeld-Jakob disease, Gerstmann-Straussler syndrome, kuru).

As it turns out, the similarities between amyloids and normal ECMs go beyond the mere fact that both are insoluble. Mounting evidence indicates that proteoglycans are common, possibly invariant, components of amyloid deposits—including those in Alzheimer's and prion diseases—and suggests that proteoglycans may be involved in the initial formation of amyloid plaques (see Snow and Wight, 1989, for review). Other typical ECM components have also been identified in amyloid deposits, as have molecules such as protease inhibitors, which tend to associate with normal ECMs as well. These findings suggest that amyloid deposits may exhibit a structural organization not unlike that of conventional ECMs.

BINDING INTERACTIONS AMONG MOLECULES OF THE NERVOUS SYSTEM ECM

Although the ECM of the brain lacks the mechanical strength of connective tissue matrices, it is nonetheless held together by a complex network of intermolecular interactions. Figure 2 summarizes what is currently known about the binding interactions that occur among the ECM molecules described above. Taking for granted that many more interactions are likely still to be discovered, it is apparent that the ECM of the nervous system—like ECMs elsewhere—is potentially very complex in its organization. One interesting point that emerges from Figure 3 is the suggestion that proteoglycans may play a pivotal role in matrix organization in the nervous system, since the vast majority of ECM molecules interact directly with them. This interaction stems from the ability of most ECM proteins to bind heparan sulfate or, in a few cases, chondroitin sulfate. Just how important a role proteoglycans play in matrix organization will only become clear after the proteoglycan composition of the nervous system ECM is better understood.

BIOLOGICAL EFFECTS OF ECM

Molecules of the ECM have been shown to regulate cellular behaviors as diverse as proliferation, differentiation, migration, and, of particular interest in the nervous system, axon growth and guidance. In many cases, it has been possible to relate particular biological activities to particular domains of ECM molecules. Some of these relationships are illustrated in Figure 3. Such associations are based on experiments in which domain-specific antibodies, peptides derived from defined regions of amino acid sequence, and proteolytic fragments of ECM molecules, are used as reagents to mimic or block biological effects. One principle

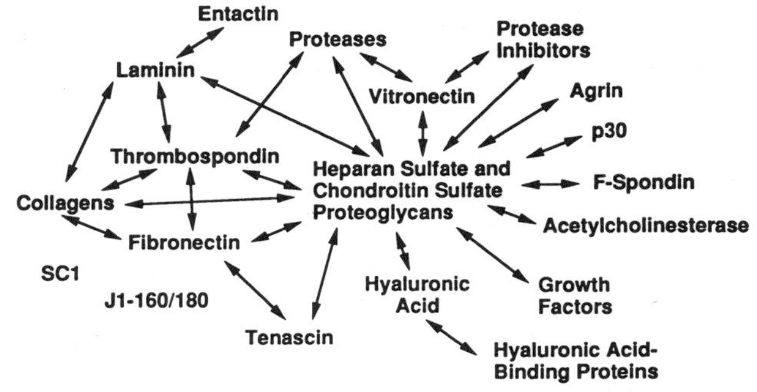

Fig. 2. Interactions among molecules of the ECM. Binding interactions that can occur between ECM molecules are indicated by arrows. Not indicated are examples of self-binding (aggregation or polymerization), which has been observed for laminin, fibronectin, thrombospondin, and collagens.

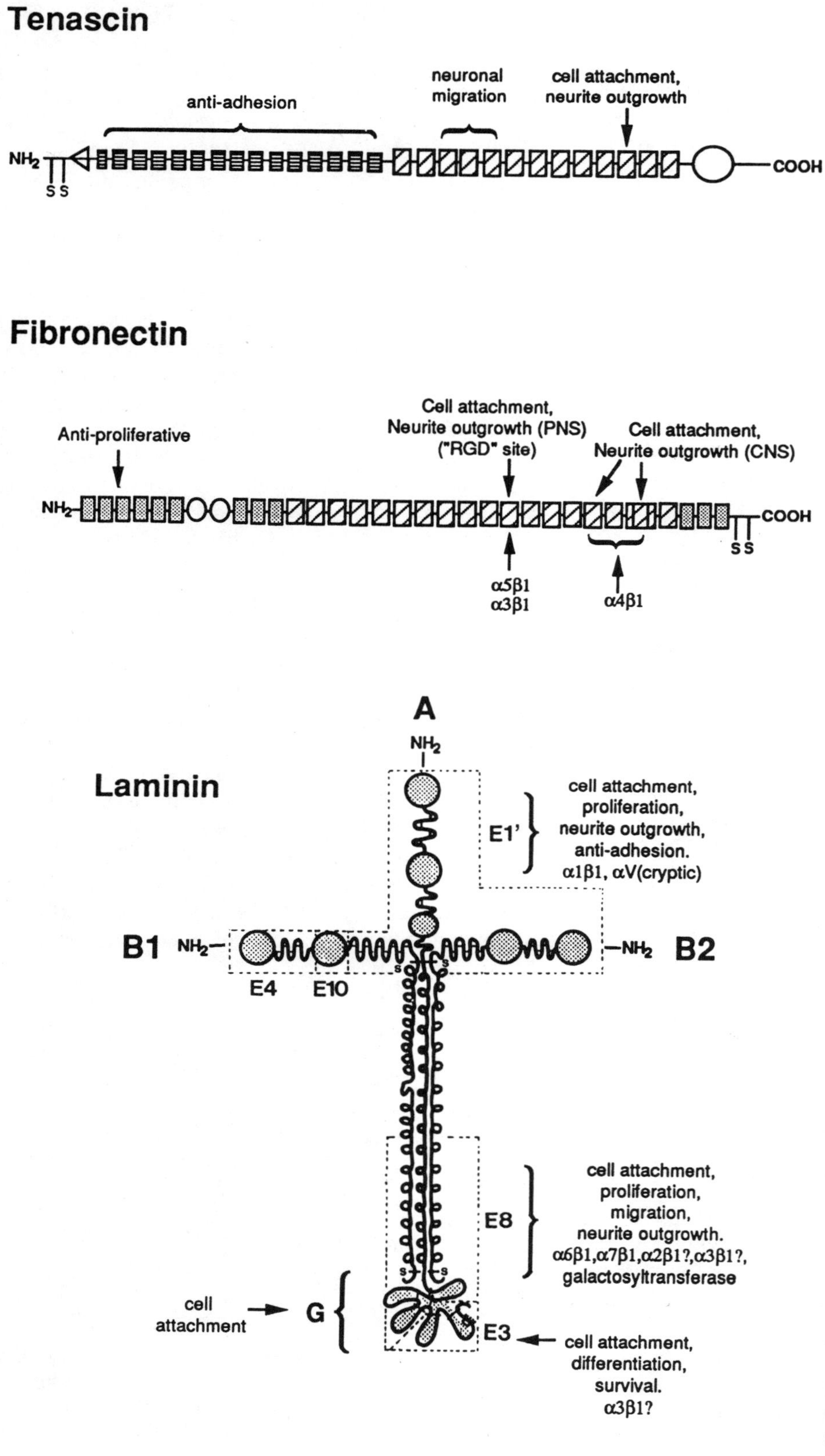

Fig. 3. Functional domains of ECM molecules. The association of biological activities and known receptor binding sites with particular domains of laminin, fibronectin, and tenascin are shown. For laminin, all three chains are named (A, B1, B2), and the regions comprised by elastase fragments (E1', E3, E8, E4, and E10) are indicated. Domain G of laminin represents the five shaded loops at the carboxyl end of the A chain, and partially overlaps the E3 and E8 fragments. For tenascin and fibronectin, only one of the polypeptide chains in each molecule is depicted. The diagram of laminin is modified from Yurchenco and Schittny (1990).

that seems to be emerging from such studies is that molecules of the ECM typically contain a number of spatially separate functional domains, and that these domains can have very different, even antagonistic, biological effects on cells. In this section, ECM effects on cells—especially on cells of the nervous system—are discussed as much as possible in terms of the molecules and molecular domains that elicit the effects. In the following section, the cellular receptors that mediate biological responses to ECM molecules are discussed.

Effects of ECM on Cell Proliferation, Differentiation, and Survival

Effects on cell proliferation, differentiation, and survival have been noted for laminin, fibronectin, thrombospondin, and tenascin. The suggestion that laminin may play a role in regulating cell proliferation has been made in several reports. In these studies, addition of soluble laminin to cell cultures, or growth of certain types of cells on laminin substrata, was shown to result in an increase in ^{3}H-thymidine incorporation and/or cell number, suggesting that laminin may have a mitogenic effect on some cells. There is as yet no consensus on the molecular domain(s) that mediate laminin's putative mitogenic effect, and there may be more than one: A peptide from the laminin A chain that contains the amino acid sequence IKVAV and is found in the E8 proteolytic fragment of laminin appears to be mitogenic for PC12 pheochromocytoma cells (Kubota et al., 1992), whereas a proteolytic fragment (E1-4) containing regions from the amino-termini of all three chains (A, B1, B2) stimulates growth of 3T3 fibroblasts (Panayotou et al., 1989)(see Fig. 3 for structures of laminin fragments). Laminin has also been shown to stimulate neuronal differentiation (i.e., the expression by cells of proteins that are characteristic of a neuronal phenotype) in adrenal chromaffin cells, a neural crest-derived cell type that possesses some biochemical characteristics of catecholaminergic neurons (Acheson et al., 1986). In this case, the biological effect has not been completely mapped, but sequences in the E3 fragment (part of the G-domain; see Fig. 3) are necessary. A similar requirement for the E3 domain has been noted for laminin's ability to promote neuronal survival under conditions where trophic factor support is limiting or absent (e.g., Edgar et al., 1984).

Fibronectin, in contrast to laminin, has been shown to have a growth inhibitory effect on some cells of the nervous system. This activity of fibronectin appears to be "cryptic", i.e., a proteolytic fragment of fibronectin, but not the intact molecule, possesses it. Growth control of normal Schwann cells by fibronectin appears to be part of an autocrine feedback loop, in which mitogen-stimulated cells secrete a metalloprotease that generates an amino-terminal fragment of the fibronectin molecule; it is this fragment (the 29 kd amino-terminal, heparin-binding fragment of fibronectin) that inhibits Schwann cell proliferation (Muir and Manthorpe, 1992). Interestingly, transformed Schwann cells do not secrete the activated form of the metalloprotease, suggesting one mechanism by which these cells might escape regulated growth control. A different domain of the fibronectin molecule appears to mediate effects on cell differentiation. This has been studied most extensively in cultures of human epidermal keratinocytes, where fibronectin has been shown to block gene expression characteristic of differentiated cells. This effect is mediated by the cell-binding fragment of fibronectin, and by a peptide (GRGDSP) contained within this fragment that is known to bind cellular ECM receptors of the integrin class (Adams and Watt, 1989). Fibronectin may be able to modulate expression of differentiated characteristics by neural crest cells as well (cf. Rogers et al., 1989).

Although both thrombospondin and tenascin have been shown to have mitogenic effects on cells (e.g., Chiquet-Ehrismann et al., 1986; Majack et al., 1986), there are as yet no reports of such effects on cells of the nervous system. It has been suggested that thrombospondin may function as part of an autocrine growth control pathway for vascular smooth muscle cells (Majack et al., 1986). A possible function for thrombospondin as an autocrine growth-stimulating factor would be particularly interesting in the developing nervous system, since widespread expression of thrombospondin has been observed at a time (E13 in the mouse embryo) when neuronal proliferation is proceeding at a high rate (cf. O'Shea and Dixit, 1988).

Effects of ECM on Axon Growth and Guidance

Laminin is well known as a potent stimulator of process (neurite) outgrowth by many types of cultured neurons, and its presence *in vivo* along pathways that growing axons follow (described above) supports the notion that it plays an important role in promoting axon growth during nervous system development (reviewed in Reichardt and Tomaselli, 1991; Sanes, 1989). *In vitro*, laminin is capable not only of stimulating neurite growth but also of precisely guiding it. Growth cones, the growing tips of neurites, accurately follow substratum pathways along which laminin has been deposited (cf. Gundersen, 1987). The neurite outgrowth effects of laminin appear to involve different molecular domains, depending on the cell type tested. For many neurons, this activity apparently resides in the E8 proteolytic fragment; however, for some the major neurite outgrowth-promoting activity maps to the E1-4 region (Fig. 3; reviewed by Beck et al., 1990).

Fibronectin's role in mediating axon growth and guidance during nervous system development is less well understood than that of laminin. Those CNS neurons that have been tested so far do not readily extend neurites on fibronectin substrata, but many PNS neurons do. Oddly, some CNS neurons (embryonic spinal cord) do extend neurites on a 33 kd proteolytic fragment of fibronectin, implying that this region of the molecule contains "cryptic" neurite outgrowth-promoting activity (Rogers et al., 1989). Unfortunately, much of the analysis of the effects of fibronectin on CNS neurons preceded studies of fibronectin expression in the brain (described above). In view of fibronectin's expression in the subplate of the cerebral cortex (Stewart and Pearlman, 1987), it would seem particularly valuable to test fibronectin for effects on those neurons that send axons into that region, namely thalamic and cortical projection neurons.

Tenascin, thrombospondin, and vitronectin, like laminin and fibronectin, have all recently been shown to stimulate neurite outgrowth by some classes of neurons (reviewed by Tomaselli and Neugebauer, 1991). The fact that so many ECM molecules can promote neurite outgrowth has raised the possibility that mere cell attachment to an ECM molecule might be all that is required for neurite extension. While this might be true for a few types of neurons, it does not appear to be the case generally. Indeed, degree of cell–substratum adhesion and degree of stimulation or guidance of neurite growth by ECM molecules do not correlate well (Gundersen, 1987; see Lander, 1990, for discussion). In fact, some neurons readily extend neurites in response to ECM molecules to which the cells themselves don't attach appreciably (Calof and Lander, 1991; Wehrle and Chiquet, 1990). Moreover, for some neurons, molecules that stimulate neurite outgrowth actually antagonize cell–substratum adhesion. This property has been termed anti-adhesiveness, and has been observed with neurons interacting with both laminin (Calof and Lander, 1991) and tenascin. Laminin's anti-adhesive activity appears to map to the E1' region of the molecule (Fig. 3; Calof et al., 1992). Tenascin's separable biological activities include not only neurite outgrowth-promoting activity (which maps to a site within its fibronectin-type III repeat units) but cell attachment activity (which maps to a potentially different site in the fibronectin type III repeats), anti-adhesive activity (which maps to the epidermal growth factor-like repeats near the amino-terminus), and a "neurite-repulsive" activity, i.e., the ability to render a substratum nonpermissive for neurite growth that has not yet been mapped (Fig. 3; Lochter et al., 1991; Spring et al., 1989). Interestingly, the ability to be either adhesive or "repulsive" (for cells or neurites) is also a characteristic of the tenascin-related molecule J1-160/180 (Morganti et al., 1990).

The neurite outgrowth-promoting activities of thrombospondin and vitronectin have not yet been mapped to specific domains, although there has been one report that thrombospondin's activity can be found in a 140 kd fragment containing the carboxy-terminal and cell-binding domains (O'Shea et al., 1991).

Understanding the role(s) that proteoglycans play in modulating axon outgrowth and guidance is made difficult by the tremendous complexity of this class of molecules and the large number of them expressed in the nervous system (cf. Herndon and Lander, 1990). No unifying concepts have yet emerged, and it is probably naive to expect proteoglycans all to exert similar effects on neurons. Several groups have reported that chondroitin sulfate and keratan sulfate proteoglycans inhibit neuronal attachment and neurite outgrowth, and in this light it has been suggested that proteoglycans may function in forming barriers to axon outgrowth during nervous system development (e.g., Cole and McCabe, 1991). At least one report, however, has suggested that the core protein of a brain-derived proteoglycan may stimulate neurite outgrowth by cortical neurons (Iijima et al., 1991).

Effects of ECM on Cell Migration

Neural crest cells, the mesenchymal cell type that gives rise to the peripheral and autonomic nervous systems, constitute the neural cell type whose migratory response to ECM has been studied most extensively. Fibronectin, collagens, laminin, and tenascin—all of which can be found in areas where neural crest cells migrate *in vivo*—all support neural crest cell migration in culture to some extent (e.g., Halfter et al., 1989; Perris and Bronner-Fraser, 1989; Reichardt and Tomaselli, 1991). It is possible that neural crest cells are intrinsically highly motile, and will migrate on any substratum that supports adequate attachment. The pathways they follow *in vivo* might then be determined more by the locations of molecules that inhibit their migration than by those that stimulate it. Some studies suggest that proteoglycans play such a migration-inhibiting role for these cells (cf. Perris et al., 1991).

In contrast to the behavior of neural crest cells, the migratory responses of other neural cells suggest that molecules of the ECM can function in an instructive manner, to specifically stimulate and guide cell migration. An interesting example of this specificity is the case of neurons and neuronal precursors of the nasal olfactory epithelium. *In vivo*, neuronal cells derived from the developing nasal epithelium migrate along nerve fiber tracts into the brain, where they give rise to the luteinizing hormone-releasing neurons of the hypothalamus. *In vitro*, neuronal cells from the embryonic olfactory epithelium are highly motile on substrata coated with laminin or its A-chain isoform, merosin, and migration of these cells is stimulated and guided by laminin but not by fibronectin, collagens, or nonspecifically adhesive substrata such as polylysine. Recent work suggests that laminin's effect on migration of these cells maps to the E8 fragment of the molecule (Fig. 3; Calof et al., 1992). It is interesting that the migration-stimulating activity of laminin maps neither to the molecule's anti-adhesive site (see above) nor to its most adhesive site for these cells, which is found in the G-domain, and involves the E3 region (Calof et al., 1992).

Migration of granule cells during histogenesis of the cerebellum may prove to be an example of ECM molecules acting sequentially to regulate neuronal migration through a complex and changing cellular environment (reviewed in Reichardt and Tomaselli, 1991). Thrombospondin appears to play an important role in facilitating migration of granule cells from the external granule layer (EGL) into the molecular layer of the developing cerebellum, since antibodies to thrombospondin prevent granule cells from entering the molecular layer in cultured cerebellar slices. Tenascin apparently functions at the later stage of promoting granule cell migration through the molecular layer: Antibodies to tenascin cause granule cells to accumulate in the molecular layer, failing to complete their migration through it and into the internal granule layer. A recent study utilizing domain-specific, function-blocking monoclonal antibodies in cerebellar slice cultures indicates that the migration-promoting activity of tenascin maps to a domain between the third and fifth fibronectin type III repeats in the carboxy-terminal portion of the molecule (Fig. 3; Husmann et al., 1992). This domain of tenascin is close to, but separate from, the domain that contains neurite outgrowth-promoting, anti-adhesive, and attachment activity.

CELLULAR RECEPTORS FOR ECM

Studies of the mechanisms of cell response to the ECM have concentrated in recent years on the search for cellular receptors for ECM molecules. Several of the classes of cell-surface molecules that have the ability to bind ECM components are described below, and are summarized in Table 1.

TABLE 1. Known Interactions of ECM Proteins with Cell Surface Molecules

	Integrins	Proteo-glycans	Glyco-lipids	Others
Laminin	+	+	+	Lectins, galactosyltransferase, other proteins
Fibronectin	+	+	+	Other proteins
Thrombospondin	+	+	+	
Tenascin	+	+		
Vitronectin	+	+		
Collagens	+	+		Other proteins
Hyaluronic acid		+		CD44/H-CAM

Structure and Function of the Integrin Family of Receptors

The most widely studied class of cellular receptors for ECM molecules is the superfamily of integral membrane proteins known as integrins. A recent review describes the structure and function of the members of this large (20 known members and still growing) and complex group of receptors, and the reader is referred to that paper for a detailed discussion of integrins (Hynes, 1992). For the purposes of this discussion, some basic features of integrin structure are summarized.

The integrins are heterodimers of noncovalently linked subunits, referred to as α and β Both subunits are transmembrane glycoproteins, and both appear to contribute to ligand binding by the dimer. The structural and functional complexity of the integrins appears to be the result of two factors: formation of different heterodimers, resulting from different pairings of the eight β subunits and the 13 α subunits, and structural variation in the subunits themselves, resulting from alternative splicing of their cytoplasmic domains. Integrins, the majority of which appear to function as receptors for ECM molecules (but see Hynes, 1992, and Hynes and Lander, 1992), are expressed by virtually every cell type, and most cells express more than one integrin. So far as is known, each of the 20 or so heterodimers has a distinct ligand specificity. However, it is important to point out that this does not mean that each integrin heterodimer can bind only one ligand. Many integrins can bind more than one ECM molecule, and a given ECM molecule can be recognized by more than one integrin. For example, there is evidence for at least six integrin "laminin receptors"—$\alpha 1\beta 1$, $\alpha 2\beta 1$, $\alpha 3\beta 1$, $\alpha 6\beta 1$, $\alpha 7\beta 1$, and $\alpha 6\beta 4$—and the first three of these are known to function as receptors for other ECM molecules as well (cf. Hynes, 1992).

Integrin receptors appear to be important in mediating attachment, migration, and axon outgrowth by neural cells in response to ECM. The avian retinotectal pathway provides an example of how regulation of integrin expression and function may be important in governing axon outgrowth during nervous system development: At a time in development when retinal ganglion cells are sending axons along a laminin-rich pathway to the brain (embryonic day 6 in the chick), they are highly responsive to laminin, as shown by neurite outgrowth assays on laminin substrata *in vitro*. Later in development, at the time when most retinal ganglion cell axons have reached their targets in the laminin-poor optic tectum (around embryonic day 11 or 12), the ability of retinal ganglion cells to extend neurites on laminin substrata is lost (Cohen et al., 1986). The time course of loss of responsiveness to laminin correlates with decreased expression of the α6 integrin subunit by retinal ganglion cells, suggesting that regulation of integrin laminin receptor ($\alpha 6\beta 1$) expression is one means by which neuronal responsiveness to the ECM is regulated (de Curtis et al., 1991). Interestingly, other retinal neurons also lose the ability to attach to and extend neurites on laminin at embryonic day 11–12, but α6 expression is not down-regulated in these cells. This suggests that an integrin's function may be suppressed, even when it is present at normal levels on the cell surface, and that this may be a second means by which neuronal responsiveness to the ECM is regulated during development. This idea is supported indirectly by the existence of a monoclonal antibody that binds the β1 integrin subunit and effectively restores the ability of embryonic day 11 retinal neurons to respond to laminin (Neugebauer and Reichardt, 1991). Since this antibody recognizes a site on the β1 integrin subunit that is not involved in ligand binding, its activity appears to represent a case of allosteric modulation of integrin function. In other words, it appears that integrins can exist in active and inactive conformations that can be interconverted through the action of other integrin-binding molecules. Evidence suggests that signals inside cells, as well as molecules outside cells, can control integrin activation (Hynes, 1992).

Although multiple integrins can act as receptors for the same ECM protein, the locations of binding sites for integrins on proteins tend to vary (Fig. 3). The locations of integrin binding sites are in turn directly related to the biological activities of some of the molecular domains that were discussed in the previous section. For example, the neurite outgrowth-promoting activity displayed by the E1' fragment of laminin requires the receptor function of integrin α1β1, which binds to E1', while the neurite outgrowth-promoting activity of the E8 fragment requires either integrin α3β1 or α6β1, which interact with sites in the E8 fragment (Tomaselli et al., 1990). It is likely that differences in the responses of different cell types to an ECM molecule and its fragments frequently result from differences in integrin expression.

Clearly, it will be important to characterize the expression of integrins by many of the cell types in the nervous system, a process that has only just begun. In theory, one might be able to predict some of the responses of a cell to its ECM simply by knowing its complement of integrins and the ECM molecules it is encountering. Unfortunately, things are unlikely to be that simple. Not only are integrins sometimes present but inactive on cell surfaces (as described above for some retinal neurons), but a single type of integrin may mediate different biological responses in different cells. For example, the integrin α6β1 mediates cell adhesion to laminin in many cells, neurite outgrowth in response to laminin in other cells (e.g., retinal cells), and stimulation of migration without stimulation of cell adhesion in other cells (e.g., olfactory receptor neurons; Calof et al., 1992).

These considerations suggest that the key to understanding how integrins mediate cellular responses to the ECM will lie in understanding the intracellular signals transduced by integrins, and how these signals converge with others to affect cell behavior. The idea that integrins transmit signals into cells has only recently gained wide acceptance, as evidence accumulates that integrin interactions with ECM molecules can elicit some of the hallmarks of signal transduction (e.g., protein tyrosine phosphorylation, elevation of cytoplasmic pH). For a recent discussion of this emerging field, the reader is referred to the review by Hynes (1992).

Nonintegrin Receptors for ECM Molecules

Several other cell-surface proteins, besides integrins, may function as receptors for ECM. A 100 kd protein isolated from brain on the basis of its ability to bind laminin is one of several candidate nonintegrin laminin receptors (Smallheiser and Schwartz, 1987). Two others may function as lectins: CBP35 (a 35 kd carbohydrate-binding protein also known as Mac 2), a laminin-binding protein expressed on macrophage cell surfaces (Woo et al., 1990), and a 67 kd laminin-binding protein isolated from tumor cells, which may be identical to an elastin receptor that functions as a galactose-specific lectin (Mecham et al., 1989; Sephel et al., 1989). Functional roles for these carbohydrate-binding proteins in mediating biological activities of laminin such as adhesion, migration, and neurite outgrowth have not yet been established. However, there is evidence that the carbohydrate moieties present on laminin do play a role in these activities, since neurite outgrowth by PC12 pheochromocytoma cells is impaired on unglycosylated laminin (reviewed by Mercurio, 1990).

Evidence from investigations of another putative laminin receptor also indicates that carbohydrate groups on laminin are important to neural cells. Cell-surface β(1→4)-galactosyltransferase is a cell-surface enzyme that appears to play some role in several cell interactions with laminin, including cell migration and neurite outgrowth. Galactosyltransferase binds to N-linked oligosaccharides on laminin, and interaction of this enzyme with oligosaccharides present on the E8 region of the molecule is apparently important for PC12 pheochromocytoma cell neurite outgrowth on laminin (Begovac et al., 1991). Neural crest cells also express galactosyltransferase, and agents that interfere with the activity of the enzyme, or antibodies directed against it, perturb neural crest migration *in vivo* and *in vitro* (Hathaway and Shur, 1992).

As described earlier, most ECM proteins possess domains that bind to the GAGs found on proteoglycans. Since proteoglycans are found not only in the ECM but also on most cell surfaces, it is not surprising that proteoglycans have been suggested to act as receptors for ECM molecules. The best studied proteoglycan in this regard is syndecan, a transmembrane protein that bears both heparan sulfate and chondroitin sulfate chains. Syndecan can mediate adhesion of epithelial cells to fibronectin, collagen, and thrombospondin (Bernfield and Sanderson, 1990; Sun et al., 1989). Although syndecan is not detectable in the CNS, other integral membrane proteoglycans are, including N-syndecan (a proteoglycan with a region of homology to syndecan; Carey et al., 1992), NG2 (a proteoglycan present on certain glial cells; Nishiyama et al., 1991), glypican (Litwack et al.,

1992), and several molecules that have been only partly characterized (Herndon and Lander, 1990). Interestingly, some of these proteoglycans are linked to the plasma membrane via phosphatidylinositol glycan linkages, providing a potential mechanism by which proteoglycans might function to transmit signals into cells.

Finally, several ECM molecules have been found to bind various types of cell-surface glycolipids, especially gangliosides and sulfated lipids (Table 1). It is not yet known whether such molecules act as true receptors for ECM molecules. However, if such a role should be demonstrated, it will most likely have important implications for the nervous system, since a high abundance and diversity of glycolipid expression is a characteristic of neural tissue.

CONCLUSIONS

In only a few decades, the ECM of the nervous system has progressed from being an entity not widely recognized to exist to being a critical controller of virtually all aspects of cellular behavior. The functions of the nervous system ECM derive not merely from its ability to act as a scaffold or a glue to bind cells together. Instead, the ECM consists of numerous biologically active components, each containing distinct functional domains capable of having a wide variety of effects on cells. The effects of ECM molecules are mediated through diverse cellular receptors, including integrins, proteoglycans, lectins, cell-surface enzymes, and other molecules. One challenge of the next few years will be to determine how the signals transmitted to cells via the ECM cooperate with developmental signals to drive the morphogenesis of the nervous system.

REFERENCES

Acheson, A., D. Edgar, R. Timpl. et al. (1986) Laminin increases both levels and activity of tyrosine hydroxylase in calf adrenal chromaffin cells. J. Cell Biol. 102:151–159.

Adams, J.C., and F.M. Watt (1989) Fibronectin inhibits the terminal differentiation of human keratinocytes. Nature 340:307–309.

Anderson, M.J., and D.M. Fambrough (1983) Aggregates of acetylcholine receptors are associated with plaques of a basal lamina heparan sulfate proteoglycan on the surface of skeletal muscle fibers. J. Cell Biol. 97:1396–1411.

Asch, A.S., L.L.K. Leung, J. Shapiro, et al. (1986) Human brain glial cells synthesize thrombospondin. Proc. Natl. Acad. Sci. USA 83:2904–2908.

Beck, K., I. Hunter, and J. Engel (1990) Structure and function of laminin: Anatomy of a multidomain glycoprotein. FASEB J. 4:148–160.

Begovac, P.C., D.E. Hall, and B.D. Shur (1991) Laminin fragment E8 mediates PC12 cell neurite outgrowth by binding to cell surface *beta*1,4 galactosyltransferase. J. Cell Biol. 113:637–644.

Bernfield, M., and R.D. Sanderson (1990) Syndecan, a developmentally regulated cell surface proteoglycan that binds extracellular matrix and growth factors. Philos. Trans. R. Soc. Lond. (B) 327:171–186.

Bornstein, P., K. O'Rourke, K. Wilkstrom, et al. (1991) A second, expressed thrombospondin gene (Thbs2) exists in the mouse genome. J. Biol. Chem. 266:12821–12824.

Brandan, E., M. Maldonado, J. Garrido, et al. (1985) Anchorage of collagen-tailed acetylcholinesterase to the extracellular matix is mediated by heparan sulfate proteoglycans. J. Cell Biol. 101:985–992.

Calof, A.L., and A.D. Lander (1991) Relationship between neuronal migration and cell-substratum adhesion: Laminin and merosin promote olfactory neuronal migration but are anti-adhesive. J. Cell Biol. 115:779–794.

Calof, A.L., P.D. Yurchenco, J.J. O'Rear, et al. (1992) Adhesion, anti-adhesion, and migration of olfactory neurons and neuronal precursors are regulated by distinct molecular domains of laminin. Soc. Neurosci. Abstr. (in press).

Carey, D.J., D.M. Evans, R.C. Stahl, et al. (1992) Molecular cloning and characterization of N-Syndecan, a novel transmembrabe heparan sulfate proteoglycan. J. Cell Biol. 117:191–201.

Chiquet, M. (1989) Tenascin/J1/Cytotactin: The potential function of hexabrachion proteins in neural development. Dev. Neurosci. 11:266–275.

Chiquet-Ehrismann, R., E.J. Mackie, C.A. Pearson, et al. (1986) Tenascin: An extracellular matrix protein involved in tissue interactions during fetal development and oncogenesis. Cell 47:131–139.

Cohen, J., J.F. Burke, C. McKinlay, et al. (1987) The role of laminin and the laminin/fibronectin receptor complex in the outgrowth of retinal ganglion cell axons. Dev. Biol. 122:407–418.

Cohen, J., J.F. Burne, J. Winter, et al. (1986) Retinal ganglion cells lose response to laminin with maturation. Nature 322:465–467.

Cole, G.J., and C.F. McCabe (1991) Identification of a developmentally regulated keratan sulfate proteoglycan that inhibits cell adhesion and neurite outgrowth. Neuron 7:1007–1018.

Daston, M.M., and N. Ratner (1991) Expression of P30, a protein with adhesive properties, in Schwann cells and neurons of the developing and regenerating peripheral nerve. J. Cell Biol. 112:1229–1239.

de Curtis, I., V. Quaranta, R.N. Tamura, et al. (1991) Laminin receptors in the retina: Sequence analysis of the chick integrin a6 subunit. J. Cell Biol. 113:405–416.

Dihanich, M., M. Kaser, E. Reinhard, et al. (1991) Prothrombin mRNA is expressed by cells of the nervous system. Neuron 6:575–581.

Eckenstein, F.P., G.D. Shipley, and R. Nishi (1991) Acidic and basic fibroblast growth factors in the nervous system: Distribution and differential alteration of levels after injury of central versus peripheral nerve. J. Neurosci. 11:412–419.

Edgar, D., R. Timpl, and H. Thoenen (1984) The heparin-binding domain of laminin is responsible for its effects on neurite outgrowth and neuronal survival. EMBO J. 3:1463–1468.

Erickson H.P., and M.A. Bourdon (1989) Tenascin: An extracellular matrix protein prominent in specialized embryonic tissues and tumors. Annu. Rev. Cell Biol. 5:71–92.

Gallagher, J.T. (1989) The extended family of proteoglycans: Social residents of the pericellular zone. Curr. Opin. Cell Biol. 1:1201–1218.

Gloor, S., K. Odink, J. Guenther, et al. (1986) A glia-derived neurite promoting factor with protease inhibitory activity belongs to the protease nexins. Cell 47:687–693.

Gundersen, R.W. (1987) Response of sensory neurites and growth cones to patterned substrata of laminin and fibronectin in vitro. Dev. Biol. 121:423–431.

Halfter, W., R. Chiquet-Ehrismann, and R.P. Tucker (1989) The effect of tenascin and embryonic basal lamina on the behavior and morphology of neural crest cells in vitro. Dev. Biol. 132:14–25.

Hathaway, H.J., and B.D. Shur (1992) Cell surface *beta*(1,4)-galactosyltransferase functions during neural crest cell migration and neurulation in vivo. J. Cell Biol. 117:369–382.

Herndon, M.E., and A.D. Lander (1990) A diverse set of developmentally regulated proteoglycans is expressed in rat central nervous system. Neuron 4:949–961.

Hoffman, S., K.L. Crossin, and G.M. Edelman (1988) Molecular forms, binding functions and developmental expression patterns of cytotactin and cytotactin-binding proteoglycan, an interactive pair of extracellular matrix molecules. J. Cell Biol. 106:519–532.

Husmann, K., A. Faissner, and M. Schachner (1992) Tenascin promotes cerebellar granule cell migration and neurite outgrowth by different domains in the fibronectin type III repeats. J. Cell Biol. 116:1475–1486.

Hynes, R.O. (1990) Fibronectins. New York: Springer Verlag, 546 pp.

Hynes, R.O. (1992) Integrins: Versatility, modulation, and signaling in cell adhesion. Cell 69:11–25.

Hynes, R.O., and A.D. Lander (1992) Contact and adhesive specificities in the associations, migrations, and targeting of cells and axons. Cell 68:303–322.

Iijima, N., A. Oohira, T. Mori, et al. (1991) Core protein of chondroitin sulfate proteoglycan promotes neurite outgrowth from cultured neocortical neurons. J. Neurochem. 56:706–708.

Iwata, M., and S.S. Carlson (1991a) A general component of adult brain extracellular matrix, the T1 chondroitin sulfate proteoglycan. Soc. Neurosci. Abstr. 17:575.

Iwata, M., and S.S. Carlson (1991b) A large chondroitin sulfate basement membrane-associated proteoglycan exists as a disulfide-stabilized complex of several proteins. J. Biol. Chem. 266:323–333.

Johnston, I.G., T. Paladino, J.W. Gurd, et al. (1990) Molecular cloning of SC1: A putative brain extracellular matrix glycoprotein showing partial similarity to osteonectin/BM40/SPARC. Neuron 2:165–176.

Klar, A., M. Baldassare, and T.M. Jessell (1992) F-spondin: A gene expressed at high levels in the floor plate encodes a secreted protein that promotes neural cell adhesion and neurite extension. Cell 69:95–110.

Kubota, S., K. Tashiro, and Y. Yamada (1992) Signaling site of laminin with mitogenic activity. J. Biol. Chem. 267:4285–4288.

Lander, A.D. (1989) Understanding the molecules of neural cell contacts: Emerging patterns of structure and function. Trends Neurosci. 12:189–195.

Lander, A.D. (1990) Mechanims by which molecules guide axons. Curr. Opin. Cell Biol. 2:907–913.

Lander, A.D. (1992) Proteoglycans. In: T. Kreis and R. Vale (eds.). Guidebook to Cytoskeleton and Structural Proteins. New York: Oxford University Press (in press).

Lawler, J., and R.O. Hynes (1986) The structure of human thrombospondin, an adhesive glycoprotein with multiple calcium-binding sites and homologies with several different proteins. J. Cell Biol. 103:1635–1648.

Letourneau, P.C., A.M. Madsen, S.L. Palm, et al. (1988) Immunoreactivity for laminin in the developing ventral longitudinal pathway of the brain. Dev. Biol. 125:135–144.

Liesi, P., S. Kaakkola, D. Dahl, et al. (1984) Laminin is induced in astrocytes of adult brain by injury. EMBO J. 3:683–686.

Litwack, E.D., C.S. Stipp, and A.D. Lander (1992) Glypican is regionally expressed in the developing rat brain. Soc. Neurosci. Abstr. (in press).

Lochter, A., L. Vaughan, A. Kaplony, et al. (1991) J1/tenascin in substrate-bound and soluble form displays contrary effects on neurite ourgrowth. J. Cell Biol. 113:1159–1171.

Majack, R.A., S.C. Cook, and P. Bornstein (1986) Control of smooth muscle cell growth by components of the extracellular matrix: Autocrine role for thrombospondin. Proc. Natl. Acad. Sci. USA 83:9050–9054.

Mecham, R.P., A. Hinek, G.L. Griffin, et al. (1989) The elastin receptor shows structural and functional similarities to the 67 kDa tumor cell laminin receptor. J. Biol. Chem. 264:16652–16657.

Mercurio, A.M. (1990) Laminin: Multiple forms, multiple receptors. Curr. Opin. Cell Biol. 2:845–849.

Morganti, M.C., J. Taylor, P. Pesheva, et al. (1990) Oligodendrocyte-derived J1-160/180 extracellular matrix glycoproteins are adhesive or repulsive depending on the partner cell type and time of interaction. Exp. Neurol. 109:98–110.

Muir, D., and M. Manthorpe (1992) Stromelysin generates a fibronectin fragment that inhibits Schwann cell proliferation. J. Cell Biol. 116:177–185.

Neugebauer, K.M., C.J. Emmett, K.A. Venstrom, et al. (1991) Vitronectin and thrombospondin promote retinal neurite outgrowth: Developmental regulation and role of integrins. Neuron 6:345–358.

Neugebauer, K.M. and L.F. Reichardt (1991) Cell surface regulation β1-integrin activity on developing retinal neurons. Nature 350:68-71.

Nishiyama, A., K.J. Dahlin, J.T. Prince, et al. (1991) The primary structure of NG2, a novel membrane-spanning proteoglycan. J. Cell Biol. 114:359–371.

O'Shea, K.S., and V.M. Dixit (1988) Unique distribution of the extracellular matrix component thrombospondin in the developing mouse embryo. J. Cell Biol. 107:2737–2748.

O'Shea, K.S., L.-H.J. Liu, and V.M. Dixit (1991) Thrombospondin and a 140 kD fragment promote adhesion and neurite outgrowth from embryonic central and peripheral neurons and from PC12 cells. Neuron 7:231–237.

O'Shea, K.S., J.S.T. Rheinheimer, and V.M. Dixit (1990) Deposition and role of thrombospondin in the histogenesis of the cerebellar cortex. J. Cell Biol. 110:1275–1283.

Panayotou, G., P. End, M. Aumailley, et al. (1989) Domains of laminin with growth-factor activity. Cell 56:93–101.

Perides, G., W.S. Lane, D. Andrews, et al. (1989) Isolation and characterization of a glial hyaluronate-binding protein. J. Biol. Chem. 264:5981–5987.

Perris, R., and M. Bronner-Fraser (1989) Recent advances in defining the role of the extracellular matrix in neural crest development. Comm. Dev. Neurobiol. 1:61–83.

Perris, R., D. Krotoski, T. Lallier, et al. (1991) Spatial and temporal changes in the distribution of proteoglycans during avian neural crest development. Development 111:583–599.

Rauch, U., P. Gao, A. Janetzko, et al. (1991) Isolation and characterization of developmentally regulated chondroitin sulfate and chondroitin/keratan sulfate proteoglycans of brain identified with monoclonal antibodies. J. Biol. Chem. 266:14785–14801.

Reichardt, L.F., and K.J. Tomaselli (1991) Extracellular matrix molecules and their receptors: Functions in neural development. Annu. Rev. Neurosci. 14:531–570.

Rogers, S.L., K.J. Edson, P.C. Letourneau et al. (1986) Distribution of laminin in the developing peripheral nervous system of the chick. Dev. Biol. 113:429-435

Rogers, S.L., P.C. Letourneau, and I.V. Pech (1989) The role of fibronectin in neural development. Dev. Neurosci. 11:248–265.

Rupp, F., D.G. Payan, C. Magill-Solc, et al. (1991) Structure and expression of a rat agrin. Neuron 6:811–823.

Sanes, J.R. (1989) Extracellular matrix molecules that influence neural development. Annu. Rev. Neurosci. 12:491–516.

Sanes, J.R., E. Engvall, R. Butkowski, et al. (1990) Molecular heterogeneity of basal laminae: Isoforms of laminin and collagen IV at the neuromuscular junction and elsewhere. J. Cell Biol. 111:1685–1699.

Sephel, G.C., B.A. Burrous, and H.K. Kleinman (1989) Laminin neural activity and binding proteins. Dev. Neurosci. 11:313–331.

Sheppard, A.M., S.K. Hamilton, and A.L. Pearlman (1991) Changes in the distribution of extracellular matrix components accompany early morphogenetic events of mammalian cortical development.

J. Neurosci. 11:3928–3942.

Smallheiser, N.R., and N.B. Schwartz (1987) Cranin: A laminin-binding protein of cell membranes. Proc. Natl. Acad. Sci. USA 84:6457–6461.

Snow, A.D., and T.N. Wight (1989) Proteoglycans in the pathogenesis of Alzheimer's Disease and other amyloidoses. Neurobiol. Aging 10:481–497.

Spring, J., K. Beck, and R. Chiquet-Ehrismann (1989) Two contrary functions of tenascin: Dissection of the active sites by recombinant tenascin fragments. Cell 59:325–334.

Stewart, G.R., and A.L. Pearlman (1987) Fibronectin-like immunoreactivity in the developing cerebral cortex. J. Neurosci. 7:3325–3333.

Sun, S., D.F. Mosher, and A. Rapraeger (1989) Heparan sulfate-mediated binding of epithelial cell surface proteoglycans to thrombospondin. J. Biol. Chem. 264:2885–2889.

Tomaselli, K.J., D.E. Hall, L.A. Flier, et al. (1990) A neuronal cell line (PC12) expresses two B1-class integrins—α1β1 and α3β1—that recognize different neurite outgrowth-promoting domains in laminin. Neuron 5:651–662.

Tomaselli, K.J., and K.M. Neugebauer (1991) 2001 interactions? An extracellular space odyssey. Curr. Opin. Neurobiol. 1:364–369.

Wehrle, B., and M. Chiquet (1990) Tenascin is accumulated along developing peripheral nerves and allows neurite outgrowth in vitro. Development 110:401–415.

Woo, H.J., L.M. Shaw, J.M. Messier, et al. (1990) The major non-integrin laminin binding protein of macrophages is identical to carbohydrate binding protein 35 (Mac-2). J. Biol. Chem. 265:7097–7099.

Yamada, K.M. (1990) Fibronectins: Structure, functions and receptors. Curr. Opin. Cell Biol. 1:956–963.

Yurchenco, P.D., and J.C. Schittny (1990) Molecular architecture of basment membranes. FASEB J. 4:1577–1590.

Zaremba, S., A. Guimaraes, R.G. Kalb, et al. (1989) Characterization of an activity dependent neuronal surface proteoglycan identified with monoclonal antibody Cat-301. Neuron 2:1207–1219.

Chapter 32

Mechanisms of Metastasis

Dario Rusciano and Max M. Burger

INTRODUCTION

Most cancer deaths occur as a consequence of the spreading of the original tumor to secondary locations in distant organs. Metastatic disease is still the leading cause of cancer lethality. Even though tumor treatment has improved in recent years, the eradication of most of an early-detected tumor may still leave enough malignant cells that have already spread to initiate new growth, diversification, and further spreading. Therefore, devising new drugs or therapies to prevent tumor spread and thus improving the long-term survival of cancer patients requires a more comprehensive understanding of the molecular mechanisms underlying the metastatic distribution of tumor cells.

The most characteristic feature of cancer is uncontrolled cell proliferation. How a normally regulated cell can lose this control has been the subject of many studies in recent years, and the quest for answers has led to the discovery of dominant acting genes (oncogenes) and tumor suppressor genes. There is strong evidence that the development of cancer is a multistage process, in which both the activation of proto-oncogenes and the inactivation of tumor suppressor genes play critical roles (Hunter, 1991).

However, a tumor cell is not necessarily a metastatic cell, since unrestrained growth does not, by itself, convey properties that result in invasion and metastasis. The neoplastic cell must possess other properties that allow it to establish a metastatic colony. These properties include: efficient growth and invasion at the primary sites; escape from the primary tumor and penetration of the circulatory system either as single cells or small emboli; survival of viable cells while circulating; arrest in the microvasculature of other organs and extravasation into the organ parenchyma; and efficient growth and invasion at the secondary site (Fig. 1). Only those malignant cells that can fulfill all these requirements will be successful in establishing metastatic colonies.

Metastases are not randomly distributed among different organs. Some tumors tend to colonize specific organs. This phenomenon was first described by Paget in 1889, who suggested that malignant cells, like the seeds of a plant, may disseminate in several directions, but only survive and grow in a congenial organ (soil). In contrast, Ewing in 1928 proposed that the vascular connections of the primary tumor determine the route of dissemination and sites of vascular arrest of malignant cells, and thus their metastatic pattern.

Both theories must be taken into consideration when attempting to explain the organ colonization pattern of most tumors. We recently reviewed the possible specific interactions that underlie the generation of organ-specific metastatic colonies (Rusciano and Burger, 1992).

ANGIOGENESIS

Growth of a solid tumor, whether a primary tumor or a metastasis, beyond 1 or 2 mm in diameter (the largest size at which passive diffusion can still efficiently provide nutrients and dispose of catabolites), requires that tumor cells be able to induce neovascularization of the tumor (Blood and Zetter, 1990). The absence of angiogenesis precludes expansion of the tumor population beyond 1–2 mm, notwithstanding the proliferative capacity of tumor cells. A tumor *in situ* is avascular, and it is possible for such small tumors to remain dormant for a long time, until something occurs to provoke further growth and vascularization. Moreover, since new capillary blood vessels have little organized extracellular matrix (Blood and Zetter, 1990), tumor cells can easily reach the vascular lumen and escape to secondary sites. This suggests that angiogenesis is relevant both for the growth of the primary tumor and for the growth of its metastases.

No relationship has been reported between site-specific metastasis and angiogenesis. The property of inducing angiogenesis seems to be intrinsic to the tumor and not related to the particular microenvironment in which it lodges. The possibility exists that the production of angiogenic factors by tumor cells may be influenced by their interactions

Molecular Genetics of Nervous System Tumors, pages 357-369

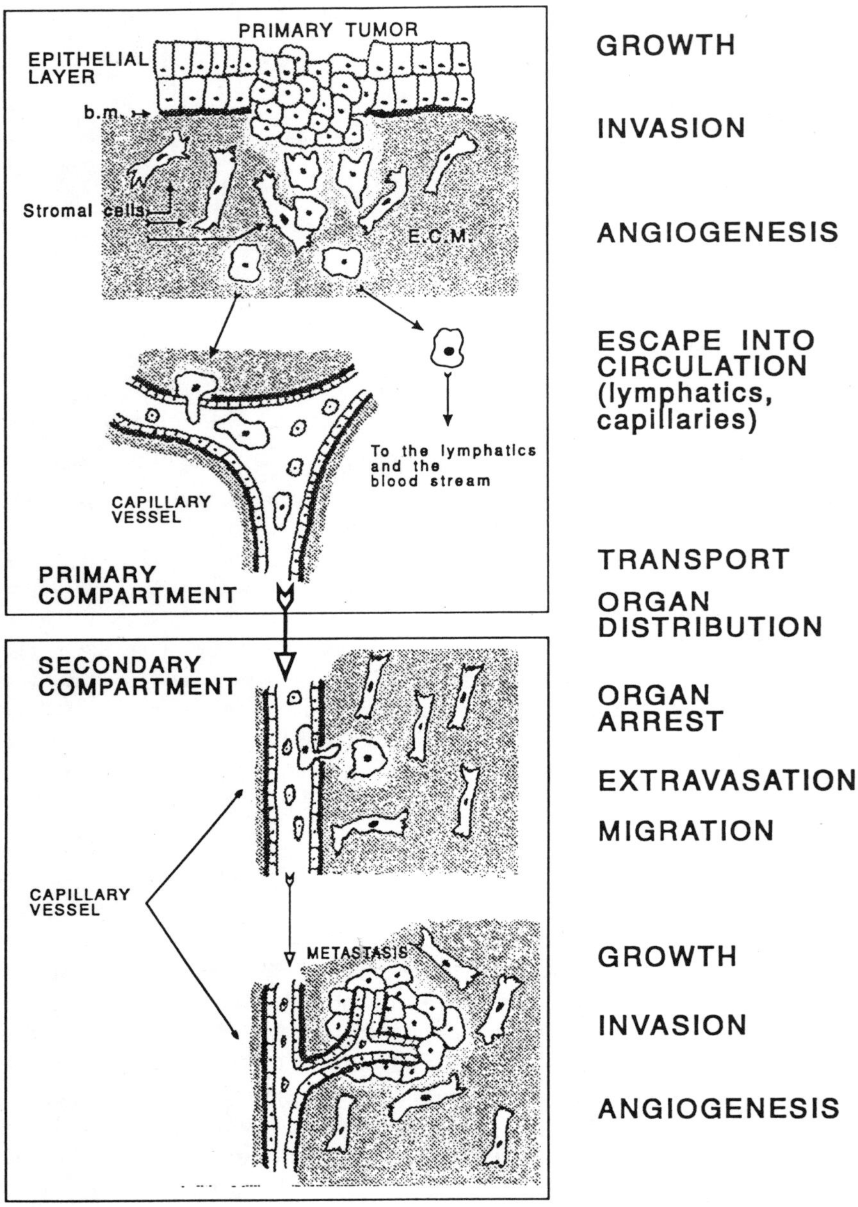

Fig. 1. The metastatic cascade of events. (The figure illustrates the case of a carcinoma, but it may also apply to other solid tumors, taking into account the obvious differences.) A successful metastatic cell has to accomplish all the depicted steps in order to give rise to a detectable metastatic colony. Such a cell may either inherently possess all the necessary requirements to go through the cascade, or it may take advantage of cooperation with other tumor cells or even with host stromal cells. b.m., basement membrane; E.C.M., extracellular matrix.

with the surrounding cells or matrix, thereby rendering some "soils" more fertile for the appropriate metastatic tumor cells to establish full-grown metastatic foci.

TUMOR INVASION AND CELL MIGRATION

Invasiveness and motility are properties of tumor cells that have great relevance for the metastatic phenotype during at least two steps of the metastatic cascade.

1. The tumor *in situ* is separated from the underlying interstitial stroma by a basement membrane. The transition to malignancy requires that the tumor cells penetrate the basement membrane and enter the stroma, where they can have further access to lymphatics and blood vessels, which are the avenues for distant migration.

2. At a further stage in the process of establishing a metastatic deposit, a second crossing of the subendothelial basement membrane is a necessary step that follows the arrest of tumor cells in the capillaries at a secondary site, and preludes metastatic colonization.

Until the 1970s it was believed that tumor invasion was a passive process caused by simple growth pressure (Eaves, 1973). We now know that the process of invasion is a complex, dynamic process that involves at least three coordinated events (Fig. 2).

1. The cells have to loosen their contacts with the original

tumor mass and the surrounding matrix to facilitate their motility and shedding into circulation. To move efficiently, these cells bind to the matrix. Cell locomotion occurs through the extension, adhesion, and contraction of cellular processes (lamellipodia and microspikes) onto the surrounding matrix.

2. The cells have to actively move through the matrix under the influence of motility and/or chemotactic factors.

3. The cells have to hydrolyze matrix components in order to ease their way through the matrix.

Weakening of Cellular Constraints

Most malignant cells lack their own extracellular matrix. Transformed cells have a lesser ability to retain fibronectin, collagen, or laminin on their surface. One explanation for this finding is that a different set of specific integrin receptors, with a lower affinity for their ligands, is expressed by transformed cells. These cells can still use basement membrane components for specific adhesion, but they are relieved of the constraints of an extracellular matrix of their own, and thus have a higher degree of freedom, making it possible for them to move and migrate more easily.

FACTORS CONTRIBUTING TO THE INVASIVE PHENOTYPE

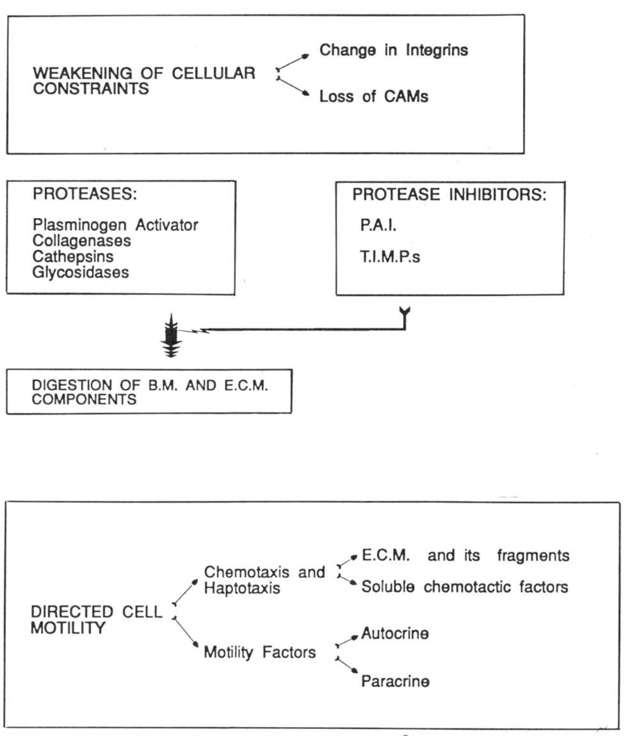

Fig. 2. A scheme of the most relevant properties leading to invasion.

In polarized endothelial cells (Lampugnani et al., 1991) and in keratinocytes (Marchisio et al., 1991), but not in dermal fibroblasts, some integrins of the β_1 family have been shown to be mainly located at cell–cell contact borders, but not at cellular free edges, suggesting a role for integrins in the maintenance of epithelial continuity. The loss of a similar function might contribute not only to loosening the contact among the tumor cells themselves but also to the loosening of contact of the tumor cells with the tissue of origin. According to this theory, analysis of the expression of specific integrin receptors in three pairs of normal and transformed rodent cell lines showed that the transformed cells expressed greatly reduced levels of the high-affinity $\alpha_5\beta_1$ fibronectin receptor, and higher levels of the $\alpha_3\beta_1$ integrin, which can function as a low-affinity fibronectin receptor (Plantefaber and Hynes, 1989). In the same study, two other integrins of the β_1 family were completely lost after transformation. Similarly, in human malignant melanoma the α_6 and β_1 subunits of the laminin receptor undergo significant down-regulation and deregulation in expression, in correlation with tumor invasiveness and metastasis (Natali et al., 1991).

To further support the link between integrin expression, cell motility, and the malignant phenotype, Giancotti and Ruoslahti (1990) reported that overexpression of the $\alpha_5\beta_1$ fibronectin receptor in transformed CHO cells leads to increased deposition of fibronectin in their extracellular matrix, reduces migratory ability, and causes a loss of tumorigenicity upon subcutaneous injection into nude mice. Recently, an inverse correlation was found between the level of expression of the $\alpha_5\beta_1$ fibronectin receptor and the rate of tumor growth in nude mice in a series of variant clonal CHO cell sublines (Schreiner et al., 1991).

Even more intriguing is the finding that a putative tumor suppressor gene for colon cancer (DCC) codes for a member of the immunoglobulin superfamily N-CAM (Fearon et al., 1990). The loss of such adhesion molecules, or of their function, may be a facilitating factor in the release of tumor cells from the primary tumor mass, promoting a more malignant, invasive phenotype (Behrens et al., 1989). Furthermore, transfection of the cell adhesion molecule E-cadherin (formerly uvomorulin) into highly invasive, uvomorulin-deficient, epithelial tumor cell lines was found to induce a remarkable loss of their *in vitro* invasive ability, which could then be reversed by the addition of specific anti-E-cadherin antibodies (Vleminckx et al., 1991). Alternatively, expression of antisense E-cadherin RNA in noninvasive, ras-transformed cells with high endogenous E-cadherin expression resulted in the induction of an invasive phenotype (Vlemnicks et al., 1991). It seems likely that E-cadherin may also positively regulate calcium-dependent gap junctional intercellular communication in mouse epidermal cells (Jongen et al., 1991), whereas a loss of intercellular junctional communication has been linked with an increased metastatic potential in a rat mammary adenocarcinoma cell line (Nicolson et al., 1988).

Cell Motility

Active locomotion of tumor cells into and through host tissue barriers is also required for successful invasion. Cell movement may occur in response to specific motility factors, which have no effect on cell growth and are related to invasive behavior of malignant cells (Warn and Dowrick, 1988).

The autocrine motility factor (AMF), produced by human melanoma, colon, and breast carcinoma cells, stimulates cell migration in the same cells producing the factor through an interaction with a gp78 cell surface receptor (Silletti et al., 1991). A second autocrine migration-stimulating factor (MSF) has also been characterized and found to be produced by fetal fibroblasts and by fibroblasts from breast tumors (Warn and Dowrick, 1988).

In contrast, the "scatter factor" acts in a paracrine fashion since it is mainly secreted by fibroblasts, but it affects epithelial cells by breaking up contiguous cell sheets, causing the loss of cell junctions and inducing random and chaotic movement (Warn and Dowrick, 1988). The purified scatter factor is able to induce MDCK epithelial cells and some human carcinoma cells to invade collagen gels. Intriguingly, the scatter factor has been proven to be identical to the hepatocyte growth factor, and perhaps also to a human lung fibroblast-derived mitogen, thereby suggesting at least a dual role for this factor (Weidner et al., 1991).

In addition to the factors that some tumor cells produce to promote their own motility, directed migration of malignant cells may also occur in response to chemotactic factors that are present in the tissue itself as soluble (chemotaxis) and/or anchored (haptotaxis) molecules (Zetter and Brightman, 1990). The presence of organ-specific factors that selectively direct cell migration into the organ itself may help tumor cells to invade at that site. Components of the extracellular matrix, or their fragments, in addition to serving as adhesive substrates, are also able to stimulate the haptotactic migration of tumor cells (Zetter and Brightman, 1990).

Chemotactic factors also play a role in the invasion and colonization of malignant tumor cells at specific sites. In the Walker 256 carcinosarcoma model system (a bone-metastasizing rat breast tumor), it was found that TGF-beta released from resorbing bones contributed to the chemotactic migration of Walker tumor cells (Orr et al., 1990). Melanoma, sarcoma, and breast carcinoma cells showed preferential migration in response to extracts from the organs for which they show metastatic predilection (brain, lung and liver, bone and brain, respectively) (Hujanen and Terranova, 1985), further suggesting that specific factors in target organs can increase specific cell recruitment to a target tissue.

ECM Solubilization

The crossing of a basement membrane and invasion into the interstitial stroma, either at the level of the primary tumor or during extravasation, also require the activity of proteolytic enzymes.

The predominant component in the basement membrane is collagen type IV. An increased expression of collagenase IV is a frequent finding in invasive tumors (Frost et al., 1990). Currently, two unique type IV collagenase enzymes, with molecular weights of 72 and 92 kd, are known. They are normally produced by a variety of cell types, and elevation of their activity shows a positive correlation with the metastatic phenotype (Frost et al., 1990).

Collagenase type IV belongs to the family of matrix metalloproteinases, which includes stromelysin and interstitial collagenase I. The stromelysin subfamily includes different enzymes with broad specificities, which are able to degrade several matrix components such as collagen types III, IV, and V, laminin, fibronectin, and proteoglycans (Frost et al., 1990). In several different animal model systems a positive correlation has been shown between stromelysin expression and tumor progression, invasion, and metastasis (Frost et al., 1990).

The expression of matrix-degrading enzymes in either tumor cells or host stromal cells may also depend on their reciprocal interaction. For example, stromelysin-3 (Basset et al., 1990) is expressed specifically in stromal cells surrounding invasive breast carcinomas but is absent from the malignant epithelial cells themselves, and is not found in fibroadenomas, which are noninvasive lesions. Stromelysin-3 may be induced by a diffusible factor secreted by tumor cells such as FGF (Basset et al., 1990). *In vitro* a similar stimulatory effect occurs between cocultured stromal fibroblasts and invasive tumor cells with respect to the secretion of interstitial collagenase I. Until recently it was believed that only stromal cells could produce interstitial collagenases in response to tumor cell factors. The cloning of a cDNA encoding interstitial collagenase from a human melanoma cell line has shown that at least some tumor cells can produce interstitial collagenase without a requirement for host stromal cell interaction (Templeton et al., 1990).

Finally, human colon carcinoma cells are invasive and metastatic to the liver, and produce high amounts of collagenase and heparanase only when implanted orthotopically in the cecal wall of nude mice, and not when injected subcutaneously (Nakajima et al., 1990).

A further link between metalloproteinase production, invasion, and metastasis was established by the finding that a tissue inhibitor of metalloproteinases (TIMP) could inhibit the proteolytic activity of these enzymes *in vitro* in a ras-transformed rat cell line as well as its metastases *in vivo* (Alvarez et al., 1990). Another confirmation of the key role played by TIMP in tumorigenicity is provided by the finding that 3T3 cells genetically modified to express an antisense RNA complementary to TIMP were down-modulated for TIMP expression, and concomitantly became invasive, tumorigenic, and metastatic (Khokha et al., 1989). The differential presence of TIMPs in different tissues and the relative amount of metalloproteinases that malignant cells can produce under different conditions may also influence the distribution of metastases.

In addition to metalloproteinases, other enzymatic activities have been positively correlated with the invasive and metastatic phenotype. Urokinase-type plasminogen activator (uPA) converts the zymogen plasminogen in plasmin, a trypsinlike enzyme with broad substrate specificity. An increased level of urokinase activity has in the past been associated with malignant behavior. More recently, several laboratories have demonstrated a direct role for uPA in aggressive behavior of tumor cells (Frost et al., 1990).

Cathepsins are a class of lysosomal acidic proteases associated with tumor progression and malignancy, although their mechanism of action in promoting metastasis is currently unknown (Frost et al., 1990). The levels of another group of lysosomal hydrolytic enzymes, glycosidases, are elevated in the interstitial fluid of tumors, in the sera of animals, in patients with tumors, and in some tumor tissue, as compared with normal adjacent tissue. A correlation was also observed between glycosidase activity and extracellular matrix degradation by human ovarian carcinoma cell lines (Niedbala et al., 1987).

THE ROLE OF THE IMMUNE SYSTEM

Natural Immunity

The ability of malignant cells to survive the tumoricidal activity exerted locally by macrophages, NK cells, and cells of the reticuloendothelial system, aside from influencing the metastatic phenotype *per se*, may also affect their organ-specific colonizing ability. NK cell activity in blood and in tissues may represent an important factor limiting survival and metastasis of neoplastic cells released from the primary tumor in the circulation.

Generally, mice with a low NK activity (whether because of genetic background or drug treatment) are more prone to develop organ metastases, while *in vivo* stimulation of NK activity increases the host resistance to the development of metastases. Furthermore, the adoptive transfer of large granular lymphocytes (LGL), which are closely associated with NK activity, into NK-depressed rats just before tumor challenge restored the ability of these animals to inhibit the development of pulmonary metastases (Barlozzari et al., 1985). A good correlation between NK cell activity and survival time without metastases has also been reported in patients with either solid tumors or leukemias (Whiteside and Herberman, 1989).

Large granular lymphocytes, as well as Kupffer cells, are involved in the defense system of the liver under various physiological and pathological conditions. In a study with nude mice it was reported that only after depletion of NK activity was a murine colon carcinoma cell line able to produce liver metastases upon intraspleen injection, indicating that liver-resident NK cells can play an important role in preventing liver metastases (Arisawa et al., 1990). More recently, a putative "immunosuppressive" molecule, able to inhibit both proliferation of spleen cells and NK cell-mediated cytotoxicity, has been specifically extracted from the membrane of the highly liver-colonizing RAW117-H10 cell subline (Joshi et al., 1991).

Following immunologic activation, pulmonary macrophages may prevent or cause regression of experimental lung metastases by mechanisms that remain largely unknown. The relationship between the ability of malignant cells to survive such killing and their metastatic potential is still a matter of debate. In fact, Fogler and Fidler (1985) failed to select *in vitro* tumor cells that were resistant to macrophage-mediated lysis, and concluded that tumor cell destruction by activated macrophages was nonselective and therefore could not be related to metastatic potential. On the contrary, Yamamura and colleagues (1984), working with a murine mammary adenocarcinoma that did not produce spontaneous metastases, found that a metastatic subline was less susceptible to both macrophage-mediated cytolysis *in vitro* and *in vivo* host antitumor mechanisms. They concluded that the *in vitro* resistance of tumor cells to macrophage cytotoxicity corresponded with their *in vivo* metastatic potential.

Active Immunity

Active humoral and cellular immune mechanisms also have a role in fighting primary and secondary tumor growths. Monoclonal antibodies can either suppress lung (Herd, 1987) or liver (Eisenthal et al., 1987) metastases by murine B16 melanoma cells.

Tumor-infiltrating lymphocytes (TIL) are thought to be specifically directed against tumor cells, but they are only weakly active in fighting cancer cells, perhaps because of the concomitant presence of specific T-suppressor cells. TIL from cancer patients can be expanded *in vitro* in the presence of interleukin-2 and reinjected into the donor after T-suppressor lymphocytes have been inhibited by cyclophosphamide. This treatment yielded promising results in a recent clinical trial with metastatic melanoma in humans (Foon, 1989).

Tumor-specific T cells may also be specifically activated by administration of a virus-modified autologous tumor cell vaccine, as shown by the increased sensibilization against autologous tumor cells measured by DTH skin reaction in treated patients (Liebrich et al., 1991).

T lymphocytes recognize foreign cell surface antigens almost exclusively in association with molecules encoded by the major histocompatibility complex (MHC). Specifically, the H-2 class I antigens appear to direct the recognition of neoplastic and virus-infected cells by cytotoxic T lymphocytes (CTL). Therefore, at least for those tumor cells that express high levels of tumor- specific antigens (TSA), it might be advantageous to hide their diversity by an unbalanced expression of the H-2 antigens. In fact, this has been a frequent finding in many murine tumors where, irrespective of their mode of induction, there was a preferential loss of the H-2K and an enhanced expression of H-2D (Gopas et al., 1990). Furthermore, the transfection of the H-$2K^b$ gene into metastatic H-$2K^b$-deficient cells caused the metastatic phenotype in both 3LL (Plaskin et al., 1988) and B16 cells to revert (De Giovanni et al., 1991).

The recognition by the host immune system of TSA in association with the MHC antigens may also be impaired by tumor cells through increased secretion of a glycocalix, which masks the tumor cell surface, as in the case of epiglycanin secretion by TA3 murine mammary carcinoma cells (Codington et al., 1979).

TUMOR CELL DEFORMABILITY

There are at least two steps during which more deformable malignant cells may gain advantage over less deformable cells. Most circulating cancer cells appear to be killed in the microcirculation, where they may be victims of mechanical damage following their passage through capillaries with a diameter as small as 5μm. Therefore, a higher deformability would allow malignant cells to better survive the stress, making possible a broader distribution to other organs downstream of the first capillary network. Since the ability of metastasizing cells to deform may depend on cytoskeletal organization, TA3 ascitic cells were treated in culture with cytochalasin B to disrupt their microfilament organization. When the cells were subsequently injected into syngeneic hosts, a higher number of extrapulmonary foci was detected, suggesting that the disruption of the microfilament system enabled the cells to traverse the lung capillary bed and lodge in other organs (Hagmar and Ryd, 1977).

During invasion and intravasation at the primary site, as well as at the moment of extravasation into a secondary site, tumor cells have to cross tissues and move through the extracellular matrix into the organ parenchyma. In order to enhance the ability of a tumor cell population to perform these steps and to better resist capillary deformation, Tullberg and Burger (1985) selected tumor cell variants able to penetrate filter pores 10 times smaller than their own diameter. These size-selected tumor cells showed cytoskeletal alterations and an increased metastatic capacity after subcutaneous implantation into mice (Tullberg and Burger, 1985).

ORGAN LODGEMENT

The arrest of circulating malignant cells in a given organ may occur by mechanical trapping into the organ microcirculation, which might be facilitated by either homotypic aggregation between tumor cells or heterotypic aggregation with host platelets. In either case the vascular connections of the primary tumor with the target organs might well explain the pattern of organ metastases. Alternatively, organ arrest may follow specific interactions with the vascular endothelium or the subendothelial matrix (Belloni and Tressler, 1989/90).

Specific Adhesion to Capillary Endothelial Cells

Capillary endothelial cells are different in different organs. In many cases the preferential adhesion of tumor cells to microvascular endothelial cells reflects their metastatic distribution. Studies on lymphomas have indicated that the metastatic spreading of malignant lymphoid cells follows the same rules that direct normal lymphocyte homing. This event is also dependent upon specific adhesive interactions between lymphocytes and endothelial cells, and appears to be mediated by the expression on endothelial cells of a category of molecules named vascular "addressins." Two of these vascular addressins (the ELAM-1 and the INCAM-110) mediate the adhesion of the human melanoma cell line Hs294T (the INCAM-110) and the human colon carcinoma cell line HT-29 (the ELAM-1) (Rice and Bevilacqua, 1989), suggesting that they might also have a role in directing the homing of some solid tumor cells.

Specific Adhesion to the Subendothelial Matrix

The contact between tumor cells and the endothelium may induce endothelial cell retraction, with exposure of the underlying basement membrane. Tumor cells may thus extravasate into the adjacent organ parenchyma by squeezing between retracted capillary endothelial cells, a step that requires both good motility and deformability of the performing cells. The exposed subendothelial matrix is usually a better adhesive substrate for tumor cells than the endothelial cell surface. The presence of specific adhesion receptors on the membrane of the metastatic cells, and the peculiar composition of the extracellular matrix at a given site, will also influence lodging and growth of malignant cells at specific sites (Belloni and Tressler, 1989/90; Rusciano and Burger, 1992).

Several basement membrane components, such as laminin, fibronectin, collagen types IV and V, heparan sulfate proteoglycan, and vitronectin, have been identified as tumor cell adhesion molecules. The interaction of malignant cells with some of these glycoproteins, such as laminin and fibronectin, may also influence their metastatic behavior. In fact, the ability of these basement membrane glycoproteins to bind cellular receptors on the one hand, and matrix proteins on the other, makes them a good candidate for involvement in the attachment/invasion stage of metastasis.

Laminin

Many tumor cells have receptors for laminin and may use this glycoprotein as an adhesion factor. Adhesion to laminin can stimulate hematogenous metastasis in the lungs by different mechanisms.

If the receptor is unoccupied, it can be used by the cells to bind directly to the basement membrane laminin. In fact, the level of laminin receptor expression on the tumor cell surface has been correlated with the ability of tumor cells to metastasize, both in mice (Alino et al., 1990) and in humans, where an increased expression of the 67K high-affinity laminin receptor in 23 patients with colon carcinoma correlated with higher tumor metastatic potential (Cioce et al., 1991). If the receptor is saturated with laminin, the cells can use it as an attachment bridge to type IV collagen in the basement membrane. Murine fibrosarcoma cells treated with intact laminin produced a higher number of lung colonies upon tail-vein inoculation (Malinoff et al., 1984), while treatment of metastatic cell lines with a cell-binding fragment of laminin or with the pentapeptide containing the essential cell binding site inhibited their adhesion to basement membrane and dramatically decreased the number of lung colonies (McCarthy et al., 1988), further suggesting a role for laminin, particularly in lung metastasis. Finally, treatment of tumor cells with a defined peptidic sequence of laminin stimulates collagenase IV production (Kanemoto et al., 1990). This might help tumor cells to cross the subendothelial matrix and locally invade the organ parenchyma. These treated murine melanoma cells have an enhanced ability to form experimental lung metastases (Kanemoto et al., 1990).

Fibronectin

Fibronectin is another pericellular matrix glycoprotein whose interaction with malignant cells plays a role in the metastatic phenotype. Treatment of tumor cells with synthetic peptides containing the RGD cell-binding domain of fibronectin inhibited experimental lung metastasis of murine melanoma cell lines (Humphries et al., 1986). Such an inhibition might result from different mechanisms. Pretreatment of tumor cells with RGD-containing peptides reduces the pulmonary retention of tail-vein-injected murine melanoma cells (Humphries et al., 1986), suggesting a direct effect on tumor cell–subendothelial matrix interaction. These peptides may also inhibit invasion through an amniotic basement membrane of human melanoma cell lines (Gehlsen et al., 1988) or prevent tumor cell-induced platelet aggregation in a murine fibrosarcoma line (Ugen et al., 1988), which suggests an indirect effect on unspecific arrest of tumor cell emboli.

Fibronectin is the most prominent component of the extracellular matrix in Disse's space of the liver, and as such might contribute to malignant cell lodgement in the liver parenchyma. For instance, we recently described that liver-colonizing F9 embryonal carcinoma cells (ECC) preferentially adhered to fibronectin and liver-derived extracellular matrix over laminin and lung-derived extracellular matrix (Rusciano et al., 1991). Most interestingly, the F9 ECC model system showed that organ-specific binding to the extracellular matrix may also negatively regulate the organ-colonizing ability of tumor cells. In fact, F9 cells bind very poorly to the extracellular matrix of the lungs, so that *in vivo* the organ retention in this organ is highly impaired after tail-vein inoculation. The cells recirculate very quickly to the liver, where they find a more adhesive substrate (Rusciano et al., 1991).

Adhesion in the Liver

The peculiar architecture of the liver, with its fenestrated endothelium directly facing the Disse's space and the absence of a classical subendothelial matrix, should facilitate penetration of blood-borne tumor cells in the organ parenchyma as well as their direct contact with hepatocytes themselves. Highly liver-metastatic ESb lymphoma cells were able to form rosettes *in vitro* with isolated autologous hepatocytes, while low-metastatic sublines of the same tumor did not (Cheingsong-Popov et al., 1983). The hepatocytes bind ESb tumor cells through lectinlike hepatic binding proteins specific for D-galactosyl and N-acetyl-D-galactosaminyl residues. More recently it was reported that in a murine colon carcinoma model system, the ability of cells to form liver metastases was proportional to the degree of tumor cell–surface galactose expression (Yeatman et al., 1989). All these findings indicate that liver-metastatic tumor cells may bind the asialoglycoprotein receptor on hepatocytes. In fact, treatment of mice with the receptor-blocking agents D-galactose and arabinogalactan completely prevented the settling of sarcoma L-1 tumor cells in the liver but did not influence their settling in the lungs (Beuth et al., 1987).

A further example of the relevance of specific adhesion to hepatocytes in liver colonization is the Lewis lung carcinoma model system, in which a liver-specific cell clone (H59) has been isolated and showed a higher adhesiveness to hepatocytes than did a lung-specific cell clone (M27). A 64 kd glycoprotein was then identified that is specifically expressed by H59 cells, and seems to be responsible for the specific adhesion to hepatocytes (Wang et al., 1991).

METASTATIC GROWTH

Lodgement of viable tumor cells in an organ is a necessary but (by itself) insufficient condition for the development of metastasis (Rusciano and Burger, 1992). Cells of the M5076 reticulum cell sarcoma, for instance, are rapidly arrested in the lungs, and retained there for three to four days. They then slowly detach and recirculate to the liver, where they find the conditions favorable for giving rise to metastatic colonies (Hart et al., 1981). A similar finding has also been reported in humans, where in most patients treated for malignant ascites by peritoneovenous shunting, the direct infusion of massive doses of viable tumorigenic cells into the jugular vein did not result in secondary tumor deposits in the lungs (Tarin et al., 1984).

The final decisive step is the growth of the lodged, viable cells into a clinically detectable colony. This critical event may depend both on intrinsic properties of the malignant cells themselves and on their interactions with the surrounding environment (Fig. 3).

Autocrine Growth Promotion

Very often the uncontrolled growth of tumor cells is a consequence of autocrine growth stimulation, an event that rarely occurs in normal tissues. The increased level of growth factor production by cancer cells may correlate with a decrease in tumor latency (Karasuyama et al., 1989). In many cancers the malignant phenotype may be associated with autonomous cell growth (Perosio and Brooks, 1989).

Increased growth factor autonomy of metastatic cells over that of nonmetastatic cells has been shown in several murine tumor systems. Metastatic, but not primary, tumor melanoma cell lines grow autonomously *in vitro*, suggesting that their metastatic behavior could be due to autostimulation (Rodeck et al., 1987). Similarly, highly metastasizing 3LL cells (but not low-metastasizing ones) can produce growth factors that stimulate their own growth according to an autocrine mechanism, and, by a paracrine mechanism, might stimulate the surrounding stromal cells (Perrotti et al., 1990). Metastatic cancer cells may help activate surrounding stromal fibroblasts (Garin-Chesa et al., 1991), leading to a reciprocal paracrine interaction between the two cell types that may facilitate both tumor progression and metastasis.

Finally, NIH-3T3 cells transformed by a signal peptide-bFGF chimeric gene, but not by bFGF in the absence of the signal peptide, acquired the ability to heavily colonize the lungs of nude mice after i.v. inoculation, strongly supporting the concept of autocrine induction of the metastatic phenotype (Egan et al., 1990).

Paracrine Growth Promotion

Autocrine growth stimulation is not the only mechanism by which malignant cells can support their growth at a metastatic site. The presence in the target organ of paracrine growth factors may also influence the outcome of metastatic colonies, and in several instances correlated with the organ-specific colonization of malignant cells, as in the case of liver-specific B16 melanoma cells (Sargent et al., 1988).

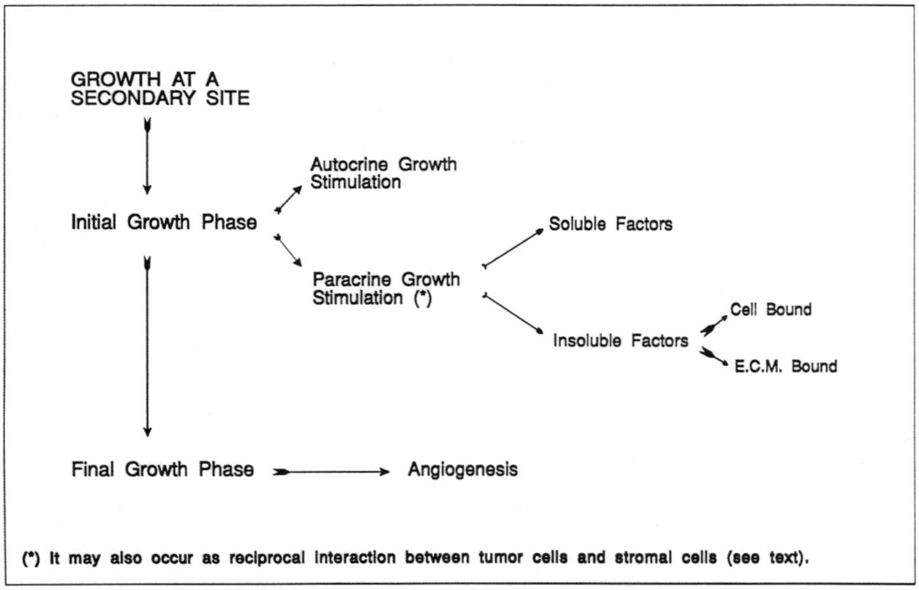

Fig. 3. A schematic illustration of the possible requirements for efficient growth of a metastatic colony (but the same may also apply to a primary tumor). The ability of a malignant cell to promote its own growth by means of autocrine growth factors, or its dependence on paracrine growth factors for efficient growth, might also contribute to determining the site(s) where metastases will develop. *This may also occur as reciprocal interaction between tumor cells and stromal cells (see text).

Furthermore, other murine B16 melanoma cell sublines, selected *in vivo* for enhanced colonization of lung and ovary, were also more responsive *in vitro* to growth-promoting effects mediated by soluble factors released by the target organ, while brain-colonizing B16 cells were virtually unresponsive to any of the organ-conditioned media. It is interesting to note in these experiments that, while lung-, kidney-, ovary-, and brain-conditioned media had either no effect or a growth-promoting effect on the tested cell lines, liver-conditioned medium always exerted a growth-inhibitory effect on all the cell lines. In contrast, a liver-colonizing subline of the RAW117 large-cell lymphoma was growth-stimulated by liver-conditioned medium, suggesting that liver may release factors, some of which are toxic and growth-inhibitory, to cells that are incapable of liver metastasis (Nicolson, 1988).

Another model system that illustrates the relevance of paracrine growth stimulation in organ-specific colonization is represented by the F9 teratocarcinoma cells. These cells, upon tail-vein inoculation, efficiently colonize the liver of syngeneic animals but are very poor lung colonizers (Terrana et al., 1987). We found that untreated F9 cells were unable to respond to lung extract-mediated growth stimulation, but that they did respond to liver-mediated growth promotion (Rusciano et al., 1991). Only after retinoic acid treatment did we observe that F9 cells were able to colonize the lungs (Terrana et al., 1987) and respond to lung-mediated growth promotion (Rusciano et al., 1992).

In another set of experiments (Nicolson et al., 1988), the highly lung-metastasizing MTLn3 cell line (derived from the rat 13762NF mammary adenocarcinoma) showed significantly higher growth stimulation by lung-conditioned medium than by medium conditioned by other tissues, as compared to the poorly metastatic MTC subline. A paracrine growth factor of 66 kd, which specifically stimulated the growth of lung-specific cell lines in three different lung-colonizing tumor systems, was then purified from lung-conditioned medium (Cavanaugh and Nicolson, 1990).

A common metastatic site for mammary cancer in humans is bone, where tumor cells produce osteolytic bone metastases and hypercalcemia. The mammary Walker 256 carcinosarcoma is a rat model system that mimics the human situation. It was found that these tumor cells respond to both chemotactic (Orr et al., 1990) and growth-stimulating (Manishen et al., 1986) factors released from the resorbing bone, which suggests that a close interaction between tumor cells and the colonized organ might be involved in the localization of secondary metastatic growths. Accordingly, bone marrow stromal cells have been shown to specifically promote the growth of mammary cancer cells (Strobel et al., 1989), while bone cells like osteoblasts or osteoclasts have not yet been shown to play any role.

A similar example is given by the carcinoma of the prostate. Prostatic adenocarcinoma is a slowly growing tumor that often metastasizes to the bone marrow, where it produces characteristic osteoblastic lesions. The growth rate of these metastases is more rapid than that observed for primary prostatic tumors, and has been linked to the presence in the bone marrow of a growth factor able to specifically promote the growth of prostatic cancer cells (Chackal-Roy et al., 1989).

The source of paracrine growth stimulation is not always a soluble molecule. A liver-specific subline (B16-L8) selected *in vivo* from the murine B16 melanoma was growth-stimulated *in vitro* by direct cell-to-cell contact with hepatocytes (Sargent et al., 1988), which suggests that B16-L8 cells are responding to some insoluble factor present on hepatocyte membranes or in their extracellular matrix.

The extracellular matrix may in fact function as a deposit for some growth factors, such as bFGF and TGF-ß. The growth factors bound to the proteoglycans of the extracellular matrix are inactive, but they may be released in their active form after proteolytic cleavage of the proteoglycan acting as a linker (Ruoslahti and Yamaguchi, 1991). In particular, bFGF binds to heparan-sulfate, which appears to protect the growth factor from degradation. It seems that only the bFGF that is released from heparan sulfate by the action of heparanase has the ability to efficiently interact with its receptor. Heparanase activity, on the other hand, correlates with the metastatic potential of various tumor cell lines, while inhibitors of heparanase reduce the incidence of lung metastasis in experimental animals (Vlodavsky et al., 1990), suggesting that heparanase might have a role in facilitating the invasion through the extracellular matrix. It might also function in liberating an active form of bFGF, which in turn may either stimulate the growth of the malignant cells or have an effect as an angiogenic promoting factor.

GENETICS OF THE METASTATIC PHENOTYPE

Somatic cell hybridization studies have shown that the metastatic phenotype of solid tumors appears to be recessive, suggesting that it is under a genetic control distinct from that causing the uncontrolled growth of primary tumor cells. The complexity of the metastatic process also suggests that it is regulated at the genetic level by the activation and/or deactivation of several specific genes (Fig. 4).

An altered expression of one or more genes directly controlling any step of the metastatic cascade described above might affect the metastatic behavior of malignant cells. Furthermore, transfection experiments have shown that some oncogenes such as ras, or the kinase family of oncogenes such as src, mos, raf, fes, and fms, are potent inducers of metastasis when transfected in certain recipient cells (Wright et al., 1990). In many cases, however, no special care was taken to distinguish between the possibility that such cells have become generally more tumorigenic, or

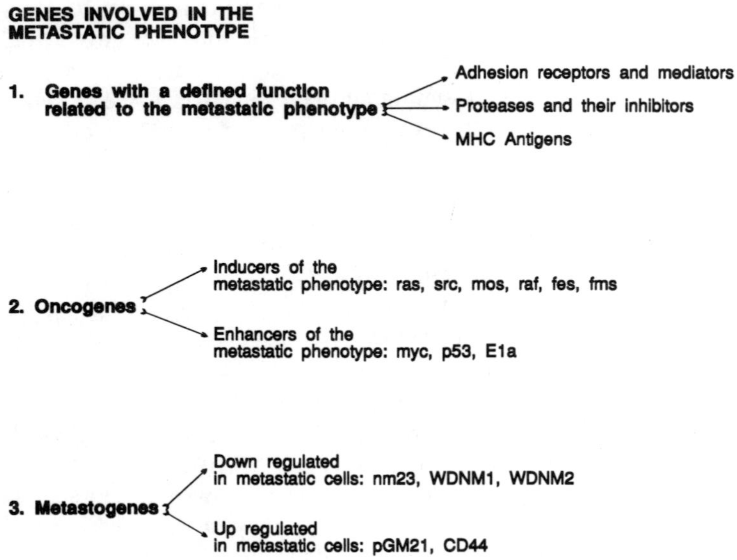

Fig. 4. A schematic report of the genes that have been shown to be involved in the metastatic phenotype, according to a functional classification.

whether they have become specifically more metastatic, since the assays used did not allow a clear distinction to be made.

Although the mechanism of action of these various oncogenes in promoting the metastatic phenotype is not clear, their expression may deregulate the membrane-associated second messenger systems.

Nuclear oncogenes such as myc, p53, or E1A might enhance the metastatic ability of certain tumorigenic cell lines. These oncogenes may exert their effect by transregulation of other genes involved in metastasis formation, such as genes coding for the MHC or proteases (Wright et al., 1990).

Finally, differential screening of cDNA libraries derived from metastatic and nonmetastatic variants of the same tumor type, coupled with subtractive hybridization, has resulted in the cloning of specific genes associated with the metastatic phenotype.

The DMBA8 rat mammary adenocarcinoma model system has been the source of three such genes. Two of them (WDNM1 and WDNM2) appear to be down-regulated in highly metastatic cells, and thereby act as metastasis-suppressor genes. More recently, sequence analysis of the WDNM1 gene suggested that it could encode a protein with proteinase-inhibiting activity, thereby influencing the invasive potential of tumor cells (Dear and Kefford, 1991). On the contrary, the third gene cloned with this model system (pGM21) appears to be positively associated with the metastatic phenotype (Phillips et al., 1990).

In 1988 Steeg reported the cloning of a gene (nm23) that was strictly associated with low metastatic ability in the K 1735 murine melanoma model system. This correlation also holds true for other rodent tumors, and is independent of the tumor-inducing agent (Steeg et al., 1991). A significant relationship has been found between expression of nm23 in human breast carcinoma and good prognosis, while in human colorectal carcinoma it is more likely an alteration in nm23 regulation that is associated with high metastatic potential. Transfection of the nm23 gene into high-metastatic K 1735 melanoma cells could reduce their metastatic potential and their ability to respond to TGF-ß. The role of the gene product in controlling the metastatic phenotype may only be inferred on the basis of its homology with nucleoside diphosphate kinase (NDPK). Since this enzyme is involved with microtubule and G-protein function, an alteration of its activity might affect the malignant phenotype at different levels. Moreover, the presence of a leucine zipper domain in the molecule might mediate nm23 binding to DNA or other proteins.

More recently, a second human nm23 gene was identified and cloned, suggesting that nm23 should be considered a family of closely related genes involved in controlling the metastatic phenotype of at least some tumor types.

A last addition to the list of metastasis-related genes is a variant form of the leukocyte adhesion molecule CD44 (Gunthert et al., 1991). The expression of its related cDNA clone, pMeta-1, has been found to correlate with metastatic propensity of rat pancreatic and mammary carcinoma cell lines. Overexpression of pMeta-1 by transfection in any one of several nonmetastasizing rat tumor cells conferred full metastatic properties in a spontaneous metastasis assay, suggesting a dominant role for this gene in the metastatic phenotype. Since CD44 has been found to mediate lymphocyte adhesion to vascular endothelial cells, it is tempting to speculate that the expression of a longer extracellular domain in the variant form of CD44 expressed in metastatic cells endows them with a more efficient adhesion mechanism, thus allowing organ colonization.

CONCLUSION

We have so far illustrated how the metastatic phenotype results indeed from a very complex series of events, involving multiple interactions, and even cooperation, between malignant and normal cells of the host. Cell cooperation during malignant progression is gaining increasing attention, underscoring the concept that all of the requirements for accomplishing the necessary steps during the metastatic cascade do not necessarily have to preexist in a metastasis-competent cell. Some may perhaps be gained at either the primary or the metastatic site by interactions with host stromal or parenchymal cells. Therefore, metastases may also be established by tumor cells that are conditionally deficient at individual steps.

There is not just a single metastatic phenotype. Malignant cells can use different strategies to achieve the same result at a certain point during metastatic progression. For instance, different types of proteases or motility factors may be used to facilitate the invasion step, or tumor-specific antigens may be hidden from the host immune system either by an unbalanced expression of MHC genes or by the secretion of a glycocalix masking the tumor cell surface.

The complexity continues when the expression of oncogenes and/or "metastogenes" is taken into account. Again, the heterogeneity of tumor cell populations is the dominant feature, either within the same tumor or in comparison with other tumors. A more comprehensive elucidation of the molecular basis of the metastatic phenotype will help identify those mechanisms and behaviors that are critical for the occurrence of cancer metastasis and that might be appropriate targets for a metastasis-oriented therapy of the cancer disease.

ACKNOWLEDGMENTS

We thank Ms. Carol Wiedmer for expert secretarial assistance, and Drs. Kurt Ballmer-Hofer and Nancy Hines for their critical reading of the manuscript.

REFERENCES

Alino, S.F., F.J. Unda, and G. Perez-Yarza (1990) Laminin surface binding sites and metastatic potential of 3LL tumor cells increased by indomethacin. Biochem. Biophys. Res. Comm. 167:731–738.

Alvarez, O.A., D.F. Carmichael, and Y.A. DeClerck (1990) Inhibition of collagenolytic activity and metastasis of tumor cells by recombinant human tissue inhibitor of metalloproteinases. J. Natl. Cancer Inst. 82:589–595.

Arisawa, Y., T. Kubota, A. Suto, et al. (1990) Nude mouse resists hepatic metastases of the allogeneic tumor, colon-26. Jpn. J. Surg. 20:487–490.

Barlozzari, T., J. Leonhardt, R.H. Wiltrout, et al. (1985) Direct evidence for the role of LGL in the inhibition of experimental tumor metastases. J. Immunol. 134:2783–2789.

Basset, P., J.P. Bellocq, C. Wolf, et al. (1990) A novel metalloproteinase gene specifically expressed in stromal cells of breast carcinomas. Nature 348:699–704.

Behrens, J., M.M. Mareel, F.M. Van Roy, et al. (1989) Dissecting tumor cell invasion: Epithelial cells acquire invasive properties after the loss of uvomorulin-mediated cell-cell adhesion. J. Cell Biol. 108:2435–2447.

Belloni, P.N., and R.J. Tressler (1989/90) Microvascular endothelial cell heterogeneity: Interactions with leukocytes and tumor cells. Cancer Metas. Rev. 8:353–389.

Beuth, J., H.L. Ko, K. Oette, et al. (1987) Inhibition of liver metastasis in mice by blocking hepatocyte lectins with arabinogalactan infusions and D-galactose. J. Cancer Res. Clin. Oncol. 113:51–55.

Blood, C.H., and B.R. Zetter (1990) Tumor interactions with the vasculature: Angiogenesis and tumor metastasis. Biochim. Biophys. Acta 1032:89–118.

Cavanaugh, P.G., and G.L. Nicolson (1990) Purification and characterization of a M_r 66,000 lung-derived (paracrine) growth factor that preferentially stimulates the in vitro proliferation of lung-metastasizing tumor cells. J. Cell. Biochem. 43:127–138.

Chackal-Roy, M., C. Niemeyer, M. Moore, et al. (1989) Stimulation of human prostatic carcinom cell growth by factors present in human bone marrow. J. Clin. Invest. 84:43–50.

Cheingsong-Popov, R., P. Robinson, P. Altevogt, et al. (1983) A mouse hepatocyte carbohydrate-specific receptor and its interaction with liver-metastasizing tumor cells. Int. J. Cancer 32:359–366.

Cioce, V., V. Castronovo, B.M. Shmookler, et al. (1991) Increased expression of the laminin receptor in human colon cancer. J. Natl. Cancer Inst. 83:29–36.

Codington, J.F., A.G. Cooper, D.K. Miller, et al. (1979) Isolation and partial characterization of an epiglycanin-like glycoprotein from a new non-strain-specific subline of TA3 murine mammary adenocarcinoma. J. Natl. Cancer Inst. 63:153–161.

De Giovanni, C., G. Palmieri, G. Nicoletti, et al. (1991) Immunological and non-immunological influence of H-2K^b gene transfection on the metastatic ability of B16 melanoma cells. Int. J. Cancer 48:270–276.

Dear, T.N., and R.F. Kefford (1991) The WDNM1 gene product is a novel member of the "four-disulphide core" family of proteins. Biochem. Biophys. Res. Commun. 176:247–254.

Eaves, G. (1973) The invasive growth of malignant tumors as a purely mechanical process. J. Pathol. 109:233–237.

Egan, S.E., L. Jarolim, S. Rogelj, et al. (1990) Growth factor modulation of metastatic lung colonization. Anticancer Res. 10:1341–1346.

Eisenthal, A., R. Lafreniere, A.T. Lefor, et al. (1987) Effect of anti-B16 melanoma monoclonal antibody on established murine B16 melanoma liver metastases. Cancer Res. 47:2771–2776.

Fearon, E.R., K.R. Cho, J.M. Nigro, et al. (1990) Identification of a chromosome 18q gene that is altered in colorectal cancers. Science 247:49–56.

Fogler, W.E., and I.J. Fidler (1985) Nonselective destruction of murine neoplastic cells by syngeneic tumoricidal macrophages. Cancer Res. 45:14–18.

Foon, K.A. (1989) Biological response modifier: The new immunotherapy. Cancer Res. 49:1621–1639.

Frost, P., R. Greig, and R. Kerbel (eds.) (1990) The Role of Cellular Proteases and Their Inhibitors in Invasion and Metastasis. Cancer Metas. Rev. (special issue) 9(4).

Garin-Chesa, P., L.J. Old, and W.J. Rettig (1991) Cell surface glycoprotein of reactive stromal fibroblasts as a potential antibody target in human epithelial cancers. Proc. Natl. Acad. Sci. USA 87:7235–7239.

Gehlsen, K.R., W.S. Argraves, M.D. Pierschbacher, et al. (1988) Inhibition of in vitro tumor cell invasion by Arg-Gly-Asp-containing synthetic peptides. J. Cell Biol. 106:925–930.

Giancotti, F.G., and E. Ruoslahti (1990) Elevated levels of the $_5\beta_1$ fibronectin receptor suppress the transformed phenotype of chinese hamster ovary cells. Cell 60:849–859.

Gopas, J., B. Rager-Zisman, M. Bar-Eli, et al. (1990) The relationship between MHC antigen expression and metastasis. Adv. Cancer Res. 53:89–115.

Gunthert, U., M. Hofmann, W. Rudy, et al. (1991) A new variant of glycoprotein CD44 confers metastatic potential to rat carcinoma cells. Cell 65:13–24.

Hagmar, B., and W. Ryd (1977) Tumor cell locomotion: A factor in metastasis formation? Influence of cytochalasin B on a tumor dissemination pattern. Int. J. Cancer 19:576–580.

Hart, I.R., J.E. Talmadge, and I.J. Fidler (1981) Metastatic behavior of a murine reticulum cell sarcoma exhibiting organ-specific growth. Cancer Res. 41:1281–1287.

Herd, Z.L. (1987) Suppression of B16 melanoma lung colonization by syngeneic monoclonal antibodies. Cancer Res. 47:2696–2703.

Hujanen, E.S., and V.P. Terranova (1985) Migration of tumor cells to organ derived chemoattractants. Cancer Res. 45:3517–3521.

Humphries, M.J., K. Olden, and K.M. Yamada (1986) A synthetic peptide from fibronectin inhibits experimental metastasis of murine melanoma cells. Science 233:467–470.

Hunter, T. (1991) Cooperation between oncogenes. Cell 64:249–270.

Jongen, W.M.F., D.J. Fitzgerald, M. Asamoto, et al. (1991) Regulation of connexin 43-mediated gap junctional intercellular communication by calcium in mouse epidermal cells is controlled by E-cadherin. J. Cell Biol. 114:545–555.

Joshi, S.S., S.J. O'Connor, D.D. Weisenburger, et al. (1991) Enhanced antiproliferative activity by metastatic RAW117 lymphoma cells. Clin. Expl. Metas. 9:27–37.

Kanemoto, T., R. Reich, L. Royce, et al. (1990) Identification of an amino acid sequence from the laminin A chain that stimulates metastasis and collagenase IV production. Proc. Natl. Acad. Sci. USA 87:2279–2283.

Karasuyama, H., N. Tohyama, and T. Tada (1989) Autocrine growth and tumorigenicity of interleukin 2-dependent helper T cells transfected with IL-2 gene. J. Exp. Med. 169:13–25.

Khokha, R., P. Waterhouse, S. Yagel, et al. (1989) Antisense RNA—induced reduction in murine TIMP levels confers oncogenicity on Swiss 3T3 cells. Nature 243:947–950.

Lampugnani, M.G., M. Resnati, E. Dejana, et al. (1991) The role of integrins in the maintenance of endothelial monolayer integrity. J. Cell Biol. 112:479–490.

Liebrich, W., P. Schlag, M. Manasterski, et al. (1991) In vitro and clinical characterisation of a Newcastle disease virus-modified autologous tumor cell vaccine for treatment of colorectal cancer patients. Eur. J. Cancer 27:703–710.

Malinoff, H.L., P.J. McCoy, Jr., J. Varani, et al. (1984) Metastatic potential of murine fibrosarcoma cells is influenced by cell surface laminin. Int. J. Cancer 33:651–655.

Manishen, W.J., K. Sivananthan, and F.W. Orr (1986) Resorbing bone stimulates tumor cell growth; a role for the host microenvironment in bone metastasis. Am. J. Pathol. 123:39–45.

Marchisio, P.C., S. Bondanza, O. Cremona, et al. (1991) Polarized expression of integrin receptors, and their relationship with the cytoskeleton and basement membrane matrix in cultured human keratinocytes. J. Cell Biol. 112:761–773.

McCarthy, J.B., A.P.N. Skubitz, S. Palm, et al. (1988) Metastasis inhibition of different tumor types by purified laminin fragments and a heparin-binding fragment of fibronectin. J. Natl. Cancer Inst. 80:108–116.

Nakajima, M., K. Morikawa, A. Fabra, et al. (1990) Influence of organ environment on extracellular matrix degradative activity and metastasis of human colon carcinoma cells. J. Natl. Cancer Inst. 82:1890–1898.

Natali, P.G., M.R. Nicotra, R. Cavaliere, et al. (1991) Tumor progression in human malignant melanoma is associated with changes in $_6\beta_1$ laminin receptor. Int. J. Cancer 49:168–172.

Nicolson, G.L. (1988) Cancer metastasis: Tumor cells and host organ properties important in metastasis to specific secondary sites. Biochim. Biophys. Acta 948:175–224.

Nicolson, G.L., K.M. Dulski, and J.E. Trosko (1988) Loss of intercellular junctional communication correlates with metastatic potential in mammary adenocarcinoma cells. Proc. Natl. Acad. Sci. USA 85:473–476.

Niedbala, M.J., R. Madiyalakan, K. Matta, et al. (1987) Role of glycosidases in human ovarian carcinoma cell mediated degradation of subendothelial extracellular matrix. Cancer Res. 47:4634–4641.

Orr, F.W., W. Millar-Book, and G. Singh (1990) Chemotactic activity of bone and platelet-derived TGF-beta for bone-metastasizing rat Walker 256 carcinosarcoma cells. Invasion Metastasis 10:241–252.

Perosio, P., and J.J. Brooks (1989) Expression of growth factors and growth factor receptors in soft tissue tumors. Implications for the Autocrine hypothesis. Lab. Invest. 60:245–253.

Perrotti, D., L. Cimino, R. Falcioni, et al. (1990) Metastatic phenotype: Growth factor dependence and integrin expression. Anticancer Res. 10:1587–1598.

Phillips, S.M., A.J. Bendall, and I.A. Ramshaw (1990) Isolation of gene associated with high metastatic potential in rat mammary adenocarcinomas. J. Natl. Cancer Inst. 82:199–203.

Plantefaber, L.C., and R.O. Hynes (1989) Changes in integrin receptors on oncogenically transformed cells. Cell 56:281–290.

Plaskin, D., C. Gelber, M. Feldman, et al. (1988) Reversal of the metastatic phenotype in Lewis lung carcinoma cells after transfection with syngeneic H-$2K^b$ gene. Proc. Natl. Acad. Sci. USA 85:4463–4467.

Rice, G.E., and M.P. Bevilacqua (1989) An inducible endothelial cell surface glycoprotein mediates melanoma adhesion. Science 246:1303–1306.

Rodeck, U., M. Herlyn, H.D. Menssen, et al. (1987) Metastatic but not primary melanoma cell lines grow in vitro independently of exogenous growth factors. Int. J. Cancer 40:687–690.

Ruoslahti, E., and Y. Yamaguchi (1991) Proteoglycans as modulators of growth factor activities. Cell 64:867–869.

Rusciano, D., and M.M. Burger (1992) Why do cancer cells metastasize into particular organs? Bioessay (in press).

Rusciano, D., P. Lorenzoni, and M.M. Burger (1991) The role of both specific cellular adhesion and growth promotion in liver colonization by F9 embryonal carcinoma cells. Int. J. Cancer 48:450–456.

Sargent, N.S.E., M. Oestreicher, H. Haidvogl, et al. (1988) Growth regulation of cancer metastases by their host organ. Proc. Natl. Acad. Sci. USA 85:7251–7255.

Schreiner, C., M. Fisher, S. Hussein, et al. (1991) Increased tumorigenicity of fibronectin receptor deficient chinese hamster ovary cell variants. Cancer Res. 51:1738–1740.

Silletti, S., H. Watanabe, V. Hogan, et al. (1991) Purification of B16-F1 melanoma autocrine motility factor and its receptor. Cancer Res. 51:3507–3511.

Steeg, P.S., K.H. Cohn, and A. Leone (1991) Tumor metastasis and nm23: Current concepts. Cancer Cells 3:257–262.

Strobel, E.S., H.G. Strobel, K.J. Bross, et al. (1989) Effects of human bone marrow stroma on the growth of human tumor cells. Cancer Res. 49:1001–1007.

Tarin, D., J.E. Price, M.G. Kettlewell, et al. (1984) Mechanisms of human tumor metastasis studied in patients with peritoneo-venous shunt. Cancer Res. 44:3584–3592.

Templeton, N.S., P.D. Brown, A.T. Levy, et al. (1990) Cloning and characterization of human tumor cell interstitial collagenase. Cancer

Res. 50:5431–5437.

Terrana, B., D. Rusciano, and L. Pacenti (1987) Organ colonization pattern of retinoic acid-treated and -untreated mouse embryonal carcinoma F9 cells. Cancer Res. 47:3791–3797.

Tullberg, K.F., and M.M. Burger (1985) Selection of B16 melanoma cells with increased metastatic potential and low intercellular cohesion using nucleopore filters. Invasion Metastasis 5:1–15.

Ugen, K.E., M. Mahalingam, P.A. Klein, et al. (1988) Inhibition of tumor cell-induced platelet aggregation and experimental tumor metastasis by the synthetic Gly-Arg-Gly-Asp-Ser peptide. J. Natl. Cancer Inst. 80:1461–1466.

Vlemincks, K., L. Vakaet, Jr., M. Mareel, et al. (1991) Genetic manipulation of E-cadherin expression by epithelial tumor cells reveals an invasion suppressor role. Cell 66:107–119.

Vlodavsky, I., G. Korner, R. Ishai-Michaeli, et al. (1990) Extracellular matrix-resident growth factors and enzymes: Possible involvement in tumor metastasis and angiogenesis. Cancer Metas. Rev. 9:203–226.

Wang, J., L. Fallavollita, and P. Brodt (1991) Identification of an Mr 64,000 plasma membrane glycoprotein mediating adhesion of tumor H59 cells to hepatocytes. Cancer Res. 51:3578–3584.

Warn, R.M., and P. Dowrick (1988) Motility factors on the march. Nature 340:186–187.

Weidner, K.M., N. Arakaki, G. Hartmann, et al. (1991) Evidence for the identity of human scatter factor and human hepatocyte growth factor. Proc. Natl. Acad. Sci. USA 88:7001–7005.

Whiteside, T.L., and R.B. Herberman (1989) The role of natural killer cells in human disease. Clin. Immunol. Immunopathol. 53:1–23.

Wright, J.A., S.E. Egan, and A.H. Greenberg (1990) Genetic regulation of metastatic progression. Anticancer Res. 10:1247–1256.

Yamamura, Y., B.C. Fischer, J.B. Harnaha, et al. (1984) Heterogeneity of murine mammary adenocarcinoma cell subpopulations. In vivo and in vitro resistance to macrophage cytotoxicity and its association with metastatic capacity. Int. J. Cancer 33:67–72.

Yeatman, T.J., K.I. Bland, E.M. Copeland, et al. (1989) Tumor cell-surface galactose correlates with the degree of colorectal liver metastasis. J. Surg. Res. 46:567–571.

Zetter, B.R., and S.E. Brightman (1990) Cell motility and the extracellular matrix. Curr. Opin. Cell Biol. 2:850–856.

Chapter 33

Metastases to the Central Nervous System

Peter A. Steck and Garth L. Nicolson

INTRODUCTION

The formation of brain metastases is usually the final, fatal phase of the progression of many solid cancers, and as such represents an intriguing and significant area of cancer biology research. The metastatic process consists of a series of very complex, sequential, and multifaceted events involving specialized subpopulations of tumor cells, and unfortunately occurs often in cancer patients with malignant disease. The essential stages and mechanisms of metastasis have been described in detail in the previous chapter. This chapter concentrates on those aspects of the metastatic process that differentiate brain metastasis from the formation of other distal metastatic lesions.

All metastatic cells that colonize distant organ sites must escape from the primary tumor by local invasion, enter the circulation, circulate in the blood, adhere to and exit from the microvasculature in a distant organ, and grow in response to the local microenvironment at the new site (Kerbel, 1990; Liotta et al., 1991; Nicolson, 1988a,b, 1991). There are several significant differences between the development of metastatic lesions in the central nervous system (CNS) and other organ sites. Metastatic cells in the CNS must adhere to brain microvessels, penetrate the blood–brain endothelial barrier, and grow in the brain.

Since the brain microvascular represents an exceedingly small fraction of the total microvascular, the process of brain metastasis formation is generally thought to be nonrandom. Indeed, from a number of clinical and animal studies, the distributions of distant metastatic lesions have been shown to be nonrandom, with certain cancers exhibiting a propensity to colonize particular organs (Nicolson, 1988a,b, 1991; Sugarbaker, 1983). Such observations lead Paget (1889) to describe the nonrandom pattern of metastatic disease as one based on the properties of the unique tumor cell "seeds" as well as the organ microenvironment or "soil." This hypothesis, in which the host environment plays an equally important role as the tumor cells in the development of distant metastatic lesions, is particularly applicable to metastatic lesions in the CNS, where different tumor cell types exhibit preferential colonization of different regions of the brain. Since many, if not most, cancers appear to favor distant metastasis formation in the first organ site encountered by blood-borne cancer cells, Ewing (1926) advanced a counterproposal. Metastatic development was proposed to be primarily due to mechanical and anatomic considerations, such as circulatory connection with the primary tumor and the ability of cancer cells to form large cell clumps in the blood that arrest nonspecifically in the microcirculation. Ewing's (1926) proposal does not explain the observation, however, that some organs and tissues are served by a very small fraction of the blood vascular, such as brain, adrenals, and bone, yet they are often involved by metastatic deposits from certain cancers. In addition, other tissues, such as heart, muscle, skin, kidney, and spleen, collectively account for a large fraction of the blood vascular but are only sporadically colonized by cancers.

The preferential colonization of particular organs, or regions of organs, implies that specific tumor–host cell interactions mediated by particular sets of properties and factors, such as those involved in malignant cell adhesion, invasion, and growth, are critical to the development of metastases at certain sites (Liotta et al., 1991; Nicolson, 1982, 1988a,b, 1991).The involvement of particular cellular traits and determinants that play a role in the development of metastases in the CNS and their consequences are the subject of this review.

CLINICAL OBSERVATIONS OF BRAIN METASTASIS

To fully appreciate the significance of CNS metastasis, a brief overview of the clinical consequences of the disease is important. A relatively large number of patients with certain advanced cancers, such as disseminated malignant melanoma, breast cancer, lung cancer, and others, will develop metastatic lesions in the brain and spinal cord, frequently resulting in severe and debilitating neurological complications. Although metastatic lesions in other organs can often be tolerated or are even asymptomatic, allowing a patient to lead a productive life, once the brain is involved with rapidly growing metastatic deposits, the quality of life declines rather quickly. The severity of symptoms represents one of the major differences between metastatic lesions in the brain and other organs. Although metastases to the CNS are a significant clinical problem, few research efforts, either clinically or experimentally, have been directed toward its origins or biology. This relatively low level of effort is at least partially due to the complexity of the necessary studies and the perception that brain metastasis represents a terminal, and thus irreversible, phase of the disease process.

The overall incidence of metastatic tumor lesions to the CNS has not been accurately defined; however, approximately 13% of the cancer population will have symptomatic complications related to brain metastases (Wright and Delaney, 1989). Autopsy studies suggest that the incidence of tumor metastasis to the CNS has increased, with about 24% of cancer patients now exhibiting CNS lesions (Posner and Chernik, 1978). Based on the current number of systemic cancers and with this estimate of metastatic frequency, about 70,000 cancer deaths will occur this year with metastatic lesions in the CNS. However, less than a third of these patients will be treated for CNS complications. In contrast, approximately 14,000 deaths from primary brain tumors are estimated to occur this year, and these will be accompanied with a significantly more extensive research effort.

The number of patients affected with brain metastases appears to be increasing. This is undoubtedly related to better therapeutic treatments of primary cancers, and thus patients are now surviving long enough to develop brain metastases. Moreover, several of the most common types of human cancers that exhibit a high frequency of brain metastases (lung, 26%; breast, 16%; renal cell carcinomas, 13%; and melanoma, 40%; Wright and Delaney, 1989) are increasing in incidence. Since many of these cancers are quite common, the vast majority of brain metastases will arise from cancers due to their high incidence. For example, almost 40% of melanoma patients will be treated for complications due to brain metastases, with between 70 and 80% having CNS lesions at autopsy. In contrast, brain metastases occur rarely in liver or prostate cancers; these cancers show only a 4 or 5% incidence of metastasis formation, respectively.

Brain metastasis is generally considered a late event in the disease pathogenesis, although for some cancers it represents the first symptomatic event and almost always indicates a poor prognosis. Once brain metastasis occurs, the medium survival time from diagnosis of CNS lesions is about six months for most cancers. Although this poor morbidity rate is often due to the effects of widely disseminated disease, some melanoma patients succumb to their disease with evidence of only CNS metastases. Brain metastases exhibit two peaks of occurrence, at 0–5 years and at 55–60 years, corresponding to the occurrence of pediatric cancers (leukemia, lymphoma, and sarcomas) and adult cancers that display a propensity to colonize the brain. The pathogenesis of brain metastasis also differentiates this process from the formation of metastatic lesions in other organs. The brain is very sensitive to alterations in local microenvironment, especially growth. This latter response is due, in part, to confinement by the skull. The brain also does not contain a lymphatic system to remove fluid buildup that accompanies tumor growth. Cerebral edema represents a major complication from CNS tumor lesions. Thus a relatively small metastasis in the brain may cause severe symptoms and signs, including paralysis, headache, seizures, and impaired cognition. Since brain metastasis is often accompanied by other complications, therapy is often only palliative. Radiation and steroids are the principal treatment regimens for maintaining the quality and duration of neurological function in patients with brain metastases. Surgery is often reserved for removal of solitary lesions if neurological functions can be restored or to relieve the devastating symptoms listed above. If radiation therapy is added to surgery for single brain metastases, the median survival times can be increased from 4–6 months to greater than 6 months. The reason these results are so poor is that tumor often recurs at the site of the resected lesion, and, in addition, multiple metastases usually form outside as well as inside the brain.

Access to the brain is tightly regulated by the blood–brain barrier (BBB) that controls the flow of many materials into the brain. The BBB is a physiologic and anatomic structure that is mainly defined by tight junctions between the brain endothelial cells, a relatively thick basement membrane, and an underlying layer of astrocytes that strictly regulate the flow of ions, nutrients, and cells into the brain. Therefore, for brain metastasis to occur, malignant cells must adhere to the luminal surface of brain microvessel endothelial cells and then invade the BBB formed by the endothelial cells, basement membrane, and astrocytes. This requires the expression of a variety of cell surface receptors and degradative enzymes (discussed below).

The distribution of metastases in the brain is divided between the leptomeninges and meninges (surface of the brain) and and the parenchymal interior (cerebrum, cerebellum, and brain stem). Thus the vast majority of metastases are established via blood-borne routes, with the rare exception of direct invasion of head and neck tumors or passage via the cerebral spinal fluid resulting in diffuse leptomeningeal metastases. The distribution of metastases in the brain roughly parallels its blood flow, with about 80% of the parenchymal lesions presenting in the supratentorial regions and 15% in the cerebellum. However, different cancers exhibit differing propensities to colonize separate regions of the brain (see below). Brain metastases occur as solitary lesions as well as multiple lesions at approximately the same incidence for most cancers. Melanoma, with its high incidence of brain metastasis, usually forms multiple lesions. Leukemias and lymphomas exhibit predominantly leptomeningeal lesions.

EXPERIMENTAL MODELS OF BRAIN METASTASIS

To study the phenomenon of organ selectivity in the metastatic process, animal metastatic models have been developed that mimic clinical disease and show preferential tumor development in particular organs. A number of model systems representing various primary cancers have been developed to examine the metastatic cascade; however, relatively few have been developed to examine brain metastasis (Nicolson, 1988a,b). A major goal of these models is to determine the biological and molecular mechanisms that are involved in metastasis to the CNS. Additionally, an important question that is still controversial in metastasis research is whether the metastatic process is the result of selective pressures or random events that lead to a rare successful metastatic cell colonizing a unique site (Kerbel, 1990). We will limit our discussion to brain metastasis. There also remains a controversy as to whether brain-metastasizing tumor cell variants are derived progressively from subpopulations of highly metastatic cells or are certain cells preexistent in a primary tumor that are capable of forming brain metastases. In most animal metastatic models, one cannot distinguish between these two possibilities. Additionally, the designation of organ-specific metastatic tumor cells generally refers to the ability of variant tumor cells to preferentially colonize certain organs. This preference is usually not strict in the sense that the tumor cells will only form tumor lesions in a target organ, but rather in the appropriate assay the tumor cells will preferentially form lesions in a specific organ or organs.

Some experimental animal models for brain metastasis have used direct inoculation of tumor cells or tumor fragments into brain. These models bear little resemblance to the mechanism of metastasis. More recent studies have utilized inoculation via the left ventricle of the heart or via one of the carotid arteries. The left ventricle predominantly supplies the carotid artery, which then supplies the blood flow for the vast majority of the brain parenchyma and meninges in humans and rodents, suggesting that this blood-borne route to initiate brain metastasis is probably valid. The first metastatic models for brain metastasis used murine Walker 256 mammary carcinoma cells or KHT sarcoma cells. Tumor cell suspensions were injected into the left ventricule of the heart or cartoid arteries (Ushio et al., 1977). It was necessary to administer cyclophosphamide or irradiation to tumor-bearing animals to permit the development of brain metastases, because without the chemotherapy or radiotherapy animals succumbed to extracerebral tumors. For example, after cyclophosphamide treatment, 90% of the surviving rats receiving Walker 256 carcinoma cells developed lethal brain metastases. This model mimics the course of certain cancers where treatment can attenuate systemic cancer, resulting in relapses at specific metastatic sites. This also illustrates the privileged location of tumor lesions in the brain.

Using a mouse B16 melanoma cell line originally developed by Fidler (1973), Brunson and colleagues (1978) selected B16 variants that exhibited a propensity to colonize the brain. This was performed by repeated cycles of *in vivo* selection by intracardic inoculation, harvesting of brain tumors, and *in vitro* proliferation of the selected cells. The selection was most successful for brain meninges colonization. The selected lines, B16-B14b and B16-15b, selected 14 or 15 times, respectively, for brain meninges tumor formation, were shown to initially colonize the dura mater (interior of the skull) and leptomeninges, eventually invading into the brain parenchyma (Kawaguchi et al., 1985). The B16-B14b cells were observed to form small tumor lesions in the brain, whereas the B16-B15b cells formed relatively large tumors that were deeply pigmented. Histological and electron microscopic evaluations of the tumors revealed that the B16-B14b cells elicited extensive infiltrations of lymphocytes that were absent in the B16-B15b cell lesions. Removal of the spleen prior to injection of the tumor cells increased the tumor growth of the B16-14b cells, suggesting that host–tumor interactions, in this case the ability to elicit an immune response, plays an important role in the development of B16 metastatic brain lesions (Kawaguchi et al., 1989). The selection for enhanced brain colonization did not result in selection of a dominant phenotypic characteristic of the cells, because the *in vitro* culture of various clones derived from the brain-selected B16-B14b line lost their preference for brain colonization upon prolonged culture *in vitro*. The loss of brain colonization properties of the B16-B14b cell clones was prevented if three of the unstable clones were cocultured together, suggesting that individual tumor cell clones are stabilized by interclonal interactions (Miner et al., 1982; Nicolson et al., 1985).

To deliver melanoma cells directly to either the brain paranchyma or the meninges, Schackert and Fidler (1988a,b) injected murine K-1735 melanoma cells via the internal or external carotid artery of C3H mice. By either route of entry, the K-1735 melanoma cells only colonized brain parenchyma, indicating that K-1735 cells do not colonize the brain by random arrest and growth of cells. They then repeated this experiment with B16 melanoma cells and found, similar to previous results (Kawaguchi et al., 1985; Miner et al., 1982; Nicolson et al., 1985), that the B16 cells only colonized the meninges and ventricles, indicating that the formation of brain metastases is not simply the result of nonspecific lodgement of tumor cells in the first microcirculatory network encountered (Schackert and Fidler, 1988a). In addition, Schackert and Fidler (1988b) showed that the colonization of particular regions of the brain was not dependent on the initial distribution of tumor cells in the brain, but rather on the interaction between host and tumor cells at specific regions. The initial arrest of all of the cell lines was demonstrated to be the same. This emphasizes that the successful brain-metastatic cell must penetrate the endothelium and BBB and proliferate in the surrounding tissue after arrest in the brain microvascular.

In metastatic melanoma patients, 49% of the brain metastases are intraparenchymal, 22% are leptomeningeal, and 32% are dural (Takakura et al., 1982), suggesting that the melanoma cells that colonized these sites probably had different phenotypic characteristics. To demonstrate that brain metastasis is a unique event requiring specialized properties, Brunson and Nicolson (1980) attempted to "adapt" B16 cells to grow in brain parenchyma by direct inoculation. After 10 cycles of "adaptation" for brain survival by direct intracerebral injection and growth, the adapted cells were assayed and found to be no more metastatic than the original cells, and they failed to colonize the brain.

In addition to animal tumor models for brain metastasis, human tumors have been introduced into athymic nude mice. Using human carcinomas of the colon, breast, kidney, and lung, Schackert et al. (1989) injected tumor cells into the internal carotid artery and found that they were able to arrest, invade, and grow in brain parenchyma. Interestingly, each tumor system formed a unique pattern of experimental metastases in the brain. Schackert et al. (1990) also inoculated human melanoma cells directly into the brains of nude mice. The various melanoma cell lines all exhibited the ability to grow in the brain parenchyma after direct intracerebral inoculation. However, each melanoma line exhibited a distinct distribution in the brain after injection into the carotid artery. Interestingly, the melanoma cell lines originally obtained from human brain parenchymal lesions exhibited a propensity to colonize the CNS, while the melanoma cells derived from lymph node or subcutaneous lesions preferentially colonized the meninges. However, each of the melanomas also showed some growth in other regions of the brain, suggesting that preferential distributions of lesions occur. Using the same melanoma cell lines, their metastatic properties were examined to determine if brain metastasis represents a terminal stage of progression of the disease, and therefore if the brain-metastatic cells represent the most malignant cells present in the host (Zhang et al., 1991). Cells derived from other metastatic lesions (lymph node or lung) were shown to be equal or greater in their malignant properties compared to the cells derived from brain metastases. The malignant properties examined included colony formation under anchorage-dependent and -independent conditions and tumorigenicity after injection subcutaneously or intravenously. The results suggested that brain metastasis does not represent a final stage of the disease process, but rather occurs by selective events due to cells with specific properties (Zhang et al., 1991). The cells that are capable of successfully completing the metastatic process may be present in the primary tumor, but they probably represent a minor subpopulation (Fidler and Balch, 1987).

TUMOR CELL PROPERTIES ASSOCIATED WITH BRAIN METASTASIS

Metastasis formation is the end result of a sequential, multistep process. There is a growing list of tumor cell and host properties that are important in each step (Kerbel, 1990; Nicolson, 1988a,b, 1991). Tumors metastasize by the growth and invasion of malignant cells at primary sites, their penetration into lymphatics, blood circulation, and body cavities; transport to distant sites; implantation at these sites; invasion; and growth and evasion of host immune and nonimmune responses. Each of these steps involves many tumor cell and host molecules. Metastasis to brain probably involves some unique characteristics of tumor cells that may not be necessary for metastasis to other organ sites.

The initial interaction of a circulating tumor cell with host organs occurs in the microcirculation. Using cell adhesion assays, the interactions between blood-borne tumor cells and microvessel endothelial cells were shown to be an important determinant for brain metastasis. For example, B16 melanoma sublines of high brain colonization potential adhered at higher rates to brain-derived microvessel endothelial cell monolayers *in vitro* than lines that failed to colonize the brain (Fig. 1) (Nicolson, 1982, 1988a). Subsequentially, several studies have examined the ability of specific organ-colonizing tumor cell variants to adhere to microvessel endothelial cells derived from various organs. These studies have shown that tumor cells of high metastatic potentials usually adhere at higher rates to target endothelial cells than poorly or nonmetastatic cells (reviewed by Nicolson, 1988a,b). Furthermore, preferential adhesion of organ-colonizing tumor cells occurs to endothelial cells derived from small but not from large blood vessels from the appropriate organ. The B16-B15b cell subline selected for brain colonization adhered at an increased rate to brain-derived microvessel endothelial cells compared to B16 sublines selected for lung or ovary colonization. Similarly, the high lung-colonizing B16 cells adhered better to lung endothelial cells (Fig. 1). Since the vast majority of brain metastases occur via the blood-borne route, the logical site of initial interaction is with the brain endothelial cells. Several studies have examined tumor cell and host endothelial cell interactions and have generally shown that increased metastatic potential and organ preference correlate with increased target organ endothelial cell adhesive properties (reviewed by Nicolson, 1988a,b). The adhesion of organ-

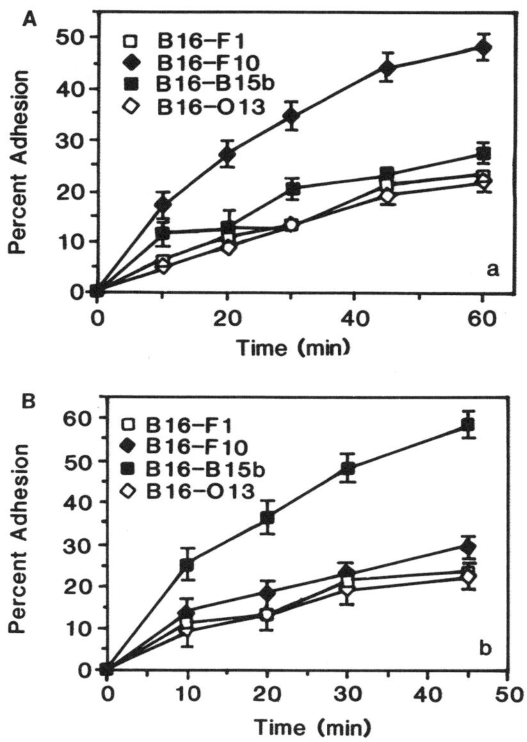

Fig. 1. Kinetics of adhesion of murine B16 melanoma sublines selected from the line B16-F1 for enhanced lung (B16-F10), brain meninges (B16-B15b), or ovary (B16-O13) colonization in mice to monolayers of (**a**) murine lung microvessel endothelial cells or (**b**) brain microvessel endothelial cells. (Reprinted from Nicolson, 1988a.)

colonizing malignant cells to the appropriate microvessel endothelial cells suggests that microvessel endothelial cells display organ-specific adhesion molecules.

Studies with animal tumor metastatic models indicate that differential quantitative expression of endothelial cell surface proteins may contribute to the altered adhesive properties of organ-colonizing tumor cells. Indeed, immunological studies have indicated that endothelial cells display organ-specific determinants on their luminal surfaces (Auerbach et al., 1985). Using *in situ* radiolabeling of the microvasculature and cultured cells isolated from various murine organs, Belloni and Nicolson (1988) showed the preferential expression of certain proteins within the microvascular of different organs. Five common proteins were observed to be expressed by all organ microvascular and cultured organ microvessel endothelial cells, whereas five alternative proteins were differentially displayed among the endothelia of various organs. For example, a 130 kd protein was observed to be expressed predominantly on brain endothelial cells. The vasculature of the brain also exhibited an increased expression of wheat germ agglutinin-binding glycoproteins, particularly a 180 kd glycoprotein, compared to the other tissues examined (heart, lung, kidney, and thyroid). These studies indicate that endothelial cells from specific organs display different cell surface determinants; whether these differentially expressed proteins play a role in organ-specific cellular adhesion or other phenomenon remains to be determined. To examine the proteins involved in cell adhesion, the specific binding of cell surface proteins from radioactively labeled cell membranes of murine lymphoma cells that exhibit organ-specific colonization were incubated with organ-derived endothelial cells, and the endothelial cell-bound proteins analyzed by polyacrylamide gel electrophoresis (Tressler and Nicolson, 1992). Only quantitative differences were seen in the amounts of particular components that bound to specific microvessel endothelial cells from highly and poorly metastatic tumor cells. Thus the differences seen in tumor cell–endothelial cell adhesive properties are probably due to quantitative, not qualitative, differences in the expression of particular tumor cell and microvessel endothelial cell surface molecules. A variety of tumor cell adhesion molecules appears to be generally associated with metastasis (Nicolson, 1991), but organ-determining molecules have not yet been isolated.

To identify the organ microvessel endothelial cell surface components involved in organ-specific metastasis, the endothelial cell surface glycoproteins were solubilized and separated by polyacrylamide gel electrophoresis, followed by transfer to nitrocellulose paper. The nitrocellulose paper was then incubated with viable, radioactively labeled organ-colonizing murine lymphoma cells to determine the binding of the cells to specific endothelial cell components (Belloni and Nicolson, 1988). Certain endothelial cell components were observed to be differentially expressed and bound differing amounts of the radiolabeled tumor cells. Although these studies used lymphoma cells and their preferential organ colonization site was liver, they demonstrate the possibility of defining specific molecules that may be involved in preferential brain colonization by metastatic tumor cells.

Another important property of brain-metastasizing tumor cells is the ability to penetrate the brain vasculature and BBB. This requires the production of degradative enzymes to solubilize the BBB basement membrane. Tumor cell lines that possess the ability to colonize the brain express high levels of basement membrane hydrolytic enzymes, such as type IV collagenase and heparanase (Liotta et al., 1991). For example, when Nakajima and associates (1983) examined a series of B16 melanoma cell lines for their abilities to degrade basement membrane proteoglycan heparan sulfate, the highly brain-metastatic B16-B15b line expressed the highest level of this enzyme.

Once malignant cells have invaded brain tissue, they can proliferate, but only if they can respond to the mitogenic environment of the brain. Alternatively, the metastatic cells could synthesize their own autocrine growth factors. Certain organs appear to contain local concentrations of growth

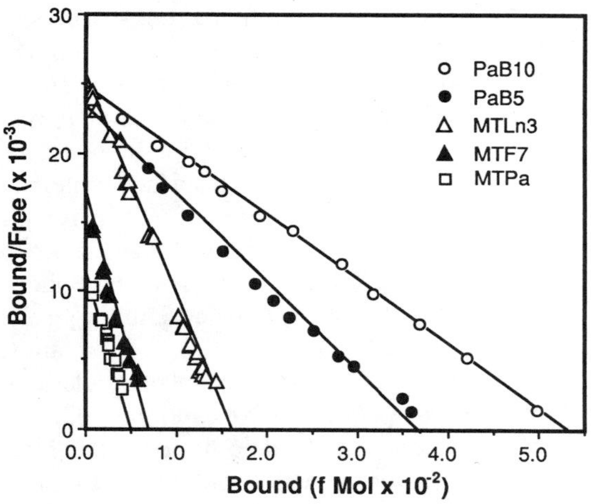

Fig. 2. Scatchard analysis of ^{125}I-transferrin binding to various rat 13762NF mammary adenocarcinoma sublines derived from the primary tumor (MTPa, MTF7), lung (MTLn3), or brain (MTPaB5, MTPaB10) metastases. Data are mean values of duplicate experiments. In all cases, bound radioactivity could be reduced to background by the addition of excess unlabeled transferrin. (Reprinted from Nicolson et al., 1992a.)

factors that may preferentially stimulate the growth of certain metastatic tumor cells. By examining the ability of specific organ-conditioned media to stimulate or inhibit the growth of organ-colonizing malignant cells, evidence for this proposal was obtained. For example, high lung-metastatic tumor cells were found to respond to lung-conditioned medium and grow at higher rates than when exposed to organ-conditioned media from other nontarget organs (Nicolson and Dulski, 1986). Using this assay, a lung-derived growth factor was subsequently isolated that differentially stimulated the growth of various metastatic cell lines that preferentially metastasized to the lung (Cavanaugh and Nicolson, 1989). Recently, this growth factor was shown to be related to transferrin (Cavanaugh and Nicolson, 1991). Interestingly, brain tissue-conditioned medium failed to differentially stimulate the growth of highly brain-colonizing B16-B15b cells, suggesting that the B16 brain-colonizing variant cells may respond to unstable or insoluble (immobilized) growth factors found in the extracellular matrix of brain.

Alternatively, they may respond to a growth factor-poor brain environment by increasing the expression of growth factor receptors that are important for tumor cell proliferation. The brain appears to contain low concentrations of free transferrinlike growth factors, because brain-conditioned medium failed to differentially stimulate the growth of brain-colonizing B16 cells. This medium did not yield high concentrations of the transferrinlike factor. To respond to this "poor growth factor environment," malignant cells may have to express increased numbers of receptors or express receptors of higher affinity for their ligands. In fact, when a series of murine melanoma cells of varying potentials to colonize brain and other organs was examined for the presence of transferrinlike receptor, the high brain-colonizing sublines, such as B16-B14b and -B15b, expressed the highest numbers of transferrin receptors (Nicolson et al., 1990). Similarly, using a rat mammary adenocarcinoma system selected to metastasize to the brain, the highly brain-metastatic cell lines expressed the highest numbers of transferrin receptors, and only the brain-metastatic lines possessed transferrin receptors of higher affinity for transferrin (Fig. 2) (Nicolson et al., 1992a). Regardless of the mechanism, the end result would be the same—that brain-metastatic cells would have a growth advantage over their nonmetastatic counterparts.

Other growth factors may also be important in brain colonization. The brain is a rich source of fibroblast growth factors (FGFs). Certain melanoma and other cancer cells have been found to use an autocrine growth loop using FGF and its receptors for cellular proliferation (Rodeck et al., 1991).

MOLECULAR AND GENETIC DETERMINANTS OF BRAIN METASTASIS

The studies described above on differential expression of metastatic tumor cell adhesion receptors, degradative enzymes, and growth factor receptors suggest that organ-specific colonization may be due to a combination of quantitative and qualitative differences of gene expression and altered responses to host environments. Presently, there is little known about the molecular genetics or specific biochemical mechanisms involved in brain metastasis, or any preferential organ metastasis, for that matter. The identification of organ-specific determinants would not only be very useful for understanding the mechanism of CNS metastasis, but it might also be useful for organ-specific delivery of therapeutic agents.

Several potential molecular markers for brain metastasis have been identified. The expression of a 90 kd cell surface glycoprotein was demonstrated to correlate with the brain colonization capabilities of B16 mouse melanoma cells (Brunson et al., 1978). This glycoprotein (gp90) was originally identified as a major cell surface protein accessible to cell surface iodination that was expressed at higher levels on the brain-colonizing B16 variants and at lower levels on lung-colonizing or poorly metastatic cells (Fig. 3). Additionally, cell clones derived from the brain-colonizing variants that lost their brain-colonization propensity had decreased expression of this glycoprotein (Miner et al., 1982). The gp90 glycoprotein, therefore, represented one of the first molecular species whose expression correlated with preferential organ colonization of metastatic cells. Antibodies were generated against gp90. The $F(ab')_2$ fragments of the antibodies failed to inhibit blood-borne implantation or adhesion to brain microvessel endothelial cells, but they did inhibit B16 cell growth (Nicolson et al., 1990). Using the antibodies and immunohistochemical techniques, gp90 was shown to be expressed on other tumor cell types and on various adult tissues (Nicolson et al., 1987). Recently, the 90 kd glycoprotein was shown to process certain properties related to the transferrin receptor (Cavanaugh and Nicolson, 1991). The 90 kd glycoprotein was shown to be a homodimer subunit of a 180 kd protein that bound transferrin in a dose-dependent manner. The B16-15b cells exhibited a greater growth stimulatory response to purified transferrin than the other variant B16 melanoma cells that displayed decreased amounts of the glycoprotein. These results suggest that brain metastasis requires alterations in the expression of growth factor receptors, particularly transferrin receptors. The overexpression of certain growth factor receptors may be required for a metastatic cell to respond to the microenvironments of various organs.

Two gene products that appear to play important roles in the metastatic process have recently been described. Neither product has been directly implicated in brain metastasis; however, they may play a general role in the metastatic process of distant organ colonization and, therefore, deserve some consideration. A cell adhesion molecule that has recently gained attention is CD44, a form of the hyaluronic acid receptor. CD44 was originally identified as an adhesion molecule that appeared to regulate the binding of lymphocytes to high endothelial venules. The process of lymphocyte homing has been used as a paradigm for blood-borne metastatic cell colonization. Subsequently, isoforms of CD44 have been isolated and shown to be different by immunological and molecular methods. These isoforms possess different binding properties for hyaluronic acid and endothelial cells (Stamenkovic et al., 1991). Recently, overexpression of a variant form of CD44 increased the metastatic potential of poorly metastatic rat pancreatic adenocarcinoma cells (Gunthert et al., 1991). Monoclonal antibodies directed toward the variant isoform of CD44 were also shown to inhibit the metastatic dissemination of these tumor cells. The results suggest that CD44 may play a causal role in the metastatic process. Additionally, the altered isoforms of CD44 expressed in metastatic clones of a rat mammary adenocarcinoma cells (Nicolson, 1988a), but not in the non- or poorly metastatic clones. To demonstrate a role for CD44 in metastasis, human melanoma cells have been selected for increased expression of CD44 (Birch et al., 1991). The high-expressing melanoma clones exhibited higher lung colonization abilities than the poorly expressing cell clones. The implantation properties could be inhibited by pretreatment of the cells with antibodies against CD44. The various clones did not, however, show significant differences when the tumor cells were injected at subcutaneous sites. The results suggest that variant forms of CD44 may play an important role in the hematogenous spread of certain cancer

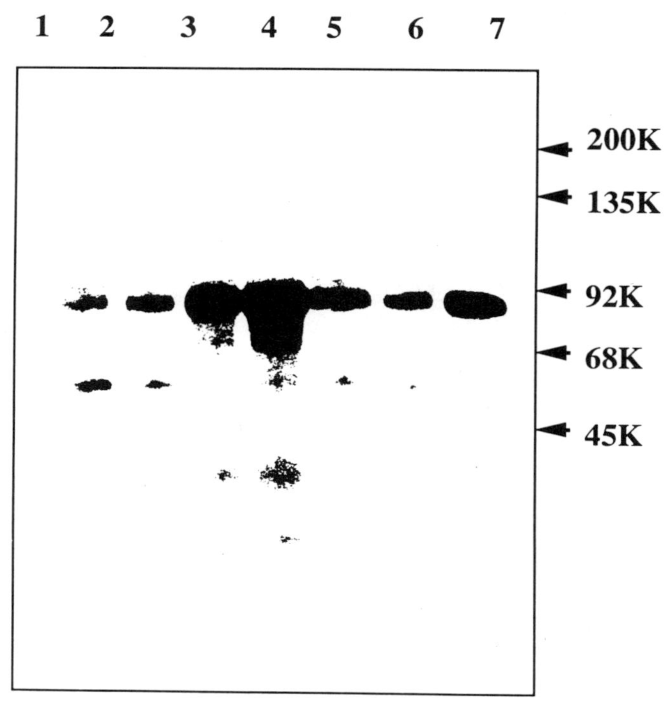

Fig. 3. Immunoprecipitation of radiolabeled B16 melanoma gp90 by anti-gp90 and its separation by polyacrylamide gel electrophoresis. **Lane 1**, B16-F1 (poorly lung metastatic); **lane 2**, B16-F10 (highly lung metastatic); **lane 3**, B16-B14b (highly brain metastatic); **lane 4**, B16-B15b (highly brain metastatic); **lane 5**, B16-O13 (highly ovary metastatic); **lane 6**, B16-F10 (highly lung metastatic); **lane 7**, human SK-MEL-23 melanoma cells. Molecular weight markers are indicated. (Reprinted from Nicolson et al., 1990.)

cells, including those that exhibit a propensity for CNS metastasis.

Another gene product that has been recently associated with metastatic potential is the product of the nm23 gene (Steeg et al., 1988). The nm23 gene was identified by subtractive hybridization of expressed genes between K1735 murine melanoma cells of low and high metastatic potentials. The poorly metastatic cells were observed to express approximately 10-fold more nm23 mRNA than did the highly metastatic cells. A similar inverse correlation with metastatic potential was shown in a series of malignant human breast carcinomas (Bevilacqua et al., 1989). The biochemical function of the nm23 gene product is currently unknown; however, this gene shares sequence homology (about 75%) with a *Drosophila* developmental gene (awd) that has nucleotide diphosphate kinase activity. This suggests that the nm23 gene product may function in microtubule assembly/disassembly or in signal transduction by regulating G-protein function. Recently, the nm23 gene was transfected into highly metastatic K1735 melanoma cells. Although the transfected cells remained metastatic, the high nm23-expressing cell clones produced lower numbers of lung tumor colonies after intravenous injection (Leone et al.,

1991). The tumorigenicity of the clones, as assessed by subcutaneous administration, was also reduced, but the growth rate of the cells was not significantly altered. Interestingly, the response of the transfected cells to transforming growth factor β (TGFβ) was altered. Thus the decreased expression of the nm23 gene may affect the ability of malignant cells to respond to TGFβ in its microenvironment. Since candidate suppressor genes, such as nm23, are altered in their expression in at least some metastatic systems, it would seem reasonable that differences in the expression of certain oncogenes would be found to be associated with metastasis (Liotta et al., 1991). Oncogene amplification has been proposed as an important mechanism for inducing tumor progression (Gitelman et al., 1987). Amplification of oncogenes has been frequently seen in some clinical cancers, but it is hardly universal (Yokota et al., 1986). Amplification of oncogenes may simply be symptomatic of additional changes in the genome, and the amplification of oncogenes (or other genes) could contribute to, without being the determinant of, tumor progression (Nicolson, 1987).

The expression of oncogenes can vary between primary tumors and their metastases (Yokota et al., 1986). Although this has also been proposed to be important in the progression of tumor cells to the metastatic phenotype, examination of a variety of primary and secondary tumors revealed that oncogenes can be overexpressed, underexpressed, or expressed equivalently in metastases compared to primary tumors (reviewed in Nicolson, 1987). Thus, the qualitative changes seen in oncogenes or the quantitative changes in their expression may contribute to progression, but they are unlikely to be the only determinants.

In tumor model experiments, the direct insertion of dominantly acting oncogenes into a suitable recipient cell resulted in acquisition of the metastatic phenotype (Liotta et al., 1991; reviewed in Nicolson, 1987). Unfortunately, this type of experiment often uses unstable, aneuploid, easily spontaneously transformable cells as recipients, like the mouse 3T3 cell lines. In other systems, two dominantly acting or activated oncogenes were necessary, an event rarely seen concurrently in spontaneous tumors. These rapid, dominantly acting changes are unlike the slow, sequential changes that characterize spontaneous transformation and tumor progression to the metastatic state *in vivo* (Nicolson, 1987, 1991). Some cells are highly resistant to oncogene-mediated conversion to the metastatic phenotype. Even within the same cell line there appears to be heterogeneity in the ability of dominantly acting activated oncogenes, such as c-H-rasEJ, to cause metastatic conversion of individual cell clones. Once transfected, these cell clones appear to diversify at rapid rates. This was easily seen in very stable nonmetastatic cell clones of a rat mammary adenocarcinoma. Soon after transfection of c-H-rasEJ, the cell clones began to diversify in their cell surface and metastatic properties (Nicolson et al., 1992b).

These *in vitro* experiments suggest that rapid cellular diversification is an important property of highly malignant cells. Qualitative events, such as an alteration in a dominantly acting oncogene, would be expected to occur at a very low rate *in vivo*, but the experiment does demonstrate a possible relationship between qualitative gene changes, cellular diversification, and malignancy, and suggests that stimulation of cellular diversification could be an important step in converting cells to more malignant phenotypes. It is usually assumed that oncogene constructs are randomly incorporated into the genome, but just the opposite appears to be the case. It is difficult to conclude that insertion of an oncogene by itself will result in a cell being converted to the metastatic phenotype. Additional changes are probably necessary, and such changes probably occur dissimilarly in each cell. Oncogene-mediated conversion to the metastatic phenotype is thought to be highly dependent on the concentration of an oncogene-encoded product, but this is also unlikely, because there was no obvious correlation between the c-H-rasEJ–encoded p21rasEJ expression and metastatic conversion (Nicolson et al., 1992b). Thus, in addition to oncogene insertion and expression, other changes are probably necessary. Finally, the insertion of c-H-rasEJ into nonmetastatic cells did convert them to the metastatic phenotype, but none of the converted cell clones was metastatic to the brain, suggesting that other changes are probably more important in determining brain-metastatic properties.

A number of studies have clearly documented that metastasis, and particularly metastasis to the CNS, requires a complex series of steps and interactions. A brain-specific metastatic determinant has not yet been identified. However, one may question if such a gene product would be expected to exist. A more likely scenario is that a series of genetic, biochemical, and biological properties would be responsible for the development of brain tumor metastases. Rather than qualitative changes, these changes are probably better reflected by quantitative changes in the expression of several genes.

REFERENCES

Auerbach, R.L., L. Alby, W. Morrissey, et al. (1985) Expression of organ specfic antigens on capillary endothelial cells. Microvasc. Res. 29:401–411.

Belloni, P.N., and G.L. Nicolson (1988) Differential expression of cell surface glycoproteins on various organ-derived microvascular endothelia and endothelial cell cultures. J. Cell. Physiol. 136:398–410.

Bevilacqua, G., M.E. Sobel, L.A. Liotta, et al. (1989) Association of low nm23 RNA levels in human primary infiltrating ductal breast carcinomas with lymph node involvement and other histopathological indicators of high metastatic potential. Cancer Res. 49:5185–5190.

Birch, M., S. Mitchell, and I.R. Hart (1991) Isolation and characterization of human cell variants expressing high and low levels of CD44. Cancer Res. 51:6660–6667.

Brunson, K.E., G. Beattie, and G.L. Nicolson (1978) Selection and altered properties of brain-colonising melanoma. Nature 272:543–545.

Brunson, K.W., and G.L. Nicolson (1980) Experimental brain metastsasis. In: L. Weiss, H.A. Gilbert, and J.B. Posner (eds.). Brain Metastasis. Boston: G.K. Hall, pp. 50–66.

Cavanaugh, P.G., and G.L. Nicolson (1989) Purification and some properties of a lung-derived growth factor that differentially stimulates the growth of tumor cell metastatic to the lung. Cancer Res. 49:3928–3933.

Cavanaugh, P.G., and G.L. Nicolson (1991) Lung-derived growth factor that stimulates the growth of lung-metastasizing tumor cells: Identification as transferrin. J. Cell. Biochem. 47:261–271.

Ewing, J. (1926) A Treatise on Tumors, 3rd ed. Philadelphia: W.B. Saunders.

Fidler, I.J., and C.M. Balch (1987) The biology of cancer metastasis and implications for therapy. Curr. Probl. Surg. 24:131–209.

Fidler, I.J. (1973) Selection of successive tumor cell lines for metastasis. Nature 242:148-149.

Gitelman, I., D.F. Dexter, and J.C. Roder (1987) DNA amplification and metastasis of the human melanoma cell line MeWo. Cancer Res. 47:3851–3855.

Gunthert, U., M. Hofmann, W. Rudy, et al. (1991) A new variant of glycoprotein CD44 confers metastatic potential to rat carcinoma cells. Cell 65:13–24.

Kawaguchi, T., M. Kawaguchi, K. Dulski, et al. (1985) Cellular behavior of metastatic B16 melanoma in experimental blood-borne implantation and cerebral invasion. Invasion Metastasis 5:16–30.

Kawaguchi, T., M. Kawaguchi, T.M. Lembo, et al. (1989) Differential tumor growth of blood-borne B16 melanoma variants in cerebral dura mater is related to tumor-host cell reactions. Clin. Expl. Metastasis 7:1–14.

Kerbel, R.S. (1990) Growth dominance of the metastatic cancer cell: cellular and molecular aspects. Adv. Cancer Res. 55:87-131.

Leone, A., U. Flatow, C.R. King, et al. (1991) Reduced tumor incidence, metastatic potential, and cytokine responsiveness of nm23-transfected melanoma cells. Cell 65:25–35.

Liotta, L.A., P.S. Steeg, and W.G. Stetler-Stevenson (1991) Cancer metastasis and angiogenesis: An imbalance of positive and negative regulation. Cell 64:327–336.

Miner, K.M., T. Kawaguchi, G.W. Uba, et al. (1982) Clonal drift of cell surface, melanogenic, and experimantal metastatic properties of in vivo-selected, brain-meninges-colonizing murine B16 melanoma. Cancer Res. 42:4631–4638.

Nakajima, M., T. Irimura, D.T. DiFerrante, et al. (1983) Heparan sulfate degradation correlates with tumor invasive and metastatic properties of B16 melanoma sublines. Science 220:611–613.

Nicolson, G.L. (1982) Cancer metastasis: Organ colonization and the cell surface properties of malignant cells. Biochim. Biophys. Acta 695:113–176.

Nicolson, G.L. (1987) Tumor cell instability, diversification, and progression to the metastatic phenotype: From oncogene to oncofetal expression. Cancer Res. 47:1473–1487.

Nicolson, G.L. (1988a) Organ specificity of tumor metastasis: Role of preferential adhesion, invasion, and growth of malignant cells at specific secondary sites. Cancer Metas. Rev. 7: 143–188.

Nicolson, G.L. (1988b) Cancer metastasis: Tumor cell and host organ properties important in colonization of specific secondary sites. Biochim. Biophys. Acta 948:175–224.

Nicolson, G.L. (1991) Tumor and host molecules important in organ preference of metastasis. Semin. Cancer Biol. 2:143–154.

Nicolson, G.L., P.G. Cavanaugh, and T. Inoue (1992a) Differential stimulation of the growth of lung-metastasizing tumor cells by lung (paracrine) growth factors: Identification of transferrin-like mitogens in lung tissue-conditioned medium. J. Natl. Cancer Inst. (in press).

Nicolson, G.L., and K.M. Dulski (1986) Organ specificity of metastatic tumor colonization is related to organ-selective growth properties of malignant cells. Int. J. Cancer 38:289–294.

Nicolson, G.L., G.E. Gallick, W.H. Spohn, et al. (1992b) Transfection of activated c-H-rasEJ/pSV2neo or pSV2neo into rat mammary cells: Rapid stimulation of clonal diversification in spontaneous metastatic and cell surface properties. Oncogene (in press).

Nicolson, G.L., T. Inoue, C.S. Van Pelt, et al. (1990) Differential expression of a Mr 90,000 cell surface transferrin receptor-related glycoprotein on murine B16 metastatic melanoma sublines selected for enhanced brain or ovary colonization. Cancer Res. 50:515–520.

Nicolson, G.L., T. Kawaguchi, M. Kawaguchi, et al. (1987) Brain surface invasion and metastasis of murine malignant melanoma variants. J. Neuro-Oncology 4:209–218.

Nicolson, G.L., C. Van Pelt, T. Irimura, et al. (1985) Stabilities and characteristics of brain meninges-colonizing murine melanoma cells. Prog. Exp. Tumor Res. 29:17–35.

Paget, S. (1889) The distribution of secondary growths in cancer of the breast. Lancet 1:571–573.

Posner, J.B., and N.L. Chernik (1978) Intracranial metastases from systemic cancer. Adv. Neurol. 19:575–587.

Rodeck, U., D. Becker, and M. Herlyn (1991) Basic fibroblast growth factor in human melanoma. Cancer Cells 3:308–312.

Schackert, G., and I.J. Fidler (1988a) Development of in vivo models for studies of brain metastasis. Int. J. Cancer 41:589–594.

Schackert, G., and I.J. Fidler (1988b) Site-specific metastasis of mouse melanomas and a fibrosarcoma in the brain or meninges of syngeneic animals. Cancer Res. 48:3478–3484.

Schackert, G., J.E. Price, C.D. Bucana, et al. (1989) Unique patterns of brain metastasis produced by different human carcinomas in athymic nude mice. Int. J. Cancer 44:892–897.

Schackert, G., J.E. Price, R. Zhang, et al. (1990) Regional growth of different human melanomas as metastases in the brain of nude mice. Am. J. Pathol. 136:95–102.

Stamenkovic, I., A. Aruffo, M. Amiot, et al. (1991) The hematopoietic and epithelial forms of CD44 are distinct polypeptides with different adhesion potentials for hyaluronate-bearing cells. EMBO J. 10:343–348.

Steeg, P.S., G. Bevilaqua, L. Kopper, et al. (1988) Evidence for a novel gene associated with low tumor metastatic potential. J. Natl. Cancer Inst. 80:200–204.

Sugarbaker, E.V. (1983) Patterns of metastasis in human malignancies. Cancer Biol. Rev. 2:235–278.

Takakura, K., K. Sano, S. Hojo, et al. (1982) Metastatic Tumors of the Central Nervous System. Tokyo: Igaku-Shoin.

Tressler, R., and G.L. Nicolson (1992) Butanol-extractable and detergent-solubilized cell surface components from murine large cell lymphoma cells associated with adhesion to organ microvessel endothelial cells. J. Cell. Biochem. 48:162–171.

Urhio, Y., J.B. Posner, and W.R. Shapiro (1977) Chemotherapy of experimental meningeal carcinomatosis. Cancer Res. 37:1232-1237.

Wright, D.C., and T.F. Delaney (1989) Treatment of metastatic cancer to the brain. In: V.T. DeVita, S. Hellman, and S.A. Rosenberg (eds.). Cancer: Principles and Practice of Oncology. New York: J.B. Lippincott, pp. 2245–2261.

Yokota, J., Y. Tsunetsegu-Yokota, H. Battifora, et al. (1986) Alterations of myc, myb, and rasHa proto-oncogens in cancers are frequent and show clinical correlation. Science 246:1265–1275.

Zhang, R.D., J.E. Price, G. Schackert, et al. (1991) Malignant potential of cells isolated from lymph node or brain metastases of melanoma patients and implications for prognosis. Cancer Res. 51:2029–2035.

Chapter 34

Viral Vectors for Experimental Brain Tumor Therapy

Robert L. Martuza

INTRODUCTION

Much of the content of the chapters in this book (Martuza, 1990, 1992)[1] documents the proliferation of knowledge over the last decade defining the molecular genetic abnormalities associated with tumorigenesis. For many tumors in the nervous system, we now know the loci of the genes that are associated with the initiation of tumor formation, as well as, in some cases, the loci of genes that are associated with tumor progression from a benign state to a malignant state. For some of these genes, only the location is known (Rouleau et al., 1990); but the genes, the gene products, and their functions have not been identified. For others such as the p53 gene (Chung et al., 1992), the gene, its product, and function have been well studied. This wealth of information will likely lead to improved diagnosis, the development of more rational prognostic indicators, and, ultimately, to the development of pharmaceutical agents that can inhibit tumorigenesis or tumor progression. Rather than these agents being compounds with a number of toxic side effects, it is also expected that some of them will have specific effects on the tumor cells themselves. However, we must realize that there are problems to be faced in the development of such new therapeutic technologies. In particular, tumors in the brain present obstacles that are not present in other systemic tumors. One such obstacle is the blood–brain barrier: Certain pharmaceuticals may not be able to cross this barrier.

While the blood–brain barrier may be disrupted in the center of a glioblastoma, it may not be disrupted in the area of the infiltrating cells, and thus larger or ionized pharmaceutical agents may not be able to cross into the areas of infiltrating glioma cells to inhibit their growth and spread. Additionally, it should be recognized that many of the genes and their products that are involved in tumorigenesis may also play a role in other tissues in the body; therefore, their inhibition or modulation may be associated with toxic side effects.

While tumors of the brain have certain features that may make therapy difficult, they also have certain features that could make therapy easier or at least should change the therapeutic design. One such feature is that primary nervous system tumors, such as glioblastoma, cause neurologic disability and death through local growth. Metastases outside the nervous system are exceedingly rare. Thus this is a local—albeit infiltrating—tumor, and local or nonsystemic therapies should be considered. Additionally, malignant primary brain tumors are a population of rapidly dividing cells contained within a population of essentially nondividing postmitotic neurons and glia. This is quite different from tumors of bone marrow or colon, where there is a rapidly dividing normal cell population as well. Thus these two features, and possibly others, could be utilized in designing therapies based on molecular genetic techniques.

Over the last several years, my colleagues and I have begun explorations into an area that we term "molecular neurosurgery" (Martuza, 1992). I have used this term in prior presentations and publications in reference to the use of engineered molecules that can be used to alter cellular function within the nervous system. These may be nucleic acids or other molecules such as porphyrins or their derivatives. They may be delivered in their molecular state, or they may be packaged in one of several ways, such as viruses, genetically engineered cells, liposomes, or as other preparations. Depending upon the molecular design and strategy chosen, therapy could be directed toward neurons, glial cells, or both. The term molecular neurosurgery accentuates the difference between those conventional types of surgery that neurosurgeons have inherited from the general surgeons. The brain does not lend itself to cutting with scalpels and other instruments. Indeed, the diseases neurosurgeons treat well are those that are *around* or *under* the brain, such as acoustic neuroma, meningioma, and aneurysms. The neurosurgeon cannot adequately address diseases *within* the brain, such as Parkinson's disease, or, in the case of tumors,

[1]Portions of this chapter have appeared in earlier reviews addressing this subject or related subjects.

infiltrating gliomas. The development of molecular neurosurgical techniques may allow the modulation of specific cells in finite loci within the nervous system without harming the normal surrounding cells.

It is interesting to note that the causes of tumors and the treatments we use for patients with tumors are often the same. For example, radiation is known to induce tumors, including tumors within the nervous system; yet one of the mainstays of tumor therapy is radiation. Radiation causes DNA breaks or mutations and it is logical to think that it can cause either a tumor-initiating mutation or, as therapy, mutations that inhibit tumor growth. Thus, since radiation acts at the level of the DNA in both instances, it may be used to create tumors, or, appropriately used, to treat tumors. Similarly, chemicals can be carcinogenic or therapeutic. Nitrosureas are potent carcinogens and cause nervous system tumors in animals. Nitrosureas are also a mainstay of tumor therapy. Their harm and their benefit likely occur at similar mechanistic levels. Viruses are a third cause of some tumors. Some viruses can integrate into the DNA, causing breaks in locations that alter cell growth. Other viruses may carry oncogenes or inhibitors of tumor suppressor genes, thus leading to tumor formation. Because viruses can alter growth at a cellular level, it seems likely that viruses could be used to cure tumors as well.

We are now in an era where the viral genome can be mapped in detail and genetically engineered. Viruses are extremely efficient at carrying their genetic information into certain cells. With appropriate engineering, viral tropisms can be altered. Viruses may be lytic and kill the cell they enter, or, alternatively, the virus may enter into latency or stable integration and form a long-lasting symbiotic relationship with the cell, in some cases expressing viral genes along with cellular genes. In this situation, the virus may be used to modulate cell growth or cell function. For these reasons, viruses appear worthy of exploration as molecular neurosurgical tools.

It is in this regard that my colleagues and I have begun experiments using a novel strategy to develop therapies for primary malignant brain tumors, namely: the use of genetically engineered viruses to destroy or modulate tumor cells within the nervous system while sparing surrounding normal neural tissue. These studies have been previously summarized in other chapters (Martuza 1990, 1992), and some of the early data has been detailed in several prior manuscripts (Ezzedine et al., 1991; Martuza et al., 1991; Short et al., 1990). This chapter reflects a review of that prior work and an expansion into additional future possibilities.

There are many viruses that could be utilized to develop therapies for tumors; however, the investigator must start somewhere, and we chose initially to explore retroviruses and herpes viruses as possible agents that could be used for tumor therapy. We chose these for specific reasons: Retroviruses have been extensively studied, the genome is mapped in detail, and they will essentially only integrate into dividing cells. Because malignant brain tumors are a dividing cell population in an organ that is otherwise nondividing and postmitotic, this last feature is very attractive. Retroviruses therefore already contain a unique feature targeting them for tumors within the nervous system. Unfortunately (as discussed later), retroviruses also contain some features that are not advantageous. For example, they are very small and thus do not allow the insertion of very much additional genetic information while still maintaining the ability to be a competent replicating virus. For this reason, as well as for safety concerns, many investigations have been done with replication-defective retroviruses. While a replication-defective viral particle can efficiently deliver a foreign gene to a dividing cell, it suffers from the ability to be able to replicate any further at that point, and thus it cannot transfer the gene to a second cell. Since most tumor therapy strategies require high-efficiency gene transfer into all or nearly all of the cells of the tumor, this represents a significant problem for the use of retroviruses.

The second virus we chose to explore overcomes this barrier. Herpes viruses can be used that are replication-competent. Herpes viruses are much larger and allow the insertion of more genetic material without inhibiting their ability to replicate. Herpes viruses also have other attractive features; for example, they can spread through the nervous system. Also, certain mutants have been identified that are associated with a decreased ability to produce encephalitis. Several antiherpetic pharmaceutical agents have been identified, allowing viral infection to be aborted if necessary. Herpes viruses are endemic in the environment, and thus safety issues are less of a problem provided that the mutant vectors are appropriately designed. However, herpes viruses have one drawback for tumor therapy relative to retroviruses: Namely, the herpes virus can enter into and grow within glial cells, neurons, and other normal cells. They do not have the built-in feature of retroviruses requiring a dividing cell population. Indeed, the common cold sore is an example of a herpes virus forming a long-term stable latent relationship (probably as an episome) within the neuron of the trigeminal ganglia. Thus, if a herpes virus is to be used for tumor therapy, some tumor cell-specific feature must be engineered into the virus so that tumor cells are killed or modulated, whereas normal neuronal cells are relatively unaffected.

RETROVIRUSES AS ANTITUMOR AGENTS

We postulated that a retrovirus could infect dividing tumor cells and carry into them a foreign gene to kill or modulate the tumor cells. Whle our initial hope was to replace *in situ* the missing or abnormal tumor suppressor gene(s) associated with a specific tumor, this hope has not yet been realized. Rather, we have focused on more basic strategies of defining the physical limitations of the use of retroviruses and of using retroviral vectors to deliver a gene that could kill tumor cells selectively. In this latter regard, several possibilities exist. For example, the virus could deliver a gene that, in and of itself, might be toxic to the cell. Diphtheria toxin, for example, could be delivered; however, this carries with it potential problems. If the virus were to infect normal cells, particularly systemic cells, toxic effects could ensue that are out of control for the investigator. This is a particular problem for the development of such therapy in humans. Therefore, we have used a different strategy: namely, using retroviruses to deliver a gene that, in and of itself, is not toxic to the cell. However, toxicity ensues when a second agent is systemically delivered under the control of the investigator or the clinician. Thus far our studies have focused on using a retrovirus to deliver the herpes thymidine kinase gene under a strong retroviral promoter (Ezzedine et al., 1991; Short et al., 1990). The expression of this gene does not affect the growth of the cell; however, it confers upon the cell a sensitivity to certain nucleoside analogues such as ganciclovir (Moolten and Wells, 1990; Moolten et al., 1990) and acyclovir. Such two-step strategies place an additional element of control under the investigator. Our initial studies demonstrated that a retroviral vector containing the thymidine kinase gene from the herpes simplex virus under the control of a strong retroviral promoter (LTR) could be integrated into growing C6 glioma cells in culture. We could further show that these cells expressed the viral kinase, and that this conferred upon them an increased sensitivity to ganciclovir. In additional studies, it could be shown that these cells were also sensitive to ganciclovir when grown *in vivo* in a rodent. Thus the concept of creating a retroviral vector containing the herpes thymidine kinase gene and using it to confer sensitivity to an exogenous agent (ganciclovir) both *in vitro* and *in vivo* was substantiated. However, it should be noted that these studies, as well as studies by others using similar strategies, utilized insertion of the foreign gene in an *in vitro* cell culture situation. This is quite different from the typical situation of a patient with a glioblastoma. In the clinical setting, the tumor is already present *in situ*, and one must transfer the foreign gene to tumor cells *in situ* rather than in a culture plate. This is a much more formidable task.

To address the efficiency of *in situ* viral infection, a virus termed "BAG" was used that was initially designed by Cepko (1989). The foreign gene in this retroviral vector is the gene for *E. coli* beta-galactosidase, which, when expressed and when the tissues are appropriately stained, turns blue in color. Gene expression can be observed readily on a cell-by-cell basis. Using this retroviral construct, my colleagues and I demonstrated that, as expected, direct inoculation of the retrovirus into an adult rodent brain was not associated with transfer of the gene into normal neuronal cells. Indeed, as noted above, the inability of a retrovirus to stably infect nondividing cells is one of its potential advantages for use in nervous system tumors. However, when we tested the ability of inoculation of an *in situ* C6 glioma to incorporate and express the foreign gene delivered by the retroviral vector, we found that this could be done only at an exceedingly low efficiency *in situ*. Thus it appeared that direct inoculation of this retroviral vector (and possibly others) would not be an efficient way to transfer a foreign gene *in situ* to a nervous system tumor. There are several reasons why this poor result may have ensued. For example, the half-life of a retrovirus in tissue is relatively short, and during that time the number of tumor cells that are actively undergoing division is probably few. Only a few cells might be expected to take up, incorporate, and express the retroviral genome.

To overcome this problem, we studied the insertion of retroviral packaging cells into the normal brain and into the tumor. Packaging cells can be constructed from various cell lines, but often fibroblasts are used. The packaging cell contains the retroviral sequences that are necessary for the construction of the virus (gag, pol, env) but does not contain the packaging sequence (psi) needed to package this nucleic acid information into the viral particle. Such packaging cells are then infected with a DNA sequence containing both the packaging sequence (psi) and the foreign gene of interest (such as HSV-TK or beta-galactosidase). The retroviral particles produced by the packaging cell are composed of the normal retroviral proteins but contain the nucleic acid of the foreign gene. These do not contain the viral nucleic acid sequences necessary to construct additional viral particles. Thus they are replication-defective. We postulated that the insertion of packaging cells into the tumor would act as a slow-release bio-pump producing viral particles over several days and increasing the efficiency of retrovirus incorporation into the tumor cells. In our first study using a packaging line termed psi-2-BAG, we found that the inoculation of the packaging cells into normal adult rodent brain was not associated with the transfer and incorporation of the foreign retroviral gene into normal brain cells. We also found that, in our system, the packaging line lasted for approximately five days. In contrast, when the packaging line was inoculated into a growing C6 intracranial glioma, we found that the retroviral particles released by the packaging line could integrate into the dividing glioma cells at a much higher efficiency than when the retrovirus was inoculated alone. In particular, we saw much higher levels of foreign gene expression in the growing periphery of the tumor. Nonetheless, the overall efficiency of gene incorporation was still a problem. In most instances, the number of tumor cells in which we could demonstrate foreign gene expression was on the order of 1%, with the maximum number being 10% in the best circumstances. For the insertion of a gene that causes tumor cell death, this may not be efficient enough to be associated with an adequate therapeutic response. In contrast, if the gene delivered was a gene associated with recruitment of lymphocytes, contained a product secreted to surrounding cells, or caused immune rejection, it may prove to have a beneficial effect. Such studies are currently underway.

There may be ways to improve the efficiency of retroviral gene transfer in nervous system tumors. Other packaging lines may have a higher efficiency, and we are currently exploring this possibility. Additionally, it may be possible to use a combination of replication-defective retroviral vectors with a wild-type retrovirus that could potentially further capture the gene of interest and transfer it to an additional cell within the tumor, thus extending this process and increasing the efficiency. The use of a wild-type virus, however, could be associated with systemic disease and the development of other tumors and thus, in a clinical setting, could potentially be a problem. Ideally, it would be best if a replication-competent retrovirus could be constructed containing the foreign gene of interest; however, because of its relatively small packaging size, the insertion of a foreign gene of any significant size thus far has not proven consistent with the retrovirus maintaining replication competence. For this reason, in addition to further developing this retroviral strategy, we have also explored a different replication-competent system: namely, the use of herpes viruses for brain tumor therapy.

HERPES VIRUSES AS ANTITUMOR AGENTS

Because herpes viruses are larger than retroviruses, they can be used to carry additional genetic material and still retain replication competence. As noted above, herpes viruses have both advantages and disadvantages relative to retroviruses. The major disadvantage is that they do not necessarily distinguish between dividing cells and nondividing cells. Thus some feature must be built into the herpes virus if it is to be used for tumor therapy. The first strategy we used takes advantage of the previously noted fact that a malignant glioma is a dividing cell population within a relatively nondividing population of normal nervous system cells. In the dividing tumor cell population, it is therefore expected that relatively high levels of enzymes necessary for DNA replication will be present, such as thymidine kinase and DNA polymerase. In contrast, in postmitotic neurons, these enzyme levels should be lower since DNA replication is not needed, and only occasional DNA repair is required. We therefore hypothesized that a herpes virus lacking the herpes thymidine kinase gene might be able to divide within the tumor cell population but not within the postmitotic neurons, nor within most normal adult glia. Indeed, prior studies demonstrated that thymidine kinase mutants and many DNA polymerase mutants have attenuated neuropathogenicity in the normal brain of rodents, possibly for this reason (Coen et al., 1982; Field and Coen, 1986; Field and Darby, 1980; Field and Wildy, 1978; Jamieson et al., 1974; Tenser and Rapp Miller, 1979). Thus our initial study was performed with a herpes simplex virus-1 (HSV1) mutant developed by my colleague Don Coen and termed *dl*sptk (Coen et al., 1989). *dl*sptk is an HSV1 mutant derived from the wild-type strain KOS and has a 360 base pair deletion in the HSV thymidine kinase gene.

We initially used the human glioma cell line U-87 obtained from the ATCC collection and demonstrated in cell culture that various strains of wild-type HSV1—as well as the mutant of interest, *dl*sptk—could infect and kill the U-87 cells (Martuza et al., 1991). Additionally, for *dl*sptk we specifically demonstrated that at low multiplicities of infection (MOI), the virus could enter one U-87 cell, make multiple copies of itself, lyse the cell, and spread to surrounding cells. Thus, even at a ratio of one viral particle per 10,000 tumor cells (MOI = 10^{-4}), *dl*sptk could mount a spreading infection in the culture plate, ultimately killing all of the glioma cells. To exclude the possibility that this effect was unique to this one human glioma line, we also tested the effects of wild-type virus and *dl*sptk on a second human glioma from the ATCC collection, T-98G. The effect was similar. We next demonstrated that this replication-competent form of tumor cell killing was not restricted to long-term cell lines. For this, we used gliomas obtained from patients undergoing surgery. Two tumors were glioblastoma multiforme; one was an anaplastic astrocytoma. These tumors were placed in cell culture and, at passage two, were tested versus *dl*sptk. In all three of these primary cultures, the effect of *dl*sptk was similar to that observed in long-term glioma cultures. In our next study we wished to demonstrate that *dl*sptk could kill tumors not only in cell culture but while growing *in situ*. For this, we used the human glioma U-87 implanted in the subrenal capsule of nude mice, an assay we and others have previously used to study nervous system and other tumors. After allowing the tumors to establish themselves, we injected either *dl*sptk in culture medium or culture medium alone containing no virus as a negative control. Tumor volume was then observed over several weeks. We noted that those tumors treated with no virus continued to grow, while the virus-treated tumors were either inhibited in growth or in some instances were apparently cured. We use the term "apparent cure" because these studies were not carried out long enough to exclude the possibility that small numbers of residual tumor cells could have regrown. Because the subrenal capsule studies use small tumors (approximately 2 mm in size), we further wished to explore the possibility that this therapy could be used on larger tumors as well. Therefore, we performed similar experiments in the subcutaneous space of nude mice using 8 mm tumors growing *in situ*. We demonstrated that *dl*sptk inhibited tumor growth relative to medium controls in this situation as well.

Finally, we wished to demonstrate this that virus would work intracranially to kill a glioma while sparing the surrounding normal brain. For this study, we inoculated U-87 human glioma cells into the right frontal lobe of a nude mouse using stereotactic techniques. In an initial series of controls with similar conditions, we demonstrated that all of the animals died from intracranial tumor growth within a period of 1½ months. In the study using *dl*sptk, the tumors were inoculated and allowed to establish. The virus was then inoculated at the same stereotactic locus at a later date at one of two doses. Negative controls were inoculated with medium alone at the same stereotactic locus. These animals were followed for several months. All of the animals inoculated with tumor and later with medium alone died as expected. In contrast, those inoculated with 1,000 particles of *dl*sptk lived significantly longer than controls. Additionally, those inoculated with 100,000 viral particles of *dl*sptk not only lived longer but several animals were apparent cures. These animals were sacrificed and the brains were serially sectioned. No definite tumor was observed. Thus we demonstrated that a virus could be genetically engineered to kill tumor cells while sparing surrounding normal brain. However, we did see evidence in the surviving animals of encephalitis. It was not the extensive hemorrhagic encephalitis that is usually seen with wild-type herpes infections. Nonetheless, this does represent a limitation of the virus *dl*sptk. Indeed, in subsequent studies (manuscript in preparation), we have demonstrated that, at an inoculation dose of 10^6 viral particles, significant toxicity from encephalitis and death is a limiting feature of *dl*sptk. For this reason we consider *dl*sptk as a beginning of such viral studies and not as an end. It demonstrates the effectiveness of this strategy and allows for further engineering of the herpes virus to increase the safety of this form of therapy.

Since this initial study of *dl*sptk on gliomas, we have asked whether this therapy would work for other nervous system tumors as well. We have now studied the effects of *dl*sptk on medulloblastoma, malignant meningioma, and neurofibrosarcoma in cell culture and have demonstrated that *dl*sptk can cause a spreading, destructive infection of these cells as well. Further, we have studied *dl*sptk in the subrenal capsule versus human malignant meningioma and have demonstrated that it can inhibit tumor growth. Additionally, using a human medulloblastoma in the subcutaneous space of a nude mouse, we have demonstrated that *dl*sptk can inhibit tumor growth and cause apparent cure. Intracranial studies are now underway. Thus we know that *dl*sptk can kill a wide variety of nervous system tumors and can be used as a starting point for further construction of safe vectors for tumor therapy.

While our initial studies focused on herpes mutants deleted in the thymidine kinase gene, we have also explored other herpes mutants with attenuated neuropathogenicity. These include herpes viruses that have mutations in the DNA polymerase gene (Coen et al., 1982; Field and Coen, 1986) and in the $_1$ 34.5 gene (Chou and Roizman, in press; Chou et al., 1990). In cell culture, we have demonstrated that mutants with deletions at these loci can cause a spreading lytic infection of human U-87 glioma cells. We are now studying the effects of these viruses *in vivo*. We find that the degree of attenuation of neuropathogenicity markedly differs. Certain DNA polymerase HSV1 mutants, while killing tumor cells *in vitro*, prove to be too toxic *in situ*. Following intracranial infection, more animals die from early encephalitis than are saved from tumor-associated death. Thus with certain DNA polymerase mutants, it may not be possible to inoculate a titer high enough for adequate tumor therapy. However, some DNA polymerase mutants have other properties that may be useful in constructing additional vectors. Among these properties are low levels of spontaneous mutation or hypersensitivity to an antiviral agent. It is possible that, by combining these effects with *dl*sptk or other mutations, a second-generation improved virus can be constructed.

In our studies of HSV1 mutated at the $_1$ 34.5 locus, we have observed promising results both in culture and *in situ*. These viruses have been able to mount a spreading infection of human glioma cells in cell culture similar to *dl*sptk; however, they demonstrate marked attenuation of neuropathogenicity when inoculated intracerebrally into nude mice. Specifically, we have not observed early encephalitis even when doses as high as 10^7 viral particles have been injected intracranially in nude mice. At this dose we have seen extended animal survival using the U-87 intracranial tumor model, as well as apparent cure. Moreover, when the animals were sacrificed, the viral encephalitis seen with *dl*sptk was not observed. Moderate inflammation and some evidence of vasculitis were seen, but no encephalitis was present. Neurologically the animals appeared normal (manuscript submitted).

Thus far, in our studies of herpes simplex viral mutants to treat brain tumors, we have explored viruses with mutations at three loci: thymidine kinase, DNA polymerase, and $_1$ 34.5. As expected, the beneficial and toxic effects have varied with each individual mutant. Nonetheless, these data, taken collectively, provide evidence that this strategy can be used to kill intracranial tumor cells of a variety of nervous system tumors with relative sparing of surrounding normal brain cells. The idea of using a replication-competent virus allows for spread of the virus through the tumor tissue in a way not possible by replication-incompetent viruses. The data generated by these initial studies are now allowing the construction of a new generation of herpes-viral vectors containing mutations at two or more of these loci. It is expected that some of these new generation of herpes mutants will provide tumor cell killing with further marked reduction in neuropathogenicity as well as a reduced possibility of reversion to wild type. Additionally, since the herpes virus is large enough, it may be possible to engineer other foreign genes and genetic elements into the virus as well. Such possibilities may include a gene to increase the sensitivity of the virus to an exogenously delivered antiviral agent. Thus the safety would be increased by allowing the investigator or clinician to abort the infection if necessary. Further, we are exploring the use of cell-specific promoters to control viral effects. With this strategy it may be possible to limit the virus to specific cell types. For example, cells may be targeted that express a particular protein or growth factor. In contrast, the virus would be ineffective in cells not expressing these products. Finally, it may be possible to engineer the herpes virus to produce an additional foreign gene product. One example of such a product is a cytokine that, in addition to viral destruction of the tumor, would cause immunologic destruction of additional tumor cells.

While our studies thus far have utilized retroviral or herpes viral strategies, other viruses are also worthy of exploration. Some may offer additional benefits to these two initial viral strategies. However, other disadvantages will also have to be recognized. While nonengineered or wild-type viruses have previously been tried as antineoplastic agents acting via lysis or immune mechanisms (Kobayashi, 1979; Southam, 1960), these earlier studies were limited by the toxicity of wild-type virus or by the randomness of spontaneous mutants. In contrast, current techniques of genetic engineering allow viruses to be created with specific strategies in mind. Thus my colleagues and I feel we have embarked upon an area of tumor therapy that is worthy of further investigation and that allows molecular strategies to be used to modulate or kill tumor cells while causing minimal problems in the surrounding normal cells. While these initial studies are very promising, an extensive amount of work still remains in order to demonstrate the safety and efficacy of this approach in the clinical setting.

ADDENDUM

Since the writing of this chapter, Culver et al. have also used a retro viral vector system expressing HSV-TK combined with ganciclovir to treat brain tumors in animals. A human therapy protocol for glioblastoma is now underway.

REFERENCES

Austin, F.C., and C.W. Boone (1979) Virus augmentation of the antigenicity of tumor cell extracts. Adv. Cancer Res. 30:301–345.

Cepko, C. (1989) Lineage analysis and immortalization of neural cells via retrovirus vectors. In: A.A. Boulton, G.B. Baker, and A.T. Campagnoni (eds.). Neuromethods: Molecular Neurobiological Techniques, vol. 16. Clifton, N.J.: Humana Press, pp. 177–218.

Chou, J., and B. Roizman. Proc. Natl. Acad. Sci. USA (in press).

Chou, J., et al. (1990) Mapping of herpes simplex virus-I neurovirulence to $\gamma_1$34.5, a gene nonessential for growth in culture. Science 250:1262–1266.

Chung, R.Y., J.M. Whaley, N. Klei, et al. (1992) TP53 gene mutations and 17p deletions in human astrocytomas. Genes Chromosomes Cancer 3:323–331.

Coen, D.M., et al. (1982) Mutations in the herpes simplex virus DNA polymerase gene can confer resistance to 9-D aminofuranosyladenine. J. Virol. 41:909–918.

Coen, D.M., et al. (1989) Thymidine kinase-negative herpes simplex virus mutants establish latency in mouse trigeminal ganglia but do not reactivate. Proc. Natl. Acad. Sci. USA 86:4736–4740.

Culver, K.W., Z. Ram, S. Wallbridge, H. Ishii, E.H. Oldfield, and R.M. Blaese (1992). In vivo gene transfer with retroviral vector-producer cells for treatment of experimental brain tumors. Science 256:1550-1552.

Ezzedine, Z.D., R.L. Martuza, M.P. Short, et al. (1991) Selective killing of glioma cells in culture and in vivo by retrovirus transfer of the herpes simplex virus thymidine kinase gene. New Biol. 3:608–614.

Field, H.J., and D.M. Coen (1986) J. Virol. 60:286.

Field, H.J., and G. Darby (1980) Pathogenicity in mice of strains of herpes simplex virus which are resistant to acyclovir in vitro and in vivo. Antimicrob. Agents Chemother. 17:209–216.

Field, H.J., and P. Wildy (1978) J. Hygiene 81:267.

Jamieson, A.T., G.A. Gentry, and J.H. Subak-Sharpe (1974) Induction of both thymidine and deoxycytidine kinase activity by herpes viruses. J. Gen. Virol. 24:465–480.

Kobayashi, H. (1979) Viral xenogenization of intact tumor cells. Adv. Cancer Res. 30:279–299.

Markert, J., D. Coen, A. Malick, et al. (1992) Expanded spectrum of viral therapy in the treatment of nervous system tumors. J. Neurosurg. 77:590-594.

Martuza, R.L. (1990) Genetic alterations of nervous system tumors. In: K. Tabuchi (ed.). Biological Aspects of Brain Tumors. Tokyo: Springer-Verlag, pp. 18–27.

Martuza, R.L. (1992) Molecular neurosurgery for glial and neuronal disorders. Stereotactic and Functional Neurosurgery 59:92-99.

Martuza, R.L., A. Malick, J.J.M. Markert, et al. (1991) Experimental therapy of human glioma by means of a genetically engineered virus mutant. Science 252:854–856.

Moolten, F.L., and J.M. Wells (1990) Curability of tumors bearing herpes thymidine kinase genes transferred by retroviral vectors. J. Natl. Cancer Inst. 82:297–300.

Moolten, F.L., J.M. Wells, R.A. Heyman, et al. (1990) Lymphoma regression induced by ganciclovir in mice bearing a herpes thymidine kinase transgene. Hum. Gene Ther. 1:125–134.

Rouleau, G.A., B.R. Seizinger, W. Wertelecki, et al. (1990) Flanking markers bracket the neurofibromatosis type 2 (NF2) gene on chromosome 22. Am. J. Hum. Genet. 46:323–328.

Short, M.P., B.B. Choi, J.-K. Lee, et al. (1990) Gene delivery to glioma cells in rat brain by grafting of a retrovirus packaging cell lines. J. Neurosci. Res. 27:427–433.

Southam, C.M. (1960) Present status of oncolytic virus studies. Trans. N.Y. Acad. Sci. 22:657–673.

Tenser, R.B., and F. Rapp Miller (1979) Trigeminal ganglion infection by thymidine kinase-negative mutants of herpes simplex virus. Science 205:915–917.

Chapter 35

Evoking an Immune Response to Glioblastoma Cells Transfected With Episome-Based Plasmid Expressing Antisense Transcripts to Insulin-Like Growth Factor I

Thomas R. Johnson, Jerzy Trojan, Susan D. Rudin, Judith Ilan, Mark L. Tykocinski, and Joseph Ilan

INTRODUCTION

Deregulated expression of growth factors and/or their receptors is associated with the growth and development of animal and human tumors, including tumors of the nervous system (Heldin and Westermark, 1989). One such growth factor is insulin-like growth factor I (IGF-I), a 70-amino acid polypeptide involved in skeletal growth and the differentiation of mesenchymal tissues (Froesch et al., 1985). IGF-I and the related polypeptide IGF-II are expressed in many tissues, including the brain, where it is possible that they are involved in the proliferation of neuronal and glial precursors (Ayer-le Lievre et al., 1991). IGF-I and -II are also expressed at high levels in some nervous system-derived tumors, e.g., astrocytomas and meningiomas (Antoniades et al., 1992) or tumor cell lines (IGF-I: C6 glioma: Kiess et al., 1989; IGF-II: SK-N-S neuroblastoma: El-Badry et al., 1991). In several instances growth factor expression in tumor cell lines has been correlated with continuous proliferation in serum-free medium by antibody-blocking studies (Blatt et al., 1984; El-Badry et al., 1991), suggesting the involvement of an autocrine stimulatory loop in cell proliferation.

The experimental work described in this chapter was performed by Trojan et al. (1992, 1993). We (Johnson et al., 1990) and others (Kiess et al., 1989) have previously reported that the rat glioma cell line C6 expresses high levels of IGF-I. IGF-I expression is enhanced in these tumor cells when they are grown in serum-free medium. Since C6 cells are also known to express both IGF-I and IGF-II receptors (Kiess et al., 1989), it is possible that an autocrine loop involving IGF-I and its receptors may exist and drive the tumorigenic phenotype. Until now, the autocrine hypothesis of malignant transformation has derived experimental support primarily from antibody-blocking and sense gene transfer studies. However, both of these experimental approaches have limitations. For instance, antibody-blocking is ineffective for the analysis of intracellular autocrine loops (Heldin and Westermark, 1989) and antibodies vary considerably in their blocking efficacy. In principle, mutagenesis strategies offer more definitive and unambiguous approaches for analyzing autocrine mechanisms in tumor cells.

Classically, in order to assign a functional role of a particular gene product, one relies on naturally occurring mutants that fail to express this particular gene in normal fashion. The *in vitro* generation of cellular mutants has limited applicability due to the diploid nature of most genes and to the lack of adequate mutant selection. Our experimental approach in this study centers around the use of antisense RNA expression to produce phenocopies of a null mutation for IGF-I. Antisense strategy bypasses inherent limitations of functional studies dependent upon natural mutant cells or artificially mutagenized cells (Izant and Weintraub, 1985).

According to the antisense concept, a nucleotide sequence is expressed in cells that is complementary to a portion of the native RNA transcript (target sequence) of the gene of interest. Binding of the antisense disrupts the function of the target sequence either by blocking its function (e.g., by binding to a splice junction), promoting its degradation, or by altering the structure of the target sequence. Examples of natural regulation by antisense RNA exist in prokaryotes (reviewed by Eguchi et al., 1991). Although

Molecular Genetics of Nervous System Tumors, pages 387–400 ©1993 Wiley-Liss, Inc.

several examples of naturally occurring antisense RNA exist in eukaryotes (reviewed in van der Krol et al., 1988; Toulme and Helene, 1988), their regulatory role is less clear. Two good candidates for gene regulation by naturally occurring antisense in eukaryotes are the presence of developmentally regulated antisense transcripts for alpha amylase in barley, which are transcribed from loci different from the sense mRNA (Rogers, 1988), and the appearance of an RNA complementary to part of the mRNA for basic fibroblast growth factor in *Xenopus*, which may be involved in the abrupt degradation of this mRNA during oocyte maturation (Bass and Weintraub, 1988; Kimelman and Kirschner, 1989).

Antisense strategy—i.e., the experimental inhibition of gene expression by introducing a source of nucleotide sequences complementary to a given endogenous mRNA species—has been applied successfully to a growing number of genes in cultured cells. Pertinent examples include growth rate reduction of human neuroblastoma cells (Negroni et al., 1991) and neuroectoderm cell lines (Whitesell et al., 1991) by antisense to the *myc* gene. Growth of transformed human astrocytes was inhibited by antisense to basic fibroblast growth factor (Morrison, 1991). A phosphoprotein, OP18, was shown by antisense inhibition to be involved in the growth regulation of neuroblastoma (Melhem et al., 1991). Antisense inhibition of glial S100 protein altered cell proliferation and cytoskeletal organization (Selinfreund et al., 1990). However, the antisense approach has frequently been complicated by incomplete inhibition of gene expression. For example, in the studies of Kim and Wold (1985), which also established the relationship between levels of antisense RNA and the inhibition/degradation of sense transcripts using a viral thymidine kinase (TK) gene, a 90% reduction in TK transcripts failed to make the cells functionally TK$^-$. While idiosyncratic gene-specific effects may account for some cases of ineffective antisense RNA-mediated inhibition of gene expression, most are probably a consequence of insufficient antisense RNA levels and inadequate suppression of the encoded protein product. Integrating vectors, used for most antisense experiments to date, do not especially provide for gene amplification and high-level accumulation of antisense RNA. Moreover, integrating vectors, which integrate randomly into the host chromosome, are subject to local regulatory effects exerted by neighboring sequences at sites of integration. Hence, transfected cells may at times express only limited amounts of antisense RNA.

We have applied the antisense strategy to study the function of IGF-I in glioma cells. In doing so, we bypassed the limitations of integrating vectors and maximized antisense RNA transcription by employing a self-amplifying episomal vector that replicates to high copy numbers extrachromosomally. The episomal expression vector incorporates Epstein-Barr virus (EBV)-replicative signals (Yates et al., 1985). We first showed that such a vector can be used in human T-cell clones for effective (>95%) inhibition of the T lymphocyte surface molecule CD8 (Hambor et al., 1988). The utility of episome-based expression vectors for effective antisense inhibition has been subsequently confirmed by others, e.g., N-*myc* in primitive human neuroectodermal cell lines (Whitesell et al., 1991) and thioredoxin in HeLa cells (Deiss and Kimchi, 1991).

In the course of this study, we have demonstrated a tumorigenic role for IGF-I. The role of IGF-I is apparently distinct from its function in certain cellular systems as a growth factor, and instead seems to relate to IGF-I's capacity to regulate a cell's differentiated (immunogenic) state. Hence, our findings highlight the fact that tumorigenic phenotypes may arise through unexpected molecular mechanisms. This, in turn, raises the prospect of novel approaches for regulating a tumorigenic phenotype for therapeutic ends.

MATERIALS AND METHODS

Detailed methodology can be found in Trojan et al. (1992, 1993). Immunocytochemical localization of antigens, *in situ* hybridization, and transfer blot hybridization were carried out as described before (Boehm et al., 1989; Johnson et al., 1990; Trojan et al., 1984). Transfection of C6 glioma and B104 neuroblastoma cells (originating from the same tumor; Benda et al., 1969) was accomplished using Lipofectin reagent (Gibco BRL) according to the supplier's instructions. A diagrammatic representation of the steps employed to assemble the antisense expression vector is shown in Figure 1. The pMK' plasmid, containing the mouse MT-I promoter fused to herpes simplex thymidine kinase, was kindly provided by Dr. Richard Palmiter (Stuart et al., 1984). pIGF-I harboring a human hepatic cDNA for IGF-I was kindly provided by Dr. Martin Jansen, Utrecht, the Netherlands (Jansen et al., 1983). pUC18 (Yanisch-Perron et al., 1985) was kindly donated by Dr. Peter Harte, Department of Genetics, Case Western Reserve University. pBluescript KS$^+$ (Stratagene) was obtained commercially. We described the construction of pREP-4 elsewhere (Groger et al., 1989). pREP-4 contributes two genes from the Epstein-Barr virus to pAnti-IGF-I, the origin of replication (EBV ori-P) and the EBNA-I nuclear antigen, which drives the EBV ori-P's replicative function. These two genes allow pAnti-IGF-I to replicate as an episome. The SV40 polyadenylation signal is present as a termination signal for transcription.

RESULTS

Stable transfectants of C6 glioma cells were derived and were transferred to serum-free medium for 24 hr. We have shown that in serum-free medium, expression of IGF-I transcripts is greatly enhanced (Johnson et al., 1990). Figure

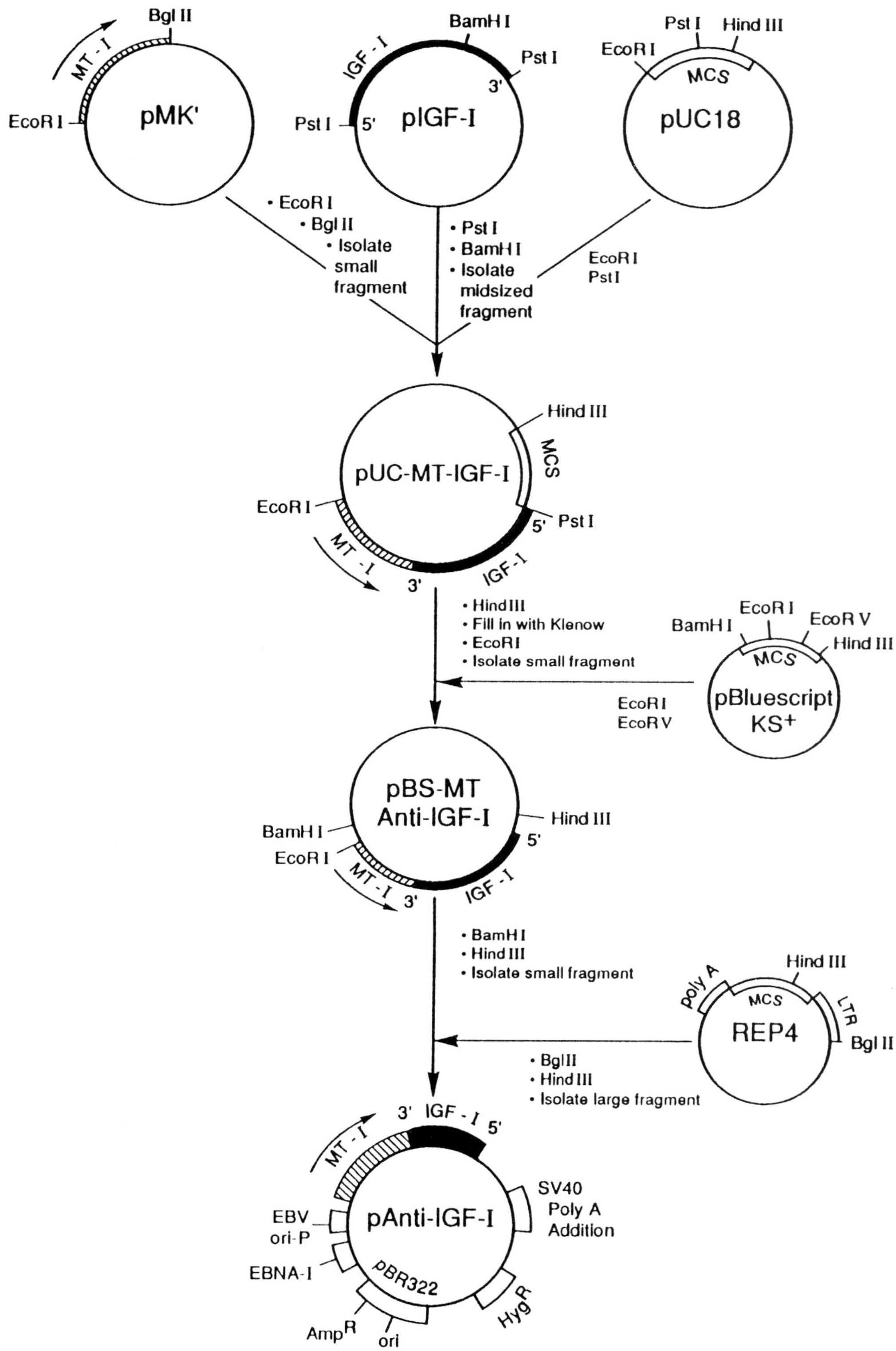

Fig. 1. Diagrammatic representation of the steps employed to construct the episomal vector pAnti-IGF-I. MT-I, metallothionein-I; MCS, multiple cloning site; LTR, long terminal repeat; EBNA-I, EBV-encoded nuclear antigen I; SV40, simian virus 40; Amp^R, ampicillin resistance gene; Hyg^R, hygromycin resistance gene (From Trojan et al., 1992).

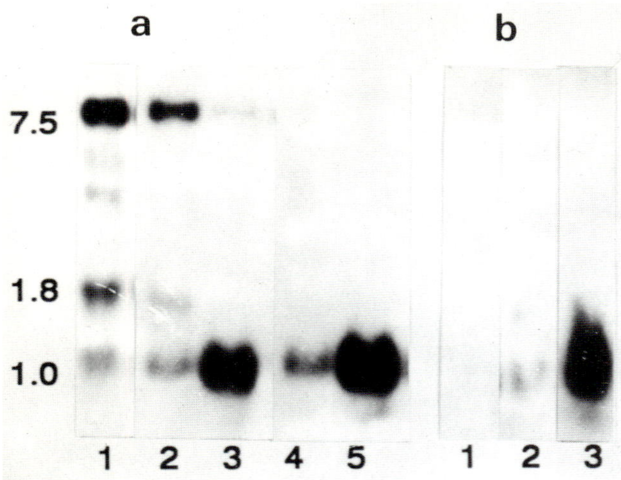

Fig. 2. Induction of antisense transcripts by $ZnSO_4$ in cultured C6 glioma and B104 neuroblastoma cells. The figure is a composite of several RNA transfer blots. Ten micrograms total RNA was used per lane. Molecular sizes (in kb) of the major IGF-I transcripts are shown on the left. **a**: Lane 1: Parental nontransfected C6 glial cells exposed to serum-free medium. Lanes 2, 3: Transfected C6 glioma cells incubated in serum-free medium in the absence (lane 2) or presence (lane 3) of $ZnSO_4$. For lanes 2 and 3, nick-translated rat IGF-I cDNA was used. Lanes 4, 5: The blot shown in lanes 2 and 3 was rehybridized to nick-translated, human IGF-I cDNA. **b**: Lane 1: Untransfected B104 neuroblastoma cells. Lanes 2, 3: Transfected B104 neuroblastoma cells incubated in the absence (lane 2) or presence (lane 3) of $ZnSO_4$ (From Trojan et al., 1992).

2 depicts an autoradiogram from transfer blots of 10 μg total RNA of transfected and untransfected C6 and B104 cells exposed to serum-free medium. The pattern of IGF-I transcripts in parental and transfected cells matches that published by us and others (Kiess et al., 1989), showing prominent bands at 7.5, 1.8, and 1 kb. Treatment with 50 μm $ZnSO_4$ (Fig. 2, lane 3) clearly induces strong expression of the 1 kb antisense transcript, as well as resulting in almost complete disappearance of the endogenous IGF-I transcripts at 7.5 and 1.8 kb. Rehybridization of the same blot with a single-stranded cRNA indicated that the endogenous 1 kb transcript also disappeared. When the same blot was probed with human nick-translated, double-stranded IGF-I cDNA, the endogenous rat IGF-I transcript is not apparent (Fig. 2a, lanes 4 and 5). Rehybridization of the blot with a single-stranded cRNA probe indicated that the 1 kb transcript in lane 4 was background antisense expression. Untransfected B104 cells show no detectable IGF-I transcripts (Fig. 2b, lane 1), while transfected B104 cells show a strong induction of antisense transcripts in cells treated with $ZnSO_4$ (compare lane 3 with lane 2).

Stable transfection resulted in change in morphology of both C6 glioma cells (Fig. 3a) and B104 neuroblastoma cells (Fig. 3b). Parental C6 cells react positively with antibody to IGF-I by the immunoperoxidase procedure in the presence (Fig. 3d) or absence (Fig. 3c) of $ZnSO_4$. In contrast, B104 neuroblastoma cells show no staining reaction with the above procedure (Fig. 3e,f). However, in the presence of $ZnSO_4$, the immunostaining disappeared almost completely (Fig. 3h) from the transfected glioma cells. *In situ* hybridization with single-stranded ^{35}S-labeled oligonucleotide showed that transfected glioma cells treated with $ZnSO_4$ express antisense transcripts (Fig. 3i,j). However, without $ZnSO_4$ no hybridization signal is apparent, using the same cells and probe at the same specific activity (Fig. 3k,l). As a control, transfected B104 neuroblastoma cells gave similar results (Fig. 3m–p).

Figure 4 further demonstrates that induction of antisense IGF-I transcripts results in disappearance of IGF-I protein. The green immunofluorescence indicating the presence of IGF-I in transfected cells (Fig. 4a) disappears when cells are treated with $ZnSO_4$ (Fig. 4b). Figure 4c–e shows the same microscopic field in different optical conditions and depicts $ZnSO_4$-treated cells with no immunofluorescent staining (Fig. 4c). The same cells hybridize strongly to ^{35}S-labeled sense oligonucleotides (Fig. 4d), which are localized in cell nuclei (Fig. 4e). Variation of the position of silver grains with respect to the microscope focal plane accounts for the apparent lack of grains over some cells in Figure 4d,e. Of special note are the results shown in Figure 4f,g, which show a mixed culture of transfected and parental C6 cells maintained together in the presence of $ZnSO_4$, stained for IGF-I protein (Fig. 4f) and also hybridized *in situ* to ^{35}S-labeled sense oligonucleotides (Fig. 4g). Only transfected cells not showing immunofluorescent staining for IGF-I protein manifest RNA antisense hybridization signals in the mixed culture (cf. cells near open arrow in Fig. 4f,g, which lack the green stain seen in the cell cluster and isolated cells near closed arrow).

Southern blots of restriction enzyme-digested DNA showed that the patterns of hybridizing fragments in the DNA from transfected cells were similar to those of the purified plasmid used for transfection. The results suggest that pAnti-IGF-I vector DNA was maintained as an episome (data not shown). Such experiments suggested an estimate of 10 episome copies per cell.

The tumorigenicity of C6 glioma cells is evident in the results shown in Figure 5. Ten rats were injected subcutaneously with 10^7 nontransfected glioma cells above the right hind leg. All 10 rats developed a large tumor. An additional 11 rats were injected subcutaneously with 10^7 transfected glioma cells in the same location. None of the rats injected with transfected cells developed a tumor during the 32 weeks of observation (an additional group of rats has been tumor-free for close to 1 year). The size of the tumor developed 3 weeks after injection of the parental glioma cells is shown in Figure 5a (arrow) and a representative rat injected with

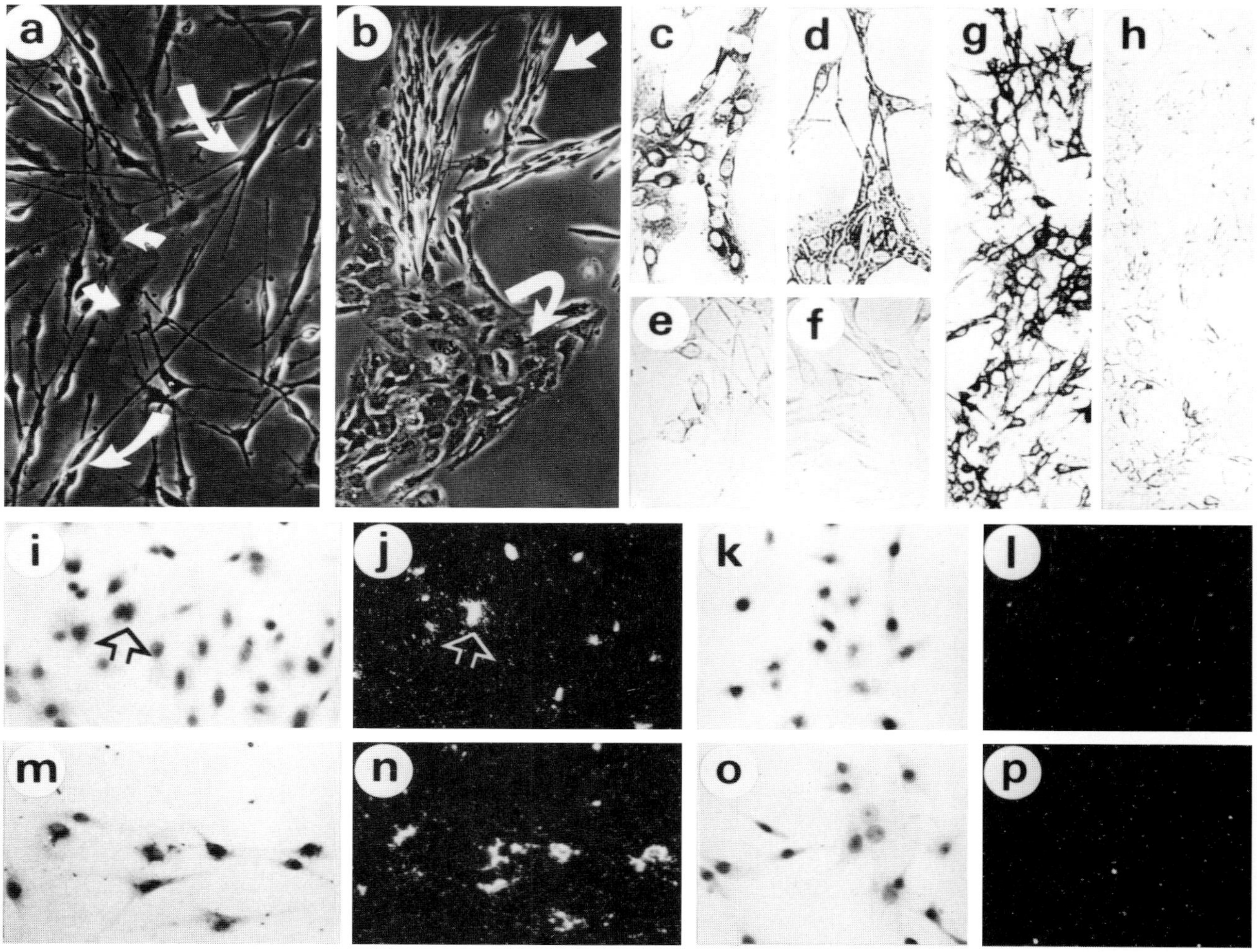

Fig. 3. Localization of IGF-I antisense transcripts and IGF-I protein in C6 glioma parental and transfected cells. **a,b:** Rat C6 glial cells (a) and B104 neuroblastoma cells (b) grown in the presence of 0.2 mg/ml or 0.5 mg/ml hygromycin, respectively, 2 weeks following transfection with pAnti-IGF-I. In panel a, the small arrows indicate the parental population of cells that gradually was eliminated during hygromycin selection over the course of 2 to 3 weeks. The large arrow represents the stable transfectants that survived the hygromycin selection. In panel b, the straight arrow represents the parental population of cells as above, while the bent arrow represents the stable transfectants surviving the hygromycin selection. ×250. **c–h:** Immunocytochemical labeling of cultured cells with antibodies to human IGF-I using quantitative immunoperoxidase technique (47). c,d: Positively stained parental C6 incubated in the absence (c) or presence (d) of 50 m $ZnSO_4$. e,f: Negative reaction obtained with B104 neuroblastoma cells incubated in the absence (e) or presence (f) of $ZnSO_4$. g,h: C6 glioma transfected cells incubated in the absence (g) or presence (h) of $ZnSO_4$. c–f: ×250; g,h: ×150. **i–p:** *In situ* hybridization of cultured cells with ^{35}S-labeled human sense oligonucleotides to detect antisense transcripts. i–l: C6 transfected cells maintained in the presence (i,j) or absence (k,l) of $ZnSO_4$. The arrow points out an example of a cell demonstrating a positive signal. In k,l, only background hybridization is apparent. m–p: B104-transfected cells maintained in the presence (m,n) or absence (o,p) of $ZnSO_4$. i,k,m,o: bright field; j,l,n,p: dark field. i–p: ×250 (From Trojan et al., 1992).

transfected glioma cells is shown on the right. An additional five rats were injected with C6 cells that had been transfected with expression vector that did not contain IGF-I sequences. All five of these rats developed tumors after 2 weeks. Figure 5b depicts a rat injected with 10^7 stably transfected B104 neuroblastoma cells as a control. In 2 weeks, a large tumor was apparent in all 10 rats injected (Fig. 5b).

Six out of 11 rats injected with the transfected C6 glioma cells developed a small lesion (<1 cm) at the point of injection a few days after injection. Three were removed for further analysis and the others disappeared after 2–3 weeks. A histological section of one of these lesions, stained with hematoxylin and eosin, is shown in Figure 4h. It shows a few glial cells (positively identified by immunostaining for glial fibrillary acid protein [GFA] in Fig. 4j) surrounded by an organizing tissue containing many mononuclear cells, mainly plasmocytes and lymphocyte-like cells. The other two lesions were histologically identical. This suggests that the

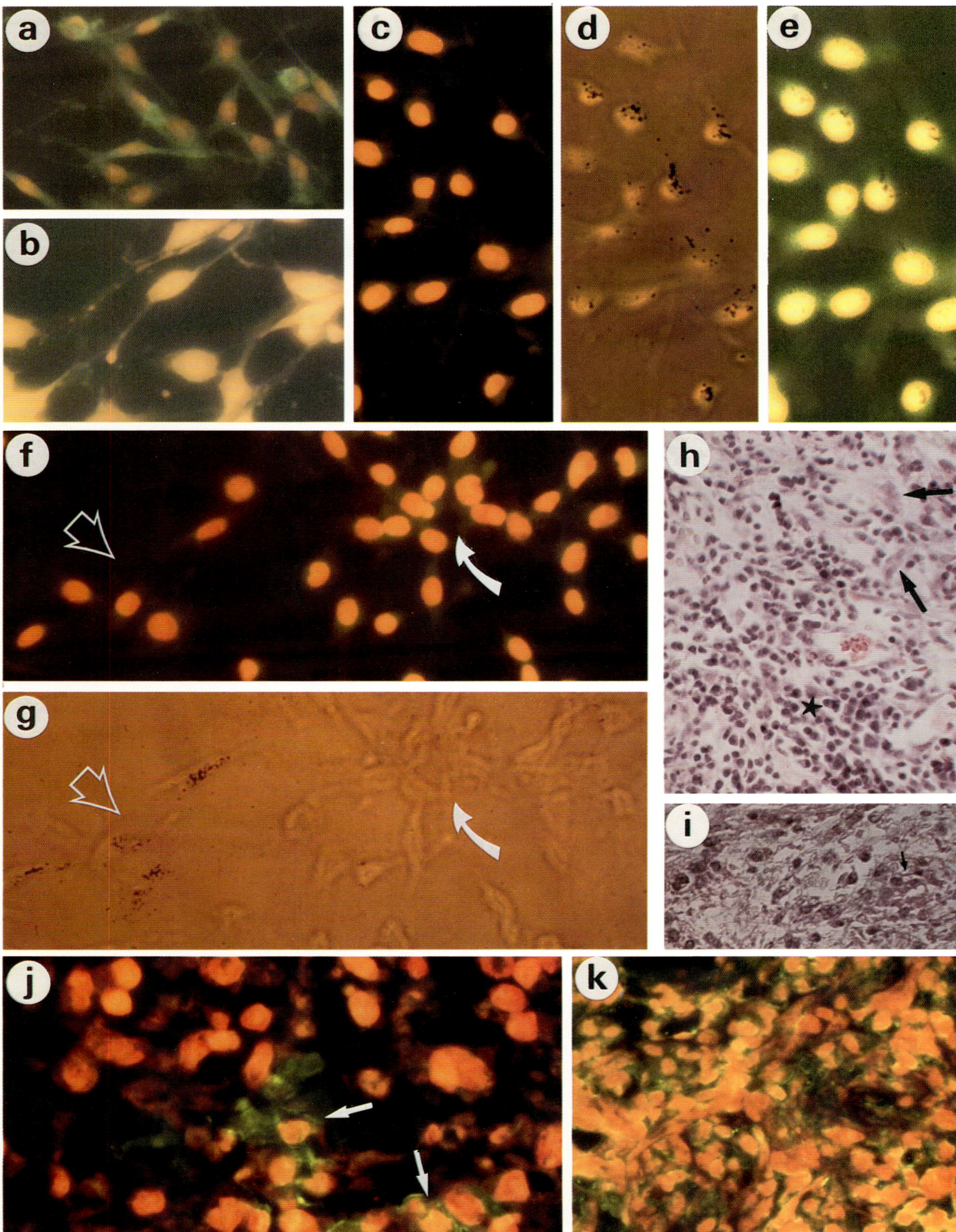
a
b
c
d
e
f
g
h
i
j
k

host immune system reacted against the transfected glioma cells. By contrast, the tumors resulting from injection of nontransfected glioma cells (which stained positively for GFA; Fig. 4k) did not show any apparent infiltration of mononuclear cells within the tumor (Fig. 4i).

The observation that the host immune system reacted against the transfected glioma cells raised the possibility that the immune response might extend to the parental cells, modifying tumor growth. In order to address this question, several experiments were carried out. In a first set of experiments, 10^7 parental glioma cells were injected subcutaneously above the left hind leg of 10 rats, as depicted in Figure 6A. When a tumor up to 1 cm in diameter was evident in each of the injected rats, a second injection was given subcutaneously above the opposite hind leg (Fig. 6B). This second injection included 10^7 transfected (antisense IGF-I) glioma cells. Within 3 weeks, complete disappearance of tumors occurred in all 10 rats (Fig. 6C). These rats are under observation and have remained tumor-free for more than 6 months. In all controls, including rats secondarily injected with glioma cells stably transfected with episomes devoid of antisense IGF-I transcriptional cassettes, tumors persisted and after 2 months grew to sizes necessitating sacrifice of the animals.

A second set of experiments further addressed the ability of transfected (antisense IGF-I) glioma cells to cure preexisting glioblastoma, and in particular addressed the issue of large (>3 cm in diameter) tumors. A flow plan of the experiments is depicted in Figure 7. Tumors were established in 10 rats as described above. After 3 to 4 weeks (when a large tumor was evident in each of the 10 animals), tumors were surgically excised from all animals. These animals were then divided into two groups. The first group of five animals did not receive any further treatment. In all of them, there was local recurrence of tumor after approximately 3 weeks, reflecting the incomplete excision of the primary tumor. The second group of five animals received an injection of 10^7 glioma cells transfected with the episome expressing antisense IGF-I immediately following surgery. None of these rats has shown tumor recurrence at the excision or other sites for 6 months, thus far.

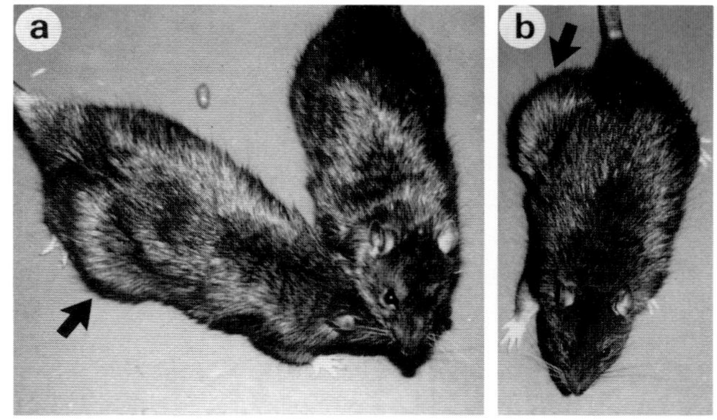

Fig. 5. Tumors derived from C6 glioma nontransfected cells (**a**, left) and transfected B104 neuroblastoma cells (**b**). Arrows indicate the position of the tumors. The rat on the right in panel a was injected with C6 glioma transfected cells and did not develop a tumor (From Trojan et al., 1992).

We next tested whether, in addition to tumor cure, tumor prevention can be achieved. Ten rats were injected for this experiment subcutaneously above the left hind leg with 10^7 parental glioma cells and simultaneously above the right hind leg with 10^7 transfected (antisense IGF-I) glioma cells (Fig. 8). No tumors have developed in any of these rats during 9 months of observation. In contrast, when only parental glioma cells were injected at the secondary site, a large tumor developed in each of the rats after 3 weeks at both sites. In some cases, tumor growth at both sites was fused into one large mass. Furthermore, when glioma cells transfected with a control episome (devoid of the IGF-I antisense component) were injected at the secondary site in five rats, all developed tumors. Therefore, the protective effect is specifically associated with transfected tumor cells

Fig. 4. Simultaneous localization of IGF-I antisense transcripts and staining of IGF-I protein in glioma cells *in vitro* and immunocytochemical identification of glioma cells *in vivo*. **a,b**: Immunocytochemical labeling of C6-transfected cells with anti-IGF-I antibodies using indirect immunofluorescence technique. Note the presence of IGF-I (green FITC fluorescence) in panel a, and absence of fluorescence in cells treated with $ZnSO_4$ in b. The nuclei are counterstained with p-phenylenediamine (p-pd)(48). a,b: ×250. **c–e**: C6-transfected cells maintained in the presence of $ZnSO_4$ for 8–10 hr. These panels represent the same microscopic field. Antisense transcripts were localized by *in situ* hybridization as in Figure 3, and were then subjected to immunofluorescent labeling of IGF-I protein. The absence of protein (c) is accompanied by a striking *in situ* hybridization signal localized primarily in the nuclei (d,e). d: Bright field; e: bright field combined with UV lamp. **f,g**: Mixed culture of C6 glioma parental and transfected cells maintained for 8–10 hr in the presence of $ZnSO_4$. The same microscopic field is visualized in f and g. Cells showing positive immunofluorescent labeling for IGF-I (f, solid arrow) show no detectable signal for IGF-I antisense transcripts (g, solid arrow); whereas cells showing a strong positive signal for antisense transcripts (g, open arrow), presumably the transfected cells, show no detectable fluorescent staining for IGF-I (f, open arrow). See also Figure 3. c–g: Nuclei are counterstained with p-pd. **h–k**: Histologic sections of injected glioma cells. h: Section (hematoxylin and eosin) of a lesion observed at the point of injection of C6 glioma transfected cells, and removed 3 weeks after cell injection. Mononuclear cells, the majority of them plasmocytes and macrophages (star), as well as lymphocytes are observed in the organizing tissue surrounding a few glial cells (arrows). i: Histological section of a small area of a tumor induced by injection of C6 glioma cells (nontransfected). Note dispersed small cells of astrocytic type, some of them forming a pseudorosette (arrow). h,i: ×200. j,k: Immunofluorescence staining of glial fibrillary acidic protein (GFA, used for identification of glial cells) in the lesion produced by injection of C6-transfected cells (j), and in the tumor developed after injection of nontransfected cells (k). j: Note GFA-positive cells (green fluorescence, arrow) surrounded by mononuclear cells (visible in h). k: An area of glioblastoma showing positive reaction with anti-GFA antibodies. Nuclei in j,k counterstained with p-pd. h,i: x200; j: x400; k: x250 (From Trojan et al., 1992).

that express antisense IGF-I. Transfected tumor cells elicit a host immune response against parental glioma cells, as well as against themselves.

The specificity of the host antitumor immune response was demonstrated in a fourth set of experiments. First, transfected (antisense IGF-I) and parental glioma cells were mixed together and the cellular mixture was coinjected above the left hind leg in each of 10 rats. No tumor developed in any of the animals for a full year of observation (Fig. 9A). A small cyst developed at the injection site in three of the six animals; two of these were surgically removed for histologic examination and the third disappeared after a short period of time. The cysts consisted mainly of lymphoid mononuclear cells and 80–90% were $CD8^+$ lymphocytes (Fig. 10a–c).

A second tumor cell-mixing experiment was performed, and in this case, transfected (antisense IGF-I) neuroblastoma cells and parental glioma cells were combined (Fig. 9B). After 3 weeks, a large tumor developed in all five rats so injected. The tumors were removed, and histologic examination in each case showed a mixed tumor consisting of neuroblastoma as well as glioma cells, as shown diagrammatically in Figure 9B. This mix of tumor cells was confirmed by histopathology (Fig. 10d,e).

A third tumor cell-mixing experiment consisted of the injection of a combination of transfected (antisense IGF-I) glioma cells and transfected (antisense IGF-I) neuroblastoma cells into each of five rats (Fig. 9C). After 3 weeks, all animals developed large tumors. Each of these tumors histologically proved to be composed solely of the neuroblastoma component (Fig. 10f,g).

These various tumor-mixing experiments together establish that a host immune response induced by transfected (antisense IGF-I) glioma cells is specific for glioma cells and can distinguish between glioma and neuroblastoma components within a mixed tumor. Furthermore, the data indicate that nonspecific bystander effects mediated by other immune cells, such as natural killer cells, are not essentially responsible for tumor regression.

During the phase of tumor involution in each of the experiments above, histologic analyses showed that tumors were abundantly infiltrated with mononuclear cells (Fig. 10a) with neovascularization. Immunoperoxidase staining for glial fibrillary acid protein (GFA) shows some glial elements among the lymphocytes (Fig. 10b). Immunocytochemistry showed that in late stages of involution 60–70% of the infiltrating cells were $CD8^+$ lymphocytes (Fig. 10c). The remainder of the infiltrating cells consisted of macrophages and $CD8^-$ lymphocytes.

Since glioblastoma is naturally a brain tumor, it was important to show that such prevention and cure can be effected at the orthotopic site in the brain. Preliminary experiments indicate that injections of $2–3 \times 10^6$ parental glioma C6 cells into rat brains bring about the development of large tumors within between 1 and 2 weeks. If such rats are injected with transfected cells subcutaneously outside the head region 3–5 days after the initial injection with parental cells, the parental cell tumors in the brain regress and show lymphocyte infiltration similar to the subcutaneous tumors described above (Fig. 10h-l). These results suggest that injection of antisense IGF-I transfected cells can cure an established glioblastoma in the orthotopic brain site in a manner analogous to the cure at the subcutaneous site.

DISCUSSION

We report here a combined *in vitro* and *in vivo* functional analysis of genetically engineered tumor cells corresponding to antisense phenocopies of an IGF-I null mutation. We obtained efficient antisense RNA-mediated inhibition of IGF-I expression in C6 cells using an *in vitro*-inducible antisense RNA transcriptional system. An apparent complete shut-off of the protein product was observed following antisense induction *in vitro*. IGF-I inhibition *in vitro* resulted in loss of tumorigenicity *in vivo*, apparently as a result of an immune response mounted against the transfected cells, which extended to the parental cells. Taken together, the *in vitro* and *in vivo* observations may support an autocrine hypothesis of tumorigenesis and demonstrate an unexpected immunologic role for IGF-I in this tumor type.

As discussed above, in order to maximize antisense expression, we employed an episome-based vector that incorporates EBV-replicative signals (Yates et al., 1985). However, we used an inducible promoter for inducible antisense transcription *in vitro*. The low basal activity *in vitro* of the episome-based MT-I promoter observed by us contrasts with the higher basal activity previously noted by us and others (Palmiter et al., 1983), and is likely to be a consequence of the serum-free environment. The MT-I promoter also includes glucocorticoid and cyclic AMP response elements. In contrast to previous studies that were performed in human cell lines, this study involved a rat cell line. A previous report failed to show EBV episome maintenance in several murine cell lines (Yates et al., 1985).

The antisense IGF-I transcripts that appeared following $ZnSO_4$ induction were prominently localized to the nuclei by *in situ* hybridization. This finding supports an intranuclear mechanism for antisense RNA-mediated inhibition, in agreement with observations made by others for several other genes (Cornelisson, 1989; Kim and Wold, 1985; Munroe, 1988; Stout and Caskey, 1990).

Since there is evidence that natural antisense RNA is involved in controlling the expression of some eukaryotic genes, our experimental strategy using genetically introduced antisense RNA may mimic natural mechanisms. Recently, antisense RNA complementary to retroviral packaging protein sequences was used to completely suppress Moloney murine leukemia virus-induced leukemia in

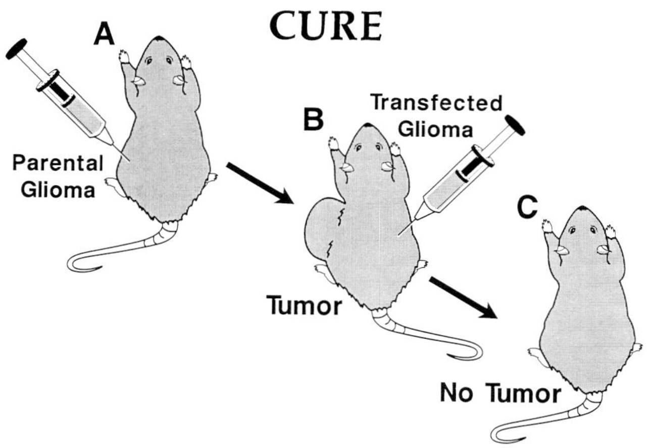

Fig. 6. Schematic representation of experiment in which an established glioma tumor is cured by injection of transfected cells.

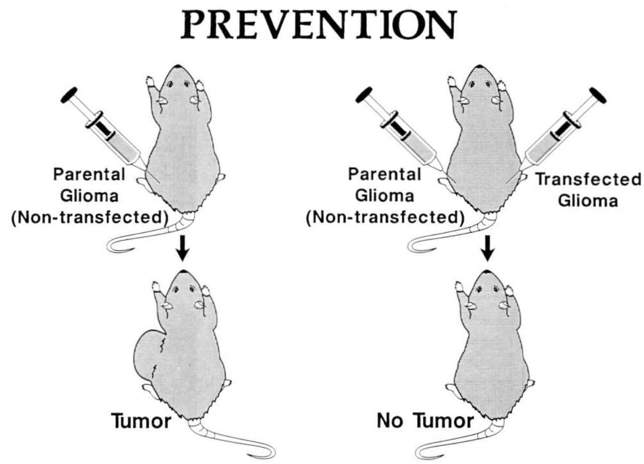

Fig. 8. Schematic representation of experiments demonstrating prevention of glioma tumor development by simultaneous injection of transfected cells at a distant site.

transgenic mice, pointing to a possible therapeutic use of antisense (Han et al., 1991). Our study differs from these previous reports in two important ways. First, we have evaluated tumorigenicity in an immunologically intact animal background. Second, our study focuses on a cytokine (IGF-I) with potential autocrine function.

A number of earlier studies support a role for autocrine stimulation in tumor formation, in that anticytokine and/or anticytokine receptor antibodies can block tumor cellular proliferation *in vitro* (Heldin and Westermark, 1989). However, antibody-blocking or oligonucleotide antisense can have limitations in the context of *in vivo* analyses. Expression of DNA constructs resulting in antisense RNA provides a direct and unambiguous experimental approach for studying both extracellular and intracellular autocrine loops. Our use of stable antisense transfectants allows analysis of their phenotypic (including tumorigenic) properties both *in vitro* and *in vivo*, and provides an *in vitro*/*in vivo* assay system that is suitable for screening transformation potential of selected genes and that is particularly useful for dissecting autocrine mechanisms.

Antibody-blocking experiments have suggested an extracellular autocrine role for IGF-I in human lung (Minuto et al., 1988) and breast (Huff et al., 1986) carcinoma cell lines, as well as a human osteosarcoma cell line (Blatt et al., 1984) *in vitro*, and for Wilms' tumor (Gansler et al., 1989) and breast carcinoma (Arteaga et al., 1989) cells *in vivo*. This has raised an intriguing

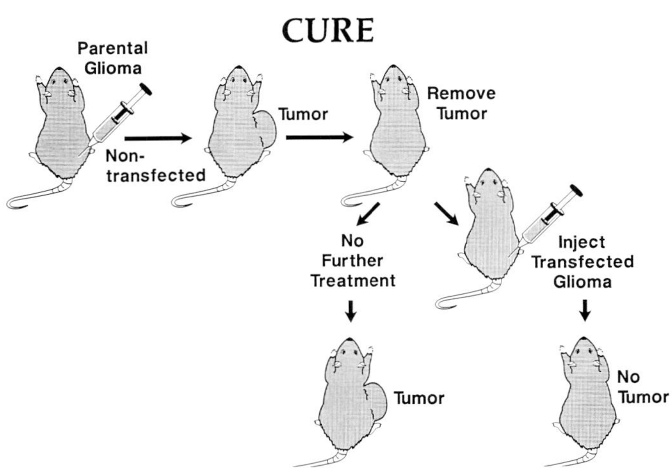

Fig. 7. Schematic representation of experiment in which large established glioma tumors are cured by a combination of surgery and injection of transfected cells.

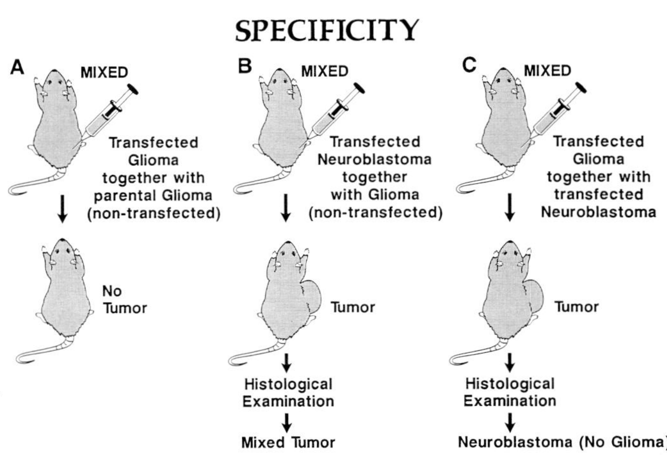

Fig. 9. Schematic representation of cell-mixing experiments demonstrating specificity of the transfected glioma cells for eliciting a response against normal glioma cells.

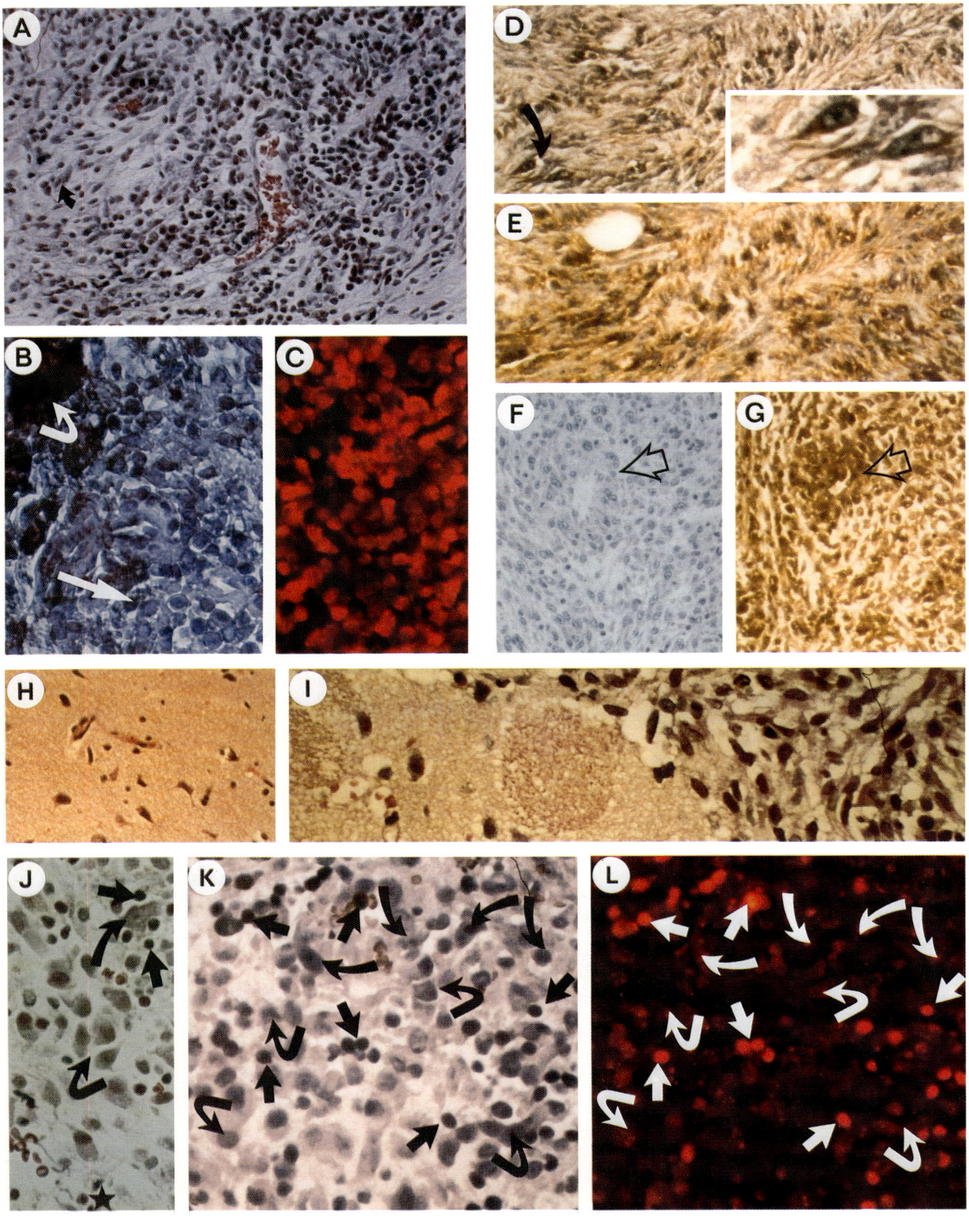
A
B
C
D
E
F
G
H
I
J
K
L

therapeutic possibility of targeting either IGF-I or its receptor with specific antibodies for the treatment of gliomas, as the tumor blood vessels are devoid of the blood–brain barrier. The antibody-mediated effect in those experiments was presumed to be based upon a change in tumor proliferative potential. However, the incomplete tumor suppression observed in the antibody-blocking experiments, as opposed to the complete suppression observed here in antisense experiments, suggests that the latter may be the more effective strategy.

We have demonstrated that the inhibition of IGF-I expression elicits a highly immunogenic phenotype in glioma cells. This immunogenicity is effective not only for purposes of tumor prevention but also for eradication of preexisting tumor at distant sites. The host immune response appears to consist predominantly of $CD8^+$ lymphocytes and is targeted specifically toward the immunogenic tumor cell type used. It can in fact recognize different glioma cell types ($IGF\text{-}I^+$ and $IGF\text{-}I^-$) admixed within a combined tumor at a single site. Moreover, the immune response distinguishes between different tumor cell types admixed in a combined tumor, as tested by an injection of mixed cells consisting of neuroblastoma and transfected glioma cells.

The presence of IGFs in placenta and the fetus has argued for a role for these cytokines in fetal development (Shen et al., 1986, 1988). Others have suggested a role for developmental genes in tumorigenesis. The direct linking of IGF-I to the tumorigenic phenotype provides support to the developmental gene: tumorigenic phenotype connection. Rat C6 glioma cells were originally derived from rat brain astrocytes (Benda et al., 1969). Normal adult astrocytes do not express IGF-I. Rat fetal glial cells, however, produce IGF-I. The production of IGF-I and its receptor in embryonic rat brain is developmentally regulated and IGF-I plays a proliferative and differentiative role (Shooter, 1991). IGF-I is expressed in other embryonic structures and fetal organs (Rotwein et al., 1987). The expression of genes in tumors that are active in embryonic or fetal structures may represent a reenactment of an early embryonic or fetal program. The salient point is that glioma cells that have escaped the host immune response have also reverted to an embryonic phenotype. Embryonic cells are known to escape immune recognition (Boehm et al., 1989). Moreover, the notion that cancer cells reenact an embryonic trait is longstanding. In fact, Otto Warburg claimed that the cause of cancer is the reversal from mitochondrial oxidation to glycolysis in malignant cells (Warburg, 1926). Embryonic and fetal cells are known to employ glycolysis rather than mitochondrial oxidation. Similarly, IGF-I may in fact be considered an "oncofetal cytokine" for glioblastoma cells at least.

Our findings are consistent with the notion that IGF-I functions as a dominant cytokine in defining not only the differentiation but, importantly, also the immunogenic phenotype of glioma cells. It can be hypothesized that a critical event in tumorigenesis may well be an IGF-I-driven phenotypic change that confers upon certain tumor types the capacity to escape the immune system. The present experiments disclose that inhibition of IGF-I in such a tumor can induce a highly immunogenic phenotype. Since some tumors express IGF-II instead of IGF-I, it will be important to learn if in such tumors inhibition of IGF-II can be correlated to an analogous induction of an immunogenic phenotype. The present experimental design provides for an efficient coupled *in vitro*/*in vivo* assay system for determining the role of other cytokines in any given tumor type that may be responsible for down-modulating tumor immunogenicity.

The prior tumor cell literature has focused on the potential role of the IGFs as growth factors driving the proliferation of tumor cells that produce them. In turn, this has led to attempts to inhibit growth of such tumor cells with anti-IGF-I or anti-IGF-II antibodies that are added exogenously. While some growth inhibition has been claimed using anti-IGF-I and anti-IGF-II antibodies *in vitro* and *in vivo* (Blatt et al., 1984; Gansler et al., 1988; Huff et al., 1986; Minuto et al., 1988), tumor development *in vivo* could not be blocked by the blocking anti-IGF antibodies. The *in vivo* experiments have been performed in immunologically deficient nude mice, and hence the issue of alterations of cell immunoge-

Figure 10. Immunocytochemical analysis of host response mounted against transfected and parental glioma cells. All panels: glioma cells, curved solid arrows; lymphocytes, straight solid arrows. **Panels A-C:** Late stage (two weeks post injection) regression of tumor resulting from subcutaneous coinjection of transfected and parental glioma cells. A: Note neovascularization (center) and lymphocyte accumulation (center and right). Hematoxylin and eosin (H&E), X150. B: Immunoperoxidase labeling with anti-GFA antibody shows clusters of persisting glioma cells among mononuclear cells. Hematoxylin, X250. C: Direct immunofluorescence staining of a section adjacent to B using an antibody to CD8 coupled to rhodamine. X250. **Panels D-G:** Tumor sections derived from subcutaneous coinjection of mixed B104 neuroblastoma and untransfected C6 glioma cells (D,E) or mixed B104 neuroblastoma and transfected C6 glioma cells (F,G). D,E: Immunoperoxidase staining with anti-GFA (D) and anti-neurofilament antibodies (E), showing both elements in mixed tumor. Hematoxylin, X200. D, Inset: higher magnification of indicated glioma cells. F,G: Neuroblastoma. Open arrows: neuroblastic rosettes. F: H&E, X200. G: A section adjacent to F. All cells show positive staining for neurofilaments. X200. **Panels H-L:** Eradication of glioblastoma in brain. H: Normal brain tissue, H&E, X250. I: Normal brain/glioblastoma (right) boundary, 1 week following injection of parental C6 cells. H&E, X40l. J: Normal brain/glioblastoma boundary, 1 week following subcutaneous injection of transfected C6 cells into a rat which 3 days previously had been injected intracranially with untransfected C6 cells. Star: normal brain. Giemsa hematoxylin, blue filter, X400. K: As J without normal brain tissue, H&E, X400. L: The same microscopic field as K, stained for CD8 antigen as in C. Fluorescence-bright field, X400 (From Trojan et al., 1993).

nicity was not even considered. Moreover, since intracrine mechanisms (Heldin and Westermark, 1989) may well be involved here, there is no certainty that anti-IGF-I antibodies will parallel antisense IGF-I RNA in altering tumor immunogenicity in an immunologically intact animal.

In general, tumors that arise *de novo* are poorly immunogenic, thereby escaping host antitumor responses (Hewett et al., 1976). The reasons for this poor immunogenicity are not well understood. This study provides one potential therapeutic approach toward enhancing tumor immunogenicity based upon antisense gene transfer. Alternative approaches based upon sense gene transfer have been reported for enhancing tumor cell immunogenicity. One involves enhancement of tumor cell immunogenicity by transfection with a foreign gene encoding an exogenous antigen such as influenza hemagglutinin (Fearon et al., 1988). However, loss of tumorigenicity of transfected tumor cells and tumor prevention using these cells were both incomplete. A second approach attempted enhancement of tumor immunogenicity by transferring into them genes expressing soluble cytokines, such as interleukin-2 (Fearon et al., 1991) and interleukin-4 (Golumbek et al., 1991; Tepper et al., 1989). Results with this approach have been more promising than those with foreign antigen transfer. Nonetheless, tumor prevention in both cases was reported to be incomplete. In contrast to these alternative approaches in which immunogenicity is being enhanced via the neo-expression in the tumor cell of a gene product not normally expressed by that cell, this study reports the enhancement of immunogenicity in an effective way by blocking expression in the tumor cell of a cytokine that is normally expressed by that cell. IGF-I is that cytokine and may function as a molecular switch in controlling the immunogenic phenotype of tumor cells.

Multiple molecular factors are known to influence the progression of somatic cells into a tumorigenic state. Among these are oncogenes, tumor suppressor genes, and angiogenic factors. Our study focuses on another aspect of the tumorigenic phenotype, namely, the capacity to elude host antitumor immune responses. We have defined IGF-I as a critical molecular factor involved in low tumor immunogenicity. In fact, many primary tumors and cell lines derived from tumors are known to produce large amounts of IGF-I or IGF-II. Hence, it is possible that IGF-I and/or IGF-II play a pivotal role in the immunogenicity of tumors other than glioblastoma of the present study. Among the tumor types that express high levels of IGF-I or IGF-II are astrocytomas and meningiomas (Antoniades et al., 1992); sarcomas (Roholl et al., 1990); thyroid adenomas (Williams et al., 1989); estradiol-regulated and nonregulated breast carcinoma (Brunner et al., 1990; Foekens et al., 1989); Wilms' tumor (Gansler et al., 1988); colonic carcinoma (Culouscou et al., 1987); squamous cell carcinoma (Brewer and Reiss, 1990); small cell lung carcinoma (Macaulay et al., 1990); teratocarcinoma (Jing et al., 1991); and hepatocarcinoma (Fu et al., 1988; Yang et al., 1991).

ACKNOWLEDGMENTS

Support for J. Trojan was provided by INSERM, Paris 75654, France. Expert assistance in cell culture was provided by K. Lewandowska, C. Williams, R. Goldschmidt, and C. Porter. We thank Dr. Steven Emancipator for his antibody reagents and for his critical advice in the immunocytochemical studies. We are also grateful to Drs. Donald Anthony and Fadi Abdul-Karim for their valuable input. We thank Dr. U. Roessmann for neurohistopathological evaluation of tumor specimens. This work was supported by National Institutes of Health grants HD-25004 and CA-43703 (Jo.I.), HD-18271 (Ju.I.), and AR-20678 (to the Northeast Ohio Arthritis Center, B.K. Blossey); and National Cancer Institute grant CA 47566 (M.L.T.). In addition, support was provided by the American Diabetes Association (Ju.I.). We also thank the Sypolt family for their generous gift toward the IGF-I line of experimentation.

REFERENCES

Antoniades, H.N., T. Galanopoulis, J. Nevile-Golden, and M. Maxwell (1992) Expression of insulin-like growth factors I and II and their receptor mRNAs in primary human astrocytomas and meningiomas: *In vivo* studies using *in situ* hybridization and immunocytochemistry. Int. J. Cancer 50:215–222.

Arteaga, C.L., L.J. Kitten, E.B. Coronado, S. Jacobs, F.C. Kull, D.C. Allred, and C.K. Oaborn (1989) Blockade of the Type I somatomedin receptor inhibits growth of human breast cancer cells in athymic mice. J. Clin. Invest. 84:1418–1423.

Ayer-le Lievre, C., P.A. Stahlbom, and V.R. Sara (1991) Expression of IGF-I and -II mRNA in the brain and craniofacial region of the rat fetus. Development 111:105–115.

Bass, B.L., and H. Weintraub (1988) An unwinding activity that covalently modifies its double-stranded RNA substrate. Cell 55:1089–1098.

Benda, P., J. Lightbody, G. Sato, L. Levine, and W. Sweet (1969) Differential rat glial cell strain in tissue culture. Science 161:370–371.

Blatt, J., C. White, S. Dienes, H. Friedman, and T. Foley (1984) Production of an insulin-like growth factor by osteosarcoma. Biochem. Biophys. Res. Commun. 123:373–376.

Boehm, K.D., M.F. Kelley, Jo. Ilan, and J. Ilan (1989) The interleukin 2 gene is expressed in the syncytiotrophoblast of the human placenta. Proc. Natl. Acad. Sci. USA 86:656–660.

Brewer, Y., and M. Reiss (1990) Insulin-like growth factor-I is an autocrine growth factor for keratinocytes and squamous carcinoma cells. Proc. Annu. Meet. Am. Assoc. Cancer Res. 31:A247.

Brunner, J., D. Yee, L. Skriver, and K.J. Cullen (1990) Estradiol regulation of IGF-I and IGF-II in the estrogen-inhibited T61 human breast cancer xenograft. Breast Cancer Res. Treat. 16:148.

Cornelissen, M. (1989) Nuclear and cytoplasmic sites for anti-sense control. Nucleic Acids Res. 17:7203–7209.

Culouscou, J.M., M. Remacle-Bonnet, F. Garrouste, J. Marvaldi, and G. Pommier (1987) Simultaneous production of IGF-I and EGF competing growth factors by HT-29 human colon cancer line. Int. J. Cancer 40:646–652.

Deiss, L.P., and A. Kimchi (1991) A genetic tool to identify thioredoxin as a mediator of a growth inhibitory signal. Science 252:117–120.

Eguchi, Y., T. Itoh, and J. Tomizawa (1991) Antisense RNA. Annu. Rev. Biochem. 60:631–652.

El-Badry, O.M., L.J. Helman, J. Chatten, S.M. Steinberg, A.E. Evans, and M.A. Israel (1991) Insulin-like growth factor II-mediated proliferation of human neuroblastoma. J. Clin. Invest. 87:648–657.

Fearon, E.R., T. Inata, B. Hunt, B. Vogelstein, and P. Frost (1988) Induction in a murine tumor of immunogenic tumor variants by transfection with a foreign gene. Cancer Res. 48:2975–2980.

Fearon, E.R., D.M. Pardoll, T. Itaya, P. Golumbek, H.I. Levitysky, J.W. Simons, H. Karasuyama, B. Vogelstein, and P. Frost (1991) Interleukin-2 production by tumor cells bypasses T helper function in the generation of an antitumor response. Cell 60:397–403.

Foekens, J.A., H. Portengen, M. Jansen, and J.G. Klijn (1989) Insulin-like growth factor I receptors and insulin-like growth factor I-like activity in human primary breast cancer. Cancer 63:2139–2147.

Froesch, C.S., J. Schwander, and J. Zapf (1985) Actions of insulin-like growth factors. Ann. Rev. Physiol. 47:443-467.

Fu, X.X., C.Y. Su, Y. Lee, L. Biempica, R. Hintz, R. Snyder, and C.E. Rogler (1988) Insulin-like growth factor II expression and oval cell proliferation associated with hepatocarcinogenesis in woodchuck hepatitis virus carriers. J. Virol. 62:3422–3430.

Gansler, T., K.D. Allen, C.F. Burant, T. Inabnett, A. Scott, M.G. Buse, D.A. Sens, and A.J. Garvin (1988) Detection of type I insulin-like growth factor (IGF) receptors in Wilms' tumor. Am. J. Pathol. 130:431–435.

Gansler, T., R. Furlanetto, T.S. Gramling, K.A. Robinson, N. Blocker, M.C. Bose, D.A. Sens, and A.J. Garvin (1989) Antibody to Type I insulin-like growth factor receptor inhibits growth of Wilms' tumor in culture and in athymic mice. Am. J. Pathol. 135:961–966.

Golumbek, P.T., A.J. Lazenby, H.I. Levitssky, L.M. Jaffee, H. Karasuyama, M. Baker, and D.M. Pardoll (1991) Treatment of established renal cancer by tumor cells engineered to secrete interleukin-4. Science 254:713–716.

Groger, R.K., D.M. Morrow, and M.L. Tykocinski (1989) Directional antisense and sense cDNA cloning using Epstein-Barr virus episomal expression vectors. Gene 81:285–294.

Hambor, J.E.R., C.A. Hauer, H.-K. Shu, R.K. Groger, D.R. Kaplan, and M.L. Tykocinski (1988) Use of an Epstein-Barr virus episomal replicon for anti-sense RNA-mediated gene inhibition in a human cytotoxic T-cell clone. Proc. Natl. Acad. Sci. USA 85:4010–4014.

Han, L., J.S. Yun, and T.E. Wagner (1991) Inhibition of Moloney murine leukemia virus-induced leukemia in transgenic mice expressing antisense RNA complementary to the retroviral packaging sequences. Proc. Natl. Acad. Sci. USA 88:4313–4317.

Heldin, C.-H., and B. Westermark (1989) Growth factors as transforming proteins. Eur. J. Biochem. 184:487–496.

Hewett, H.B., E.R. Blake, and A.S. Walder (1976) Critique of evidence for active host defense against cancer based on personal studies of 27 murine tumors of spontaneous origin. Br. J. Cancer 33:241–259.

Huff, K.K., D. Kaufman, K.H. Gabbay, E.M. Spencer, M.E. Lippman, and R.B. Dickson (1986) Secretion of an insulin-like growth factor I related protein by human breast cancer cells. Cancer Res. 46:4613–4619.

Izant, J.G., and H. Weintraub (1985) Constitutive and conditional suppression of exogenous and endogenous genes by anti-sense RNA. Science 229:345–352.

Jansen, M., F.M.A. van Schaik, A.T. Richer, B. Bullock, D.E. Woods, K.H. Gabbay, A.L. Nussbaum, J.S. Sussenbach, and J.L. Van den Brande (1983) Sequence of cDNA encoding human insulin-like growth factor I precursor. Nature 306:609–611.

Jing, N.H., R. Shiurba, H. Kitani, H. Kawakatsu, Y. Tomooka, and T. Sakakura (1991) Secretion of polypeptides related to epidermal growth factor and insulin-like growth factor I by a human teratocarcinoma cell line. In Vitro Cell. Dev. Biol. 27:864–872.

Johnson, T.R., B.K. Blossey, S.D. Rudin, and J. Ilan (1990) Regulation of IGF-I RNA transcript levels in C6 glial cells. J. Cell Biol. 111:505a.

Kaufman, K., H. Gabbay, E.M. Spencer, M.E. Lippman, and R.B. Dickson (1986) Secretion of an insulin-like growth factor I related protein by human breast cancer cells. Cancer Res. 46:4613–4619.

Kiess, W., L. Lee, D.E. Graham, L. Greenstein, L.Y. Tseng, M.M. Richler, and S.P. Nissley (1989) Rat C6 glial cells synthesize insulin-like growth factor I (IGF-I) and express IGF-I receptors and IGF-II/mannose-6-phosphate receptors. Endocrinology 124:1727–1736.

Kim, S.K., and B.J. Wold (1985) Stable reduction of thymidine kinase activity in cells expressing high levels of anti-sense RNA. Cell 42:129–138.

Kimelman, D., and M.W. Kirschner (1989) An antisense mRNA directs covalent modification of the transcript encoding fibroblast growth factor in Xenopus oocytes. Cell 59:687–696.

Macaulay, V.M., M.J. Everard, J.D. Teale, P.A. Trott, J.J. van Wyk, I.E. Smith, and J.L. Millar (1990) Autocrine function for insulin-like growth factor I in human small cell lung cancer cell lines and fresh tumor cells. Cancer Res. 50:2511–2517.

Melhem, R.F., J.R. Strahler, N. Hailat, X.X. Zhu, and S.M. Hanash (1991) Involvement of OP18 in cell proliferation. Biochem. Biophys. Res. Commun. 179:1649–1655.

Minuto, F., P. Del Monte, A. Barreca, A. Alama, G. Cariola, and G. Giordano (1988) Evidence for autocrine mitogenic stimulation by somatomedin C/insulin-like growth factor I on an established human lung cancer cell line. Cancer Res. 48:3716–3719.

Morrison, R.S. (1991) Suppression of basic fibroblast growth factor expression by antisense deoxyribonucleotides inhibits the growth of transformed human astrocytes. J. Biol. Chem. 266:728–734.

Munroe, S.H. (1988) Antisense RNA inhibits splicing of pre-mRNA *in vitro*. EMBO J. 7:2523–2532.

Negroni, A., S. Scarpa, A. Romeo, S. Ferrari, A. Modesti, and G. Raschella (1991) The role of *myc* in neuroblastoma cell proliferation. Cell Growth Diff. 2:511–518.

Palmiter, R.D., G. Norstedt, R.E. Gelinas, R.E. Hammer, and R.L. Brinster (1983) Metallothionein-human GH fusion genes stimulate growth of mice. Science 222:809–814.

Rogers, J. (1988) RNA complementary to a-amylase mRNA in barley. Plant Mol. Biol. 11:125–138.

Roholl, P.J., A. Skottner, I. Prinsen, C.J. Lips, W. Den Otter, and J.A. Van Unnik (1990) Expression of insulin-like growth factor I in sarcomas. Histopathology 16:455–460.

Rotwein, P., K.M. Pollock, M. Watson, and J.K. Milbrandt (1987) Insulin-like growth factor gene expression during rat embryonic development. Endocrinology 121:2141–2144.

Selinfreund, R.H., S.W. Barger, M.J. Welsh, and L.J. van Eldik (1990) Antisense inhibition of glial S100 beta production results in alterations in cell morphology, cytoskeletal organization, and cell proliferation. J. Cell Biol. 111:2021–2028.

Shen, S.-J., M. Daimon, C.-Y. Wang, M. Jansen, and J. Ilan (1988) Isolation of an insulin-like growth factor II cDNA with a unique 5'-untranslated region from human placenta. Proc. Natl. Acad Sci. USA 85:1947–1951.

Shen, S.-J., C.-Y. Wang, K.K. Nelson, M. Jansen, and J. Ilan (1986) Expression of insulin-like growth factor II in human placentas from normal and diabetic pregnancies. Proc. Natl. Acad. Sci. 83:9179–9182.

Shooter, E.M. (1991) In: E.M. Spencer (ed.). Modern Concepts of Insulin-like Growth Factors. New York: Elsevier, pp. 297–307.

Stout, J.T., and C.T. Caskey (1990) Antisense RNA inhibition of HPRT synthesis. Somat. Cell Mol. Genet. 16:369–382.

Stuart, G.W., P.F. Searle, H.Y. Chen, R.L. Brinster, and R.D. Palmiter (1984) A 12 base-pair DNA motif that is repeated several times in metallothionein gene promoters confers metal regulation to a heterologous gene. Proc. Natl. Acad. Sci. USA 81:7318–7322.

Tepper, R.I., P.K. Pattengale, and P. Leder (1989) Murine interleukin-4 displays potent antitumor activity *in vivo*. Cell 57:503–512.

Toulme, J.-J., and C. Helene (1988) Antimessenger oligodeoxyribonucleotides: An alternative to antisense RNA for artificial regulation of gene expression—a review. Gene 72:51–88.

Trojan, J., B.K. Blossey, T.R. Johnson, S.D. Rudin, M. Tykocinski, Ju. Ilan, and J. Ilan (1992) Loss of tumorigenicity of rat glioblastoma directed by episome-based antisense cDNA transcription of insulin-like growth factor I. Proc. Natl. Acad. Sci. USA 89:4874–4878.

Trojan, J., T.R. Johnson, S.D. Rudin, Ju. Ilan, M.L. Tykocinski, and J. Ilan (1993) Treatment and prevention of rat glioblastoma by C6 cells expressing antisense IGF-I RNA. Science Science 259: 94-97.

Trojan, J., J. Uriel, M.A. Deugnier, and J. Gaillard (1984) Immunocytochemical quantitative study of alphafetoprotein in normal and neoplastic neural development. Dev. Neurosci. 6:251–259.

van der Krol, A.R., J.N.M. Mol, and A.R. Stuitje (1988) Antisense genes in plants: An overview. Gene 72:45–50.

Wang, C.-Y., M. Daimon, S.-J. Shen, G.L. Englemann, and J. Ilan (1989) Insulin-like growth factor I messenger RNA in the developing human placenta and in term placenta of diabetics. Mol. Endocrinol. 2:217–229.

Warburg, O. (1926) Stoffwechsel der Tumoren. Berlin: J. Springer. [Tr. by F. Dickens (1930) The Metabolism of Tumors. London: Constable.]

Whitesell, L., S. Rosolen, and L.M. Neckers (1991) Episome-generated N-myc antisense RNA restricts the differentiation potential of primitive neuroectodermal cell lines. Mol. Cell Biol. 11:1360–1371.

Williams, D.W., E.D. Williams, and D. Wynford-Thomas (1989) Evidence for autocrine production of IGF-I in human thyroid adenomas. Mol. Cell Endocrinol. 61:139–143.

Yang, D.Y., and C.E. Rogler (1991) Analysis of insulin-like growth factor II (IGF-II) expression in neoplastic nodules and hepatocellular carcinomas of woodchucks utilizing in situ hybridization and immunocytochemistry. Carcinogenesis 12:1983–1901.

Yanisch-Perron, C., J. Vieira, and J. Messing (1985) Improved M13 phage cloning vectors and host strains: Nucleotide sequences of the M13mp18 and pUC19 vectors. Gene 33:103–119.

Yates, J.L., N. Warren, and B. Sugden (1985) Stable replication of plasmids derived from Epstein-Barr virus in various mammalian cells. Nature 313:812–815.

Index